A Note from the Publisher

This volume was printed directly from a typescript prepared by the author, who takes full responsibility for its content and appearance. The Publisher has not performed his usual functions of reviewing, editing, typesetting, and proofreading the material prior to publication.

The Publisher fully endorses this informal and quick method of publishing lecture notes at a moderate price, and he wishes to thank the author for preparing the material for publication.

SLOW NEUTRON SCATTERING

AND

THERMALIZATION

SLOW NEUTRON SCATTERING

AND

THERMALIZATION

With Reactor Applications

D. E. PARKS M. S. NELKIN
J. R. BEYSTER N. F. WIKNER

United States Atomic Energy Commission

W. A. BENJAMIN, INC.
New York 1970

SLOW NEUTRON SCATTERING AND THERMALIZATION

With Reactor Applications

Standard Book Number: 8053-7760-3 (cloth)
Library of Congress Catalog Card Number: 78-136509
Manufactured in the United States of America
12345R43210

W. A. BENJAMIN, INC.
New York, New York 10016

Foreword

There are occasions when the time is ripe for a book to be written on a specific subject, and this is one of them. The general area of slow neutron physics was opened in the nineteen thirties when Fermi and his collaborators discovered that absorption cross sections were much larger when neutrons were first "slowed down." It gained a tremendous impetus in the early forties when it was determined that a controlled nuclear chain reaction could be achieved with natural uranium if the fast fission neutrons were "moderated" to thermal energies.

In 1942, Wigner and Wilkins gave the first theoretical treatment of the thermalization process for the simple case of a free hydrogen gas and subsequently for monatomic heavy gases. Because of its complexity the subject lay dormant for a considerable time. Also, it was thought to be sufficient for reactor applications to describe the neutron energy spectrum by a Maxwellian, perhaps with an effective temperature somewhat different from the actual one, fitted by balance considerations to an epithermal tail. It became clear in the fifties, however, that the accurate neutronic design of reactors demanded a much more detailed knowledge of the neutron energy spectra.

The acquisition of such knowledge was made possible by advances in many areas of neutron physics. Major contributions to the general theory of slow neutron scattering were made by Placzek, Van Hove, Glauber, and Zemach. Techniques to measure scattering cross sections with consideration of energy and momentum transfer were developed by Hughes and Brockhouse. Actual neutron spectra were obtained by Poole by the time-of-flight technique, and diffusion parameters were measured by Von Dardel, who introduced the pulse decay technique. Since then there has been a continuous expansion and evolution to which many people have contributed.

Slow neutrons constitute one of the most effective probes of the dynamics of molecules, solids, and liquids since their De Broglie wavelengths are of the order of interatomic distances and their energies in the range of phonon excitations. Thus there was always a great deal of interest in this subject from the point of view of basic physics. Reactor applications, however, added considerably to the motivation of research in the slow neutron field. Today, since it turned out that the requirements of the reactor designers could be satisfied by somewhat simplified approximations, the emphasis has shifted back more and more to an investigation of fundamental problems in solid and liquid state physics. This holds true also in the area of thermalization itself.

The literature on the subject, as all literatures, has proliferated enormously in recent years. The field is dealt with briefly in many books such as *Neutron Physics* by Beckurts and Wirtz and recently in somewhat more detail in *Slow Neutrons* by Turkin, which, however, is rather short on applications. *Thermal Neutron Scattering,* edited by Egelstaff, gives a comprehensive treatment of slow neutron scattering. The emphasis is primarily experimental and little consideration is given to thermalization and reactor applications. The *Naval Reactor Physics Handbook,* edited by Radkowsky, gives many of the important formulations, but only with sketchy derivations and applications slanted toward the naval reactors. The very recent book *The Slowing Down and Thermalization of Neutrons* by M. M. R. Williams comes closest to the present volume in content and style. It gives a thorough exposition of the theory of thermalization, with perhaps more emphasis on mathematical models and less discussion of scattering theory and measurements of neutron spectra. There have been a number of conferences with extensive proceedings, such as Vienna 1960, Brookhaven 1962, Chalk River 1963, Bombay 1964, and Karlsruhe 1965. Such proceedings with their many contributors, however, serve better the initiated than the newcomer. What was missing was a thorough exposition of the whole field which provides a self-contained treatment of the underlying theory and bridges the gap to the experiments and the many applications.

Such a book could hardly have been written by a single author. The John Jay Hopkins Laboratory of the General Atomic Division of General Dynamics Corporation had brought together a group of people with competence in theory, experiment, and reactor design in order to fill its own needs. This group soon distinguished itself by outstanding contributions to slow neutron physics. In their work they in turn felt keenly the need for a suitable text and they tried to persuade one of its members to write one. His answer was: "Why don't we all write a book?" The support of the U.S. Atomic Energy Commission was solicited, a fixed-price contract was signed, and so this volume came into being.

It does not seem to be amiss to define what this book intends to be, and what it does not. It is not a *Handbuch* which covers everything. It gives a thorough and detailed exposition of the theoretical foundations. Among these is the theory of slow neutron scattering in Chapter 2. This is a demanding subject that requires some knowledge of quantum mechanics and lattice dynamics; however, the remainder of the book is so written as to be readable without going through the complex details of these developments. Another topic that is thoroughly covered consists in the modifications of neutron transport theory made necessary by the multivelocity features in the thermal region. These two subjects, scattering and transport, are, of course, at the heart of slow neutron physics and will remain so even in the presence of future advances.

With respect to applications, a more selective approach has been used. The book does not cover details of computer codes and experimental procedures since these are subject to rapid evolution. It does give fairly comprehensive results on most of the important moderators, both with respect to the scattering laws and the actually resulting spectra. On the experimental side, time-of-flight methods are used somewhat more extensively than die-away measurements on pulsed assemblies. This corresponds to the actual activities at this Laboratory. Also, in reactor applications the discussion is directed more toward the typical as compared with a complete treatment of particular reactors.

The plan of the book is explained in detail in the Introduction, Chapter 1. It should be emphasized that it represents a collective effort. It is thus not possible or desirable to give a breakdown on authorship of the various chapters and sections. It should, however, be mentioned that this work could not have been completed without the devoted and persistent labors of D. E. Parks, who coordinated the efforts of his collaborators and prodded them successfully to complete their assignments in spite of their peregrinations in space and time.

San Diego, California
May 1966

L. W. Nordheim

To The Reader

The following material was inadvertently left out of the text. For the convenience of the reader, the new material, its position and page number are shown below:

76. To follow equation, mid-page.

    Thus by Eq. (2.71), if the neutron scattering is to be approximated as in Eq. (2.119), the initial neutron energy must be at least large enough to satisfy

    $$E_0/\hbar \gg A\omega_0 e^{\hbar\omega_0 k_B T}$$

665. To follow "the" in line 8:

    energy variation of the α ($\underline{E}$). At high energy, Σ_s and Σ become

708. To follow "clear" in line 14:

    that the continuous spectrum is related to the unboundedness of the time

Three pages were misnumbered by the typist and are shown as blanks in the text. No text is missing. The pages are 636, 677 and 688.

Contents

Chapter 1

INTRODUCTION

Neutron thermalization is concerned with the energy, space, and time distribution of thermal neutrons in moderating materials. The distinguishing feature with respect to other problems of neutron scattering and migration is the importance of the chemical binding and thermal motion of the moderating nuclei. As in the other parts of nuclear reactor theory, there is the quantum mechanical problem of the individual neutron collisions and the essentially classical neutron transport problem dealing with the distribution resulting from multiple collisions. Although these subjects are coupled in the analysis of any actual experiment or reactor design, they each have a substantial theoretical base that is of interest in its own right. It is the objective of this book to develop the theoretical basis of both subjects and to combine them in quantitative applications. We consider in some detail the experiments on neutron distributions which test the combined theories and we present selected applications to specific nuclear reactor systems.

In considering the individual scattering events, the basic problem is to understand the interchange of energy between the neutron and the states of excitation of the atomic motions in the moderator. This is a problem in molecular or solid state physics rather than in nuclear physics. It was shown by Fermi[1] in 1936 that the true neutron-nucleus interaction could be replaced, to a good approximation, by a pseudo-potential interaction in which all of the nuclear physics is combined into a single parameter, the scattering length. The essential physical point is that the scattering by free nuclei at rest is very simply described, and there is no new nuclear physics introduced when the scattering nucleus is contained in a molecule or crystal. Our degree of understanding of the slow neutron scattering process is therefore dependent on our degree of understanding of the relevant molecular or crystalline excitations. For crystals these are the phonons, or quantized lattice vibrations; and for molecular gases they are the vibrations, rotations, and translations of the molecule. For liquids we do not have a completely satisfactory description. The relation between neutron scattering and the dynamics of atomic motions has its basis in the validity of the first Born approximation when the neutron-nucleus interaction is expressed by means of the Fermi pseudo-potential. For this reason the formal content of the theory is not essentially different from that for the scattering of x-rays or other quanta that interact weakly with a system of interacting particles. The distinguishing feature that gives a neutron its largely singular capability for sensing the behavior of matter at the microscopic level has its origin in the neutron kinematics. Like x-rays, the neutrons have a de Broglie wavelength that is comparable to interatomic spacings. For this reason the measurement of the angular distribution of scattered neutrons can give information about the static structure of matter. Unlike x-rays, however, thermal neutrons on scattering undergo energy changes that are comparable to or larger than their initial energies. This unique aspect of slow neutrons has led to the extensive development of slow-neutron inelastic scattering experiments as a method for studying the dynamical behavior of solids, liquids, and polyatomic gases.

In the scattering of monoenergetic neutrons by a single crystal, a sharp peak in the energy distribution of neutrons scattered at a given angle corresponds to the emission or absorption of a single phonon. The energy and momentum transfer associated with these peaks give the frequency versus wave number for the normal modes of lattice vibration. A similar result is obtained when neutrons emit or absorb a single elementary excitation in liquid helium at low temperatures. In this type of experiment unambiguous results on the behavior of matter at the microscopic level are obtained directly from the location of peaks in the scattered distribution without the need of measuring the related cross section.

In polycrystalline solids or in liquids at higher temperature the dynamical information is contained in the full energy and angle distribution of the scattered neutrons. In such cases experiments are much more difficult but have already yielded specific information about the diffusive motions in liquids and about other aspects of molecular dynamics, some of which are inaccessible to other experimental techniques. Experiments of this type are rapidly becoming more quantitatively definitive, and the theory needed to interpret them is highly developed. Much of this theoretical development is given in Chaps. 2 and 3 of this book.

Although the energy-transfer cross sections for slow neutrons are sensitive to the atomic motions in the moderator, the space and energy distributions of neutrons in a reactor are moderately insensitive to these cross sections. If this were not so, we could not have had a reasonably successful theory of thermal-neutron reactors during the many years when no adequate understanding of neutron thermalization was available. The essential point is that neutrons tend toward thermal equilibrium with the moderator. If the number of scattering collisions before absorption is sufficiently large, the thermal-neutron flux spectrum is approximately a Maxwellian distribution,

$$\phi_M(E) = (k_B T)^{-2} E \exp\left(-\frac{E}{k_B T}\right), \tag{1.1}$$

where k_B is Boltzmann's constant and T is the absolute temperature of the moderator. That this is the distribution for thermodynamic equilibrium between neutron and moderator follows directly from the dilute nature of the neutron gas and is independent of the mechanism by which equilibrium is reached. It is when we examine the departures from equilibrium produced by absorption or leakage or when we study the approach to equilibrium that we encounter the full complexities of the problem of neutron thermalization.

The neutron thermalization problem was first formulated by Wigner and Wilkins[2] during World War II. They set up the balance equation for neutrons in an infinite homogeneous medium moderated by an ideal monatomic gas and reduced this equation to a second-order differential equation for the case of monatomic hydrogen gas. This work was later extended by Wilkins,[3] who derived the differential equation appropriate to a heavy monatomic ideal gas. The subject lay rather dormant until about 1954, when it began to be intensively studied. By 1954 it was clear that the lack of understanding of thermal-neutron energy spectra was a major unknown in the neutronic design of reactors. This awareness was greatly enhanced by the emergence of reactor physics into the open literature at the 1955 Geneva Conference on the peaceful uses of atomic energy.

There were many developments making it possible to satisfy the practically motivated demand for solving the problem of neutron thermalization. These included

(1) the development of a general theoretical understanding of slow-neutron scattering,

(2) a general theoretical understanding of neutron thermalization,

(3) measurements of neutron flux spectra sufficiently precise to allow meaningful comparisons of theory and experiment,

(4) the development of digital computers to allow predictions of theory to be quantitatively evaluated, and

(5) experimental determination of the cross sections for slow-neutron inelastic scattering in moderators.

In this book, we develop the first three of these areas in detail and illustrate their application to specific reactor systems. We make extensive use of developments in the fourth and fifth areas, but we do this without extensive discussion of the techniques involved. This choice is dictated primarily by the areas in which the authors have special competence.

We have already briefly discussed the first of these areas, and will shortly discuss the second. Concerning the third, we note that the path from basic scattering theory to the behavior of a reactor is much too long and too full of unknowns to proceed entirely by theory. The development of experimental techniques that could give information at intermediate stages was very important. Particularly important was the Poole technique[4] for measuring spectra in homogeneous mixtures of absorber and moderator, and the pulsed-neutron technique for measuring thermal-neutron diffusion parameters developed by von Dardel[5] and

Antonov.[6]

Although the theory of neutron thermalization developed primarily in response to the needs of a particular application, it has yielded significant knowledge of a more fundamental nature. The approach of a "neutron gas" to thermal equilibrium with a moderator provides one of the most intensively studied and best understood examples of the approach of a system to equilibrium. The theoretical description of the neutron distribution in terms of the linearized Boltzmann equation of neutron transport theory is well founded. Recent theoretical developments have led (see Chap. 8) to a rather good understanding of the subtle nature of the eigenvalue problem defining the relaxation lengths and times for this equation. The advances in this area are closely paralleled by similar advances in understanding the mathematical structure of the linearized Boltzmann equation that is used in the kinetic theory of gases (for example, to describe high-frequency sound propagation). The neutron problem has the considerable advantage, however, that detailed experimental information on the neutron distribution is available from pulsed-neutron experiments. This had led to many calculations of the quantitative predictions of the theory and to the beginnings of a critical examination of the experimental relevance of our recently acquired mathematical sophistication.

In a more practical context, the importance of an accurate calculation of thermalization varies with the over-all spectrum of the reactor. A very heavily loaded reactor has very few thermal fissions, and thus the thermal neutrons are not of great interest. For a well-thermalized reactor, such as a swimming pool research reactor, the thermal-neutron spectrum is almost Maxwellian and follows the water temperature. The thermal reaction rates of importance for reactor design can be calculated in this limit by averaging the relevant energy-dependent cross sections over the Maxwellian flux spectrum of Eq. (1.1). In the early days of reactor design, the energy dependence of these cross sections was not known with great accuracy so that there was little incentive to calculate the flux spectrum more accurately.

We now have much more accurate basic cross section data and much more demanding criteria for accurate prediction of spatial and energy distributions of slow neutrons. Considerations of neutron thermalization are of greatest importance for reactors with about two thirds of the fissions below 0.5 eV. In these reactors the thermal-neutron distribution is quite far from Maxwellian. Most of the power reactors now being designed have spectra of this general type. There is also the exceptional case of moderation by metallic hydrides, where the energy transfer is nearly quantized in units of about 0.1 eV. For reactors using these moderators, thermalization must be considered in detail, even for very dilute fuel loadings, if the temperature dependence of the reactivity is to be accurately calculated.

The contents of this book fall in three parts. Chapters 2 and 3 deal with slow-neutron scattering, Chaps. 4, 5, 6, and 8 with the theory and measurement of thermal-neutron distributions, and Chaps. 7 and 9 with selected problems in the physics of reactor design.

Chapters 2 and 3 can be used as a detailed source of slow-neutron scattering theory. Our principal objectives in the first five sections of Chap. 2 are to set forth the basic concepts underlying the theory of slow-neutron scattering, to develop the general theoretical formalism, to state the physical content of the fully developed theory in terms of the time dependence of the motions of the atoms in the scattering system, and to illustrate this formalism by means of simple examples. We try to make the treatment self-contained by giving an elementary exposition of the quantum theory of scattering as it applies to the neutron problem.

Our treatment of slow-neutron scattering by real systems, or perhaps it is better to say nearly real systems, is accompanied by a brief description of the dynamical behavior of crystal lattices and molecular gases. We remain within the well-understood framework of the harmonic approximation in discussing the atomic vibrations. The main justification for this is the great success the harmonic approximation already enjoys in accounting for many of the observed properties of crystalline and molecular vibrations.

We have chosen to treat the scattering of neutrons by crystals and molecular gases almost entirely within the framework of the time-dependent formalism. Though not the only possible approach to the problem, it is certainly the most concise. The unique power of the time-dependent formalism is first exhibited in Sec. 2-8, where we develop Wick's short-collision-time method for obtaining cross sections in the limit of neutron energies large compared with $k_B T$ and with the characteristic frequencies of atomic motions.

The theory of slow-neutron scattering has its most elegant form when cast in the language of space-time correlations between the particles composing the macroscopic scattering systems. In this form, the theory is particularly suitable for the study of poorly understood dynamical systems such as liquids. In Sec. 2-9, the theoretical utility of expressing the dynamical behavior of statistical systems in terms of space-time correlation functions is illustrated by applying it to the study of kinetic models of atomic motions in the liquid state.

Although to a large extent theoretical, the third chapter has a much less fundamental tone than the second. In the first two sections of Chap. 3 we develop approximate methods for calculating slow-neutron scattering cross sections. The cross section of primary interest is the differential cross section $\sigma(E_0,E,\theta)$ for energy transfer from an initial energy E_0 to a final energy E with scattering through an angle θ in the laboratory system. Whereas in Chap. 2 we establish the general formalism connecting $\sigma(E_0,E,\theta)$ and the microscopic behavior of the scattering system, in Chap. 3 we proceed from a calculational basis that is less rigorous but more tractable than that of the general formalism. This basis follows from the general formalism on making the incoherent and Gaussian approximations, which are already mentioned in Chap. 2. The Gaussian and incoherent approximations have been extensively used in the analysis of slow-neutron scattering experiments and in applications of slow-neutron scattering theory to problems in neutron thermalization. In Sec. 3-1 we discuss in some further detail the validity of these approximations. Section 3-2 contains detailed treatments of the phonon expansion and short-collision-time methods.

Neutron scattering by materials that are important as moderators is studied in the last three sections of Chap. 3. We do not discuss extensively the experimental techniques of slow-neutron scattering, but we do make extensive use of the experimental results. Our primary objectives are to show how these results compare with predictions based on specific models for atomic motions and to set forth the physical models in terms of which many of the neutron thermalization calculations of subsequent chapters are carried out.

The behavior of thermal-neutron distributions in bulk matter forms the subject matter of Chaps. 4 through 9. An understanding of the general aspects of the theory of thermal-neutron distributions requires very little detailed familiarity with the theory of slow-neutron scattering. In most of the general theoretical developments following Chap. 3, the only properties of the differential-energy-transfer cross section $\sigma(E_0,E,\theta)$ that play a significant role are the detailed balance condition and the form $\sigma(E_0,E,\theta)$ in the limit of neutron scattering by atoms that are free and at rest. Further, some of the more specific theoretical developments, such as the Wigner-Wilkins and the Wilkins models for the neutron spectrum in an infinite homogeneous medium, require only a knowledge of the scattering in the relatively simple case where the effects of interatomic binding are absent. Quantitative calculations of thermalization in real moderators are carried out, however, with differential-energy-transfer cross sections that are derived from molecular and crystalline models that include the effects of interatomic binding.

Although a complete understanding of thermalization effects in reactors requires a quantitative specification of the energy-transfer cross section $\sigma(E_0,E,\theta)$, the reader who is interested only in reactor applications may completely bypass Chaps. 2 and 3. He would thus consider the energy-transfer cross section, like any other cross section, to be part of the input information that is required to solve a specific design problem. Considering the degree of authomation that has been achieved in the neutronic aspects of reactor design, such a view of the cross sections is compatible with a quantitative capability for assessing the role played by thermalization in reactor design problems, even though it results at best in only a qualitative understanding of the fundamentals of neutron thermalization.

Chapter 4, on thermal-neutron transport theory, is intended to provide much of the conceptual and formal basis for work in the subsequent chapters. The fundamental quantities of thermal-neutron transport theory are defined, together with the limitations of the theory and the basic assumptions underlying it. The primary emphasis of the chapter is on those aspects of neutron transport theory that are most directly connected with the energy distribution of neutrons near thermal energies. In particular, we extend those concepts, such as thermal diffusion length, migration area, neutron lifetime, and extrapolation distance, which are well understood for neutrons diffusing with a single speed, to a theory that includes energy exchange between thermal neutrons and the moderator. The thermal-neutron spectrum in an infinite homogeneous medium, which is the simplest area of thermal-neutron transport theory, is the subject of Chap. 5. In order to have a physical understanding of the general features of the spectrum, we study the general properties of the infinite-medium neutron balance in some detail and, in particular, in the limit of weak absorption. On completing this study, we turn to the specific examples provided by the spectrum in monatomic hydrogen gas and in a monatomic heavy gas. These are not moderators of practical interest, but they do constitute examples that submit more readily to mathematical analysis than the general problem. Further, these examples exhibit many of the same mathematical features as the problem associated with the spectrum in a real moderator.

Although it is difficult to obtain explicit solutions for the spectrum at all energies for an arbitrary moderator, it is possible to obtain explicit approximate solutions for energies that are large in relation to the characteristic energies of atomic motion. In carrying out this program in Sec. 5-6, we make extensive use of Wick's short-collision-time method, which is developed in Sec. 2-8. Section 5-6 is the only part of Chap. 5 in which detailed aspects of the general scattering formalism are indispensable to the development of the subject matter.

The effects of chemical binding are usually large enough that we cannot use the differential equations of the monatomic hydrogen or heavy gas models to calculate reactor spectra. It is unlikely, however, that the full information contained in $\sigma(E_0,E\theta)$ is needed for most applications. It is therefore of interest to look for approximate differential equations that correct for chemical binding effects. In Sec. 5-7 we outline one method by means of which this program has been carried out. The last section in Chap. 5 deals with effects that are induced by departures from the ideal infinite homogeneous system. In this section, we have in mind mainly the experimental determination of a thermal-neutron spectrum that closely approximates that in the ideal system. The treatment of effects induced by the inhomogeneous distribution of absorbers in the form of thin foils and by the nonuniformity of the source providing thermal neutrons in the assembly forms the basis for many of the considerations of experimental design that are discussed in Chap. 6.

After the construction of the first multiplying assemblies, many techniques were developed for measuring neutron spectra. The earliest spectral studies were performed in reactor environments by means of thermal activation of foils and by beam-absorption techniques. For the purpose of detailed comparisons with calculated spectra, we require measurements of much higher resolution than can be conveniently achieved by such integral techniques. Thus, in Chap. 6 we deal exclusively with measurements by neutron time-of-flight techniques. Two different approaches to the time-of-flight measurement of neutron spectra are the direct, pulsed-source time-of-flight method (the Poole technique), and the chopped-beam, steady-source time-of-flight method.

In discussing these methods, we emphasize both the experimental design and the problem of correcting experimental data for undesirable effects that are associated in an inseparable way with the design. Such effects include, for example, the spectral distortions and energy uncertainties associated with the finite time required for the emission of a neutron from a pulsed assembly.

The time-of-flight methods have been applied extensively to the measurement of thermal-neutron spectra in the common moderators. In Sec. 6-4 we compare the results of some measurements of neutron spectra with theoretical results, emphasizing in particular the spectra of neutrons thermalized by light water, zirconium hydride, or graphite. We also give results for heavy water, polyethylene, and benzene, the latter being considered as a prototype of the organic moderators. These comparisons demonstrate the significant effect of chemical binding on thermal-neutron spectra and provide the most direct evidence of the importance of a proper account of neutron thermalization for accurate predictions of the neutronic behavior of neuclear reactors.

Our treatment of neutron thermalization in practical reactor configurations is not exhaustive; it is hoped that the analysis of a few problems will be useful as an illustration of the application of accurate thermalization considerations to the problems of reactor analysis and design. For such illustration, we use as practical examples reactor configurations wherein the effects of thermalization and of chemical binding in particular are demonstrably significant and have been thoroughly investigated. The results of such investigations are contained in Secs. 7-2 and 9-2. In Sec. 7-1 we discuss some aspects of reactor cross sections which are particularly relevant to subsequent parts of Chap. 7, which deals exclusively with problems in homogeneous reactors. Problems of thermalization in nonuniform reactors are the subject of Chap. 9.

In homogeneous or nearly homogeneous reactors, the practically significant effects of thermalization are manifested in two ways. First, the increased hardening of the spectrum in the presence of interatomic binding leads to an increase in the thermal-neutron migration area and, thereby, to an increase in the leakage from the reactor. Second, when a significant fractions of thermal-neutron absorptions occur in a resonance absorber, there is a significant effect of thermalization on reactivity even in a very large system. In homogeneous water reactors, although an accurate prediction of reaction rates depends on a proper treatment of chemical binding effects, an accurate calculation of reactivity does not. Thermal neutrons in a homogeneous water reactor are absorbed at a much faster rate than they leak from the system, and thus the reactivity, which depends on the ratios of reaction rates, is not very sensitive to the shifts in the thermal spectrum, at least as long as significant concentrations of thermal resonance absorbers are not present.

In homogeneous solid-moderator reactors, on the other hand, the reactivity and its temperature dependence may be significantly affected by leakage of thermal neutrons, and the latter is sensitive to the shift of the thermal spectrum toward higher energies that is produced by the effects of interatomic binding. Because of the very tight binding of carbon atoms in the graphite lattice, such spectral shifts are particularly large in graphite systems. Although the quantitative results in Sec. 7-2 are specific to moderation by graphite, they are nevertheless strongly indicative of the role played by chemical binding effects in reactors moderated by beryllium and beryllium oxide.

In the presence of absorbers with resonances at low energies, the shift in the thermal spectrum induced by changes of temperature may lead to appreciable effects on reactivity even when there is no appreciable leakage from the system. In practical solid-moderated reactors, the effect is significant at temperatures high enough that the spectrum in the vicinity of the resonance energy shifts significantly with temperature. At such high temperatures the effect of thermalization is important, but the effect of chemical binding is not; i.e., it is important to calculate the spectrum accurately, but this may be done as if neutrons were scattering from a gas of free atoms. The practical importance to temperature coefficients of a large spectral shift in the vicinity of the resonance energy of a strong resonance absorber in a homogeneous reactor is considered in detail in Sec. 7-3.

Significant problems occur in all reactor spectra where there is a strong spatial dependence of the thermal-neutron spectrum. These problems, where both the thermalization and migration of neutrons play a significant role, are the subject of Chap. 9. We seek principally to determine the importance of the physical description of atomic motions and scattering for the neutron balance in a reactor. By doing this, we establish that our capability for performing accurate reactor design calculations is not severely limited by our imperfect capability for calculating neutron scattering in the thermal energy range.

In Chap. 9 we make extensive use of multigroup methods in our study of thermalization in nonuniform reactors, but we do not make a detailed study of such methods. We do, however, describe the methods of overlapping neutron distributions. These methods circumvent many of the difficulties associated with the direct solution of the multigroup transport equation. Much of the early research that led to the use of overlapping neutron distributions is described in Chap. 8.

The principal topics in Chap. 8, however, are the characteristic lengths and times in neutron thermalization. We begin with a systematic consideration of the diffusion length considered as the eigenvalue defining the asymptotic relaxation length of the thermal-neutron flux far from sources. This asymptotic situation can be realized experimentally to a good approximation so that a meaningful comparison between theory and experiment can be given. The same is true for the asymptotic relaxation time for the flux in a finite moderating medium, although the theory here usually deals with too highly idealized a problem to make a precise comparison with experiment.

We also consider the higher eigenvalues describing the space and time variation of the neutron spectrum. Although these are important to the physical understanding of the theory, they are not readily measurable. The comparison between theory and experiment in the area of spatial and temporal variations of the spectrum relies almost entirely on direct numerical solutions of the transport equation.

Except for the numerical calculations that are used in the comparison of theoretical and experimental eigenvalues, the emphasis in Chap. 8 is primarily analytical. This chapter is intended to synthesize recent advances in our understanding of the transport equation for thermal neutrons in a way that digests the principal physical content. In a subject whose subtleties are still being uncovered, this can be, at most, only partially successful.

REFERENCES

1. E. Fermi, On the Motion of Neutrons in Hydrogenous Substances, Translation from *Ricerca Scientifica, VII*, **2**: 13 (1936), available from OTS as PB-5922332 (AEC File No. NP-2385); also in Enrico Fermi, *Collected Papers*, Vol. I, p. 980, E. Segrè, Chairman of Editorial Board, University of Chicago Press, Chicago, 1962.

2. E. P. Wigner and J. E. Wilkins, Jr., *Effect of the Temperature of the Moderator on the Velocity Distribution of Neutrons with Numerical Calculations for H as Moderator*, USAEC Report AECD-2275, Oak Ridge National Laboratory, September 1944.

3. J. E. Wilkins, Jr., *Effect of the Temperature of the Moderator on the Velocity Distribution of Neutrons for a Heavy Moderator*, USAEC Report CP-2481, November 1944.

4. J. B. Sykes, *Feasibility of a Proposed Method of Measurement of Thermal Neutron Spectra in a Thermal Pile*, British Report AERE-T/R-1367, 1954.

5. G. F. von Dardel, The Interaction of Neutrons with Matter Studied with a Pulsed Neutron Source, *Trans. Roy. Inst. Technol., Stockholm*, **75** (1954); and G. F. von Dardel and N. G. Sjöstrand, Diffusion Parameters of Thermal Neutrons in Water, *Phys. Rev.*, **96**: 1245 (1954).

6. A. V. Antonov *et al.*, A Study of Neutron Diffusion in Beryllium, Graphite, and Water by the Impulse Method, in *Proceedings of the First United Nations International Conference on the Peaceful Uses of Atomic Energy, Geneva, 1955*, Vol. 5, p. 3, United Nations, New York, 1956.

Chapter 2

THE THEORY OF SLOW-NEUTRON SCATTERING

In 1934, Fermi observed that the high initial energies of neutrons emitted in a nuclear reaction are reduced to energies of the order of those of thermal agitation by means of successive elastic collisions with nuclei of hydrogen or other light elements. The energies of these slow neutrons are of the order of 0.025 eV for a temperature T of 300°K, and their de Broglie wave length λ is of the order of $h/(2mk_BT)^{\frac{1}{2}} \approx 1.8 \times 10^{-8}$ cm. Here, h is Planck's constant, k_B is Boltzmann's constant, and m is the mass of the neutron. These energies are comparable to the characteristic rotational and vibrational energies of atoms in molecules and crystals, and the de Broglie wave length is comparable to the interatomic distances. In view of these facts, it is not surprising tha that subsequent investigations of the properties of these slow neutrons revealed that their interaction with matter as manifested by the results of scattering experiments is subject to the following influences:

The thermal agitation of the atoms of the scatterer.

The interference between waves scattered from different nuclei.

The forces which bind the atoms together in the scatterer (chemical binding).

One of the earliest experiments that demonstrated the existence of and some of the properties of neutrons with energies comparable to thermal energies was performed by the Columbia University Group.[1] Their apparatus consisted of a source of fast neutrons, surrounded by a block of paraffin. The paraffin block was surrounded on all sides except one by thick layers of cadmium, a strong absorber of slow neutrons. The side which was free of cadmium faced a neutron detector at a distance d from it. A mechanical velocity selector was placed between the paraffin block and the neutron detector. The velocity of neutrons transmitted by the selector was determined from their time of flight t through the relation v = d/t. The crude velocity selector used by the Columbia group was the first step in the development of the high-resolution neutron spectrometers that are prevalent today.

In spite of the very poor resolution of the early Columbia spectrometer, it gave very significant results. In addition to showing that neutrons slowed down to thermal energies, it enabled a crude determination of their velocity distribution in the paraffin block. With the paraffin block at room temperature (T = 300°K), the flux spectrum of emitted neutrons, $\phi(v) = n(v)v$, was determined, at least roughly, to be a Maxwellian flux spectrum at the temperature of the paraffin; i.e.,

$$\phi(v) \approx v^2 \exp\left(-\frac{mv^2}{2k_BT}\right),$$

where m and v are the mass and speed, respectively, of a neutron, and T is the temperature of the paraffin block.

The necessity of taking a wave-mechanics point of view in describing the interactions of slow neutrons with matter became apparent at a very early stage in the history of neutron physics. Early experiments on the absorption of slow neutrons, although highly qualitative in nature, were sufficient to show that in many elements the absorption cross sections were enormously large compared with the geometrical cross sections of nuclei. This result, which constitutes a serious difficulty from a classical point of view, is easily understandable in terms of the wave-mechanical description of the neutron. In 1936, von Halban and Preiswerk[2] provided a more striking verification of the wave character of the neutron when they showed experimentally that neutrons could be diffracted by a crystal.

That chemical-binding forces should have a significant influence on the scattering of slow neutrons was first pointed out by Fermi[3] in 1936. The manner in which the chemical-binding effect is manifested in the cross section for the scattering of slow neutrons by protons bound in paraffin is indicated in Fig. 2.1. Above about 1 eV, the neutron energy is much larger than the characteristic energies (of the order of a few tenths of an electron volt) of hydrogen atom vibrations. Since the proton recoil energy resulting from the impact is much greater than these characteristic energies, the proton can be considered as if it were free. As the neutron energy decreases to values comparable to or less than the characteristic vibrational energies, the cross section increases monotonically to a value of approximately four times that of the free-proton value.

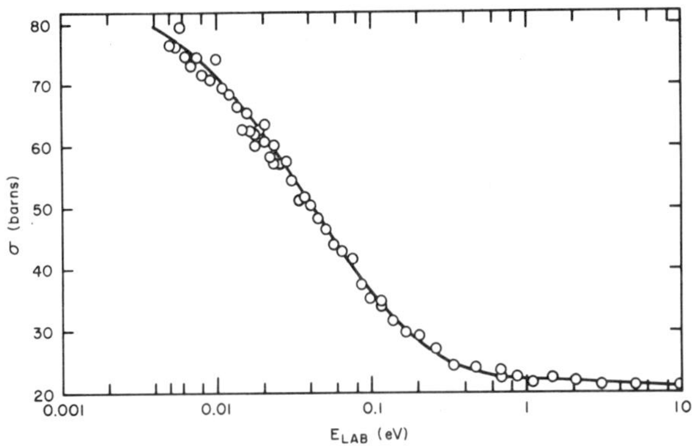

Figure 2.1 The cross section for the scattering of neutrons by hydrogen atoms bound in paraffin. (From Sachs, based on Rainwater's daca, Nuclear Theory, Ref. 4.)

Actually, the value of four is rigorously correct for the case of the scattering of very slow neutrons by a proton bound in an otherwise-inert heavy molecule maintained at a very low temperature. More specifically, the energy of the neutron must be less than the energy separation between the ground state and the first excited state of the scattering system, and the temperature must be sufficiently low that there is no appreciable thermal excitation of scatterers (molecules or crystals) to their first excited state. If these conditions do not apply, the neutron may lose energy by exciting the molecule to an excited state, or it may gain energy by de-excitation of a thermally excited level. Furthermore, if the molecule in which the proton is bound is not heavy compared to the proton, the translation of the molecule as a unit and the recoil that the molecule experiences in a collision have an effect on the scattering and on the exchange of energy between the neutron and molecule. Finally, the fact that the molecule contains more than one nucleus which scatters neutrons leads, in general, to complicated interference effects in the scattering.

In the remainder of this chapter we present in considerable detail the physical principles and mathematical techniques which are relevant to an understanding of the phenomena which we have just considered.

In order to establish a fundamental basis for a discussion of the scattering of slow neutrons by chemically bound nuclei, we first review the principal concepts that occur in the scattering of slow neutrons by free nuclei. Our important objective is to determine those properties of nuclear forces which influence the scattering of slow neutrons.

For a given state of spin angular momentum of the neutron-nucleus system, an essentially complete description of the scattering is determined by a knowledge of a single parameter--the Fermi scattering length of the neutron-nucleus system. The spin-dependence of the neutron-nucleus force implies that the scattering length is different for different spin states of the system. We approach the problem by first disregarding the spin-dependence of the nuclear forces. Following this, we introduce the spin-dependent forces and discuss their relevance for the coherence properties of the scattered wave.

2-1 ELEMENTS OF SLOW-NEUTRON SCATTERING THEORY

The wave function Ψ describing the scattering of neutrons which interact with a free nucleus by means of a central potential U(r) satisfies, in center of mass coordinates, the Schrödinger equation

$$-\frac{\hbar^2}{2\mu}\nabla^2\Psi(\mathbf{r}) + U(r)\Psi(\mathbf{r}) = E\Psi(\mathbf{r}), \qquad (2.1)$$

where $\hbar = h/2\pi$, $\mathbf{r}$ is the displacement vector[a] of the neutron relative to the nucleus, and E is the energy in the center-of-mass system. The reduced mass, μ, of the neutron-nucleus system is related to the neutron

[a] In this book, all vectors appear in bold-face type.

mass, m, and the nuclear mass, M, by

$$\mu = \frac{mM}{m + M} . \tag{2.2}$$

The important physical fact that leads to a useful relationship between the scattering cross section and the potential function U(r) is that the range of the neutron-nucleus interaction is extremely small compared with the wave length of slow neutrons. In order to utilize this fact, it is convenient to convert the differential equation (2.1) to an integral equation containing the boundary condition that the solution at large separations approaches the sum of an incident plane wave $\Psi_{in}(\mathbf{r}) = e^{ikz}$ and an outgoing spherical wave. Thus,[5]

$$\Psi(\mathbf{r}) = e^{ikz} - \frac{\mu}{2\pi\hbar^2} \int \frac{e^{ik|\mathbf{r}-\mathbf{r}'|}}{|\mathbf{r} - \mathbf{r}'|} U(r')\Psi(\mathbf{r}')d\mathbf{r}' , \tag{2.3}$$

where the positive z-direction is parallel to the direction of the incident neutron beam. The center-of-mass number k and the corresponding de Broglie wave length λ are given by

$$k = \frac{2\pi}{\lambda} = \frac{(2\mu E)^{\frac{1}{2}}}{\hbar} . \tag{2.4}$$

After noting that for large separations between the neutron and nucleus the factor $\exp(ik|\mathbf{r} - \mathbf{r}'|)/|\mathbf{r} - \mathbf{r}'|$ does not vary appreciably over the range of r' where U(r) is nonvanishing, we may write

$$|\mathbf{r} - \mathbf{r}'| = r\left(1 - \frac{\mathbf{r} \cdot \mathbf{r}'}{r^2} + \ldots\right)$$

and obtain the asymptotic solution

$$\Psi(\mathbf{r}) \approx e^{ikz} - \frac{\mu}{2\pi\hbar^2} \frac{e^{ikr}}{r} \int e^{-i\mathbf{k}\cdot\mathbf{r}'} U(r')\Psi(\mathbf{r}')d\mathbf{r}' , \tag{2.5}$$

where $\mathbf{k} = k(\mathbf{r}/r)$ is the propagation vector of the scattered wave

$$\Psi_{sc}(\mathbf{r}) = -\frac{e^{ikr}}{r} f(\mathbf{k}),$$

and

$$f(\mathbf{k}) = \frac{\mu}{2\pi\hbar^2} \int e^{-i\mathbf{k}\cdot\mathbf{r}'} U(r')\Psi(\mathbf{r}')d\mathbf{r}' \tag{2.6}$$

is the scattering amplitude. The scattering amplitude is simply related to $d\sigma/d\Omega$, the differential cross section for the scattering of neutrons into the solid angle $d\Omega$ about the direction $\mathbf{\Omega} = \mathbf{k}/k$. The quantity $d\sigma/d\Omega$ is defined so that $d\sigma$ is the total flux of neutrons through the surface element of area $(r^2 d\Omega)$, at the position r, divided by the incident neutron flux. This definition[b] can be expressed quantitatively by

$$\frac{d\sigma}{d\Omega} = \frac{\frac{\hbar k}{m} \Psi^*_{sc} \Psi_{sc} r^2}{\frac{\hbar k}{m} \Psi^*_{in} \Psi_{in}} = |f(\mathbf{k})|^2 . \tag{2.7}$$

We are interested in the scattering only for the case $k\rho \ll 1$, i.e., for wave lengths much greater than the range ρ of nuclear forces. In this case, an extremely good approximation is to set k equal to zero in Eq. (2.6). Then, it is immediately clear that $f(\mathbf{k})$ is independent of the angle of scattering and that the scattering is spherically symmetric in the center-of-mass system. In the terminology of the method of partial waves, the only scattering that one obtains in this approximation is s-wave scattering. The characteristic property of s-waves is that they are spherically symmetric or, equivalently, that they correspond to a quantum-mechanical state with a zero value for the orbital angular momentum quantum number ℓ. Furthermore, we note that, since U(r) is spherically symmetric, we could calculate the s-wave scattering correctly by considering only the s-wave component of the incident plane wave. The corrections to the result of Eq. (2.6) arising from the higher terms in the Taylor expansion of $\exp(-i\mathbf{k} \cdot \mathbf{r})$ correspond to states of orbital angular momentum with $\ell \neq 0$. These higher terms are of the order of $k\rho$ compared with the result we obtain by neglecting the exponential factor in Eq. (2.6).

It is a very good approximation to replace $\Psi(\mathbf{r})$ in Eq. (2.6) by $\phi(\mathbf{r})$, its value for the limiting case of zero energy in the center-of-mass system. This follows from Eq. (2.1) and from the smallness of E compared with the average depth, V_0, of the potential well that is represented by U(r). The correction is of the order of E/V_0, which is about 10^{-5} for a kinetic energy of one electron volt. Thus, any results

[b] In Eq. (2.7), and elsewhere in this book, the * denotes the complex conjugate.

on the scattering of slow neutrons by free nuclei which are obtained at zero energy also apply with a high degree of accuracy even at several tens of electron volts.

The value of $f(\mathbf{k})$ for zero energy is a real quantity called the Fermi scattering length, a, of the neutron-nucleus system; i.e.,

$$a = \frac{\mu}{2\pi\hbar^2} \int U(r)\phi(\mathbf{r})d\mathbf{r} \ . \tag{2.8}$$

That a is a real quantity follows from the reality of $U(r)$ and the form of Eq. (2.3) for $k = 0$. A geometrical interpretation of a can be inferred from Eq. (2.5). For zero energy, this equation gives

$$r\Psi \approx r - a \ ,$$

so that a is the radius at which the asymptotic wave function (in this case, the exact s-wave function outside the range of nuclear forces) extrapolates to zero. The magnitude of the Fermi scattering length is fixed by experiment through its relation to the total scattering cross section at zero energy,

$$\sigma_0 = 4\pi a^2 \ , \tag{2.9}$$

as determined by integrating Eq. (2.7) over all solid angles. Some methods for determining the sign of the scattering length are discussed in Sec. 2-1.2.

Following Blatt and Weisskopf,[6] we observe that the sign of the scattering length may be related to the existence or nonexistence of a bound state of the system. Provided that the potential is not very different in strength from one which would give a binding energy of exactly zero, a positive value of a implies the existence of a bound state, while a negative value implies no bound state at all. This is illustrated, in terms of the behavior of the wave function, in Figs. 2.2(a) and 2.2(b), which show typical wave functions for zero energy giving positive and negative scattering lengths, respectively. The behavior of the wave function for $r < \rho$ is to a first approximation independent of the energy E; hence, it is the same for a bound state with a not too large binding energy. By making E slightly negative, the interior part of the wave function of Fig. 2.2(a) is only slightly changed, but in such a way that it can join smoothly to a decreasing exponential in the exterior region. Thus, for a positive value of a, a bound state exists. On the other hand, a wave function with a negative scattering length does not have sufficient curvature in the interior region to join on smoothly to a decreasing exponential in the exterior region, and hence no bound state can exist. If the potential is so strong that there is more than one bound state, the above arguments do not apply and, in general, the scattering length may be either positive or negative.

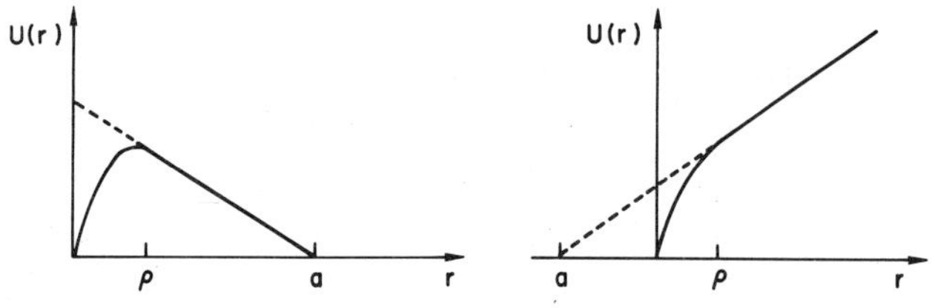

Figure 2.2 (a) A positive scattering length implies that a bound state exists; (b) a negative scattering length implies that the system has no bound state close to zero energy

The potential U(r) uniquely determines the scattering length through Eq. (2.8) and the solution of the Schrödinger equation (2.1). The reverse is not true, however. There are an infinite number of potentials which give the same scattering length. In particular, we would obtain the same scattering length if we replaced the true potential U(r) by the "localized-impact pseudo-potential"

$$V'(r) = \frac{2\pi\hbar^2 a}{\mu}\,\delta(\mathbf{r})\left[\frac{\partial}{\partial r}\,r\right], \tag{2.10}$$

where $\delta(\mathbf{r}) = \delta(x)\delta(y)\delta(z)$, and $\delta(x)$ is the Dirac delta function. With this replacement, the solution of the Schrödinger equation for $k = 0$ becomes

$$\Psi(r) = 1 - \frac{a}{r}, \tag{2.10a}$$

as can be verified by substituting Eqs. (2.10) and (2.10a) into Eq. (2.3). The potential has been chosen so that the wave function at all values of r is replaced by the asymptotic wave function. Since the asymptotic wave function determines the scattering, the wave function

(2.10a) gives the same scattering length as the wave function of the true potential U(r).

This may also be verified by direct substitution of Eqs. (2.10) and (2.10a) into Eq. (2.8), since $(\partial/\partial r)[r(1 - ar^{-1}] = 1$. For scattering by a single particle, the localized-impact pseudo-potential is equivalent to the potential

$$\tilde{V}(r) = \frac{2\pi\hbar^2 a}{\mu}\,\delta(\mathbf{r}) = \frac{2\pi\hbar^2 a_b}{m}\,\delta(\mathbf{r}) \qquad (2.10b)$$

provided that the first Born approximation for $\phi(\mathbf{r})$ is used to calculate a by Eq. (2.8). If higher-order approximations to the wave function are to be used, this equivalence no longer applies, and the full pseudo-potential (2.10) must be used. The pseudo-potential (2.10) has been constructed as that potential with scattering length a for which the first Born approximation is exact. The contribution to the scattering from higher-order perturbations is zero because of the property that $[(\partial/\partial r)r]$ gives zero when operating on the scattered wave. If Eq. (2.10b) were used to calculate beyond the first Born approximation, corrections to the cross section would be obtained, but these corrections would be spurious.

The device of using Eq. (2.10), or equivalently (2.10b) plus the first Born approximation, is quite artificial in the present context. For neutron scattering by a system of chemically bound nuclei, however, the replacement of the true potential by Eq. (2.10) becomes an extremely accurate approximation rather than an exact result. The further step of using (2.10b) and the first Born approximation is no longer equivalent to Eq. (2.10), but constitutes the almost universally used Fermi approximation, to which small corrections have been calculated in a few special cases.[7]

The quantity $a_b = am/\mu$ is called the bound-scattering length. The significance of a_b is determined by noting that

$$\sigma_b = 4\pi a_b^2 \qquad (2.11)$$

is the zero-energy cross section for neutron scattering by a nucleus rigidly bound to a fixed position. We can see this from the arguments presented below.

More generally, we consider the scattering by a nucleus bound in an otherwise neutronically inert molecule of mass $\mathcal{M}$. In this case, the Schrödinger equation describing the relative motion of the neutron-molecule system is given by

$$-\frac{\hbar^2}{2}\frac{m+\mathcal{M}}{m\mathcal{M}}\nabla^2\Psi + U\Psi = E\Psi \ . \tag{2.12}$$

Rewriting Eq. (2.12) in the form

$$-\frac{\hbar^2}{2\mu}\nabla^2\Psi + \frac{1}{\mu}\frac{m\mathcal{M}}{m+\mathcal{M}}U\Psi = \frac{1}{\mu}\frac{m\mathcal{M}}{m+\mathcal{M}}E\Psi \ , \tag{2.12a}$$

we see that $(m\mathcal{M}/\mu)(m+\mathcal{M})^{-1}U(r)$ is an effective potential for scattering by the nucleus bound in the molecule. Introducing this effective potential in place of U(r) in Eq. (2.8) and using Eq. (2.9), the zero-energy scattering cross section is found to be

$$\sigma_0' = 4\pi\left(\frac{\mathcal{M}}{M}\frac{m+M}{m+\mathcal{M}}\right)^2 a^2 \ , \tag{2.13}$$

where, again, M is the mass of the scattering nucleus. In the limit of an infinitely heavy molecule, Eq. (2.13) becomes the bound-atom cross section

$$\sigma_b = 4\pi\left(\frac{M+m}{M}\right)^2 a^2 = 4\pi a_b^2 \ . \tag{2.14}$$

For the case of neutron scattering by a proton bound in an infinitely heavy molecule, Eq. (2.14) gives $\sigma_b = 4\sigma_0$. Now we understand the origin of Fermi's factor of four which entered the discussion at the beginning of this chapter.

2-1.1 THE EFFECTS OF SPIN. That nuclear forces are spin-dependent is most directly seen by comparing the properties of the deuteron with the low-energy cross section for neutron-proton scattering. A theoretical estimate of the low-energy cross section for neutron-proton scattering can be made by relating the scattering length to the known properties of the ground state of the deuteron.[6] The estimate for the cross section that one obtains by this procedure is

$$\sigma = 2.4 \text{ barns.}$$

This result is in violent disagreement with the experimental result of 20.36 barns. The experiment was performed by measuring the scattering at neutron energies between 1 and 10 eV, where molecular binding would have an insignificant effect. In this energy range, the discrepancy is much larger than the inaccuracies involved in going from Eq. (2.1) to Eq. (2.9).

This discrepancy was cleared up by Wigner, who noted that there is a significant difference between the ground state of the deuteron and the scattering states of the neutron-proton system. Experimentally, the total angular momentum of the deuteron in the ground state is unity. It is predominantly a state of zero orbital angular momentum ($\ell = 0$), and the spins of the neutron and proton are in a triplet spin state (their total spin angular momentum is unity). In a scattering experiment, the situation is different. A beam of neutrons is sent against an assembly of stationary protons. No attempt is made to order either the spin directions of the neutrons in the incident beam or the spin directions of the protons in the scatterer; therefore, the spins are uncorrelated in a scattering experiment.

On the average, a pair of uncorrelated particles, each with a spin of one-half, are found in the triplet state with a probability of three-fourths and in the singlet state with a probability of one-fourth. Since the nuclear forces depend on the spin orientation of the neutron-proton system, the ground state of the deuteron provides information only about the scattering of particles in the triplet state, and not at all about the scattering in the singlet state.

Now we shall expand the mathematical formalism so that proper account is taken of the spin-dependence of the nuclear forces. We consider the case of slow-neutron scattering by a nucleus of spin I. For this system, since $\ell = 0$, there are only two possible definite values for the total angular-momentum quantum number J; namely, $J = I \pm \frac{1}{2}$. The neutron-nucleus interaction is different for each of these states. To account for this difference, we introduce the operator[4]

$$\underset{\sim}{a} = \frac{I + 1 + \sigma \cdot I}{2I + 1} a_+ + \frac{I - \sigma \cdot I}{2I + 1} a_- \quad . \tag{2.15}$$

The quantities $(\hbar/2)\sigma$ and $\hbar I$ are the angular-momentum operators of the neutron and nucleus, respectively, and a_+ and a_- are the Fermi scattering lengths corresponding to the angular-momentum states with quantum numbers $J = I \pm \frac{1}{2}$, respectively. The useful property of $\tilde{\mathbf{a}}$ is that it reduces to a_+ or a_- according as $J = I + \frac{1}{2}$ or $J = I - \frac{1}{2}$. This is easily proved by noting that

$$\begin{aligned} \sigma \cdot I &= (I + \tfrac{1}{2}\sigma)^2 - I^2 - \frac{1}{4}\sigma^2 \\ &= J(J+1) - I(I+1) - \frac{3}{4} . \end{aligned} \tag{2.16}$$

Thus,

$$\boldsymbol{\sigma} \cdot \mathbf{I} = I \qquad J = I + \tfrac{1}{2} \tag{2.17}$$

$$\boldsymbol{\sigma} \cdot \mathbf{I} = -(I+1) \qquad J = I - \tfrac{1}{2} . \tag{2.18}$$

We may regard $\tilde{\mathbf{a}}$ as an operator which operates on the spin state of the incident neutron and nucleus to yield the spin state of the system after scattering. Thus, if the initial spin state corresponding to the incident plane wave and target nucleus is $|\chi_i\rangle$, then the scattered wave is given by

$$\Psi_{sc} = \frac{e^{ikr}}{r} \tilde{\mathbf{a}} |\chi_i\rangle . \tag{2.19}$$

2-1.2 THE COHERENT AND INCOHERENT CROSS SECTIONS. The differential cross section in the center-of-mass system corresponding to the transition from an initial spin state $|\chi_i\rangle$ to a final spin state $|\chi_f\rangle$ is given[c] by

---

[c] It is interesting to observe that part of the operator $\tilde{\mathbf{a}}$, namely,

$$(\sigma_x I_x + \sigma_y I_y) \frac{a_+ - a_-}{(2I+1)}$$

induces transitions in which the magnetic quantum numbers of the neutron and nucleus are changed by one unit. If

$$|\chi_i\rangle = |I, \sigma, M_I, M_\sigma\rangle ,$$

that is, if the initial state has well defined values, M_I and M_σ, of the magnetic quantum numbers of the nucleus and neutron, respectively,

$$\frac{d\sigma_{if}}{d\Omega} = \left|\langle\chi_f|\,\tilde{\mathbf{a}}\,|\chi_i\rangle\right|^2 . \tag{2.20}$$

Since we are interested in experiments that do not distinguish the final spin orientations of the particles, we must sum over final spin states to obtain the total cross section corresponding to a given spin before scattering. Thus,

$$\frac{d\sigma_i}{d\Omega} = \sum_f \frac{d\sigma_{if}}{d\Omega} = \sum_f \langle\chi_i|\,\tilde{\mathbf{a}}\,|\chi_f\rangle\langle\chi_f|\,\tilde{\mathbf{a}}\,|\chi_i\rangle$$

$$= \langle\chi_i|\,\tilde{\mathbf{a}}^2\,|\chi_i\rangle . \tag{2.21}$$

In addition, when the incident neutron beam and target are unpolarized, it is necessary to average the above result over the $2(2I + 1)$ equally probable initial spin states of the entire system. Performing this averaging process, and utilizing the facts that $\sum_i \langle\chi_i|\boldsymbol{\sigma}\cdot\mathbf{I}|\chi_i\rangle = 0$ and that $\sum_i \langle\chi_i|(\boldsymbol{\sigma}\cdot\mathbf{I})^2|\chi_i\rangle = 2I(I + 1)(2I + 1)$,[d] we find

$$\frac{d\sigma}{d\Omega} = \frac{1}{2(2I + 1)} \sum_i \langle\chi_i|\,\tilde{\mathbf{a}}^2\,|\chi_i\rangle = a_c^2 + a_{inc}^2 , \tag{2.22}$$

where

---

then the only final states $|\chi_f\rangle$ for which $\langle\chi_f|\sigma_x I_x + \sigma_y I_y|\chi_i\rangle$ can be nonvanishing are $|I,\sigma, M_I \pm 1, M_\sigma \pm 1\rangle$. Thus, the part of $\tilde{\mathbf{a}}$ that we are concerned with here causes transitions in which the spin flips. On the other hand, the remaining part of $\tilde{\mathbf{a}}$, namely,

$$\frac{I + 1}{2I + 1} a_+ + \frac{I}{2I + 1} a_- + \frac{\sigma_z I_z}{2I + 1} (a_+ - a_-)$$

has nonvanishing matrix elements only between identical initial and final states.

[d] To prove these results, we utilize a representation for which, in addition to I and σ, J and M_J are good quantum numbers. Here, the magnetic quantum number takes $2J + 1$ values between J and $-J$, while J takes the values $I \pm \frac{1}{2}$. We thus have

$$a_c = \frac{I + 1}{2I + 1} a_+ + \frac{I}{2I + 1} a_- \tag{2.23}$$

and

$$a_{inc} = \frac{[I(I + 1)]^{\frac{1}{2}}}{2I + 1} (a_+ - a_-) \tag{2.24}$$

are the coherent and incoherent scattering lengths, respectively. The separation of the scattering cross section of a single nucleus into coherent and incoherent parts is motivated by considerations of the scattering by an assembly of nuclei with randomly oriented spins (discussed in Sec. 2-1.3). In a sample for which the nuclear spins are randomly oriented, the coherent scattering cross section

$$\sigma_c = 4\pi a_c^2 \tag{2.25}$$

corresponds to the cross section for the type of scattering which produces waves capable of interfering with waves scattered from other nuclei in the sample. The incoherent cross section

$$\sigma_{inc} = 4\pi a_{inc}^2 \tag{2.26}$$

corresponds, on the other hand, to waves which cannot give rise to interference with waves scattered by other nuclei.

Another essentially different type of incoherent scattering, called strictly incoherent, occurs when a change of the internal state of a single scatterer is involved in the collision process. In such a

---

$$\sum_i \langle \chi_i | \sigma \cdot \mathbf{I} | \chi_i \rangle = \sum_{J,M_J} \langle \chi_{J,M_J} | \sigma \cdot \mathbf{I} | \chi_{J,M_J} \rangle$$

$$= \sum_{M_J=-(I+\frac{1}{2})}^{I+\frac{1}{2}} \langle \chi_{J+\frac{1}{2},M_J} | \sigma \cdot \mathbf{I} | \chi_{I+\frac{1}{2},M_J} \rangle$$

$$+ \sum_{M_J=-(I-\frac{1}{2})}^{I+\frac{1}{2}} \langle \chi_{I-\frac{1}{2},M_J} | \sigma \cdot \mathbf{I} | \tilde{\chi}_{I-\frac{1}{2},M_J} \rangle$$

$$= I(2I + 2) - (I + 1)2I = 0,$$

process, the scattered wave cannot interfere with the incident wave. An example of such a process occurs in neutron scattering when the scattering nucleus flips its spin in the course of a collision. Another example of strictly incoherent scattering is provided by the inelastic scattering of neutrons by a single scatterer. From this, however, we must not be led to conclude that interference effects are absent in inelastic scattering. In the scattering of neutrons by a crystal, although the inelastically scattered neutron wave cannot interfere with the incident wave, there can still be interference between waves of the same frequency originating from different scattering centers.

The refraction of neutrons is the result of interference between the incident and scattered waves, and hence is an example of a strictly coherent phenomenon. The measurement of the index of neutron refraction of a medium constitutes a method for determining the coherent scattering length of nuclei. This method has been used to fix the absolute signs of the scattering lengths of some nuclei and to obtain the coherent cross section of the proton. The discussion below is taken from Sachs' book.[8]

The most practical experimental method for obtaining the index of refraction is to determine the angle, θ_c, of critical reflection from a plane surface of material. If the angle θ_c is measured from the reflecting surface, then to an extremely good approximation for neutrons of wave number k

$$\theta_c = \frac{2}{k}\left(\frac{\pi \mathfrak{N} m a_c}{\mu}\right)^{\frac{1}{2}},$$

where $\mathfrak{N}$ is the number of nuclei per unit volume. This result follows from Snell's law and the very accurate

$$1 - \frac{2\pi \mathfrak{N} m a_c}{\mu k^2}$$

---

the last line following on using Eqs. (2.17) and (2.18). Similarly, we deduce that

$$\sum_i \langle \chi_i | (\sigma \cdot \mathbf{I})^2 | \chi_i \rangle = I^2(2I + 2) + (I + 1)^2 2I$$

$$= 2I(I + 1)(2I + 1) .$$

for the index of refraction ν. Critical reflection from the surface will occur only if ν is less than 1, and thus only for a positive coherent scattering length. Determination of the mere existence of critical reflection is a means of establishing that a scattering length is positive. This was used by Fermi and Zinn[9] to fix the signs of the coherent scattering lengths of several elements. Their results can be combined with the measurements of relative sign by crystal diffraction to obtain the signs for many elements.[10]

When the scattering material is made up of a mixture of two nuclear species with coherent scattering lengths $a_c^{(1)}$ and $a_c^{(2)}$, the critical angle is

$$\theta_c = \frac{2}{k}\left[\pi m\left(\frac{\mathcal{n}_1 a_c^{(1)}}{\mu_1} + \frac{\mathcal{n}_2 a_c^{(2)}}{\mu_2}\right)\right]^{1/2},$$

where $\mathcal{n}_1$ and $\mathcal{n}_2$ are the numbers per unit volume of each nuclear type, and μ_1 and μ_2 are the reduced masses of the system consisting of the neutron and each type of nucleus. This result provides the basis for a method that has been applied by Hughes, Burgy, and Ringo[11] to obtain an accurate value of the coherent scattering length of the proton in terms of that of carbon. They compare the critical angles at a given neutron energy for mirrors consisting of various liquid hydrocarbons. Since the proton and carbon have scattering lengths of opposite sign, the hydrocarbons can be chosen to have an average coherent scattering length near zero. The square of θ_c is then a linear function of the carbon-to-hydrogen ratio, and only a slight extrapolation of the observed θ_c is required to determine the ratio corresponding to $\theta_c = 0$. This gives directly and accurately the ratio of the proton scattering length to that of carbon. The latter value is known from the scattering cross section of the spin-zero isotope C^{12}, which, from Eq. (2.24), produces no spin incoherent scattering. When the results of the reflection experiment are combined with the value of the neutron-proton scattering cross section, the triplet and singlet scattering lengths of the proton are found to be

$$a_+ = 0.538 \times 10^{-12}\ \text{cm}$$

$$a_- = -2.37 \times 10^{-12}\ \text{cm}.$$

This completes the discussion of the concepts of strict coherence and strict incoherence. Henceforth, the concept of coherence will be used as the property of a nucleus to scatter neutron waves which can interfere with waves scattered from other nuclei rather than with the incident radiation.

2-1.3 SPIN INCOHERENCE. Since slow neutrons, like X-rays, have wave lengths comparable to atomic dimensions, diffraction effects are expected when neutrons fall on an assembly of nuclei. For the present, attention is restricted to the case of neutron scattering by nuclei which are fixed at definite positions. This relatively simple example is considered since it provides the required understanding of the properties of the scattered neutrons. In particular, we are led to a better understanding of the concepts of spin-coherent and spin-incoherent scattering, and the distinction between them. At the same time, we are brought to a position from which we can comment on the scattering of neutrons by hydrogen molecules and the manner in which this experiment was used to determine the Fermi scattering length for the neutron-proton singlet state.

Consider a system of N infinitely heavy nuclei which are fixed at positions $\mathbf{R}_1, \mathbf{R}_2, \ldots, \mathbf{R}_N$. Each nucleus is assumed to have a spin I, and the scattering lengths corresponding to the neutron-nucleus states with spins $I + \frac{1}{2}$ and $I - \frac{1}{2}$ are denoted by a_+ and a_-, respectively. When a monoenergetic incident wave hits the system, the total scattered wave is the sum of the waves arising separately from the individual scattering centers. Allowing for the spin, the scattered wave is

$$\Psi_{sc} = \sum_{n=1}^{N} \frac{e^{ik|\mathbf{r}-\mathbf{R}_n|}}{|\mathbf{r} - \mathbf{R}_n|}\left(\frac{I + 1 + \boldsymbol{\sigma} \cdot \mathbf{I}_n}{2I + 1}\, a_+ + \frac{I - \boldsymbol{\sigma} \cdot \mathbf{I}_n}{2I + 1}\, a_-\right) e^{i\mathbf{k}_0 \cdot \mathbf{R}_n} |\chi_\iota\rangle\,, \tag{2.27}$$

where $|\chi_\iota\rangle$ is the initial spin state of the entire system of neutrons and nuclei, $\mathbf{I}_n$ is the spin operator of the *nth* nucleus, and the amplitude of the incident neutron wave at $\mathbf{R}_n$ is $\exp(i\mathbf{k}_0 \cdot \mathbf{R}_n)$. At positions $\mathbf{r}$ which are far from any of the scattering centers, this becomes

$$\Psi_{sc} = \frac{e^{ikr}}{r}\sum_n e^{i\kappa\cdot R_n}\left(\frac{I + 1 + \sigma \cdot I_n}{2I + 1} a_+ + \frac{I - \sigma \cdot I_n}{2I + 1} a_-\right)|\chi_\iota> . \quad (2.28)$$

where $\kappa = k_0 - k$.

The differential cross section for the case of a given initial spin state $|\chi_\iota>$ is given by

$$\frac{d\sigma_\iota}{\Omega} = \sum_f \left|<\chi_f| \sum_n e^{i\kappa\cdot R_n}\left[a_c + \frac{\sigma \cdot I_n}{2I + 1}(a_+ - a_-)\right]|\chi_\iota>\right|^2$$

$$= <\chi_\iota|\left|\sum_n e^{i\kappa\cdot R_n}\left[a_c + \frac{\sigma \cdot I_n}{2I + 1}(a_+ - a_-)\right]\right|^2|\chi_\iota> . \quad (2.29)$$

Since the neutron beam is assumed to be unpolarized, it is necessary to average $d\sigma_\iota/d\Omega$ over initial states of the neutron spin. Further, if the nuclear spins are randomly oriented and if there is no spin-orientation correlation between nuclei at different positions, then all of the cross terms which appear as a result of having performed the squaring operation indicated in Eq. (2.29) vanish upon averaging over the initial spin states of the scattering nuclei. Performing the indicated averaging and remembering that all of the initial spin states of the system are equally probable, gives

$$\frac{d\sigma}{d\Omega} = a_c^2\left|\sum_n e^{i\kappa\cdot R_n}\right|^2 + Na_{inc}^2 . \quad (2.30)$$

The scattered intensity, which is proportional to the result obtained by summing $|\Psi_{sc}|^2$ over final spin states and averaging over initial spin states is proportional to $d\sigma/d\Omega$. The scattered intensity of neutrons, as well as the scattering cross section, separate naturally into a coherent part and an incoherent part. The coherent part proportional to a_c^2 exhibits the effects of the interference between the waves scattered from the different scattering centers. For neutrons of infinite wave length, with $\kappa = 0$, the coherent scattered intensity takes the maximum possible value $N^2a_c^2$. This is expected, since the condition of complete constructive interference, namely, that there is no difference in phase between the waves scattered from different centers, is satisfied for $\mathbf{k} = 0$. The incoherent part is proportional to the number of particles and is obtained by adding the incoherent contributions from the individual scattering centers.

In general, the neutron scattering from any material consists both of coherent and incoherent components; however, there are important cases where the magnitude of one of these components is negligibly small relative to the other. It is clear that a material consisting of a single spin-zero isotope scatters in a completely coherent manner, since σ_{inc} vanishes. The isotopes C^{12} and O^{16} are spin-zero nuclei which are almost 100% abundant in the naturally occurring elements. Thus, samples of graphite or of oxygen gas provide examples for which the neutron scattering is almost completely coherent.

Hydrogen and vanadium provide examples of materials which scatter in an almost completely incoherent way. Naturally occurring hydrogen and vanadium consist almost exclusively of the isotopes H^1 and V^{51}, respectively. For these isotopes, the scattering lengths a_+ and a_- are of opposite sign and have a magnitude such that a_c given by Eq. (2.23) is very small compared with a_{inc}. More generally, spin incoherence can exist only for nuclei which contain an odd number of either protons or neutrons, since I is zero for stable even-even nuclei.

2-1.4 SPIN CORRELATIONS. It is the property of the separation of the scattering from a many-body system into a coherent and an incoherent part that motivates the separation of the cross section for scattering from a single nucleus into coherent and incoherent parts in the manner indicated in Eq. (2.22). In order to accomplish this separation in the scattering by many body systems, we required that there exist no correlation between the spin orientation of nuclei at different positions. Practical conditions exist, however, for which this assumption is not valid. These conditions occur for molecules which contain identical nuclei and whose wave functions are therefore subject to symmetry requirements which imply nuclear spin correlations.

An important example where the requirements imposed by symmetry are significant is the case of slow-neutron scattering by molecular hydrogen at low temperatures. In order to understand how the requirements of symmetry affect the scattering, we first note that the states of the H_2 molecule are divided into two groups, those for which the proton spins form a triplet (orthohydrogen) and those for which they form a singlet (parahydrogen).

In practice, it is possible to obtain an essentially pure parahydrogen sample by cooling H_2 to low temperatures. For values of k_BT which are small compared with the energy difference between the lowest two rotational states of the H_2 molecule, the statistical equilibrium mixture of ortho- and parahydrogen consists predominantly of parahydrogen. To understand this phenomenon, one must consider the role played by symmetry. Because of the Pauli principle, the total wave function of H_2 must be antisymmetric with respect to interchange of the two hydrogen atoms. At low temperatures, the only vibrational and electronic state of the molecule which is necessary to consider is the ground state, for which the electronic and vibrational wave functions are symmetric. Since the para- and orthohydrogen spin states are antisymmetric and symmetric, respectively, the rotational wave function must be symmetric for parahydrogen and antisymmetric for orthohydrogen. Furthermore, since the ground rotational state is the symmetric one of zero angular momentum, H_2 in statistical equilibrium at low temperatures consists predominantly of parahydrogen.[e]

If for the moment we ignore the fact that the neutron can induce transitions between the ortho- and parahydrogen states, then the cross section for scattering by ortho- or parahydrogen can be deduced from Eq. (2.29), wherein the initial state $|\chi_i\rangle$ is a product of a neutron spin function and the spin function $\chi(J)$ of the two protons in the molecule. The summation in Eq. (2.29) is over all final states which are consistent with a given spin angular momentum J of the molecule. The result of Eq. (2.29) still has to be averaged over the initial spin states of the neutron to give the cross section for scattering of an unpolarized neutron beam. Thus, we must evaluate

---

[e] Since the ortho-para transition rate is extremely slow a catalyst is often used to achieve the low-temperature statistical equilibrium distribution.

$$\frac{d\sigma}{d\Omega} = \frac{16}{9} \langle\chi(J)| \left\langle \left| \sum_{n=1}^{2} e^{i\kappa\cdot\mathbf{R}_n} \times \right.\right.$$

$$\left.\left. \times \left[a_c + \frac{\sigma \cdot \mathbf{I}_n}{2I+1}(a_+ - a_-) \right] \right|^2 \right\rangle_\nu |\chi(J)\rangle , \qquad (2.31)$$

where $\langle\ldots\rangle_\nu$ denotes an average over spin states of the unpolarized neutron beam. The factor 16/9 appears in accordance with Eq. (2.13) because of the binding of the protons in a molecule of mass $\mathcal{M} = 2m = 2M$. Upon performing the squaring operation indicated in Eq. (2.31), we find that we need to evaluate $\langle \sigma \cdot \mathbf{I}_n \rangle_\nu$, $\langle (\sigma \cdot \mathbf{I}_n)^2 \rangle_\nu$, and $\langle (\sigma \cdot \mathbf{I}_1) \times (\sigma \cdot \mathbf{I}_2) \rangle_\nu$. From the well-known properties of the Pauli spin operators,[4] it follows that $\langle \sigma \cdot \mathbf{I}_n \rangle_\nu = 0$, and $\langle (\sigma \cdot \mathbf{I}_n)^2 \rangle_\nu = I(I+1)$.

Finally, using the relationship

$$\langle [\sigma \cdot (\mathbf{I}_1 + \mathbf{I}_2)]^2 \rangle_\nu = \langle (\sigma \cdot \mathbf{J})^2 \rangle_\nu = J(J+1)$$

$$= 2I(I+1) + 2(\sigma \cdot \mathbf{I}_1)(\sigma \cdot \mathbf{I}_2) ,$$

we obtain

$$\frac{d\sigma}{d\Omega} = \frac{16}{9} a_c^2 \left| \sum_{n=1}^{2} e^{ik\cdot\mathbf{R}_n} \right|^2 + \frac{32}{9} a_{inc}^2 \times$$

$$\times \left\{ 1 + \left[\frac{J(J+1)}{2I(I+1)} - 1 \right] \cos \kappa \cdot (\mathbf{R}_2 - \mathbf{R}_1) \right\} \qquad (2.32)$$

for the differential scattering cross section per hydrogen molecule. Hence, when spin correlations are present there is a further contribution to the cross section. This takes the form of an additional interference effect between the waves scattered by identical nuclei, and contributes to $d\sigma/d\Omega$ the term

$$\frac{32}{9} a_{inc}^2 \left[\frac{J(J+1)}{2I(I+1)} - 1 \right] \cos \kappa \cdot (\mathbf{R}_2 - \mathbf{R}_1) ,$$

which depends on the incoherent rather than the coherent scattering amplitude. For the case of scattering of zero-energy neutrons by para- $(J = 0)$ or ortho- $(J = 1)$ hydrogen, Eq. (2.32) yields, after setting $I = \frac{1}{2}$ and integrating over solid angles

$$\sigma_{para} = \frac{16}{9}\pi(a_- + 3a_+)^2 \tag{2.33}$$

$$\sigma_{ortho} = \frac{16}{9}\pi[(a_- + 3a_+)^2 + 2(a_+ - a_-)^2] . \tag{2.34}$$

Teller[12] pointed out that a measurement of the para- and orthohydrogen cross sections allows a definite determination of the sign of a_-/a_+. The value of $a_+ = 0.526 \times 10^{-12}$ cm has been estimated from the known properties of the ground state of the deuteron. From measurements in the energy region above a few electron volts, where interference, chemical binding, and temperature effects are completely insignificant, the experimental cross section for scattering by a single proton is

$$\sigma = \pi(3a_+^2 + a_-^2) = 20.36 \text{ barns.}$$

From the preceding information we can estimate that $a_- \approx \pm 4a_+$. Since the ratio $|a_-/a_+|$ is so close to 3, the ratio $\sigma_{ortho}/\sigma_{para}$ is nearly equal to or much larger than unity, according as a_- is positive or negative. Measurements of the kind first made by Sutton et al.[13] give $\sigma_{ortho}/\sigma_{para} \sim 30$, thus demonstrating that a_- must be negative. The best values of a_- and a_+ are given by[4]

$$a_- = -2.38 \times 10^{-12} \text{ cm}$$

$$a_+ = 0.526 \times 10^{-12} \text{ cm.}$$

These values are not as accurate as the corresponding values obtained by neutron reflection from liquid mirrors (see Sec. 2-1.2).

A precise theoretical explanation of the orthohydrogen cross section must take into account an effect which we have so far neglected. The physical origin of this effect is that the lowest energy state of orthohydrogen is higher than that of parahydrogen. Therefore, it is energetically possible for the neutron to gain energy from a process in which ortho- is converted to parahydrogen. The operator affecting this transition is given by Eq. (2.28) as

$$\sum_{n=1}^{2} e^{i\kappa\cdot R_n}\left(\frac{I + 1 + \sigma \cdot I_n}{2I + 1} a_+ + \frac{I - \sigma \cdot I_n}{2I + 1} a_-\right). \tag{2.35}$$

For a neutron wave length which is long compared with the separation of the protons in H_2, the expansion, to the first order, of Eq. (2.35) in powers of κ gives

$$\sum_{n=1}^{2}\left(\frac{I + 1 + \sigma \cdot I_n}{2I + 1} a_+ + \frac{I - \sigma \cdot I_n}{2I + 1} a_-\right)$$

$$+ \frac{i\kappa \cdot b}{2} \frac{\sigma \cdot (I_1 - I_2)}{2I + 1} (a_+ - a_-) , \tag{2.36}$$

where **b** is the displacement between protons. In writing the above result, we have taken the origin of coordinates to be at the center of mass of the H_2 molecule. In the zero-energy limit, the only significant term is the one involving the sum over n. This term, however, does not produce transitions between the ortho- and parahydrogen states, because the rotational states involved in the transition are states of opposite parity. For shorter neutron wave lengths, however, the term proportional to $\kappa\cdot b$ in (2.36) gives a nonzero probability for the ortho-para transition in which the neutron gains energy.

2-1.5 ISOTOPIC INCOHERENCE. There remains one type of incoherence, called isotopic incoherence, which we have yet to consider. To illustrate isotopic incoherence, let us consider the hypothetical case of neutron scattering by a sample consisting of N spin-zero isotopes of a single element. The isotopes with scattering lengths a_i are present with the fractional abundances c_i which satisfy the condition $\sum_i c_i = 1$. Since the isotopes are not distributed regularly, the sample cannot scatter neutrons in a completely coherent way. The disorder introduced by the random distribution of the isotopes gives rise to another scattering of incoherent type. The scattering cross section is given by

$$\frac{d\sigma}{d\Omega} = \left\langle \left| \sum_{n=1}^{N} a^{(n)} e^{i\kappa\cdot R_n} \right|^2 \right\rangle_{is} , \tag{2.37}$$

where $a^{(n)}$ is the scattering length of the nucleus at $\mathbf{R}_n$ and the average, denoted by $\langle \ldots \rangle_{is}$, is to be taken over the distribution of isotopes in the scattering sample. Before performing the indicated averaging process, it is convenient to rewrite Eq. (2.37) in the form

$$\frac{d\sigma}{d\Omega} = \sum_{n=1}^{N} \left\langle \left[a^{(n)}\right]^2 \right\rangle_{is} + \sum_{n,n'=1}^{N}{}' \left\langle a^{(n)} a^{(n')} \right\rangle_{is} e^{i\kappa\cdot(\mathbf{R}_n - \mathbf{R}_{n'})},$$

where the prime on the summation sign indicates the omission of terms with n' equal to n. Let $P[a^{(n)}a^{(n')} = a_i a_j]$ for i,j = 1,2 be the probability that $a^{(n)}a^{(n')}$ is $a_i a_j$. Then,

$$P\left[a^{(n)}a^{(n')} = a_i^2\right] = c_i^2 ,$$

$$P\left[a^{(n)}a^{(n')} = a_i a_j\right] = 2c_i c_j \qquad i \neq j .$$

Using these results in Eq. (2.37), we find that

$$\frac{d\sigma}{d\Omega} = N \sum_{\iota} c_{\iota} a_{\iota}^2 + \left(\sum_{\iota} c_{\iota} a_{\iota}\right)^2 \sum_{n,n'=1}^{N}{}' e^{i\kappa\cdot(\mathbf{R}_n - \mathbf{R}_{n'})} .$$

Since

$$\sum_{n,n'=1}^{N} e^{i\kappa\cdot(\mathbf{R}_n - \mathbf{R}_{n'})} = \left|\sum_{n=1}^{N} e^{i\kappa\cdot\mathbf{R}_n}\right|^2 - N ,$$

we find that

$$\frac{d\sigma}{d\Omega} = \frac{\sigma_c}{4\pi} \left|\sum_{n=1}^{N} e^{i\kappa\cdot\mathbf{R}_n}\right|^2 + N\frac{\sigma_{inc}}{4\pi} , \tag{2.38}$$

where, in the present case,

$$\sigma_c = 4\pi\left(\sum_i c_i a_i\right)^2 = 4\pi\langle a\rangle_{is}^2 , \tag{2.39}$$

$$\sigma_{inc} = 4\pi\left[\sum_i c_i a_i^2 - \left(\sum c_i a_i\right)^2\right] = 4\pi\left(\langle a^2\rangle_{is} - \langle a\rangle_{is}^2\right). \tag{2.40}$$

As in the case of the scattering of neutrons by a system of nuclei with randomly oriented spins, it is convenient to associate with each nucleus a coherent and an incoherent scattering cross section.

In the more general case of scattering by a sample consisting of a single atomic species, but with a completely disordered distribution of spins and isotopic species, one obtains

$$\sigma_c = 4\pi\langle a\rangle^2_{sp,is} \tag{2.41}$$

and

$$\sigma_{inc} = 4\pi\left(\langle a^2\rangle_{sp,is} - \langle a\rangle^2_{sp,is}\right), \tag{2.42}$$

where now the symbol $\langle\ldots\rangle_{sp,is}$ denotes an average over the distribution of spin orientations and isotopes. (The reader who is not interested in the derivation of this result should skip to Sec. 2-2.) If the relative abundance of the *ith* isotopic component with spin I_i and scattering lengths $a_{+,i}$ and $a_{-,i}$ is $c_i\left(\sum_i c_i = 1\right)$, then

$$\langle a\rangle_{sp,is} = \sum_i c_i\left(\frac{I_i + 1}{2I_i + 1}\, a_{+,i} + \frac{I_i}{2I_i + 1}\, a_{-,i}\right) \tag{2.43}$$

$$\langle a^2\rangle_{sp,is} = \sum_i c_i\left(\frac{I_i + 1}{2I_i + 1}\, a^2_{+,i} + \frac{I_i}{2I_i + 1}\, a^2_{-,i}\right). \tag{2.44}$$

To obtain Eqs. (2.41) and (2.42), we observe that

$$\frac{d\sigma}{d\Omega} = \left\langle <\chi_\iota\left|\sum_{n=1}^{N} \tilde{a}_n e^{i\boldsymbol{\kappa}\cdot\mathbf{R}_n}\right|^2 |\chi_\iota>\right\rangle_{sp,is} =$$

$$= \left\langle <\chi_\iota\left|\sum_{n=1}^{N} e^{i\boldsymbol{\kappa}\cdot\mathbf{R}_n}\left[a_c^{(n)} + \frac{\boldsymbol{\sigma}\cdot\mathbf{I}_n}{2I_n + 1}\left(a_+^{(n)} - a_-^{(n)}\right)\right]\right|^2 |\chi_\iota>\right\rangle_{sp,is} \tag{2.45}$$

We explicitly perform the squaring operations indicated in Eq. (2.45) and carry out the average over spin orientations. Neglecting spin correlations and remembering that spin-orientation averages of terms of the form $\langle(\sigma\cdot\mathbf{I}_n)(\sigma\cdot\mathbf{I}_{n'})\rangle_{sp}$ vanish for $n \neq n'$, we see that

$$\frac{d\sigma}{d\Omega} = \left\langle \left| \sum_{n=1}^{N} e^{i\boldsymbol{\kappa}\cdot\mathbf{R}_n} a_c^{(n)} \right|^2 \right\rangle_{is} +$$

$$+ \sum_{n=1}^{N} \left\langle \frac{I_n(I_n + 1)}{(2I_n + 1)^2} \left[a_+^{(n)} - a_-^{(n)} \right]^2 \right\rangle_{is} , \qquad (2.46)$$

where

$$a_c^{(n)} = \frac{I_n + 1}{2I_n + 1} a_+^{(n)} + \frac{I_n}{2I_n + 1} a_-^{(n)} ,$$

and hence

$$\frac{d\sigma}{d\Omega} = \sum_{n,n'=1}^{N} e^{i\boldsymbol{\kappa}\cdot(\mathbf{R}_n - \mathbf{R}_{n'})} \left\langle a_c^{(n)} a_c^{(n')} \right\rangle_{is} +$$

$$+ \sum_{n=1}^{N} \left\langle \frac{I_n(I_n + 1)}{(2I_n + 1)^2} \left(a_+^{(n)} - a_-^{(n)} \right)^2 \right\rangle_{is} . \qquad (2.47)$$

The first term on the right-hand side of Eq. (2.47) may be written in the form

$$\sum_{n=1}^{N} \left\langle \left(a_c^{(n)} \right)^2 \right\rangle_{is} + \sum_{n,n'=1}^{N}{}' e^{i\boldsymbol{\kappa}\cdot(\mathbf{R}_n - \mathbf{R}_{n'})} \left\langle a_c^{(n)} a_c^{(n')} \right\rangle_{is} ,$$

Now,

$$P\left[a_c^{(n)} a_c^{(n')} = a_{c,i}^2 \right] = c_i^2 ,$$

$$P\left[a_c^{(n)} a_c^{(n')} = a_{c,i} a_{c,i'} \right] = 2c_i c_{i'} , \qquad i \neq i .$$

Thus,

$$\sum_{n,n'} e^{i\boldsymbol{\kappa}\cdot(\mathbf{R}_n - \mathbf{R}_{n'})} \left\langle a_c^{(n)} a_c^{(n')} \right\rangle_{is}$$

$$= N \sum_{\iota} c_{\iota} a_{c,\iota}^2 + \sum_{n,n'}{}' e^{i\boldsymbol{\kappa}\cdot(\mathbf{R}_n - \mathbf{R}_{n'})} \left(\sum_{\iota} c_{\iota} a_{c,\iota} \right)^2 =$$

$$= \left(\sum_l c_l a_{c,l} \right)^2 \sum_{n,n'} e^{i\boldsymbol{\kappa}\cdot(\mathbf{R}_n - \mathbf{R}_{n'})} +$$

$$+ N\left[\sum_l c_l a_{c,l}^2 - \left(\sum_l c_l a_{c,l} \right)^2 \right].$$

Substituting into Eq. (2.47),

$$\frac{d\sigma}{d\Omega} = \left(\sum_l c_l a_{c,l} \right)^2 \sum_{n,n'=1}^{N} e^{i\boldsymbol{\kappa}\cdot(\mathbf{R}_n - \mathbf{R}_{n'})} + N\left[\sum_l c_l a_{c,l}^2 - \right.$$

$$\left. - \left(\sum_l c_l a_{c,l} \right)^2 + \sum_l c_l \frac{I_l(I_l + 1)}{(2I_l + 1)^2} (a_{+,l} - a_{-,l})^2 \right]$$

$$= \langle a \rangle_{sp,is}^2 \sum_{n,n'=1}^{N} e^{i\boldsymbol{\kappa}\cdot(\mathbf{R}_n - \mathbf{R}_{n'})} +$$

$$+ N\left(\langle a^2 \rangle_{sp,is} - \langle a \rangle_{sp,is}^2 \right), \qquad (2.48)$$

where $\langle a \rangle_{sp,is}$ and $\langle a^2 \rangle_{sp,is}$ are defined by Eqs. (2.43) and (2.44). From Eq. (2.48), the validity of Eqs. (2.41) and (2.42) follows immediately.

2-2 SLOW-NEUTRON SCATTERING BY SYSTEMS IN THERMODYNAMIC EQUILIBRIUM

Despite the great strength of the interaction between slow neutrons and nuclei, the range of the nuclear forces is so short that the perturbation of a neutron collision on the atomic states of the scattering system is small. Fermi[3] was the first to point out that it is in fact possible to replace the true potential between neutron and nucleus by the auxiliary potential of Eq. (2.10) and then to calculate the scattering in the first Born approximation. The use of the first Born approximation is an essential simplification since it gives the scattering in terms of a dynamical response function $S(\boldsymbol{\kappa},\omega)$ which does not depend on the neutron variables but only on the dynamics of the nuclear motion in the scattering system. In this section we develop the formalism for slow-neutron scattering by a many-particle system and derive the dynamical response functions $S(\boldsymbol{\kappa},\omega)$ which determine the desired cross section.

The scattering system may be a molecule, a crystal, or a fluid. We refer to it as a molecule in the general discussion. We assume a molecular Hamiltonian

$$H_0 = -\sum_{\ell=1}^{N} \frac{\hbar^2}{2M_\ell} \nabla_\ell^2 + V(\mathbf{r}_1,\mathbf{r}_2,\ldots,\mathbf{r}_N) , \tag{2.49}$$

where M_ℓ is the mass of the ℓth nucleus and $\mathbf{r}_\ell$ is its position. The interatomic potential V is the change in the electronic energy with a change in the nuclear coordinates. We do not explicitly consider the electronic coordinates, but assume that the potential V is known. The assumption that V depends only on the nuclear coordinates is the Born-Oppenheimer, or adiabatic, approximation. It corresponds to assuming that the electronic wave functions follow the changes in nuclear position so rapidly that the appropriate Hamiltonian for the dynamical problem is the same as for a static distortion of the molecule. This assumption is basic to almost all molecular and lattice dynamics, and is expected to be accurately applicable to the problems considered in this book. We emphasize that the adiabatic approximation pertains to the molecular dynamics. The scattering formalism which we develop below is independent of this approximation and applies even when the molecular Hamiltonian cannot be expressed in the form (2.49). It does, however, depend on the assumption that the molecular Hamiltonian has a complete orthonormal set of eigenstates $u_s(\mathbf{r}_1,\ldots,\mathbf{r}_N)$ satisfying

$$H_0 u_s(\mathbf{r}_1,\ldots,\mathbf{r}_N) = \epsilon_s u_s(\mathbf{r}_1,\ldots,\mathbf{r}_N) .$$

We consider the collision of a neutron of momentum $\hbar\mathbf{k}_0$ with the molecule in an initial state u_n of energy ϵ_n. For simplicity, the spin-dependence of the neutron-nucleus interaction is neglected. After deriving the results for this case, we make the generalization to systems with spin-dependent forces. The wave equation for the system of interest is

$$\left[-\frac{\hbar^2}{2m}\nabla_\nu^2 + H_0 + \sum_{\ell=1}^{N} U_\ell(|\mathbf{r}_\nu - \mathbf{r}_\ell|) \right] \Psi = \epsilon\Psi , \tag{2.50}$$

where $\mathbf{r}_\nu$ is the position vector of the neutron, H_0 is the molecular Hamiltonian,

$$\epsilon = \frac{\hbar^2 k_0^2}{2m} + \epsilon_n ,$$

and $U_\ell(|\mathbf{r}_\nu - \mathbf{r}_\ell|)$ is the interaction energy between the neutron and the ℓth nucleus. The range of U_ℓ is not only small compared with the wave length of the incident neutron, but is also small compared with the amplitude of the nuclear vibrations.

We begin by converting the partial differential equation (2.50) into an integral equation with the boundary condition that the wave function, at large separations between the neutron and the scatterer, consists of an incident plane wave and outgoing spherical waves. More explicitly, for large separations we require

$$\psi(\mathbf{r}_\nu, \mathbf{r}_1, \ldots, \mathbf{r}_N) = e^{i\mathbf{k}_0 \cdot \mathbf{r}_\nu} u_n(\mathbf{r}_1, \ldots, \mathbf{r}_N) - \sum_s f_{ns} \frac{e^{ik_s r_\nu}}{r_\nu} u_s(\mathbf{r}_1, \ldots, \mathbf{r}_N) . \tag{2.51}$$

Here f_{ns} is the scattering amplitude for a process in which the molecule makes a transition to a state u_s with energy ϵ_s. The wave number k_s of the scattered neutron is determined by

$$\frac{\hbar^2}{2m}(k_0^2 - k_s^2) = \epsilon_s - \epsilon_n \tag{2.52}$$

and is thus dependent on s in Eq. (2.51).

The derivation of the integral equation commences by expanding the scattered wave in the complete set of eigenfunctions $u_s(\mathbf{r}_1, \ldots, \mathbf{r}_N)$ belonging to the Hamiltonian H_0:

$$-e^{i\mathbf{k}_0\cdot\mathbf{r}_\nu}u_n(\mathbf{r}_1,\ldots,\mathbf{r}_N) + \Psi(\mathbf{r}_\nu,\mathbf{r}_1,\ldots,\mathbf{r}_N)$$

$$= \sum_s \phi_s(\mathbf{r}_\nu)u_s(\mathbf{r}_1,\ldots,\mathbf{r}_N) \ . \qquad (2.53)$$

The sum in Eq. (2.53) is interpreted to extend over continuum, as well as over discrete eigenstates. From Eq. (2.49) and the orthogonality of the states u_s, there follows

$$\nabla_\nu^2\phi_s(\mathbf{r}_\nu) + k_s^2\phi_s(\mathbf{r}_\nu) = \frac{2m}{\hbar^2}\int d\mathbf{r}_1\ldots\int d\mathbf{r}_N u_s^*(\mathbf{r}_1,\ldots,\mathbf{r}_N) \times$$

$$\times \sum_{\ell=1}^{N} U_\ell(|\mathbf{r}_\nu - \mathbf{r}_\ell|)\Psi(\mathbf{r}_\nu,\mathbf{r}_1,\ldots,\mathbf{r}_N) \ . \qquad (2.54)$$

The solution of this equation corresponding to outgoing neutron waves is

$$\phi_s(\mathbf{r}_\nu) = -\frac{2m}{\hbar^2}\frac{1}{4\pi}\int d\mathbf{r}'_\nu \frac{e^{ik_s|\mathbf{r}_\nu-\mathbf{r}'_\nu|}}{|\mathbf{r}_\nu - \mathbf{r}'_\nu|}\int d\mathbf{r}_1\ldots\int d\mathbf{r}_N u_s^*(\mathbf{r}_1,\ldots,\mathbf{r}_N) \times$$

$$\times \sum_{\ell=1}^{N} U_\ell(|\mathbf{r}'_\nu - \mathbf{r}_\ell|)\Psi(\mathbf{r}'_\nu,\mathbf{r}_1,\ldots,\mathbf{r}_N) \ , \qquad (2.55)$$

which, for $\mathbf{r}_\nu$ much greater than the dimensions of the scatterer, becomes

$$\phi_s(\mathbf{r}_\nu) = -\frac{m}{2\pi\hbar^2}\frac{e^{ik_s r_\nu}}{r_\nu}\int d\mathbf{r}'_\nu \times$$

$$\times \sum_{\ell=1}^{N}\int d\mathbf{r}_1\ldots\int d\mathbf{r}_N e^{-i\mathbf{k}_s\cdot\mathbf{r}_\ell}u_s^*(\mathbf{r}_1,\ldots,\mathbf{r}_N) \times$$

$$\times\ U(|\mathbf{r}'_\nu - \mathbf{r}_\ell|)\Psi(\mathbf{r}'_\nu,\mathbf{r}_1,\ldots,\mathbf{r}_N) \ . \qquad (2.56)$$

Here we have utilized the short range of the nuclear potential U by replacing $\mathbf{r}'_\nu$ by $\mathbf{r}_\ell$ in the factor $\exp[(ik_s|\mathbf{r}_\nu - \mathbf{r}'_\nu|)]/(|\mathbf{r}_\nu - \mathbf{r}'_\nu|)$. The coefficient of $-\exp(ik_s r_\nu)/r_\nu$ is the scattering amplitude f_{ns}; that is

$$f_{ns}(\mathbf{k}_0,\mathbf{k}_s) = \frac{m}{2\pi\hbar^2}\int d\mathbf{r}_\nu \times$$

$$\times \sum_{\ell=1}^{N}\int d\mathbf{r}_1 \ldots \int d\mathbf{r}_N e^{-i\mathbf{k}_s\cdot\mathbf{r}_\ell} u_s^*(\mathbf{r}_1,\ldots,\mathbf{r}_N) \times \tag{2.57}$$

$$\times\, U_\ell(|\mathbf{r}_\nu - \mathbf{r}_\ell|)\Psi(\mathbf{r}_\nu,\mathbf{r}_1,\ldots,\mathbf{r}_N) \;.$$

The differential scattering cross section of the molecule for the process in which the molecule goes from the state u_n to the state u_s is

$$\frac{d\sigma_{n,s}}{d\Omega}(\mathbf{k}_0,\mathbf{k}_s) = \frac{k_s\phi_s^*\phi_s r^2}{k_0\left|e^{i\mathbf{k}_0\,\mathbf{r}_\nu}\right|^2} = \frac{k_s}{k_0}\,|f_{ns}(\mathbf{k}_0,\mathbf{k}_s)|^2 \;. \tag{2.58}$$

That is, $d\sigma_{n,s}(\mathbf{k}_0,\mathbf{k}_s)$ is the flux of neutrons with the wave vector $\mathbf{k}_s$ through the surface element $r^2 d\Omega$ at $\mathbf{r}$, divided by the incident flux.

The derivation of Eqs. (2.57) and (2.58) is exact, but the calculation of f_{ns} requires a knowledge of the wave function Ψ of the coupled neutron-molecule system. The problem is greatly simplified by first replacing the neutron-nucleus interaction U_ℓ by the localized-impact pseudo-potential of Eq. (2.10) and then calculating in first Born approximation. For a single free particle, both steps in this procedure are exact. In the present case they are both approximate, but very accurate.

Consider first the replacement of U_ℓ of the pseudo-potential of Eq. (2.10). This is a very good approximation because of the great strength and short range of the nuclear force. Inside the range of the nuclear force, the effects of the interatomic potential are negligible, so that the dependence of the wave function on the neutron-nucleus separation $r = |\mathbf{r}_\nu - \mathbf{r}_\ell|$ is the same as for a free nucleus. Furthermore, the range of the nuclear force is so short that this dependence can be replaced by its asymptotic behavior (2.10a).

For a molecule in which only one nucleus scatters neutrons, the scattering due to the pseudo-potential (2.10) is again exactly calculable in first Born approximation, as in the free-atom case. The operator $[(\partial/\partial r)r]$ in Eq. (2.10) selects only the incident plane wave part of the wave function. If more than one nucleus scatters, however, an exact calculation using Eq. (2.10) gives corrections of order a^2 and higher to the scattering amplitude. These corrections arise from the rescattering from one nucleus of a wave originally scattered by another nucleus. The wave scattered by one nucleus has a finite amplitude at the position of another nucleus, and the $[(\partial/\partial r)r]$ term in Eq. (2.10) does not give a null result when operating on these terms.

Since the scattered wave amplitude is of order (a/d) at the internuclear separation d, we expect the relative magnitude of the rescattering corrections to the scattered amplitude to be of this order. Typically, this is a correction of order 10^{-4} to the scattering amplitude and to the cross section. Neglecting these corrections, we use, for the remainder of our discussion, the very accurate Fermi approximation of replacing $U_\ell(\mathbf{r}_\nu - \mathbf{r}_\ell)$ in Eq. (2.57) by

$$\tilde{V}_\ell(|\mathbf{r}_\nu - \mathbf{r}_\ell|) = \frac{2\pi\hbar^2}{m} a_b^\ell \delta(\mathbf{r}_\nu - \mathbf{r}_\ell) \tag{2.59}$$

and then using the first Born approximation.

In Eq. (2.59), a_b^ℓ is the bound-atom scattering length of the ℓth nucleus. The Born approximation is realized by replacing the exact wave function $(\mathbf{r}_\nu,\mathbf{r}_1,\ldots,\mathbf{r}_N)$ in Eq. (2.57) by the wave function

$$\exp(i\mathbf{k}_0 \cdot \mathbf{r}_\nu)u_n(\mathbf{r}_1,\ldots,\mathbf{r}_N)$$

for the incident channel. The scattering amplitude in this approximation is given by

$$f_{ns}(\mathbf{k}_0,\mathbf{k}_s) = \sum_{\ell=1}^{N} a_b^\ell \int d\mathbf{r}_1 \cdots \int d\mathbf{r}_N u_s^*(\mathbf{r}_1\ldots,\mathbf{r}_N) e^{i\boldsymbol{\kappa}_s\cdot\mathbf{r}_\ell} u_n(\mathbf{r}_1,\ldots,\mathbf{r}_N)$$

$$= \sum_{\ell=1}^{N} a_b^\ell \langle s|e^{i\boldsymbol{\kappa}_s\cdot\mathbf{r}_\ell}|n\rangle , \tag{2.60}$$

where $\hbar\boldsymbol{\kappa}_s = \hbar\mathbf{k}_0 - \hbar\mathbf{k}_s$ is the momentum transferred to the molecule in the process under consideration. It is a characteristic feature of a Born approximation scattering amplitude that it depends only on the momentum transfer in a collision and not separately on the initial and final momenta.

Except at absolute zero temperature the molecule is not initially in a definite quantum state. The experimentally measurable differential cross section is an average over a thermal equilibrium distribution of initial state of the molecule. The probability that the molecule will be in the state u_n is given by

$$P_n = [Z(\beta)]^{-1} e^{-\beta\epsilon_n} ,$$

where $\beta = (k_B T)^{-1}$ and

$$Z(\beta) = \sum_n e^{-\beta\epsilon_n} \tag{2.61}$$

is the partition function of the molecule in the canonical ensemble. This is a general result of quantum statistical mechanics. It applies for free or interacting particles and applies whether the particles obey Fermi or Bose statistics. The statistics of the particles enter only in the enumeration of the states n of the N-particle system.

The quantity of experimental interest is the double differential cross section $d^2\sigma/dEd\Omega$ for a neutron of initial energy

$$E_0 = \frac{\hbar^2 k_0^2}{2m} \tag{2.62a}$$

to scatter into the solid angle $d\Omega$ and the energy interval dE at final energy

$$E = \frac{\hbar^2 k^2}{2m} . \tag{2.62b}$$

The momentum transfer is fixed at the value

$$\hbar\boldsymbol{\kappa}_s = \hbar\boldsymbol{\kappa} = \hbar(\mathbf{k}_0 - \mathbf{k}) .$$

The cross section for this process is obtained by summing Eq. (2.58) over all final states u_s consistent with the conservation of energy, and averaging over a thermal-equilibrium distribution of initial states. Using Eq. (2.60) this gives for the cross section per atom

$$\frac{d^2\sigma}{dEd\Omega}(\mathbf{k}_0,\mathbf{k}) = \frac{k}{Nk_0}[Z(\beta)]^{-1}\sum_{n,s} e^{-\beta\epsilon_n}\left|\sum_{\ell=1}^{N} a_b^{\ell}\left\langle s\left|e^{i\boldsymbol{\kappa}\cdot\mathbf{r}_\ell}\right|n\right\rangle\right|^2 \times$$

$$\times\ \delta(E_0 + \epsilon_n - E - \epsilon_s)\ . \tag{2.63a}$$

Let us now consider the case where all the atoms of the scatterer belong to a single atomic species. For this case, we wish to generalize Eq. (2.63) to a result that applies when the scatterer consists of a completely disordered distribution of nuclear spins and isotopic species. The generalization to systems with different atomic species is straightforward.

To obtain the cross section for the scattering of neutrons, we require

$$\left\langle\left|\sum_{\ell=1}^{N}\tilde{a}_b^{\ell}\left\langle s\left|e^{i\boldsymbol{\kappa}\cdot\mathbf{r}_\ell}\right|n\right\rangle\right|^2\right\rangle_{sp,is}, \tag{2.63b}$$

where $\tilde{a}_b^{\ell}$ is to be regarded as a spin-dependent operator of the kind introduced in Eq. (2.15). The average is to be taken over the spin-orientation and isotopic distributions of the scatterer. The problem of averaging in (2.63a) is precisely the same problem of averaging that occurs in Eq. (2.45), if we assume in (2.63a) that the matrix elements $\langle s|\exp(i\boldsymbol{\kappa}\cdot\mathbf{r}_\ell)|n\rangle$ do not vary from one isotope to another.[f] Consequently, we may use Eqs. (2.41) and (2.42) to immediately generalize Eq. (2.63) to the case of scattering by a monatomic system with a completely disordered arrangement of nuclear spins and isotopic species. The result is conveniently written as

$$\frac{d^2\sigma}{dEd\Omega}(\mathbf{k}_0,\mathbf{k}) = \frac{k}{4\pi\hbar k_0}\left[\sigma_{c,b}S_c(\boldsymbol{\kappa},\omega) + \sigma_{inc,b}S_{inc}(\boldsymbol{\kappa},\omega)\right], \tag{2.64}$$

where the scattering functions S_c and S_{inc} are given by

---

[f] This assumption is extremely good in practice. Naturally occurring samples of light elements are very nearly monoisotopic. For heavy elements, the variation of $\langle s|e^{i\boldsymbol{\kappa}\cdot\mathbf{r}_\ell}|n\rangle$ from one isotope to the next is small.

$$S_c(\kappa,\omega) = \frac{[Z(\beta)]^{-1}}{N} \sum_{n,s} e^{-\beta\epsilon_n} \left| \sum_{\ell=1}^{N} s|e^{i\kappa\cdot r_\ell}|n \right|^2 \times$$

$$\times\ \delta\left(\omega + \frac{\epsilon_n - \epsilon_s}{\hbar}\right), \tag{2.65}$$

$$S_{inc}(\kappa,\omega) = \frac{[Z(\beta)]^{-1}}{N} \sum_{n,s} e^{-\beta\epsilon_n} \sum_{\ell=1}^{N} \left| \langle s|e^{i\kappa\cdot r_\ell}|n\rangle \right|^2 \times$$

$$\times\ \delta\left(\omega + \frac{\epsilon_n - \epsilon_s}{\hbar}\right), \tag{2.66}$$

In Eq. (2.64) $\omega = (E_0 - E)/\hbar$, and $\sigma_{c,b}$ and $\sigma_{inc,b}$ are the bound coherent and incoherent cross sections, respectively.

In writing Eqs. (2.64), (2.65), and (2.66), we have explicitly ignored spin-correlation effects. In almost all applications, these effects are either absent or of little importance. An important exception occurs in the case of slow-neutron scattering by liquid hydrogen;[14] however, this case is not considered beyond the approximate treatment given in Sec. 2-1.4.

In complete analogy to the case of scattering by static point nuclei treated in Sec. 2-1.5, we have expressed the scattering cross section as the sum of a coherent and an incoherent part:

$$\frac{d^2\sigma}{dEd\Omega} = \frac{d^2\sigma_c}{dEd\Omega} + \frac{d^2\sigma_{inc}}{dEd\Omega}, \tag{2.67}$$

where

$$\frac{d^2\sigma_c}{dEd\Omega} = \frac{k}{4\pi\hbar k_0}\,\sigma_{c,b}S_c(\kappa,\omega), \tag{2.68}$$

and

$$\frac{d^2\sigma_{inc}}{dEd\Omega} = \frac{k}{4\pi\hbar k_0}\,\sigma_{inc,b}S_{inc}(\kappa,\omega). \tag{2.69}$$

The incoherent scattering is obtained by adding the scattering contributions from individual atoms, whereas the coherent part contains not only interference effects between waves scattered by different atoms but also direct terms. It is frequently convenient to separate the direct terms from the interference terms according to

$$\frac{d^2\sigma}{dEd\Omega} = \frac{k}{4\pi\hbar k_0}\{(\sigma_{c,b} + \sigma_{inc,b})S_{inc}(\boldsymbol{\kappa},\omega) + $$

$$+ \sigma_{c,b}[S_c(\boldsymbol{\kappa},\omega) - S_{inc}(\boldsymbol{\kappa},\omega)]\} . \tag{2.70}$$

This separation is important for many applications, but for now we are more interested in the separation of the scattering into coherent and incoherent parts.

The functions $S_c(\boldsymbol{\kappa},\omega)$ and $S_{inc}(\boldsymbol{\kappa},\omega)$ defined by Eqs. (2.65) and (2.66) are the dynamical response functions mentioned at the beginning of this section. These functions depend only on the properties of the "molecule" and not on the particular probe which interacts weakly with the molecule. The fact that one is dealing with neutron scattering appears only in the bound-atom cross sections and in the connection of $\boldsymbol{\kappa}$ and ω with the kinematics of neutron scattering. Except for an atomic form factor $|F(\kappa)|^2$ associated with the longer-range interaction, the function $S_c(\boldsymbol{\kappa},\omega)$ is the same as that which describes the scattering of electrons or X-rays in the Born approximation.[g]

In this book we normally deal with an average over orientations of randomly oriented molecules or microcrystals. After averaging over orientation, the functions $S(\boldsymbol{\kappa},\omega)$ depend only on

$$\left.\begin{aligned} \omega &= (E_0 - E)/\hbar \\ \text{and}\quad \kappa^2 &= \frac{2m}{\hbar^2}[E_0 + E - 2(EE_0)^{\frac{1}{2}}\cos\theta] , \end{aligned}\right\} \tag{2.71}$$

where θ is the angle of scattering in the laboratory coordinate system. In this case the double differential cross section is a function only of E_0, E, and θ and is denoted by $\sigma(E_0,E,\theta)$. In this and the following chapter we are concerned with the calculation and measurement of $\sigma(E_0,E,\theta)$ as it relates to the dynamics of the scattering system through

[g] There is no spin or isotope incoherence in X-ray scattering. The equivalence of neutron and X-ray scattering applies only to the scattering in which no electronic excitation occurs.

$S(\boldsymbol{\kappa},\omega)$. In later chapter, we use this cross section and other quantities derived from it. Some of the quantities of interest are the energy-transfer cross section.

$$\sigma(E_0,E) = 2\pi \int_{-1}^{1} \sigma(E_0,E,\theta)d(\cos\theta) , \tag{2.72}$$

the differential cross section

$$\frac{d\sigma(E_0)}{d\Omega} = \int_0^{\infty} \sigma(E_0,E,\theta)dE , \tag{2.72a}$$

the total scattering cross section

$$\sigma_s(E_0) = \int_0^{\infty} \sigma(E_0,E)dE = 2\pi \int_{-1}^{1} \frac{d\sigma(E_0)}{d\Omega} d(\cos\theta) , \tag{2.72b}$$

and the transport cross section

$$\sigma_{tr}(E_0) = 2\pi \int_{-1}^{1} \frac{d\sigma(E_0)}{d\Omega} (1 - \cos\theta)d(\cos\theta) . \tag{2.72c}$$

2-3 SOME PROPERTIES OF THE SCATTERING FUNCTIONS

The scattering function $S(\boldsymbol{\kappa},\omega)$ depends in general on both the structure and dynamics of the scattering system.[h] This dependence enters through the appearance of the matrix elements of $\exp(i\boldsymbol{\kappa}\cdot\mathbf{r}_\ell)$ between the energy eigenstates of the scattering system and through the appearance of the energy eigenvalues ϵ_s in the expressions defining $S(\boldsymbol{\kappa},\omega)$. The scattering function $S(\boldsymbol{\kappa},\omega)$ has some general properties, however, which are independent of the structure and dynamics of the scattering system. The first property of interest is the detailed

---

[h] In the remainder of this chapter, the omission of the subscript c or inc implies that the statement is true for both the coherent and incoherent scattering functions.

balance condition

$$S(-\boldsymbol{\kappa},-\omega) = e^{-\beta\hbar\omega} S(\boldsymbol{\kappa},\omega) \; . \tag{2.73}$$

In a general context, the property of detailed balance is a consequence of the principle of microscopic reversibility and of the thermal equilibrium of the scattering system.[i] In the present context, we may derive it in a straightforward manner. For the coherent scattering function,

$$S_c(-\boldsymbol{\kappa},-\omega) = \frac{[Z(\beta)]^{-1}}{N} \sum_{n,s} e^{-\beta\epsilon_n} \left| \left\langle s \right| \sum_{\ell=1}^{N} e^{-i\boldsymbol{\kappa}\cdot\mathbf{r}_\ell} \left| n \right\rangle \right|^2 \times$$

$$\times\, \delta\left(-\omega + \frac{\epsilon_n - \epsilon_s}{\hbar}\right) . \tag{2.74}$$

Using the result

$$\left| \left\langle s \right| \sum_{\ell=1}^{N} e^{-i\boldsymbol{\kappa}\cdot\mathbf{r}_\ell} \left| n \right\rangle \right|^2 = \left| \left\langle n \right| \sum_{\ell=1}^{N} e^{i\boldsymbol{\kappa}\cdot\mathbf{r}_\ell} \left| s \right\rangle \right|^2 \tag{2.75}$$

and interchanging the summation indices in Eq. (2.74) yields

$$S_c(-\boldsymbol{\kappa},-\omega) = \frac{[Z(\beta)]^{-1}}{N} \sum_{n,s} e^{-\beta\epsilon_s} \left| \left\langle s \right| \sum_{\ell} e^{i\boldsymbol{\kappa}\cdot\mathbf{r}_\ell} \left| n \right\rangle \right|^2 \times$$

$$\times\, \delta\left(\omega + \frac{\epsilon_n - \epsilon_s}{\hbar}\right) .$$

Because of the δ-function, ϵ_s is related to ${}_n$ by

$$\epsilon_s = \epsilon_n + \hbar\omega \; .$$

---

[i] When the scattering is not described by the Born approximation we

When the scattering is not described by the Born approximation we must use microscopic reversibility to relate the appropriate transition probability to its inverse (after summing over spin states). This is discussed in Appendix A of the article by H. Hurwitz, Jr., M. S. Nelkin, and G

Thus, we finally obtain

$$S_c(-\boldsymbol{\kappa},-\omega) = e^{-\beta\hbar\omega}\frac{[Z(\beta)]^{-1}}{N}\sum_{n,s} e^{-\beta\epsilon_n}\left|\left\langle s\right|\sum_{\ell=1}^{N} e^{i\boldsymbol{\kappa}\cdot\mathbf{r}}\left|n\right\rangle\right|^2 \times$$

$$\times\ \delta\left(\omega + \frac{\epsilon_n - \epsilon_s}{\hbar}\right) = e^{-\beta\hbar\omega}S_c(\boldsymbol{\kappa},\omega) \ .$$

The same method of derivation yields the equivalent result for $S_{inc}(-\boldsymbol{\kappa},-\omega)$. The principle of microscopic reversibility entered the present derivation as Eq. (2.75). The thermal equilibrium of the scattering system is explicit in the use of the Boltzmann factors $e^{-\beta\epsilon_n}$ in performing the average over the initial states of the scattering system.

The detailed balance condition allows one to define a symmetric function:

$$S_0(\boldsymbol{\kappa},\omega) = e^{-\frac{1}{2}\beta\hbar\omega}S(\boldsymbol{\kappa},\omega) = S_0(-\boldsymbol{\kappa},-\omega) \ . \tag{2.76}$$

Furthermore, if the scattering system contains an ensemble of randomly oriented subsystems (molecules or crystallites), the neutron scattering depends only on the magnitude of the momentum transfer. Then the scattering can be described in terms of the symmetric function

$$S_0(\kappa^2,\omega) = e^{-\frac{1}{2}\beta\hbar\omega}\left\langle S(\boldsymbol{\kappa},\omega)\right\rangle_R = S_0(\kappa^2,-\omega) \ , \tag{2.77}$$

where the symbol $\langle\ldots\rangle_R$ denotes an average over the random orientations of the subsystems.

For an ensemble of randomly oriented subsystems the quantity

$$\frac{d^2\sigma}{dEd\Omega}\,(\mathbf{k}_0,\mathbf{k})$$

depends only on the initial and final energies and the angle of scattering θ of the neutron. Thus, we may replace Eq. (2.64) by

$$\sigma(E_0,E,\theta) = \frac{k}{4\pi\hbar k_0}\,e^{\frac{1}{2}\beta\hbar\omega}[\sigma_{c,b}S_{c,0}(\kappa^2,\omega) +$$

$$+\ \sigma_{inc,b}S_{inc,0}(\kappa^2,\omega)] \ . \tag{2.78}$$

The use of the property (2.77) for the analysis of scattering data has been emphasized by Egelstaff[15] and is discussed further in Chap. 3.

The condition of detailed balance is extremely important for problems of neutron thermalization. The relation (2.77) between the scattering function for positive and negative ω implies the following constraint on the differential energy transfer cross section for neutrons:

$$\phi_M(E_0)\sigma(E_0,E,\theta) = \phi_M(E)\sigma(E,E_0,\theta) \ , \tag{2.79}$$

where $\phi_M(E) = \beta^2 E e^{-\beta E}$ is the Maxwellian flux distribution function. The implications of this relation for neutron thermalization are considered in subsequent chapters.

Two other general properties of interest are the sum rule

$$\int_{-\infty}^{\infty} \omega \left\langle S(\boldsymbol{\kappa},\omega)\right\rangle_R d\omega = \frac{\hbar\kappa^2}{2} M^{-1} \tag{2.80}$$

and the following result for the zero*th* moment of the incoherent scattering function:

$$\int_{-\infty}^{\infty} S_{inc}(\boldsymbol{\kappa},\omega)d\omega = 1 \ . \tag{2.80a}$$

The validity of Eq. (2.80a) does not depend on an average of $S_{inc}(\boldsymbol{\kappa},\omega)$ over orientations of the scattering system. It follows on integrating Eq. (2.66) over ω and using the completeness and normalization of the states $|s\rangle$ to obtain

$$\sum_s \sum_{\ell=1}^{N} \left|\left\langle s\right| e^{i\boldsymbol{\kappa}\cdot\mathbf{r}_\ell}\left|n\right\rangle\right|^2 = \sum_{\ell=1}^{N}\sum_s \left\langle n\right| e^{i\boldsymbol{\kappa}\cdot\mathbf{r}_\ell}\left|s\right\rangle\left\langle s\right| e^{i\boldsymbol{\kappa}\cdot\mathbf{r}_\ell}\left|n\right\rangle$$

$$= \sum_{\ell=1}^{N} \left\langle n|n\right\rangle = N \ .$$

We may define *nth*-order moments of the scattering function by

$$\int_{-\infty}^{\infty} \omega^n \left\langle S(\boldsymbol{\kappa},\omega)\right\rangle_R d\omega \ .$$

In contrast to the first moment, and the zeroth moment of the incoherent scattering function, all other moments depend explicitly on the dynamics and structure of the scattering system. Specific expressions for some of these moments are derived in Sec. 2-8. Equation (2.80), which holds for both the coherent and incoherent scattering functions, was first derived by Placzek,[16] and will be derived here after the following general comments are made.

The sum rule (2.80) does not depend on the thermal equilibrium of the scattering system since, as we can deduce from the following derivation, it holds for any initial state. Its applicability to neutron-scattering cross sections does depend, however, on the validity of the Born approximation result, Eq. (2.64), and the velocity independence of interatomic forces. The detailed balance condition (2.79), on the other hand, is valid independent of the Born approximation but depends on the thermal equilibrium of the system. To prove these last two statements requires a more complicated analysis than we present here.

The general results (2.64), (2.79), (2.80), and (2.80a) serve as internal-consistency criteria on experimental data. The most generally valid is Eq. (2.79), which states that the cross section $(E,E_0, \;)$ is simply related to the inverse cross section $\sigma(E_0,E,\theta)$ through

$$\sigma(E,E_0,\theta) = \sigma(E_0,E,\theta)(E_0/E)\exp[\beta(E - E_0)] \;.$$

It is thus a matter of convenience whether one measures a cross section by energy-gain or energy-loss scattering.

The Born approximation result (2.64) is more accurate than any current experimental work. The number of variables on which $(E_0,E, \;)$ depends nontrivially is thus reduced to two: κ^2 as given by Eq. (2.71) and $\omega = (E_0 - E)\hbar^{-1}$. This reduction along with (2.79) leads to the symmetric function (2.77) as a logical method for the presentation of experimental results. The internal-consistency criteria among data obtained at various incident energies and scattering angles are usefully employed in this method of data presentation.

The sum rule (2.80) depends further on the assumption of velocity-independence for the potential energy of the scattering system. This is not a serious limitation, but the sum rule is nevertheless not a very useful internal-consistency criterion for experimental data. This is true primarily since the full range of ω contributing to the integral in Eq. (2.80) is not accessible experimentally. The sum rule is, however, an important constraint on any theoretical model. The sum rule (2.80) is closely related to the fact that the scattering approaches free-atom slowing down at high energies. This property is clearly required of any model to be used in thermalization calculations, but is also important in connection with an understanding of the molecular dynamics. A particular example where this sum rule is a useful check on theoretical models is that of liquid helium at very low temperatures. Similar sum rules occur in many other problems in physics where the dynamical response of a system to a weakly perturbing probe is considered.

We now proceed to a proof of Eq. (2.80). The reader who is not interested in the mathematical details of the theory may wish to go on to Sec. 2-4. To prove Eq. (2.80), we first multiply both sides of Eq. (2.65) by ω, average over the orientations of the subsystems, and integrate over all ω. We obtain

$$\int_{-\infty}^{\infty} \omega \left\langle S_c(\boldsymbol{\kappa},\omega) \right\rangle_R d\omega = \frac{[Z(\beta)]^{-1}}{N} \sum_{n,s} e^{-\beta\epsilon_n} \sum_{\ell,j=1}^{N} \frac{(\epsilon_s - \epsilon_n)}{\hbar} \times$$

$$\times \left\langle \left\langle n \middle| e^{-i\boldsymbol{\kappa}\cdot\mathbf{r}_j} \middle| s \right\rangle \left\langle s \middle| e^{i\boldsymbol{\kappa}\cdot\mathbf{r}_\ell} \middle| n \right\rangle \right\rangle_R =$$

$$= - \frac{[Z(\beta)]^{-1}}{N\hbar} \sum_{n} e^{-\beta\epsilon_n} \left\langle \sum_{\ell,j=1}^{N} \sum_{s} \times \right.$$

$$\times \left(\left\langle n \middle| e^{-i\boldsymbol{\kappa}\cdot\mathbf{r}_j} \middle| s \right\rangle \left\langle s \middle| e^{i\boldsymbol{\kappa}\cdot\mathbf{r}_\ell} H_0 \middle| n \right\rangle - \right.$$

$$\left. \left. - \left\langle n \middle| e^{-i\boldsymbol{\kappa}\cdot\mathbf{r}_j} H_0 \middle| s \right\rangle \left\langle s \middle| e^{i\boldsymbol{\kappa}\cdot\mathbf{r}_\ell} \middle| n \right\rangle \right) \right\rangle_R$$

$$= - \frac{[Z(\beta)]^{-1}}{N\hbar} \sum_{n} e^{-\beta\epsilon_n} \sum_{\ell,j=1}^{N} \times$$

$$\times \left\langle \left\langle n \middle| e^{-i\boldsymbol{\kappa}\cdot\mathbf{r}_j} \left[e^{i\boldsymbol{\kappa}\cdot\mathbf{r}_\ell}, H_0 \right] \middle| n \right\rangle \right\rangle_R . \qquad (2.81)$$

In Eq. (2.81), we have utilized the fact that $H_0|s\rangle = \epsilon_s|s\rangle$; that is, that the states $|s\rangle$ are eigenstates of the Hamiltonian H_0 of the scattering system. To obtain the final expression on the right-hand side of Eq. (2.81), we have used the completeness of the states $|s\rangle$ and introduced the commutator operator

$$[A,B] = AB - BA .$$

For a velocity independent potential, we now find that

$$e^{-i\boldsymbol{\kappa}\cdot\mathbf{r}_\ell}\left[e^{i\boldsymbol{\kappa}\cdot\mathbf{r}_\ell},H_0\right] = H_0 - e^{-i\boldsymbol{\kappa}\cdot\mathbf{r}_\ell}H_0 e^{i\boldsymbol{\kappa}\cdot\mathbf{r}_\ell} = -\frac{\hbar^2\kappa^2}{2M} - \frac{\hbar\boldsymbol{\kappa}\cdot\mathbf{p}_\ell}{M} . \qquad (2.82)$$

Here, $\mathbf{p}_\ell$ denotes the momentum operator $-i\hbar\nabla_\ell$ conjugate to the displacement $\mathbf{r}_\ell$. Substituting this result into Eq. (2.81) gives

$$\int_{-\infty}^{\infty} \omega\left\langle S_c(\boldsymbol{\kappa},\omega)\right\rangle_R d\omega = \int_{-\infty}^{\infty} \omega\left\langle S_{inc}(\boldsymbol{\kappa},\omega)\right\rangle_R d\omega - \frac{[Z(\beta)]^{-1}}{N\hbar}\sum_n e^{-\beta\epsilon_n} \times$$

$$\times \sum_{\ell\neq j}\left\langle\left\langle n|e^{-i\boldsymbol{\kappa}\cdot\mathbf{r}_j}[e^{i\boldsymbol{\kappa}\cdot\mathbf{r}_\ell},H_0]|n\right\rangle\right\rangle_R , \qquad (2.83)$$

where

$$\int_{-\infty}^{\infty} \omega\left\langle S_{inc}(\boldsymbol{\kappa},\omega)\right\rangle_R d\omega = \frac{[Z(\beta)]^{-1}}{N\hbar}\left[\sum_{n,\ell} e^{-\beta\epsilon_n}\frac{(\hbar\kappa)^2}{2M} + \right.$$

$$\left. + \sum_{n,\ell} e^{-\beta\epsilon_n}\frac{\hbar}{M}\left\langle\left\langle n|\boldsymbol{\kappa}\cdot\mathbf{p}_\ell|n\right\rangle\right\rangle_R\right] . \qquad (2.84)$$

The average over orientations of the subsystems which compose the scatterer can be accomplished by averaging over directions of $\boldsymbol{\kappa}$ for a particular subsystem of fixed orientation. Clearly then,

$$\left\langle\left\langle n|\boldsymbol{\kappa}\cdot\mathbf{p}_\ell|n\right\rangle\right\rangle_R = 0 .$$

Actually, for systems which are spatially bounded, this vanishes even without the average over orientations. This follows from

$$\langle n|\mathbf{p}_\ell|n\rangle = \frac{iM}{\hbar}\langle n|H_0\mathbf{r}_\ell - \mathbf{r}_\ell H_0|n\rangle = 0 ,$$

which holds for eigenstates $|n\rangle$ of H_0. Finally, using Eq. (2.61), we obtain Eq. (2.80) for the incoherent case.

To establish Eq. (2.80) for the coherent case, we must show that the interference terms in Eq. (2.83) vanish. Since $|n\rangle$ is an eigenstate of H_0,

$$\left\langle\left\langle n \right| e^{-i\boldsymbol{\kappa}\cdot(\mathbf{r}_j-\mathbf{r}_\ell)} H_0 - e^{-i\boldsymbol{\kappa}\cdot\mathbf{r}_\ell} H_0 e^{i\boldsymbol{\kappa}\cdot\mathbf{r}_\ell} \left| n \right\rangle\right\rangle_R = \tag{2.85}$$

$$= \left\langle\left\langle n \right| H_0 - e^{-i\boldsymbol{\kappa}\cdot\mathbf{r}_j} H_0 e^{i\boldsymbol{\kappa}\cdot\mathbf{r}_j} e^{-i\boldsymbol{\kappa}\cdot(\mathbf{r}_j-\mathbf{r}_\ell)} \left| n \right\rangle\right\rangle_R .$$

Using the relation (2.82)

$$\left(-\frac{(\hbar\kappa)^2}{2M} - \frac{\hbar\boldsymbol{\kappa}\cdot\mathbf{p}_j}{M}\right) e^{-i\boldsymbol{\kappa}\cdot(\mathbf{r}_j-\mathbf{r}_\ell)} |n\rangle$$

$$= e^{-i\boldsymbol{\kappa}\cdot(\mathbf{r}_j-\mathbf{r}_\ell)} \left[\frac{(\hbar\kappa)^2}{2M} - \frac{\hbar\boldsymbol{\kappa}\cdot\mathbf{p}_j}{M}\right] |n\rangle ,$$

we can write Eq. (2.85) in the form

$$- \left\langle\left\langle n \right| e^{-i\boldsymbol{\kappa}\cdot(\mathbf{r}_j-\mathbf{r}_\ell)} \left[\frac{(\hbar\kappa)^2}{2M} + \frac{\hbar\boldsymbol{\kappa}\cdot\mathbf{p}_\ell}{M}\right] \left| n \right\rangle\right\rangle_R$$

$$= \left\langle\left\langle n \right| e^{-i\boldsymbol{\kappa}\cdot(\mathbf{r}_j-\mathbf{r}_\ell)} \left[\frac{(\hbar\kappa)^2}{2M} - \frac{\hbar\boldsymbol{\kappa}\cdot\mathbf{p}_j}{M}\right] \left| n \right\rangle\right\rangle_R \tag{2.86}$$

$$= \left\langle\left\langle n \right| e^{i\boldsymbol{\kappa}\cdot(\mathbf{r}_j-\mathbf{r}_\ell)} \left[\frac{(\hbar\kappa)^2}{2M} + \frac{\hbar\boldsymbol{\kappa}\cdot\mathbf{p}_j}{M}\right] \left| n \right\rangle\right\rangle_R ,$$

where the last equation follows by replacing $\boldsymbol{\kappa}$ with $-\boldsymbol{\kappa}$ in the right-hand side of the first of Eq. (2.86). That it is valid to make this replacement follows, because of the average over orientations implied by the symbol $\langle\ldots\rangle_R$. We now combine into pairs the terms in Eq. (2.83) which differ from one another by an interchange of the indices ℓ and j and show that each such pair of terms for $\ell \neq j$ gives a null contribution in Eq. (2.83). We require

$$\left\langle\left\langle n\left|e^{-i\kappa\cdot(\mathbf{r}_j-\mathbf{r}_\ell)}\left(H_0 - e^{-i\kappa\cdot\mathbf{r}_\ell}H_0 e^{i\kappa\cdot\mathbf{r}_\ell}\right)\right|n\right\rangle\right\rangle_R +$$

$$+\left\langle\left\langle n\left|e^{i\kappa\cdot(\mathbf{r}_j-\mathbf{r}_\ell)}\left(H_0 - e^{-i\kappa\cdot\mathbf{r}_j}H_0 e^{i\kappa\cdot\mathbf{r}_j}\right)\right|n\right\rangle\right\rangle_R$$

$$= -\left\langle\left\langle n\left|e^{-i\kappa\cdot(\mathbf{r}_j-\mathbf{r}_\ell)}\left[\frac{(\hbar\kappa)^2}{2M} + \frac{\hbar\kappa\cdot\mathbf{p}_\ell}{M}\right]\right|n\right\rangle\right\rangle_R -$$

$$-\left\langle\left\langle n\left|e^{i\kappa\cdot(\mathbf{r}_j-\mathbf{r}_\ell)}\left[\frac{(\hbar\kappa)^2}{2M} + \frac{\hbar\kappa\cdot\mathbf{p}_j}{M}\right]\right|n\right\rangle\right\rangle_R .$$

That we obtain the claimed null result follows from Eq. (2.86). Thus, we have proved Eq. (2.80) for both the coherent and incoherent cases. Note that the proof of Eq. (2.80) required Eq. (2.82), which is valid only for a system whose potential energy is independent of the velocities of the particles composing the system. Note also that the proof of (2.80) did not require the thermal equilibrium of the scattering system.

2-4 THE TIME-DEPENDENT FORMALISM

We note that $S(\kappa,\omega)$ is defined in terms of a sum over quantities involving the exact eigenstates of the scattering system. In the few special cases when these are known, and the number of initial and final states contributing is small, $S(\kappa,\omega)$ can be calculated directly from Eqs. (2.65) and (2.66). Usually, the exact eigenstates are not known. Even when they are known, the number of initial and final states contributing to the sum over states is often so large that explicit summation is not feasible. It is therefore preferable to eliminate the appearance of the exact eigenstates. This can be done by introducing a Fourier transform of $S(\kappa,\omega)$ with respect to ω.

We can define the intermediate scattering functions $\chi_c(\kappa,t)$ and $\chi_{inc}(\kappa,t)$ by

$$\chi(\kappa,t) = \int_{-\infty}^{\infty} e^{i\omega t} S(\kappa,\omega)\,d\omega . \qquad (2.87)$$

Throughout the remainder of this chapter we work primarily with these functions. This had the advantage that the dynamics of the scattering system are exhibited more explicitly, enabling approximations to be made on a sound physical basis. Furthermore, for the important case of a harmonically vibrating system, the exact calculation is considerably simplified by working with $\chi(\boldsymbol{\kappa},t)$. After the function $\chi(\boldsymbol{\kappa},t)$ has been calculated, $S(\boldsymbol{\kappa},\omega)$ can be obtained by using the inverse Fourier transformation

$$S(\boldsymbol{\kappa},\omega) = \frac{1}{2\pi}\int_{-\infty}^{\infty} e^{-i\omega t}\chi(\boldsymbol{\kappa},t)dt \ . \tag{2.88}$$

We consider first the properties of $\chi_{inc}(\boldsymbol{\kappa},t)$. Carrying out the integration in Eq. (2.87) using Eq. (2.66) we obtain

$$\chi_{inc}(\boldsymbol{\kappa},t) = [NZ(\beta)]^{-1}\sum_{n,s} e^{-\beta\epsilon_n} \times \sum_{\ell=1}^{N}\left|\left\langle s\left|e^{i\boldsymbol{\kappa}\cdot\mathbf{r}_\ell}\right|n\right\rangle\right|^2 e^{i(\epsilon_s-\epsilon_n)t/\hbar} \ . \tag{2.88a}$$

We could perform the sum over final states in Eq. (2.88a) were it not for the factor $\exp(i\epsilon_s t/\hbar)$ which depends on s. Since this is a numerical factor, we can bring it inside the matrix element and use

$$e^{i\epsilon_s t/\hbar}u_s = e^{iH_0 t/\hbar}u_s \ . \tag{2.88b}$$

This follows since u_s is an eigenstate of H_0 with energy ϵ_s. Equation (2.88a) then reduces to

$$\chi_{inc}(\boldsymbol{\kappa},t) = [NZ(\beta)]^{-1}\sum_{n,s} e^{-(\beta+it\hbar^{-1})\epsilon_n} \times \sum_{\ell=1}^{N}\left\langle n\left|e^{-i\boldsymbol{\kappa}\cdot\mathbf{r}_\ell}\right|s\right\rangle\left\langle s\left|e^{iH_0 t/\hbar}e^{i\boldsymbol{\kappa}\cdot\mathbf{r}_\ell}\right|n\right\rangle . \tag{2.88c}$$

The sum over final states s can now be carried out using the completeness of the set of states u_s to give

$$\chi_{inc}(\boldsymbol{\kappa},t) = [NZ(\beta)]^{-1} \sum_n e^{-(\beta+it\hbar^{-1})\epsilon_n} \times$$

$$\times \sum_{\ell=1}^{N} \left\langle n \middle| e^{-i\boldsymbol{\kappa}\cdot\mathbf{r}_\ell} e^{iH_0 t/\hbar} e^{i\boldsymbol{\kappa}\cdot\mathbf{r}_\ell} \middle| n \right\rangle . \tag{2.88d}$$

Next we bring the factor $\exp(-it\epsilon_n/\hbar)$ inside the matrix element and use Eq. (2.88b). We introduce the Heisenberg position operator

$$\mathbf{r}_\ell(t) = e^{iH_0 t/\hbar} \mathbf{r}_\ell e^{-iH_0 t/\hbar} , \tag{2.88e}$$

which is obtained from the Schrödinger position operator $\mathbf{r}_\ell$ by a unitary transformation. The same transformation applies to any analytic function of $\mathbf{r}_\ell$ so that

$$e^{i\boldsymbol{\kappa}\cdot\mathbf{r}_\ell(t)} = e^{iH_0 t/\hbar} e^{i\boldsymbol{\kappa}\cdot\mathbf{r}_\ell} e^{-iH_0 t/\hbar} .$$

We thus finally obtain

$$\chi_{inc}(\boldsymbol{\kappa},t) = [NZ(\beta)2^{-1} \sum_n e^{-\beta\epsilon_n} \times$$

$$\times \sum_{\ell=1}^{N} \left\langle n \middle| e^{-i\boldsymbol{\kappa}\cdot\mathbf{r}_\ell(0)} e^{i\boldsymbol{\kappa}\cdot\mathbf{r}_\ell(t)} \middle| n \right\rangle . \tag{2.89}$$

We may write Eq. (2.89) in the more compact form

$$\chi_{inc}(\boldsymbol{\kappa},t) = \frac{1}{N} \sum_{\ell=1}^{N} \left\langle e^{-i\boldsymbol{\kappa}\cdot\mathbf{r}_\ell(0)} e^{i\boldsymbol{\kappa}\cdot\mathbf{r}_\ell(t)} \right\rangle_T , \tag{2.90}$$

where, for any operator $\mathbf{A}$,

$$\langle \mathbf{A} \rangle_T = \frac{\mathrm{Tr}[\mathbf{A} \exp(-\beta H_0)]}{\mathrm{Tr}\,[\exp(-\beta H_0)]}$$

and the trace of an operator $\mathbf{B}$ is defined by

$$\mathrm{Tr}\mathbf{B} = \sum_s \langle s|\mathbf{B}|s \rangle .$$

For example,

$$Z(\beta) = \mathrm{Tr}\ [\exp(-\beta H_0)] \ .$$

By a procedure similar to that which led to Eq. (2.90), we may express the coherent intermediate scattering function as

$$\chi_c(\boldsymbol{\kappa},t) = \frac{1}{N} \sum_{\ell,\ell'=1}^{N} \left\langle e^{-i\boldsymbol{\kappa}\cdot\mathbf{r}_\ell(0)} e^{i\boldsymbol{\kappa}\cdot\mathbf{r}_{\ell'}(t)} \right\rangle_T . \tag{2.91}$$

An alternative and equally concise form for the intermediate scattering function may be obtained from Eq. (2.88d) by defining

$$e^{-i\boldsymbol{\kappa}\cdot\mathbf{r}_\ell} H_0 e^{i\boldsymbol{\kappa}\cdot\mathbf{r}_\ell} = H_\ell(\boldsymbol{\kappa}) \ .$$

From this result we are led to

$$\chi_c(\boldsymbol{\kappa},t) = \frac{1}{N} \sum_{\ell,\ell'=1}^{N} \left\langle e^{-i\boldsymbol{\kappa}\cdot[\mathbf{r}_\ell(0)-\mathbf{r}_{\ell'}(0)]} e^{iH_{\ell'}(\boldsymbol{\kappa})t/\hbar} e^{-iH_0 t/\hbar} \right\rangle_T \tag{2.92}$$

$$\chi_{inc}(\boldsymbol{\kappa},t) = \frac{1}{N} \sum_{\ell=1}^{N} \left\langle \exp[iH_\ell(\boldsymbol{\kappa})t/\hbar]\ \exp[-iH_0 t/\hbar] \right\rangle_T \ . \tag{2.92a}$$

The appearance of the trace of an operator in Eqs. (2.90), (2.91), and (2.92) shows that the desired result does not depend on the exact set of states used to describe the scattering system. This type of simplification was first used in a similar context by Lamb.[17] It has since been used in the theoretical literature by a number of authors.[18-21] It has been emphased by Van Hove <u>et al</u>.[22] that the lack of dependence of $S(\boldsymbol{\kappa},\omega)$ on the exact eigenstates is a typical feature of many observables of interest in many-body problems. This independence is fortunate, since we can calculate the exact eigenstates only for very idealized systems. The approximate calculation of $S(\boldsymbol{\kappa},\omega)$ may be possible, however, for many real systems. The form of Eq. (2.91) is well suited to calculation by the diagrammatic methods now being employed for many-body problems.

In addition to its formal utility for expressing the scattering function in a manner more amenable to calculation, the Fourier transform in Eq. (2.87) introduces a time-dependent description of the dynamics of the scattering system. That it is physically legitimate to interpret the Fourier-transform variable as <u>time</u> follows from the appearance of the properly defined time-dependent operators in Eqs. (2.90) through (2.92). Furthermore, in Sec. 2-7, where we discuss the physical meaning of the χ-functions in more detail, we see that it can be interpreted as the duration of the interaction between the neutron and the scattering system.

The expressions for the sum rule and the detailed balance condition assume a particularly simple form in terms of the intermediate function. From Eqs. (2.80) and (2.87) it is easily proved that

$$-i \left. \frac{d\langle\chi(\boldsymbol{\kappa},t)\rangle_R}{dt} \right|_{t=0} = \frac{\hbar\kappa^2}{2M} .$$

Using (2.73), we obtain

$$\chi(-\boldsymbol{\kappa},-t) = \int_{-\infty}^{\infty} e^{-i\omega t} S(-\boldsymbol{\kappa},\omega)d\omega = \int_{-\infty}^{\infty} e^{i\omega t} S(-\boldsymbol{\kappa},-\omega)d\omega$$

$$= \int_{-\infty}^{\infty} e^{i\omega t} e^{-\beta\hbar\omega} S(\boldsymbol{\kappa},\omega)d\omega$$

$$= \chi(\boldsymbol{\kappa},t + i\beta\hbar) .$$

An alternative time-dependent formulation, suggested by Schofield,[23] involves the introduction of the Fourier transform of the symmetric function defined by Eq. (2.76):

$$\chi_0(\boldsymbol{\kappa},t) = \int_{-\infty}^{\infty} e^{i\omega t} S_0(\boldsymbol{\kappa},\omega)d\omega .$$

Utilizing Eq. (2.76), the relation

$$\chi_0(\boldsymbol{\kappa},t) = \chi(\boldsymbol{\kappa},t + \tfrac{1}{2}i\beta\hbar)$$

follows easily. Utilizing the symmetry and reality of $S_0(\boldsymbol{\kappa},\omega)$ inside the above integral we find

$$\chi_0(-\kappa,t) = \chi_0(\kappa,-t) \ .$$

In addition for real values of κ,t it also follows that

$$\chi_0(\kappa,-t) = \chi_0^*(\kappa,t) \ .$$

If the scattering is such that $S_0(\kappa,\omega) = S_0(-\kappa,\omega)$, as it is for liquids and polycrystals, then

$$\chi_0(\kappa,t) = \chi_0^*(\kappa,t) \ ,$$

so that $\chi_0(\kappa,t)$ is real for real values of t. The condition of detailed balance is then automatically satisfied if one works with real values of $\chi_0(\kappa,t)$. We refer to this formulation as the symmetric formulation.

2-5 NEUTRON SCATTERING BY SOME SIMPLE SYSTEMS

In this section, we emphasize the utility of the time-dependent formulation by using it to carry out the calculation of the scattering function for some especially simple dynamical systems. The scatterer is considered to consist of a single atom. The $\chi_{inc}(\kappa,\omega)$ and $\chi_c(\kappa,\omega)$ are then identical, and represent an average of $\langle s|\ \exp[-i\kappa\cdot r(0)] \times \exp[i\kappa\cdot r(t)]|s\rangle$ over an ensemble of these single-particle systems.

The simplest example that we can consider is the case of neutron scattering by an atom of mass M which is free and at rest in the laboratory system. The atom then is in the state $|0\rangle$, which is the eigenstate of the momentum operator belonging to the momentum eigenvalue $\mathbf{p} = 0$. Strictly speaking, this case does not correspond to a state of thermodynamic equilibrium. It is, however, precisely the situation that one assumes in the theory of neutron moderation at energies where the effects of thermal agitation and chemical binding are negligible.

In this case, the average over the initial states in Eq. (2.92a) reduces to

$$\chi_{inc}(\kappa,t) = \langle 0|\ \exp\{i[H_0(\mathbf{p} + \hbar\kappa)t/\hbar]\}\ \exp\{-[H_0(\mathbf{p})t/\hbar]\}|0\rangle \ ,$$

where $H_0(\mathbf{p})$ is the Hamiltonian for a free particle of mass M,

$$H_0(\mathbf{p}) = \frac{p^2}{2M} . \tag{2.93}$$

Use of this expression gives

$$\chi_{inc}(\boldsymbol{\kappa},t) = \exp[i(\hbar\kappa^2/2M)t] . \tag{2.94}$$

This result is different from the one we would obtain if we were to proceed from Eq. (2.89) and therein mistakenly set $\mathbf{r}_\ell(t) = \mathbf{r}_\ell(0)$. Our initial assumption that the scattering particles are free and at rest is different from the classical statement that they are free and at definite positions. Quantum mechanically, a particle which is initially at a given position cannot remain localized. This is further manifested by the fact that the operator $\mathbf{r}(t)$ does not commute with H_0.

The scattering function is obtained by substituting Eq. (2.94) into Eq. (2.88):

$$S(\boldsymbol{\kappa},\omega) = \delta\left(\omega - \frac{\hbar\kappa^2}{2M}\right). \tag{2.95}$$

Then, from Eqs. (2.70) and (2.71), using the fact that $S_c = S_{inc}$ for the present problem, we find

$$\begin{aligned}\sigma(E_0,E,\theta) &= \frac{k\sigma_b}{4\pi\hbar k_0}\,\delta\left(\omega - \frac{\hbar\kappa^2}{2M}\right)\\ &= \frac{\sigma_b}{4\pi}\frac{k}{k_0}\,\delta\left(E - E_0 + \frac{E + E_0 - 2(EE_0)^{1/2}\cos\theta}{A}\right) ,\end{aligned} \tag{2.95a}$$

where $A = M/m$ and $\sigma_b = \sigma_{inc} + \sigma_c$ is the familiar bound-atom cross section.

Integration of this result over all solid angles for fixed values of E and E_0 gives the scattering kernel of slowing-down theory. The result is most easily obtained by transforming the variables of integration from $\cos\theta$ to

$$x = E - E_0 + \frac{E + E_0 - 2(EE_0)^{1/2}\cos\theta}{A}$$

and writing

$$\sigma(E_0,E) = \frac{A\sigma_b}{4E_0}\int_{x_-}^{x_+} dx\delta(x) ,$$

where

$$x_\pm = E - E_0 + \frac{\left(E^{1/2} \pm E_0^{1/2}\right)^2}{A} .$$

Completing the integration and using Eqs. (2.9) and (2.14) to obtain

$$\sigma_0 = \frac{\sigma_b}{[1 + (1/A)]^2} ,$$

we arrive at the familiar result

$$\left.\begin{aligned} \sigma(E_0,E) &= \frac{(A+1)^2}{4AE_0}\sigma_0 \quad E_0 \geq E \geq \left(\frac{A-1}{A+1}\right)^2 E_0 \\ \sigma(E_0,E) &= 0 \qquad\qquad \text{for all other } E . \end{aligned}\right\} \tag{2.96}$$

We now consider the case that the atom of mass M is one from an ensemble of such atoms which are in statistical equilibrium at the temperature T. Using the free particle Hamiltonian, Eq. (2.93), as well as Eq. (2.92a),

$$\chi(\boldsymbol{\kappa},t) = \frac{\mathrm{Tr}\left\{\exp\left(-\frac{p^2}{2Mk_BT}\right)\exp\left[i\,\frac{(\mathbf{p}+\hbar\boldsymbol{\kappa})^2}{2M\hbar}\,t\right]\exp\left(-i\,\frac{p^2}{2M\hbar}\,t\right)\right\}}{\mathrm{Tr}\left[\exp\left(-\frac{p^2}{2Mk_BT}\right)\right]} .$$

If we evaluate the traces in the momentum representation the $\mathbf{p}$ becomes an ordinary vector momentum, and $\mathrm{Tr} \to \int \ldots d\mathbf{p}$. Thus,

$$\chi(\boldsymbol{\kappa},t) = \exp\left(i\,\frac{\hbar\kappa^2}{2M}\,t\right)\frac{\int \exp\left(-\frac{p^2}{2Mk_BT} + i\,\frac{\boldsymbol{\kappa}\cdot\mathbf{p}}{M}\,t\right)d\mathbf{p}}{\int \exp\left(-\frac{p^2}{2Mk_BT}\right)d\mathbf{p}} .$$

From this we find

$$\chi(\boldsymbol{\kappa},t) = \exp\left[\frac{\hbar\kappa^2}{2M}\left(it - \frac{k_BT}{\hbar}\,t^2\right)\right]. \tag{2.97}$$

Finally, from Eq. (2.88),

$$S(\boldsymbol{\kappa},\omega) = \left(\frac{M}{2\pi\kappa^2 k_B T}\right)^{\frac{1}{2}} \exp\left[\frac{-M\left(\hbar\omega - \dfrac{\hbar^2\kappa^2}{2M}\right)^2}{2\hbar^2\kappa^2 k_B T}\right] . \qquad (2.98)$$

The scattering kernel $\sigma(E_0,E)$ and the total scattering cross section $\sigma(E_0)$, which occur implicitly in Eq. (5.1) for the spectrum of neutrons in an infinite homogeneous gaseous moderator, can be obtained by using Eqs. (2.98) and (2.70) in Eqs. (2.72) and (2.72b), respectively. In using Eq. (2.70), we must set $S_c = S_{inc} = S$. The slightly tedious integrations that must be performed are carried out in Appendix I. The results are

$$\sigma(E_0,E) = A\frac{\sigma_b}{8E_0}\{\mathrm{erf}(\eta x - \rho x_0) - \mathrm{erf}(\eta x + \rho x_0) +$$

$$+ \exp[-(x^2 - x_0^2)][\mathrm{erf}(\rho x + \eta x_0) - \mathrm{erf}(\rho x - \eta x_0)]\} \qquad (2.99)$$

$$E \geq E_0 ,$$

$$\sigma(E_0,E) = A\frac{\sigma_b}{8E_0}\{\mathrm{erf}(\eta x + \rho x_0) + \mathrm{erf}(\eta x - \rho x_0) +$$

$$+ \exp[-(x^2 - x_0^2)][\mathrm{erf}(\eta x_0 - \rho x) - \mathrm{erf}(\eta x_0 + \rho x)]\} \qquad (2.99a)$$

$$E \leq E_0 ,$$

and

$$\sigma(E_0) = \sigma_0\{(1 + 1/2z)[\mathrm{erf}(z)^{\frac{1}{2}}] + (\pi z)^{-\frac{1}{2}} \exp(-z)\} , \qquad (2.100)$$

where

$$\eta = \frac{A + 1}{2A^{\frac{1}{2}}} ,$$

$$\rho = \frac{A - 1}{2A^{\frac{1}{2}}} ,$$

$$x = (\beta E)^{\frac{1}{2}} ,$$

$$x_0 = (\beta E_0)^{\frac{1}{2}} ,$$

$$z = Ax_0^2 ,$$

and

$$\operatorname{erf} x = \frac{2}{\pi^{\frac{1}{2}}} \int_0^x e^{-u^2} du . \tag{2.100a}$$

Equation (2.100) is familiar from the kinetic theory of gases where it defines the collision mean free path in a gas of hard spheres as a function of the speed of an initially specified molecule.[24]

We note the behavior of the cross section for low and high values of the incident neutron energy. For $z = A\beta E_0 << 1$, the cross section varies inversely as the velocity of the neutron:

$$\sigma(E_0) \approx 2\sigma_0(\pi A\beta E_0)^{-\frac{1}{2}} ; \tag{2.101}$$

while, for $A\beta E_0 >> 1$,

$$\sigma(E_0) \sim \sigma_0[1 + \tfrac{1}{2}(A\beta E_0)^{-1}] . \tag{2.102}$$

The manner in which the neutron cross section varies at low energies, namely, as $E_0^{-\frac{1}{2}}$, is a property common to all substances at nonvanishing temperatures. This follows from Eqs. (2.65) and (2.66) after observing that in general

$$\int k\, S\left(\mathbf{k}_0 - \mathbf{k}, \frac{\hbar k_0^2 - \hbar k^2}{2m}\right) d\mathbf{k}$$

approaches a finite limit as $k_0 \to 0$. In Sec. 2-8 we show that it is also generally true that for any substance the leading effects of finite temperature and chemical binding vary as $1/E_0$ at high energies.

We next consider a particle of mass M which is harmonically and isotropically bound to a fixed center of force. The Hamiltonian for this system is

$$H_0 = \frac{p^2}{2M} + \tfrac{1}{2}M\omega_0^2 r^2 . \tag{2.103}$$

The solution of the equations of motion gives the time-dependent displacement operator

$$\mathbf{r}(t) = \mathbf{r}(0) \cos \omega_0 t + \frac{\mathbf{p}(0)}{M\omega_0} \sin \omega_0 t \ . \tag{2.104}$$

In order to evaluate $\chi(\boldsymbol{\kappa},t)$, we make use of the rule

$$e^A e^B = e^{A+B+\frac{1}{2}[A,B]} \ . \tag{2.105}$$

This relation, which holds for any operators **A** and **B** which commute with their commutator [**A**,**B**], is proved in Appendix II. Since

$$[\boldsymbol{\kappa} \cdot \mathbf{r}(0), \boldsymbol{\kappa} \cdot \mathbf{r}(t)]$$

$$= \left[\boldsymbol{\kappa} \cdot \mathbf{r}(0), \frac{\boldsymbol{\kappa} \cdot \mathbf{p}(0)}{M}\right] \frac{\sin \omega_0 t}{\omega_0} = \frac{i\hbar\kappa^2}{M} \frac{\sin \omega_0 t}{\omega_0}$$

is a c-number, i.e., since it commutes with all operators, we may write

$$\left\langle e^{-i\boldsymbol{\kappa}\cdot\mathbf{r}(0)} e^{i\boldsymbol{\kappa}\cdot\mathbf{r}(t)} \right\rangle_T =$$

$$= \left\langle e^{i\boldsymbol{\kappa}\cdot[\mathbf{r}(t)-\mathbf{r}(0)]} \right\rangle_T \exp\left(+ \frac{i\hbar\kappa^2}{2M\omega_0} \sin \omega_0 t \right) . \tag{2.106}$$

The evaluation of a thermal average of the type required in Eq. (2.106) may be performed by use of a corollary of a theorem due to Bloch,[25] which we prove in Appendix III. This theorem states that if Q is any linear combination of harmonic oscillator coordinates, then

$$\left\langle \exp(Q) \right\rangle_T = \exp\left(\tfrac{1}{2} \left\langle Q^2 \right\rangle_T \right) . \tag{2.107}$$

Employing this evaluation of the thermal average in Eq. (2.106), we find

$$\left\langle \exp\{i\boldsymbol{\kappa} \cdot [\mathbf{r}(t) - \mathbf{r}(0)]\} \right\rangle_T$$

$$= \exp\left(- \frac{\kappa^2}{2} \left\langle [r_\kappa(\cos \omega_0 t - 1) + \frac{p_\kappa}{M\omega_0} \sin \omega_0 t]^2 \right\rangle_T \right), \tag{2.108}$$

where

$$r_\kappa = \frac{\boldsymbol{\kappa} \cdot \mathbf{r}(0)}{\kappa} \qquad p_\kappa = \frac{\boldsymbol{\kappa} \cdot \mathbf{p}(0)}{\kappa} \ .$$

To complete the evaluation of the thermal average, we recall that[26]

$$\frac{\langle n|p_\kappa^2|n\rangle}{2M} = \langle n|\,\frac{M\omega_0^2 r_\kappa^2}{2}\,|n\rangle = \frac{\epsilon_{\kappa,n}}{2} = \frac{\hbar\omega_0}{2}\,(n+\tfrac{1}{2}) \qquad n = 0, 1, 2, \ldots, \tag{2.109}$$

and that

$$\langle n|r_\kappa p_\kappa + p_\kappa r_\kappa|n\rangle = 0 \;, \tag{2.110}$$

when $|n\rangle$ and $\epsilon_{\kappa,n}$ are the eigenstates and eigenvalues of the Hamiltonian

$$\mathcal{H}_\kappa = \frac{p_\kappa^2}{2M} + \frac{M\omega_0^2}{2}\,r_\kappa^2 \;.$$

Using Eqs. (2.109) and (2.110) in Eq. (2.108), we may write

$$\left\langle \exp\{i\boldsymbol{\kappa}\cdot[\mathbf{r}(t)-\mathbf{r}(0)]\}\right\rangle_T =$$

$$= \exp\left\{-\frac{\kappa^2}{2M\omega_0^2}\left[(\cos\omega_0 t - 1)^2 + \sin^2\omega_0 t\right]\frac{\mathrm{Tr}\left(\mathcal{H}_\kappa e^{-\beta H_0}\right)}{\mathrm{Tr}\,e^{-\beta H_0}}\right\} \cdot \tag{2.111}$$

A straightforward calculation gives

$$\frac{\mathrm{Tr}\,\mathcal{H}_\kappa e^{-\beta H_0}}{\mathrm{Tr}\,e^{-\beta H_0}} = \hbar\omega_0\left[\frac{\sum_{n=0}^{\omega}(n+\tfrac{1}{2})e^{-(n+\frac{1}{2})\beta\hbar\omega_0}}{\sum_{n=0}^{\infty}e^{-(n+\frac{1}{2})\beta\hbar\omega_0}}\right] = (\bar{n}+\tfrac{1}{2})\hbar\omega_0 \;, \tag{2.112}$$

where

$$\bar{n} = \frac{\sum_{n=0}^{\infty} n e^{-n\beta\hbar\omega_0}}{\sum_{n=0}^{\infty} e^{-n\beta\hbar\omega_0}} = \left(e^{\beta\hbar\omega_0} - 1\right)^{-1} . \tag{2.112a}$$

Equation (2.112a) is familiar from the theory of black-body radiation as the average number of photons of energy $\hbar\omega_0$ in a given normal mode at temperature $T = (k_B\beta)^{-1}$. Collecting our results and expressing trigonometric functions in terms of exponentials, we obtain

$$\chi(\boldsymbol{\kappa},t) = \exp\left\{\frac{\hbar\kappa^2}{2M\omega_0}\left[(\bar{n}+1)e^{i\omega_0 t} + \bar{n}e^{-i\omega_0 t} - (2\bar{n}+1)\right]\right\}. \quad (2.113)$$

We still require the time Fourier transform of Eq. (2.113) in order to obtain $S(\boldsymbol{\kappa},\omega)$ and the differential-energy-transfer cross section. The result of this integration cannot be expressed in a simple closed form, as was possible for the case of freely moving atoms. Nevertheless, the Fourier transform of Eq. (2.113) is easily evaluated if we first expand $\chi(\boldsymbol{\kappa},t)$ as a complex Fourier series. This expansion consists in writing (2.113) in the form

$$\chi(\boldsymbol{\kappa},t) = \exp\left[-\frac{\hbar\kappa^2}{2M\omega_0}(2\bar{n}+1)\right]\exp\left\{\frac{\hbar\kappa^2}{4M\omega_0\sinh(\beta\hbar\omega_0/2)}\times\right.$$
$$\left.\times\left[\exp(\tfrac{1}{2}\beta\hbar\omega_0 + i\omega_0 t) + \exp(-\tfrac{1}{2}\beta\hbar\omega_0 - i\omega_0 t)\right]\right\}$$

and noting that the second exponential[27] has the form of the generating function for modified Bessel functions I_n:

$$e^{\frac{1}{2}x(y+y^{-1})} = \sum_{n=-\infty}^{\infty} y^n I_n(x) \;.$$

Carrying out the expansion as indicated yields

$$\chi(\boldsymbol{\kappa},t) = \exp\left[-\frac{\hbar\kappa^2}{2M\omega_0}(2\bar{n}+1)\right]\times$$
$$\times\sum_{n=-\infty}^{\infty} e^{in\omega_0 t}e^{\frac{1}{2}n\beta\hbar\omega_0} I_n\left(\frac{\hbar\kappa^2}{2M\omega_0\sinh\frac{1}{2}\beta\hbar\omega_0}\right).$$

Eq. (2.88) then gives

$$S(\boldsymbol{\kappa},\omega) = \exp\left[-\frac{\hbar\kappa^2}{2M\omega_0}(2\bar{n}+1)\right]\times$$
$$\times\sum_{n=-\infty}^{\infty} e^{\frac{1}{2}n\beta\ \omega_0}\delta(\omega - n\omega_0) I_n\left(\frac{\hbar\kappa^2}{2M\omega_0\sinh\frac{1}{2}\beta\hbar\omega_0}\right)$$

Energy transfers are confined to the discrete values $\hbar\omega = n\hbar\omega_0$. This is expected since the energies available to the harmonic oscillator are similarly confined.

It is instructive to consider the total cross section for neutron scattering by a system of nuclei at zero temperature executing isotropic simple harmonic vibrations about fixed equilibrium positions. In this case, $\bar{n} = 0$. There are no oscillators in excited states, so that a neutron can only lose energy in a collision. This simple model was first used by Fermi[3] in his study of the moderation of neutrons in hydrogenous substances. For this simple model, the separate terms in the expansion of Eq. (2.114) have such a simple dependence on κ and ω that they can be integrated to give their contribution to the total scattering cross section. Utilizing Eq. (2.114), we can express the cross section for T = 0 in the form

$$\sigma(E_0) = \sum_{n=0}^{\infty} \sigma_n(E_0) ,$$

where

$$\sigma_n(E_0) = \frac{1}{4\pi k_0} \frac{\sigma_b}{n!} \int_0^{E_0/\hbar} k \, d\omega \int d\Omega \, \exp\left(-\frac{\hbar\kappa^2}{2M\omega_0}\right) \times \left[\left(\frac{\hbar\kappa^2}{2M\omega_0}\right)^n \delta(\omega - n\omega_0)\right] . \tag{2.115}$$

In this expression, κ^2 is to be regarded as a function of $\omega = [(E_0 - E)/\hbar]$, E_0, and the angle of scattering θ. The limits of integration arise from the fact that the energy lost by the neutron cannot be greater than the energy it had before the scattering occurred. On transforming from an integration over solid angle in Eq. (2.115) to integration over the variable

$$x = \frac{\hbar\kappa^2}{2M\omega_0} ,$$

we can write

$$\sigma_n(E_0) = \frac{\sigma_b}{4n!} A \frac{1}{y_0} \int_0^{E_0/\hbar} \delta(\omega - n\omega_0) d\omega \int_{x_-}^{x_+} x^n e^{-x} \, dx ,$$

where, by Eq. (2.71), with cos θ = ±1,

$$x_\pm = \frac{\hbar}{2M\omega_0} (k \pm k_0)^2$$

and

$$y_0 = \frac{E_0}{\hbar\omega_0} .$$

Integrating over ω, we obtain finally

$$\sigma_n(E_0) = \frac{\sigma_b}{4n!} A \frac{1}{y_0} \int_{a_-(y_0,n)}^{a_+(y_0,n)} x^n e^{-x} dx \qquad y_0 \geq n , \qquad (2.116)$$

$$\sigma_n(E_0) = 0 \qquad y_0 \leq n . \qquad (2.116a)$$

where

$$a_\pm(y_0,n) = A^{-1}[(y_0 - n)^{\frac{1}{2}} \pm y_0^{\frac{1}{2}}]^2 .$$

For $y_0 \leq 1$, the only nonvanishing contribution to the cross section comes from the elastic scattering (n = 0):

$$\sigma_0(E_0) = \frac{\sigma_b}{4} A \frac{1}{y_0} [1 - \exp(-4y_0/A)] .$$

Figure 2.3 shows schematically the elastic cross section, the inelastic cross sections for n = 1,2,3,..., and the total cross section as a function of y_0. For $E_0 = 0$, the cross section is equal to the bound-atom cross section. For $y_0 > 4$, the total cross section is within a few percent of the free-atom value.

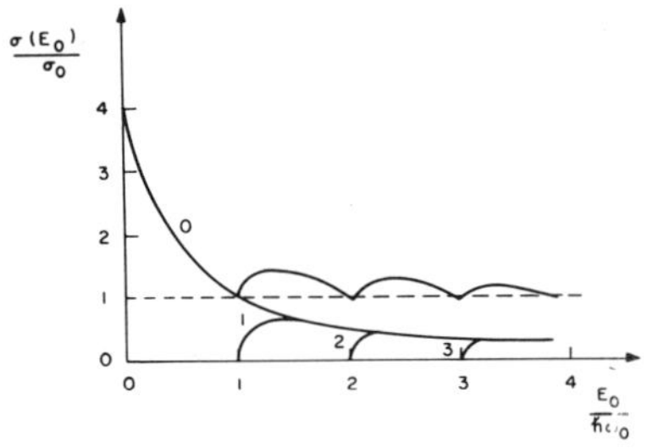

Figure 2.3 Elastic and inelastic scattering cross sections for a neutron interacting with a harmonic isotropic oscillator of protonic mass. (From Fermi, Ricerca Scientifica, Ref. 3.)

We can gain further insight into the scattering of neutrons by an oscillator by considering Eq. (2.114) in the limit that

$$\left.\begin{aligned} p &= \omega/\omega_0 \\ x &= \frac{\hbar\kappa^2}{2M\omega_0 \sinh \frac{1}{2}\hbar\beta\omega_0} \end{aligned}\right\} \gg 1 . \qquad (2.117)$$

In this limit we show that if

$$\Delta = \frac{\omega - (\hbar\kappa^2/2M)}{2k_B\bar{T}\kappa^2/M} \qquad (2.118)$$

satisfies

$$\Delta \ll (\hbar^2\kappa^2/8Mk_B\bar{T})^{\frac{1}{2}} ,$$

then from Eq. (2.114) we obtain the approximate result in Eq. (2.119) by integrating over any one of the delta-function peaks in the scattered neutron beam:

$$\frac{1}{\omega_0}\int_{(n-\frac{1}{2})\omega_0}^{(n+\frac{1}{2})\omega_0} S(\boldsymbol{\kappa},\bar{\omega})\,d\bar{\omega} = \left(\frac{M}{2\pi\kappa^2 k_B\bar{T}}\right)^{\frac{1}{2}} \times$$

$$66 \times \exp\left\{\frac{-[\omega - (\hbar\kappa^2/2M)]^2}{2k_B\bar{T}\kappa^2/M}\right\}, \tag{2.119}$$

where

$$\omega = n\omega_0$$

and

$$k_B\bar{T} = \tfrac{1}{2}\hbar\omega_0 \coth \tfrac{1}{2}\beta\hbar\omega_0 \ . \tag{2.119a}$$

In the course of deriving this result, it becomes clear that it is most accurate in the vicinity of the values of ω and κ^2 for which the intensity of scattered neutrons is a maximum. We discuss the physical significance of Eq. (2.119) after first presenting its derivation.

To prove Eq. (2.119), we make use of the approximate result[28]

$$(2\pi)^{\frac{1}{2}} I_p(x) \approx (p^2 + x^2)^{-1/4} \times \\ \times \exp\left[(p^2 + x^2)^{\frac{1}{2}} - p \sinh^{-1}\frac{p}{x}\right], \tag{2.120}$$

which is accurate if $p >> 1$ and $x >> 1$. We use Eq. (2.118) in (2.117) to express p in terms of x and Δ:

$$p = x\left(\sinh \tfrac{1}{2}\beta\hbar\omega_0\right)\left[1 + 2\left(\frac{k_B\bar{T}}{\hbar\omega_0}\right)^{\frac{1}{2}} \frac{\Delta}{(x \sinh \tfrac{1}{2}\beta\hbar\omega_0)^{\frac{1}{2}}}\right].$$

On expanding $(p^2 + x^2)^{\frac{1}{2}}$ and $p \sinh^{-1}(p/x)$ for small values of $\delta = \Delta[x \sinh \tfrac{1}{2}\beta\hbar\omega_0]^{-\frac{1}{2}}$ we find

$$I_{\bar{\omega}/\omega 0}\frac{\hbar\kappa^2}{2M\omega_0 \sinh \tfrac{1}{2}\beta\hbar\omega_0} = \left(2\pi k_B\bar{T}\kappa^2/M\omega_0^2\right)^{-\frac{1}{2}}[1 + O(\delta)] \times \\ \times \exp\left[\frac{\hbar\kappa^2}{M\omega_0}\frac{k_B\bar{T}}{\hbar\omega_0} - \frac{\hbar\bar{\omega}\beta}{2} - \Delta^2 + O(\delta)\right]. \tag{2.120a}$$

From Eqs. (2.120a) and (2.114) one easily infers that Eq. (2.119) is a good approximation for $\delta << 1$, a condition which is equivalent to the inequality (2.118).

The right-hand side of Eq. (2.119) is formally identical to the result obtained in the case of scattering by a monatomic gas at an effective temperature $\bar{T}$ which is equal to two thirds of the average kinetic energy of any oscillator in the system. Thus, when the inequalities (2.117) are satisfied, the energy distribution of singly scattered neutrons is predominantly determined by the momentum distribution of particles in the scattering system. The first of these inequalities is valid when the energy loss of the neutron is large compared with the harmonic oscillator level spacing. For values of $\hbar\omega_0 \lesssim kT$, the second inequality can be satisfied even the free-atom recoil energy $\hbar^2\kappa^2/2M$ is also large compared with this level spacing. For $kT \ll \hbar\omega_0$, on the other hand, this inequality is satisfied when this recoil energy is small compared with $\hbar\omega_0$.

The range of values of κ^2 for which Eq. (2.119) is valid is large if $k_BT/\hbar\omega_0 >> 1$. For this case, by (2.119a) $\bar{T} \to T$, so that the scattering function given by Eq. (2.119) is very nearly equal to that of a free gas at physical temperature T of the scatterer, even for energy transfers $\hbar\omega \lesssim k_BT$. If, on the other hand, $\hbar\omega_0 >> k_BT$, Eq. (2.117) shows that the validity of Eq. (2.119) requires very large momentum transfers:

$$\frac{\hbar\kappa^2}{2M\omega_0} >> e^{\hbar\omega_0/k_BT} .$$

An early set of experiments by Brockhouse and Hurst[29] clearly illustrate the effects of interatomic binding on the energy distribution of scattered neutrons. Their experiments consisted in scattering a monoenergetic beam of 0.35-eV neutrons from thin, solid samples and measuring the transmission through a cadmium filter of the neutrons scattered through an angle of 90°. Since the cadmium cross section falls rapidly from over 6000 barns to less than 150 barns as the neutron energy increases from 0.2 eV to 0.5 eV, the transmission through the filter is quite sensitive to the energy distribution of the scattered neutrons. On the other hand, we do not expect the energy distribution to be sensitive to the effects of crystal structure since the wavelength of 0.35-eV neutrons is small compared with interatomic spacings. The materials whose neutron-scattering properties were investigated are lead, aluminum, diamond, and graphite. Figures 2.4(a), (b), and (c) compare experimental results with calculations based on the free gas and

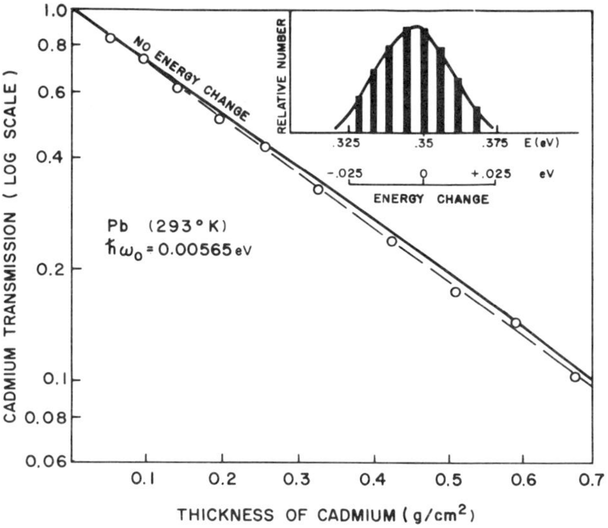
CADMIUM TRANSMISSION (LOG SCALE)
THICKNESS OF CADMIUM (g/cm2)
NO ENERGY CHANGE
RELATIVE NUMBER
.325
.35
.375
E (eV)
-.025
0
+.025
eV
ENERGY CHANGE
Pb (293° K)
ħω0 = 0.00565 eV

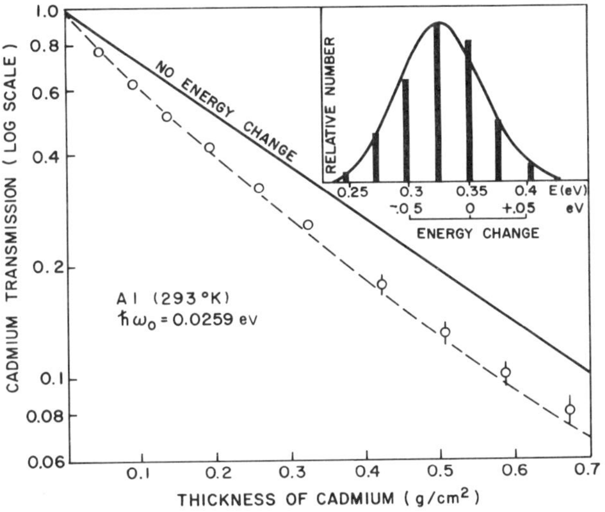
CADMIUM TRANSMISSION (LOG SCALE)
THICKNESS OF CADMIUM (g/cm2)
NO ENERGY CHANGE
RELATIVE NUMBER
0.25
0.3
0.35
0.4
E(eV)
-0.5
0
+.05
eV
ENERGY CHANGE
Al (293 °K)
ħω0 = 0.0259 eV

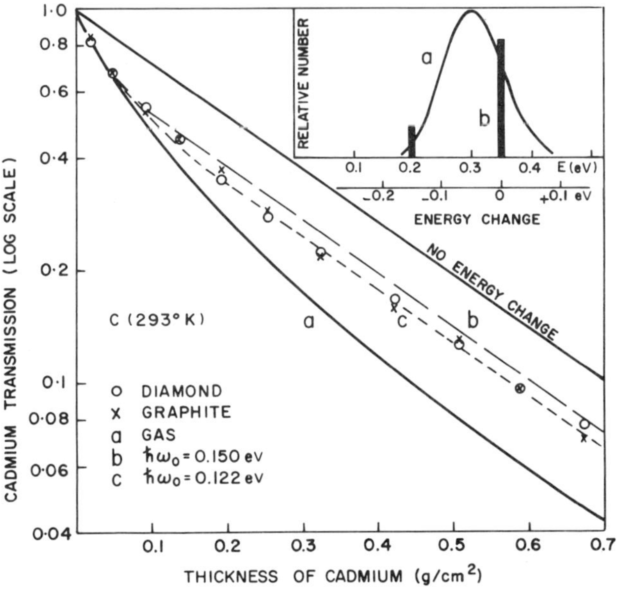
CADMIUM TRANSMISSION (LOG SCALE)
THICKNESS OF CADMIUM (g/cm2)
RELATIVE NUMBER
a
b
E (eV)
-0.2
-0.1
0
+0.1 eV
ENERGY CHANGE
NO ENERGY CHANGE
C (293° K)
a
c
b
o DIAMOND
x GRAPHITE
a GAS
b ħω0 = 0.150 eV
c ħω0 = 0.122 eV

Einstein models. The results for Al and for Pb agree well with predictions of the gas model at the physical temperature T. In view of our previous discussion and the large ratio of initial neutron energy to $\hbar\omega_0$, this is an expected result. Of particular significance are the results for graphite. For this important moderator the cadmium transmission is very different from what would be expected if graphite scattered neutrons as does a perfect gas. At room temperature, the same conclusion applies for other commonly used moderators, inasmuch as in the Einstein approximation they are all characterized by values of $\hbar\omega_0/k_BT > 1$.

2-6 SLOW-NEUTRON SCATTERING BY CRYSTALLINE SOLIDS

2-6.1 ELASTIC SCATTERING. When a neutron of wavelength comparable to interatomic didstances moves through a medium composed of regular arrays of atoms, interference effects become important. These effects are particularly pronounced for the scattering of slow neutrons by non-absorbing single-crystals and, to a lesser extent, for poly-crystalline substances. The following treatment deals in considerable detail with the scattering of slwo neutrons by monatomic crystals. Some of the results obtained are generalized to the case of crystals containing more than one type of atom. Since we wish to emphasize the effects of crystal structure on neutron scattering, we at first make a simplification in the dynamical behavior of the nuclei composing the crystal.

This simplification consists in assuming that the nuclei of the crystal do not interact with one another and that all of them execute isotropic, simple harmonic vibrations with the same frequency about their positions of equilibrium. This is called the Einstein model. As a consequence of these assumptions, we see that interference effects occur only in the coherent elastic scattering, and not at all in the inelastic scattering. (Recall that there are no interference effects in the incoherent scattering.) The validity of this assertion follows from Eq. (2.91) and is proved below.

We first review briefly some of the elementary concepts of crystalline structures and introduce some of the notation to be used in this section. The sums in Eq. (2.91) are over all nuclei in the crystal. The position of any nucleus in a crystal can be specified by giving both its equilibrium position and its displacement from equilibrium; the equilibrium position of a nucleus is specified by giving the location of the unit cell which contains it as well as its position relative to some corner of the unit cell.

Suppose that $\mathbf{a}_i (i = 1,2,3)$ are the fundamental translation vectors of the crystal lattice; the $\mathbf{a}_1$ are not necessarily orthogonal to one another. The unit cell may be taken to be a rhombohedral parallelepiped, the edges of which are determined by the fundamental translation vectors. Although the crystal may have an arbitrary shape, any results that we obtain will be valid if we assume that it is a rhombohedral parallelepiped whose faces are parallel to the faces of the unit cell and whose edges have lengths L_1, L_2, L_3 which are large integer multiples of the edge lengths of a unit cell; this is

$$L_1 = N_1 a_1, \quad L_2 = N_2 a_2, \quad L_3 = N_3 a_3 \ .$$

The crystal contains $N_c = N_1 N_2 N_3$ unit cells. Any given cell may be specified relative to one corner of the crystal by the vector

$$\mathbf{R}_p = \sum_{i=1}^{3} p_i \mathbf{a}_i \ , \tag{2.121}$$

where p_1, p_2, p_3 are integers which run from zero to N_1, N_2, N_3, respectively. We call the cell that is specified by the integers p_1, p_2, p_3 the *p*th unit cell in the lattice. In addition, we assume that there are n atoms in a unit cell and denote the positions of these relative to the points $\mathbf{R}_p$ by $\boldsymbol{\rho}_s$ $(s = 1,2,\ldots n)$. Then, denoting the displacement of the *s*th nucleus in the *p*th unit cell from its equilibrium position by $\mathbf{u}_p^s$, we may express its position $\mathbf{r}_p^s$ relative to the origin of coordinates at the corner of the crystal by

$$\mathbf{r}_p^s = \mathbf{R}_p + \boldsymbol{\rho}_s + \mathbf{u}_p^s \ . \tag{2.122}$$

Let us now return to the calculation of the neutron scattering by an Einstein crystal. It is clear that the treatment of the scattering by an Einstein oscillator given in Sec. 2-5 applies with little modification to the incoherent scattering by an Einstein crystal. The only difference is that in Eqs. (2.115) and (2.116) the bound incoherent cross section $\sigma_{inc,b}$ should occur in place of σ_b.

The coherent scattering can be obtained from Eqs. (2.91), (2.88), and (2.68). First, note that (2.91) becomes proportional to the coherent part of Eq. (2.30) if $\mathbf{r}_{\ell'}(t) = \mathbf{r}_{\ell'}(0) = \mathbf{R}_{\ell'}$, that is, if the nuclei are at rest at their equilibrium positions. To account for the dynamical features of the problem, we use our assumption that the particles of the system do not interact and write Eq. (2.91) for the intermediate coherent scattering function in the form

$$\chi_c(\boldsymbol{\kappa},t) = \chi_{inc}(\boldsymbol{\kappa},t) + \frac{1}{N}\sum_{\ell,\ell'}^{N}{}' \left\langle e^{-i\boldsymbol{\kappa}\cdot\mathbf{r}_\ell(0)}\right\rangle_T \left\langle e^{i\boldsymbol{\kappa}\cdot\mathbf{r}_{\ell'}(t)}\right\rangle_T , \tag{2.123}$$

where we recall that the prime on the summation sign indicates the omission of terms with $\ell' = \ell$. The thermal expectation value in Eq. (2.90) for $\chi_{inc}(\boldsymbol{\kappa},t)$ cannot be decomposed into a product of expectation values as can be done with the interference part of (2.123): whereas $\mathbf{r}_\ell(0)$ and $\mathbf{r}_{\ell'}(t)$ commute for $\ell' \neq \ell$, they do not for $\ell' = \ell$ unless $t = 0$. Nor is this separation possible for systems of particles which interact with one another. On the other hand, digressing briefly, we note that such a separation is possible for any system of noninteracting particles, and not just for the case of harmonically vibrating particles which we are discussing here. For noninteracting systems of particles, we can show that $\left\langle e^{i\boldsymbol{\kappa}\cdot\mathbf{r}_{\ell'}(t)}\right\rangle_T$ is independent of t, and hence, from Eq. (2.88), that interference scattering is completely elastic. In fact,

$$\begin{aligned}\left\langle e^{i\boldsymbol{\kappa}\cdot\mathbf{r}_{\ell'}(t)}\right\rangle_T &= [Z(\beta)]^{-1}\sum_n \left\langle n\left| e^{-\beta H_0} e^{i(H_0/\hbar)t} e^{i\boldsymbol{\kappa}\cdot\mathbf{r}_{\ell'}(0)} e^{-iH_0/\hbar)t}\right| n\right\rangle \\ &= \frac{\sum_n e^{-\beta\epsilon_n}\left\langle n\left| e^{-i\boldsymbol{\kappa}\cdot\mathbf{r}_{\ell'}(0)}\right| n\right\rangle}{\sum_n e^{-\beta\epsilon_n}} \\ &= \left\langle e^{i\boldsymbol{\kappa}\cdot\mathbf{r}_{\ell'}(0)}\right\rangle_T .\end{aligned} \tag{2.124}$$

Let us now use Eq. (2.122) to cast $\chi_c(\kappa,t)$ into a form that is convenient for discussing neutron scattering by crystals. Using Eqs. (2.122) and (2.124) in Eq. (2.123) there results

$$\chi_c(\kappa,t) = \chi_{inc}(\kappa,t) - \frac{1}{nN_c}\sum_{\ell=1}^{nN_c}\left|\left\langle e^{-i\kappa\cdot\mathbf{r}_\ell(0)}\right\rangle_T\right|^2 + $$

$$+ \chi_c^{(0)}(\kappa,t) , \qquad (2.125)$$

where

$$\chi_c^{(0)}(\kappa,t) = \frac{1}{nN_c}\sum_{p,p'} e^{i\kappa\cdot(\mathbf{R}_{p'}-\mathbf{R}_p)} \sum_{s,s'} e^{i\kappa\cdot(\rho_{s'}-\rho_s)}$$

$$\left\langle e^{-i\kappa\cdot\mathbf{u}_p^s(0)}\right\rangle_T \left\langle e^{i\kappa\cdot\mathbf{u}_p^{s'}(0)}\right\rangle_T . \qquad (2.126)$$

The sums in Eq. (2.126) now go over all values of the indices. This is made possible by the appearance of the second member on the right-hand side of Eq. (2.125). From Eq. (2.107)

$$\left\langle e^{-i\kappa\cdot\mathbf{u}_p^s(0)}\right\rangle_T = \left\langle e^{i\kappa\cdot\mathbf{u}_p^s(0)}\right\rangle_T = e^{-\frac{1}{2}\left\langle[\kappa\cdot\mathbf{u}_p^2(0)]^2\right\rangle_T}$$

$$= e^{-W_p^s(\kappa)} . \qquad (2.127)$$

Furthermore, since the different particles of the system are assumed to be dynamically equivalent,

$$2W_p^s(\kappa) = \left\langle[\kappa\cdot\mathbf{u}_p^s(0)]^2\right\rangle_T = \left\langle[\kappa\cdot\mathbf{u}]^2\right\rangle_T = 2W(\kappa) \qquad (2.127a)$$

independent of s and p. Also,

$$\left|\left\langle e^{-i\kappa\cdot\mathbf{r}_\ell(0)}\right\rangle_T\right|^2 = \left|\left\langle e^{-i\kappa\cdot\mathbf{u}_p^s(0)}\right\rangle_T\right|^2 = e^{-2W(\kappa)} ,$$

Thus,

$$\chi_c(\kappa,t) = \chi_{inc}(\kappa,t) - e^{-2W(\kappa)} + \chi_c^{(0)}(\kappa) ,$$

with

$$\chi_c^{(0)}(\boldsymbol{\kappa}) = \chi_c^{(0)}(\boldsymbol{\kappa},t) = e^{-2W(\boldsymbol{\kappa})} \left| \sum_{s=1}^{n} e^{i\boldsymbol{\kappa}\cdot\boldsymbol{\rho}_s} \right|^2 \frac{1}{nN_c} \times$$

$$\times \sum_{p,p'} e^{i\boldsymbol{\kappa}\cdot(\mathbf{R}_p - \mathbf{R}_{p'})} \; . \tag{2.128}$$

Since the elastic part of $\chi_{inc}(\boldsymbol{\kappa},t)$ is equal to $e^{-2W(\boldsymbol{\kappa})}$, $\chi_c^{(0)}(\boldsymbol{\kappa})$ is the only part of $\chi_c(\boldsymbol{\kappa},t)$ that contributes to the coherent elastic scattering. From Eqs. (2.88) and (2.68), this contribution is given by

$$\frac{d^2\sigma_{el}}{dEd\Omega} = \frac{1}{4\pi}\sigma_{c,b}\,\delta(E_0 - E)\chi_c^{(0)}(\boldsymbol{\kappa}) \; . \tag{2.129}$$

Let us now examine the various terms in Eq. (2.128). We consider first the lattice sum

$$\frac{1}{N_c}\sum_p e^{i\mathbf{f}\cdot\mathbf{R}_p} \; ,$$

where $\mathbf{f}$ is some vector having the dimensions of a reciprocal length. For this purpose, it is convenient to introduce a set of vectors τ by the requirement that

$$e^{i\tau\cdot\mathbf{R}_p} = 1 \tag{2.130}$$

for any p. The set of vectors τ form a lattice known as the reciprocal lattice. It depends only on the translation group of the original lattice and not at all on the structure of the unit cell.

The reciprocal lattice consists of the ensemble of points, the positions of which are given by

$$\tau(n_1', n_2', n_3') = 2\pi \sum_{i=1}^{3} n_i' \mathbf{b}_i \; , \tag{2.131}$$

where n_1', n_2', n_3' are integers and the $\mathbf{b}_i$ are the primitive translations of the reciprocal lattice defined by the relations

$$\mathbf{b}_i \cdot \mathbf{a}_j = \delta_{ij} \; . \tag{2.132}$$

There are nine equations for the three unknown vectors $\mathbf{b}_i$. Their solution is

$$\mathbf{b}_1 = \frac{\mathbf{a}_2 \times \mathbf{a}_3}{\mathbf{a}_1 \cdot \mathbf{a}_2 \times \mathbf{a}_3}$$

$$\mathbf{b}_2 = \frac{\mathbf{a}_3 \times \mathbf{a}_1}{\mathbf{a}_2 \cdot \mathbf{a}_3 \times \mathbf{a}_1} \qquad (2.133)$$

$$\mathbf{b}_3 = \frac{\mathbf{a}_1 \times \mathbf{a}_2}{\mathbf{a}_3 \cdot \mathbf{a}_1 \times \mathbf{a}_2}$$

Two useful properties of the reciprocal lattice vectors are (1) that $\boldsymbol{\tau}(n_1', n_2', n_3')$ is normal to the planes in the direct lattice space defined by Eq. (2.121), which intersect the edges of the crystal (recall that the edges are parallel to $\mathbf{a}_1$, $\mathbf{a}_2$, $\mathbf{a}_3$) at intervals of a_1/n_1', a_2/n_3', and (2) that the length of $\boldsymbol{\tau}$ is equal to 2π times the reciprocal of the spacing between these planes

$$|\boldsymbol{\tau}(n_1', n_2', n_3')| = \frac{2\pi}{d(n_1', n_2', n_3')} . \qquad (2.134)$$

Dividing n_1', n_2', n_3' by their largest common integral factor m gives three integers: $h = n_1'/m$, $k = n_2'/m$, and $l = n_3'/m$, which are the Miller indices of the set of lattice planes whose normal is parallel to $\boldsymbol{\tau}(n_1', n_2', n_3')$. The set of planes which intersect the crystal edges at ma_1/n_1', ma_2/n_2', ma_3/n_3' have a spacing equal to

$$d = md(n_1', n_2', n_3') = \frac{2\pi m}{|\boldsymbol{\tau}(n_1', n_2', n_3')|} . \qquad (2.135)$$

We discuss later the significance of the reciprocal lattice for problems of neutron diffraction.

Now, consider

$$S(\mathbf{f}) = \lim_{N_c \to \infty} \frac{1}{N_c} \left| \sum_p e^{i(\mathbf{f} \cdot \mathbf{R}_p)} \right|^2 , \qquad (2.136)$$

where by $N_c \to \infty$ we mean that all of N_1, N_2, N_3 approach infinity. It is easy to show that

$$\frac{1}{N_l}\left|\sum_{P_l=0}^{N_l} e^{i(\mathbf{f}\cdot\mathbf{a}_l)p_l}\right|^2 = \frac{1}{N_l}\,\frac{\sin^2 \tfrac{1}{2}(N_l + 1)\mathbf{f}\cdot\mathbf{a}_l}{\sin^2 \tfrac{1}{2}\mathbf{f}\cdot\mathbf{a}} .$$

From this result and from the representation of the Dirac delta function

$$\sum_n \delta(x - 2\pi n) = 2\pi \lim_{N\to\infty} \frac{1}{N}\,\frac{\sin^2 \tfrac{1}{2}Nx}{\sin^2 \tfrac{1}{2}x} ,$$

it follows that

$$S(\mathbf{f}) = (2\pi)^3 \sum_{\boldsymbol{\tau}} \prod_{i=1}^{3} \delta(\mathbf{f}\cdot\mathbf{a}_i - \boldsymbol{\tau}\cdot\mathbf{a}_i)$$

$$= \frac{(2\pi)^3}{V_0} \sum_{\boldsymbol{\tau}} \delta(\mathbf{f} - \boldsymbol{\tau}) . \tag{2.137}$$

The volume per unit cell of the crystal, $V_0 = \mathbf{a}_1 \cdot \mathbf{a}_2 \times \mathbf{a}_3$, enters as the Jacobian of the transformation from the oblique coordinate axes $\mathbf{a}_i$ to Cartesian coordinates. The differential coherent elastic cross section $d\sigma_{c,el}/d\Omega$ can therefore be written

$$\frac{d\sigma_{c,el}}{d\Omega} = \int_0^\infty \frac{d^2\sigma_{c,el}}{dE\,d\Omega}\,dE = \sum_{\boldsymbol{\tau}} \frac{d\sigma_{c,el}(\boldsymbol{\tau})}{d\Omega} , \tag{2.138}$$

where

$$\frac{d\sigma_{c,el}(\boldsymbol{\tau})}{d\Omega} = \frac{2\pi^2\sigma_{c,b}}{nV_0}\, e^{-2W(\boldsymbol{\tau})}\,|F(\boldsymbol{\tau})|^2\,\delta(\boldsymbol{\kappa} - \boldsymbol{\tau}) , \tag{2.139}$$

and

$$F(\boldsymbol{\tau}) = \sum_{s=1}^{n} e^{i\boldsymbol{\tau}\cdot\boldsymbol{\rho}_s} . \tag{2.140}$$

Thus, the coherent elastic scattering vanishes unless $\boldsymbol{\kappa}$ is equal to a reciprocal lattice vector.

The physical significance of the reciprocal lattice space can now easily be understood. Because of the δ-function behavior of (2.139), coherent elastic scattering will occur only if

$$\kappa = \mathbf{k}_0 - \mathbf{k} = \boldsymbol{\tau}(n_1', n_2', n_3'). \tag{2.141}$$

This is the Bragg condition. The momentum transfer must be parallel to one of the $\boldsymbol{\tau}(n_1', n_2', n_3')$, that is, perpendicular to the set of lattice planes whose normal is parallel to $\boldsymbol{\tau}(n_1', n_2', n_3')$. The magnitude of the momentum transfer and the angle of scattering θ between $\mathbf{k}_0$ and $\mathbf{k}$ are related by (see Eq. (2.135))

$$|\mathbf{k} - \mathbf{k}_0| = 2k_0 \sin \tfrac{1}{2}\theta = \frac{4\pi}{\lambda} \sin \tfrac{1}{2}\theta = |\boldsymbol{\tau}| = \frac{2\pi m}{d} \tag{2.142}$$

or

$$2d \sin \tfrac{1}{2}\theta = m\lambda \ . \tag{2.142a}$$

The argument $\boldsymbol{\tau}$ attached to the symbol for the differential cross-section in Eq. (2.139) reminds us that it represents the contribution from τ reflection, i.e., from the reflection of order m at the set of lattice planes [h,k,l] with Miller indices h,k,l.

The Bragg condition plays a prominent role in the theory of diffraction by crystals. It forms the theoretical basis which enables the determination of crystal structures by means of X-ray and neutron techniques. It is seen from Eq. (2.139) that only for certain values of θ is a strongly scattered beam found. Figure 2.5 is a diffraction pattern that Wollan and Shull,[30] as cited by Hughes,[31] obtained by irradiating a crystal of NaCl with neutrons from a pile at the Oak Ridge National Laboratory. In the experiment of Wollan and Shull, a single-crystal of NaCl scatters a beam of neutrons having a wide range of wavelengths; the diffracted neutrons then form a pattern of Laue spots. By observing the values of θ, the directions of the crystal planes and the spacing between them can be found. From the variation of intensity of reflections from the various planes, information can be obtained about the density of the atoms which occupy them, and thus the crystal structure can be determined.

In their early experiments Fermi and Marshall[10] wanted to measure scattering amplitudes by diffracting essentially monoenergetic neutrons from a single-crystal. Their apparatus, shown in Fig. 2.6, contained a first crystal used as a monochromator. The almost monochromatic beam from this first crystal was scattered by the single-crystal under investigation. The neutrons diffracted by the second crystal for different incident angles were detected by a BF_3 counter, which was rotated about the scattering crystal to record the scattered beam as the crystal was rotated. The intensities of the Bragg maxima in the diffraction pattern provided information on the coherent scattering amplitudes of the nuclei in the single-crystal.

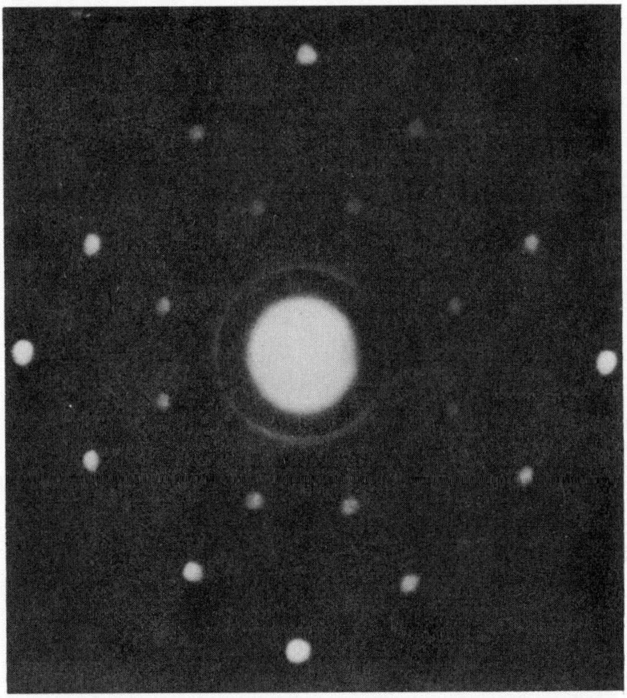

Figure 2.5 A Laue photograph of NaCl obtained by Wolland Shull[30] at the Oak Ridge pile. (From Hughes, Pile Neutron Research, Ref. 31.)

The intensities of the various Bragg maxima are proportional to the square of the form factor $F(\tau)$, which represents the sum of the amplitudes of the waves scattered from each nucleus in a unit cell. The structure factor given by Eq. (2.140) is the appropriate one only for monatomic crystals. For a polyatomic Einstein crystal, Eq. (2.139) is still valid if we replace $\sigma_{c,b}|F(\tau)|^2$ by

$$(4\pi)^{-1}\left|\sum_{s}^{n} a_{c,b}^{s} e^{i\tau\cdot\rho_s}\right|^2 ,$$

where $a_{c,b}^{s}$ is the bound coherent scattering length of the *s*th nucleus in the unit cell. In certain of the diffraction maxima of a crystal consisting of two atomic species, the sum or the difference of the amplitudes of the elements enters, depending on the relative phases of the waves received at the detector from different points in the unit cell. The intensity of different peaks is determined by the relative scattering lengths. For example, a low intensity in a peak that corresponds to the sum of amplitudes of two elements shows that their scattering lengths have opposite signs.

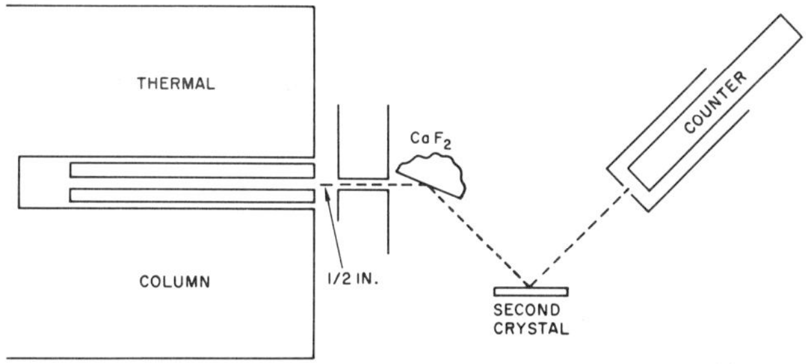

Figure 2.6 Crystal diffraction apparatus. (From Fermi and Marshall, Phys. Rev., Ref. 10.)

Fermi and Marshall made no attempt to obtain the absolute values of scattering lengths, but only the algebraic sign. They found that most scattering lengths have the same sign as carbon, which has been determined to be positive by means of the mirror-reflection method, and that only a few have a negative sign. This was to be expected since a negative amplitude occurs only when there is a resonance near to and above thermal energies.

The only factor in Eq. (2.139) which we have not discussed is the Debye-Waller factor $e^{-2W(\kappa)}$. For an isotropic Einstein oscillator of frequency ω_0,

$$2W(\boldsymbol{\kappa}) = \frac{\hbar\kappa^2}{M\omega_0}(\bar{n} + \tfrac{1}{2}) ,$$

with $\bar{n}$ given by (2.112a). The Debye-Waller factor in Eq. (2.139) reflects the influence of the atomic motions on the coherent elastic scattering. The zero-point vibrations and thermal agitation of the atoms which make up the crystal lead to departures from the ideal lattice structure, and to a corresponding reduction in the intensity of the Laue spots. This factor is considered in detail in Sec. 2-6.3.

Let us now turn our attention to the scattering of neutrons by polycrystals. The scatterer is assumed to be composed of single-crystals which are randomly oriented with respect to one another throughout the sample. To obtain the coherent elastic scattering cross section for polycrystalline materials, it is necessary to average both sides of Eq. (2.139) over all possible orientations of the single-crystals, or equivalently, over all directions of $\boldsymbol{\tau}$. For this purpose, we require

$$\frac{1}{4\pi}\int \delta(\boldsymbol{\kappa} - \boldsymbol{\tau})d\hat{\tau} ,$$

where $\hat{\tau} = \boldsymbol{\tau}/\tau$. To evaluate this, we first express the three-dimensional Dirac δ-function in spherical coordinates as follows:

$$\delta(\boldsymbol{\kappa} - \boldsymbol{\tau})d\boldsymbol{\tau} = \frac{1}{\kappa^2}\,\delta(\hat{\boldsymbol{\tau}} - \hat{\boldsymbol{\kappa}})\ (\tau - \kappa)\tau^2 d\tau d\hat{\boldsymbol{\tau}}$$

$$= \frac{1}{\tau^2}\,\delta(\hat{\boldsymbol{\tau}} - \hat{\boldsymbol{\kappa}})\delta(\tau - \kappa)d\boldsymbol{\tau} ,$$

where $\delta(\hat{\boldsymbol{\tau}} - \hat{\boldsymbol{\kappa}})$ is a two-dimensional delta function. Using Eq. (2.142) there follows

$$\frac{1}{4\pi}\int \delta(\boldsymbol{\kappa} - \boldsymbol{\tau})d\hat{\tau} = \frac{1}{4\pi\tau^2}\,\delta(\tau - \kappa)$$

$$= \frac{1}{4\pi\tau^2}\,(\tau - 2k_0 \sin\tfrac{1}{2}\theta) . \qquad (2.143)$$

Thus,

$$\left\langle \frac{d\sigma_{d,el}}{d\Omega} \right\rangle_R = \frac{1}{4\pi} \sum_{\boldsymbol{\tau}} \int \frac{d\sigma_{c,el}(\boldsymbol{\tau})}{d\Omega} \, d\hat{\boldsymbol{\tau}}$$

$$= \frac{\pi\sigma_{c,b}}{2nV_0} \sum_{\boldsymbol{\tau}} \frac{1}{\tau^2} e^{-2W(\boldsymbol{\tau})} |F(\boldsymbol{\tau})|^2 \times$$

$$\times \, \delta(\tau - 2k_0 \sin \tfrac{1}{2}\theta) \, , \qquad (2.144)$$

where $\langle \ldots \rangle_R$ implies an average over crystal orientations. Integrating Eq. (2.144) over all angles of scattering gives the total cross section for a polycrystal:

$$\sigma_{c,el} = \frac{\pi^2 \sigma_{c,b}}{nV_0 k_0^2} \sum_{\boldsymbol{\tau}}^{\tau \leq 2k_0} \frac{1}{\tau} e^{-2W(\boldsymbol{\tau})} |F(\boldsymbol{\tau})|^2 \, , \qquad (2.145)$$

where the summation extends over all reciprocal lattice vectors whose magnitude is not greater than $2k_0$. If the crystal has only one atom per unit cell, $F(\boldsymbol{\tau})$ is equal to 1, and if further, $W(\boldsymbol{\tau})$ depends only on the magnitude of τ, then the sum over reciprocal lattice vectors in Eqs. (2.144) and (2.145) can be expressed in the form

$$\sum_{\tau \leq 2k_0} \frac{M_\tau}{\tau} e^{-2W(\tau)} \, ,$$

where M_τ denotes the multiplicity, that is, the number of $\boldsymbol{\tau}$ vectors whose magnitudes are equal to τ.

Equations (2.144) and (2.145) apply with little modification to the coherent elastic scattering by any harmonically vibrating crystal. In general, the atoms in a unit cell of a crystal are not dynamically equivalent. As a consequence, a different Debye-Waller factor $e^{-2W_s(\boldsymbol{\tau})}$ is associated with each of the s atoms in the unit cell. Furthermore, if the unit cell contains more than one type of atom, then the correct expression for the coherent elastic scattering can be obtained from Eqs. (2.144) and (2.145) by replacing

$$\sigma_{c,b} e^{-2W(\boldsymbol{\tau})} |F(\boldsymbol{\tau})|^2$$

with

$$\frac{1}{4\pi}\left|\sum_{s=1}^{n} a^{s}_{c,b} e^{-W_s(\tau)} e^{i\tau\cdot\rho_s}\right|^2 .$$

For the case of elastic scattering by a polycrystal, the neutrons scattered by a certain set of lattice planes fall into circular cones. The diffraction pattern consists of concentric circles with their centers on the line defined by the incident neutron beam. These circles are the so-called Debye-Scherrer rings.

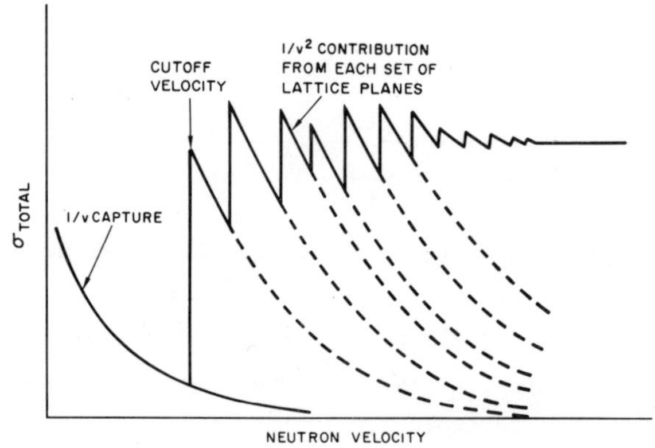

Figure 2.7 The transmission cross section in the crystal region showing the contribution of various sets of crystal planes. (From Hughes, Pile Neutron Research, Ref. 31.)

The manner in which neutrons reflected from different sets of lattice planes contribute to the cross section is shown in Fig. 2.7. For wavelengths longer than twice the largest lattice spacing, Eq. (2.145) shows that there exists no coherent eleastic scattering. When the neutron energy increases until $\lambda = 2d$, the planes with largest spacing d which are perpendicular to the direction of incidence $\mathbf{k}_0$ scatter neutrons through an angle of 180°. At slightly higher energies, the angle of scattering from these particular planes decreases. At the same time, planes which are not oriented at right angles to $\mathbf{k}_o$ but which have the spacing d, begin to reflect neutrons. The contribution to the scattering by the set of planes with the spacing d then decreases with increasing energy, being proportional to E_0^{-1}. As E_0 increases further, more sets of planes can reflect neutrons, and the cross section at any energy is the sum of the contributions from all the reflecting planes, with the contribution from each plane varying as E_0^{-1}. At energies greater than about 0.1 eV, so many planes contribute to the scattering that the cross section appears as a smooth function of energy on any except very fine energy scales. As the energy approaches 1 eV, the coherent elastic scattering (and also the incoherent elastic scattering) becomes very small because of the Debye-Waller factor; however, the total scattering changes very little with energy, because the decreased elastic scattering is compensated by an increase in inelastic scattering. When this state of affairs is reached, the scattering is not very different from that by free atoms at rest.

2-6.2 LATTICE DYNAMICS. In the previous section, we discussed the scattering of slow neutrons by a crystal composed of atoms harmonically bound to their positions of equilibrium, but which do not interact with one another. The unrealistic assumption that the atoms are noninteracting does not affect the important features of the elastic scattering by crystals. On the other hand, for very slow neutrons, the inelastic scattering of both the coherent and incoherent types is strongly influenced by the interactions between atoms. This is one reason that slow neutrons provide such a powerful probe for investigating the dynamical properties of condensed states of matter.

Since thermal neutrons, like X-rays, have wavelengths of the order of the interatomic distances in crystals, they can be used for diffraction studies. However, unlike X-rays, a thermal neutron has an energy which is of the same order as, or smaller than, the characteristic energies of lattice vibrations. In general, the relative energy change of a thermal neutron which interacts with the lattice is large and readily observable. Observation of the energy distribution of scattered neutrons thus provides a direct means of studying the dynamics of the vibrating crystal lattice.

So far we have emphasized only the usefulness of slow neutrons for investigating the atomic motions in crystalline materials. Of more direct interest to the reactor physicist is the question of how the energy distributions of neutrons in bulk matter are affected by the atomic motions in moderators. A rigorous treatment of the thermalization properties of the solid moderators such as graphite, beryllium, beryllium oxide, and zirconium hydride requires a description in terms of crystal-lattice vibrations. Even in light and heavy water, each molecule, considered as a unit, executes small vibrations and small torsional oscillations during time intervals comparable to the time of interaction between a molecule and a typical thermal neutron. It is only on a time scale which is long compared with these interaction times that the molecular motions tend to depart from bounded oscillations. In this respect, a liquid resembles a solid, and as we see in Sec. 3-4.1, this similarity is reflected in a quantitative as well as in a qualitative way. Furthermore, as in the case of atomic vibrations in a crystal, the internal vibrations of a molecule, which are extremely important for a quantitative description of the thermalization of neutrons, can be described to a very good approximation in terms of the theory of harmonic vibrations.

In the rest of this section we review briefly the dynamics of lattice vibrations; for a more comprehensive treatment of the field the reader is referred to the texts of Born and Huang,[32] Peierls,[33] and Maradudin, Montroll, and Weiss.[34]

The scattering of neutrons by an isotropic Einstein oscillator, already discussed in Sec. 2-5, constitutes the simplest example of the general problems to be reviewed. We consider first, in classical mechanics, the motion of nuclei within the immediate neighborhood of a configuration of stable equilibrium. A rigorous theory of the thermal vibrations about the stable equilibrium is very difficult and, besides, has little bearing on the problems considered in this book. Therefore, we consider the crystal as perfect, apart from the small vibrations about equilibrium, and we neglect all surface effects. In addition, we assume that the displacements of atoms from their positions of equilibrium are small compared with interatomic distances, so that the motions can be described in terms of a vibration potential V which is a homogeneous quadratic form in the nuclear coordinate:

$$V = \frac{1}{2} \sum_{\substack{p',s',\alpha' \\ p,s,\alpha}} V^{ss'}_{\alpha\alpha'}(R_p, R_{p'}) u^{s}_{p,\alpha} u^{s'}_{p',\alpha'} , \tag{2.146}$$

where $u^s_{p,\alpha}$ ($\alpha = 1,2,3$) are the Cartesian components of $\mathbf{u}^s_p$. Then, the equation of motion for the vibration of the atom of mass M_s near $\mathbf{R}_p + \boldsymbol{\rho}_s$ is

$$M_s \ddot{u}^s_{p,\alpha} = - \sum_{s',\alpha',p'} V^{ss'}_{\alpha\alpha'}(R_p,R_{p'}) u^{s'}_{p',\alpha'} , \qquad (2.147)$$

where $\ddot{u} = d^2u/dt^2$. The force constants $V^{ss'}_{\alpha\alpha'}(R_p,R_{p'})$ are symmetric in their indices, and depend only on the relative positions of the unit cells at $\mathbf{R}_p$ and $\mathbf{R}_{p'}$; that is,

$$V^{ss'}_{\alpha\alpha'}(R_p,R_{p'}) = V^{ss'}_{\alpha\alpha'}(R_p - R_{p'})$$

$$= V^{s's}_{\alpha'\alpha}(R_{p'} - R_p) . \qquad (2.148)$$

We consider first Eqs. (2.147) as classical mechanical equations of motion and look for normal vibrations

$$\mathbf{u}^s_p = (N_c M_s)^{-\frac{1}{2}} \mathbf{C}^s(R_p) q(t) , \qquad (2.149)$$

where $\mathbf{C}^s(R_p)$ are the normal-mode amplitude vectors and $q(t)$ satisfies

$$\ddot{q} + \omega^2 q = 0 . \qquad (2.150)$$

In Eq. (2.149) the factor $(N_c M_s)^{-\frac{1}{2}}$ is not essential, but will lead to a convenient normalization of the $\mathbf{C}^s$. From Eqs. (2.147), using Eqs. (2.149) and (2.150), there result $3nN_c$ equations for the Cartesian components C^s_α of $\mathbf{C}^s$:

$$\omega^2 C^s(\mathbf{R}_p) = \sum_{s',\alpha',p'} \frac{V^{ss'}_{\alpha\alpha'}(R_p,R_{p'})}{(M_s M_{s'})^{\frac{1}{2}}} C^{s'}_{\alpha'}(\mathbf{R}_{p'}) . \qquad (2.151)$$

The property (2.148) allows us to take the vectors $\mathbf{C}^s(R_p)$ in the form

$$\mathbf{C}^s(\mathbf{R}_p) = \text{constant } \mathbf{C}^s \exp(i\mathbf{f} \cdot \mathbf{R}_p) , \qquad (2.152)$$

where $\mathbf{C}^s$ is independent of $\mathbf{R}_p$. To prove this, observe that if $\mathbf{C}^s(\mathbf{R}_p)$ is a solution of Eq. (2.151), then so is $\mathbf{C}^s(\mathbf{R}_p + \mathbf{a})$, where $\mathbf{a}$ is any lattice vector. Thus, we may choose the solutions so that $\mathbf{C}^s(\mathbf{R}_p + \mathbf{a})$ differs from $\mathbf{C}^s(\mathbf{R}_p)$ by a constant factor

$$\mathbf{C}^s(\mathbf{R}_p + \mathbf{a}) = e^{i\mathbf{f}\cdot\mathbf{a}}\mathbf{C}^s(\mathbf{R}_p) \ ,$$

which is true for arbitrary values of $\mathbf{f}$. Now the validity of Eq. (2.152) is easily deduced.

We must now consider the permissible values of $\mathbf{f}$. Strictly speaking, Eq. (2.148) is not valid for a crystal which is finite in extent, for then there exist complicated surface effects. We can avoid the difficulties arising from surface effects if first we imagine the crystal which we are considering to be a large subsystem of a crystal which extends to infinity in all directions. Then we impose the cyclic condition of Born and von Karman, namely, that an atom at the position $\mathbf{R}_p + \boldsymbol{\rho}_s + N_1\mathbf{a}_1 + N_2\mathbf{a}_2 + N_3\mathbf{a}_3$ execute the same motion as the atom at $\mathbf{R}_p + \boldsymbol{\rho}_s$. The motion of an atom in the interior of a large crystal is not significantly affected by surface effects, so that we may choose mathematical boundary conditions somewhat arbitrarily. The cyclic condition is the most convenient one.

From Eqs. (2.152) and (2.149), it follows that the cyclic boundary condition can be quantitatively expressed as

$$\left.\begin{aligned} & e^{i(\mathbf{f}\cdot\mathbf{a}_l)N_l} = 1 \\ \text{or}\quad & \mathbf{f}\cdot\mathbf{a}_l = n\frac{2\pi}{N_l} \qquad 0 \le n_l \le N_l \ . \end{aligned}\right\} \tag{2.153}$$

An important consequence of Eq. (2.153) is that there are N_c permitted values of $\mathbf{f}$ and that these are uniformly distributed over $\mathbf{f}$ space with the density $N_cV_0/(2\pi)^3$. This follows on noting that Eq. (2.153) implies that

$$\mathbf{f} = 2\pi\sum_{i=1}^{3}\frac{n_i'\mathbf{b}_i}{N_i} \qquad n_1' = 0, \pm1, \pm2, \ldots, \tag{2.153a}$$

where $\mathbf{b}_i$ are the basis vectors of the reciprocal lattice. Thus, the volume per point in $\mathbf{f}$ space is

$$\frac{(2\pi)^3}{N_1N_2N_3}(\mathbf{b}_1 \cdot \mathbf{b}_2 \times \mathbf{b}_3) = \frac{(2\pi)^3}{N_cV_0} .$$

Here, we have used the fact that the volumes of the unit cells in the direct and reciprocal lattices are the inverse of one another. Another useful property in the space of the **f** vectors is that

$$\frac{1}{N_c}\sum_p{}' e^{i(\mathbf{f}-\mathbf{f}')\cdot\mathbf{R}_p} = 1 \qquad \mathbf{f} - \mathbf{f}' = 0 \quad \text{or} \quad \tau ,$$

$$\frac{1}{N_c}\sum_p e^{i(\mathbf{f}-\mathbf{f}')\cdot\mathbf{R}_p} = \frac{1}{N_c} \qquad \mathbf{f} - \mathbf{f}' \neq 0 \quad \text{or} \quad \tau . \tag{2.154}$$

The first equation follows from Eq. (2.130) and the fact that p assumes N_c values. The second equation follows on noting that, for $\mathbf{f} - \mathbf{f}' \neq 0$ or τ,

$$\frac{1}{N_c}\sum_p e^{i\mathbf{f}\cdot\mathbf{R}_p} = \prod_{j=1}^{3}\frac{1}{N_j}\sum_{p_j=0}^{N_j}\exp(2\pi i n_j' p_j \mathbf{b}_j \cdot \mathbf{a}_j/N_j)$$

$$= \prod_{j=1}^{3}\frac{1}{N_j}\sum_{p_j=0}^{N_j}\exp(2\pi i n_j' p_j/N_j) ;$$

and that, for $n_j' \neq N_j$,

$$\sum_{p_j=0}^{N_j}\exp(2\pi i n_j' p_j/N_j) = \frac{1 - \exp[2\pi i n_j'(N_j + 1)/N_j]}{1 - \exp(2\pi i n_j'/N_j)} = 1 .$$

For a crystal containing a large number of atoms, it is an extremely good approximation to write

$$\frac{1}{N_c}\sum_p e^{i(\mathbf{f}-\mathbf{f}')\cdot\mathbf{R}_p} = \Delta(\mathbf{f},\mathbf{f}') , \tag{2.154a}$$

where $\Delta(\mathbf{f},\mathbf{f}')$ equals one if **f** and **f'** differ by either zero or a reciprocal lattice vector and equals zero otherwise.

Returning to Eq. (2.152), we see that two vectors **f'** and **f** describe the same normal mode if $(\mathbf{f} - \mathbf{f}') \cdot \mathbf{R}_p/2\pi$ is an integer for any lattice vector $\mathbf{R}_p$, that is, if $\mathbf{f}' - \mathbf{f}$ is equal to a reciprocal lattice vector $\boldsymbol{\tau}$. When this is the case, **f'** and **f** are said to be equivalent. Of all these equivalent vectors, we choose the one of smallest magnitude. All of the $3N_c$ values of **f** so selected will cover a region of **f** space known as the first Brillouin zone.[35] This region of **f** space has inversion symmetry about $\mathbf{f} = 0$, it contains all equivalent points and no nonequivalent points, and its volume is equal to that of the unit cell of the reciprocal lattice. We could equally well have chosen a unit cell as our basic region in **f** space. This, however, is not convenient, because it would restrict **f** to values having one sign only. A Brillouin zone, on the other hand, contains $-\mathbf{f}$ if it contains **f**, which is convenient when dealing with the traveling waves in a solid.

It still remains to solve the $3nN_c$ set of equations for the quantities $C^s_\alpha(\mathbf{R}_p)$. Substitution of Eq. (2.152) into Eq. (2.151) reduces the latter set of $3nN_c$ equations in $3nN_c$ unknowns to 3n equations in 3n unknowns for each of the N_c values of **f**:

$$\omega^2 C^s_\alpha = \sum_{s',\alpha'} \sum_{p'} \frac{V^{ss'}_{\alpha\alpha'}(R_p - R_{p'})}{(M_s M_{s'})^{\frac{1}{2}}} \exp[i\mathbf{f} \cdot \mathbf{R}_{p'} - \mathbf{R}_p)]C^{s'}_{\alpha'}$$

$$= \sum_{s',\alpha'} G^{ss'}_{\alpha\alpha'}(-\mathbf{f})C^{s'}_{\alpha'} , \tag{2.155}$$

where

$$G^{ss'}_{\alpha\alpha'}(\mathbf{f}) = \sum_p \frac{V^{ss'}_{\alpha\alpha'}(\mathbf{R}_p)}{(M_s M_{s'})^{\frac{1}{2}}} e^{i\mathbf{f}\cdot\mathbf{R}_p} . \tag{2.156}$$

The 3n solutions of Eq. (2.155) for each of the N_c values of **f** give a total of $3nN_c$ independent normal modes, corresponding to $3nN_c$ degrees of freedom of the vibrating crystal.

The system of Eq. (2.155) has a solution only if the determinant of coefficients vanishes, i.e., only if

$$\| G^{ss'}_{\alpha\alpha'}(-\mathbf{f}) - \omega^2 \delta_{ss'} \delta_{\alpha\alpha'} \| = 0. \tag{2.157}$$

This is a function of degree 3n in ω^2, and, in general, there exist 3n roots. These must be real and positive; otherwise, the equilibrium configuration from which we started could not be stable. For each **f**, we label the 3n values of ω^2 by $\omega_j^2(\mathbf{f})$, for j = 1,...,3n. For $\omega^2 = \omega_j^2(\mathbf{f})$, the 3n components C_α^s can be obtained apart from an arbitrary multiplicative constant. We label the solution for the $\mathbf{C}^s$ corresponding to $\omega^2 = \omega_j^2(\mathbf{f})$ by $\mathbf{C}_j^s$ and choose the arbitrary constant so that

$$\sum_s \mathbf{C}_j^{s*}(\mathbf{f}) \cdot \mathbf{C}_j^s(\mathbf{f}) = \sum_s \left|\mathbf{C}_j^s(\mathbf{f})\right|^2 = 1 . \tag{2.158}$$

It is now convenient to obtain a number of mathematical relations which we shall need later. First, from Eqs. (2.156) and (2.148) it is evident that

$$G_{\alpha\alpha'}^{ss'}(-\mathbf{f}) = G_{\alpha\alpha'}^{ss'*}(\mathbf{f}) = G_{\alpha'\alpha}^{s's}(\mathbf{f}) . \tag{2.159}$$

Replacing **f** by -**f** in Eq. (2.157), taking the complex conjugate, and using Eq. (2.159), one finds that

$$\omega_j^2(\mathbf{f}) = \omega_j^2(-\mathbf{f}) . \tag{2.160}$$

Operating on Eq. (2.155) in precisely the same manner, we find that we may choose

$$\mathbf{C}_j^s(-\mathbf{f}) = \mathbf{C}_j^{s*}(\mathbf{f}) . \tag{2.160a}$$

From Eqs. (2.155), (2.156), and (2.130), it is also clear that

$$\begin{aligned} \omega_j^2(\mathbf{f} + \tau) &= {}_j^2(\mathbf{f}) , \\ \mathbf{C}_j^s(\mathbf{f} + \tau) &= \mathbf{C}_j^s(\mathbf{f}) . \end{aligned} \tag{2.161}$$

Finally, writing the general solution of the equations of motion as the superposition

$$\mathbf{u}_p^s = \sum_{\mathbf{f}} \sum_{j=1}^{3n} \frac{1}{(N_c M_s)^{1/2}} e^{i\mathbf{f}\cdot\mathbf{R}_p} \mathbf{C}_j^s(\mathbf{f}) q_j(\mathbf{f},t) , \tag{2.162}$$

one notes that the reality of the atomic displacements, together with

Eq. (2.160a) implies that

$$q_j^*(-\mathbf{f},t) = q_j(\mathbf{f},t) \; . \tag{2.163}$$

A useful orthogonality relation can be obtained by considering Eq. (2.155) for the normal-mode vectors, $\mathbf{C}_j^s(\mathbf{f})$, and $\omega_j^2(\mathbf{f})$:

$$\sum_{s,\alpha} C_{j',\alpha}^{s*}(\mathbf{f})C_{j,\alpha}^{s}(\mathbf{f}) = \delta_{j'j} \; . \tag{2.164}$$

The validity of this result for j = j' follows immediately from Eq. (2.158). For $j \neq j'$, the proof of orthogonality proceeds in the usual way. One first multiplies Eq. (2.155) for $C_{j,\alpha}^s(\mathbf{f})$ by $C_{j',\alpha}^{s*}(\mathbf{f})$ and then sums over s and α. Next, take the complex conjugate of Eq. (2.155) for $C_{j',\alpha}^s(\mathbf{f})$, multiply both sides of the equation by $C_{j,\alpha}^s(\mathbf{f})$, and sum over s and α. Subtracting this summation from the previous one and using Eq. (2.159), there results

$$[\omega_j^2(\mathbf{f}) - \omega_{j'}^2(\mathbf{f})] \sum_{s,\alpha} C_{j',\alpha}^{*s}(\mathbf{f})C_{j,\alpha}^{s}(\mathbf{f}) = 0 \; .$$

If $\omega_{j'}^2(\mathbf{f}) \neq \omega_j^2(\mathbf{f})$, then Eq. (2.163) for $j' \neq j$ is satisfied. If $\omega_{j'}^2(\mathbf{f}) = \omega_j^2(\mathbf{f})$, then Eq. (2.155) has more than one independent solution. These may be chosen in such a way that Eq. (2.164) is satisfied.

Other useful relations are the completeness relations

$$\sum_{j=1}^{3n} C_{j,\alpha'}^{s'*}(\mathbf{f})C_{j,\alpha}^{s}(\mathbf{f}) = \delta_{\alpha\alpha'}\delta_{ss'} \; . \tag{2.165}$$

To prove this result, consider

$$\mathbf{F}_s(\mathbf{f},t) = \sum_p (N_c M_s)^{\frac{1}{2}} e^{-i\mathbf{f}\cdot\mathbf{R}_p} \mathbf{u}_p^s(t) \; . \tag{2.166}$$

Using Eq. (2.162) and Eq. (2.154), we can show that

$$\mathbf{F}_s(\mathbf{f},t) = \sum_{j=1}^{3n} \mathbf{C}_j^s(\mathbf{f})q_j(\mathbf{f},t) \; .$$

This result, and the orthogonality relation, Eq. (2.164), lead to

$$q_j(\mathbf{f},t) = \sum_s \mathbf{C}_j^{s*}(\mathbf{f}) \cdot \mathbf{F}_s(\mathbf{f},t) . \qquad (2.167)$$

Combining the last two equations and noting that $\mathbf{u}_p^s$ and therefore $\mathbf{F}_s$ are arbitrary functions, we conclude that Eq. (2.165) is valid.

Using Eqs. (2.160) through (2.164), and Eq. (2.154), one can show in a straightforward way that the kinetic energy of the crystal

$$T = \tfrac{1}{2} \sum_{s,p} M_s \dot{\mathbf{u}}_p^s(t)^2$$

is given in terms of the normal coordinates $q_j(\mathbf{f},t)$ by

$$T = \tfrac{1}{2} \sum_{j,\mathbf{f}} |\dot{q}_j(\mathbf{f},t)|^2 . \qquad (2.168)$$

We can similarly transform the potential energy, but its form follows more readily from the fact that the equation of motion must be Eq. (2.150), which requires that

$$V = \tfrac{1}{2} \sum_{j,\mathbf{f}} [\omega_j(\mathbf{f})]^2 |q_j(\mathbf{f})|^2 . \qquad (2.169)$$

For a system of harmonic oscillators, the transition from classical mechanics to quantum mechanics is straightforward. The important Eqs. (2.150) and (2.162) still apply, but now u and q are to be regarded as operators. Being the coordinates of a particle, the $\mathbf{u}_p^s$ satisfy the commutation relations

$$[M_s \dot{u}_{p,\alpha}^s(t), u_{p',\alpha'}^{s'}(t)] = 1\hbar\delta_{pp'}\delta_{ss'}\delta_{\alpha\alpha'} . \qquad (2.170)$$

In addition, all displacements u commute with one another and all velocities $\dot{u}$ commute with one another. Now, using Eqs. (2.167) and (2.166),[j]

[j] The dagger † is used as a superscript to denote the Hermitian conjugate of an operator.

$$[\dot{q}^{\dagger}_{j'}(\mathbf{f}',t), q_j(\mathbf{f},t)] = \frac{1}{N_c}\sum_{pp'} e^{i(\mathbf{f}'\cdot\mathbf{R}_{p'}-\mathbf{f}\cdot\mathbf{R}_p)} \sum_{ss'} (M_s M_{s'})^{\frac{1}{2}} \times$$

$$\times \sum_{\alpha\alpha'} C^{s'}_{j',\alpha'}(\mathbf{f}')C^{s*}_{j,\alpha}(\mathbf{f})[\dot{u}^{s'}_{p',\alpha'}(t), u^{s}_{p,\alpha}(t)] .$$

Finally, taking account of Eqs. (2.170), (2.164), and (2.154a), we obtain

$$[\dot{q}^{\dagger}_{j'}(\mathbf{f}',t), q_j(\mathbf{f},t)] = -i\hbar\delta_{jj'}\Delta(\mathbf{f},\mathbf{f}') . \tag{2.171}$$

Further, it is evident that the $q_j(\mathbf{f},t)$ commute with one another, as do the $\dot{q}_j(\mathbf{f},t)$.

It is convenient for some purposes to separate the $q_j(\mathbf{f},t)$ into their positive and negative frequency parts;

$$q_j(\mathbf{f},t) = \left(\frac{\hbar}{2\omega_j(\mathbf{f})}\right)^{\frac{1}{2}} \left[A_j(\mathbf{f})e^{-i\omega_j(\mathbf{f})t} + A^{\dagger}_j(-\mathbf{f})e^{i\omega_j(\mathbf{f})t}\right] , \tag{2.172}$$

where

$$A_j(\mathbf{f}) = i[2\hbar\omega_j(\mathbf{f})]^{-\frac{1}{2}}[\dot{q}_j(\mathbf{f},0) - i\omega_j(\mathbf{f})\, q_j(\mathbf{f},0)] . \tag{2.173}$$

From Eq. (2.171) and the statement which follows it, we can determine that the $A_j(\mathbf{f})$ satisfy the commutation relations

$$\left.\begin{aligned} [A_j(\mathbf{f}), A^{\dagger}_{j'}(\mathbf{f}')] &= \delta_{jj'}\Delta(\mathbf{f},\mathbf{f}') \\ [A_j(\mathbf{f}), A_{j'}(\mathbf{f}')] &= 0 . \end{aligned}\right\} \tag{2.174}$$

The most general displacement of an atom in a crystal can now be written

$$u^s_p(t) = \sum_{j\mathbf{f}} \left(\frac{\hbar}{2N_c M_s \omega_j(\mathbf{f})}\right)^{\frac{1}{2}} C^s_j(\mathbf{f}) e^{i\mathbf{f}\cdot\mathbf{R}_p} \times$$

$$\times \left[A_j(\mathbf{f})e^{-i\omega_j(\mathbf{f})t} + A^{\dagger}_j(-\mathbf{f})e^{i\omega_j(\mathbf{f})t}\right], \tag{2.175}$$

and the Hamiltonian for the crystal vibration takes the form

$$H_0 = T + V = \sum_{jf} [A_j^\dagger(\mathbf{f})A_j(\mathbf{f}) + \tfrac{1}{2}]\hbar\omega_j(\mathbf{f}) \ . \tag{2.176}$$

In complete analogy to the case of a single isotropic harmonic oscillator, the energy of the normal mode with the polarization index j and the propagation vector **f** corresponding to the frequency $\omega_j(\mathbf{f})$ is given by

$$\epsilon_j(\mathbf{f}) = [n_j(\mathbf{f}) + \tfrac{1}{2}]\hbar\omega_j(\mathbf{f}) \ , \tag{2.177}$$

where $n_j(\mathbf{f})$ is a positive integer or zero. The total energy of vibration when the crystal is in an eigenstate of the Hamiltonian is

$$\epsilon = \sum_{jf} [n_j(\mathbf{f}) + \tfrac{1}{2}]\hbar\omega_j(\mathbf{f}) \ .$$

The term

$$\sum_{jf} \tfrac{1}{2}\hbar\omega_j(\mathbf{f})$$

represents the zero-point energy. Its appearance in the energy is a consequence of the uncertainty principle, which requires that the atoms cannot be at rest at their positions of equilibrium. The term $n_j(\mathbf{f})\ \omega_j(\mathbf{f})$ represents the increase in energy above that of the ground state owing to the excitation of an oscillator j to its $n_j(\mathbf{f})$*th* excited state. Instead of this description, however, one may say that there are $n_j(\mathbf{f})$ quanta of vibration with the wave number **f** and polarization j. Such quanta are called phonons. Associated with each phonon are its frequency $\omega_j(\mathbf{f})$, its propagation vector **f**, and the n polarization vectors $\mathbf{C}_j^s(\mathbf{f})$.

In this same context, it is useful to recall some properties of $A_j(\mathbf{f})$ and $A_j^\dagger(\mathbf{f})$. Let $|\ldots n_j(\mathbf{f})\ldots n_{j'}(\mathbf{f}')\ldots$ denote a state of the crystal in which there are $n_j(\mathbf{f})$ phonons with frequency $\omega_j(\mathbf{f})$ and propagation vector **f** and $n_{j'}(\mathbf{f}')$ phonons with frequency $\omega_{j'}(\mathbf{f}')$ and propagation vector **f'**. Then

$$A_j^\dagger(\mathbf{f})\left|\ldots n_j(\mathbf{f})\ldots n_{j'}(\mathbf{f}')\ldots\right\rangle = [n_j(\mathbf{f})+1]^{\frac{1}{2}}\left|\ldots n_j(\mathbf{f})+1\ldots n_{j'}(\mathbf{f}')\ldots\right\rangle ,$$

$$A_j(\mathbf{f})\left|\ldots n_j(\mathbf{f})\ldots n_{j'}(\mathbf{f}')\ldots\right\rangle = [n_j(\mathbf{f})]^{\frac{1}{2}}\left|\ldots n_j(\mathbf{f})-1\ldots n_{j'}(\mathbf{f}')\ldots\right\rangle . \tag{2.178}$$

Thus the operators $A_j^\dagger(\mathbf{f})$ and $A_j(\mathbf{f})$, which are defined in connection with the quantum mechanics of a simple harmonic oscillator, may be considered in the present context as phonon creation and annihilation operators, respectively. From Eq. (2.178) and the orthonormality condition

$$\left\langle\ldots n_j'(\mathbf{f})\ldots n_{j'}'(\mathbf{f}')\ldots\middle|\ldots n_j(\mathbf{f})\ldots n_{j'}(\mathbf{f}')\ldots\right\rangle = \ldots\delta_{n_j'(\mathbf{f}),n_j(\mathbf{f})}\cdots\delta_{n_{j'}'(\mathbf{f}'),n_{j'}(\mathbf{f}')}\cdots\Big\rangle$$

one can immediately obtain the expectation values

$$\left\langle\ldots n_j(\mathbf{f})\ldots n_{j'}(\mathbf{f}')\ldots\middle|A_j(\mathbf{f})A_{j'}(\mathbf{f}')\middle|\ldots n_j(\mathbf{f})\ldots n_{j'}(\mathbf{f}')\ldots\right\rangle = 0 ,$$

$$\left\langle\ldots n_j(\mathbf{f})\ldots n_{j'}(\mathbf{f}')\ldots\middle|A_j^\dagger(\mathbf{f})A_{j'}^\dagger(\mathbf{f}')\middle|\ldots n_j(\mathbf{f})\ldots n_{j'}(\mathbf{f}')\ldots\right\rangle = 0 ,$$

$$\left\langle\ldots n_j(\mathbf{f})\ldots n_{j'}(\mathbf{f}')\ldots\middle|A_j^\dagger(\mathbf{f})A_{j'}(\mathbf{f}')\middle|\ldots n_j(\mathbf{f})\ldots n_{j'}(\mathbf{f}')\ldots\right\rangle = n_j(\mathbf{f})\delta_{jj'}\Delta(\mathbf{f},\mathbf{f}') . \tag{2.179}$$

Here occupation numbers not explicitly indicated are to be regarded as identical.

The determination of the normal modes by solving Eq. (2.155) is a very laborious task, even when all the force constants $V_{\alpha\alpha'}^{ss'}(R_p)$ are known. For the simplest case of just one atom per unit cell, it requires the solution of three simultaneous equations for each **f**. Nevertheless, such solutions have been obtained for many substances, even for crystals containing more than one atom per unit cell. There will be occasion to illustrate the results of some of these investigations, together with experimental results that have been obtained by means of the techniques of slow-neutron scattering.

Usually a quantitative solution of the problem of lattice vibrations has as its object the determination of the dispersion relations

$$\omega = \omega_j(\mathbf{f}) \ . \tag{2.180}$$

Various approximate procedures adopted for the description of crystal properties differ from one another mainly in the assumption made about $\omega_j(\mathbf{f})$. In the simple Einstein model of a crystal, the frequency $\omega_j(\mathbf{f})$ is assumed to be a constant independent of j and **f**. In this very rough model, a crystal composed of N atoms is described as 3N - 6 3N harmonic oscillators all of the same frequency ω_0. In most cases, a better approximation is provided by the simple Debye model which (1) replaces the dispersion relation by the form that it would have in a continuous isotropic medium

$$\omega_j(\mathbf{f}) = c|\mathbf{f}| \qquad |\mathbf{f}| \leq f_0 \ ,$$

where the proportionality constant c is the velocity of sound, and (2) replaces the cell of the reciprocal lattice by a sphere of radius

$$\frac{f_0}{2\pi} = \left(\frac{3}{4\pi V_0}\right)^{\frac{1}{3}} ,$$

where V_0 is the volume of the unit cell of the crystal. The simple Einstein and Debye models, although good approximations for many purposes, give only a very rough description of the dynamical behavior of any real solid.

Let us now consider some general features of the lattice vibrations of a real crystal. First, suppose that all atoms of the crystal are uniformly displaced by the same amount in all directions; that is,

$$\mathbf{u}_p^s(t) = \mathbf{c} \ ,$$

where **c** is a constant vector. This displacement corresponds to a mode in which **f** = 0 and $\omega = 0$. Since the direction of **c** is arbitrary, there are three independent modes of this kind, to which we assign the polarization indices j = 1, 2, 3. For a simple lattice with one atom per unit cell (n = 1), these are the only modes with **f** = 0.

For n > 0, there are 3n - 3 other modes with **f** = 0. For these,

$\omega_j(0) \neq 0$. They represent vibrations in which the displacement of corresponding atoms in different unit cells, that is, atoms with the same ρ_s but different $\mathbf{R}_p$, is the same, but the atoms in the same cell move relatively to one another. In the case where n = 2, e.g., for the alkali halides, this fact, together with the fact that the total momentum associated with the vibrations must be zero, shows that the two atoms in a unit cell must move in opposite directions.

For arbitrary values of n, those 3n - 3 branches of the dispersion curves $\omega = \omega_j(\mathbf{f})$ whose frequency $\omega_j(\mathbf{f}) \neq 0$ when $\mathbf{f} = 0$ are called optical branches. The corresponding modes of vibration are called optical modes. Their name derives from the fact that these modes, and more specifically the optical modes with $\mathbf{f} \approx 0$, are the ones that are excited when infrared radiation is absorbed by an ionic crystal.

The three branches of the dispersion curves for which $\omega_j = \omega_j(\mathbf{f})$ approaches zero when $\mathbf{f} \to 0$ are the acoustical branches, and the corresponding modes are called acoustical modes. Expanding $\omega_j^2(\mathbf{f})$ about $\mathbf{f} = 0$ and retaining terms through the second order in $\mathbf{f}$, one can show that

$$\omega_j(\mathbf{f}) = c_j|\mathbf{f}| \qquad j = 1, 2, 3 . \tag{2.181}$$

In arriving at this result, we have used $\omega_j(0) = 0$ for j = 1, 2, 3 and $\omega_j^2(\mathbf{f}) = \omega_j^2(-\mathbf{f})$. Equation (2.181) is similar to the Debye approximation, except that now the velocity of sound c_j is dependent on the direction of polarization, since we have not specialized to the case of an isotropic continuum. From the way Eq. (2.181) was obtained, it is clear that the Debye approximation is generally a good one only for small values of $|\mathbf{f}|$ of, more specifically, for elastic waves, where the wavelength is long compared with interatomic distances.

Models for lattice vibrations are often constructed by assuming that only a few of the interatomic force constants $V_{\alpha\alpha'}^{ss'}(\mathbf{R}_p)$ appearing in Eq. (2.156) are different from zero. These are then adjusted to give a best fit to the experimentally determined elastic constants of the crystal, or better, to dispersion relations which have been measured by means of slow-neutron-scattering techniques (see Sec. 2-6.3). Figure 2.8, for example, shows measured dispersion relations for aluminum for two symmetric directions in the basic cell of the reciprocal lattice, namely, the cube edge and face diagonal.[36] The dashed curves were computed by Brockhouse using measured elastic constants and a force model of Begbie and Born.[37] Dispersion curves for aluminum were measured about the same time by Brockhouse and Stewart at Chalk River,[38] by Carter, Palevsky, and Hughes at Brookhaven,[39] and somewhat later by Larsson, Dahlborg, and Holmryd at Stockholm.[40] These investigators employed the techniques of slow-neutron scattering. Figure 2.8 also shows dispersion curves deduced by Walker[41] from X-ray intensity measurements, which constitute an entirely different though less direct method than that of neutron spectroscopy. The upper and lower branches of the dispersion curves in Fig. 2.8 correspond to longitudinal and transverse lattice vibrations, respectively. In the longitudinal mode $\mathbf{c}_j^s(\mathbf{f})$, and therefore the direction in which the atoms vibrate, is parallel to the propagation vector **f** of the lattice wave. In a transverse mode, the atoms vibrate in a plane perpendicular to **f**.

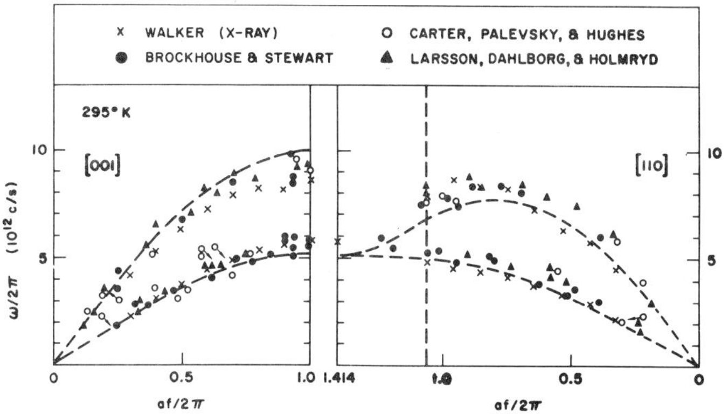

Figure 2.8 Dispersion relations in aluminum. (From Brockhouse, MMPP-NPC-1-4, Ref. 36.)

In general, the lattice vibrations are neither transversely nor longitudinally polarized. A real normal mode usually describes an elliptically polarized wave. In such a case, the atoms describe harmonic motion in an elliptical path, with the normal to the plane of the ellipse not necessarily parallel to **f**. It is only for particular directions of wave propagation that atoms vibrate either parallel or perpendicular to **f**.

In the [0,0,1] direction in aluminum, there exist transverse lattice vibrations. These modes are doubly degenerate, i.e., for each **f** there are two transverse modes with the same frequency. This is the reason that we see only two branches of the dispersion curves in Fig. 2.8 instead of the three that one generally expects for a crystal with one atom per unit cell. Figure 2.9 shows results for germanium obtained in 1957 by Brockhouse and Iyengar.[42] Germanium has two atoms per unit cell, so that there are six branches of the dispersion relations. In the symmetry directions shown, these are a longitudinal optical branch, a longitudinal acoustical branch, a doubly degenerate transverse optical branch, and a doubly degenerate transverse acoustical branch.

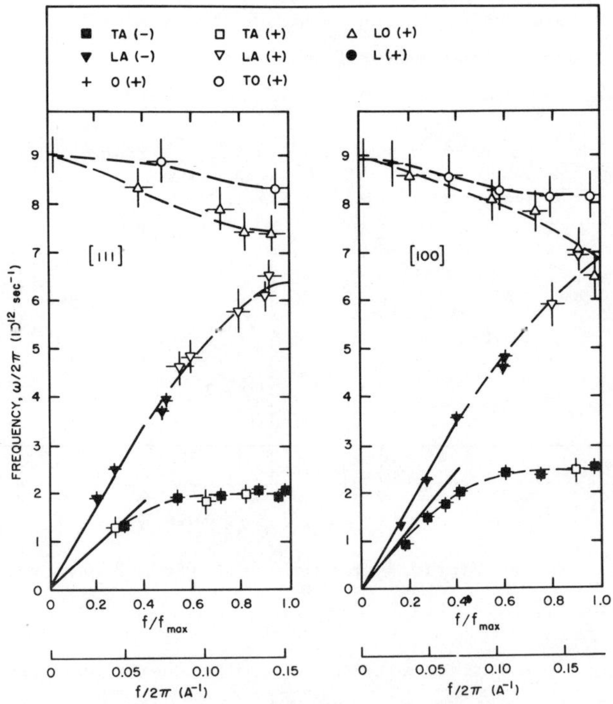

Figure 2.9 Dispersion relations in germanium. (From Brockhouse and Iyengar, Phys. Rev., Ref. 42.)

These results for germanium and aluminum are among the earliest measurements of dispersion relations. Since the time of these measurements, the methods of slow-neutron spectroscopy have been improved considerably and have been more extensively used in the study of lattice vibrations. In the next section we consider the theoretical basis underlying the efficacy of slow neutrons as a tool for studying atomic motions in solids.

2-6.3 INELASTIC SCATTERING. The angular and energy distribution of inelastically scattered neutrons with an initially low energy depend sensitively on the exact form of the dispersion relation and the details of the vibration spectrum. In fact, by observing the energy and angle distributions for the case of scattering by single crystals, one can determine the dispersion relations $\omega = \omega_j(\mathbf{f})$ between the frequency and the wave vector and, in certain cases, the frequency distribution function $\rho(\omega)$, defined as the fraction of normal vibrations per unit frequency interval.[43,44]

The sensitivity of the distributions in energy and angle to crystal dynamics is most strongly exhibited in coherent one-phonon processes. For an incident wavelength of the order of a lattice constant, the changes of absolute energy in a given direction of scattering are of the same order of magnitude as the incident energy. This is in distinct contrast to X-ray scattering by a crystal, in which case the energy change of a photon, still of the order of average phonon energies, is an extremely small fraction of the energy of the incident X-ray. The small relative size of the wavelength shift for X-ray scattering is not readily observable, and hence $\omega_j(\mathbf{f})$ must be inferred from measurements of scattered intensity. The coherent one-phonon scattering of neutrons results in peaks in the neutron energy distribution at energies corresponding to energy transfers of $\hbar\omega = \hbar\omega_j(\mathbf{f})$. To determine the $\omega_j(\mathbf{f})$, one is not required to know the scattered intensity, but only the energies at which the peaks occur.

In the above remarks we have been concerned only with the coherent one-phonon scattering. The energy distribution of neutrons incoherently scattered by one-phonon processes is also of considerable interest in connection with crystal dynamics, especially for cubic crystals, for which the scattered energy distribution of neutrons and the frequency distribution function of the crystal are simply related. For multiphonon processes, coherent as well as incoherent, the neutron scattering is related in a very complicated and less sensitive way to the crystal dynamics.[45] For the purpose of determining the dispersion relations and the frequency spectrum, these processes are therefore of less interest than the one-phonon processes. On the other hand, multiphonon processes dominate the approach of a burst of initially nonthermal neutrons to equilibrium. For the present, we shall concentrate on the one-phonon processes, taking up the calculation of multiphonon processes in Sec. 3-2.1.

Our first objective is to cast the general equation (2.91) for $\chi_c(\kappa,t)$ into a form which is more convenient for discussing neutron scattering by a crystal. Using Eq. (2.122),

$$\chi_c(\kappa,t) = \frac{1}{nN_c} \sum_{pp'ss'} e^{-i\kappa\cdot(R_{p,s}-R_{p',s'})} \times$$

$$\times \left\langle e^{-i\kappa\cdot u_p^s(0)}\, e^{i\kappa\cdot u_{p'}^{s'}(t)} \right\rangle_T , \tag{2.182}$$

where

$$R_{p,s} = R_p + \rho_s \ .$$

From Eq. (2.105) and Bloch's theorem, Eq. (2.107),

$$\left\langle e^{-i\kappa\cdot u_p^s(0)}\, e^{i\kappa\cdot u_{p'}^{s'}(t)} \right\rangle_T$$

$$= \exp\left\{\tfrac{1}{2}\left[\kappa\cdot u_p^s(0), \kappa\cdot u_{p'}^{s'}(t)\right]\right\} \times$$

$$\times \exp\left\{-\tfrac{1}{2}\left\langle\left[\kappa\cdot u_{p'}^{s'}(t) - \kappa\cdot u_p^s(0)\right]^2\right\rangle_T\right\}.$$

Here, we have used the fact that the commutator of the displacement operators $u_p^s(0)$ and $u_{p'}^{s'}(t)$ is a c-number for the potential (2.146); that is, the commutator commutes with any displacement or momentum operator. Indeed, from the commutation relations (2.174) and Eq. (2.175) it is easy to show that for the case when all atoms have the same mass $M = M_s$

$$\left[\kappa \cdot u_p^s(0), \kappa \cdot u_{p'}^{s'}(t)\right] = \frac{\hbar i}{MN_c} \sum_{jf} \left[\kappa \cdot C_j^s(f)\right]^* \left[\kappa \cdot C_j^{s'}(f)\right] \times$$

$$\times\ e^{if\cdot(R_{p'}-R_p)} \left[\frac{\sin \omega_j(f)t}{\omega_j(f)}\right], \qquad (2.183)$$

Here we have also used Eqs. (2.160) and (2.160a). Again using Eqs. (2.174) and (2.175), in addition to Eq. (2.179), we find that

$$\left\langle \kappa \cdot u_{p'}^{s'}(t')\kappa \cdot u_p^s(t)\right\rangle_T = \frac{\hbar}{2MN_c} \sum_{jf} \kappa \cdot C_j^{s'}(f) \left[\kappa \cdot C_j^s(f)\right]^* \times$$

$$\times\ e^{if\cdot(R_{p'}-R_p)} g_j(f,t - t'), \qquad (2.184)$$

where

$$g_j(f,t - t') = \frac{\left\{\bar{n}_j(f)e^{-i\omega_j(f)(t-t')} + \left[\bar{n}_j(f) + 1\right]e^{i\omega_j(f)(t-t')}\right\}}{\omega_j(f)}, \qquad (2.185)$$

and, using Eq. (2.179),

$$\bar{n}_j(f) = \bar{n}\left[\omega_j(f)\right] = \left\langle A_j^\dagger(f)A_j(f)\right\rangle_T$$

$$= \frac{\sum_{n_j(f)=0}^{\infty} n_j(f) \exp\left[-\beta n_j(f)\hbar\omega_j(f)\right]}{\sum_{n_j(f)=0}^{\infty} \exp\left[-\beta n_j(f)\hbar\omega_j(f)\right]}$$

$$= \left\{\exp\left[\beta\hbar\omega_j(f)\right] - 1\right\}^{-1} \qquad (2.186)$$

is the average number of phonons with the frequency $\omega_j(f)$ and wave vector **f**. This result for $\bar{n}_j(f)$ is equivalent to Eq. (2.112a), and shows the formal similarity between phonons as quantized excitations of the lattice vibrations and photons as quantized excitations of the electromagnetic field. In both cases, the average population is given by Bose-Einstein statistics. This arises entirely from the properties of the harmonic oscillator in quantum mechanics, and has no connection with the symmetry properties of the many-particle wave function. Equation (2.184) with t' = t, p' = p, and s' = s gives the Debye-Waller factor

$$2W_s(\kappa) = \left\langle [\kappa \cdot \mathbf{u}_p^s(t)]^2 \right\rangle_T = \left\langle [\kappa \cdot \mathbf{u}_p^s(0)]^2 \right\rangle_T$$

$$= \frac{\hbar}{2MN_c} \sum_{jf} \frac{\left[\kappa \cdot \mathbf{C}_j^s(\mathbf{f})\right]^2}{\omega_j(\mathbf{f})} \left\{ 2\bar{n}\left[\omega_j(\mathbf{f})\right] + 1 \right\} . \tag{2.187}$$

Combining all of our results, we finally obtain

$$\left\langle \exp\left[-i\kappa \cdot \mathbf{u}_p^s(0)\right] \exp\left[i\kappa \cdot \mathbf{u}_{p'}^{s'}(t)\right] \right\rangle_T$$

$$= \exp[-W_s(\kappa) - W_{s'}(\kappa)] \times$$

$$\times \exp\left\{ \frac{\hbar}{2MN_c} \sum_{jf} [\kappa \cdot \mathbf{C}_j^s(\mathbf{f})]^* [\kappa \cdot \mathbf{C}_j^{s'}(\mathbf{f})] \times \right.$$

$$\left. \times \exp[i\mathbf{f} \cdot (\mathbf{R}_{p'} - \mathbf{R}_p)] g_j(\mathbf{f},t) \right\} . \tag{2.188}$$

The coherent scattering by a harmonically vibrating crystal consisting of one atomic species follows from Eqs. (2.68), (2.88), (2.182), and (2.188):

$$\frac{d^2\sigma_c}{dE d\Omega} = \sigma_{c,b} \frac{k}{8\pi^2 \hbar k_0} \sum_{s,s'} \exp[-i\kappa \cdot (\rho_s - \rho_{s'})] \times$$

$$\times \exp[-W_s(\kappa) - W_{s'}(\kappa)] \sum_p e^{i\kappa \cdot \mathbf{R}_p} \int_{-\infty}^{\infty} dt e^{-i\omega t} \times$$

$$\times \exp\left\{ \frac{\hbar V_0}{2M} \left(\frac{1}{2\pi}\right)^3 \sum_j \int_{\text{cell}} d\mathbf{f} [\kappa \cdot \mathbf{C}_j^s(\mathbf{f})]^* \times \right.$$

$$\left. \times [\kappa \cdot \mathbf{C}_j^{s'}(\mathbf{f})] e^{-i\mathbf{f} \cdot \mathbf{R}_p} g_j(\mathbf{f},t) \right\} . \tag{2.189}$$

Here we have used the periodicity of the lattice to express

$$\frac{1}{N_c}\sum_{p,p'} e^{i\mathbf{f}\cdot(\mathbf{R}_{p'}-\mathbf{R}_p)} Q(\mathbf{R}_{p'} - \mathbf{R}_p) = \sum_p e^{-i\mathbf{f}\cdot\mathbf{R}_p} Q(\mathbf{R}_p) \;, \qquad (2.190)$$

which holds for any Q defined on the lattice, and we have replaced the summation over **f** by

$$\frac{1}{N_c}\sum_{\mathbf{f}} (\cdots) = \frac{V_0}{(2\pi)^3}\int_{\text{cell}} (\cdots)d\mathbf{f} \;, \qquad (2.191)$$

where the integration goes over the first Brillouin zone. The validity of this replacement in the limit of large N_c follows, on recalling that the density of points in **f** space is equal to $N_c V_0/(2\pi)^3$, where V_0 is the volume of a unit cell of the direct lattice.

It is now a simple matter to express the incoherent intermediate scattering function in terms of normal-mode vectors and frequencies. On comparing Eq. (2.90) with Eq. (2.91), we notice that χ_{inc} follows from χ_c by ignoring the terms in Eq. (2.182) for which p' ≠ p and s' ≠ s. Thus,

$$\begin{aligned} {}_{inc}(\varkappa, t) &= \frac{1}{nN_c}\sum_{p,s}\left\langle \exp\left[-i\varkappa \cdot \mathbf{u}_p^s(0)\right] \exp\left[i\varkappa \cdot \mathbf{u}_p^s(t)\right]\right\rangle_T \\ &= \frac{1}{n}\sum_{s=1}^{n} \exp[-2W_s(\varkappa)] \times \\ &\quad \times \exp\left[\frac{\hbar}{2MN_c}\sum_{j\mathbf{f}}\left|\boldsymbol{\kappa}\cdot\mathbf{c}_j^s(\mathbf{f})\right|^2 g_j(\mathbf{f},t)\right] , \end{aligned} \qquad (2.192)$$

and

$$\begin{aligned} \frac{d^2\sigma_{inc}}{dE d\Omega} &= \sigma_{inc,b}\,\frac{k}{8\pi^2\hbar n k_0}\sum_s \exp[-2W_s(\boldsymbol{\kappa})]\int_{-\infty}^{\infty} dt e^{-i\omega t} \times \\ &\quad \times \exp\left[\frac{\hbar V_0}{2M}\left(\frac{1}{2\pi}\right)^3 \times \right. \\ &\quad \left. \times \sum_j \int_{\text{cell}} d\mathbf{f}\left|\boldsymbol{\kappa}\cdot\mathbf{c}_j^s(\mathbf{f})\right|^2 g_j(\mathbf{f},t)\right] . \end{aligned} \qquad (2.193)$$

The expressions for the intermediate scattering functions are still very complicated; to evaluate the cross sections from them by means of Eqs. (2.88) and (2.64) presents a formidable task even when the normal-mode frequencies and polarizations are known. Nevertheless, it is possible to extract much of the physical content of Eqs. (2.189) and (2.193) without becoming involved in too many mathematical details.

To illustrate some of the physical content of Eqs. (2.189) and (2.193) we consider the first few terms in the phonon expansion of the coherent and incoherent scattering cross sections. The phonon expansion consists in expanding the functions

$$\exp\left[\frac{\hbar V_0}{2M}(\cdots)\right]$$

in Eqs. (2.189) and (2.193) in powers of its argument. The first two terms in the expansion correspond to elastic scattering and to scattering in which one phonon is exchanged between the neutron and the crystal. Let

$$\frac{d^2\sigma}{dEd\Omega} = \frac{d^2\sigma_0}{dEd\Omega} + \frac{d^2\sigma_1}{dEd\Omega} + \cdots ,$$

where $d^2\sigma_r/dEd\Omega$ represents the contribution to the cross section from processes which involve the exchange of r-phonons between the neutron and the crystal. The omission of a subscript c or inc means that the expansion applies for both the coherent and incoherent scattering. The coherent zero-phonon or elastic term is given by

$$\frac{d^2\sigma_{c,0}}{dEd\Omega} = \frac{\sigma_{c,b}}{4\pi\hbar n}\sum_{s,s'} \exp[-i\kappa \cdot (\rho_s - \rho_{s'})] \times$$

$$\times \exp[-W_s(\kappa) - W_{s'}(\kappa)]\sum_p \exp[i\kappa \cdot R_p]\delta(\omega)$$

$$= \frac{(2\pi)^3}{V_0}\frac{\sigma_{c,b}}{4\pi n}\sum_{s,s'} \exp[-i\tau(\rho_s - \rho_{s'})] \times$$

$$\times \exp[-W_s(\tau) - W_{s'}(\tau)]\sum_\tau \delta(\kappa - \tau)\delta(E_0 - E) , \quad (2.194)$$

where we have used Eqs. (2.190) with Q = 1, (2.136), and (2.137). This expression should be compared with Eq. (2.139) in Sec. 2-6.1. Since the elastic scattering is discussed at length there, it will not be treated any further here.

The coherent one-phonon term is given by

$$\frac{d^2\sigma_{c,1}}{dE d\Omega} = \frac{\sigma_{c,b}}{2M} \frac{k}{8\pi^2 \hbar n k_0} \sum_{s,s'} \exp[i\kappa \cdot (\rho_s - \rho_{s'})] \times$$

$$\times \exp[-W_s(\kappa) - W_{s'}(\kappa)] \frac{\hbar V_0}{(2\pi)^3} \sum_p e^{i\kappa \cdot R_p} \int_{-\infty}^{\infty} dt e^{-i\omega t} \times$$

$$\times \sum_j \int_{\text{cell}} df e^{if \cdot R_p} [\kappa \cdot C_j^s(f)]^* [\kappa \cdot C_j^{s'}(f)] \times$$

$$\times \frac{[\bar{n}_j(f) + 1] \exp[i\omega_j(f)t] + \bar{n}_j(f) \exp[-i\omega_j(f)t]}{\omega_j(f)}$$

$$= \frac{\sigma_{c,b}}{2M} \frac{k}{4\pi n k_0} \times$$

$$\times \sum_{s,s'} \exp[-i\kappa \cdot (\rho_s - \rho_{s'})] \exp[-W_s(\kappa) - W_{s'}(\kappa)] \times$$

$$\times \sum_{\tau} \sum_{j,\epsilon} \int_{\text{cell}} df [\kappa \cdot C_j^s(f)]^* [\kappa \cdot C_j^{s'}(f)] \times$$

$$\times \frac{[n_j(f) + \frac{1}{2}(1 + \epsilon)] \delta[\omega - \epsilon\omega_j(f)]}{\omega_j(f)} \delta(\kappa - f - \tau), \quad (2.195)$$

where $\epsilon = \pm 1$, that is, $\epsilon = +1$ and $\epsilon = -1$, corresponding to the emission and absorption, respectively, by the neutron of a phonon of frequency $\omega_j(f)$. The summation on ϵ is performed over both of its values.

Consider a coherent one-phonon scattering process in which an initially monoenergetic neutron with momentum $\hbar k_0$ emerges with a final momentum $\hbar k$. Experimentally these processes are manifested as discrete peaks in the energy distribution of scattered neutrons. These peaks stand out above the smooth background distribution arising from incoherent processes and coherent multiphonon processes. The wave vector of the phonon which is exchanged between the lattice and the neutron must be given by

$$\mathbf{f} = \mathbf{k}_0 - \mathbf{k} - \boldsymbol{\tau} , \tag{2.196}$$

and its energy must be

$$\hbar\omega_j(\mathbf{f}) = \epsilon(\hbar^2/2m)(k_0^2 - k^2) . \tag{2.197}$$

These two equations form the quantitative basis for the investigation of lattice dynamics by means of neutron scattering. Equation (2.196) resembles a conservation law but is not an expression of conservation of momentum. The quantity $\hbar\mathbf{f}$ is not the momentum of a phonon, since the physical momentum corresponding to any normal vibration is zero. The second equation, on the other hand, expresses conservation of energy in a completely physical sense.

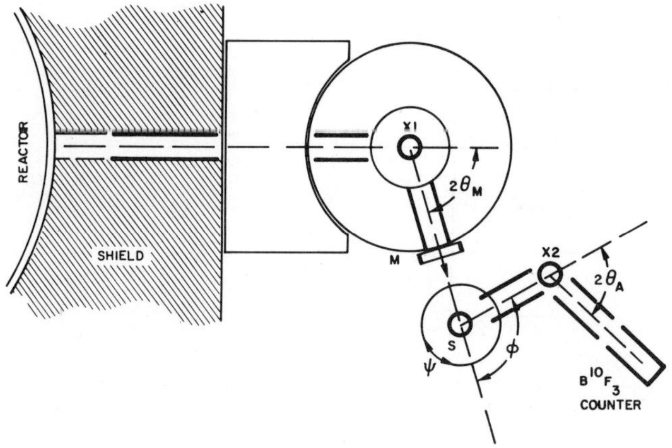

Figure 2.10 Apparatus for measuring dispersion relations. (From Brockhouse, Report MMPP-NPC-1-4, Ref. 36.)

From Eqs. (2.196) and (2.197), it is easy to understand how measurements of neutron scattering can be used to determine dispersion relations in solids. Figure 2.10 shows an actual experimental arrangement.[36] Neutrons from the pile fall on the monochromating crystal X1. The neutrons of the desired wavelength are Bragg-reflected from the monochromator and impinge on the single crystal at S. Scattered neutrons are observed at some angle ϕ, and their energies are measured by Bragg reflection from the analyzing crystal X2. The wavelengths are determined from the angles $2\theta_M$ and $2\theta_A$, using the Bragg law.

Suppose that experiments are carried out at a fixed angle of scattering and for a fixed orientation of the single-crystal sample with respect to the almost monochromatic neutron beam incident on it. With the analyzing crystal set to select a particular wavelength, the final momentum of a neutron which can enter the detector is completely determined. Now, from Eq. (2.196) we can find the wave vector **f** of the phonon which was exchanged between the neutron and the sample. First, $\boldsymbol{\tau}$ must be known. To find $\boldsymbol{\tau}$ we must remember that **f** is restricted to the first Brillouin zone of the reciprocal lattice. Thus, $\boldsymbol{\tau}$ is the reciprocal lattice vector which gives the smallest value of $|\mathbf{k} - \mathbf{k}_0 + \boldsymbol{\tau}|$. Having found $\boldsymbol{\tau}$, **f** is uniquely determined. Now, Eq. (2.195) shows that a scattering process involving one phonon of wave vector **f** will not correspond to a detectable event unless one of the frequencies $\omega_j(\mathbf{f})$ corresponding to **f** satisfies the energy-conservation condition, Eq. (2.197). This does not happen in general, but, as the energy selected by the analyzing crystal is varied, one can show that a condition is eventually reached when both Eq. (2.196) and Eq. (2.197) are satisfied. Consider, for example, the case of incident neutrons of very small energy, $k_0 \to 0$. In this case the neutron can only absorb phonons. For any direction of scattering, that is, for any direction of **f**, one can show[43,44] that the scattering surfaces $\omega = \omega_j(\mathbf{f})$ are intersected a finite number of times by any straight line extending from $\mathbf{f} = 0$ to a point on a sphere of radius

$$k_{max} = (2m\omega_{max}/\hbar)^{\frac{1}{2}}$$

centered at $\mathbf{f} = 0$. Here ω_{max} is the maximum of $\omega_j(\mathbf{f})$ for all j and **f**. The final values of the neutron momentum are restricted to lie in the interval

$$0 \le k \le k_{max} \ . \qquad (2.198)$$

Physically, therefore, when zero-energy neutrons coherently absorb a single phonon, they are scattered in every direction; for each direction, the energy of the scattered neutron has a finite number of discrete values in the interval defined by (2.198). In principle, there are three times as many discrete outgoing energies as there are atoms in a unit cell, thus yielding 3n points on the dispersion curve. (Frequently, however, fewer than 3n peaks in the energy distribution are observed experimentally because of the degeneracy of some branches of the dispersion relations for particular directions of **f**, or because of low scattered intensities and high backgrounds.) By repeating the experiments for different crystal orientations, for different angles of scattering, and perhaps for different incident energies, the dispersion relations can be constructed. For a more detailed discussion of the energy distributions of neutrons which undergo one-phonon coherent scattering, including the case of finite wavelengths, the reader is referred to the article of Placzek and Van Hove.[43]

Figures 2.11 and 2.12 show results of the first experiments of Brockhouse made in 1955.[38] Distributions of the counting rate versus the setting of the analyzing spectrometer[46] are plotted in Fig. 211(a), in which is incorporated an auxiliary scale of the outgoing energy. The incident energy was 0.063 eV. The experiments were performed for a fixed angle of scattering, the five distributions (a through e in Fig. 2.11) corresponding to five orientations of the crystal samples with respect to the incident neutron beam, as illustrated in the reciprocal lattice diagram in Fig. 2.12. The positions of the centers of the neutron groups are plotted as indicated by the code numbers. The triads of numbers show the positions of the relevant reciprocal lattice points. The dispersion relations in Fig. 2.8 for aluminum were constructed from experimental results such as those described here.

So far, we have not considered the effects of incoherent scattering, nor of coherent scattering with the emission and absorption of two or more phonons. The first term in the phonon expansion of the incoherent cross section, the elastic term

$$\frac{d^2\sigma_{inc,0}}{dEd\Omega} = \frac{\sigma_{inc,b}}{4\pi\hbar n}\sum_s \exp[-2W_s(\kappa)]\,\delta(\omega) \qquad (2.199)$$

contains no features which are essentially different from the case of

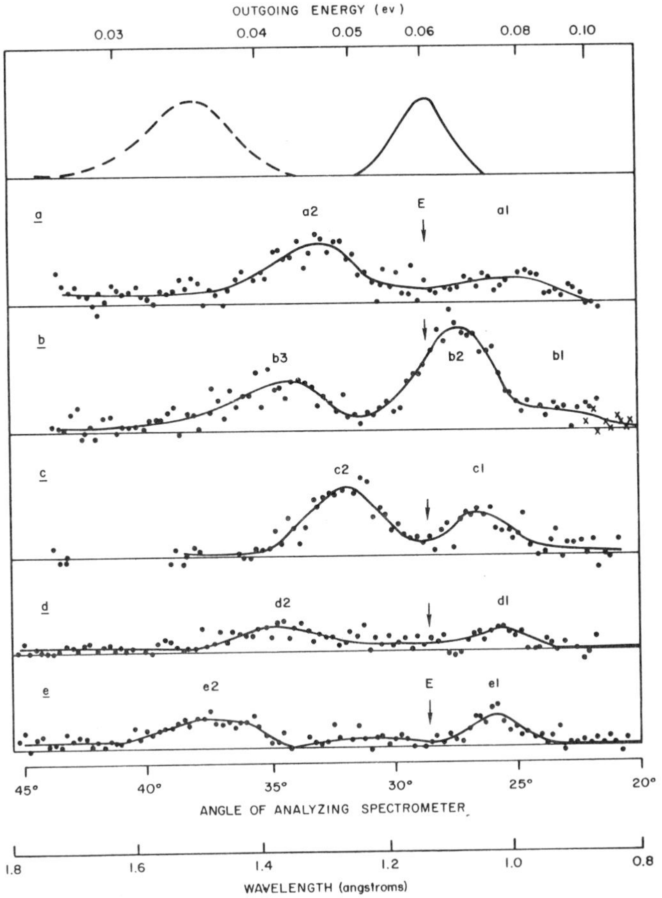

Figure 2.11 Counting rate versus setting of the analyzing spectrometer for five orientations of the crystalline sample with respect to the incident neutron beam. (From Brockhouse, AECL Report No. 211, Ref. 46.)

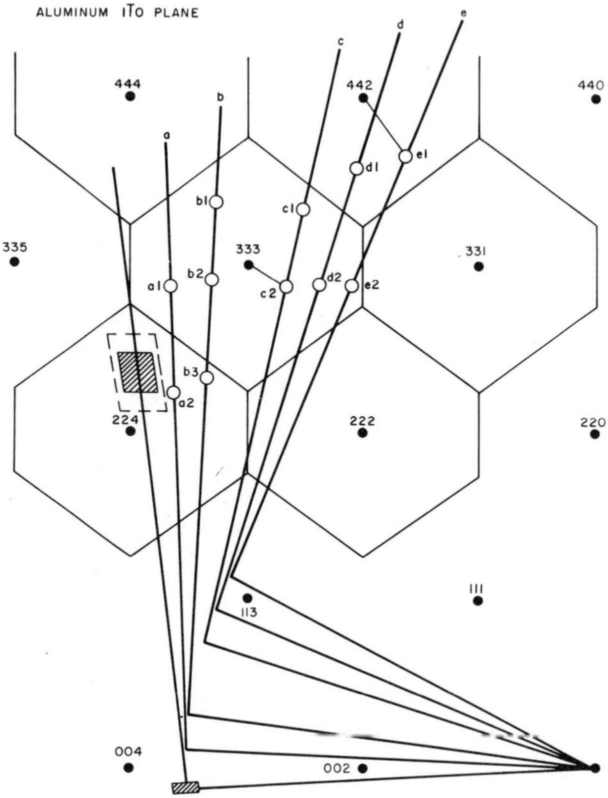

Figure 2.12 Relevant reciprocal lattice points for the experimental results shown in Fig. 2.11. (From Brockhouse, AECL Report No. 221, Ref. 46.)

elastic scattering by an Einstein crystal, and is not discussed further here. The incoherent one-phonon term is given by

$$\frac{d^2\sigma_{\text{inc},1}}{dEd\Omega}(\mathbf{k}_0 \to \mathbf{k}) = \sigma_{\text{inc},b}\,\frac{k}{4\pi\hbar n k_0}\sum_s \exp[-2W_s(\boldsymbol{\kappa})] \times$$

$$\times \sum_{j,\epsilon} \frac{1}{2M}\,\frac{\hbar V_0}{(2\pi)^3}\int_{\text{cell}} d\mathbf{f}\;\frac{\left|\boldsymbol{\kappa}\cdot \mathbf{C}_j^s(\mathbf{f})\right|^2}{{}_j(\mathbf{f})} \times$$

$$\times\,[\bar{n}_j(\mathbf{f}) + \tfrac{1}{2}(1+\epsilon)]\,\delta[\omega - \epsilon\omega_j(\mathbf{f})]\,. \qquad (2.200)$$

In this expression the δ-function corresponding to the conservation condition, Eq. (2.196), does not appear. As a consequence, for each direction of scattering, the energy distribution of neutrons resulting from incoherent one-phonon scattering does not contain the discrete spectrum characteristic of coherent one-phonon scattering. Nevertheless, the neutron energy distributions associated with incoherent one-phonon scattering contain considerable information about the crystal dynamics, though less than that which is contained in the discrete distributions associated with coherent one-phonon scattering. The case of the incoherent scattering by a cubic crystal having one atom per unit cell is of particular interest. In this case, the three polarization vectors are mutually orthogonal to one another and are of equal magnitude. Thus, from Eq. (2.158),

$$|c_{j,\alpha}(\mathbf{f})|^2 = 1/3 \qquad j = 1,2,3.$$

It is useful now to introduce the partial frequency-distribution functions $\rho_j(\omega)$, defined as the number of normal vibrations in the j*th* branch per unit frequency interval divided by the total number of vibrations in the j*th* branch,

$$\left.\begin{aligned} \rho_j(\omega)d\omega &= \frac{V_0}{(2\pi)^3} \int_{\omega \le \omega_j(\mathbf{f}) \le \omega + d\omega} d\mathbf{f} \\ \rho_j(\omega)d\omega &= 0 \qquad\qquad \omega_j(\mathbf{f}) \ge \omega_{j\max} \end{aligned}\right\} \tag{2.201}$$

for $j = 1,\ldots,3n$ and the total frequency distribution function

$$\rho(\omega) = \frac{1}{3n} \sum_{j=1}^{3n} \rho_j(\omega) . \tag{2.202}$$

The frequency distribution functions are normalized; i.e.,

$$\int_0^\infty \rho_j(\omega)d\omega = \int_0^\infty \rho(\omega)d\omega = 1 . \tag{2.203}$$

Clearly, $\rho(\omega)d\omega$ is proportional to the volume in $\mathbf{f}$ space contained between two infinitesimally separated surfaces of constant frequency. For a cubic crystal with one atom per unit cell, we now find

$$\frac{d^2\sigma_{inc,1}}{dEd\Omega} = \sigma_{inc,b} \frac{k}{8\pi k_0 M} e^{-2W(\boldsymbol{\kappa})} \sum_{\epsilon} \frac{\kappa^2}{3n} \sum_{j=1}^{3} \int_0^{\omega_{jmax}} \frac{\rho_j(\omega')}{\omega'} \times$$

$$\times [\bar{n}(\omega') + \tfrac{1}{2}(1+\epsilon)]\delta(\omega - \epsilon\omega')d\omega'$$

$$= \sigma_{inc,b} \frac{k}{4\pi k_0} e^{-2W(\boldsymbol{\kappa})} \frac{\kappa^2}{2M} \sum_{\epsilon} \int_0^{\omega_{max}} \frac{\rho(\omega')}{\omega'} \times$$

$$\times [n(\omega') + \tfrac{1}{2}(1+\epsilon)]\delta(\omega - \epsilon\omega')d\omega' \ . \qquad (2.204a)$$

Similarly, the general expression (2.193) in this case becomes

$$\frac{d^2\sigma_{inc}}{dEd\Omega} = \sigma_{inc,b} \frac{k}{8\pi^2 \hbar n k_0} e^{-2W(\boldsymbol{\kappa})} \int_{-\infty}^{\infty} dt e^{-i\omega t} \times$$

$$\times \exp\left(\frac{\hbar\kappa^2}{2M} \int_0^{\omega_{max}} \frac{\rho(\omega')}{\omega'} \times \right.$$

$$\left. \times \{[\bar{n}(\omega') + 1]e^{i\omega' t} + \bar{n}(\omega')e^{-i\omega' t}\}d\omega'\right) \ . \qquad (2.204b)$$

Defining $\rho(\omega')$ for negative value of ω' by

$$\rho(-\omega') = \rho(\omega') \ , \qquad (2.205)$$

we may extend the lower limit of integration from 0 to $-\omega_{max}$. Doing this and performing the integration gives

$$\frac{d^2\sigma_{inc,1}}{dEd\Omega} = \sigma_{inc,b} \frac{k}{8\pi k_0} e^{-2W(\boldsymbol{\kappa})} \frac{\kappa^2}{2M} \frac{\rho(\omega)}{\omega \sin h(\hbar\omega/2k_B T)} \times$$

$$\times \exp(\hbar\omega/2k_B T) \ . \qquad (2.206)$$

Thus, the one-phonon incoherent scattering is simply related to the frequency distribution function of the crystal. More generally, one can obtain precisely the same result for any monatomic polycrystal, provided that the Debye-Waller factor does not vary from atom to atom in the unit cell and does not depend strongly on the direction of $\boldsymbol{\kappa}$. These conditions apply to a very high accuracy if $W_s(\boldsymbol{\kappa}) << 1$. To show that Eq. (2.206) is valid for monatomic polycrystals when $W_s(\boldsymbol{\kappa}) << 1$, average Eq. (2.200) over all crystal orientations or, equivalently, over all directions of κ for a crystal of fixed orientation. Since

$$\left\langle \left| \kappa \cdot \mathbf{C}_j^s(\mathbf{f}) \right|^2 \right\rangle_R = (\kappa^2/3) \left| \mathbf{C}_j^s(\mathbf{f}) \right|^2 , \tag{2.207}$$

the normalization condition (2.158) together with Eqs. (2.201) and (2.202) lead immediately to the desired result.

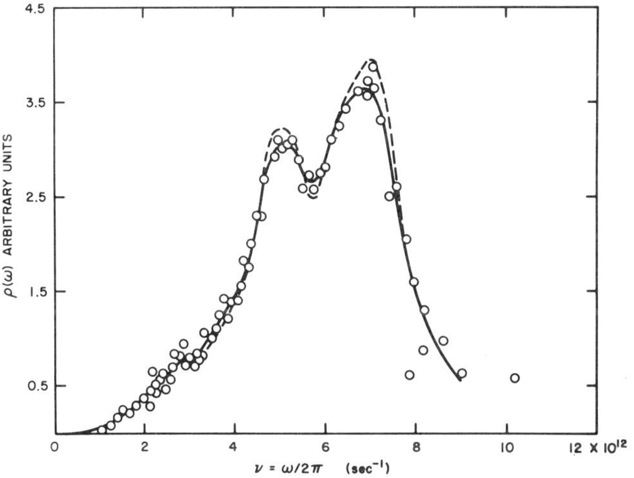

Figure 2.13 The lattice vibration spectrum of vanadium. The solid line is the spectrum obtained by assuming a monochromatic incident neutron beam, and the dashed line shows the correction for the actual energy distribution of the incident beam. (From Eisenhauer et al., Phys. Rev., Ref. 48.)

The frequency distribution of vanadium has been studied experimentally by many investigators.[47-51] Figure 2.13 shows the $\rho(\omega)$ for vanadium which Eisenhauer, et al.,[48] have deduced from the results of slow-neutron scattering experiments. Since vanadium is a body-centered cubic crystal having only one atom per unit cell, and since it is an almost perfectly incoherent scatterer of neutrons, the methods of neutron spectroscopy are very well suited for determining its vibration spectrum.

There remains to be discussed the neutron energy distributions associated with coherent and incoherent multiphonon processes. We do not study these processes here. In Sec. 3-2.1, multiphonon scattering is discussed in greater detail. It suffices to state that for initial neutron energies which are small compared with $\hbar\omega_{max}$, and for sample temperatures which are small compared with $\hbar\omega_{max}/k_B$, the multiphonon terms constitute a negligible contribution to the neutron inelastic scattering. Furthermore, as we have already noted, the discrete peaks in the one-phonon coherent scattering stand out above the background distribution arising from incoherent processes and coherent multiphonon processes.

The background distribution is not completely smooth. For a cubic crystal with one atom per unit cell, the energy distribution of neutrons associated with incoherent one-phonon processes reflects the discontinuities which are present in the derivative of the frequency distribution $\rho(\omega)$.[43] Sjölander[45] has shown that the energy distribution of neutrons scattered in coherent two-phonon processes, for certain directions of scattering, can have logarithmic singularities at the energies corresponding to those of the one-phonon peaks. Furthermore, this distribution exhibits singularities of the kind that occur in incoherent one-phonon scattering.

This concludes our discussion of neutron scattering as it relates to the determination of the lattice dynamics of crystalline solids. In Chap. 3 we discuss further neutron scattering by solids, emphasizing those aspects of the problem which are most relevant for understanding the thermalization of neutrons by moderators.

2.7 NEUTRON SCATTERING BY MOLECULAR GASES

Now we consider the problem of neutron scattering by molecular gases. Section 2-1 already contains an approximate treatment of this problem. Equation (2.13) is valid provided (1) that only one nucleus in the molecule scatters neutrons, (2) that the temperature is so low that there is no appreciable thermal excitation of molecules to excited states, and (3) that the energy of the neutron is much greater than k_BT but less than that which is required to raise the molecule to an excited state. In practice, however, these restrictions are so severe that the simple result (2.13) is seldom applicable.

In neutron scattering from most molecules there exist effects due to interference between waves scattered from different atoms in the molecule. Although appreciable, these effects are not so conspicuous as in the case of coherent scattering from solids. This is to be expected, since the number of atoms in a molecule is small in comparison with the number of atoms having statically correlated positions in a crystal.

As in solids, the inelastic scattering of slow neutrons is strongly influenced by the dynamics of the scattering system. The molecular dynamics is conveniently described in terms of the motion of the molecular center of mass, the rotation of the rigid equilibrium configuration, and the small vibrations about the positions of equilibrium. In practical applications involving molecules that contain a small number of atoms, the characteristic frequencies of vibration are often much greater than both $k_B T$ and the characteristic frequencies of rotation. Thus, the translations and rotations of a quasi-rigid molecule can be studied experimentally. The departures from rigidity arise from motions in the vibrational ground state. Since numerous measurements have been made of slow-neutron scattering by molecules in their vibrational ground state,[52-55] we consider this case in some detail. First we discuss the general formalism as it applies to molecules with vibrational degress of freedom which can be excited either thermally or by means of the neutron.

The formal result, Eq. (2.70), which applies to the case of scattering by a sample consisting of a single atomic species, is now generalized to include the case where more than one type of atom is present in the scatterer. Since this generalization is easy to make, we only write down the results.[56] In doing this it is convenient to express the scattering per molecule as the sum of a direct part and an interference part.

$$\frac{d^2\sigma}{dEd\Omega} = \frac{d^2\sigma_d}{dEd\Omega} + \frac{d^2\sigma_{int}}{dEd\Omega} ,$$

where

$$\frac{d^2\sigma_d}{dEd\Omega} = \sum_{s=1}^{n}\left[\left(a^s_{c,b}\right)^2 + \left(a^s_{inc,b}\right)^2\right]\frac{k}{\hbar k_0}\frac{1}{2\pi} \times$$

$$\times \int_{-\infty}^{\infty} e^{-i\omega t}\chi_{ss}(\boldsymbol{\kappa},t)dt \tag{2.208}$$

and

$$\frac{d^2\sigma_{int}}{dEd\Omega} = \sum_{s,s'=1}^{n}{}' a^s_{c,b}a^{s'}_{c,b}\frac{k}{\hbar k_0}\frac{1}{2\pi}\int_{-\infty}^{\infty} e^{-i\omega t}\chi_{ss'}(\boldsymbol{\kappa},t)dt \ . \tag{2.209}$$

The $a^s_{c,b}$ and $a^s_{inc,b}$ are the bound atom coherent and incoherent scattering lengths, respectively, of the *sth* nucleus. The summation in Eq. (2.209) extends over all n nuclei in the molecule, s and s' taking values which are different from one another. In writing Eq. (2.209) we have ignored spin correlations between any possibly identical nuclear constituents of the molecule. The intermediate scattering functions $\chi_{ss'}(\boldsymbol{\kappa},t)$, which contain the effects of the molecular dynamics, are given by

$$\chi_{ss'}(\boldsymbol{\kappa},t) = \left\langle e^{-i\boldsymbol{\kappa}\cdot\mathbf{r}_s(0)} e^{i\boldsymbol{\kappa}\cdot\mathbf{r}_{s'}(t)}\right\rangle_T \ . \tag{2.210}$$

2-7.1 MOLECULAR DYNAMICS. In order to treat the molecular dynamics and to correlate it with neutron scattering, it is convenient to separate the nuclear coordinates into a sum of terms corresponding to the translational, vibrational, and rotational degrees of freedom. The instantaneous position of the *sth* nucleus in the molecule may be expressed as

$$\mathbf{r}_s(t) = \mathbf{R}(t) + \mathbf{b}^s(t) + \mathbf{u}^s(t) \ , \tag{2.211}$$

where $\mathbf{R}(t)$ is the position vector of the molecular center of mass, $\mathbf{b}^s(t)$ is the displacement of the equilibrium position of the *sth* nucleus from the center of mass, and $\mathbf{u}^s(t)$ is the instantaneous displacement of the nucleus from its equilibrium position due to vibration. By neglecting the interaction between vibrations and rotations, the molecular

Hamiltonian can be expressed as the sum of three parts,

$$H_0 = H_{tr} + H_r + H_v ,$$

which are, separately, the Hamiltonians for the translational, rotational, and vibrational motions of the molecule. Ignoring the spin state, the molecular wave functions assume the form

$$\Psi = \Psi_{tr}\Psi_r\Psi_v ,$$

in which the effects of the different dynamical modes are explicitly separated. We now consider briefly the vibrational and rotational motions. (For a more detailed discussion of molecular dynamics, see the text by Wilson, Decius, and Cross.[57])

The vibrations are studied in a system of coordinates which are rigidly attached to the equilibrium molecule, i.e., the rigid molecule which exists in the absence of vibrational effects; the origin of the coordinates is chosen to coincide with the position of the center of mass. In this coordinate system, the Hamiltonian describing small vibrations about the equilibrium configuration takes the form

$$H_v = \sum_{s,\alpha} \frac{\left(p_\alpha^s\right)^2}{2M_s} + \tfrac{1}{2} \sum_{s',\alpha'} V_{\alpha\alpha'}^{ss'} u_\alpha^s u_{\alpha'}^{s'} ,$$

where u_α^s ($\alpha = 1,2,3$) is the component of the displacement of the *sth* particle in the α direction of the rotating coordinate system. The force constants $V_{\alpha\alpha'}^{ss'}$ have the symmetry expressed by Eq. (2.148), but without the $\mathbf{R}_p$ and $\mathbf{R}_{p'}$. The momentum components p_α^s are defined with respect to an observer fixed in the rotating system. For the moving observer, the equilibrium vectors $\mathbf{b}^s$ are constants, and the linear and angular momenta associated with the vibrations vanish. Since the origin of coordinates is at the center of mass, these conditions imply that

$$\sum_s M_s \mathbf{u}^s = 0 , \tag{2.212}$$

and

$$\sum_s M_s (\mathbf{b}^s \times \mathbf{u}^s) = 0 . \tag{2.213}$$

The equations of motion for the molecular vibrations are

$$M_s \ddot{u}_\alpha^s = -\sum_{s',\alpha'} V_{\alpha\alpha'}^{ss'} u_{\alpha'}^{s'} ,$$

where the components of acceleration are the ones that the moving observer would measure. The treatment of the vibrations is very similar to that which was given for the lattice vibrations of a crystal, only now we are dealing with a system with few degrees of freedom.

We again look for normal-mode solutions of the form

$$\mathbf{u}^s = (M_s)^{-\frac{1}{2}} \mathbf{C}^s q(t) ,$$

where q(t) satisfies Eq. (2.150). This leads to an eigenvalue equation like Eq. (2.151). The secular equation for the frequencies becomes

$$\| (M_s M_{s'})^{-\frac{1}{2}} V_{\alpha\alpha'}^{ss'} - \omega^2 \delta_{ss'} \delta_{\alpha\alpha'} \| = 0 . \tag{2.214}$$

This is an equation of degree 3n × 3n which in general possesses 3n roots, ω_j^2, j = 1,...,3n for the eigenvalues ω^2. The arguments leading to Eqs. (2.164) and (2.165) are still valid, so that these equations also apply to the case of molecular vibrations. Now we choose the $\mathbf{C}_j^s$ to be real vectors. Had we defined the normal modes in a crystal as standing waves rather than traveling waves, the **C**'s would also have been real in that case.

Six of the roots of the secular equation (2.214) are equal to zero.[57] These roots correspond to the free translation and rotation of the molecule. Let the normal modes corresponding to free translation of the molecule be numbered by the indices j = 3n, 3n-1, and 3n-2, and those corresponding to free rotations by j = 3n-3, 3n-4, 3n-5. Because we have chosen a coordinate system whose origin is at the center of mass, and since this system rotates with the molecule, Eqs. (2.212) and (2.213) must be satisfied in general and in particular for each normal-mode j. These conditions fix the normal-mode coordinates with j = 3n-5,...,3n to be $q_j(t) = 0$. The most general vibration of the molecule can thus be written

$$u_\alpha^s(t) = \sum_{j=1}^{3n-6} (M_s)^{-\frac{1}{2}} C_{j,\alpha}^s q_j(t) .$$

If we introduce creation and annihilation operators as in Eq. (2.173), this becomes

$$u_\alpha^s(t) = \sum_{j=1}^{3n-6} \left(\frac{\hbar}{2M_s\omega_j}\right)^{\frac{1}{2}} C_{j,\alpha}^s \left(A_j e^{-i\omega_j t} + A_j^\dagger e^{i\omega_j t}\right). \qquad (2.215)$$

The commutation relations, Eq. (2.174), still apply, and the vibrational Hamiltonian has the form of Eq. (2.176). For the molecular case, however, these equations do not require the argument **f**.

The total displacement due to vibrations can now be written as

$$\tilde{\mathbf{u}}^s(t) = \sum_{\alpha=1}^{3} \mathbf{i}'_\alpha u_\alpha^s(t) ,$$

where $\mathbf{i}'_\alpha$ are three mutually orthogonal vectors which rotate with the molecule. The $\mathbf{i}'_\alpha$ are obviously time-dependent with respect to a stationary observer. Expressed in terms of stationary orthonormalized basis vectors $\mathbf{i}_\alpha$, the displacement due to vibrations is

$$\mathbf{u}^s(t) = \sum_{j=1}^{3n-6} \left(\frac{\hbar}{2M_s\omega_j}\right)^{\frac{1}{2}} \mathbf{C}_j^s(t) \left(A_j e^{-i\omega_j t} + A_j^\dagger e^{i\omega_j t}\right), \qquad (2.216a)$$

where

$$\mathbf{C}_j^s(t) = \sum_{\alpha=1}^{3} \mathbf{i}_\alpha \sum_{\alpha'=1}^{3} (\mathbf{i}_\alpha \cdot \mathbf{i}'_{\alpha'}) C_{j,\alpha'}^s .$$

The polarizations $\mathbf{C}_j^s(t)$ thus depend on the instantaneous orientation of the molecule in space, the time-dependence being contained in the factors $\mathbf{i}_\alpha \cdot \mathbf{i}'_{\alpha'}$, representing the projections of the rotating vectors $\mathbf{i}'_{\alpha'}$ on the fixed system.

Quantum mechanically, the vector $\tilde{\mathbf{u}}^s(t)$ whose components are given in Eq. (2.215) is the Heisenberg operator

$$\tilde{\mathbf{u}}^s(t) = \exp(iH_v t/\hbar)\mathbf{u}^s(0) \exp(-iH_v t/\hbar) , \qquad (2.216b)$$

and $\mathbf{u}^s(t)$ is

$$\mathbf{u}^s(t) = \exp[i(H_r + H_v)t/\hbar]\mathbf{u}^s(0)\exp[-i(H_r + H_v)t/\hbar]$$

$$= \exp(iH_r t/\hbar)\tilde{\mathbf{u}}^s(0)\exp(-iH_r t/\hbar) \; . \qquad (2.216c)$$

It follows that the $\mathbf{C}_j^s(t)$ in Eq. (2.216a) are operators related to the polarization vectors $\mathbf{C}_j^s$ in the rotating system

$$\mathbf{C}_j^s(t) = \exp(iH_r t/\hbar)\mathbf{C}_j^s \exp(-iH_r t/\hbar) \; . \qquad (2.216d)$$

As a consequence of the rotation of the vibration polarization vectors, it is impossible to make a rigorous separation of the effects of vibration and rotation on the scattering of neutrons. This becomes clear below.

2-7.2 THE INTERMEDIATE SCATTERING FUNCTION FOR MOLECULES. Now let us concern outselves with the general formulation of the scattering functions for molecular gases. In this, we follow closely the work of Zemach and Glauber.[56] Using Eq. (2.211) and noting that $\mathbf{R}(t)$ commutes with $\mathbf{b}^s(t)$ and $\mathbf{u}^s(t)$, Eq. (2.210) can be written

$$\chi_{ss'}(\kappa,t) = \chi^{tr}(\kappa,t)\Big\langle \exp\{-i\kappa\cdot[\mathbf{b}^s(0) + \mathbf{u}^s(0)]\} \times \exp\{i\kappa\cdot[\mathbf{b}^{s'}(t) + \mathbf{u}^{s'}(t)]\}\Big\rangle_T \; , \qquad (2.217)$$

where, from Eq. (2.97),

$$\chi^{tr}(\kappa,t) = \Big\langle e^{-i\kappa\cdot\mathbf{R}(0)}e^{i\kappa\cdot\mathbf{R}(t)}\Big\rangle_T$$

$$= \exp\left[\frac{\hbar\kappa^2}{2\mathcal{M}}(it - k_B T t^2/\hbar)\right] \qquad (2.218)$$

represents the effect of the Maxwellian velocity distribution of the mass centers of the molecules, each of whose mass is $\mathcal{M}$. The expectation value in Eq. (2.217) is taken with respect to the vibrational and rotational states of the molecule. In an obvious but somewhat more explicit notation, the rotation-vibration term, which multiplies χ_{tr} in Eq. (2.217), is

$$\chi_{ss'}^{vr}(\kappa,t) = \left\langle \Psi_r \middle| e^{-i\kappa\cdot\mathbf{b}^s(0)} \left\langle \Psi_v \middle| e^{-i\kappa\cdot\mathbf{u}^s(0)} \exp[i(H_r + H_v)t/\hbar] \times \right.\right.$$

$$\left.\left. \times\, e^{i\kappa\cdot\mathbf{u}^{s'}(0)} e^{i\kappa\cdot\mathbf{b}^{s'}(0)} \exp[-i(H_r + H_v)t/\hbar] \middle| \Psi_v \right\rangle_T \middle| \Psi_r \right\rangle_T$$

$$= \left\langle \Psi_r\, e^{-i\kappa\cdot\mathbf{b}^s(0)} \exp(iH_r t/\hbar)\chi_{ss'}^{v}(\kappa,t) \exp(-iH_r t/\hbar) \times \right.$$

$$\left. \times\, e^{i\kappa\cdot\mathbf{b}^{s'}(t)} \middle| \Psi_r \right\rangle_T , \tag{2.219}$$

where

$$\chi_{ss'}^{v}(\kappa,t) = \left\langle \Psi_v \middle| e^{-i\kappa\cdot\tilde{\mathbf{u}}^s(0)} e^{i\kappa\cdot\tilde{\mathbf{u}}^{s'}(t)} \middle| \Psi_v \right\rangle_T \tag{2.220}$$

is the vibrational intermediate scattering function in the rotating system. In obtaining Eq. (2.220), we have used Eq. (2.216b) and the fact that H_v commutes with H_r and $\mathbf{b}^s(0)$. The vibrational scattering function $\chi_{ss'}^{v}(\kappa,t)$ expressed in terms of ω_j and $\mathbf{C}_j^s$ can be easily obtained by referring to Eq. (2.188) and its derivation

$$\chi_{ss'}^{v}(\kappa,t) = \exp[-W_s(\kappa) - W_{s'}(\kappa)] \exp\left[\frac{\hbar}{2}\left(M_s M_{s'}\right)^{-\frac{1}{2}} \times \right.$$

$$\left. \times \sum_{j=1}^{3n-6} (\kappa \cdot \mathbf{C}_j^s)(\kappa \cdot \mathbf{C}_j^{s'}) g(\omega_j, t) \right] , \tag{2.221}$$

where

$$W_s(\kappa) = \frac{\hbar}{2M_s} \sum_{j=1}^{3n-6} \frac{\left(\kappa \cdot \mathbf{C}_j^s\right)^2}{\omega_j} [2\bar{n}(\omega_j) + 1] . \tag{2.222}$$

$$g(\omega,t) = \frac{[\bar{n}(\omega) + 1]e^{i\omega t} + \bar{n}(\omega)e^{-i\omega t}}{\omega} , \tag{2.223}$$

and

$$\bar{n}(\omega) = [e^{\beta\hbar\omega} - 1]^{-1} . \tag{2.224}$$

Note the form of Eq. (2.219). Because of the noncommutativity of H_r and $\mathbf{C}_j^s$, the effects of rotation and vibration are not separable in the intermediate scattering function.

Measurements of neutron scattering by molecules are frequently carried out for neutron energies and for values of $k_B T$ which are much less than any characteristic energy ω_j of the molecule. In these cases, the vibrational motions are not appreciably excited by thermal agitation, nor can they be excited by the neutron. The function $\chi^{v}_{ss'}(\boldsymbol{\kappa},t)$ may then be replaced by that part of it which represents elastic vibrational transitions:

$$\chi^{v}_{ss'}(\boldsymbol{\kappa},t) \approx \exp[-W_s(\boldsymbol{\kappa}) - W_{s'}(\boldsymbol{\kappa})] \; .$$

For the conditions described, this quantity is nearly equal to one. However, in practical calculations, some authors[58,59] have found that a better approximation is to set

$$\exp(iH_r t/\hbar)\chi^{v}_{ss'}(\boldsymbol{\kappa},t)\; \exp(-iH_r/t\hbar) \approx$$

$$\approx \exp[-W_s(\boldsymbol{\kappa}) - W_{s'}(\boldsymbol{\kappa})] \; , \qquad (2.225)$$

and then to approximate $\chi^{vr}_{ss'}(\boldsymbol{\kappa},t)$ by

$$\chi^{vr}_{ss'}(\boldsymbol{\kappa},t) = \left\langle \exp[-W_s(\boldsymbol{\kappa}) - W_{s'}(\boldsymbol{\kappa})]\right\rangle_R \chi^{r}_{ss'}(\boldsymbol{\kappa},t) \; , \qquad (2.226)$$

where $\langle \ldots \rangle_R$ represents an average over molecular orientations and

$$\chi^{r}_{ss'}(\boldsymbol{\kappa},t) = \left\langle \Psi_r \middle| e^{-i\boldsymbol{\kappa}\cdot\mathbf{b}^s(0)} e^{i\boldsymbol{\kappa}\cdot\mathbf{b}^s(t)} \middle| \Psi_r \right\rangle_T \qquad (2.227)$$

is the intermediate scattering function corresponding to the rotations of a rigid molecule. By means of the approximation (2.226), some account is made for the effects of molecular vibrations in the vibrational ground state.

Many authors have treated the problem of slow-neutron scattering by rigid molecules.[21,58-61] The approach employed by some of these authors consists in explicitly summing over those rotational transitions which make a significant contribution to the scattering.[59-61] This procedure is the most natural one to use for low temperatures and small neutron energies. At higher energies and temperatures, the number of transitions which contribute significantly to the cross section becomes large. Then it is advantageous to find approximate expressions for calculating the scattering. To this end, the time-dependent formalism has been applied with considerable success. In Sec. 2-7.4 we consider the use of the time-dependent formalism for developing useful approximations.

2-7.3 SUMMATION OVER ROTATIONAL TRANSITIONS. First, consider the calculation by explicit summation over the possible rotational transitions. We express Eq. (2.227) in a form which explicitly displays the rotational transitions involved in the scattering

$$\chi^{r}_{ss'}(\kappa,t) = \left\langle \Psi_r \right| \exp[-i\boldsymbol{\kappa}\cdot\mathbf{b}^{s}(0)]\exp(iH_r t/\hbar)\exp[i\boldsymbol{\kappa}\cdot\mathbf{b}^{s'}(0)]\times$$
$$\times\exp(-iH_r t/\hbar)\left|\Psi_r\right\rangle_T$$
$$= [Z(\beta)]^{-1}\sum_{\iota}\exp(-\beta\epsilon_\iota)\sum_f\left\langle \iota\right|\exp[-i\boldsymbol{\kappa}\cdot\mathbf{b}^{s}(0)]\left|f\right\rangle\times$$
$$\times\left\langle f\right|\exp[i\boldsymbol{\kappa}\cdot\mathbf{b}^{s'}(0)]\left|\iota\right\rangle\exp[\iota(\epsilon_f-\epsilon_\iota)t/\hbar] ,$$

where $|\iota\rangle$ and $|f\rangle$ denote the initial and final rotational states, respectively, $Z(\beta) = \sum_\iota \exp(-\beta\epsilon_\iota)$ is the rotational partition function, and the summations are over all initial and final rotational states.

To obtain $d^2\sigma/dEd\Omega$ we require

$$S_{ss'}(\boldsymbol{\kappa},\omega) = \frac{1}{2\pi}\int_{-\infty}^{\infty} e^{-i\omega t}\chi^{tr}(\boldsymbol{\kappa},t)\chi^{r}_{ss'}(\boldsymbol{\kappa},t)dt$$

$$= [Z(\beta)]^{-1}\sum_{\iota,f}\exp(-\beta\epsilon_\iota)\left\langle\iota\right|\exp[-i\boldsymbol{\kappa}\cdot\mathbf{b}^{s}(0)]\left|f\right\rangle \times$$

$$\times\left\langle f\right|\exp[i\boldsymbol{\kappa}\cdot\mathbf{b}^{s'}(0)]\left|\iota\right\rangle\frac{1}{2\pi}\int_{-\infty}^{\infty}\times$$

$$\times\exp\left[\frac{i(\epsilon_f-\epsilon_\iota-\hbar\omega)t}{\hbar}+\frac{\hbar\kappa^2}{2\mathcal{M}}\frac{(i\hbar t-k_BTt^2)}{\hbar}\right]dt$$

$$= [Z(\beta)]^{-1}\sum_{\iota,f}\exp(-\beta\epsilon_\iota)\left\langle\iota\right|\exp[-i\boldsymbol{\kappa}\cdot\mathbf{b}^{s}(0)]\left|f\right\rangle \times$$

$$\times\left\langle f\right|\exp[i\boldsymbol{\kappa}\cdot\mathbf{b}^{s'}(0)]\left|\iota\right\rangle\left(\frac{2\pi\kappa^2k_BT}{\mathcal{M}}\right)^{-\frac{1}{2}}\times$$

$$\times\exp\left\{\frac{-\mathcal{M}[\epsilon_f-\epsilon_\iota-\hbar\omega+(\hbar\kappa^2/2\mathcal{M})]^2}{2\hbar^2\kappa^2k_BT}\right\}. \qquad (2.228)$$

Note that the effect of the thermal distribution of center-of-mass velocities is to Doppler-broaden the discrete line spectrum that would be observed at a temperature of zero degrees absolute.

For a complete calculation of the cross section, it still remains to evaluate the matrix elements appearing in Eq. (2.228). We do not go into this problem in very much detail, but refer the reader to the works of Rahman,[60] who gives expressions for the matrix elements for symmetric and asymmetric rotators, and of Griffing,[59,62] who considers in detail the scattering of neutrons by methane (CH_4), a spherical top molecule. Here, we consider a methane molecule in its ground state, the purpose being to illustrate the procedure for computing the matrix elements.

The methane molecule is a rather simple one to consider. In this case, the rotational Hamiltonian is

$$H_r = \frac{L^2}{2I}$$

where L is the angular momentum of the molecule and I its moment of inertia. The eigenfunctions of H_r are the orthonormal spherical harmonics $Y_{Jm}(\theta,\phi)$; for these states the energy is

$$E_J = BJ(J+1) \qquad J = 0,1,2,\ldots,$$

where $B = \hbar^2/2I$ is the rotational constant, and J and m are the rotational quantum numbers. For methane, B is about 6×10^{-4} eV. The angle θ measures the inclination between any one of the principal axes of inertia and a fixed space direction, which we choose to be the direction of $\boldsymbol{\kappa}$. The azimuthal angle ϕ measures the rotation of the molecule around $\boldsymbol{\kappa}$.

For the case when the molecule is initially in the ground state, the required matrix elements are

$$\left\langle Jm\middle|e^{i\boldsymbol{\kappa}\cdot\mathbf{b}^{s'}}\middle|00\right\rangle = \int Y^*_{Jm}(\theta,\phi)e^{i\boldsymbol{\kappa}\cdot\mathbf{b}^{s'}}Y_{00}(\theta,\phi)d\Omega ,$$

where for convenience we abbreviate $\mathbf{b}^{s'}(0)$ to $\mathbf{b}^{s'}$. For a nucleus which lies on the principal axis whose polar angle is θ,

$$e^{i\boldsymbol{\kappa}\cdot\mathbf{b}^{s'}} = (4\pi)^{\frac{1}{2}}\sum_{\ell=0}^{\infty}(2\ell+1)^{\frac{1}{2}}i^{\ell}j_{\ell}(\kappa b^{s'})Y_{\ell 0}(\theta) ,$$

where $b^{s'}$ is the magnitude of $\mathbf{b}^{s'}$, and the j_ℓ are spherical Bessel functions. It now follows from the orthonormality of the spherical harmonics that

$$\left\langle J_m\middle|e^{i\boldsymbol{\kappa}\cdot\mathbf{b}^{s'}}\middle|00\right\rangle = (2J+1)^{\frac{1}{2}}i^J j_J(\kappa b^{s'})\delta_{m0} . \tag{2.229}$$

Since all atoms in methane lie on a principal axis and since θ may be the polar angle between any such axis and $\boldsymbol{\kappa}$, Eq. (2.229) is valid for all atoms in methane. In particular, for any pair of protons, $b^s = b^{s'}$, so that

$$\left\langle 00\middle|e^{-i\boldsymbol{\kappa}\cdot\mathbf{b}^{s}}\middle|Jm\right\rangle\left\langle Jm\middle|e^{i\boldsymbol{\kappa}\cdot\mathbf{b}^{s'}}\middle|00\right\rangle = (2J+1)[j_J(\kappa b^s)]^2\delta_{m0} .$$

It is interesting to consider the case where s and s' are equal and where they refer to the carbon atom, which is located at the center of mass of the molecule. In this case, $b^s = 0$ and the matrix elements in Eq. (2.228) are nonvanishing only when the initial and final states are the same, i.e., only when the neutron does not induce a rotational transition. This is expected, since an impulse applied at the center of mass of the molecule cannot change its angular momentum.

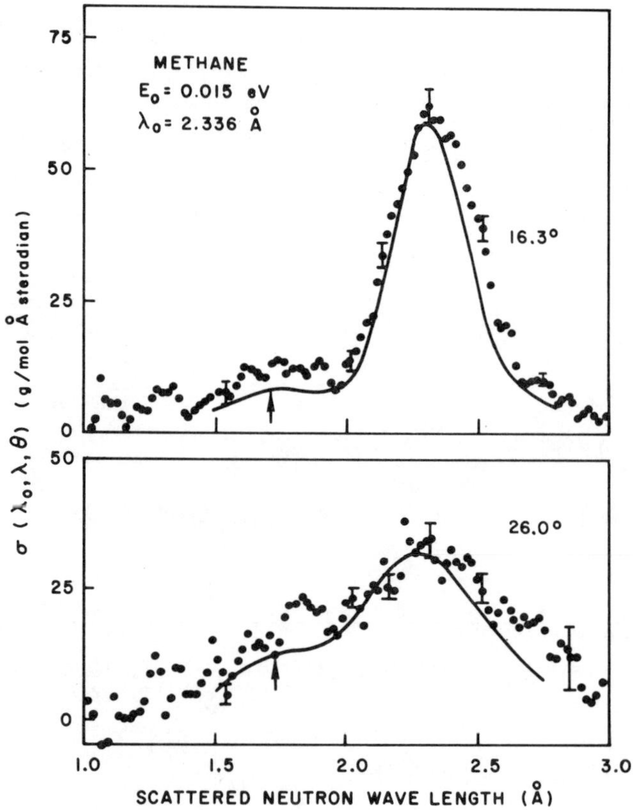

Figure 2.14 The experimental and computed partial differential cross sections shown as a function of the scattered neutron wavelength. The experimental data has been normalized to the theoretical results so that the magnitude of the main peak is approximately equal. The arrow indicates the maximum of that portion of the curve that is influenced by rotations. The temperature of the gas is 21°C. (From Griffing, Chalk River Symposium, Vol. I, Ref. 59.)

The procedure for carrying out the complete calculation of the cross section is now clear. We do not proceed any further with this problem but close this part of the discussion by showing in Fig. 2.14 a comparison of Griffing's calculations[59] with experimental results for gaseous methane.[63] The intensities of scattered neutrons as a function of wavelength are plotted for an incident energy of 0.015 eV, a temperature of 21°C, and scattering angles of 16.3° and 26.0°. The case of gaseous methane is of interest since the molecular dynamics is well understood and accurate calculations are feasible. Agreement between theory and experiment is thus expected and is obtained except for the low incident neutron energy and small scattering angles shown. The disagreement here is due largely to angular resolution and other experimental effects.

2-7.4 THE QUASI-CLASSICAL APPROXIMATION. For high incident neutron energies or high temperatures, the sum over states in Eq. (2.228) converges very slowly. In this case, the full time-dependent formalism can be applied to the development of approximate methods for calculating the scattering. This formalism has been applied by Zemach and Glauber,[21] by Krieger and Nelkin,[58] and by Volkin[64] to the problem of scattering by rotating molecules. These authors either neglect vibrational effects completely or treat them in the approximation of Eq. (2.226).

Another and much earlier approach to the problem of slow-neutron scattering by molecules is that of Sachs and Teller,[65] in which only the total cross section is considered and interference and vibrational effects are neglected. The basis of the Sachs-Teller method is the concept of a nuclear-mass tensor whose properties replace the effects of chemical binding. The Sachs-Teller method has been extended by Messiah[66] to include molecular vibrations.

In the remainder of this section we derive the mass-tensor concept by means of the time-dependent formalism, following closely the treatment in Ref. 58. As a first step, we express Eq. (2.227) for $s = s'$ in a form which is similar to Eq. (2.92a). (For reasons discussed in Sec. 2-8.3, this treatment applies only to the direct scattering. The extension to interference scattering in Ref. 58 is incorrect. The limitation to direct scattering is not a serious restriction, however, for hydrogenous molecules, where we apply the results.) Thus, instead of Eq. (2.227) we have

$$\chi_{ss}^{r}(\kappa,t) = \left\langle \Psi_r \right| \exp[iH_r^s(\kappa)t/\hbar)] \exp(-iH_r t/\hbar) \left| \Psi_r \right\rangle_T , \qquad (2.230)$$

where

$$H_r^s(\kappa) = e^{-i\kappa \cdot \mathbf{b}^s} H_r e^{i\kappa \cdot \mathbf{b}^s} . \qquad (2.230a)$$

The quasi-classical approximation consists essentially in treating the operators $H_r^s(\kappa)$ and H_r in Eq. (2.230) as classical commuting variables and replacing the rigid rotator wave functions Ψ_r by rotational wave packets characterized by simultaneously well-defined values of orientation and angular momentum. Such wave packets exist to a good approximation when the rotator is sufficiently excited, i.e., when $k_B T$ is much greater than the rotational constant B, and when $E_0 >> (Bk_B T)^{\frac{1}{2}}$, the level spacing in the neighborhood of the most probable level.

The quasi-classical approximation requires the quantity $H_r^s(\kappa) - H_r$. The rigid-molecule Hamiltonian is

$$H_r = (L_x^2/2I_x) + (L_y^2/2I_y) + (L_z^2/2I_z) , \qquad (2.231)$$

where the x, y, z directions lie along the principal axes, which rotate with the molecule, **L** is the angular momentum

$$\mathbf{L} = \sum_{s'} \mathbf{b}^{s'} \times \mathbf{p}^{s'} ,$$

and I_x, I_y, and I_z are the moments of inertia about the principal axes. Now, for any function $f(\mathbf{p}_1,\ldots\mathbf{p}_s,\ldots\mathbf{p}_n)$ of the momentum operators,

$$e^{-i\kappa \cdot \mathbf{b}^s} f(\mathbf{p}_1,\ldots\mathbf{p}_s,\ldots\mathbf{p}_n) e^{i\kappa \cdot \mathbf{b}^s} = f(\mathbf{p}_1,\ldots\mathbf{p}_s + \hbar\kappa,\ldots\mathbf{p}_n) . \qquad (2.231a)$$

Thus, in order to evaluate $H_r^s(\kappa)$ we see that we need

$$[L_{s,z}(\kappa)]^2 = \left[\sum_{s'} (\mathbf{b}^{s'} \times \mathbf{p}^{s'})_z + \hbar(\mathbf{b}^s \times \kappa)_z\right]^2$$

$$= L_z^2 + 2\hbar L_z(b^s \times \kappa)_z + \hbar^2(b^s \times \kappa)_z^2 \qquad (2.232)$$

and similar expressions involving the x and y components of $L + \hbar\mathbf{b}^s \times \kappa$. In accordance with the classical approximation which is being made, the noncommutativity of $\mathbf{L}$ and $\mathbf{b}^s$ is completely ignored. Using Eqs. (2.231) and (2.232) in Eq. (2.230) leads to the approximate result

$$\chi^r_{ss}(\kappa,t)$$

$$= \left\langle \exp\left[\frac{L_x(\mathbf{b}^s \times \kappa)_x}{I_x} + \frac{L_y(\mathbf{b}^s \times \kappa)_y}{I_y} + \frac{L_z(\mathbf{b}^s \times \kappa)_z}{I_z}\right] it \times \right.$$

$$\left. \times \exp\left\{\frac{\hbar}{2}\left[\frac{(\mathbf{b}^s \times \kappa)_x^2}{I_x} + \frac{(\mathbf{b}^s \times \kappa)_y^2}{I_y} + \frac{(\mathbf{b}^s \times \kappa)_z^2}{I_z}\right] it\right\}\right\rangle_{T,R}, \qquad (2.233)$$

where $\langle \ldots \rangle_{T,R}$ now implies a classical average over the Boltzmann distribution of angular momenta and over the orientations of the molecule.

Before carrying out these averages, let us consider the physical sense of Eq. (2.233). For this purpose, consider a stationary rigid dumbbell molecule. In this case, $\mathbf{L} = 0$ and no orientation average is required. The translational χ function is simply

$$\chi^{tr} = \exp\left(i\,\frac{\hbar\kappa^2}{2\mathcal{M}}\,t\right). \qquad (2.234)$$

Figure 2.15 illustrates the two possible results of a classical collision with a molecule. Consider first the case illustrated in Figure 2.15(a), where as a result of the neutron impact the molecule recoils in a direction parallel to its axis. Then $\underset{\sim}{\kappa} \times \underset{\sim}{b} = 0$, $\chi^r_{ss}(\underset{\sim}{\kappa},t) = 1$ and the total effect of translations and rotations is represented by $\chi^{rt}_{ss}(\underset{\sim}{\kappa},t) = \exp(i\frac{\hbar\kappa^2}{2\mathcal{M}}t)$

Comparison with Eq. (2,294) shows this to be a sensible result; namely, the scattering is like that from a particle at rest which is free to recoil and whose mass is equal to the molecular mass, $\mathcal{M} = 2M$. For the case...

the scattering is like that from a particle at rest which is free to recoil and whose mass is equal to the molecular mass, $\mathcal{M} = 2M$. For the case shown in Fig. 2.15(b), on the other hand,

$$\chi_{ss}^{rt}(\kappa,t) = \exp\left(\frac{\hbar\kappa^2}{2M}\, it\right)$$

and the scattering is like that from a particle of mass M which is free to recoil.

The simple example considered here illustrates simply but adequately the mass-tensor concept introduced by Sachs and Teller.[65] The mass tensor $\mathbf{M}_s$ is obtained after first expressing

$$F_s = \frac{(\mathbf{b}^s \times \kappa)_x^2}{I_x} + \frac{(\mathbf{b}^s \times \kappa)_y^2}{I_y} + \frac{(\mathbf{b}^s \times \kappa)_z^2}{I_z}$$

in the form

$$F_s = \kappa \cdot \mathbf{K}_s \cdot \kappa \, ,$$

where $\mathbf{K}_s$ is the tensor

$$\mathbf{K}_s = \mathbf{B}_s^\dagger \cdot \mathbf{I}^{-1} \cdot \mathbf{B}_s \, ,$$

the matrices $\mathbf{I}^{-1}$ and $\mathbf{B}_s$ are defined by

$$\mathbf{I}^{-1} = \begin{bmatrix} I_x^{-1} & 0 & 0 \\ 0 & I_y^{-1} & 0 \\ 0 & 0 & I_z^{-1} \end{bmatrix}$$

$$\mathbf{B}_s = \begin{bmatrix} 0 & b_z^s & -b_y^s \\ -b_z^s & 0 & b_x^s \\ b_y^s & -b_x^s & 0 \end{bmatrix}$$

and $\mathbf{B}_s^\dagger$ is the transpose of $\mathbf{B}_s$. In terms of these quantities

Figure 2.15 The classical scattering of neutrons by a dumbbell molecule.

$$\chi_{ss}^{rt}(\kappa,t) = \exp\left[\frac{\hbar\kappa^2}{2\mathcal{M}}\left(it - \frac{k_BT}{\hbar}t^2\right)\right] \times$$
$$\times \left\langle \exp\left[\left(-\mathbf{L}\cdot\mathbf{I}^{-1}\cdot\mathbf{B}_s\cdot\kappa + \tfrac{1}{2}\hbar\kappa\cdot\mathbf{K}_s\cdot\kappa\right)it\right]\right\rangle_{T,R} .$$

The evaluation of the thermal average requires the result

$$\frac{\int \exp(-H_r/k_BT)\,\exp(-\mathbf{L}\cdot\mathbf{I}^{-1}\cdot\mathbf{B}_s\cdot\kappa)d\mathbf{L}}{\int \exp(-H_r/k_BT)d\mathbf{L}}$$
$$= \exp[-k_BT(\kappa\cdot\mathbf{K}_s\cdot\kappa)t^2/2] .$$

On observing the form of Eq. (2.233), we see that this result follows simply as the product of terms of the form

$$\frac{\int_{-\infty}^{\infty} \exp\{(-L_z^2/2I_zk_BT) + [L_z(\mathbf{b}_s\times\kappa)_z it/I_z]\}dL_z}{\int_{-\infty}^{\infty} \exp(-L_z^2/2I_zk_BT)dL_z}$$
$$= \exp[-k_BT(\mathbf{b}\times\kappa)_z^2t^2/2I_z] .$$

Finally, after introducing the inverse mass tensor

$$\mathcal{M}_s^{-1} = \mathbf{K}_s + \mathcal{M}^{-1}\mathbf{1} , \qquad (2.235)$$

where $\mathbf{1}$ is the unit tensor, there results

$$\chi_{ss}^{rt}(\kappa,t) = \left\langle \exp\left[\frac{\hbar}{2}\kappa\cdot\mathcal{M}_s^{-1}\cdot\kappa(it - k_BT^2/\hbar)\right]\right\rangle_R . \qquad (2.236)$$

To perform the averaging over orientations, we note that $\chi_{ss}(\boldsymbol{\kappa},t)$ appears as the average of a product of factors. The first of these,

$$\exp\left[-\frac{\hbar}{2}\,\boldsymbol{\kappa}\cdot\mathcal{M}_s^{-1}\cdot\boldsymbol{\kappa}(it - k_B T t^2/\hbar)\right], \tag{2.237}$$

represents, in the quasi-classical approximation, the combined effects of molecular rotation and translation on the scattering from the *sth* nucleus. The second factor represents the dynamical effects of ground-state vibrations and may be approximated in the manner indicated by Eq. (2.226).

An approximate average of the Sachs-Teller factor is obtained by averaging in the exponent

$$\left\langle \chi_s^{ST}(\boldsymbol{\kappa},t)\right\rangle_R \approx \exp\left[\frac{\hbar}{2}\left\langle \boldsymbol{\kappa}\cdot\mathcal{M}_s^{-1}\cdot\boldsymbol{\kappa}\right\rangle_R (it - k_B T^2/\hbar)\right]$$

$$= \exp\left[\frac{\hbar\kappa^2}{2M_s^{(0)}}(it - k_B T t^2)/\hbar\right], \tag{2.238}$$

where

$$\left[M_s^{(0)}\right]^{-1} = \frac{1}{3}\,\mathrm{Tr}(\mathcal{M}_s^{-1})$$

$$= \frac{1}{\mathcal{M}} + \frac{1}{3}\times$$

$$\times\left(\frac{b_{2,y}^2 + b_{s,z}^2}{I_x} + \frac{b_{s,x}^2 + b_{s,z}^2}{I_y} + \frac{b_{s,x}^2 + b_{s,y}^2}{I_z}\right). \tag{2.239}$$

The question of an improved choice for $M_s^{(0)}$ in Eq. (2.238) is considered further in Ref. 58. It turns out that Eq. (2.239) is a reasonable choice, but can be improved slightly by using the Sachs-Teller result for the total cross section in which the average over orientations is performed exactly.

The accuracy of the Krieger-Nelkin formulae have been investigated in considerable detail by McMurry, Gannon, and Hestir.[67] Figure 2.16 compares some of their calculations with measured results for methane.[63] In the figure, the symbol Q.M. means that an exact calculation was made in the manner indicated by Eq. (2.228). The curves obtained on the basis of Eq. (2.238) are labeled K.N. The letters E.A. imply the use of Eq. (2.236) and "exact" numerical calculations of averages over molecular orientations. A significant feature in these figures is the improved accuracy of the quasi-classical approximation at the larger incident energy and angle of scattering.

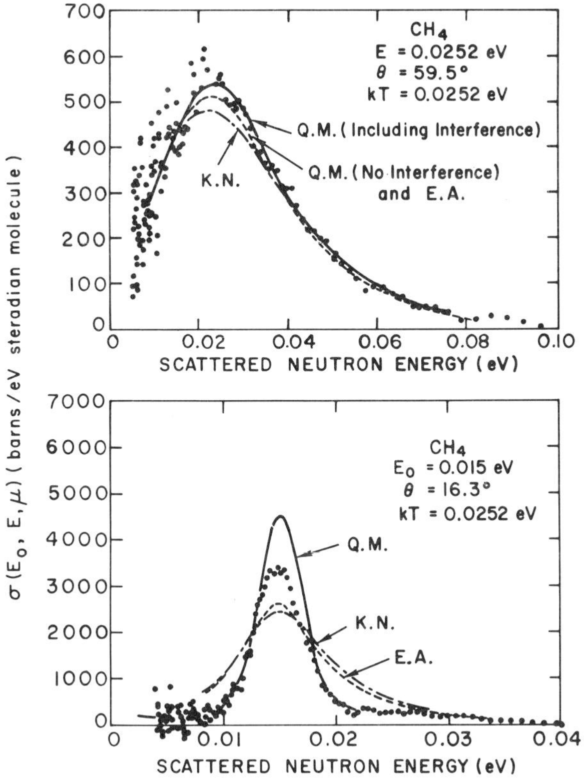

Figure 2.16 Partial differential scattering cross sections of methane for low neutron energies. (From McMurry, Gannon, and Hestir, Nucl. Sci. Eng., Ref. 67.)

2-8 THE SHORT-COLLISION-TIME APPROXIMATION

2-8.1 THE PHYSICAL MEANING OF $\chi(\kappa,t)$. Placzek[16] and Wick[19] have considered the scattering of neutrons by weakly bound systems. In particular, they obtain formulae for the total scattering cross section which are valid when the neutron energy is large compared with energies that are characteristic of the dynamical behavior of the scattering system.

Placzek's method consists in expanding $S(\kappa,\omega)$ in powers of $m/M = A^{-1}$. His results, therefore, require that the mass of the scattering nucleus be large in comparison with the neutron mass. For a scatterer composed of heavy nuclei, the energy transfer is, in general, a small fraction (of the order of A^{-1}) of the initial kinetic energy of the neutron and, at the same time, is large compared with the characteristic energies of atomic motions in the scattering system. Under these conditions, Placzek shows that the cross section is a smooth function of the neutron energy and can be represented by simple asymptotic formulae. For example, the direct part of the scattering from a system of nuclei can be given as an expansion in inverse powers of the neutron energy.

Wick's calculation is based on the fact that for large energies the duration of a collision is short compared with the natural periods of atomic motions. Besides its greater formal elegance, Wick's method is more deeply rooted in the physical nature of a collision between a neutron and the scatterer than is the method of Placzek. In addition, it removes the restriction that A^{-1} be small, but it is equivalent to Placzek's expansion for large values of A.

We shall examine Wick's method in some detail, concerning ourselves initially with its underlying intuitive physical basis. Of fundamental importance in this context is the recognition that the variable t appearing in Eq. (2.88) may be interpreted as the duration of the collision between the neutron and the scatterer. We have already seen that it is proper to regard t as a time. It still remains to show that it is indeed a collision time. For this purpose, consider the expression

$$\chi_{\ell\ell'}^{(\iota)}(\boldsymbol{\kappa},t) = \langle\iota|\ \exp[-i\boldsymbol{\kappa}\cdot\mathbf{r}_\ell(0)]\ \exp(iH_0t/\hbar)\ \times$$

$$\times\ \exp[i\boldsymbol{\kappa}\cdot\mathbf{r}_{\ell'}(0)]\ \exp(-iH_0t/\hbar)|\iota\rangle$$

$$= \langle\iota|\ \exp(iH_0t/\hbar)\ \exp[-i\boldsymbol{\kappa}\cdot\mathbf{r}_{\ell'}(0)]\ \times$$

$$\times\ \exp(-iH_0t/\hbar)\ \exp[i\boldsymbol{\kappa}\cdot\mathbf{r}_\ell(0)]|\iota\rangle\ .$$

The sum over ℓ and ℓ' of the time Fourier transform of expressions of this kind relate the neutron scattering to the dynamical state of a system whose wave function is $|\iota\rangle$ in the absence of the perturbation produced by the neutron.

Suppose that $(\mathbf{p}_\ell)_{Av}$ denotes the expectation value of the momentum $\mathbf{p}_\ell$ of the ℓth particle at $t = 0$ in the state $|\iota\rangle$. Then it is easy to show that in the state $\exp[i\boldsymbol{\kappa}\cdot\mathbf{r}_\ell(0)]|\iota\rangle$ the expectation value of $\mathbf{p}_\ell$ is $(\mathbf{p}_\ell)_{Av} + \hbar\boldsymbol{\kappa}$, while the expectation values of the momenta of all other particles are unchanged. Thus, we may say that at $t = 0$ the neutron imparted an impulse $\hbar\boldsymbol{\kappa}$ to the ℓth nucleus, leaving the system in the state $\exp[i\boldsymbol{\kappa}\cdot\mathbf{r}_\ell(0)]|\iota\rangle$. After the impulse, the perturbed system evolves under the influence of interatomic forces until, at the time t, the wave function of the system is

$$e^{-iH_0t/\hbar}\ e^{i\boldsymbol{\kappa}\cdot\mathbf{r}_\ell(0)}\ |\iota\rangle\ .$$

Finally, at t, the $\ell' th$ nucleus receives an impulse $-\hbar\boldsymbol{\kappa}$ which changes the expectation value of its momentum from

$$[p_{\ell'}(t)]_{Av} = \langle\iota|\ \exp[-i\boldsymbol{\kappa}\cdot\mathbf{r}_\ell(0)]\ \exp(iH_0t/\hbar)p_{\ell'}(0)\ \times$$

$$\times\ \exp(-iH_0t/\hbar)\ \exp[i\boldsymbol{\kappa}\cdot\mathbf{r}_\ell(0)]|\iota\rangle$$

to $[\mathbf{p}_{\ell'}(t)]_{Av} - \hbar\boldsymbol{\kappa}$, leaving the system immediately after the impulse in the state

$$e^{-i\boldsymbol{\kappa}\cdot\mathbf{r}_{\ell'}(0)}\ e^{-iH_0t/\hbar}\ e^{i\boldsymbol{\kappa}\cdot\mathbf{r}_\ell(0)}\ |\iota\rangle\ .$$

Thus, t is the time that has elapsed between the first and second impulses, and is properly interpreted as a collision time.

Now we note that $\exp(-iH_0t/\hbar)|\iota\rangle$ is the wave function of the unperturbed system at time t. Therefore, the quantity $\chi_{\ell\ell'}^{(\iota)}(\boldsymbol{\kappa},t)$ is the probability amplitude for finding the system in the unperturbed state immediately after the completion of the interaction between the neutron and the atomic system. It may be considered as a function which expresses the correlation between equal and opposite momentum transfers applied t seconds apart.

It may seem strange that the scattering of a neutron in a single elementary process with an accompanying momentum transfer $\hbar\boldsymbol{\kappa}$ should be describable in terms of a second-order process in which the total momentum transfer is zero. Wick explains this circumstance by appealing to the optical theorem, according to which the total cross section is related linearly to the imaginary part of the amplitude of the forward elastic scattered wave. To obtain the first-order Born approximation to the scattering cross section from the forward elastic scattered wave, the latter must be calculated in the second Born approximation. The second-order process leading to the forward elastic scattering is just the successive transfer of equal and opposite momenta to the scatterer. These two elementary interactions are now physically related parts of the single process in which the neutron scatters by exchanging the momentum $\hbar\boldsymbol{\kappa}$ with the atomic system.

The appearance of the collision time in Eq. (2.87) suggests at once an approximation based on the plausible physical idea that for sufficiently large values of the incident neutron energy the duration of the collision must be short. The resulting assumption is that large values of t do not give a significant contribution to the integral over t. This assumption is completely reasonable in view of the interpretation given to $\chi_{\ell\ell'}^{(\iota)}(\boldsymbol{\kappa},t)$ as a probability amplitude. Because of this meaning, one expects that the probability of finding a system in its unperturbed state after the successive application of two impulses of amounts $\hbar\boldsymbol{\kappa}$ and $-\hbar\boldsymbol{\kappa}$ will be appreciable for large $\hbar\boldsymbol{\kappa}$ only when the time-lag between these impulses is short compared with the characteristic periods of the system.

2-8.2 THE DIRECT SCATTERING. We now turn to the formal development of the discussion given in the preceding paragraphs. In so doing, we first examine the noninterference part of the scattering, reserving the discussion of the interference terms for Sec. 2-8.3. For simplicity in notation, we drop any indices which are required to indicate the nucleus whose contribution to the scattering is being considered. This causes no difficulty since the noninterference part of the scattering by a system of particles is merely the sum of the contributions from the single particles.

As a first approximation, we expect that the intermediate scattering function would be given by Eq. (2.94), which applies in the case of scattering by atoms which are free and at rest. We further expect that for large κ this factor is the most strongly time-dependent one in the expression for $\chi_{inc}(\boldsymbol{\kappa},t)$. Therefore, it is convenient to isolate the factor, Eq. (2.94), from the time-dependence of $\chi_{inc}(\boldsymbol{\kappa},t)$. This may be done easily using the form (2.92a) for χ_{inc}. The quantity $H_\ell(\boldsymbol{\kappa}) = H(\boldsymbol{\kappa})$ is given by (see Eq. (2.231a))

$$H(\boldsymbol{\kappa}) = H_0 + \frac{\hbar \mathbf{p} \cdot \boldsymbol{\kappa}}{M} + \frac{\hbar^2 \kappa^2}{2M} .$$

Thus

$$\begin{aligned}\chi_{inc}(\boldsymbol{\kappa},t) &= \left\langle e^{-i\boldsymbol{\kappa}\cdot\mathbf{r}(0)} e^{i\boldsymbol{\kappa}\cdot\mathbf{r}(t)} \right\rangle_T \\ &= \exp(i\hbar\kappa^2 t/2M) \left\langle \exp\left[i\left(H_0 + \frac{\hbar\boldsymbol{\kappa}\cdot\mathbf{p}}{M}\right)t/\hbar\right] \times \right. \\ &\quad \left. \times \exp(-iH_0 t/\hbar) \right\rangle_T , \qquad (2.240)\end{aligned}$$

which accomplishes the desired separation of the factor corresponding to the recoil of a free nucleus from the intermediate scattering function. On the basis of previous considerations, we now expect that for large $\boldsymbol{\kappa}$ the expectation value of

$$g(\boldsymbol{\kappa},t) = \exp[i(H_0 + \hbar\,\frac{\boldsymbol{\kappa}\cdot\mathbf{p}}{M})t/\hbar]\,\exp(-iH_0 t/\hbar) \qquad (2.241)$$

appearing in Eq. (2.240) is weakly time-dependent compared with the free-atom recoil factor. Neglecting the time-dependence of $g(\boldsymbol{\kappa},t)$ altogether, one obtains the result corresponding to the case of a collision of a neutron with a free atom at rest.

At this point, we note that Placzek's expansion is the one that is obtained by expanding Eq. (2.240), including the free-nucleus recoil factor, in powers of t. As a consequence, the first term in Placzek's expansion for the total cross section is the cross section corresponding to a rigid binding of the nucleus to its equilibrium position. It is only by summing the infinite number of terms corresponding to the expansion of the recoil factor in powers of t that Placzek obtains the free-atom result as a first approximation. In Wick's calculation, the effects of recoil are taken into account from the beginning.

We now expand $g(\boldsymbol{\kappa},t)$ in powers of t:

$$g(\boldsymbol{\kappa},t) = \sum_{n=0}^{\infty} \left(\frac{it}{\hbar}\right)^n \frac{g_n(\boldsymbol{\kappa})}{n!} \tag{2.242}$$

The coefficients $g_n(\boldsymbol{\kappa})$ can be expressed in terms of the incoherent scattering function. For this purpose, we assume that the scattering system is randomly oriented. Using Eqs. (2.88), (2.240), and (2.241),

$$\left\langle g(\boldsymbol{\kappa},t)\right\rangle_{T,R} = \int_{-\infty}^{\infty} \exp\left[i\left(\omega - \frac{\hbar\kappa^2}{2M}\right)t\right]\left\langle S_{inc}(\boldsymbol{\kappa},\omega)\right\rangle_R d\omega \; . \tag{2.243}$$

Expanding the exponential in powers of t and comparing the result with Eq. (2.242) shows that

$$s_n(\kappa) = \left\langle g_n(\boldsymbol{\kappa})\right\rangle_{T,R} = \int_{-\infty}^{\infty} \left(\hbar\omega - \frac{h^2\kappa^2}{2M}\right)^n \left\langle S_{inc}(\boldsymbol{\kappa},\omega)\right\rangle_R d\omega \; . \tag{2.244}$$

In particular, $s_0(\kappa) = 1$, and from Eq. (2.80) $s_1(\kappa) = 0$. Thus, the $s_n(\kappa)$ are the central moments of the scattering function

The quantities

$$p_n(\kappa) = \int (\hbar\omega)^n \left\langle S_{inc}(\boldsymbol{\kappa},\omega)\right\rangle_R d\omega \tag{2.245}$$

are called Placzek moments. They admit a simple physical interpretation. If a momentum transfer $\hbar\boldsymbol{\kappa}$ is imparted to the system, the average energy transfer accompanying this fixed momentum transfer is given by $p_n(\kappa)$. The Placzek moments are of course different from the moments of the energy transfer occurring in an actual collision. For the latter, the momentum transfer is not fixed. The moments of energy transfer in an actual collision are given by

$$[(\Delta E)^n]_{Av} = \frac{\int_0^\infty \sigma(E_0,E)(E_0 - E)^n dE}{\int_0^\infty \sigma(E_0,E)dE}$$

$$= \frac{\iint_{-\infty}^{E_0/\hbar} \left(\frac{E_0 - \hbar\omega}{E_0}\right)^{1/2} \left\langle S_{inc}(\boldsymbol{\kappa},\omega)\right\rangle_R (\hbar\omega)^n d\omega d\Omega}{\iint_{-\infty}^{E_0/\hbar} \left(\frac{E_0 - \hbar\omega}{E_0}\right)^{1/2} \left\langle S_{inc}(\boldsymbol{\kappa},\omega)\right\rangle_R d\omega d\Omega} , \quad (2.246)$$

where $d\Omega$ is the element of solid angle into which neutrons are scattered. For heavy atoms and large incident energies, the energy transfers are much less than E_0. In this case, one makes a very small error (1) if $\hbar\omega$ in the square root factors in Eq. (2.246) and in

$$\kappa^2 = \frac{2m}{\hbar^2} \{2E_0 - \hbar\omega - 2[E_0(E_0 - \hbar\omega)]^{1/2} \cos\theta\}$$

is neglected (in this approximation $\boldsymbol{\kappa}$ remains constant for a fixed angle of scattering and a fixed initial energy), and (2) if the upper limit of integration is replaced by infinity. Then

$$[(\Delta E)^n]_{Av} = (4\pi)^{-1} \int p_n(\kappa) d\Omega \ ;$$

the mean of the n*th* power of the energy transfer in a collision is the n*th* Placzek moment averaged over solid angles.

The $s_n(\kappa^2)$ for $n > 1$ depend explicitly on expectation values of functions of the coordinates and momenta of the particles of the scattering system. To relate the $s_n(\kappa^2)$ to the dynamical properties of the scatterer, we require the coefficients $g_n(\boldsymbol{\kappa})$ appearing in Eq. (2.242). These can be obtained by first differentiating Eq. (2.241) with respect to time:

$$\frac{d}{dt} g(\boldsymbol{\kappa},t) = i/\hbar \exp\left[i\left(H_0 + \frac{\hbar\boldsymbol{\kappa}\cdot\mathbf{p}}{M}\right) t/\hbar\right] L(0) \exp(-iH_0 t/\hbar)$$

$$= (i/\hbar) g(\boldsymbol{\kappa},t) L(t) , \quad (2.247)$$

where

$$L(t) = \frac{\hbar\boldsymbol{\kappa} \cdot \mathbf{p}(t)}{M}$$

Using Eq. (2.242) and the expansion

$$L(t) = \sum_{n=0}^{\infty} \frac{1}{n!}\left(\frac{\hbar}{i}\right)^n \left(\frac{d^n L}{dt^n}\right)_{t=0} \left(\frac{it}{\hbar}\right)^n$$

in Eq. (2.247), and equating coefficients of like powers of $(it/\hbar)$, gives the recursion relation

$$g_{n+1}(\boldsymbol{\kappa}) = g_n(\boldsymbol{\kappa})L + (\hbar/i)ng_{n-1}(\boldsymbol{\kappa})\dot{L} +$$

$$+ \tfrac{1}{2}(\hbar/i)^2 n(n-1)g_{n-2}(\boldsymbol{\kappa})\ddot{L} + \dots , \qquad (2.248)$$

where

$$L = L(0) ,$$

$$\dot{L} = \left(\frac{dL}{dt}\right)_{t=0} ,$$

$$\ddot{L} = \left(\frac{d^2L}{dt^2}\right)_{t=0} , \qquad \text{etc.}$$

Since $g_0(\boldsymbol{\kappa}) = 1$, it follows that $g_n(\boldsymbol{\kappa})$ is a polynomial of degree n in $\boldsymbol{\kappa}$. Terms in these polynomials which involve $\boldsymbol{\kappa}$ raised to an odd power vanish when averaged over orientations of a randomly oriented system. Using this fact, and the recursion relation Eq. (2.248), we can express the first few central moments, Eq. (2.244), as

$$s_0(\kappa) = 1 ,$$

$$s_1(\kappa) = 0 ,$$

$$s_2(\kappa) = \left\langle L^2 \right\rangle_{T,R} ,$$

$$s_3(\kappa) = (\hbar/i)\left\langle \dot{L}L + 2L\dot{L} \right\rangle_{T,R} ,$$

$$s_4(\kappa) = \left\langle L^4 \right\rangle_{T,R} + (\hbar/i)^2\left\langle \ddot{L}L + 3L\ddot{L} + 3\dot{L}^2 \right\rangle_{T,R} .$$

We show below how these results can be written in the more explicit form given by Wick:

$$s_0(\kappa) = 1 \ ,$$

$$s_1(\kappa) = 0 \ ,$$

$$s_2(\kappa) = \tfrac{2}{3}(\hbar^2\kappa^2/M)K_{Av} = (\hbar^2\kappa^2/M)k_B\bar{T} \ , \qquad (2.249)$$

$$s_3(\kappa) = (\hbar^2\kappa^2/2M)B_{Av} \ ,$$

$$s_4(\kappa) = \tfrac{4}{5}(\hbar^2\kappa^2/M)^2(K^2)_{Av} + (\hbar^2\kappa^2/M)C_{Av} \ ,$$

where

$$K_{Av} = \tfrac{3}{2}\, k_B\bar{T} = \left\langle p^2/2M \right\rangle_T \ ,$$

$$B_{Av} = \tfrac{1}{3}(\hbar^2/M)\left\langle \nabla^2 V \right\rangle_T \ ,$$

$$C_{Av} = \tfrac{1}{3}(\hbar^2/M)\left\langle (\nabla V)^2 \right\rangle_T \ , \qquad (2.250)$$

$$(K^2)_{Av} = \left\langle (p^2/2M)^2 \right\rangle_T \ ,$$

and V is the potential energy of the system of particles. The spatial derivatives are taken with respect to the coordinates of the particle whose contribution to the cross section is being calculated. The leading correction to the cross section for scattering by free atoms at rest is proportional to $s_2(\kappa)$, and, therefore, to the average kinetic energy of the particle being considered. In general, because of quantum-mechanical effects, $\bar{T}$ is greater than the physical temperature T of the medium. At zero temperature, for example, $\bar{T}$ is in general non-zero, and is a measure of the energy of zero-point vibrations.

We now derive Eq. (2.249), beginning with $s_2(\kappa)$. In this case, the result follows simply by averaging over directions of κ:

$$s_2(\kappa) = \left\langle L^2 \right\rangle_{T,R} = \hbar^2\left\langle (\boldsymbol{\kappa}\cdot\mathbf{p}/M)^2 \right\rangle_{T,R} = \tfrac{2}{3}(\hbar^2\kappa^2/M)\left\langle p^2/2M \right\rangle_T \ .$$

To obtain s_3 and s_4 we use the fact that for an eigenstate $|j\rangle$ of the Hamiltonian the expectation of the time-derivative of any operator **A** vanishes:

$$\langle j|\frac{d\mathbf{A}}{dt}|j\rangle = \frac{i}{\hbar}\langle j|H_0\mathbf{A} - \mathbf{A}H_0|j\rangle = 0 \ .$$

Thus, for example

$$s_3(\kappa) = \frac{\hbar}{i}\left\langle \dot{L}L + 2L\dot{L}\right\rangle_{T,R} = \frac{\hbar}{i}\left\langle \tfrac{1}{2}\frac{d}{dt}L^2 + L\dot{L}\right\rangle_{T,R} = \frac{\hbar}{i}\left\langle L\dot{L}\right\rangle .$$

Since $\langle L\dot{L} + \dot{L}L\rangle = 0$ we can express s_3 in the form

$$s_3(\kappa) = \tfrac{1}{2}\frac{\hbar}{i}\left\langle L\dot{L} - \dot{L}L\right\rangle_{T,R} = \tfrac{1}{2}\left\langle [L,[H_0,L]]\right\rangle_{T,R} .$$

Finally, since $[H_0,L] = -(\hbar^2/i)(\boldsymbol{\kappa}\cdot\nabla V)/M$,

$$s_3(\kappa) = \frac{\hbar^2}{2}(\hbar/i)^2\left\langle -\frac{\boldsymbol{\kappa}\cdot\nabla}{M}\left(\frac{\boldsymbol{\kappa}\cdot\nabla V}{M}\right) + \left(\frac{\boldsymbol{\kappa}\cdot\nabla V}{M}\right)\frac{\boldsymbol{\kappa}\cdot\nabla}{M}\right\rangle_{T,R}$$

$$= (\hbar^4\kappa^2/6M^2)\left\langle\nabla^2 V\right\rangle_T .$$

A similar procedure gives the result in Eq. (2.249) for $s_4(\kappa)$.

Let us now apply the results of Eq. (2.249) to the calculation of the scattering cross section and the energy-transfer moments. The scattering cross section is given by

$$\sigma(E_0) = \frac{\sigma_b}{4\pi\hbar}\int\frac{k}{k_0}\left\langle S_{inc}(\boldsymbol{\kappa},\omega)\right\rangle_R dE\,d\Omega$$

$$= \frac{\hbar\sigma_b}{4\pi m k_0}\int\left\{\frac{1}{2\pi}\int\exp\left[-i\left(\omega - \frac{\hbar\kappa^2}{2M}\right)t\right]\times\right.$$

$$\left.\times\left\langle g(\boldsymbol{\kappa},t)\right\rangle_{T,R}dt\right\}d\boldsymbol{\kappa} . \tag{2.251}$$

By transforming the integration variable from $\boldsymbol{\kappa}$ to

$$\mathbf{q} = \boldsymbol{\kappa} - \frac{A}{A+1}\mathbf{k}_0 = \boldsymbol{\kappa} - \mathbf{k}_r , \tag{2.252}$$

and using the series expansion for $\left\langle g(\boldsymbol{\kappa},t)\right\rangle_{T,\Omega}$, there results

$$\frac{1}{2\pi}\int\exp\left[-i\left(\omega - \frac{\hbar\kappa^2}{2M}\right)t\right]\left\langle g(\boldsymbol{\kappa},t)\right\rangle_{T,R}dt =$$

$$= \frac{2m}{\hbar}\frac{A}{A+1}\sum_{n=0}^{\infty}\frac{1}{n!}\left(\frac{2m}{\hbar 2}\right)^n\left(\frac{A}{A+1}\right)^n\delta^{(n)}(q^2 - k_r^2)s_n(\kappa) .$$

Here we have used the definition of the Dirac δ-function and of its derivatives,

$$a^{-(n+1)} \frac{d^n\delta(x)}{dx^n} = a^{-(n+1)}\delta^{(n)}(x) = \frac{1}{2\pi}\int_{-\infty}^{\infty} e^{iaxt}(it)^n dt \ . \qquad (2.253)$$

These have the property that

$$\int_{-\epsilon_1}^{\epsilon_2} f(x)\,\delta^{(n)}(x)dx = (-1)^n f^{(n)}(0) \ , \qquad (2.253a)$$

where

$$f^{(n)}(0) = \left(\frac{d^n f}{dx^n}\right)_{x=0} \ ,$$

and ϵ_1, ϵ_2 are any positive numbers. The expression for the cross section now becomes

$$\sigma(E_0) = \frac{\sigma_b}{2\pi k_0}\,\frac{A}{A+1}\sum_{n=0}^{\infty}\frac{1}{n'}\left(\frac{2m}{\hbar 2}\right)^n\left(\frac{A}{A+1}\right)^n \Gamma_n^{(0)}(k_r) \ , \qquad (2.254)$$

where

$$\Gamma_n^{(0)}(k_r) = \int \delta^{(n)}(q^2 - k_r^2)s_n(\kappa)d\mathbf{q} \ .$$

In this equation κ is understood to have the value determined from Eq. (2.252). The purpose of the transformation (2.252) is now clear. As a result of it, the argument of $\delta^{(n)}$ depends only on the magnitude of $\mathbf{q}$ and not on its direction. Consequently, it is easy to perform the integration over $\mathbf{q}$. Noting that

$$q^2 d^q d\Omega_q = \tfrac{1}{2} q d(q^2)\ \frac{2\pi}{2k_r q}\ d(\kappa^2) \ ,$$

there follows

$$\Gamma_n^{(0)}(k_r) = \pi\int qd(q^2)\delta^{(n)}(q^2 - k_r^2)\frac{1}{2k_r q}\int s_n(\kappa)d\kappa^2$$

$$= (-1)^n \frac{\pi}{2k_r}\left(\frac{\partial^n}{\partial x^n}\int_{y_-}^{y_+} s_n(\kappa)d(\kappa^2)\right)_{x=k_r^2} , \qquad (2.255)$$

where $y_{\pm}(x) = (x^{\frac{1}{2}} \pm k_r)^2$. Using Eq. (2.249), we obtain explicit expressions for the first few $\Gamma_n^{(0)}$:

$$\Gamma_0^{(0)} = 2\pi k_r ,$$

$$\Gamma_1^{(0)} = 0 ,$$

$$\Gamma_2^{(0)} = \pi\hbar^2 k_B\bar{T}/Mk_r ,$$

$$\Gamma_3^{(0)} = 0 ,$$

$$\Gamma_4^{(0)} = -\frac{3}{4}\pi\hbar^2 C_{Av}/Mk_r^5 .$$

The cross section is given by

$$\sigma(E_0) = \sigma_0\left[1 + \tfrac{1}{2}\frac{k_B\bar{T}}{AE_0} - \frac{1}{32}\left(\frac{A+1}{A}\right)^2\frac{C_{Av}}{AE_0^3} + O(E_0^{-4})\right] . \qquad (2.256)$$

We note that the inclusion of higher terms in the collision-time expansion would result in contributions involving higher powers of E_0^{-1}, starting with E_0^{-4}. The Doppler effect, represented by the term containing $\bar{T}$, gives the dominant correction for chemical binding effects at high energies. The increase in cross section produced by the Doppler effect is verified by measurements of $\sigma_0(E)$ in hydrogenous substances. These measurements are discussed in Sec. 3-4.1. Except for hydrogenous substances, however, Eq. (2.256) is not very useful. The reason for this is that for most materials the elastic scattering interference effect at high energies, which we have so far neglected, is of the same order of magnitude as the Doppler effect.

We can also apply the method which leads to Eq. (2.256) to derive expansions for the moments of the energy transfer.[68] To obtain

$$M^{(\ell)}(E_0) = \int_0^\infty \sigma(E_0 \to E)(E_0 - E)^\ell dE \ , \tag{2.257}$$

we can use the series on the right-hand side of Eq. (2.254), with $\Gamma_n^{(0)}(k_r)$ replaced by

$$\Gamma_n^{(\ell)}(k_r) = \int \delta^{(n)}(q^2 - k_r^2)\left\{\frac{\hbar^2}{2m}\left[k_0^2 - \left(\frac{\mathbf{k}_0}{A+1} - \mathbf{q}\right)^2\right]\right\}^\ell s_n(\kappa^2)d\mathbf{q} \ .$$

The first energy-transfer moment is given by

$$M^{(1)}(E_0) = \sigma_0 \frac{2AE_0}{(A+1)^2}\left[1 - \frac{2A-1}{A}\frac{k_B\bar{T}}{E_0} + \frac{1}{24}\frac{A+1}{A}\frac{3A+1}{A}\frac{B_{Av}}{E_0^2} - \right.$$

$$\left. - \frac{1}{15}\frac{4A+1}{A^2}\frac{(K^2)_{Av}}{E_0^2} + O(E_0^{-3})\right] \ . \tag{2.258}$$

The terms $\sigma_0 2AE_0/(A+1)^2$ is the first moment of energy loss for collisions against free nuclei at rest. The bracketed term represents corrections for chemical-binding effects at high energies. The leading correction, which is proportional to $\bar{T}$, represents the decrease in average energy loss arising from the Doppler effect.

We have stated that the validity of the short-collision-time approximation rests on the assumption that only values of t much less than the periods of atomic motions contribute to the time integral in Eq. (2.251). There is one case in which this assumption breaks down, namely, in the case of a harmonically and isotropically vibrating nucleus with A = 1. That one may expect such a difficulty for an isotropic harmonic oscillator of any mass M follows directly from Eq. (2.113), from which it is clear that the χ-function is unity for $t = 2\pi n/\omega_0$ (where n is an integer) as well as for t = 0. Wick has examined the contributions to the cross section for t near $2\pi n/\omega_0$ for the case of a harmonic oscillator with arbitrary mass in its ground state. When the scattering cross section is expressed in the form

$$\frac{\sigma(E_0)}{\sigma_0} = 1 + n_0^{-1}c(n_0) \ ,$$

where $n_0 = E_0/\hbar\omega_0$, the effect of the long-collision-time contributions is manifested by a $c(n_0)$ which has a damped oscillatory behavior. The existence of this oscillatory behavior was first pointed out by Placzek.[16] For A = 1, the damping is zero; the damping increases with increasing A, so that, for very large A, $c(n_0)$ rapidly approaches a constant value with increasing n_0.

The nature of the process which leads to the undamped oscillatory behavior of $c(n_0)$ for A = 1 can be described in terms of the double-impulse picture discussed in the first part of this section. In terms of this picture, the collision time can be long if and only if the neutron is left with almost zero velocity after the first impulse. This is possible for a head-on collision with a nucleus whose mass is m and whose speed is only a very small fraction of that of the neutron. The second collision can occur either immediately or after 1,2,3... periods, when the nucleus passes near its initial position (where it left the neutron at rest) with the right direction of motion to kick the neutron again in the forward direction. For an oscillator with A > 1, there is the difference that after the first impulse, the neutron is left with an appreciable velocity, and the second impulse occurs in an interval of time which is short compared with the period of vibration of the atom.

2-8.3 THE INTERFERENCE SCATTERING. The interference scattering can be calculated by formal techniques very similar to those employed for calculating the direct scattering. There is of course an essential physical difference between the two cases. There is no good reason for extracting a factor like Eq. (2.94) from the interference part of $\chi_c(\boldsymbol{\kappa},t)$. First, if the nuclei with coordinates $\mathbf{r}_\ell$ and $\mathbf{r}_{\ell'}$ have different masses, we do not know which mass to use in Eq. (2.94). The separation of a term like (2.94), although it can be accomplished, is without significance. Such a strongly oscillating factor is compensated for by terms which occur in the expectation value of the remaining operators. More precisely, one can show that when $\chi_{\ell\ell'}^{(\iota)}(\boldsymbol{\kappa},t)$ is averaged over the thermal distribution of initial states and then expressed in the form $e^{if(t)}$, no linear term appears[19] in the expansion of f(t) about t = 0. (It is for this reason that the treatment of interference terms given in Ref. 58 is incorrect.) This means that a factor expressing the effects of free-atom recoil does not apply to the interference terms.

Then it is natural to expand the entire interference part of the intermediate scattering function in powers of t. When this is done, the expansion that one obtains is exactly the same as the expansion given by Placzek.[16]

The calculation of the interference terms requires the short-time expansion of Eq. (2.92), with the omission of terms with $\ell' = \ell$. We denote this quantity by $\chi_{int}(\kappa,t)$. Averaging over orientations of the system gives

$$\left\langle \chi_{int}(\boldsymbol{\kappa},t)\right\rangle_R = \frac{1}{N}\sum_{\ell,\ell'} \frac{G_n^{\ell\ell'}}{n!}\left(\frac{it}{\hbar}\right)^n , \tag{2.259}$$

where

$$G_0^{\ell\ell'} = \left\langle \exp\left\{-i\boldsymbol{\kappa}\cdot[\mathbf{r}_\ell(0) - \mathbf{r}_{\ell'}(0)]\right\}\right\rangle_{T,R} = \left\langle \exp(-i\boldsymbol{\kappa}\cdot\mathbf{r}_{\ell\ell'})\right\rangle_{T,R}$$

$$= \left\langle \frac{\sin \kappa r_{\ell\ell'}}{\kappa r_{\ell\ell'}}\right\rangle_T ,$$

$$G_1^{\ell\ell'} = 0,$$

$$G_2^{\ell\ell'} = \frac{(\hbar\kappa)^4}{4M^2} G_0^{\ell\ell'} + \frac{\hbar^2}{M^2}\left\langle \exp(-i\boldsymbol{\kappa}\cdot\mathbf{r}_{\ell\ell'})[(\boldsymbol{\kappa}\cdot\mathbf{p}_\ell)(\boldsymbol{\kappa}\cdot\mathbf{p}_{\ell'})]\right\rangle_{T,R}$$

and $r_{\ell\ell'} = |r_\ell(0) - r_{\ell'}(0)|$.

The interference contribution to the cross section can be expressed in the form[16]

$$\sigma_{int}(E_0) = \frac{\sigma_{c,b}}{N}\sum_{\ell\ell'}{}' \left\{ \frac{1}{2k_0^2}\left\langle\frac{1}{r_{\ell\ell'}^2}\right\rangle_T - \frac{1}{2k_0^2}\left\langle\frac{\cos 2k_0 r_{\ell\ell'}}{r_{\ell\ell'}^2}\right\rangle_T + \right.$$

$$+ A^{-2}\left\langle(\cos 2\mathbf{k}_0\cdot\mathbf{r}_{\ell\ell'} - 2\mathbf{k}_0\cdot\mathbf{r}_{\ell\ell'} \times\right.$$

$$\left.\times \sin 2\mathbf{k}_0\cdot\mathbf{r}_{\ell\ell'})\,\frac{(\mathbf{p}_\ell\cdot\mathbf{k}_0)(\mathbf{p}_{\ell'}\cdot\mathbf{k}_0)}{\hbar^2 k_0^4}\right\rangle_T +$$

$$\left. + A^{-2}\left\langle\frac{\sin 2k_o r_{\ell\ell'}}{k_0 r_{\ell\ell'}} + \cos 2k_0 r_{\ell\ell'}\right\rangle_T + \ldots \right\} . \tag{2.260}$$

The first line in this equation represents the static approximation for the interference scattering, and is obtained when the time-dependence of $\chi_{int}(\boldsymbol{\kappa},t)$, or equivalently the time-dependence of the nuclear motion, is neglected. In the static approximation the calculated cross section for interference scattering is purely elastic. The terms proportional to A^{-2} represent part of the correction to the static approximation. For short wavelength, all the terms involving trigonometric functions fluctuate rapidly, and their magnitude goes to zero with decreasing wavelength.[69] The cross section is then represented by

$$\sigma_{int}(E_0) = \frac{\sigma_{c,b}}{2Nk_0^2}\sum_{\ell\ell'}\left\langle\left(r_{\ell\ell'}^2\right)^{-1}\right\rangle_T . \tag{2.261}$$

The exact wavelength at which Eq. (2.260) goes over into Eq. (2.261) depends on the nature of the scattering system, and its determination requires rather careful consideration.[69]

Wick has pointed out that the series given in Eq. (2.260) is not an asymptotic expansion in the same sense as Eq. (2.256). Proceeding further in the expansion would yield other terms which are simple powers of E_0^{-1}, i.e., terms which would not appear multiplied by rapidly oscillating trigonometric functions. At high energies, these terms would be more important than the strongly oscillating terms in Eq. (2.260). For large masses and moderate energies, however, the oscillating terms in Eq. (2.260) are more important than those which do not explicitly appear.

So far, we have applied the short-collision-time approximation to the calculation of integral properties of the scattering, as, for example, the total scattering cross section and the energy-change moments. This approximation is not satisfactory for calculating $\sigma(E_0,E,\theta)$, for in this case it leads to an expansion in terms of singular functions. In Sec. 3-2.2 we shall develop short-collision-time approximations which express $\sigma(E_0,E,\theta)$ in terms of nonsingular functions. Actually the quasi-classical approximation for rigid rotators given in Sec. 2-7.3 is an example of such an approximation. Another application of the short-collision-time method is made in Chap. 5, where it is applied to calculate the leading corrections for the effects of chemical binding on the neutron spectrum in the energy region where these effects are small.

2-9 SPACE-TIME CORRELATION FUNCTIONS AND THE DYNAMICS OF ATOMIC MOTION IN LIQUIDS

2-9.1 DEFINITIONS AND SIMPLE PROPERTIES. We have so far considered the functions $S(\boldsymbol{\kappa},\omega)$ and $\chi(\boldsymbol{\kappa},t)$ as concise ways of stating the neutron scattering properties of systems with known dynamical behavior. Our focus has been on the neutron, and on the calculation of the scattering cross sections needed for considerations of neutron thermalization. We now turn our attention to the complementary problem of interpreting neutron scattering experiments on systems of poorly understood dynamical behavior. In particular, we consider liquids and dense gases. For these systems we need to express the scattering in a way that more clearly exhibits the qualitative physical aspects of the atomic motions. The basic idea here was given by Van Hove,[18] who noted that neutron scattering could be expresssed in terms of space-time correlation functions defined as the Fourier transforms with respect to momentum transfer of the intermediate scattering functions given in Section 2-4. These space-time correlation functions can then be given simple probabilistic definitions for systems whose atomic motions can be described classically.

It is worth recalling that the functions $S(\boldsymbol{\kappa},\omega)$ and $\chi(\boldsymbol{\kappa},t)$ that we have been studying are well-defined descriptions of the dynamical behavior of the scattering system, independently of their relation to neutron scattering. We may consider $\boldsymbol{\kappa}$ and ω as dynamical variables of the system without any reference to their relation to momentum and energy transfer in neutron scattering. Functions very similar to $S(\boldsymbol{\kappa},\omega)$ are frequently the focus of calculations in the theory of many-particle systems. That the theory can be directly tested at the microscopic level by neutron scattering experiments has been a great impetus to the development of the theory, but the tendency of solid-state theory and statistical mechanics to move toward this type of description is inevitable in any case as our knowledge becomes more refined.

Restricting our attention to systems with a single atomic species, we introduce the space-time correlation function $G(\mathbf{r},t)$ defined by

$$G(\mathbf{r},t) = (2\pi)^{-3}N^{-1}\sum_{\ell,\ell'}\int e^{-i\boldsymbol{\kappa}\cdot\mathbf{r}} \times$$

$$\times \left\langle e^{-i\boldsymbol{\kappa}\cdot\mathbf{r}_\ell(0)}\, e^{i\boldsymbol{\kappa}\cdot\mathbf{r}_{\ell'}(t)}\right\rangle_T d\boldsymbol{\kappa}\ , \tag{2.262a}$$

$$= (2\pi)^{-3}\int e^{-i\boldsymbol{\kappa}\cdot\mathbf{r}}\chi(\boldsymbol{\kappa},t)d\boldsymbol{\kappa} \tag{2.262b}$$

$$= (2\pi)^{-3}\iint e^{i(\omega t-\boldsymbol{\kappa}\cdot\mathbf{r})}S_c(\boldsymbol{\kappa},\omega)d\boldsymbol{\kappa}d\omega\ . \tag{2.262c}$$

In order to obtain a clearer picture of the physical content of $G(\mathbf{r},t)$, we express it in the form

$$G(\mathbf{r},t)$$

$$= N^{-1}\sum_{\ell,\ell'}\left\langle\int\delta[\mathbf{r}+\mathbf{r}_\ell(0)-\mathbf{r}']\delta[\mathbf{r}'-\mathbf{r}_{\ell'}(t)]d\mathbf{r}'\right\rangle_T\ , \tag{2.263}$$

where the δ-function of an operator is to be understood as

$$\delta(\mathbf{r}) = (2\pi)^{-3}\int e^{i\boldsymbol{\kappa}\cdot\mathbf{r}}d\boldsymbol{\kappa}$$

and

$$\delta[\mathbf{r}'-\mathbf{r}_\ell(t)] = e^{iH_0t/\hbar}\,\delta[\mathbf{r}-\mathbf{r}_\ell(0)]\,e^{-iH_0t/\hbar}\ . \tag{2.264}$$

The result (2.263) follows from the convolution formula for a Fourier transform of an ordered product.

When calculated quantum mechanically, $G(\mathbf{r},t)$ is a complex function with no simple physical interpretation. For many substances it is a good approximation to calculate $G(\mathbf{r},t)$ as if the system obeyed classical mechanics. In that case, we can interpret the thermal average in Eq. (2.263) as an average over a classical equilibrium ensemble, and we can neglect the fact that $\mathbf{r}_{\ell'}(t)$ does not commute with $\mathbf{r}_\ell(0)$. We thus obtain

$$G^{cl}(\mathbf{r},t) = N^{-1}\sum_{\ell,\ell'}\left\langle\delta[\mathbf{r}+\mathbf{r}_\ell(0)-\mathbf{r}_{\ell'}(t)\right\rangle_T\ . \tag{2.265}$$

The function $G^{cl}(\mathbf{r},t)$ can be interpreted as the probability (unnormalized) of finding an atom in a unit volume at $\mathbf{r}$ at time t given that an atom was at the origin at time zero. We thus have a direct association between the energy and momentum distribution of scattered neutrons and the development of the atomic motions in space and time.

We emphasize that $G^{cl}(\mathbf{r},t)$ is the classical limit of the space-time correlation function defined by Eq. (2.263). This is not simply related to the classical limit of the neutron scattering process. It turns out, however, that the classical correlation function (2.265) is useful in describing neutron scattering experiments from substances such as liquid argon that obey classical statistical mechanics to a good approximation. We give more careful consideration to the nature of the classical limit in Section 2-9.3.

In considering neutron scattering we have found it convenient to split $S(\boldsymbol{\kappa},\omega)$ into a direct part representing the motion of individual atoms, and an interference part representing the correlated motion of pairs of atoms. A similar breakup of $G(\mathbf{r},t)$ is useful for our subsequent considerations. Following Van Hove, we introduce a self-correlation function $G_s(\mathbf{r},t)$ defined by

$$\begin{aligned} G_s(\mathbf{r},t) &= (2\pi)^{-3} \iint e^{-i\boldsymbol{\kappa}\cdot\mathbf{r}} e^{i\omega t} S_{inc}(\boldsymbol{\kappa},\omega)\, d\boldsymbol{\kappa}\, d\omega \\ &= (2\pi)^{-3} \int \left\langle e^{-i\boldsymbol{\kappa}\cdot\mathbf{r}_1(0)}\, e^{i\boldsymbol{\kappa}\cdot\mathbf{r}_1(t)} \right\rangle_T e^{-i\boldsymbol{\kappa}\cdot\mathbf{r}}\, d\boldsymbol{\kappa} \\ &= \left\langle \int \delta\left(\mathbf{r} + \mathbf{r}_1(0) - \mathbf{r}'\right) \delta\left(\mathbf{r}' - \mathbf{r}_1(t)\right) d\mathbf{r}' \right\rangle_T . \end{aligned} \tag{2.266}$$

In Eq. (2.266) we have made use of the fact that we are dealing with a single monatomic species, so that the ensemble average of the motion is the same for any atom. Thus atom number one in Eq. (2.266) could equally well be any other atom in the system. Similarly, we introduce the time-dependent pair correlation function $G_d(\mathbf{r},t)$ describing the motion of distinct atoms:

$$G_d(\mathbf{r},t) = (N-1) \int \left\langle e^{-i\boldsymbol{\kappa}\cdot\mathbf{r}_1(0)} e^{i\boldsymbol{\kappa}\cdot\mathbf{r}_2(t)} \right\rangle_T e^{-i\boldsymbol{\kappa}\cdot\mathbf{r}} d\boldsymbol{\kappa}$$

$$= (N-1) \left\langle \int \delta\left(\mathbf{r} + \mathbf{r}_1(0) - \mathbf{r}'\right) \times \right.$$

$$\left. \times\, \delta\left(\mathbf{r}' - \mathbf{r}_2(t)\right) d\mathbf{r}' \right\rangle_T . \tag{2.267}$$

Again we have used the identity of the atoms so that atoms 1 and 2 can be any two distinct atoms in the system.[k] These new functions are related to $G(\mathbf{r},t)$ through

$$G(\mathbf{r},t) = G_s(\mathbf{r},t) + G_d(\mathbf{r},t) . \tag{2.268}$$

The notation here is somewhat unfortunate, but is dictated by common usage. The subscript s when attached to $G(\mathbf{r},t)$ carries the same meaning as the subscript inc when attached to $S(\boldsymbol{\kappa},\omega)$ or to $\chi(\boldsymbol{\kappa},t)$. The sum of $G_s(\mathbf{r},t)$ and $G_d(\mathbf{r},t)$ is the double Fourier transform of the coherent scattering function $S_c(\boldsymbol{\kappa},\omega)$.

The function $G(\mathbf{r},t)$ was originally introduced as a generalization of the radial distribution function $g(\mathbf{r})$, which is familiar from X-ray and neutron diffraction. In ordinary diffraction experiments the measured quantity is the differential scattering cross section integrated over the energy of the scattered particle. For neutron diffraction, the desired result would be obtained by integrating Eq. (2.68) over energy at a fixed scattering angle. If we set the factor k/k_0 in Eq. (2.68) equal to one and assume that the integral at fixed scattering angle is the same as the integral at fixed momentum transfer, we obtain

$$\frac{d\sigma}{d\Omega} = \frac{\sigma_{c,b}}{4\pi} \int S_c(\boldsymbol{\kappa},\omega)\,d\omega = \frac{\sigma_{c,b}}{4\pi}\,\chi_c(\boldsymbol{\kappa},0) . \qquad (2.269$$

This is the so-called static approximation alluded to in Section 2-8.3. It is very accurate for most cases for which neutron diffraction is used

---

[k] Equations (2.266) and (2.267) are rigorously correct if the system is confined in a box and satisfies periodic boundary conditions. This is equivalent to ignoring surface effects.

for structure determination.

For X-ray diffraction, a similar result is obtained, except that the constant $\sigma_{c,b}$ is replaced by an atomic form factor whose κ-dependence must be determined independently. For X-rays, the static approximation is of course extremely accurate since the change in energy on scattering is a very small fraction of the initial energy.

The quantity measured in an ordinary diffraction experiment is thus

$$\chi_c(\boldsymbol{\kappa},0) = \int e^{i\boldsymbol{\kappa}\cdot\mathbf{r}}[G_s(\mathbf{r},0) + G_d(\mathbf{r},0)]d\boldsymbol{\kappa} \ . \qquad (2.270)$$

From the last line of Eq. (2.266) we see that

$$G_s(\mathbf{r},0) = \delta(\mathbf{r}) \ , \qquad (2.271)$$

and from Eq. (2.267) we see that

$$G_d(\mathbf{r},0) = (N - 1)\left\langle\delta(\mathbf{r} + \mathbf{r}_1(0) - \mathbf{r}_2(0))\right\rangle_T = \frac{N - 1}{V} g(\mathbf{r}) \ . \qquad (2.272)$$

In Eq. (2.272), V is the volume of the system and the definition of $g(\mathbf{r})$ is the usual one for the radial distribution function of a fluid. Combining Eqs. (2.270), (2.271), and (2.272) and introducing

$$\rho = \frac{N - 1}{V} \ ,$$

we can write Eq. (2.270) in the form

$$\frac{d\sigma}{d\Omega} = \frac{\sigma_{c,b}}{4\pi} [1 = \Gamma(\boldsymbol{\kappa})] \ , \qquad (2.273)$$

where

$$\Gamma(\boldsymbol{\kappa}) = \rho\int e^{i\boldsymbol{\kappa}\cdot\mathbf{r}} g(\mathbf{r})d\mathbf{r} \ . \qquad (2.274)$$

Actually, for a fluid, $g(\mathbf{r})$ depends only on $r = |\mathbf{r}|$, and thus $\Gamma(\boldsymbol{\kappa})$ depends only on $\kappa = |\boldsymbol{\kappa}|$ and is given by

$$\Gamma(\kappa) = \rho\int_0^\infty g(r)\left(\frac{\sin \kappa r}{\kappa r}\right) 4\pi r^2 dr \ . \qquad (2.275)$$

Similar relations hold for all the Fourier transforms with respect to $\mathbf{r}$

and κ that occur for the time-dependent functions.

Radial distribution function for liquids have been determined for a number of substances by X-ray and neutron techniques.[70-73] Figure 2.17 shows the typical appearance of g(r) for a monatomic fluid consisting of atoms like argon and krypton. The appearance of this curve is quite consistent with the interpretation of $G_d(\mathbf{r},0)$ given by Eq. (2.272), that is, that $\rho g(\mathbf{r}) = G_d(\mathbf{r},0)$ describes the average density distribution of all particles except the one at $\mathbf{r} = 0$, from which the density is observed. The near-zero value of g(r) for small r extends to a radius that is comparable to the radius of an atom. This reflects the nearly perfect impenetrability of atoms resulting from the Pauli principle and electrostatic forces. The peaks at larger radii represent the local ordering of atoms which disappears at large distances, where g(r) approaches unity.

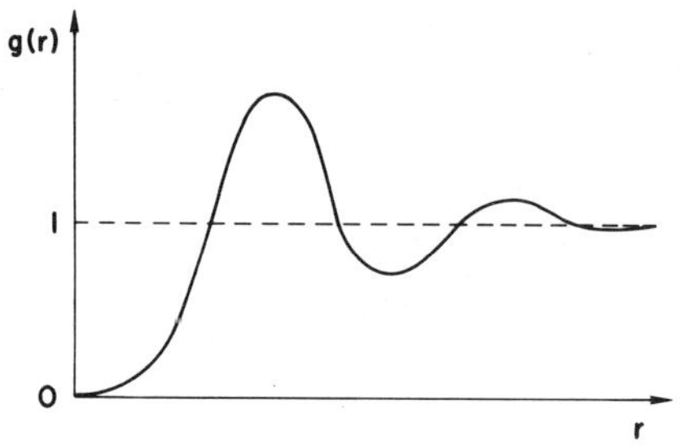

Figure 2.17 The static pair correlation function g(r).

The radial distribution function g(r) is of great interest in equilibrium statistical mechanics.[74] It can be shown that a knowledge of g(r) determines completely the thermodynamic properties of the system. For example, the internal energy is given by

$$\frac{U}{Nk_BT} = \frac{3}{2} + \frac{\rho}{2k_BT}\int_0^\infty V(r)g(r)4\pi r^2 dr \ , \tag{2.276}$$

where V(r) is the interaction potential between atoms. In writing Eq. (2.276) we have assumed that the atoms in the system interact through a two-body velocity-independent potential. A second general result that we will need later relates the volume integral of [g(r) - 1] to the isothermal compressibility. We turn to this result in Sec. 2-9.5, where we discuss more general relations among fluctuations, correlations, and $G(r,t)$.

2-9.2 THE BEHAVIOR OF $G_s(r,t)$ FOR CLASSICAL SYSTEMS. Returning again to the space-time correlation functions, we consider first the motion of individual atoms as described by $G_s(r,t)$. The simplest property of $G_s(r,t)$ that is of physical interest is the mean-square displacement

$$\left\langle r^2(t)\right\rangle = \frac{\int_0^\infty r^2 G_s(\mathbf{r},t)4\pi r^2 dr}{\int_0^\infty G_s(\mathbf{r},t)4\pi r^2 dr} \equiv 6\gamma(t) \ . \tag{2.277}$$

The denominator of Eq. (2.277) is just $\chi_{inc}(0,t)$ and is equal to one. The numerator of Eq. (2.277) is obtained by expanding $\left\langle \chi_{inc}(\boldsymbol{\kappa},t)\right\rangle_R$ in powers of κ^2:

$$\chi_{inc}(\boldsymbol{\kappa},t) \equiv \left\langle \chi_{inc}(\boldsymbol{\kappa},t)\right\rangle_R = 1 - \kappa^2\gamma(t) + O(\kappa^4) \ . \tag{2.278}$$

In general, $\gamma(t)$ and $G_s(r,t)$ are complex functions. For certain simple physical systems we can use our calculations of $\chi_{inc}(\boldsymbol{\kappa},t)$ to write $\gamma(t)$. For an ideal gas,

$$\gamma(t) = (2M\beta)^{-1}[t^2 - i\hbar\beta t] \ , \tag{2.279}$$

while for a crystal with one atom per unit cell and a normal-mode frequency distribution $\rho(\omega)$,

$$\gamma(t) = \frac{\hbar}{2M}\int_0^\infty \rho(\omega)\omega^{-1}\,\{\cot h(\tfrac{1}{2}\beta\hbar\omega)[1 - \cos\omega t] - - i \sin \omega t\}d\omega \ . \tag{2.280}$$

This result also applies for a monatomic polycrystal. To get a better physical feeling for $\gamma(t)$, consider first the classical limit of $G_s(\mathbf{r},t)$. Then $G_s(\mathbf{r},t)$ and $\gamma(t)$ become real functions. In this limit we have

$$\chi_{\text{inc}}(\kappa,t) \to \left\langle e^{i\kappa\cdot[\mathbf{r}_1(t)-\mathbf{r}_1(0)]} \right\rangle_T$$

so that Eq. (2.278) implies

$$\lim_{\hbar\to 0} \gamma(t) = \tfrac{1}{2} \left\langle [z_1(t) - z_1(0)]^2 \right\rangle_{cl} \equiv \tfrac{1}{2}w(t) , \qquad (2.281)$$

where $z_1(t)$ is the component of the displacement $\mathbf{r}_1(t)$ in the direction of κ. Again, for the case of an ideal gas

$$w(t) = \frac{k_B T t^2}{M} , \qquad (2.282)$$

and for a crystal

$$w(t) = \frac{2k_B T}{M} \int_0^\infty \rho(\omega)\omega^{-2}[1 - \cos \omega t]d\omega . \qquad (2.283)$$

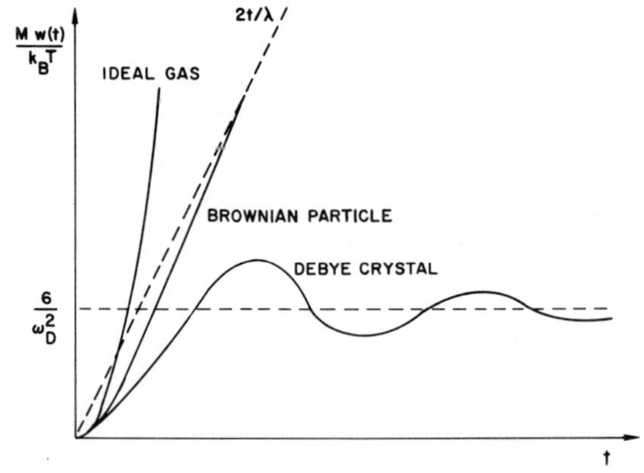

Figure 2.18 The width function for an ideal gas, Brownian motion, and a Debye crystal.

Figure 2.18 shows w(t) for an ideal gas, for a Debye crystal, and for a Brownian particle whose displacement satisfies the Langevin equation. For the Debye crystal,

$$\rho(\omega) = 3\omega^2/\omega_D^3 \qquad 0 < \omega < \omega_D$$
$$= 0 \qquad \omega > \omega_D \ . \tag{2.283a}$$

We discuss the Brownian particle in more detail shortly. In an ideal gas, an atom initially at the origin streams outward indefinitely without collision; thus the mean square displacement is proportional to t^2 for all times. In a solid, the initial behavior is the same as in a gas, at least at high temperatures where the classical limit of Eq. (2.283) is valid. At long times, however, the mean square displacement saturates since the atom is bound to its lattice site. Because of the large number of frequencies involved, the oscillatory behavior of w(t) characteristic of harmonic motion is damped at long times. In fact, the asymptotic value of w(t) at long times is just the mean square amplitude of vibration that determines the Debye-Waller factor at temperatures high compared with the Debye temperature.

If we were to consider an Einstein crystal with

$$\rho(\omega) = \delta(\omega - \omega_0) \ ,$$

we would find that w(t) oscillates in an undamped manner for long times. Actually the long-time behavior of w(t) for a harmonic system has been the subject of some interesting studies.[75] If we take a system with a finite but large number of degrees of freedom, then w(t) is given by a sum over frequencies rather than an integral. Such a sum is by definition a quasi-periodic function of time and does not have a constant asymptotic value. The recurrence of oscillations at very long times is closely related to the well-known Poincaré cycles. If we let the number of degrees of freedom in the system become infinite <u>before</u> we consider the limit of long times, however, we get an integral such as (2.283) that approaches a constant behavior at long times. We emphasize that this apparent approach to a smooth asymptotic behavior requires taking the limit of an infinite system first. Although the harmonic system is a somewhat peculiar special case, it turns out in general[76] that correlation functions of the type of $\gamma(t)$ have quasi-periodic behavior at long times for finite systems, and that our usual intuitive ideas about long-time behavior for the correlation functions require that we consider the limit of an infinite system.

In a liquid or a non-ideal gas, an atom that is initially moving away from the origin makes collisions with other atoms in the fluid. At long times the mean square displacement is proportional to the time. In fact, the coefficient of self-diffusion of the fluid can be defined by

$$D = \lim_{t\to\infty} \frac{w(t)}{2t} \tag{2.284}$$

The mean square displacement starts out proportional to t^2 at short times and is proportional to t at long times. A simple model for such behavior is the motion of a Brownian particle as described by the Langevin equation or equivalently by the Fokker-Planck equation.[77] This gives a mean square displacement

$$w(t) = \frac{2k_B T}{M\lambda^2} [\lambda|t| - (1 - e^{-\lambda|t|})] \ , \tag{2.285}$$

where λ is the so-called friction constant. This is similar to the mean square displacement for an atom in a dilute gas where the random binary collisions can be described by the Boltzmann equation. Recent work[78,79] indicates that a linearized Boltzmann equation of the same form as the neutron transport equation is appropriate to this problem. A simplified solution of this equation[78] yields (2.285) with the friction constant to be interpreted as $1/\tau$, where τ is the mean time between collisions.

In a liquid, we expect that w(t) will have a solid-like behavior at short times and go over into a diffusive behavior at long times. This might include a fairly well-defined region at intermediate times in which w(t) is almost constant. Since we have no satisfactory theory of the liquid state, we must rely on experiment to tell us about this intermediate-time behavior. It is just this time region that is effectively explored by inelastic neutron scattering experiments.

The analysis of neutron scattering experiments to determine w(t) is rather complicated, and our knowledge of this function from experiment is still quite incomplete. As one might expect, the behavior of w(t) for intermediate times is different for different types of liquids. The solid-like behavior is quite apparent, for example, in water, while the motion in liquid argon is almost gas-like. There is an excellent recent review of our knowledge in this area by Larsson and Sjölander[80] to which the reader is referred for more detailed information.

Proceeding with our theoretical discussion of the mean square displacement, we can rewrite w(t) in terms of the velocity correlation function for the substance. If we then introduce the spectrum of this correlation function, it turns out that this spectrum is a natural generalization of the frequency spectrum of normal modes $\rho(\omega)$ that appears in Eqs. (2.280) and (2.283) and that with this substitution these two equations remain true. This idea is originally expressed by Egelstaff[15] and has been developed more fully by several authors. A rather complete set of references is contained in the review by Larsson and Sjölander mentioned above.

The physics is somewhat more transparent if we first consider a classical system, for which we can write

$$z(t) = z(0) = \int_0^t v_z(t_1)dt_1 \ .$$

It follows that

$$w(t) = \int_0^t \int_0^t \left\langle v_z(t_1) v_z(t_2) \right\rangle_{cl} dt_1 dt_2 \ . \qquad (2.286)$$

Since the system is in equilibrium, the velocity correlation function appearing in Eq. (2.286) in the form

$$w(t) = 2 \int_0^t (t - t_1) \left\langle v_z(0) v_z(t_1) \right\rangle_{cl} dt_1 \ . \qquad (2.287)$$

In particular, if we make use of the definition (2.284) of the diffusion coefficient, we obtain

$$D = \int_0^\infty \left\langle v_z(0) v_z(t) \right\rangle_{cl} dt \ . \qquad (2.288)$$

The general character of Eq. (2.288) is of interest. It expresses a transport coefficient in terms of a time-dependent correlation function of the equilibrium system. Actually, our definition of D was through Eq. (2.284), so that the diffusion coefficient has been defined in terms of an equilibrium correlation function rather than through Fick's law. The two definitions must be the same, however, since Fick's law plus the equation of continuity imply that the mean square displacement is 2Dt.

There has been a great deal of recent interest in equations like (2.288) since they place the calculation of transport coefficients on a basis comparably sound to the calculation of thermodynamic functions from the partition function. That is to say, the transport coefficients characteristic of non-equilibrium phenomena can be expressed in terms of time-dependent correlations at equilibrium.

Such relations are usually known as Kubo formulae, and they are known for all the transport coefficients of interest. For more information on this subject, the reader is referred to the review article by Chester.[81] We see in Sec. 2-9.5 that a relation between G(r,t) and transport coefficients can also be established but this relation is considerably more complicated.

To illustrate the physical meaning of (2.287) and (2.288) for a simple case, we consider the Brownian particle diffusing according to the Langevin equation. For this case,

$$\left\langle v_z(0)v_z(t)\right\rangle = \frac{k_B T}{M}\, e^{-\lambda|t|} , \tag{2.289}$$

and (2.288) gives a relation between the diffusion coefficient and the friction constant

$$D = \frac{k_B T}{M\lambda} . \tag{2.290}$$

Equations (2.289) and (2.290) are not true for an atom moving in a liquid. (We could of course use (2.290) as the definition of the friction coefficient, but (2.289) would still not be true.)

In writing (2.287) we have expressed the mean square displacement of an atom in a fluid in terms of the correlation function of the stationary stochastic process v(t). In describing such a process it is often more convenient to work with the Fourier transform of the correlation function. This transform is known as the spectrum of the process.[77] (If the reader is not familiar with the language of stochastic processes, he can omit reading those sentences phrased in this language.) We define the spectrum $f_0(\omega)$ by

$$\left\langle v_z(0)v_z(t)\right\rangle_{cl} = \frac{k_B T}{M}\int_0^\infty f_0(\omega)\cos\omega t\,d\omega\ . \qquad (2.291)$$

Since the distribution of velocities at $t = 0$ is Maxwellian, it follows that

$$\int_0^\infty f_0(\omega)d\omega = 1\ . \qquad (2.292)$$

The diffusion coefficient is simply related to the value of $f_0(\omega)$ at zero frequency. Substituting (2.291) into (2.288), we obtain

$$D = \frac{\pi k_B T f_0(0)}{2M}\ . \qquad (2.293)$$

Finally, $w(t)$ can be written in terms of $f_0(\omega)$ by substituting (2.291) into (2.287):

$$w(t) = \frac{2k_B T}{M}\int_0^\infty f_0(\omega)\omega^{-2}[1 - \cos\omega t]d\omega\ . \qquad (2.294)$$

We note that (2.294) is identical with (2.283), except that $f_0(\omega)$ replaces $\rho(\omega)$. Thus the form (2.294) for $w(t)$ does not depend on the motion being harmonic but can be generalized to arbitrary motion. In the special case of a harmonic crystal, the frequency distribution of the normal modes coincides with the spectrum of the velocity correlation function. For a more general motion, the frequency distribution of the normal modes is not a well-defined concept, but we still have the relation (2.294) between the mean square displacement and the spectrum of the velocity correlation function.

It should be emphasized that the analogy between these two functions should not be carried too far. For a harmonic solid, for example, $\rho(\omega)$ is independent of temperature and determines all the thermodynamic properties of the solid.

For a more general motion $f_0(\omega)$ depends on temperature and does not determine the thermodynamic properties. In particular, since the diffusion coefficient may have a quite complicated dependence on temperature, we see from (2.288) that $f_0(0)$ may also vary in a complicated way with temperature.

As originally emphasized by Egelstaff, the spectrum $f_0(\omega)$ gives a convenient way of characterizing the nature of the atomic motions. For example, the systems we have discussed so far give the following results:

<u>Ideal Gas</u> $$f_0(\omega) = \delta(\omega) \ , \qquad (2.295)$$

<u>Einstein Crystal</u> $$f_0(\omega) = \delta(\omega - \omega_0) \ , \qquad (2.296)$$

<u>Debye Crystal</u> $$f_0(\omega) = \frac{3\omega^2}{\omega_D^3} \qquad 0 < \omega < \omega_D \ , \qquad (2.297)$$

<u>Langevin Diffusion</u> $$f_0(\omega) = \frac{2\lambda}{\pi}\left(\omega^2 + \lambda^2\right)^{-1} . \qquad (2.298)$$

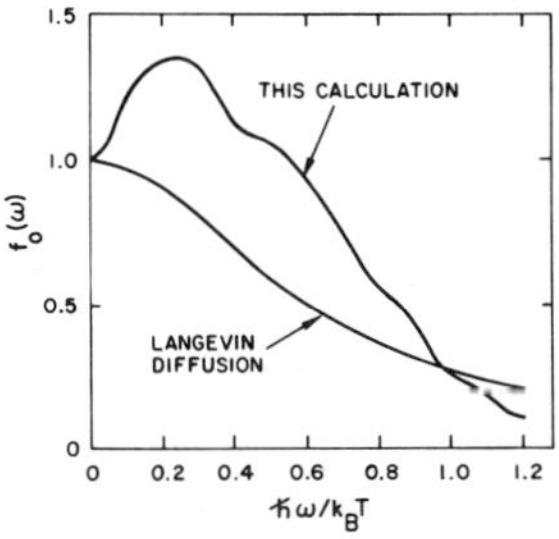

Figure 2.19 Spectrum of the velocity auto-correlation function. The Lorentzian spectrum of a Langevin-type correlation is also shown. (From Rahman, <u>Phys. Rev</u>., Ref. 82.)

We expect the spectrum for a liquid to exhibit some mixture of diffusive and solid-like behavior, but we cannot say more without the help of experiment or detailed calculations. Recently, Rahman[82] carried out an extensive computer calculation solving the classical equations of motion for 864 particles interacting with a potential appropriate to argon atoms. His calculation should be a good simulation of molecular dynamics in liquid argon at 94.4°K and a density of 1.374 g/cm^3. In Fig. 2.19 we show his calculated result for the spectrum of the velocity correlation function; for comparison, the result (2.298) for Langevin diffusion is also given. The way in which this spectrum varies with temperature and density has not been calculated. It is also not very well known how similar this result is to what one would obtain for other liquids. We should thus consider this result as an encouraging start toward the quantitative understanding of this very interesting and difficult subject.

We have so far considered the mean square displacement $w(t)$ and the spectrum $f_0(\omega)$ as dynamical quantities of interest in their own light without much thought as to their relevance to neutron scattering. We know, however, for the particular cases of the ideal gas and of the cubic crystal with a single atom per unit cell, that the scattering is determined entirely by $\gamma(t)$. In these special cases the intermediate scattering function is given by

$$\chi_{inc}(\boldsymbol{\kappa},t) = \exp[-\kappa^2\gamma(t)] \ . \tag{2.299}$$

Since $\chi_{inc}(\boldsymbol{\kappa},t)$ is a Gaussian function of κ, its Fourier transform $G_s(\mathbf{r},t)$ is a Gaussian function of r and is given by

$$G_s(\mathbf{r},t) = [4\pi\gamma(t)]^{-\frac{3}{2}} \exp\left[\frac{-r^2}{4\gamma(t)}\right] \ . \tag{2.300}$$

In the classical limit this becomes

$$\chi_{inc}^{cl}(\boldsymbol{\kappa},t) = \exp[-\tfrac{1}{2}\kappa^2 w(t)] \ , \tag{2.301}$$

$$G_s^{cl}(\mathbf{r},t) = [2\pi w(t)]^{-\frac{3}{2}} \exp\left[\frac{-r^2}{2w(t)}\right] . \tag{2.302}$$

It has been shown on quite general grounds that the Gaussian form of Eqs. (2.299) through (2.302) is correct in the limit of short times[83] and in the limit of long times.[84] Furthermore, in addition to the systems mentioned above, it is true at all times for the Brownian particle diffusing according to the Langevin equation. However, we do in general expect deviations from Gaussian behavior at intermediate times, and therefore $\gamma(t)$ is not a complete description of the incoherent neutron scattering.

Equations (2.299) through (2.302) are often assumed to hold at all times. This is known as the Gaussian approximation. This approximation can be checked directly by experiment. If Eq. (2.299) holds, the

Fourier transform of $S_{inc}(\kappa,\omega)$ with respect to ω can be determined from measurements at a single momentum transfer. If we make measurements of $S_{inc}(\kappa,\omega)$ for several values of κ, we can thus determine whether or not the Gaussian approximation is valid.

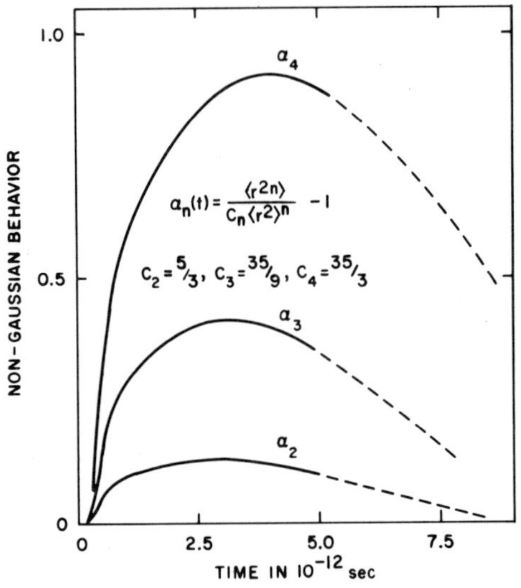

Figure 2.20 The non-Gaussian character of $G_s(r,t)$ showing an initial Gaussian behavior lasting about 0.15×10^{12} sec and, on extrapolating to the right, a return to a Gaussian form at about 10^{-11} sec. Maximum departure of r^4 from its Gaussian value is only about 13%. (From Rahman, Phys. Rev., Ref. 82.)

On the theoretical side, we have estimates of non-Gaussian effects in several cases. In a molecular system, the Gaussian approximation consists in using Krieger-Nelkin averaging over the molecular orientations, as discussed in Sec. 2-7. This is probably quite good for the molecular rotations but is less accurate for the vibrations of a planar molecule such as H_2O. Corrections associated with a better averaging over orientation have been calculated for H_2O by Koppel and Young[85] and are discussed in Chap. 3. In graphite, also, the proper averaging over orientation of the microcrystals gives appreciable non-Gaussian effects. These are also discussed further in Chap. 3. For monatomic fluids there have been two calculations of non-Gaussian effects. The kinetic model considered by Nelkin and Ghatak[78] gave a mean square displacement that was the same as one would obtain for a Brownian particle. The higher moments of the displacement, however, are larger than one would obtain from the Gaussian approximation (2.302). Similar results were obtained by Rahman[82] in his molecular dynamics calculation for liquid argon. Rahman's results are shown in Fig. 2.20. The actual importance of these non-Gaussian effects to the analysis of experiments in liquid argon cannot be determined until the double Fourier transforms of the space-time correlation functions have been calculated and compared with the available experiments on liquid argon.[86]

Finally, we consider the behavior of $S_{inc}(\boldsymbol{\kappa},\omega)$ in the limit of small κ and ω. In this limit, we are looking at the motion of an atom on a spatial scale that is large compared with atomic spacings, and for frequencies low compared with characteristic frequencies of thermal vibration. We expect this limit to be governed by the macroscopic laws appropriate to the average motion of an atom in the fluid. The appropriate description is in terms of simple diffusion. For small values of κ, $\chi_{inc}(\boldsymbol{\kappa},t)$ decays slowly with time. In taking the Fourier transform, the time integral will have its major contributions from large values of t. Thus we can to a good approximation let

$$\chi_{inc}(\boldsymbol{\kappa},t) \approx \exp(-\kappa^2 D|t|) \tag{2.303}$$

since w(t) approaches 2Dt for long times and non-Gaussian corrections can be shown to be small in this limit. The corresponding result for $G_s(\mathbf{r},t)$ can be obtained by replacing w(t) by $2D|t|$ in Eq. (2.302). This is the familiar solution of the diffusion equation in an infinite homogeneous medium. The corresponding result for $S_{inc}(\boldsymbol{\kappa},\omega)$ is

$$S_{inc}(\boldsymbol{\kappa},\omega) = \frac{D\kappa^2}{\pi} \frac{1}{\omega^2 + \left(D\kappa^2\right)^2} . \tag{2.304}$$

In the limit of small momentum transfer, we thus expect that the energy distribution of scattered neutrons should be Lorentzian, with a half-width at half-maximum given by

$$\Delta\omega = D\kappa^2 \ . \tag{2.305}$$

The diffusion coefficients determined from neutron inelastic scattering at small momentum transfer are consistent with those measured by other techniques, but neutron scattering is not a particularly advantageous technique for determining diffusion coefficients. Equation (2.304), however, is quite soundly based theoretically in the limit of small κ. Neutrons are of great value in determining just how this limit is approached. From the fact that Eq. (2.304) can be approximately applied for the smallest κ available in neutron experiments (about $2 \times 10^7\ \text{cm}^{-1}$), we see that the macroscopic law of diffusion has some relevance even on a spatial scale as small as 10 Å. In Sec. 2-9.5 we consider the macroscopic equations governing the behavior of coherent scattering. Although the equations and the related macroscopic phenomena are more complicated, the basic physical idea of associating the macroscopic equations with the correlation functions remains the same. One of the most important contributions of neutron scattering experiments is to establish the limiting space and time scales on which these macroscopic equations apply. The experimental information available is still too incomplete to do this in a definitive way, but it is possible to do more than we have done here. The interested reader is referred to the review by Larsson and Sjölander[80] for a more complete description.

2-9.3 EXTENSION TO QUANTUM MECHANICAL SYSTEMS. Before discussing more fully the physical content of $G(\mathbf{r},t)$, let us consider some of its general formal properties. First we recall that $G(\mathbf{r},t)$ is in general a complex function. It does, however, possess the Hermitian symmetry

$$G^*(-\mathbf{r},-t) = G(\mathbf{r},t) \ . \tag{2.306}$$

This follows directly from Eq. (2.262), using the reality of $S(\boldsymbol{\kappa},\omega)$. Next, observe that

$$F(\mathbf{r},t) = G(\mathbf{r},t + i\hbar\beta/2) \tag{2.307}$$

is a real and even function of $\mathbf{r}$ and t; that is,

$$F^*(\mathbf{r},t) = F(-\mathbf{r},-t) = F(\mathbf{r},t) \ . \tag{2.308}$$

This follows directly from the relation between $G(r,t + i\beta\hbar/2)$ and the even function $S_0(\boldsymbol{\kappa},\omega)$ given by Eq. (2.76):

$$F(\mathbf{r},t) = G(\mathbf{r},t + i\beta\hbar/2)$$

$$= (2\pi)^{-3} \int d\boldsymbol{\kappa} d\omega e^{i(\omega t-\boldsymbol{\kappa}\cdot\mathbf{r})} S_{0c}(\boldsymbol{\kappa},\omega) \ . \tag{2.309}$$

Finally, we want to derive the relation

$$\text{im } G(\mathbf{r},t) = -\tan\left(\tfrac{1}{2}\beta\hbar \frac{\partial}{\partial t}\right) \text{re } G(\mathbf{r},t) \tag{2.310}$$

between the real (re) and imaginary (im) parts of G. To do this, observe that, by Eq. (2.73),

$$G^*(\mathbf{r},t) = (2\pi)^{-3} \int d\boldsymbol{\kappa} d\omega e^{-i(\omega t-\boldsymbol{\kappa}\cdot\mathbf{r})} S_c(\boldsymbol{\kappa},\omega)$$

$$= (2\pi)^{-3} \int d\boldsymbol{\kappa} d\omega e^{i(\omega t-\boldsymbol{\kappa}\cdot\mathbf{r})} e^{-\beta\hbar\omega} S_c(\boldsymbol{\kappa},\omega) \ .$$

Now we proceed in a straightforward way by constructing

$$\text{re } G = \tfrac{1}{2}(G + G^*)$$

$$= \tfrac{1}{2}(2\pi)^{-3} \int d\boldsymbol{\kappa} d\omega e^{i(\omega t-\boldsymbol{\kappa}\cdot\mathbf{r})} (1 + e^{-\beta\hbar\omega}) S_c(\boldsymbol{\kappa},\omega)$$

$$\text{im } G = \tfrac{1}{2}i(G - G^*)$$

$$= \tfrac{1}{2}i(2\pi)^{-3} \int d\boldsymbol{\kappa} d\omega e^{i(\omega t-\boldsymbol{\kappa}\cdot\mathbf{r})} (1 - e^{-\beta\hbar\omega}) S_c(\boldsymbol{\kappa},\omega) \ .$$

Next, we express re G in the form

$$\text{re } G = \tfrac{1}{2}(G + G^*)$$

$$= \tfrac{1}{2}(2\pi)^{-3} \int d\boldsymbol{\kappa} d\omega e^{i(\omega t-\boldsymbol{\kappa}\cdot\mathbf{r})} (1 - e^{-\beta\hbar\omega}) \cot h \frac{\beta\hbar\omega}{2} S_c(\boldsymbol{\kappa},\omega)$$

and operate on both sides of this expression with

$$\tan h\left(-\tfrac{1}{2}\beta\hbar i \frac{\partial}{\partial t}\right) = \frac{1}{i}\left[\tan\left(\frac{\beta\hbar}{2}\frac{\partial}{\partial t}\right)\right] .$$

Noting that

$$\tan h\left(-\tfrac{1}{2}\beta\hbar i \frac{\partial}{\partial t}\right) e^{i\omega t} = \tan h\left(\tfrac{1}{2}\beta\hbar\omega\right) ,$$

Eq. (2.310) follows immediately.

Each of the relations from Eq. (2.306) through (2.310) holds for $G_s(\mathbf{r},t)$ and $G_d(\mathbf{r},t)$ separately. Relations of this kind between real and imaginary parts of correlation functions are known as fluctuation-dissipation theorems. We can also obtain such relations between the real and imaginary parts of the velocity correlation function. This allows us to generalize the analogy between Eq. (2.283) and Eq. (2.294) to the quantum-mechanical case. We follow the paper of Rahman <u>et al.</u>,[84] giving the most important results without proof.

We can in general write

$$\ln\chi_{inc}(\kappa,t) = -\kappa^2\gamma(t) + \kappa^4\gamma_2(t) + \dots . \tag{2.311}$$

Neglecting the second and higher terms on the right-hand side of (2.311) corresponds to the Gaussian approximation. It can be shown that

$$\gamma(t) = \frac{i\hbar t}{2M} + \int_0^t (t - t_1)\left\langle v_z(0)v_z(t_1)\right\rangle dt_1 , \tag{2.312}$$

where the Heisenberg operator $v_z(t)$ is defined by

$$v_z(t) = \frac{1}{M} \exp\left(\frac{iH_0 t}{\hbar}\right) p_z \exp\left(\frac{-iH_0 t}{\hbar}\right),$$

and

$$p_z = \frac{\hbar}{i}\frac{\partial}{\partial z}$$

is the momentum operator for the atom whose motion we are considering. Now we introduce two frequency spectra related to the real and imaginary parts of the velocity correlation function:

$$\mathrm{im}\left\langle v_z(0)v_z(t)\right\rangle_T = \frac{\hbar}{2M}\int_0^\infty \omega f(\omega) \sin \omega t d\omega , \tag{2.313}$$

$$\mathrm{re}\left\langle v_z(0)v_z(t)\right\rangle_T = \frac{k_B T}{M}\int_0^\infty g(\omega) \cos \omega t d\omega . \tag{2.314}$$

The functions $f(\omega)$ and $g(\omega)$ are simply related by virtue of the thermal equilibrium of the scattering system. This relation is

$$g(\omega) = (\tfrac{1}{2}\beta\hbar\omega) \cot h(\tfrac{1}{2}\beta\hbar\omega) f(\omega) \tag{2.315}$$

and is another special case of a general fluctuation dissipation theorem. The proof is very similar to that given in Sec. 2-3 for the detailed balance relation.

Substituting Eqs. (2.313, (2.314), and (2.315) into Eq. (2.312), we get

$$\gamma(t) = \frac{\hbar}{2M}\int_0^\infty f(\omega) \{\cot h(\tfrac{1}{2}\beta\hbar\omega)[1 - \cos \omega t] - i \sin \omega t\}\omega^{-1} d\omega \ . \tag{2.316}$$

We note that Eq. (2.316) is exactly of the same form as Eq. (2.280), but is not restricted to harmonic lattice vibrations. We again have the spectrum of the velocity correlation function as a generalization of the normal-mode frequency distribution, but now in the quantum-mechanical case. Furthermore, it can be shown that

$$\int_0^\infty f(\omega) d\omega = 1 \tag{2.317}$$

and that $f(\omega)$ is non-negative.

We can thus reduce the discussion of the quantum-mechanical mean square displacement to a single real positive and normalized frequency distribution. Whatever such frequency distribution we use, the function $\gamma(t)$ calculated from Eq. (2.316) will have the correct short time behavior and the correct relationship between real and imaginary parts. Furthermore, from examining the long-time behavior it can be shown that

$$D = \int_0^\infty \text{re} \left\langle v_z(0) v_z(t) \right\rangle_T dt = \frac{\pi k_B T}{2M} f(0) \ . \tag{2.318}$$

For a system whose motions are almost classical we can thus expect $f(\omega)$ and $f_0(\omega)$ to be very similar. If we restrict ourselves to the Gaussian approximation, we can combine a classical description of the atomic motions with a quantum-mechanical scattering calculation by replacing $f(\omega)$ by $f_0(\omega)$ in Eq. (2.316) and using Eq. (2.299) to calculate the scattering. By doing this we satisfy the detailed balance condition (2.73) and the first moment sum rule (2.80) that must hold for the true scattering.

A very similar formulation of the problem has been given by Egelstaff and Schofield,[87] who work in the symmetric formulation discussed in Sec. 2-4. Combining this formulation with the Gaussian approximation, one obtains

$$e^{-\frac{1}{2}\beta\hbar\omega} S_{inc}(\boldsymbol{\kappa},\omega) = \frac{1}{\pi} \int_0^\infty \cos \omega t \exp[-\kappa^2 \zeta(t)] dt , \qquad (2.319)$$

where

$$\zeta(t) = \gamma(t + \tfrac{1}{2} i\beta\hbar)$$

$$= \frac{\hbar}{2M} \int_0^\infty \frac{f(\omega)}{\omega} \frac{\cos h(\frac{1}{2}\beta\hbar\omega) - \cos \omega t}{\sin h(\frac{1}{2}\beta\hbar\omega)} d\omega . \qquad (2.320)$$

This formulation is entirely equivalent to the one we have discussed and has the advantage that it deals only with manifestly real quantities.

Examining Eqs. (2.319) and (2.320) we see that there are two reasons why the classical description of Sec. 2-9.2 may not be adequate. The first is that $f(\omega)$ as defined by Eq. (2.313) (or preferably by Eqs. (2.314) and (2.315)) may differ from $f_0(\omega)$ as defined by Eq. (2.291). This is associated with the adequacy of a classical description of the dynamics. The second source of deviation from classical behavior arises from the explicit appearance of Planck's constant in Eqs. (2.319) and (2.320).

Actually, in replacing $f(\omega)$ by $f_0(\omega)$ we are not making a consistently classical calculation, inasmuch as such a calculation would be irrelevant to atomic phenomena. We first reduce the problem to a Hamiltonian containing a potential energy depending only on the nuclear coordinates. The origin of this potential energy is certainly quantum mechanical. The reduction to this form of Hamiltonian is known as the Born-Oppenheimer approximation, and has been discussed briefly in Sec. 2-2. Having made this reduction, we then calculate the atomic motions classically to obtain $f_0(\omega)$. It is difficult to know how accurate an approximation it is to replace $f(\omega)$ by $f_0(\omega)$ for a liquid. We note, however, that it is precisely the correct procedure in the case of a solid. Once we have made the Born-Oppenheimer approximation in a solid, the frequency distribution of normal modes $\rho(\omega)$ is calculated by solving the classical normal-mode problem. The analogy between $\rho(\omega)$ for a solid and $f(\omega)$ for a liquid cannot be taken too seriously, but it suggests that the analogous procedure of calculating $f(\omega)$ from classical equations of motion should be quite good. This is particularly true when one realizes that greater uncertainties are probably associated with obtaining an appropriate Hamiltonian to describe the atomic motions in the liquid than in solving the equations of motion classically. (Of course, the above argument does not apply to liquid helium, where a completely quantum-mechanical treatment is manifestly required.)

Even if we replace $f(\omega)$ by $f_0(\omega)$, there are quantum-mechanical corrections associated with the explicit appearance of Planck's constant in Eqs. (2.319) and (2.320). For values of $\hbar\omega \gtrsim k_BT$, these corrections are formally identical to those we must make for phonons in a solid. Even if $f(\omega)$ has no high-frequency components, however, there is a quantum-mechanical term in Eq. (2.320). We recall that for an ideal gas, Eq. (2.98) gives

$$S(\kappa,\omega) = \left(\frac{M\beta}{2\pi\kappa^2}\right)^{\frac{1}{2}} \exp\left(-\frac{\beta M\omega}{2\kappa^2}\right) \exp(\tfrac{1}{2}\beta\hbar\omega) \exp\left(-\frac{\beta\hbar^2\kappa^2}{8M}\right). \quad (2.321)$$

The second exponential in Eq. (2.321) comes from Eq. (2.319). The third exponential can be thought of as a correction associated with the uncertainty principle that does not allow the atom to be precisely localized at $t = 0$. The condition for this correction to be small is

$$\frac{\hbar^2\kappa^2}{2M} \ll k_BT ,$$

or, in other words, the product of κ and the De Broglie wavelength of an atom with thermal kinetic energy should be much less than one. For liquid argon at 90°K, this condition becomes

$$\kappa \ll 12 \times 10^8 \text{ cm}^{-1}$$

and is well satisfied by the experiments of interest. The ideal gas limit is relevant to large values of κ. Quantum-mechanical corrections become less important as κ becomes smaller and longer time intervals become dominant.

To summarize, if we make the Gaussian approximation and assume that the spectrum of the velocity correlation function can be calculated from a classical description of the molecular dynamics, then all the needed quantum-mechanical corrections to calculate the incoherent neutron scattering are contained explicitly in Eqs. (2.319) and (2.320). We have given some qualitative discussion of these corrections, but it is just as easy and preferable to work directly with Eqs. (2.319) and (2.320) rather than to make further approximations. If we go beyond the Gaussian approximation, we do not have an automatic procedure. The work of Aamodt et al.[88] indicates, however, that a reasonable approximation is to replace the first exponential in Eq. (2.321) with the double Fourier transform of the classical Van Hove correlation function, and to retain the second and third exponential as quantum-mechanical corrections. For the time-dependent pair-correlation function $G_d(\mathbf{r},t)$ there is essentially nothing known about the quantum-mechanical corrections for almost classical systems, and rather little known even for the classical case. We now turn our attention to this function for classical fluids.

2-9.4 COHERENT SCATTERING FROM CLASSICAL LIQUIDS. All of the available experimental data on neutron inelastic scattering by monatomic liquids is for elements for which the scattering is primarily coherent. Except for the case of argon, where the incoherence is isotopic in origin and can in principle be varied by varying the isotopic composition, it is not possible to experimentally separate $G_s(\mathbf{r},t)$ from $G_d(\mathbf{r},t)$. It is therefore necessary to understand something about coherent inelastic scattering even if the motion of a single atom is the subject that we want to study. The time-dependent pair correlations contained in $G_d(\mathbf{r},t)$ are of course also of considerable fundamental interest, but they are considerably more complicated and less well understood than is the motion of a single atom. It is just because we understand so little about the coherent effects that are mixed into the scattering data that the rather well-developed theory of the preceding sections cannot be ambiguously compared with experiment.

We can make a few exact statements about $G_d(\mathbf{r},t)$. At $t = 0$ it is given by Eq. (2.272). In the limit of very long times we expect the probability of finding an atom at r not to depend on where some other atom was at $t = 0$, so that

$$\lim_{t\to\infty} G_d(\mathbf{r},t) = \frac{N - 1}{V} .$$

We discuss this point further in the following section in connection with the relation between $G(\mathbf{r},t)$ and density correlations. The basic behavior of $G_d(\mathbf{r},t)$ with time is expected to be a relaxation of the short-range spatial order that is present at $t = 0$. To illustrate this behavior, the behavior of $G_d(\mathbf{r},t)$ for liquid argon is shown in Fig. 2.21. This figure is taken from Rahman's molecular dynamics computation,[82] which we discussed earlier.

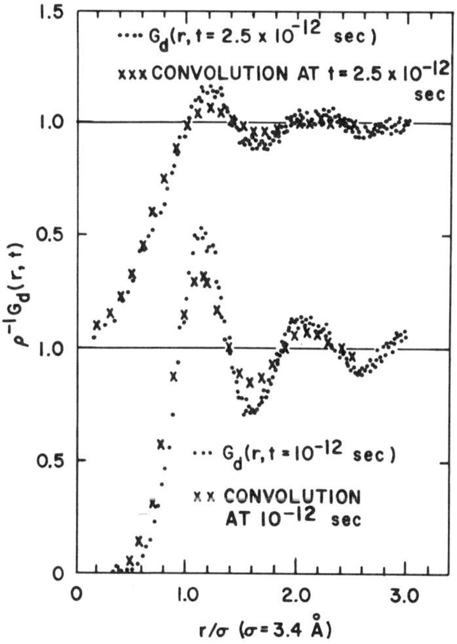

Figure 2.21 Time-dependent pair-correlation function $G_d(r,t)$ shown at two values of t. The convolution approximation of Vineyard[86] gives a too rapid decay of G_d. (From Rahman, Phys. Rev., Ref. 82.)

The crosses in Fig. 2.21 correspond to the convolution approximation introduced by Vineyard.[89] This approximation assumes that $G_d(\mathbf{r},t)$ can be written as

$$G_d(\mathbf{r},t) = \int G_d(\mathbf{r}',0)G_s(\mathbf{r} - \mathbf{r}',t)d\mathbf{r}' \ . \tag{2.322}$$

We see from Fig. 2.21 that this gives qualitatively reasonable results but that it overestimates the rate at which the spatial correlations decay. If we take the Fourier transform of Eq. (2.322) with respect to $\mathbf{r}$ and t, we obtain

$$S_c(\boldsymbol{\kappa},\omega) = [1 + \Gamma(\boldsymbol{\kappa})]S_{inc}(\boldsymbol{\kappa},\omega) \ ; \tag{2.323}$$

thus, Vineyard's approximation says that the energy distribution of coherently scattered neutrons is the same as that of incoherently scattered neutrons for the same momentum transfer.

The slower decay of correlations in $G_d(\mathbf{r},t)$ in the neighborhood of the principal maximum of $g(\mathbf{r})$ implies that the energy distribution of coherent scattering will be narrower than that for incoherent scattering for values of $\boldsymbol{\kappa}$ near the first diffraction maximum. This narrowing effect was first predicted by de Gennes[90] on quite general grounds, and has been observed experimentally for several liquids. The argument here goes back to the short-time behavior as expressed through the sum rules discussed in Sec. 2-8. Although de Gennes started from a classical theory, the same results can be obtained from the classical limit of the results of Placzek and Wick, which we have discussed in Sec. 2-8. The essential point is that the interference scattering gives no contribution to the second moment of the energy transfer for a classical system. We thus have

$$\int \omega^2 S_{inc}(\boldsymbol{\kappa},\omega)d\omega = \int \omega^2 S_c(\boldsymbol{\kappa},\omega)d\omega = \frac{\kappa^2 k_B T}{M} \ . \tag{2.324}$$

The corresponding integrals over ω without the factor ω^2 in the integrand are quite different, however:

$$\left.\begin{aligned}\int S_{inc}(\boldsymbol{\kappa},\omega)d\omega &= 1\\ \int S_c(\boldsymbol{\kappa},\omega)d\omega &= [1+\Gamma(\boldsymbol{\kappa})]\ .\end{aligned}\right\}\qquad(2.325)$$

We thus have

$$\left.\begin{aligned}\left\langle\omega^2\right\rangle_{inc} &= \frac{\int \omega^2 S_{inc}(\boldsymbol{\kappa},\omega)d\omega}{\int S_{inc}(\boldsymbol{\kappa},\omega)d\omega} = \frac{\kappa^2 k_B T}{M}\\ \left\langle\omega^2\right\rangle_c &= \frac{\int \omega^2 S_c(\boldsymbol{\kappa},\omega)d\omega}{\int S_c(\boldsymbol{\kappa},\omega)d\omega} = \frac{\kappa^2 k_B T}{M[1+\Gamma(\boldsymbol{\kappa})]}\ .\end{aligned}\right\}\qquad(2.326)$$

Making some reasonable assumptions about the dependence of $S(\boldsymbol{\kappa},\omega)$ on ω, it follows from Eq. (2.326) that the energy distribution of coherently scattered neutrons is narrower than that of incoherently scattered neutrons for κ values such that $\Gamma(\boldsymbol{\kappa})$ is in the neighborhood of its principal maximum. Since this maximum is closely related to the principal maximum in $g(\mathbf{r})$, the slower decay of correlations seen in $G_d(\mathbf{r},t)$ is directly related to the narrower energy distribution of coherently scattered neutrons.

So far in our discussions we have implicitly or explicitly considered $S(\boldsymbol{\kappa},\omega)$ as a function of ω, with $\boldsymbol{\kappa}$ as a parameter. This point of view emphasizes the dynamics of atomic motion and is particularly appropriate for incoherent scattering. For coherent scattering, however, it is sometimes preferable to think in terms of the structure as modified by the atomic motions. This is what we have done in Fig. 2.21 in plotting $G_d(\mathbf{r},t)$ as a function of r with t as a parameter. The corresponding treatment of experimental data is to plot $S(\boldsymbol{\kappa},\omega)$ as a function of κ for fixed ω. An example of such a display of data is given in Fig. 2.22, where we show Randolph's results[91] for liquid sodium at 150°K. (We recall that ordinary diffraction measures the integral of $S(\boldsymbol{\kappa},\omega)$ over all ω.) The characteristic feature of Fig. 2.22 is the broadening of the diffraction maxima with increased energy transfer.

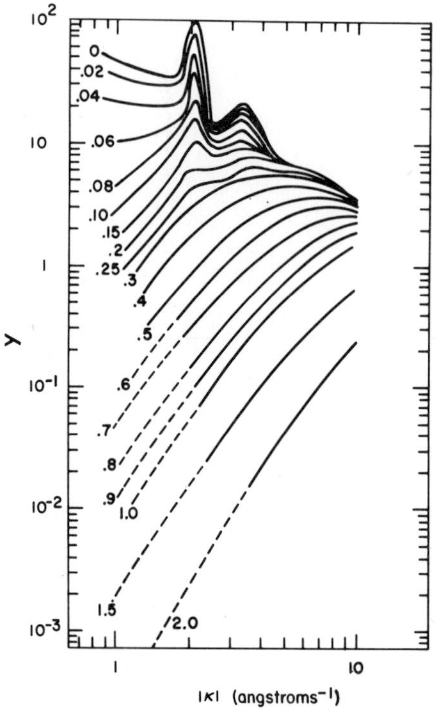

Figure 2.22 Reduced partial differential cross section $(E_0/E)^{\frac{1}{2}}\sigma(E_0, E, \theta)$ $\exp[E_0 - E)/2k_BT] = y$ as a function of κ for liquid sodium at 150°C. The parameter labeling the curves is the dimensionless energy transfer $(E_0 - E)/k_BT$. (From Randolph, Phys. Rev., Ref. 91.)

Similar effects have been observed in liquid argon[92] and have been interpreted in terms of a solid-like model to give phonon dispersion relations for the liquid. Such a model is at best crudely applicable to liquid argon, but our understanding of coherent inelastic scattering in liquids is sufficiently primitive that new qualitative ideas are still welcome. So far we have a few good experiments with neutrons, and one good "experiment" on a digital computer.[82] Improved phenomenological theory is also becoming available.[93] It seems likely that our knowledge in this area will become much more quantitative in the next few years and will yield considerable insight into the nature of the liquid state.

2-9.5 DENSITY FLUCTUATIONS, $G(\mathbf{r},t)$, AND HYDRODYNAMICS. We have so far considered $G(\mathbf{r},t)$ in specifically atomic terms and have emphasized the individual-particle aspects of the motion. It is often useful, however, to emphasize the collective aspects of the system. This is most effectively done by exploiting the formal connection between $G(\mathbf{r},t)$ and the propagation of density fluctuations in the system. To do this we introduce the quantum-mechanical density operator

$$\rho(\mathbf{r},t) \sum_{\ell} \delta[\mathbf{r} - \mathbf{r}_{\ell}(t)] \ . \tag{2.327}$$

The average number density

$$n(\mathbf{r}) = \left\langle \rho(\mathbf{r},t) \right\rangle_T \tag{2.328}$$

is a real quantity; it is independent of the time since the expectation value of a Heisenberg operator is independent of time for a stationary state. The quantity $n(\mathbf{r})$ has a probabilistic interpretation in the sense that $n(\mathbf{r})d\mathbf{r}/N$ is the probability for finding a particle in the volume $d\mathbf{r}$ at $\mathbf{r}$.

Following Glauber,[94] we introduce a two-point density function

$$n(\mathbf{r}',t';\ \mathbf{r}'',t'') = \left\langle \rho(\mathbf{r}',t')\rho(\mathbf{r}'',t'') \right\rangle_T \ . \tag{2.329}$$

The usefulness of the two-point density function in the present context arises from the simple relation that exists between it and the Van Hove correlation function

$$\begin{aligned} G(\mathbf{r},t) &= N^{-1} \int \left\langle \rho(\mathbf{r}' - \mathbf{r},0)\rho(\mathbf{r}',t) \right\rangle_T d\mathbf{r}' \\ &= N^{-1} \int n(\mathbf{r}' - \mathbf{r},0;\ \mathbf{r}',t)d\mathbf{r}' \ . \end{aligned} \tag{2.330}$$

An important property of $n(\mathbf{r}',t';\ \mathbf{r}'',t'')$ is its invariance with respect to time translations. This follows from

$$n(\mathbf{r}',t';\ \mathbf{r}'',t'') = [Z(\beta)]^{-1}\mathrm{Tr}\Big\langle e^{-\beta H_0} e^{iH_0t'/\hbar}\rho(\mathbf{r}',0)e^{-iH_0t'/\hbar} \times$$
$$\times\ e^{iH_0t''}\rho(\mathbf{r}'',0)e^{-iH_0t''/\hbar}\Big\rangle$$
$$= [Z(\beta)]^{-1}\mathrm{Tr}\Big\langle e^{-\beta H_0}\rho(\mathbf{r}',0)e^{iH_0(t''-t')/\hbar} \times$$
$$\times\ \rho(\mathbf{r}'',0)e^{-iH_0(t''-t')/\hbar}\ ,$$

which gives

$$n(\mathbf{r}',t';\ \mathbf{r}'',t'') = n(\mathbf{r}',0;\ \mathbf{r}'',t''-t')\ . \tag{2.331}$$

This property is a completely general one for systems in statistical equilibrium. A property of less general validity is the invariance with respect to space translations

$$n(\mathbf{r}'-\mathbf{R},t';\ \mathbf{r}''-\mathbf{R},t'') = n(\mathbf{r}',t';\ \mathbf{r}'',t'')\ , \tag{2.332}$$

for any $\mathbf{R}$. This holds only for spatially uniform systems.

Since the density operators $\rho(\mathbf{r}',t')$ and $\rho(\mathbf{r}'',t'')$ do not commute unless $t' = t''$, the function $n(\mathbf{r}',t';\ \mathbf{r}'',t'')$ is in general complex, its values being restricted only by the condition imposed by the Hermitian property of the density operators.

$$n^*(\mathbf{r}',t';\ \mathbf{r}'',t'') = \Big\langle\rho(\mathbf{r}'',t'')\rho(\mathbf{r}',t')\Big\rangle_T$$
$$= n(\mathbf{r}'',t'';\ \mathbf{r}',t')\ . \tag{2.333}$$

The imaginary part of the two-point density function is related to the commutator of the density operators by

$$\mathrm{im}\ n(\mathbf{r}',t';\ \mathbf{r}'',t'') = \frac{1}{2i}[n(\mathbf{r}',t';\ \mathbf{r}'',t'') - n^*(\mathbf{r}',t';\ \mathbf{r}'',t'')]$$
$$= \frac{1}{2i}\Big\langle[\rho(\mathbf{r}',t'),\rho(\mathbf{r}'',t'')]\Big\rangle_T\ . \tag{2.334}$$

In the limit of a classical system the density operators in Eq. (2.333) commute, and thus the two-particle density is a real function. In this case there exists a simple physical interpretation, namely, that $n(\mathbf{r}',t';\ \mathbf{r}'',t'')d\mathbf{r}'d\mathbf{r}''/N^2$ is the joint probability that there is a particle in $d\mathbf{r}'$ at t' and one in $d\mathbf{r}''$ at t''. The particles observed in the two-volume elements at different times may be one and the same particle or may be a distinct pair of particles. To further develop the connection between neutron scattering and the two-point density function, it is convenient to factor $n(\mathbf{r}',t';\ \mathbf{r}'',t'')$ into the product of the average single-particle density function and a conditional density function $n(0,0|\mathbf{r},t)$ that represents the density of particles at $\mathbf{r}$ at the time t given that a particle was at the origin at $t = 0$. Thus,

$$n(\mathbf{r}',t';\ \mathbf{r}'',t'') = n(\mathbf{r}')n(0,0|\mathbf{r}'' - \mathbf{r}',t'' - t') \ , \qquad (2.335)$$

where we have used the property of time-translational invariance and have assumed that the microscopic properties of the medium are invariant under spatial translations. From Eqs. (2.330), (2.335), and

$$\int n(\mathbf{r}')d\mathbf{r}' = N \ ,$$

it follows that

$$G(\mathbf{r},t) = n(0,0|\mathbf{r},t) \ . \qquad (2.336)$$

Thus, $G(\mathbf{r} - \mathbf{r}',t - t') = N(\mathbf{r}',t'|\mathbf{r},t) = n(0,0|\mathbf{r} - \mathbf{r}',t - t')$represents the density of particles at $\mathbf{r}$ at time t given that there is a particle at $\mathbf{r}'$ at the time t'. In the classical case where the density operators appearing in Eq. (2.330) are commuting c-numbers, G is a real function and admits the probabilistic interpretation that $N^{-1}(\mathbf{r} - \mathbf{r}',t - t')d\mathbf{r}$ is the probability that at time t there will be an atom at $\mathbf{r}$ given that there is an atom at $\mathbf{r}'$ at the time t'.

The two-point density function also enters into the description of the fluctuations of the density of a system in its equilibrium state. The local density fluctuations are related to the two-point density function in the sense that

$$\left\langle[\rho(\mathbf{r}'',t'') - \rho(\mathbf{r}',t')]^2\right\rangle_T = \left\langle\rho^2(\mathbf{r}'',t') + \rho^2(\mathbf{r}',t')\right\rangle_T - 2\,\mathrm{re}\ n(\mathbf{r}',t';\ \mathbf{r}'',t'') \ . \qquad (2.337)$$

For a system whose microscopic properties are invariant under space translations, this reduces to

$$\left\langle[\rho(\mathbf{r}'',t'') - \rho(\mathbf{r}',t')]^2\right\rangle_T = 2(N/V)^2 - 2\,\frac{N}{V}\,\mathrm{re}\ G(\mathbf{r}'' - \mathbf{r}',t'' - t') \ . \qquad (2.338)$$

Having determined that the real part of $G(\mathbf{r},t)$ is related to the fluctuations in the density of the equilibrium system, it still remains to determine the meaning of its imaginary part. Van Hove[95] has pointed out that the imaginary part of $G(\mathbf{r},t)$ divided by $\hbar$ describes the local disturbance produced by a neutron in the density of the scattering system. Equation (2.310) tells us that this local disturbance is related to the equilibrium density fluctuations of the scattering system in the absence of the neutron. This is a typical form for a fluctuation dissipation theorem[96] in which the dissipative response of a system to a perturbation is related to the equilibrium fluctuations of the system in the absence of the perturbation.

We recognize that the one- and two-particle density functions, and consequently $G(\mathbf{r},t)$, are statistical mechanical properties of the system. Discussion of these functions does not require the simultaneous discussion of neutron scattering. The same is true of $S(\boldsymbol{\kappa},\omega)$, which we need not view as a scattering law in which $\hbar\boldsymbol{\kappa}$ and $\hbar\omega$ are the momentum and energy, respectively, transferred by the neutron to the system; we consider it simply as the Fourier transform of $G(\mathbf{r},t)$ with respect to space and time, with $\boldsymbol{\kappa}$ and ω being the transform variables. The neutron-scattering experiment is merely the method by which $G(\mathbf{r},t)$ can be observed.

For large values of $|t|$ or $\mathbf{r}$, the pair distribution $G(\mathbf{r},t)$ for the systems of large numbers of particles studied in statistical mechanics has an especially simple form. For such systems, the particles in regions widely separated in space or in time are statistically independent. Thus, for sufficiently large $\mathbf{r}$ or t, the two-particle density function $n(\mathbf{r}',t'; \mathbf{r}'',t'')$ approaches $n(\mathbf{r}')n(\mathbf{r}'')$ and, from Eq. (2.330), the pair distribution $G(\mathbf{r},t)$ approaches the asymptotic value

$$G_{as}(\mathbf{r},t) = N^{-1} \int n(\mathbf{r}' - \mathbf{r})n(\mathbf{r}')d\mathbf{r}' , \tag{2.339}$$

which is called the autocorrelated density. In particular, for a homogeneous system, $n(\mathbf{r}')$ is constant, and $G(\mathbf{r},t)$ reduces to

$$G_{as}(\mathbf{r},t) = n(\mathbf{r}) = \frac{N}{V} , \tag{2.340}$$

where V is the volume of the system. Except for systems that exhibit long-range order, the convergence of $G(\mathbf{r},t)$ to G_{as} takes place over a length R_0 of the order of intermolecular distances and a time of the order of that needed for an average particle of the system to travel over a distance R_0.

Since there is no correlation between the positions of the particles at widely separated space points, or after sufficiently long time intervals, it is really the difference between $G(\mathbf{r},t)$ and $G_{as}(\mathbf{r},t)$ which expresses the correlations in the positions of particles. The separation of $G(\mathbf{r},t)$ into its asymptotic value and a correlation term $G'(\mathbf{r},t) = G(\mathbf{r},t) - G_{as}(\mathbf{r},t)$ has a simple significance for scattering. Expressing $S_c(\boldsymbol{\kappa},\omega)$ in terms of $G'(\mathbf{r},t)$ and $G_{as}(\mathbf{r},t)$ gives for a system extending to infinity in all directions

$$
\begin{aligned}
S_c(\boldsymbol{\kappa},\omega) &= N^{-1}\delta(\omega) \int e^{i\boldsymbol{\kappa}\cdot\mathbf{r}} d\mathbf{r} \int n(\mathbf{r}' - \mathbf{r})n(\mathbf{r}')d\mathbf{r}' + \\
&\quad + \frac{1}{2\pi} \iint e^{i(\boldsymbol{\kappa}\cdot\mathbf{r}-\omega t)} G'(\mathbf{r},t)d\mathbf{r}dt \\
&= N^{-1}\delta(\omega) \left| \int e^{i\boldsymbol{\kappa}\cdot\mathbf{r}} n(\mathbf{r})d\mathbf{r} \right|^2 + \\
&\quad + \frac{1}{2\pi} \iint e^{i(\boldsymbol{\kappa}\cdot\mathbf{r}-\omega t)} G'(\mathbf{r},t)d\mathbf{r}dt \,.
\end{aligned}
\tag{2.341}
$$

For a target that is homogeneous as well as infinite in extent, the first term reduces to $(2\pi)^3 (N/V)\delta(\omega)\delta(\boldsymbol{\kappa})$. For a target of finite but large dimensions, the first term is highly peaked for small $\boldsymbol{\kappa}$; i.e., it corresponds to nearly forward scattering. In practice, the scattering angles are so small that the scattered neutrons are indistinguishable from the unscattered ones. The first term in (2.341) makes no contribution to the scattering observed in conventional experiments and is usually omitted from the expression for the differential cross section.

The interpretation of $G(r,t)$ in terms of density operators can be exploited to calculate $S(\boldsymbol{\kappa},\omega)$ in the limit of small κ and small ω. By virtue of Eq. (2.336), this limit is determined by the long-wavelength and low-frequency behavior of the density disturbance induced by localizing a particle at the origin at $t = 0$. For sufficiently slow space and time variation this behavior can be calculated from the macroscopic equations appropriate to the propagation of small density disturbances in the fluid. These equations are the linearized Navier-Stokes equations with appropriate initial conditions. Since we localize an atom at the origin at t = 0 without affecting its velocity distribution, these initial conditions correspond to an initially localized density disturbance with no temperature disturbance. The double Fourier transformed equations can be solved analytically. The results in the limit of small κ and small ω are of greatest physical interest and have been given by Kadanoff and Martin[97]

$$AS_c(\boldsymbol{\kappa},\omega) = \left(1 - \frac{C_v}{C_p}\right)\frac{D_T\kappa^2}{\omega^2 + \left(D_T\kappa^2\right)^2} + \frac{C_v}{C_p}\frac{c^2\kappa^2\Gamma}{\left(\omega^2 - c^2\kappa^2\right)^2 + \left(\omega^2\Gamma\kappa^2\right)^2} - \left(1 - \frac{C_v}{C_p}\right)\frac{D_T\kappa^2(\omega^2 - c^2\kappa^2)}{\left(\omega^2 - c^2\kappa^2\right)^2 + \left(\omega\kappa^2\Gamma\right)^2} . \tag{2.342}$$

In Eq. (2.342), C_v and C_p are the specific heats at constant volume and constant pressure, respectively, c is the speed of sound in the fluid,

$$D_T = \frac{K}{mnC_p}$$

and

$$\Gamma = \left(\xi + \frac{4}{3}\eta\right)(mn)^{-1} + D_T\left(\frac{C_p}{C_v} - 1\right) . \tag{2.343}$$

In Eq. (2.343), mn is the mass density, K is the thermal conductivity, D_T is the thermal diffusivity, η is the shear viscosity, and ξ is the second or bulk viscosity. The quantity Γ in Eq. (2.343) is familiar in hydrodynamics[98] where it gives the classical sound absorption of a fluid.

The constant A in Eq. (2.342) is given by

$$A = -\frac{1}{\pi}\frac{k_BTN}{V^2}\left(\frac{\partial V}{\partial P}\right)_T = \frac{1}{\pi}\frac{k_BT}{\left(\frac{\partial P}{\partial \rho}\right)_T} . \tag{2.344}$$

We thus satisfy the well-known Ornstein-Zernike[99] relation,

$$1 + \Gamma(0) = 1 + \rho \int_0^\infty [g(r) - 1]4\pi r^2 dr = \frac{-k_B TN}{V^2}\left(\frac{\partial V}{\partial P}\right)_T , \qquad (2.345)$$

which was mentioned at the end of Sec. 2-9.1.

Equation (2.342) for coherent scattering is the analog of Eq. (2.304) for incoherent scattering. It depends only on the thermodynamic and transport properties of the fluid, but is more complicated than the simple diffusive behavior appropriate to incoherent scattering. It is not known whether the hydrodynamic limit of Eq. (2.342) can be approximately reached for the smallest κ values available in neutron scattering, but it is almost certainly applicable to the closely related problem of light scattering by a fluid. As pointed out by Van Hove,[18] whenever the first Born approximation can be used to calculate the scattering, the distribution of the scattered radiation is determined by $S(\boldsymbol{\kappa},\omega)$. The physical phenomena contained in Eq. (2.342) have in fact long been familiar in the scattering of light by liquids.

Perhaps the most familiar term in Eq. (2.342) is the second term, which gives two peaks in the frequency distribution of the scattered radiation. The displacement of these peaks is determined by the sound velocity in the medium, and their width is determined by the attenuation of sound of wave number κ. These two peaks are often known as the Brillouin doublet. Their measurement has become much easier with the advent of laser technology, and the study of Brillouin scattering by liquids is of considerable current interest. (The third term in Eq. (2.342) is not important for the small values of κ appropriate to light scattering.)

It was first pointed out by Landau and Placzek that the undisplaced component in Eq. (2.342) must also be present.[100] Physically, this term arises because our original density disturbance was made isothermally. A fraction C_v/C_p of this disturbance propagates as an isentropic sound wave; the remaining entropy fluctuations do not propagate but diffuse with the thermal diffusivity of the fluid. We can deduce the necessity of the undisplaced component from sum rule arguments. Suppose we replace C_v/C_p in Eq. (2.342) by some undetermined fraction, f. In the limit of small κ, only the second term in Eq.

(2.342) contributes to the second moment sum rule, and this term gives

$$\int \omega^2 S_c(\kappa,\omega)d\omega = \frac{fc^2\kappa^2 k_B T}{\left(\frac{\partial P}{\partial \rho}\right)_T} = \frac{f\kappa^2 k_B T}{M} \frac{\left(\frac{\partial P}{\partial \rho}\right)_S}{\left(\frac{\partial P}{\partial \rho}\right)_T}$$

If we then make use of Eq. (2.324), we obtain

$$f = \frac{\left(\frac{\partial P}{\partial \rho}\right)_T}{\left(\frac{\partial P}{\partial \rho}\right)_S} = \frac{C_v}{C_p} .$$

The validity of Eq. (2.342) requires that the propagation of a spontaneous density fluctuation be governed by the same laws as the propagation of a small externally induced density disturbance. For an external disturbance induced by a perturbing Hamiltonian coupled to the density, this can be proven to be the case.[94] A more important question from the point of view of applying Eq. (2.342) is the range of κ and ω for which it applies. For monatomic liquids very little is known about this question, and it is of considerable fundamental interest. With the combination of light scattering and ultrasonic measurements, this point should become less obscure in the next few years. From the theoretical point of view, one needs a theory that has the correct hydrodynamic limit. As emphasized by Kadanoff and Martin,[97] most approximate microscopic theories fail badly in this limit. An important exception is the kinetic theory of gases based on the Boltzmann equation. This applies only to a dilute gas, but applies over the entire range of κ from the collision-dominated hydrodynamic limit to the free streaming characteristic of an ideal gas without collision (Knudsen gas). The approach of $S(\kappa,\omega)$ to the hydrodynamic limit for a dilute gas has been calculated by Yip and Nelkin,[101] and discussed in a more careful way by Nelkin, Van Leeuwen, and Yip,[102] but there are no direct comparisons of theory and experiment yet available. To extend this kinetic description to fluids of higher density is a difficult problem that is likely to be the subject of much theoretical work over the next several years.

REFERENCES

1. E. Amaldi, The Production and Slowing Down of Neutrons, p. 39, in *Handbuch der Physik, Neutrons and Related Gamma Ray Problems,* Vol. 38, Pt. 2, Springer-Verlag, Berlin, 1959.

2. H. von Halban, Jr., and P. Preiswerk, Experimental Proof of Diffraction of Neutrons, *Compt. Rend.*, **203**: 73 (1936); and Slow Neutrons, *J. Phys. Radium,* **8**: 29 (1937).

3. E. Fermi, On the Motion of Neutrons in Hydrogenous Substances, Translation from *Ricerca Scientifica*, VII, **2**: 13 (1936), available from OTS as PB-5922332 (AEC File No. NP-2385); also in Enrico Fermi, *Collected Papers*, Vol. I, p. 980, E. Segrè, Chairman of Editorial Board, University of Chicago Press, Chicago, 1962.

4. R. G. Sachs, *Nuclear Theory*, Addison-Wesley Publishing Company, Inc., Cambridge, Mass., 1953. Figure 2.1 in this book was replotted by Sachs from data originally published by L. J. Rainwater *et al.*, Slow Neutron Velocity Spectrometer Studies of H, D, F, Mg, S, Si, and Quartz, *Phys. Rev.*, **73**: 733 (1948).

5. David Bohm, *Quantum Theory*, Chap. 21, Prentice-Hall, New York, 1951.

6. J. M. Blatt and V. F. Weisskopf, *Theoretical Nuclear Physics*, pp. 48-86, John Wiley and Sons, New York, 1952.

7. G. C. Summerfield, On the Fermi Approximation in Thermal Neutron Scattering, *Ann. Phys. (N.Y.)*, **26**: 72 (1964).

8. R. G. Sachs, *op. cit.*, p. 110.

9. E. Fermi and W. H. Zinn, Reflection of Neutrons on Mirrors, *Phys. Rev.*, **70**: 103 (1946).

10. E. Fermi and L. Marshall, Interference Phenomena of Slow Neutrons, *Phys. Rev.*, **71**: 666 (1947).

11. D. J. Hughes, M. T. Burgy, and G. R. Ringo, Coherent Neutron-Proton Scattering by Liquid Mirror Reflection, *Phys. Rev.*, **77**: 291 (1950); and M. T. Burgy, G. R. Ringo, and D. J. Hughes, Coherent Neutron-Porton Scattering by Liquid Mirror Reflection, *Phys. Rev.*, **84**: 1160 (1951).

12. Edward Teller, Interference of Neutron Waves in Ortho- and Parahydrogen, *Phys. Rev.*, **49**: 420 (1936).

13. R. B. Sutton *et al.*, Scattering of Slow Neutrons by Ortho- and

Parahydrogen, *Phys. Rev.*, **72**: 1147 (1947).

14. G. Sarma, Diffusion des neutrons lents par l'hydrogene liquide, in *Inelastic Scattering of Neutrons in Solids and Liquids: Proceedings of the Symposium held at Vienna, Austria, October 11-14, 1960*, p. 397, International Atomic Energy Agency, Vienna, 1961.

15. P. A. Egelstaff, The Theory of the Thermal Neutron Scattering Law, in *Inelastic Scattering of Neutrons in Solids and Liquids* (*op. cit.*, Ref. 14), p. 25, International Atomic Energy Agency, Vienna, 1961.

16. G. Placzek, The Scattering of Neutrons by Systems of Heavy Nuclei, *Phys. Rev.*, **86**: 377 (1952).

17. W. E. Lamb, Jr., Capture of Neutrons by Atoms in a Crystal, *Phys. Rev.*, **55**: 190 (1939).

18. L. Van Hove, Correlations in Space and Time and Born Approximation Scattering in Systems of Interacting Particles, *Phys. Rev.*, **95**: 249 (1954).

19. G. C. Wick, Scattering of Neutrons by Systems Containing Light Nuclei, *Phys. Rev.*, **94**: 1228 (1954).

20. R. J. Glauber, Time-dependent Displacement Correlations and Inelastic Scattering by Crystals, *Phys. Rev.*, **98**: 1692 (1955).

21. A. C. Zemach and R. J. Glauber, Dynamics of Neutron Scattering by Molecules, *Phys. Rev.*, **101**: 118 (1956).

22. L. Van Hove, N. M. Hugenholtz, and L. P. Howland, Interactions of Elastic Waves in Solids, in *Problems in the Quantum Theory of Many-Particle Systems*, p. 1, W. A. Benjamin, New York, 1961.

23. P. Schofield, Space-Time Correlation Function Formalism for Slow Neutron Scattering, *Phys. Rev. Letters*, **4**: 239 (1960).

24. J. H. Jeans, *The Dynamical Theory of Gases*, 4th ed., Chap. 10, Dover Publications, New York, 1954.

25. F. Bloch, Theory of Exchange Problem and Remanence Phenomena of Ferromagnetic Substances, *Z. Physik*, **74**: 295 (1932).

26. L. I. Schiff, *Quantum Mechanics*, 2nd ed., pp. 60-69, McGraw-Hill Book Company, Inc., New York, 1955.

27. G. N. Watson, *A Treatise on the Theory of Bessel Functions*, 2nd ed., The MacMillan Company, New York, 1944.

28. A. Erdélyi *et al.*, *Higher Transcendental Functions*, Vol. 2, p. 86 (Bateman Manuscript Project, California Institute of Technology), McGraw-Hill Book Company, New York, 1953-1955.

29. B. N. Brockhouse and D. G. Hurst, Energy Distribution of Neutrons

Scattered from Solids, *Phys. Rev.*, **88**: 542 (1952).

30. E. O. Wollan, C. G. Shull, and M. C. Marney, Laue Photography of Neutron Diffraction, *Phys. Rev.*, **73**: 527 (1948).

31. D. J. Hughes, *Pile Neutron Research*, p. 267, Addison Wesley Publishing Company, Inc., Cambridge, Mass., 1953.

32. M. Born and K. Huang, *Dynamical Theory of Crystal Lattices*, Clarendon Press, Oxford, 1954.

33. R. E. Peierls, *Quantum Theory of Solids*, Chap. 1, Clarendon Press, Oxford, 1955.

34. A. A. Maradudin, E. W. Montroll, and G. H. Weiss, *Theory of Lattice Dynamics in the Harmonic Approximation*, Solid State Physics Series, Suppl. 3, Academic Press, New York, 1963.

35. R. A. Smith, *Wave Mechanics of Crystalline Solids*, John Wiley and Sons, Inc., New York, 1961.

36. B. N. Brockhouse, *Lattice Waves, Spin Waves, and Neutron Scattering*: A paper based on lectures presented at the Neutron Physics Conference, Mackinac Island, Michigan, June 12-17, 1961, Report MMPP-NPC-1-4, Michigan Memorial Phoenix Project, University of Michigan.

37. G. H. Begbie and M. Born, Thermal Scattering of X-rays by Crystals. I. Dynamical Foundation, *Proc. Roy. Soc. (London)*, **A188**: 179 (1946).

38. B. N. Brockhouse and A. T. Stewart, Scattering of Neutrons by Phonons in an Aluminum Single Crystal, *Phys. Rev.*, **100**: 756 (1955); and Normal Modes of Aluminum by Neutron Spectrometry, *Rev. Mod. Phys.*, **30**: 236 (1958).

39. R. S. Carter, H. Palevsky, and D. J. Hughes, Inelastic Scattering of Slow Neutrons by Lattice Vibrations in Aluminum, *Phys. Rev.*, **106**: 1168 (1957).

40. K.-E. Larsson, U. Dahlborg, and S. Holmryd, A Study of Some Temperature Effects on the Phonons in Aluminum by Use of Cold Nuetrons, *Arkiv Fysik*, **17**: 369 (1960).

41. C. B. Walker, X-Ray Study of Lattice Vibrations in Aluminum, *Phys. Rev.*, **103**: 547 (1956).

42. B. N. Brockhouse and P. K. Iyengar, Normal Modes of Germanium by Neutron Spectrometry, *Phys. Rev.*, **111**: 747 (1958).

43. G. Placzek and L. Van Hove, Crystal Dynamics and Inelastic Scattering of Neutrons, *Phys. Rev.*, **93**: 1207 (1954).

44. L. S. Kothari and K. S. Singwi, Interaction of Thermal Neutrons with Solids, in *Solid State Physics: Advances in Research and*

Applications, Vol. 8, pp. 110-190, F. Seitz and D. Turnbull, Eds., Academic Press, New York and London, 1959.

45. A. Sjölander, On Two-Phonon Processes in Neutron Diffraction Against Crystals, *Arkiv Fysik*, **13**: 215 (1958).

46. B. N. Brockhouse, Slow Neutron Spectrometry--A New Tool for Study of the Dynamics of Condensed Systems, in *Papers Presented by the A.E.C.L. Staff at the Meeting of the Royal Society of Canada Held at Toronto, June 6-8, 1955*, Canadian Report AECL No. 221.

47. K. C. Turberfield and P. A. Egelstaff, The Phonon Frequency Distribution in Vanadium at Several Temperatures, in *Inelastic Scattering of Neutrons in Solids and Liquids* (*op. cit.*, Ref. 14), p. 581, International Atomic Energy Agency, Vienna, 1961.

48. C. M. Eisenhauer, I. Pelah, D. J. Hughes, and H. Palevsky, Measurement of Lattice Vibrations in Vanadium by Neutron Scattering, *Phys. Rev.*, **109**: 1046 (1958).

49. M. G. Zemlianov, Yu. M. Kagan, N. A. Tchernoplekov, and A. G. Chicherin, A Study of the Phonon Spectrum and Dispersion Curves for Vanadium, in *Inelastic Scattering of Neutrons in Solids and Liquids: Proceedings of the Symposium held at Chalk River, Canada, September 10-14, 1962*, Vol. II, p. 125, International Atomic Energy Agency, Vienna, 1963.

50. R. Haas, W. Kley, K. H. Krebs, and R. Rubin, The Phonon Frequency Distribution of Vanadium, in *Inelastic Scattering of Neutrons in Solids and Liquids* (*op. cit.*, Ref. 49), Vol. II, p. 145.

51. I. Pelah, R. Haas, W. Kley, K. H. Krebs, J. Peretti, and R. Rubin, Observation of a Low Energy Peak in the Phonon Frequency Distribution of Vanadium, in *Inelastic Scattering of Neutrons in Solids and Liquids* (*op. cit.*, Ref. 49), Vol. II, p. 155.

52. E. Melkonian, A Precise Determination of the Slow Neutron Cross Section of the Free Proton, *Phys. Rev.*, **76**: 1744 (1949).

53. E. Melkonian, Slow Neutron Velocity Spectrometer Studies of O_2, N_2, A, H_2, H_2O, and Seven Hydrocarbons, *Phys. Rev.*, **76**: 1750 (1950).

54. N. Z. Alcock and D. G. Hurst, Neutron Diffractions by the Gases N_2, CF_4, and CH_4, *Phys. Rev.*, **83**: 1100 (1951).

55. H. L. Anderson, E. Fermi, and L. Marshall, Production of Low Energy Neutrons by Filtering through Graphite, *Phys. Rev.*, **70**: 815 (1946).

56. A. C. Zemach and R. J. Glauber, Neutron Diffraction by Gases, *Phys. Rev.*, **101**: 129 (1956).

57. E. B. Wilson, Jr., J. C. Decius, and P. C. Cross, *Molecular Vibrations: The Theory of Infrared and Raman Vibrational Spectra*,

McGraw-Hill Book Company, New York, 1955.

58. T. J. Krieger and M. S. Nelkin, Slow Neutron Scattering by Molecules, *Phys. Rev.*, **106**: 290 (1957).

59. G. W. Griffing, Scattering of Slow Neutrons by Gaseous Methane, in *Inelastic Scattering of Neutrons in Solids and Liquids* (*op. cit.*, Ref. 49), Vol. I, p. 435.

60. A. Rahman, Scattering of Slow Neutrons by Molecules, *J. Nucl. Energy, Pt. A*, **13**: 128 (1961).

61. H. L. McMurry, Calculation of Partial Differential Scattering Cross Sections for Slow Neutrons, *Nucl. Sci. Eng.*, **15**: 429 (1963).

62. G. W. Griffing, Influence of the Rotational Levels on the Scattering of Slow Neutrons by Gaseous Methane, *Phys. Rev.*, **124**: 1489 (1961).

63. P. D. Randolph, R. M. Brugger, K. A. Strong, and R. E. Schmunk, Inelastic Scattering of Slow Neutrons from Methane, *Phys. Rev.*, **124**: 460 (1961).

64. H. C. Volkin, Slow-Neutron Scattering by Rotators, *Phys. Rev.*, **113**: 866 (1959).

65. R. G. Sachs and E. Teller, The Scattering of Slow Neutrons by Molecular Gases, *Phys. Rev.*, **60**: 18 (1941).

66. A. M. L. Messiah, Scattering of Slow Neutrons by H_2 and CH_4, *Phys. Rev.*, **84**: 204 (1951).

67. H. L. McMurry, L. J. Gannon, and W. A. Hestir, Evaluation of Techniques for Computing Partial Differential Scattering Cross Sections, *Nucl. Sci. Eng.*, **15**: 438 (1963).

68. N. Corngold, Chemical Binding Effects in the Thermalization of Neutrons, *Ann. Phys. (N.Y.)*, **11**: 338 (1960).

69. G. Placzek, B. R. A. Nijboer, and L. Van Hove, Effect of Short Wavelength Interference on Neutron Scattering by Dense Systems of Heavy Nuclei, *Phys. Rev.*, **82**: 392 (1951).

70. B. R. Orton, G. I. Williams, and B. A. Shaw, An X-ray Structure Investigation of the Liquids of Sodium, Potassium, and Sodium-Potassium Alloys, *Acta Met.*, **8**: 177 (1960).

71. K. Furukawa, B. R. Orton, J. Hamor, and G. I. Williams, The Structure of Liquid Tin, *Phil. Mag.*, **8**: 141 (1963).

72. G. Caglioti and F. P. Ricci, The Structure of Liquid Bromine, *Nuovo Cimento*, **24**: 103 (1962).

73. G. E. Coote and B. C. Haywood, Scattering of Neutrons by Liquid Bromine, in *Inelastic Scattering of Neutrons in Solids and Liquids*

(*op. cit.*, Ref. 49), Vol. I, p. 249.

74. T. L. Hill, *Statistical Mechanics*, Chap. 30, McGraw-Hill Book Company, New York, 1956.

75. P. Mazur and E. M. Montroll, Poincaré Cycles, Ergodicity, and Irreversibility in Assemblies of Uncoupled Harmonic Oscillators, *J. Math. Phys.*, **1**: 70 (1960).

76. I. C. Percival, Almost Periodicity and the Quantal H Theorem, *J. Math. Phys.*, **2**: 235 (1961).

77. Ming Chen Wang and G. E. Uhlenbeck, On the Theory of Brownian Motion II, *Rev. Mod. Phys.*, **17**: 323 (1945) (reprinted in *Selected Papers on Noise and Stochastic Processes*, Nelson Wax, Ed., Dover Publications, New York, 1954).

78. M. Nelkin and A. Ghatak, Simple Binary Collision Model for Van Hove's $G_s(r,t)$, *Phys. Rev.*, **135**: A4 (1964).

79. J. M. J. van Leeuwen and Sidney Yip, Derivation of Kinetic Equations for Slow-Neutron Scattering (unpublished manuscript, 1964).

80. Karl-Eric Larsson and A. Sjölander, Scattering of Slow Neutrons in Liquids, in *Thermal Neutron Scattering*, P. A. Egelstaff, Ed., p. Academic Press, New York, 1965.

81. G. V. Chester, The Theory of Irreversible Processes, p. 411 in *Reports on Progress in Physics*, Vol. 26, The Institute of Physics and the Physical Society, London, 1963.

82. A. Rahman, Correlations in the Motion of Atoms in Liquid Argon, *Phys. Rev.*, **136**: A405 (1964).

83. P. Schofield, Some Properties of the Space-Time Correlation Function, in *Inelastic Scattering of Neutrons in Solids and Liquids* (*op. cit.*, Ref. 14), p. 39.

84. A. Rahman, K. S. Singwi, and A. Sjölander, Theory of Slow-Neutron Scattering by Liquids. I, *Phys. Rev.*, **126**: 986 (1962).

85. J. U. Koppel and J. A. Young, Neutron Scattering by Water Taking into Account the Anisotropy of the Molecular Vibration, *Nucl. Sci. Eng.*, **19**: 412 (1964).

86. B. A. Dasannacharya and K. R. Rao, Neutron Scattering from Liquid Argon, *Phys. Rev.*, **137**: B1318 (1965).

87. P. A. Egelstaff and P. Schofield, On the Evaluation of the Thermal Neutron Scattering Law, *Nucl. Sci. Eng.*, **12**: 260 (1962).

88. R. Aamodt, K. M. Case, M. Rosenbaum, and P. F. Zweifel, Quasi-Classical Treatment of Neutron Scattering, *Phys. Rev.*, **126**: 1165 (1962).

89. G. H. Vineyard, Scattering of Slow Neutrons by a Liquid, *Phys. Rev.*, **110**: 999 (1958).

90. P. G. de Gennes, Liquid Dynamics and Inelastic Scattering of Neutrons, *Physica*, **25**: 825 (1959).

91. P. D. Randolph, Slow-Neutron Inelastic Scattering from Liquid Sodium, *Phys. Rev.*, **134**: A1238 (1964).

92. N. Kroo, G. Borgonovi, K. Sköld, and K.-E. Larsson, Inelastic Scattering of Cold Neutrons by Condensed Argon, *Phys. Rev. Letters*, **12**: 721 (1964).

93. K. S. Singwi, Coherent Scattering of Slow Neutrons by a Liquid, *Phys. Rev.*, **136**: A 969 (1964).

94. R. J. Glauber, Scattering of Neutrons by Statistical Media, p. 571 in *Lectures in Theoretical Physics*, Vol. IV (Lectures delivered at the Summer Institute for Theoretical Physics, University of Colorado, Boulder, 1961), Interscience Publishers, New York and London, 1962.

95. L. Van Hove, A Remark on the Time-Dependent Pair Distribution, *Physica*, **24**: 404 (1958).

96. H. B. Callen and T. A. Welton, Irreversibility and Generalized Noise, *Phys. Rev.*, **83**: 34 (1951).

97. L. Kadanoff and P. C. Martin, Hydrodynamic Equations and Correlation Functions, *Ann. Phys. (N.Y.)*, **24**: 419 (1963).

98. L. D. Landau and E. M. Lifshitz, *Fluid Mechanics*, p. 300, translated from the Russian by J. B. Sykes and W. H. Reid, Pergamon Press, London, 1959.

99. L. D. Landau and E. M. Lifshitz, *Statistical Physics*, p. 365, translated from the Russian by E. Peierls and R. F. Peierls, Pergamon Press, London, 1958.

100. L. D. Landau and E. M. Lifshitz, *Electrodynamics of Continuous Media*, Chap. 14, translated from the Russian by J. B. Sykes and J. S. Bell, Pergamon Press, London, 1960.

101. S. Yip and M. Nelkin, Application of a Kinetic Model to Time-Dependent Density Correlations in Fluids, *Phys. Rev.*, **135**: A1241 (1964).

102. M. Nelkin, J. M. J. van Leeuwen, and S. Kip, A Kinetic Description of the Van Hove Correlation Functions in *Inelastic Scattering of Slow Neutrons, Proceedings of a Symposium held in Bombay, 15-19 December 1964*, Vol. II, p. 35, International Atomic Energy Agency, Vienna, 1965.

Chapter 3

CALCULATION AND MEASUREMENT OF SLOW NEUTRON-SCATTERING CROSS SECTIONS

The objective of the theory of neutron thermalization is a quantitative prediction of the slowing down and diffusion of neutrons in the energy region below a few electron volts. In this energy region the cross section for a given momentum change depends not only on the nuclear mass and scattering amplitude but also on the atomic structure, the interatomic binding, and the temperature of the moderator atoms. In order to calculate the neutron distribution in energy and position, it is necessary to have experimental information about the relevant cross sections in addition to a theoretical basis that relates these cross sections to the dynamical behavior of the scattering system.

The general theory of slow-neutron scattering has been extensively developed and forms an essentially exact basis for the study of neutron thermalization by chemically bound systems. The fundamental cross section of interest for neutron thermalization is the differential cross section $\sigma(E_0,E,\theta)$ for energy transfer from E_0 to E with scattering through an angle θ in the laboratory system. This cross section is related to the detailed dynamics of atomic motion through its dependence on the scattering law $S(\boldsymbol{\kappa},\omega)$ as expressed by Eq. (2.64). The dynamics of atomic motions is extremely complicated and, for most substances, especially liquids, is not known in detail. The problem of most fundamental interest is the determination of these dynamical properties from experimental measurements of slow-neutron scattering. This problem is emphasized in Chap. 2.

There is no sharp separation between the fundamental physical problem of determining the atomic motions in solids and liquids and the applied problem of neutron thermalization in reactors. In many cases, the features of atomic motions that are of greatest fundamental interest are also of primary interest for reactor applications. For example, the discovery of the optical vibrational level in zirconium hydride was motivated by reactor physics considerations but was made possible by the development of equipment for fundamental studies.[1,2] The work of Schmunk *et al.*[3] on the dispersion relations in beryllium has been applied to the study of the spectrum of neutrons moderated by this substance.[4] A similar application has been made of the work of Yoshimori and Kitano,[5] who calculated the dispersion relations and the frequency distribution of lattice vibrations in graphite for the purpose of understanding the anomalous temperature dependence of the specific heat of graphite at low temperatures.[6,7] The works of Egelstaff *et al.*[8] and Larsson *et al.*[9] are motivated by considerations that are relevant to both the fundamental physics of atomic motions in water and the problem of neutron thermalization.

Even if the dynamics of moderator-atom motions were known in complete detail, it would be extremely difficult to incorporate all of this information into neutron-thermalization calculations for reactors. Reactor spectra and reactor cross sections, however, can at most be only moderately sensitive to the detailed features of the atomic motions; therefore, it is both necessary and justified to work with somewhat simplified models for the atomic motions. Even with these simplified models, it is often necessary to make a number of mathematical and physical approximations in order to calculate $\sigma(E_0,E,\theta)$. In the absence of a complete experimental knowledge of $\sigma(E_0,E,\theta)$, it is imperative to compare calculations, using these models and approximations, with precise experimental information on the neutron distribution in position and energy for configurations of interest in reactor physics. Much of the content of subsequent chapters is devoted to this end.

The results plotted in Fig. 2.4 are somewhat indicative of the extent to which the details of atomic motions must be considered. The data points represent the cadmium transmission of initially monoenergetic neutrons that have been scattered through an angle of 90° by thin samples of lead, aluminum, and carbon. For lead and aluminum, these results indicate that an ideal gas model satisfactorily explains the experimental results, at least for the neutron energy of 0.35 eV used in the experiment. On the basis of Eq. (2.119) and the discussion following its derivation we expect that the gas model would enable an accurate correlation of experimental results with theory over a range of energies extending as low as about 0.01 eV in lead and 0.05 eV in aluminum. This statement applies only to the results of integral-type measurements and not, for example, to the results of high-resolution experiments on the energy-angle distributions of singly scattered, initially monoenergetic neutrons.

The conclusion obtained for the case of carbon scatterers is entirely different from that which applies for lead and aluminum. Because of the presence of a large fraction of modes with high frequencies of vibration, carbon does not scatter like an ideal gas of carbon atoms at the physical temperature of the scattering sample, even for neutrons having an initial energy as large as 0.35 eV. Nor can we expect the Einstein model with a single frequency ω_0 to apply at all energies, for then we would have the unrealistic circumstance that neutrons with an energy $E_0 < \hbar\omega_0$ could not lose energy to the lattice. For graphite, then, and in fact for most of the commonly used moderators, one is compelled to consider scattering calculations in terms of dynamical descriptions that are more complete than those given in Sec. 2-5.

3-1 BASIC ASSUMPTIONS AND APPROXIMATIONS

The essential problem is to relate the neutron scattering to the atomic dynamics in a way that is sufficiently simple for numerical calculations of $\sigma(E_0,E,\theta)$ to be feasible. At the same time, the description of the atomic motions that enters into the formulation should reflect with reasonable accuracy all features of the moderator-atom motions to which the scattering of thermal neutrons is most sensitive. Actually, this last requirement is more demanding than is necessary since the quantities of primary interest in thermalization and in reactor design are far less sensitive to the dynamical behavior of a system at the microscopic level than is the scattering law.

The principal approximations that we shall make in order to calculate the scattering are

1) the incoherent approximation for inelastic scattering (but not for elastic scattering), and
2) the Gaussian approximation

$$\chi_{inc}(\kappa,t) = \exp[-\kappa^2\gamma(t)] , \tag{3.1}$$

where $\gamma(t)$ is given by Eq. (2.316).

First, let us discuss the validity of the incoherent approximation. For single crystals, the cross section for interference scattering $\sigma_{int}(\mathbf{k}_0)$ is a highly singular function of the wave vector $\mathbf{k}_0$ of the incident neutron. Delta-function singularities occur, both in the elastic and inelastic scattering, at values of k_0 satisfying the Bragg condition, Eq. (2.142). For polycrystals the order of these singularities is greatly reduced. The interference part of the scattering cross section of polycrystals, $\sigma_{int}(E_0)$, is finite for all E_0 but has discontinuities for neutron wave vectors of magnitude $k_0 = \tau/2$, where τ is the magnitude of a reciprocal lattice vector (see Sec. 2-6.1). The sizes of these discontinuities decrease exponentially with increasing energy and are of little importance for E_0 greater than about 0.1 eV.

Placzek and Van Hove[10] have made general estimates of the magnitude of the interference effect in the scattering by polycrystals. In the case of short wave lengths they show that σ_{int} is represented by its eleastic part with a relative error of order $(u_0/d)^2$, where u_0 is the root-mean-square displacement of a nucleus from its equilibrium position and d is the lattice constant. They also show that the elastic and one-phonon parts approximate σ_{int} with a relative error of the order of $(u_0/d)^4$. For polycrystalline materials at ordinary temperatures these relative errors are extremely small, being of the order of 10^{-2} and 10^{-4}, respectively. The representation of the interference scattering by its elastic part is accurate for all neutron wave vectors $\mathbf{k}_0$ that satisfy $k_0 d \gg 1$.

For neutron wavelengths such that $k_0 d \lesssim 1$, the conclusions stated above are questionable. For the limiting case of $k_0 \to 0$, Placzek and Van Hove[10] have shown that the interference part of the one-phonon scattering cross section in polycrystals is in general comparable to its noninterference part. Since for $k_0 d \lesssim 1$ and at ordinary temperatures the main contribution to the inelastic scattering is from one-phonon processes, the inelastic cross section in polycrystals is significantly affected by interference effects. Nevertheless, this fact is of little consequence in assessing the significance of interference effects on thermal-neutron spectra, inasmuch as there are an extremely small number of neutrons in the very-long-wavelength part of a reactor spectrum. The important question that remains to be answered concerns the magnitude of these effects at energies where significant numbers of thermal neutrons are present.

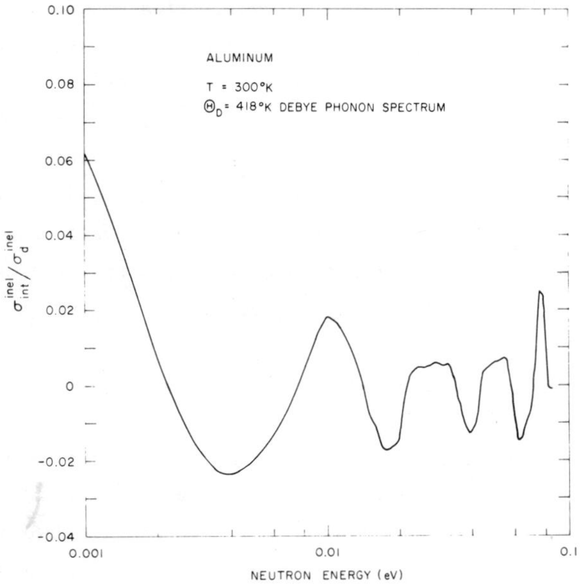

Figure 3.1 The ratio $\sigma_{int}^{inel}/\sigma_d^{inel}$ of the interference and non-interference parts of the total inelastic scattering cross section of aluminum. (From Marshall and Stuart, USAEC Report UCRL-5568, Ref. 12.)

Marshall and Stuart[11,12] have calculated interference effects on inelastic scattering of slow neutrons in the energy region extending from about 10^{-3} eV to 10^{-1} eV. This region overlaps the long-wavelength region, where the interference and noninterference parts of the one-phonon scattering are in general comparable, and the $k_0d >> 1$ region, where the effects of interference are negligible. They give specific results for the cases of scattering by polycrystalline aluminum and polycrystalline magnesium. Figure 3.1 shows the results of their calculations of the ratio of the interference part of the inelastic cross section to the noninterference inelastic cross section of aluminum, which scatters neutrons only coherently. These calculations are based on the Debye approximation described in Sec. 2-6.2. Although the Debye approximation may lead to a poor estimate of the detailed features of interference scattering, the calculation of the magnitude of the interference effect is significant. For energies greater than about 0.001 eV the magnitude of the interference effect is practically negligible. The same conclusion undoubtedly applies for the interference part of the inelastic scattering by other polycrystalline materials. It should be even more accurate to use the incoherent approximation to calculate inelastic scattering by liquids, inasmuch as dynamical correlations, which give rise to interference effects in inelastic scattering, are weaker and of much shorter range in liquids than in solids.

So far, the only conclusion that we have made concerns the insignificance of interference effects for the total inelastic scattering cross section. We expect the effects of interference to be considerably greater on the inelastic part of the scattering kernel $\Sigma_0(E_0,E)$ than on the total inelastic cross section. Nevertheless, since this kernel appears integrated in the equation of neutron balance, it is safe to presume that the effects of interference on the neutron spectrum will be as small as the effects on the inelastic cross section. To our knowledge, however, there have been no estimates of the magnitude of these effects on the neutron spectrum. On the basis of the preceding considerations, we calculate inelastic slow-neutron scattering in the incoherent approximation. Except where explicitly stated to the contrary, full account is taken of the interference phenomena in the calculation of elastic scattering. Quantitatively, for moderators consisting of a single atomic species the approximation that we are making takes the form

$$\sigma(E_0,E,\theta) = \frac{k}{4\pi\hbar k_0}\left\{\sigma_b\left\langle S_{inc}(\boldsymbol{\kappa},\omega)\right\rangle_R + \right.$$

$$\left. + \sigma_{c,b}\left[S_c^{el}(\boldsymbol{\kappa},\omega)_R - \left\langle S_{inc}^{el}(\boldsymbol{\kappa},\omega)\right\rangle_R\right]\right\}, \tag{3.2}$$

where $S^{el}(\boldsymbol{\kappa},\omega)$ is that part of the scattering function that corresponds to elastic scattering, $\langle\ldots\rangle_R$ indicates an average over orientations of the scattering system, and $\hbar\kappa$ and $\hbar\omega = E_0 - E$ are the momentum and energy, respectively, lost by the neutron. The quantities σ_b and $\sigma_{c,b}$ are the total and coherent bound atom cross sections, respectively (see Sec. 2-1.2). The result (3.2) should be compared with Eq. (2.70).

The second fundamental approximation that we make is the Gaussian approximation, Eq. (3.1). This particular form for $\chi_{inc}(\boldsymbol{\kappa},t)$ is verified exactly by some of the simple systems that we have so far considered, namely by an ideal gas, an isotropic Einstein oscillator, a simple cubic crystal whose lattice vibrations are harmonic, and an atom diffusing according to the Langevin equation. In general, however, Eq. (3.1) is not an exact result, and we can determine the corrections to it only by quantitative calcuations. For monatomic liquids these corrections are rooted in the dynamics of atomic motions. Quantitative results for non-Gaussian effects in liquid argon have already been presented in Sec. 2-9.

For harmonically vibrating systems, non-Gaussian effects can also occur, but these are associated entirely with the anisotropy of the vibrations. In general, the effect of the anisotropy on $S_{inc}(\boldsymbol{\kappa},\omega)$ is relatively large for large values of κ^2. For these large values of κ^2 the magnitude of $S_{inc}(\boldsymbol{\kappa},\omega)$ tends to be small but not insignificant. Thus, we can determine the importance of anisotropic motions on thermal-neutron scattering and spectra only by quantitative calculations. The results of such calculations are presented later in this chapter and also in Chap. 6. It is found that the anisotropy of the motions, although significant for $S_{inc}(\boldsymbol{\kappa},\omega)$, affects only slightly the neutron spectrum in a reactor. The largest effects of anisotropic molecular vibrations occur in H_2O-moderated systems; as we see in Sec. 6-4, accounting for this effect gives rise to differences in calculated neutron spectra of about 5%.

3-2 METHODS FOR CALCULATING SLOW-NEUTRON SCATTERING

We now undertake the development of analytical methods for computing the differential-energy-transfer cross section for neutron scattering by moderators, or equivalently, the scattering functions $\langle S_{inc}\rangle_R$, $\langle S_c^{el}\rangle_R$ and $\langle S_{inc}^{el}\rangle_R$ appearing in Eq. (3.2). We have already calculated the coherent elastic scattering by crystals in Sec. 2-6.1. We do not consider this component of the scattering in any further detail here, but we do give specific results in Secs. 3-4.3 and 3-4.4 for graphite and beryllium, respectively. In the remainder of this section we are concerned only with the calculation of $\langle S_{inc}\rangle_R$.

With the Gaussian approximation, Eq. (3.1) and Eq. (2.88), we have

$$S_{inc}(\kappa,\omega) \equiv \langle S_{inc}(\boldsymbol{\kappa},\omega)\rangle_R$$

$$= \frac{1}{2}\int_{-\infty}^{\infty} [\exp(-i\omega t)]\ \exp[-\kappa^2\gamma(t)]dt\ , \qquad (3.3)$$

where, by Eq. (2.316), $\gamma(t)$ is related to a real, positive, and normalized frequency distribution $f(\omega)$. We have two problems. First, we must determine $f(\omega)$, and then we must perform the Fourier transform in Eq. (3.3). For the present, we consider the latter problem, supposing that we already know $f(\omega)$.

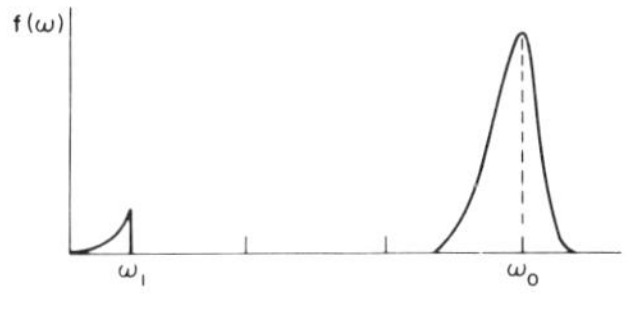

Figure 3.2 Schematic representation of distribution associated with hydrogen-atom motions in zirconium hydride.

Before proceeding, we want to indicate our motivation for making some of the formal developments that follow below. Let us suppose that the distribution $f(\omega)$ is as shown in Fig. 3.2. This distribution represents schematically the $f(\omega)$ associated with the hydrogen-atom motions in zirconium hydride. For this important moderator there is a distribution of frequencies extending from zero frequency up to a frequency ω_1 corresponding to about 0.02 eV. The narrow high-frequency band is centered around $\omega_0 \approx 0.14$ eV and has an area that is a few orders of magnitude greater than that below ω_1. Physically, it is clear that neutrons that have an energy sufficiently great to excite one or two quanta or "phonons" in the high-frequency band may also excite large numbers of quanta of low frequencies. Thus, an approximate method of calculation that converges rapidly for excitation of a few quanta but poorly for excitation of a large number of quanta cannot be used conveniently for the calculation of the $S_{inc}(\kappa,\omega)$ corresponding to the composite $f(\omega)$ shown in Fig. 3.2. On the other hand, neutron excitation of the high-frequency vibrations in Fig. 3.2 can be conveniently treated by a "phonon expansion," while neutron excitation of the low-frequency vibrations can be treated by means of one of the short-collision-time approximations we discuss below, which, roughly speaking, are accurate for initial neutron energies $E_0 >> \hbar\omega_1$.

Returning now to the general case, we express $f(\omega)$ as

$$M^{-1}f(\omega) = \sum_{j=1}^{K} M_j^{-1} f_j(\omega) \ , \tag{3.4}$$

where M is the mass of the scattering nucleus, and introduce

$$\gamma_j(t) = \frac{\hbar}{2M_j}\int_0^\infty f_j(\omega) \ \{\cot h(\tfrac{1}{2}\beta\hbar\omega)[1 - \cos(\omega t)] - i(\sin \omega t)]\}\omega^{-1}d\omega \ . \tag{3.5}$$

The $f_j(\omega)$ and M_j so far are arbitrary; in subsequent sections we consider how they are obtained.

We impose the conditions that $f_j(\omega)$ be finite and non-negative, that it vanish for $\omega_j \geq \omega_{jm}$, and that

$$\int_0^\infty f_j(\omega)d\omega = \int^{\omega_{jm}} f_j(\omega)d\omega = 1 \ . \tag{3.6}$$

Similar conditions are also satisfied by f( ). Since f( ) is normalized to unity, Eqs. (3.4) and (3.6) imply that the effective masses M_j

satisfy

$$M^{-1} = \sum_{j=1}^{K} M_j^{-1} . \tag{3.7}$$

We further require that each $f_j(\omega)$ save one satisfy

$$\lim_{\omega \to 0} \omega^{-2} f_j(\omega) = b_j > 0 , \tag{3.8a}$$

where b_j is finite and positive. The remaining $f_j(\omega)$ satisfies either Eq. (3.8a) or one of the following conditions:

$$\lim_{\omega \to 0} f_j(\omega) = f(0) \neq 0 \tag{3.8b}$$

$$f_j(\omega) = \delta(\omega) . \tag{3.8c}$$

The conditions on $f_j(\omega)$ together with Eq. (3.4) determine the behavior of $f(\omega)$ in the limit $\omega \to 0$. The behavior of $f(\omega)$ in the limit $\omega \to 0$ describes the behavior of the mean-square displacement of an atom at long times and the character of the neutron scattering for small energy transfers. In case (3.8a) the atomic motion is bounded at all times and there is a delta-function peak in $S_{inc}(\kappa,\omega)$ at zero-energy transfer. This peak corresponds to strictly elastic scattering, in which process the quantum state of the scatterer is not changed. In cases (3.8b) and (3.8c) the mean-square displacement at long times is proportional to $|t|$ and t^2, respectively, and there is no strictly elastic scattering. For classical fluids, $f(\omega)$ satisfies Eq. (3.8b). In this case, $S_{inc}(\kappa,\omega)$ has a peak with a finite width, depending on κ, centered about zero-energy transfer. This is called the quasi-elastic peak. In the limit of simple diffusion the quasi-elastic scattering is determined by Eq. (2.304). For an ideal gas, where $f(\omega)$ is given by Eq. (3.8c), every collision changes the quantum state of the scatterer and there is no strictly elastic scattering.

The quantity $f(\omega)$ and its components $f_j(\omega)$ contain the information in terms of which the neutron scattering is to be computed. How $f(\omega)$ is obtained in practice and how we may choose the $f_j(\omega)$ to facilitate the calculation of $S_{inc}(\kappa,\omega)$ are described in subsequent sections of this chapter. For the present, we examine the mathematical consequences of the decomposition in Eq. (3.4).

We introduce

$$\chi_j(\kappa,t) = \exp[-\kappa^2\gamma_j(t)] \tag{3.9}$$

and

$$S_j(\kappa,\omega) = \frac{1}{2\pi}\int_{-\infty}^{\infty} [\exp(-i\omega t]\chi_j(\kappa,t)dt \ , \tag{3.10}$$

where j = 1, 2,...K. Let us now define

$$S^J(\kappa,\omega) = \frac{1}{2\pi}\int_{-\infty}^{\infty} \exp(-i\omega t) \prod_{j=1}^{J} \chi_j(\kappa,t)dt \tag{3.11}$$

for $J < K$. From Eqs. (3.1), (3.3), (3.4), and (2.316)

$$S^K(\kappa,\omega) = S_{inc}(\kappa,\omega) \ . \tag{3.12}$$

The convolution theorem for Fourier transforms applied to the product of the functions χ_j enables us to express $S^J(\kappa,\omega)$ as a (J-1)-fold convolution of all $S_j(\kappa,\omega)$ having $j \leq J$. This is equivalent to

$$S^J(\kappa,\omega) = \int_{-\infty}^{\infty} S_J(\kappa,\omega')S^{J-1}(\kappa,\omega - \omega')d\omega' \ , \tag{3.13}$$

a result that forms a convenient basis for a recursive numerical calculation of $S_{inc}(\kappa,\omega) = S^K(\kappa,\omega)$. First, however, we must know $S_j(\kappa,\omega)$.

In the following section we develop approximate methods for calculating $S_j(\kappa,\omega)$. For this purpose it is convenient to introduce the notation

$$K_j(\omega) = \frac{f_j(\omega)\ \exp(\hbar\omega/2k_BT)}{2\omega\ \sin h(\hbar\omega/2k_BT)} \ , \tag{3.14}$$

in terms of which we may write[a]

$$\chi_j(\kappa,t) = \exp\left\{-\frac{\kappa^2}{2M_j}\int_{-\infty}^{\infty} K_j(\omega')[1 - \exp(i\omega't)]d\omega'\right\}. \qquad (3.15)$$

This form for $\chi_j(\kappa,t)$ follows after we define

$$f_j(-\omega) = f_j(\omega) \ , \qquad (3.16)$$

and recognize that the integrand in Eq. (3.5) is an even function of its argument.

The special cases where $f_j(\omega)$ is equal to $\delta(\omega - \omega_j)$ have already been considered in Sec. 2-5. When $\omega_j = 0$, $S_j(\kappa,\omega)$ is as given in Eq. (2.98). When $\omega_j \neq 0$, Eq. (2.114) applies. In the following section we do not consider these special cases any further.

The approximations to be discussed in Sec. 3-2 do not exhaust all possible methods by means of which we may evaluate $S_j(\kappa,\omega)$. They are nevertheless the most extensively used approximations; together with Eqs. (2.98) and (2.114) they form the basis for codes like LEAP,[13] SUMMIT,[14] GAKER[15,16] and KERNEL,[17] which are designed for the high-speed digital computation of thermal-neutron scattering kernels.

3-2.1 PHONON EXPANSION

(a) Exact Calculation. We consider first the phonon expansion of $S_j(\kappa,\omega)$. As we shall see in the development given below, this expansion exists if $f_j(\omega)$ satisfies the condition (3.8a), but not if it satisfies either Eq. (3.8b) or (3.8c). Thus the phonon expansion exists for harmonic or anharmonic lattice vibrations of crystals but not for the diffusive or streaming motion of an atom.

The designation of the expansion considered in this section as a phonon expansion is suggested by its formal equivalence to the expansion that applies in the case of neutron scattering by harmonic lattice vibrations. The use of the term phonon in a more general context should not be construed as suggesting the existence in the general case of

---

[a] When Eq. (3.8b) holds, $K_j(\omega) \sim \omega^{-2}$ for ω 0. Equation (3.15), however, is equivalent to Eq. (3.9) with $\gamma_j(t)$ given by Eq. (3.5) if we take the principal value of the integral occurring in Eq. (3.15).

excitations with well-defined energies and momenta.

To proceed with the mathematical development, it is convenient to introduce

$$G_{j1}(\omega) = \frac{K_j(\omega)}{\alpha_j} \tag{3.17}$$

where

$$\alpha_j = \int_{-\infty}^{\infty} K_j(\omega)\,d\omega \tag{3.18}$$

exists by virtue of Eqs. (3.6) and (3.8a). Equation (3.18) corresponds to the normalization

$$\int_{-\infty}^{\infty} G_{j1}(\omega)\,d\omega = 1 \;. \tag{3.19}$$

The phonon expansion consists in retaining the time-independent term in the argument of the exponent in Eq. (3.15) and expanding the rest of the exponent in powers of its argument. If we use Eq. (3.17), the result is

$$\chi_j(\kappa,t) = \left[\exp\left(-\frac{\hbar\kappa^2}{2M_j}\alpha_j\right)\right]\sum_{n=0}^{\infty}\frac{1}{n!}\left(\frac{\hbar\kappa^2}{2M_j}\alpha_j\right)^n \times$$

$$\times \left[\int_{-\infty}^{\infty} d\omega' G_{j1}(\omega')e^{i\omega' t}\right]^n \;. \tag{3.20}$$

We define

$$G_{jn}(\omega) = \frac{1}{2\pi}\int_{-\infty}^{\infty} e^{-i\omega t}dt\left[\int_{-\infty}^{\infty} G_{j1}(\omega')e^{i\omega' t}d\omega'\right]^n \tag{3.21}$$

and notice that the G_{j1} given by Eq. (3.21) is consistent with the definition of Eq. (3.17). When we substitute Eq. (3.20) into Eq. (3.10), we find

$$S_j(\kappa,\omega) = \left[\exp\left(-\frac{\kappa^2}{2M_j}\alpha_j\right)\right]\left[\sum_{n=0}^{\infty}\frac{1}{n!}\left(\frac{\kappa^2}{2M_j}\alpha_j\right)^n G_{jn}(\omega)\right] \;. \tag{3.22}$$

From Eq. (3.21) we further determine that

$$\int_{-\infty}^{\infty} G_{jn}(\omega)d\omega = \frac{1}{2\pi}\int_{-\infty}^{\infty} d\omega \int_{-\infty}^{\infty} e^{-i\omega t}dt\left[\int_{-\infty}^{\infty} G_{j1}(\omega')e^{i\omega' t}d\omega'\right]^n$$

$$= \int_{-\infty}^{\infty} \delta(t)dt\left[\int_{-\infty}^{\infty} G_{j1}(\omega')e^{i\omega' t}d\omega'\right]^n$$

$$= \left[\int_{-\infty}^{\infty} G_{j1}(\omega')d\omega'\right]^n = 1 \ . \qquad (3.23)$$

It still remains to calculate the quantities $G_{jn}(\omega)$. We can do this in a numerical but essentially exact way. For values of $n \gtrsim 4$, where the exact calculation is cumbersome, there exist approximate but very accurate analytical forms for $G_{jn}(\omega)$. We shall discuss these after first considering the exact calculation of $G_{jn}(\omega)$. For convenience, we omit the subscript j. This should lead to no confusion since the results that follow apply for any $f(\omega)$ for which the condition (3.8a) is satisfied. The $n = 0$ term follows in a straightforward way from Eq. (3.21):

$$G_0(\omega) = \delta(\omega) \ . \qquad (3.25)$$

From Eqs. (3.14) and (3.17),

$$G_1(\omega) = \frac{f(\omega)\exp\dfrac{\hbar\omega}{2k_BT}}{2\alpha\omega \sin h\dfrac{\hbar\omega}{2k_BT}} \ . \qquad (3.26)$$

The application of the convolution theorem for Fourier transforms to Eq. (3.21) leads to the recursion relations

$$G_n(\omega) = \int_{-\infty}^{\infty} G_1(\omega')G_{n-1}(\omega - \omega')d\omega' \ . \qquad (3.27)$$

A form of these relations that is somewhat more convenient for numerical calculations is obtained by introducing the quantities

$$g_n(\omega) = G_n(\omega) \exp - \frac{\hbar\omega}{2k_BT} . \tag{3.28}$$

Then, from Eq. (3.27),

$$g_n(\omega) = \int_{-\infty}^{\infty} g_1(\omega')g_{n-1}(\omega - \omega')d\omega' . \tag{3.29}$$

That the functions $g_n(\omega)$ satisfy the symmetry property

$$g_n(-\omega) = g_n(\omega) \tag{3.30}$$

follows from the symmetry of

$$g_1(\omega) = \frac{f(\omega)}{2\alpha\omega \sin \frac{\hbar\omega}{2k_BT}} \tag{3.31}$$

and, by induction, from Eq. (3.29). Equation (3.29) forms a particularly convenient basis for the iterative numerical calculation of the low-order phonon terms ($n \lesssim 4$) by means of a high-speed computer.

(b) Approximations for Multiphonon Terms. The time required for the digital computation of $G_n(\omega)$ increases roughly as n^2. Thus, for large n (in practice for $n \gtrsim 4$) it is desirable to obtain approximate analytical forms for $G_n(\omega)$ or $g_n(\omega)$. To do this we follow closely the treatment of multiphonon terms given by Sjölander.[18]

It is well known in mathematical statistics that a function of the type $G_n(\omega)$ defined by Eq. (3.21) tends to a Gaussian function as n increases.[19] The same is also true of $g_n(\omega)$. To account for the deviation of $G_n(\omega)$ from a Gaussian function, we can expand it in a series of Hermite polynomials, a development that is known in mathematical statistics as the Edgewood series.[19] Since the development of this series is thoroughly described in textbooks on statistics, only a heuristic derivation of it is given here.

To begin, we introduce the function

$$B(t) = \int_{-\infty}^{\infty} G_1(\omega)e^{i\omega t}d\omega = \int_{-\omega_m}^{\omega_m} G_1(\omega)e^{i\omega t}d\omega . \tag{3.32}$$

If we compare Eq. (3.32) with Eq. (3.19) and remember that $G_1(\omega)$ is non-negative, it is easy to show that

$$\left.\begin{aligned} |B(t)| &< 1 \qquad t \neq 0, \\ B(0) &= 1 . \end{aligned}\right\} \tag{3.33}$$

On integrating Eq. (3.32) by parts it follows that we can find some positive constant K such that

$$|B(t)| < \frac{K}{|t|} \qquad t \to \infty . \tag{3.34}$$

(The inequality (3.34) holds whenever G is piecewise-continuous and piecewise-differentiable in the interval $-\omega_m \le \omega \le \omega_m$. These conditions on G apply for any well-behaved $f(\omega)$ that satisfies the condition (3.8a). We are not concerned here with the case $f(\omega) = \delta(\omega - \omega_0)$ since it has already been treated in Sec. 2-5). For large values of n, a small region around t = 0 makes the main contribution to

$$G_n(\omega) = \frac{1}{2\pi}\int_{-\infty}^{\infty} \exp(-i\omega t)[B(t)]^n dt , \tag{3.35}$$

except possible for large values of ω, in which case, however, G_n is very small. It seems reasonable, therefore, to expand B(t) in powers of t. To avoid the appearance of the singular functions that result from such an expansion, we first express Eq. (3.35) in the form

$$G_n(\omega) = \frac{1}{2\pi}\int_{-\infty}^{\infty} dt\ [\exp(-i\omega t)]\ \exp[n\ln B(t)] \tag{3.36}$$

and make the expansion of $\ln B(t)$ in powers of t;

$$\ln B(t) = \sum_{\nu=0}^{\infty} \frac{\beta_\nu}{\nu!}(it)^\nu . \tag{3.37}$$

This expansion does not in general converge, but holds only as an asymptotic expansion for small values of t (see discussion after Eq. (3.48) below). The coefficients β_ν are related to the coefficients b_ν of the expansion

$$B(t) = \sum_{\nu=0}^{\infty} \frac{b_\nu}{\nu!} (it)^\nu \tag{3.38}$$

by

$$\begin{aligned}
\beta_0 &= 0, \\
\beta_1 &= b_1 , \\
\beta_2 &= b_2 - b_1^2 , \\
\beta_3 &= b_3 - 3b_1b_2 + 2b_1^3 , \\
\beta_4 &= b_4 - 3b_2^2 - 4b_1b_3 + 12b_1^2 - 6b_1 .
\end{aligned} \tag{3.39}$$

In turn the b_ν are related to the moments of $G_1(\omega)$ by

$$b_\nu = \overline{\omega^\nu} = \int_{-\infty}^{\infty} \omega^\nu G_1(\omega)d\omega . \tag{3.40}$$

In particular,

$$\begin{aligned}
b_0 &= 1 , \\
b_1 &= \frac{1}{\alpha} , \\
b_2 &= \frac{1}{\alpha}\int_0^{\infty} f(\omega) \cot h\left(\frac{\hbar\omega}{2k_BT}\right)d\omega = \frac{2k_B\overline{T}}{\hbar\alpha} .
\end{aligned} \tag{3.41}$$

The equivalence of $\overline{T}$ in Eq. (3.41) and that defined by Eq. (2.250) follows by setting $t = 0$ in Eqs. (2.313) and (2.314) and then using Eq. (2.315). For convenience we introduce the notation σ^2 for β_2; thus,

$$\sigma^2 = \frac{2k_B\overline{T}}{\hbar\alpha} - \frac{1}{\alpha^2} . \tag{3.42}$$

One easily verifies that σ^2 is a positive quantity by noting that

$$\sigma^2 = \overline{\omega^2} - \bar{\omega}^2 = \int_{-\infty}^{\infty} (\omega - \bar{\omega})^2 G_1(\omega)d\omega .$$

We now expand $\ell n B(t)$ in Eq. (3.36) and express $G_n(\omega)$ as follows:

$$G_n(\omega) = \frac{1}{2\pi}\int_{-\infty}^{\infty} dt\ \exp\left[-\frac{n\sigma^2 t^2}{2} - it(\omega - n\bar{\omega})\right] \times$$

$$\times \exp\left[\sum_{\nu=3}^{\infty} \frac{\beta_\nu}{\nu!}(it)^\nu\right]. \tag{3.43}$$

Transforming the variable of integration to $z = n^{\frac{1}{2}}\sigma t$ and expanding the second exponential factor in the last part of Eq. (3.43) in powers of its argument gives

$$G_n(\omega) = \left(4\pi^2 n\sigma^2\right)^{-\frac{1}{2}}\int_{-\infty}^{\infty} dz\ \exp\left[-\frac{z^2}{2} - \frac{iz(\omega - n\bar{\omega})}{n^{\frac{1}{2}}\sigma}\right]\times$$

$$\times \sum_{\nu=0}^{\infty} C_\nu^{(n)}(iz)^\nu , \tag{3.44}$$

where

$$C_0^{(n)} = 1 ,$$

$$C_1^{(n)} = C_2^{(n)} = 0,$$

$$C_3^{(n)} = \frac{\beta_3}{3!\sigma^3}\frac{1}{n} = \frac{A_1}{n} ,$$

$$C_4^{(n)} = \frac{\beta_4}{4!\sigma^4}\frac{1}{n} = \frac{A_2}{n} , \tag{3.45}$$

$$C_5^{(n)} = \frac{\beta_5}{5!\sigma^5}\frac{1}{n} ,$$

$$C_6^{(n)} = \frac{\beta_6}{6!\sigma^6}\frac{1}{n^2} + \frac{\beta_3^2}{2(3!\sigma^3)^2}\frac{1}{n} .$$

To integrate the series in Eq. (3.44) we use

$$\int_{-\infty}^{\infty} dz(iz)^{\nu} \exp\left(-\frac{z^2}{2} - izx\right) = (2\pi)^{\frac{1}{2}} H_{\nu}(x) \left[\exp\left(\frac{-x^2}{2}\right)\right]. \tag{3.46}$$

where

$$H_{\nu}(x) = (-1)^{\nu} \left[\exp\left(\frac{x^2}{2}\right)\right] \frac{d^{\nu}}{dx^{\nu}} \left[\exp\left(\frac{-x^2}{2}\right)\right] \tag{3.47}$$

is the Hermite polynomial of order ν. Retaining only terms through order n^{-1} in the $C_{\nu}^{(n)}$, the resulting expression for $G_n(\omega)$, arranged in ascending powers of $n^{-\frac{1}{2}}$, is

$$G_n(\omega) = (2\pi n\sigma^2)^{-\frac{1}{2}} \left[\exp\left(-\frac{x_n^2}{2}\right)\right] \left\{1 - \frac{A_1}{n} H_3(x_n) + \right.$$

$$\left. + \frac{1}{n}\left[A_2 H_4(x_n) + A_1^2 H_6(x_n)\right] + O(n^{-\frac{3}{2}}) \right\}, \tag{3.48}$$

where

$$x_n = \frac{\omega - n\bar{\omega}}{n^{\frac{1}{2}}\sigma}. \tag{3.49}$$

The series in Eq. (3.48) does not in general converge, but it applies only as an asymptotic expansion for large values of n.[19] An asymptotic series has the property that the error is of smaller order than that of the last term retained. The accuracy of Eq. (3.48) cannot always be improved by including more terms. As x_n increases, the optimum number of terms decreases and the minimum error increases. For a fixed number of terms, the error increases as x_n increases.

Now we give the results of numerical calculations which illustrate the accuracy of the approximate equation, Eq. (3.48). For this purpose, we use Sjölander's calculations,[18] which are based on the frequency spectrum shown in Fig. 3.3. This is the spectrum Walker[20] obtained for aluminum from measurements of diffuse X-ray scattering. The frequencies are given in units of the maximum vibrational frequency of aluminum, $\omega_m = 5.91 \times 10^{13}\ \text{sec}^{-1}$. Figure 3.4 shows $G_1(\omega)$ for a temperature of 300°K. Figures 3.5, 3.6, and 3.7 compare the $G_n(\omega)$ obtained from Eq. (3.48) with the $G_n(\omega)$ that have been determined numerically to a high degree of accuracy from the exact equations (3.28) and (3.29). In the figures, Approximation A refers to Eq. (3.48) and Approximation B refers to the first term of Eq. (3.48). The constants σ^2, $\bar{\omega}$, A_1, and A_2 in Eq. (3.48) are determined numerically starting from the f(ω) shown in Fig. 3.3. In Fig. 3.7, the curve corresponding to Approximation A is indistinguishable from the exact result. Sjölander makes similar comparisons between exact and approximate calculations of $G_n(\omega)$ at 74.8°K, a temperature that is considerably less than $\hbar\omega_m/k_B$.

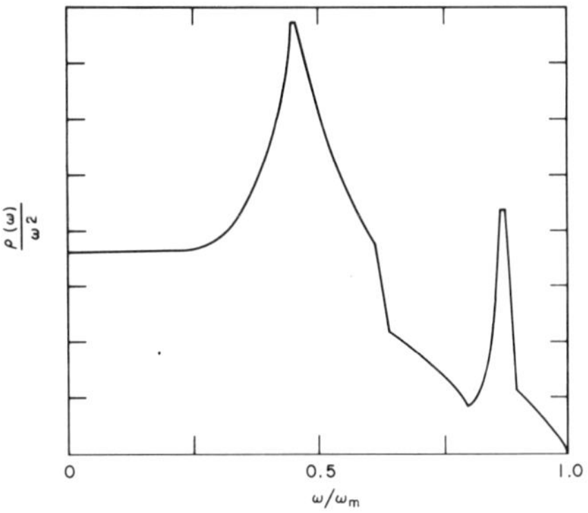

Figure 3.3 Walker's[20] phonon spectrum for aluminum. (From Sjölander, Arkiv Fysik, Ref. 18.)

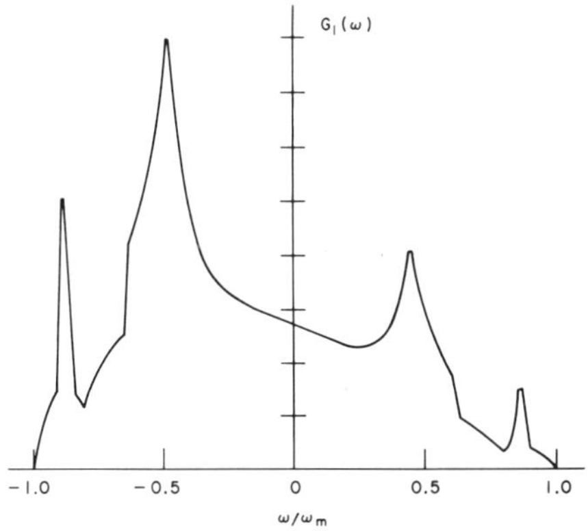

Figure 3.4 $G_1(\omega)$ for aluminum at T = 300°K. (From Sjölander, Arkiv Fysik, Ref. 18.)

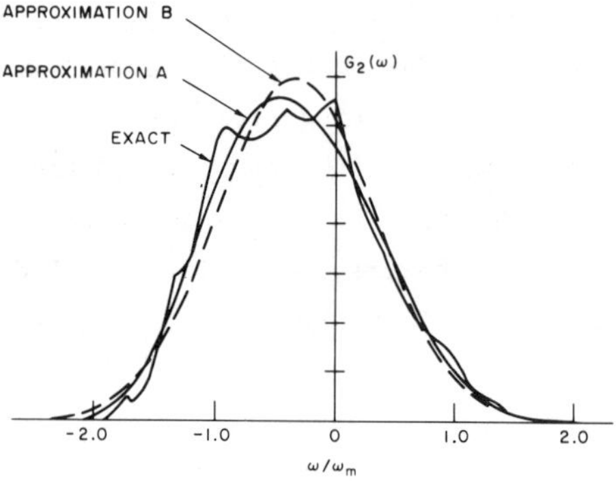

Figure 3.5 $G_2(\omega)$ for aluminum at T = 300°K. (From Sjölander, Arkiv Fysik, Ref. 18.)

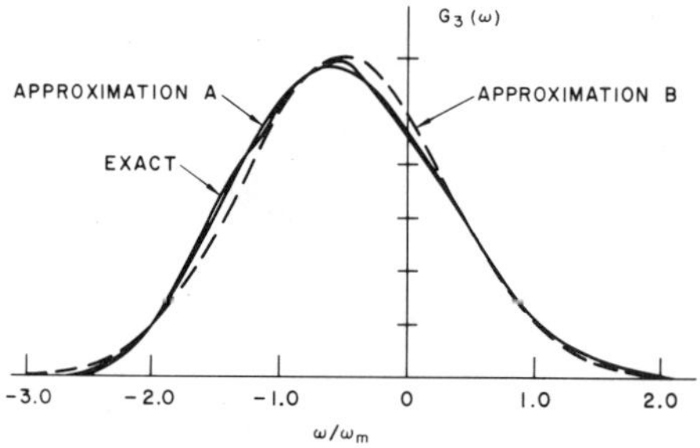

Figure 3.6 $G_3(\omega)$ for aluminum at T = 300°K. (From Sjölander, Arkiv Fysik, Ref. 18.)

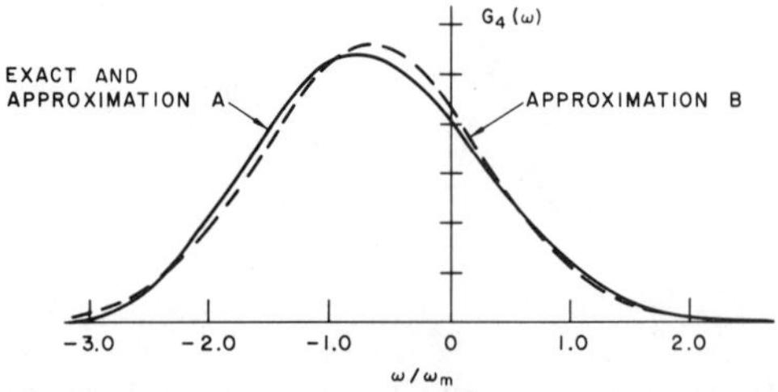

Figure 3.7 $G_4(\omega)$ for aluminum at T = 300°K. (From Sjölander, Arkiv Fysik, Ref. 18.)

From these comparisons, we conclude that, except where $G_n(\omega)$ is very small, the approximation (3.48) gives very accurate values of $G_n(\omega)$ for $n > 2$ at both high and low temperatures. Retaining only the leading term in Eq. (3.48) is less accurate but should be satisfactory for most applications, in particular for problems in neutron thermalization. To obtain the two-phonon term, one should use Eqs. (3.28) and (3.29) with $n = 2$, wherein $g_1(\omega)$, Eq. (3.31), is determined from a realistic frequency spectrum.

In the approximate formulation given above, which leads to Eq. (3.48), the detailed balancing condition is not automatically satisfied except in the limit of high temperatures. In practice, this means that energy-gain scattering ($\omega < 0$) at low temperatures ($k_BT \ll \hbar\omega_n$) is not very accurately given by Eq. (3.48). In this case, however, only a few phonon processes contribute significantly to the energy-gain scattering, and the contribution of these processes can be conveniently calculated recursively using the exact expression, Eq. (3.29), in which case the detailed balancing property is satisfied.

Alternatively, with an accuracy that is sufficient for neutron-thermalization calculations, we can determine $G_n(-\omega)$ from the exact equation

$$G_n(-\omega) = G_n(\omega)\ \exp\left(-\frac{\hbar\omega}{k_BT}\right) \tag{3.50}$$

after we first obtain $G_n(\omega)$ for $\omega \geq 0$ from the approximate equation (3.48). Equation (3.50), which follows from Eqs. (3.28) and (3.29), is another statement of the condition of detailed balancing.

It is possible to formulate an approximation for the multiphonon terms in such a way that the condition of detailed balancing is exactly satisfied. This formulation, which is called the symmetric formulation, has been considered by Sjölander[18] and by Schofield and Hassitt.[21] The symmetric formulation, however, for reasons that we indicate below, leads to approximations that are of much more restricted practical utility than the formulation leading to Eq. (3.48).

The symmetric formulation results on changing the variable of integration in Eq. (3.21) from t to

$$z = t - \frac{i\hbar}{2k_BT}$$

and shifting the path of integration so that it runs along the real z-axis. Using Eq. (3.34) the reader can verify the equivalence of integrating in Eq. (3.21) along the real t-axis and the real z-axis. If we use Eq. (3.28), Eq. (3.21) becomes[b]

$$g_n(\omega) = \frac{1}{2\pi}\int_{-\infty}^{\infty} dz e^{-i\omega z}\left[\int_{-\infty}^{\infty} d\omega' g_1(\omega')e^{i\omega' z}\right]^n , \tag{3.51}$$

and, from Eq. (3.22),

$$S(\kappa,\omega) = \left[\exp\left(-\frac{\hbar\kappa^2}{2M}\alpha + \frac{\hbar\omega}{2k_BT}\right)\right]\sum_{n=0}^{\infty}\frac{1}{n!}\left(\frac{\hbar\kappa^2}{2M}\alpha\right)^n g_n(\omega) . \tag{3.52}$$

The term "symmetric formulation" stems from the property (3.30) of the $g_n(\omega)$. The entire asymmetry of $S(\kappa,\omega)$, for fixed κ, comes from the factor $\exp(\hbar\omega/2k_BT)$.

The development of approximate expressions for $g_n(\omega)$ now proceeds from Eq. (3.51) in a manner that is precisely analogous to the development that leads from Eq. (3.21) to the approximate result given by Eq. (3.48). The result is

$$g_n(\omega) = \left(\frac{\bar{\alpha}}{\alpha}\right)^n (2\pi n\bar{\sigma}^2)^{-\frac{1}{2}}\left[\exp\left(-\frac{\omega^2}{2n\bar{\sigma}^2}\right)\right]\times$$

$$\times\left[1 + \frac{\bar{\alpha}_4 - 3\bar{\alpha}_2^2}{4!\bar{\sigma}^4}\frac{1}{n}H_4\left(\frac{\omega}{n^{\frac{1}{2}}\bar{\sigma}}\right) + O\left(\frac{1}{n^2}\right)\right] , \tag{3.53}$$

where

$$\bar{\alpha} = \int_{-\infty}^{\infty}\frac{f(\omega)}{2\omega\sin h\frac{\hbar\omega}{2k_BT}}d\omega ,$$

$$\bar{\sigma}^2 = \bar{\alpha}_2 ,$$

$$\bar{\alpha}_n = \frac{1}{2}\int_{-\infty}^{\infty}\frac{f(\omega)}{2\omega\sin h\frac{\hbar\omega}{2k_BT}}\omega^n d\omega , \tag{3.54}$$

---

[b] We continue to omit the subscript j.

and α is given by Eq. (3.18). In the limit of high temperatures it is easy to verify that Eqs. (3.53) and (3.48) are identical. This verification is most easily accomplished by noting that according to Eq. (3.28), $g_n(\omega)$ and $G_n(\omega)$ are identical when k_BT is much larger than ω_m.

We can also see the reason for the rather limited utility of Eq. (3.53). This approximation for $g_n(\omega)$ is most accurate in a neighborhood centered around $\omega = 0$. For $|\omega| >> n^{\frac{1}{2}}\bar{\sigma}$, $g_n(\omega)$ is small and is not very accurately given by Eq. (3.53). To obtain $S_{inc}(\kappa,\omega)$, however, and hence the scattering, we must multiply each $g_n(\omega)$ by the factor $\exp(\hbar\omega/2k_BT)$, which is large except for $\hbar\omega < k_BT$. The effect of multiplying by this factor is that $S_{inc}(\kappa,\omega)$ has its maximum at a value of ω that lies outside the region around $\omega = 0$ where Eq. (3.53) is a good approximation. For problems in thermalization in which the moderators are characterized by Debye temperatures that are large compared with T, the approximation corresponding to Eq. (3.48) is more useful than the one based on the symmetric formulation.

This nearly completes our discussion of the calculation of $S_j(\kappa,\omega)$ by means of the phonon expansion, Eq. (3.22). When we take advantage of Eqs. (3.27) and (3.48), the computation and summation of terms in Eq. (3.22) and the numerical convolution of the different $S_j(\kappa,\omega)$, Eq. (3.13), is an easy task for modern high-speed computers. For most moderators, the results of this section permit a rapid computation of $\sigma_0(E',E)$, Eq. (2.72), for all values of E' and E on a closely spaced energy mesh extending from zero to a few tenths of an electron volt. On this mesh, provided that $[(\hbar\kappa^2/2)(\alpha_j/M_j)] \lesssim 4$ for each j, convergence to within a few percent of the exact answer for $\sigma_0(E',E)$ is obtainable by summing fewer than ten terms in Eq. (3.22).

3-2.2 SHORT-COLLISION-TIME APPROXIMATIONS

(a) <u>An Asymptotic Expansion for Large κ^2</u>. For $(\kappa^2/2M_j)\alpha_j >> 1$, the phonon expansion for $S_j(\kappa,\omega)$ does not converge very rapidly. In this case, however, it is possible to obtain expansions that are asymptotically valid for large values of κ^2. Here we develop two different expansions for $S_{inc}(\kappa,\omega)$, both of which are expressed as series in inverse powers of the momentum transfer $\hbar\kappa$. The first of these is valid in the region of κ,ω-space where the scattered neutron distribution is most intense.[22] The second one has a greater range of validity but it requires more work to apply it to problems of neutron thermalization.[23] We do not require the condition (3.8a) for either of these expansions. Within the framework of the Gaussian approximation they apply for any system.

To begin the development of the short-collision-time approximation we express Eq. (3.15) in the form

$$\chi_j(\kappa,t) = [Q_j(t)]^{\overline{N}_j} , \tag{3.55}$$

where

$$Q_j(t) = \exp\left\{-\frac{k_B\overline{T}_j}{\hbar}\int_{-\infty}^{\infty} K_j(\omega)[1 - \exp(i\omega t)]d\omega\right\}. \tag{3.56}$$

$$\overline{N}_j = \frac{\hbar^2\kappa^2}{2M_j k_B\overline{T}_j} , \tag{3.57}$$

and

$$\frac{k_B\overline{T}_j}{\hbar} = \frac{1}{2}\int_{-\infty}^{\infty} \omega^2 K_j(\omega)d\omega = \frac{1}{2}\int_0^{\infty} \omega f_j(\omega) \cot h \frac{\hbar\omega}{2k_BT} d\omega . \tag{3.58}$$

The quantity $\overline{N}_j$ is a measure of the excitation of the system by a neutron that loses a momentum $\hbar\kappa$. In particular, for a crystal, $\overline{N}_j$ is a measure of the average number of phonons of type j emitted in the transfer of momentum $\hbar\kappa$ from the neutron to the lattice vibrations. The expansions that we develop below apply when $\overline{N}_j \gg 1$.

For notational convenience we again omit the subscript j. Now observe the properties[c]

$$Q(0) = 1$$

$$|Q(t)| < Q(0) \qquad t \neq 0 .$$

The inequality follows from

---

[c] We continue to exclude the case $f(\omega) = \delta(\omega - \omega_0)$. In that case $|Q(t)| < Q(0)$ does not apply when t is an integral multiple of $2\pi/\omega_0$.

$$|Q(t)| = \left| \exp\left[-\frac{k_B\bar{T}}{\hbar}\int_{-\infty}^{\infty} d\omega K(\omega)(1 - \cos\omega t)\right] \times \right.$$

$$\left. \times \exp\left[i\frac{k_B\bar{T}}{\hbar}\int_{-\infty}^{\infty} d\omega K(\omega)\sin\omega t\right] \right|$$

$$= \left| \exp\left[-\frac{k_B\bar{T}}{}\int_{-\infty}^{\infty} d\omega K(\omega)(1 - \cos\omega t)\right] \right|$$

$$< 1 \qquad t \neq 0 .$$

Thus, for large values of $\bar{N}$, $\chi(\kappa,t)$ is small except for a small neighborhood around $t = 0$, and it seems reasonable to make an expansion of $Q(t)$ about $t = 0$.

If we express Q as $\exp(\ln Q)$ and expand $\ln Q$ about $t = 0$, then the expansion is formally equivalent to the one made for $B(t)$ in the previous section. In fact, the entire development of $S(\kappa,\omega)$ for $\bar{N} >> 1$ is formally equivalent to that of $G_n(\omega)$ for large n. To see this, we have only to identify $\bar{N}$ with n and $Q(t)$ with $B(t)$, and to compare the result of inserting Eq. (3.55) into Eq. (3.10) with Eq. (3.35).

To continue, we expand the argument of the exponent in Eq. (3.56) in powers of t, substitute the result in Eq. (3.55), and obtain

$$\chi(\kappa,t) = [Q(t)]^{\bar{N}} = \left\{\exp\left[\frac{\hbar\kappa^2}{2M}\left(it - \frac{k_B\bar{T}}{\hbar}t^2\right)\right]\right\} \times$$

$$\times \left\{\exp\left[\frac{\hbar\kappa^2}{2M}\sum_{\nu=3}^{\infty}\frac{x_\nu}{\nu!}(it)^\nu\right]\right\} , \qquad (3.59)$$

where

$$x_\nu = \int_{-\infty}^{\infty} \omega^\nu K(\omega)d\omega = \int_{-\infty}^{\infty} d\omega \frac{\omega^\nu f(\omega) \exp\left(\frac{\hbar\omega}{2k_BT}\right)}{2\omega \sin h\left(\frac{\hbar\omega}{2k_BT}\right)}$$

$$= \int_0^{\infty} d\omega\omega^{\nu-1} f(\omega) \cot h\left(\frac{\hbar\omega}{2k_BT}\right) \qquad \nu \text{ even}, \tag{3.60}$$

$$= \int_0^{\infty} d\omega\omega^{\nu-1} f(\omega) \qquad \nu \text{ odd} .$$

We next expand the second exponential factor in Eq. (3.59) in powers of it and substitute the resulting expression for $\chi(\kappa,t)$ into Eq. (3.10). Utilizing the definition (3.46) of the Hermite polynomials, we finally obtain

$$S(\kappa,\omega) = (2\pi\Delta^2)^{-\frac{1}{2}} \left[\exp\left(-\frac{x^2}{2}\right)\right]\left\{1 + \frac{1}{12\sqrt{2}}\left(\frac{k_B\overline{T}}{R}\right)^{\frac{1}{2}} \frac{\hbar^2 x_3}{(k_B\overline{T})^2} H_3(x) + \right.$$

$$\left. + \frac{1}{96}\frac{k_B\overline{T}}{R}\left[\frac{\hbar^3 x_4}{(k_B\overline{T})^3} H_4(x) + \frac{1}{6}\frac{\hbar^4 x_3^2}{(k_B\overline{T})^4} H_6(x)\right] + 0\left[\left(\frac{k_BT}{R}\right)^{\frac{3}{2}}\right]\right\}, \tag{3.61}$$

where

$$R = \frac{\hbar^2\kappa^2}{2M} \tag{3.62}$$

is the free atom recoil energy,

$$\hbar^2\Delta^2 = \frac{\hbar^2\kappa^2 k_B\overline{T}}{M} \tag{3.63}$$

is the square of the Doppler width, and

$$x = \frac{\hbar\omega - R}{\hbar\Delta} . \tag{3.64}$$

At this point we remind the reader that a subscript j may be associated with each of the quantities S, Δ, x, x_ν, $\overline{T}$, M, and R appearing in Eq. (3.61). If this is not the case, then in Eq. (3.61) $S(\kappa,\omega) = S_{inc}(\kappa,\omega)$.

The term outside the brace-enclosed term in Eq. (3.61) is the result for an ideal gas of mass M and temperature $\bar{T}$. The occurrence of such a leading term may be expected on physical grounds or on the basis of the discussion of scattering by an Einstein oscillator given in Sec. 2-5. Since the expansion in Eq. (3.61) is known to be asymptotic,[22] the unavoidable minimum error inherent in the use of the series decreases as $(k_B\bar{T}/R) = \bar{N}^{-1}$ increases. The correction terms vanish both in the limit of large recoil and in the limit of weak binding for arbitrary recoil. For any assumed frequency distribution, the terms of order and higher can be calculated in a straightforward manner.

For the correction terms in Eq. (3.61) to be small, we require that

$$\frac{k_B\bar{T}}{R} << 1 ,$$

and

$$x^2 = \frac{(\hbar\omega - R)^2}{\hbar^2\Delta^2} = \frac{(\hbar\omega - R)^2}{2Rk_B\bar{T}} << 1 . \tag{3.65}$$

Thus, for large free-atom recoil energies, the approximation (3.61) is best when the energy loss $\hbar\omega$ is in the neighborhood of the energy loss that would occur when neutrons are scattered by atoms that are free and at rest. This is just the region of ω,R-space where the intensity of scattered neutrons is greatest. Equation (3.61) results in a significant error only in the wings of the scattered distribution, where both $S(\kappa,\omega)$ and the right-hand side of Eq. (3.61) are small. Further, using Eq. (3.61) as the approximation for $S_{inc}(\kappa,\omega)$, we can obtain a result for the total scattering cross section which differs from Wick's expansion, Eq. (2.256), only by terms that vanish like $\exp(-AE/k_B\bar{T})$ when E approaches infinity.[22]

In the approximate equation (3.61) the property of detailed balancing is not preserved. We could just as well use the above methods to develop a short-collision-time approximation starting from the symmetric scattering function in Eq. (2.319), in which case the detailed-balancing property would be preserved. In the limit of high temperatures, the result would be identical with the high-temperature limit of Eq. (3.61). For low temperatures, on the other hand, the utility of the approximate formula would be very limited for precisely the reason that the utility of Eq. (3.53) is limited. Furthermore, the leading terms in such an approximation would lead to a result for the cross section $\sigma_s(E)$ which would contain neither the correct free-atom limit nor the correction terms given in Eq. (2.256).

In the next section, we give an expansion, derived by Egelstaff and Schofield,[23] that automatically satisfies the condition of detailed balance. Furthermore, their expansion is accurate over a considerably greater range of κ,ω-space than is Eq. (3.61). On the other hand, to obtain the increased accuracy one is forced to do considerably more numerical work to generate the expansion for any given $f(\omega)$ than is required by Eq. (3.61).

(b) Calculation by the Method of Steepest Descents. Egelstaff and Schofield[23] have applied the method of steepest descents for evaluating $S(\kappa,\omega)$ for ranges of κ and ω where the phonon expansion converges slowly. As was the case in the previous section, their expansion is not tied to the condition (3.8a). The basis of the method is to deform the contour of integration from the real t-axis to pass through a saddle point in the complex t-plane and then to evaluate the integral along a path that in the neighborhood of the saddle point is a path of steepest descent.[24]

Consider the symmetric function

$$S_{0j}(\kappa,\omega) = \exp(-\tfrac{1}{2}\beta\hbar\omega)S_j(\kappa,\omega) = \frac{1}{2\pi}\int_{-\infty}^{\infty} e^{-i\omega t}\chi_{0j}(\kappa,t)dt\ . \tag{3.66}$$

By Eqs. (2.319), (2.320), (3.14), and (3.15),

$$\chi_{0j}(\kappa,t) = \exp[-\kappa^2\zeta_j(t)] = \chi_j\left(\kappa, t + \frac{i\hbar}{2k_BT}\right)\ . \tag{3.67}$$

For easy reference we write χ_{0j} as (omitting the subscript j)

$$\chi_0(\kappa,t) = \exp\left[-\frac{\hbar\kappa^2}{2M}\int_0^{\infty} d\omega' \frac{f(\omega')}{\omega' \sin h \dfrac{\hbar\omega'}{2k_BT}} \times \left(\cos h \frac{\omega'}{2K_BT} - \cos \omega' t\right)\right]\ . \tag{3.68}$$

The saddle point occurs at the point $t = i\tau$, where

$$\frac{d}{dt}[e^{-i\omega t}\chi_0(\kappa,t)] = 0 . \tag{3.69}$$

Thus, $\tau = \tau(\omega/\kappa^2)$ is determined by

$$\omega - \frac{\hbar\kappa^2}{2M}\int_0^\infty f(\omega')\frac{\sin h\ \omega'\tau}{\sin h\ \frac{\hbar\omega'}{2k_BT}}\,d\omega' = 0 \tag{3.70}$$

as a function of ω/κ^2, which satisfies the condition

$$\tau\left(-\frac{\omega}{\kappa^2}\right) = -\tau\left(\frac{\omega}{\kappa^2}\right). \tag{3.71}$$

We next expand $\{-i\omega t + \ell n[\chi_0(\kappa,t)]\}$ in a Taylor series about the point $t = -i\tau$, obtaining

$$e^{-i\omega t}\chi_0(\kappa,t) = \{\exp[-\omega\tau - Rf_0(\tau)]\}\,\{\exp[-\tfrac{1}{2}Rf_2(\tau)t'^2]\ \times$$

$$\times\left[\exp R\sum_{n=3}^{\infty}\frac{f_n(\tau)}{n!}\,(it')^n\right], \tag{3.72}$$

where $t' = t + i\tau$, R is given by Eq. (3.62),

$$\hbar f_0(\tau) = \int_0^\infty d\omega'\ \frac{f(\omega')}{\omega'\ \sin h\ \frac{\hbar\omega'}{2k_BT}}\quad \cos h\ \frac{\hbar\omega'}{2k_BT} - \cos h\ \omega'\tau\ , \tag{3.73}$$

and, for $n > 1$,

$$\hbar f_n(\tau) = \int_0^\infty d\omega'\ \frac{f(\omega')\omega'^{\,n-1}}{\sin h\ \frac{\hbar\omega'}{2k_BT}}\ \sin h\ \omega'\tau \qquad n \text{ odd}\ ,$$

$$= \int_0^\infty d\omega'\ \frac{f(\omega')\omega'^{\,n-1}}{\sin h\ \frac{\hbar\omega'}{2k_BT}}\ \cos h\ \omega'\tau \qquad n \text{ even}\ . \tag{3.74}$$

Now we deform the path of integration in Eq. (3.66) from the axis of real t to the contour C indicated by the arrow in Fig. 3.8. From $-\infty$ to $-t_0$ and from t_0 to ∞, C is coincident with the real axis. In the interval $-t_0 \leq \text{Re}\ t \leq t_0$, C is parallel to the real axis and passes through the saddle point at $t = -i\tau$, in the neighborhood of which C is a path of steepest descent. The integrals along the vertical segments do not in general vanish in the limit that t_0 goes to infinity.[d] For $\omega_m t_0 \gg 1$ and for $R/\omega_m \gg 1$, however, the contribution from these segments is small compared with the contribution from the neighborhood of $t = -i\tau$. In fact, using arguments similar to those following Eq. (3.55), it follows for large R that the main contribution to the integral over C comes from the immediate neighborhood of $t = -i\tau$.

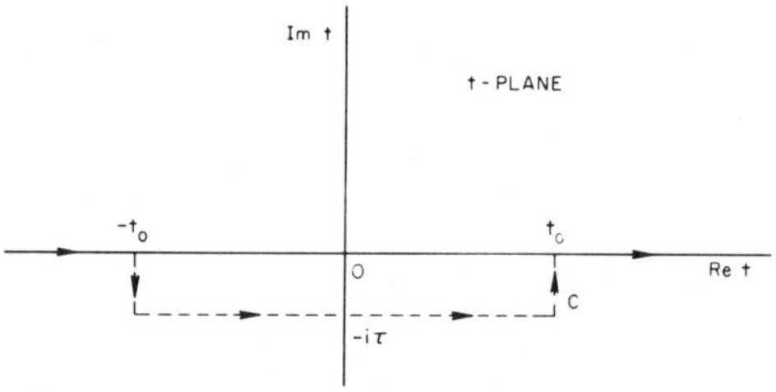

Figure 3.8 Deformed contour through the saddle point at $t = -i\tau$.

Expanding the last exponential factor on the right-hand side of Eq. (3.72) in powers of t', neglecting the contribution from the vertical segments of C, and letting $t_0 \to \infty$, we find

$$S_0(\kappa,\omega) = [2\pi R f_2(\tau)]^{-\frac{1}{2}} \{\exp[-\omega\tau - Rf_0(\tau)]\} \times$$
$$\times \left\{1 + \left[\frac{1}{Rf_2(\tau)}\right] C_1(\tau) + \left[\frac{1}{Rf_2(\tau)}\right]^2 C_2(\tau) + \ldots\right\}, \qquad (3.75)$$

where

$$C_1(\tau) = \frac{1}{8} g_4 - \frac{5}{24} g_3^2 ,$$
$$C_2(\tau) = \frac{385}{1152} g_3^4 - \frac{35}{64} g_3^2 g_4 + \frac{35}{384} g_4^2 + \frac{7}{48} g_3 g_5 - \frac{1}{48} g_6 , \qquad (3.76)$$

with g_n defined by

---

[d] When t_0 approaches infinity, the contribution from the vertical segments vanishes for gases and liquids but not for solids.

$$g_n \equiv g_n(\tau) = \frac{f_n(\tau)}{f_2(\tau)} . \tag{3.77}$$

Like Eq. (3.61), Eq. (3.75) is an asymptotic expansion and does not in general converge for any value of R. Nevertheless, provided R is sufficiently large, it should give an accurate value for $S_0(\kappa,\omega)$.

The result for $S_0(\kappa,\omega)$ in Eq. (3.75) satisfies the detailed balance theorem by virtue of its invariance with respect to a change of sign of τ, which, by Eq. (3.71), is equivalent to changing the sign of ω. It does not allow us, however, to satisfy exactly any of the moment theorems, as, for example, Eq. (2.80). Nevertheless, it has been shown[23] that the moments are given correctly to some order in $(k_B\bar{T}/R)$ depending on the number of terms that are included in the series for $S_0(\kappa,\omega)$.

Egelstaff and Schofield[23] have considered the range of validity of the asymptotic expansion of the case of (1) a solid in which $f(\omega)$ vanishes for $\hbar\omega > k_B$ and (2) a diffusive system with the width function given by Eq. (3.84) in Sec. 3-2.3. Here we present their arguments only for the first case.

If $f(\omega')$ is zero for $\hbar\omega' > k_B$, then, from Eqs. (3.74), we have the inequalities

$$f_n < \frac{k_B}{\hbar} f_{n-1} \qquad n \geq 3 , \tag{3.78}$$

from which we deduce that

$$g_n = \frac{f_n}{f_2} < \left(\frac{k_B}{\hbar}\right)^{n-2} \qquad n \geq 3 , \tag{3.79}$$

$$\begin{aligned} C_1 &\lesssim \left(\frac{k_B}{\hbar}\right)^2 \\ C_2 &\lesssim \left(\frac{k_B}{\hbar}\right)^4 \end{aligned} \tag{3.80}$$

and, in general,

$$C_n \lesssim \left(\frac{k_B}{\hbar}\right)^{2n} . \tag{3.81}$$

Thus, provided

$$\frac{(k_B\ /\hbar)^2}{Rf_2} << 1 \ , \tag{3.82}$$

the asymptotic expansion is valid.

In the limit that $R \to 0$ for a fixed ω, it follows from Eq. (3.70) that $\tau \to \infty$. Furthermore, for $\tau \to \infty$

$$g_n(\tau) \sim \quad {\frac{k_B}{\hbar}}^{n-2}$$

and

$$f_2(\tau) \sim \frac{k_B}{\hbar}\ f_1(\tau) = \frac{k_B}{\hbar^2}\ \frac{\hbar\omega}{R} \ ,$$

so that

$$C_1 \sim \quad {\frac{k_B}{\hbar}}^{2}$$

and

$$C_2 \sim \quad {\frac{k_B}{\hbar}}^{4}$$

The expansion parameter in the limit of small R thus is

$$\frac{1}{Rf_2(\tau)}\ {\frac{k_B}{\hbar}}^{2} \sim \frac{k_B}{\hbar\omega} \ . \tag{3.83}$$

Egelstaff and Schofield determine that for $\hbar\omega > 0.3k_B$ the term $[Rf_2(\tau)]^{-4}C_4(\tau)$ in the expansion (3.75) should be less than about 0.01 for all values of R. Since the expansion is asymptotic, however, this is not the true error in the approximate equation (3.75). This is borne out by the calculations of Egelstaff and Schofield for a Debye model. The results are shown in Fig. 3.9, where we plot the dimensionless quantity $\bar{S} = (k_B T/\hbar)S_0$ as a function of $\hbar\omega/k_B T$. The asymptotic form of $S_0(\kappa,\omega)$ passes through the centers of the various disconnected sections of the multiphonon contribution and, when the discontinuities become small, represents with great accuracy the true values of $S_0(\kappa,\omega)$. Of course, the discontinuities become small only for large R.

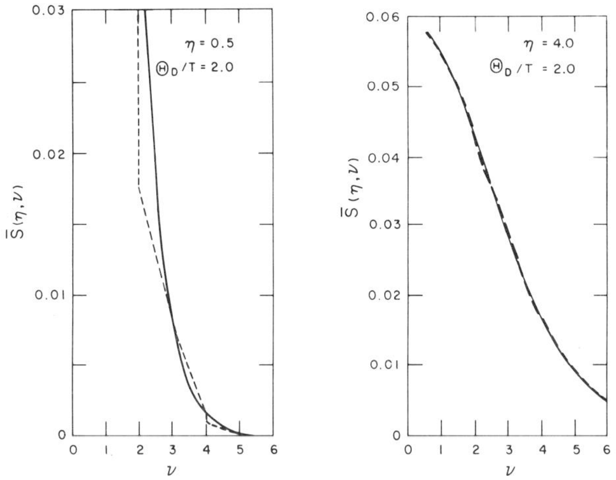

Figure 3.9 Comparison of the short-collision-time and phonon-expansion calculations for a Debye solid. Agreement between these methods is obtained at high values of $\nu = \hbar\omega/k_B T$ for any value of $\eta = \hbar^2\kappa^2/2Mk_B T$. (From Egelstaff and Schofield, Nucl. Sci. Eng., Ref. 23.)

Finally, in the limit of high temperatures

$$\tau \to \frac{\hbar}{2k_B T}\,\frac{\hbar\omega}{R}$$

Using this result, we find it easy to show that Eq. (3.75) becomes identical to $S_0(\kappa,\omega)$ for a free gas. For high temperatures, thus, Eqs. (3.75) and (3.61) become equivalent approximations.

3-2.3 AN APPROXIMATION FOR DIFFUSIVE MOTIONS. For diffusive motions, $f(\omega)$ goes to a finite limit when $\omega \to 0$ and thus the phonon expansion does not apply. The short-collision-time approximations of the previous two sections do exist for the diffusive motions, but such approximations do not constitute an accurate result for small values of the momentum and energy transfers. We therefore require a separate discussion in order to account satisfactorily for the effects of molecular diffusion in neutron scattering.

Egelstaff and Schofield[23] give a prescription that permits a two-parameter fit of data obtained by scattering neutrons from fluids. They represent the diffusive motions by a $\gamma(t)$ that is given by

$$\gamma_d(t) = D\left[\left(t^2 - i\beta\, t + c^2\right)^{\frac{1}{2}} - c\right]. \tag{3.84}$$

At long times this approaches $D|t|$, a result that is in accord with Eqs. (2.303) and (3.1) and describes simple diffusion. For short times,

$$\gamma_d(t) \xrightarrow[t\to 0]{} -\frac{D\beta\hbar}{2c}\, it\ .$$

When we expand either Eq. (2.312) or (2.316) for small t, we see that this must equal $[-(\ /2M_d)it]$, where M_d is the effective mass that is associated with the motions represented by γ_d. Thus,

$$\frac{D}{c} = \frac{k_B T}{M_d}\ . \tag{3.85}$$

If we should attempt to represent the motions in a monatomic liquid by Eq. (3.84) at all times, then M_d would be equal to the atomic mass. In general, however, Eq. (3.84) is not a good approximation for all times but only in the limit of long times. To accurately approximate $\gamma(t)$ for all times we must combine other $\gamma_j(t)$ with $\gamma_d(t)$, in which case M_d appears as one of the M_j in Eq. (3.7) and therefore is greater than the atomic mass M. The ratio $M_d/M > 1$ represents, in rough approximation, the number of atoms in a cluster of atoms that move together for a short time.

The parameter c is a measure of the time that elapses before the motion of an atom achieves its asymptotic behavior. In other words, it is a measure of time that atoms remain localized near their initial positions before diffusing away.

The $f(\omega)$ corresponding to Eq. (3.84) is[23]

$$f_d(\omega) = \frac{4MD}{\pi\hbar}\left[c^2 + \frac{1}{4}\left(\frac{\hbar}{k_B T}\right)^2\right]^{\frac{1}{2}} \sin h\, \frac{\hbar\omega}{2k_B T}\, K_1 \times$$

$$\times \left\{\left[c^2 + \frac{1}{4}\left(\frac{\hbar}{k_B T}\right)^2\right]^{\frac{1}{2}} \omega\right\}, \tag{3.86}$$

where, in the notation of Erdélyi et al.,[25] $K_1(x)$ is a modified Bessel function of the third kind.[e] For $\omega = 0$, this gives the result in Eq. (2.293).

The $S_0(\kappa,\omega)$ corresponding to Eq. (3.84) is given by

$$S_d(\kappa,\omega) = \frac{\kappa^2 D}{\pi}\left[c^2 + \frac{1}{4}\left(\frac{\hbar}{k_B T}\right)^2\right]^{\frac{1}{2}} \times$$

$$\times \frac{K_1\left\{\left[c^2 + \frac{1}{4}\left(\frac{\hbar}{k_B T}\right)^2\right]^{\frac{1}{2}} \left(\omega^2 + D\kappa^2\right)^{2\ \frac{1}{2}}\right\}}{\left(\omega^2 + D\kappa^2\right)^{2\ \frac{1}{2}}} \times$$

$$\times \exp \frac{\hbar\kappa^2 D}{k_B T} \quad . \tag{3.87}$$

In the classical limit ($\hbar \to 0$), and for $c^2(\omega^2 + D\kappa^2) << 1$, this reduces to the classical result for simple diffusion given by Eq. (2.304). As we expect, it is only in the wings of the quasi-elastic peak observed in slow-neutron scattering experiments that the description corresponding to Eq. (3.87) gives results that differ significantly from those that we obtain from simple diffusion theory. In Sec. 3-4.1 we compare results based on Eq. (3.87) with results based on observation of quasi-elastic scattering from water.

3-2.4 OTHER METHODS. The Placzek mass expansion,[26] which we mentioned in Sec. 2-8, has also been applied to calculate the scattering kernel and its integral properties. We do not discuss these calculations in this chapter, but we mention Marshall and Stuart's[11] calculations of the total inelastic scattering cross section and Purohit's[27] calculations of energy-transfer moments.

---

[e] In Ref. 23, $K_1(x)$ is referred to as a modified Bessel function of the second kind, but it is identical to one of the third kind as defined on page 5 of Ref. 25.

3-3 THE SCATTERING LAW AND ITS MEASUREMENT

If $\chi_{inc}(\kappa,t)$ has the form given by Eq. (3.1), then a knowledge of $\gamma(t)$, Eq. (2.316), is sufficient to determine the correlation function $G_s(\mathbf{r},t)$ and the noninterference part of the scattering. Conversely, the measurement of $S_{inc}(\kappa,\omega)$ for all energy transfers at a fixed value of κ determines $\gamma(t)$ and hence $f(\omega)$. For most moderators, however, there is a significant interference component of the scattering so that the direct observation of S_{inc} is in general not possible. Even if the interference scattering were negligible, it would still be difficult to carry out the experiment for a sufficiently large range of values of ω. Further, $\chi_{inc}(\kappa,t)$ contains non-Gaussian effects, and these become more important as κ^2 increases.

Egelstaff and Schofield[23] have given a method for the interpretation of slow-neutron scattering measurements which, in principle, permits a direct, but in practice only rough, determination of $f(\omega)$. They express $\left\langle S_{inc}(\kappa,\omega)\right\rangle_R$ as a sum of terms:

$$S_{inc}(\kappa,\omega) \equiv \left\langle S_{inc}(\kappa,\omega)\right\rangle_R = S_g(\kappa,\omega) + S_n(\kappa,\omega) .$$

In the Gaussian approximation, S_n vanishes and $S_{inc} - S_g$ is the Gaussian function given by Eq. (3.3). For small values of κ, $S_n(\kappa,\omega)$ is of the order of κ^4 (cf. Secs. 2-9.2 and 2-9.3). Using Eq. (3.22), which applies for bounded atomic motions, and Eqs. (3.14) and (3.17), it then follows that

$$\lim_{\kappa\to 0} \frac{S_{inc}(\kappa,\omega)}{\kappa^2} = \lim_{\kappa\to 0} \frac{S_g(\kappa,\omega)}{\kappa^2} = \frac{\hbar}{2M} \frac{f(\omega)\exp\dfrac{\hbar\omega}{2k_BT}}{2\omega\sin h\dfrac{\hbar\omega}{2k_BT}} . \qquad (3.88)$$

These results also obtain for fluids, where the motion of an atom is unbounded.[23]

Frequently, the experimental data are presented by giving the scattering law as a function of the dimensionless variables

$$\eta = \frac{\hbar^2\kappa^2}{2Mk_BT} \qquad (3.89a)$$

$$\nu = \frac{\hbar\omega}{k_B T} \tag{3.89b}$$

For an isotropic system containing only one atomic species, the scattering law $\bar{S}$ is defined by

$$\sigma(E_0, E, \theta) = \frac{\sigma_b}{4\pi}\left(\frac{E}{E_0}\right)^{\frac{1}{2}}(k_B T)^{-1} e^{\nu}\bar{S}(\eta,\nu) \ . \tag{3.90}$$

Thus, $\bar{S}$ is a dimensionless quantity and it contains in general both a coherent and an incoherent part. For molecules consisting of more than one atomic species, Eq. (3.90) is still valid, but the values of η in Eq. (3.89a) and σ_b in Eq. (3.90) must be defined in terms of the mass and bound-atom cross section, respectively, of only one of the constituent atoms. For this purpose, the lightest atom in the molecule is usually chosen. Once this choice is made, the precise meaning of $\bar{S}(\eta,\nu)$ follows when Eq. (3.90) is compared with Eqs. (2.208) and (2.209), the latter equations being considered for the case of an isotropic system.

If neutrons are scattered essentially by only one type of atom, we can obtain Eq. (3.88). Expressed in terms of $\bar{S}_{inc}$, the incoherent part of $\bar{S}$, and the dimensionless variables η and ν, Eq. (3.88) becomes

$$\lim_{n\to 0} \frac{\bar{S}_{inc}(\eta,\nu)}{\eta} \equiv \frac{p(\nu)}{\nu^2} = \frac{\bar{f}(\nu)}{2\nu \sin h(\nu/2)} \ , \tag{3.91}$$

where

$$\bar{f}(\nu) = \frac{k_B T}{\hbar} f \ \frac{k_B T}{\hbar} \nu \tag{3.92}$$

satisfies

$$\int_0^\infty \bar{f}(\nu)d\nu = 1 \ . \tag{3.93}$$

If more than one type of atom has a significant cross section for neutron scattering, we cannot determine the $f(\nu)$ associated with each of these atoms without making some rather specific assumptions; the D_2O molecule provides one example where this is the case. Unless stated to the contrary, we continue to assume that only one species of atom in the system scatters neutrons significantly.

Equation (3.88), or equivalently Eq. (3.91), suggests how, in the absence of interference effects, $f(\omega)$ may be obtained from measurements of neutron scattering. If there exists around $\kappa^2 = 0$ a range of values of κ^2 for which the measured curve of $S_{inc}(\kappa,\omega)/\kappa^2$ is essentially a linear function of κ^2, the extrapolation of this curve back to $\kappa = 0$ enables us to obtain $f(\omega)$. The interference effects, which dominate the scattering for small values of κ^2, vanish strongly when κ is large compared with the inverse distance between atoms. If, as κ increases, they become insignificant before the terms in S_{inc}/κ^2 that are proportional to κ^4 become important, we can still perform the extrapolation to $\kappa^2 = 0$ and thus obtain $f(\omega)$. The available evidence, however, indicates that, apart from purely incoherent scatterers, such favorable cases do not exist. This has been demonstrated in particular for neutron scattering by beryllium powder.[28] We discuss this case further in Sec. 3-4.4. In spite of the difficulties originating from interference effects, the extrapolation procedure has been used to obtain an $f(\omega)$ for graphite,[29,30] beryllium,[31] H_2O and D_2O.[32,33] (In H_2O, however, the interference effects are not significant.)

It has been attempted in some cases to make the extrapolation to $\kappa^2 = 0$ after first correcting the data for interference effects and for terms in S_{inc}/κ^2 that are proportional to κ^4. Egelstaff[34] has considered these corrections in general, and they have been applied to the analysis of measurements in H_2O and D_2O.[35] Although we consider that the results from these difficult experiments usually provide only a rough determination of $f(\omega)$, they often provide a useful guide to its real form. This is particularly true for those substances where more definitive experiments from which we can determine $f(\omega)$ do not exist.

Egelstaff[36] and Brockhouse[37] have described some of the experiments for measuring the differential-energy-transfer cross section $\sigma(E_0,E,\theta)$, or equivalently $\bar{S}(\eta,\nu)$. Figure 3.10 illustrates schematically one apparatus that permits a rapid rate of data collection. Instruments of this type have been installed at the NRU reactor at Chalk River, Canada, and at the MTR reactor at the National Reactor Testing Station in Idaho.

In Fig. 3.10 the neutron beam from the reactor passes through four choppers (for a discussion of choppers see Secs. 6-1.1 and 6-3).

Choppers A and C separate thermal neutrons from the entire neutron spectrum leaking from the reactor. Choppers B and D perform as a mechanical velocity selector, passing bursts of neutrons with a well-defined velocity. The sample scatters approximately ten percent of these neutrons and they are then counted by detectors that are arranged in a ring as shown in the figure. The distribution in energy at each angle of scattering is then determined by the time-of-flight method (see Chap. 6).

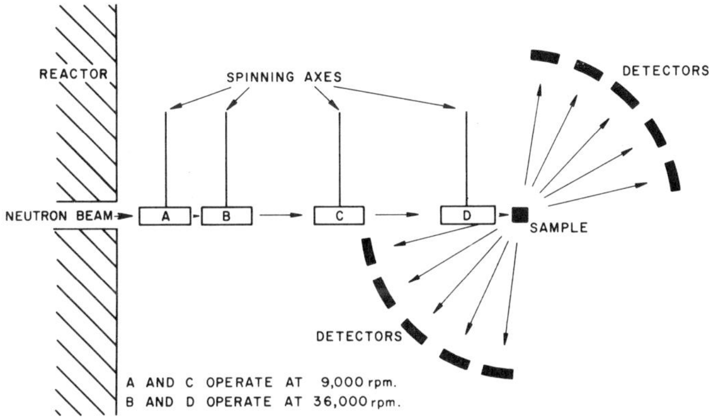

Figure 3.10 General layout of thermal-neutron scattering law experiment at the NRU reactor, Chalk River, Canada. (From Egelstaff, Nucl. Sci. Eng., Ref. 36.)

The ranges of momentum and energy transfers that are experimentally accessible in high-resolution measurements of neutron scattering depend strongly on the type of source that is used. Currently available reactor sources provide sufficiently intense beams only for neutrons having energies less than about 0.15 eV. Recently, neutrons produced with the aid of electron linear accelerators have been used to extend the accessible range of energy transfers to about 0.5 eV. For energy transfers larger than about 0.1 eV, one usually observes neutrons that have lost energy on scattering, their initial energies being of the order of a few tenths of an electron volt.

In the following sections we examine some of the results obtained from neutron-scattering experiments using neutrons produced both by reactors and indirectly with the aid of a linear accelerator. We emphasize those experiments in which the neutrons are scattered by H_2O, D_2O, zirconium hydride, graphite, and beryllium.

3-4 SLOW-NEUTRON SCATTERING BY MODERATORS

The direct measurement of $\sigma(E_0,E,\theta)$ at all energies and angles of interest for thermalization studies is quite impractical. Apart from the enormous amount of data that is required, there are experimental difficulties associated with the low intensity of neutrons that is available at neutron energies greater than about 0.3 eV. We must thus rely heavily on theoretical methods, those of the previous sections, for example, for extrapolating from the region of energies and angles of scattering that are accessible in neutron experiments to those regions that are not so accessible but at which the effects of atomic motions are still important.

Two complementary approaches have been taken in the application of results from neutron-scattering experiments to the problem of thermalization. These consist in comparing the theoretical results expected from prescribed dynamical models with the results of scattering experiments or, as discussed in Sec. 3-3, in inferring from the experiments a dynamical description of the scattering system in terms of which $\sigma(E_0,E,\theta)$ may be calculated.

For many moderators the information from neutron scattering experiments is not by itself sufficient to infer a dynamical description that is sufficiently complete for describing thermalization phenomena. This is true, for example, for H_2O, where measurements with currently available reactor sources do not give information on $f(\omega)$ for $\hbar\omega \gtrsim 0.2$ eV. As we mentioned earlier, however, high-current electron linear accelerators show promise of filling this gap.

There are experiments that do not employ neutrons which, in many cases, are more naturally suited to the study of high-frequency motions in moderators. Much information that is relevant for thermalization calculations has been obtained from experiments involving optical radiation. The techniques of Raman and infrared spectroscopy have led to an assignment of the fundamental frequencies of internal vibrations in the water molecule and have also given information relating to the frequencies of hindered rotations in water. Information less sensitively related to atomic motions but still relevant for the thermalization problem is contained in the specific heat as a function of temperature. The application of some of these results is discussed in the following sections.

Table 3.1

BOUND-ATOM CROSS SECTIONS FOR MODERATORS*

	σ_b (barns)	$\sigma_{c,b}$ (barns)	Sign of Coherent Amplitude
H	81.5 ± 0.4	1.79 ± 0.02	(-)
D	7.6 ± 0.1	5.4 ± 0.3	(+)
Be^9	7.54 ± 0.07	7.53 ± 0.07	(+)
C	5.53 ± 0.03	5.50 ± 0.04	(+)
O	4.24 ± 0.02	4.2 ± 0.3	(+)
Zr	6.3 ± 0.3	5.0 ± 0.3	(+)

*From Hughes and Schwartz, USAEC Report BNL-325, 2nd ed., Ref. 38.

The only quantities of a specifically nuclear character that enter into the thermal-neutron scattering problem are the coherent and incoherent cross sections. In Table 3.1 we list the bound-atom coherent cross sections $\sigma_{c,b}$ and $\sigma_b = \sigma_{c,b} + \sigma_{inc,b}$ for the elements that occur in substances commonly used as moderators. We also give the sign of the coherent scattering length. Using this table, we can determine whether there is a significant interference contribution to the scattering at low energies. (For thermal energies above about 0.01 eV, the interference has a negligible effect for quantities of interest in neutron thermalization.) For example, in H_2O, the $\sigma_{c,b}$ for hydrogen and the square root of the product of $\sigma_{c,b}$ for hydrogen and that for oxygen are both small compared with σ_b for hydrogen. Thus, the interference is negligible.

3-4.1 LIGHT WATER AND HEAVY WATER. The geometric structure and energy states of H_2O are well known from spectroscopic investigations. The two light atoms in H_2O and the oxygen atom form an isosceles triangle as shown in Fig. 3.11. The oxygen atom is at the vertex of the $\gamma = 104°$ angle formed by the two equal sides of length $\ell = 0.96$ Å. There are three normal internal vibrations of the molecule. Table 3.2 shows these modes and their fundamental vibration energies in the vapor phase. The frequencies corresponding to these vibrations appear in the infrared absorption and Raman spectra of the H_2O molecule.

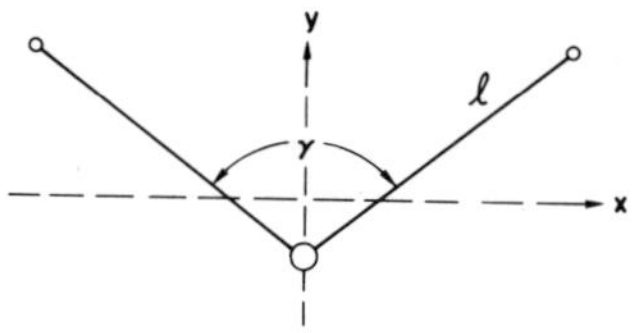

Figure 3.11. Schematic diagram of a water molecule.

The x and y axes in Fig. 3.11, which intersect at the center of mass of the molecule, are two principal axes. The third principal axis is in a direction perpendicular to the x-y plane. The moments of inertia about the x, y, and z axes are

$$I_x = 1.02 \times 10^{-40} \text{ g-cm}^2 ,$$

$$I_y = 1.92 \times 10^{-40} \text{ g-cm}^2 , \qquad (3.94)$$

$$I_z = 2.95 \times 10^{-40} \text{ g-cm}^2 .$$

Since no two of these moments are equal, the H_2O molecule is an asymmetric rotator. Because of this asymmetry, the rotation spectrum of H_2O is very complex, having many closely spaced energies. The smallest transition energies are about 10^{-4} eV.

The structure and normal vibrations of D_2O are similar to those of H_2O. In D_2O the bond length ℓ and the bond angle γ are 1.01 Å and 106°, respectively. Because the deuteron has a greater mass than the proton, the internal vibration frequencies of D_2O are less than for H_2O (see Table 3.2). The principal moments of inertia of D_2O are greater than those of H_2O, and the free D_2O molecule also has a complicated rotation spectrum of many closely spaced levels.

Table 3.2

FUNDAMENTAL VIBRATIONS, $\hbar\omega$, OF H_2O AND D_2O MOLECULES*

(In electron volts)

Type		H_2O Vapor	D_2O Vapor
Symmetric stretching	ω_1 	0.474	0.342
Bending	ω_2 	0.205	0.150
Asymmetric stretching	ω_3 	0.488	0.356

*From G. Herzberg, Infrared and Raman Spectra, Ref. 39.

NOTE: The frequencies in this table are for vibrations of infinitesimal amplitude. Anharmonic effects shift the fundamental frequencies to lower values. Intermolecular effects in the liquid and solid phases also reduce slightly the fundamental frequencies.

Although H_2O and D_2O are dynamically similar, they scatter neutrons very differently. Table 3.1 shows that the scattering from H_2O comes predominantly from incoherent scattering by hydrogen, with a small but generally significant direct contribution from the oxygen. For D_2O, on the other hand, the deuterons and the oxygen all give large contributions to the cross section. Further, since the oxygen and the deuteron have coherent scattering lengths that are greater than the incoherent scattering length of the deuteron, there are significant interference effects in the scattering for neutrons having a wavelength greater than or comparable to the length of the OD bond. These effects have been calculated in detail by Butler;[40] we consider some of his results later in this section.

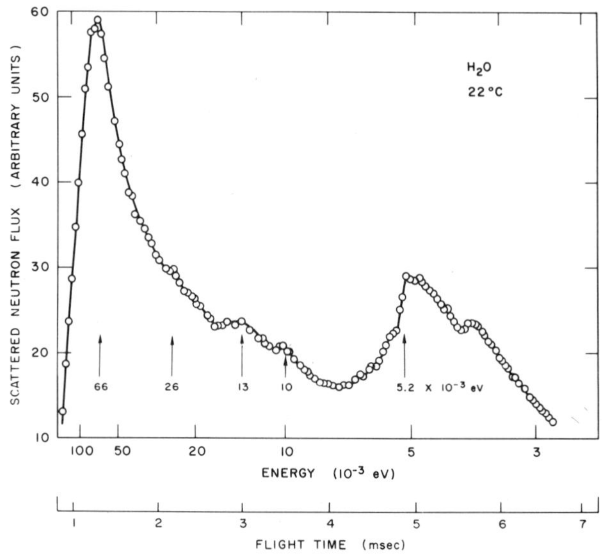

Figure 3.12 The time-of-flight spectrum of neutrons scattered by a 0.12-mm water sample. The arrow at 5.2×10^{-3} eV marks the edge of the elastic peak and the others show the inelastic peaks for energy gains. The energy scale and the values at the peaks refer to the scattered neutrons; the energy gains are less by 5.0×10^{-3} eV. The structure near 4×10^{-3} eV results from passage of the thermal-neutron beam through aluminum in the reactor structure. (From Hughes et al., Phys. Rev., Ref. 42.)

In the liquid phase the molecular dynamics of water is not as well understood as in the vapor phase. For a comprehensive review of the prevailing theoretical ideas relating to the intermolecular bonds in liquid and solid H_2O, we refer the reader to an article by Springer.[41] Except for the internal molecular vibrations, most of our knowledge of the dynamics of the H_2O molecule comes from experiments with slow neutrons. Figure 3.12 shows the spectrum of neutrons after scattering of an incident beam of cold neutrons selected by filtration through a refrigerated block of polycrystalline beryllium.[42,43] The largest energy in the incident beam is 0.005 eV, corresponding to the Bragg cutoff in beryllium. A prominent feature of Fig. 3.12 is the peak at 0.066 eV, corresponding to an energy change of 0.061 eV. The motions responsible for this peak are the same ones that produce the band observed at about 0.07 eV in the Raman spectrum of H_2O.[44] Figure 3.13 shows that peaks similar to the one at 0.066 eV in room temperature H_2O also appear in ice at -30°C and in D_2O at 22°C, but at different transition energies.

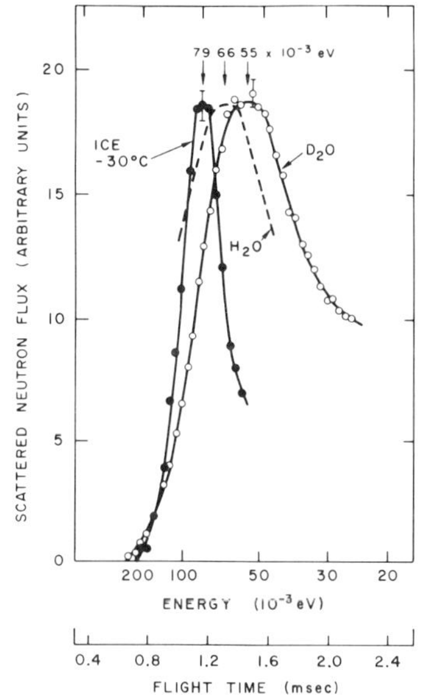

Figure 3.13 The high-energy transition measured for H_2O, ice at -30°C, and for D_2O at 22°C; the position of the peak for H_2O at 22°C is shown by the dashed line. The transition energies are less than the quoted energies by 5.0×10^{-3} eV. (From Hughes et al., Phys. Rev., Ref. 42.)

The appearance of the peaks at transition energies of 0.061 eV in H_2O and at 0.050 eV in D_2O is attributed to the hindered rotational oscillation of a molecule in the field of its neighbors. The interaction between the molecules in the liquid arises from the coupling between the molecular dipole moment and the local electric field produced by its neighbors. Yip and Osborn[45] have given a theory of neutron scattering by linear and symmetric molecules in an electric field with random orientation, but the theory has not been extended to include the case of scattering by asymmetric molecules. Considering the interpretation that the peak arises from a hindered rotational oscillation, the shift of the peak in Fig. 3.13 to energies higher and lower than 0.061 eV in ice and in D_2O, respectively, is an expected result.

Figure 3.12 shows that there are also three distinct peaks corresponding to energy gains of 0.021, 0.008, and 0.005 eV. The first two of these appear as barely resolved peaks in the Raman spectrum of H_2O.[46] Magat[46] attributes the 0.008-eV transition, like the 0.061-eV one, to a hindered rotation of a molecule in the field of its neighbors, and the 0.021-eV transition to the vibration of an entire molecule about its average position.

The curves in Fig. 3.12 reveal two barely discernible peaks, one corresponding to an energy gain and the other to an energy loss of about 5×10^{-4} eV. The occurrence of these peaks and the general shape of the scattered flux near 5×10^{-3} eV is not consistent with the diffusion-broadened distribution that is expected.[42] Hughes et al.[42] associate the two peaks in the nearly elastic distribution with very-low-energy quantum transitions in liquid water. Larsson et al.,[9] using an experimental technique similar to that employed by Hughes et al., and Stiller and Danner,[47] using a different experimental technique but still one that involves a beryllium filter, also see evidence for a discrete transition energy of about 5×10^{-4} eV.

Other experimentalists have investigated the quasi-elastic scattering from H_2O.[48-52] In particular, in the measurements of Brockhouse[48,49] and of Cribier and Jacrot,[51] which employ a line spectrum of incident neutrons selected by crystal reflection, the energy resolution is about the same as in the experiments of Hughes et al. The measured spectrum, however, does not exhibit the very-low-energy transitions observed by the authors mentioned in the previous paragraph. Instead, the spectrum can be explained in terms of a classical conception of the motions in H_2O.

To our knowledge, the discrepancies between the different experiments have not been explained. However interesting the fundamental problem posed by these discrepancies may be, they have no relevance for the thermalization problem in reactors, and henceforth we treat the motions associated with the low-energy transfer scattering in H_2O from a completely phenomenological point of view. (Before abandoning the discussion of very-low-energy quantum transitions entirely, however, we point out that they are relevant for thermalization at very low temperatures. The principal application is to refrigerated materials that are used in conjunction with research reactors or accelerators to produce beams of very cold neutrons.)

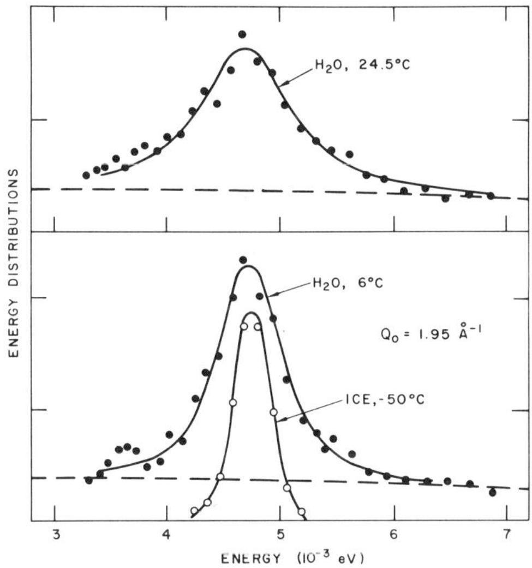

Figure 3.14 Energy distributions from H_2O at 6° and 24.5°C and from ice at -50°C. (From Brockhouse et al., in Inelastic Scattering of Neutrons in Solids and Liquids, Ref. 49.)

Figure 3.14 shows distributions measured by Brockhouse for water at 24.5°C and 6°C, and for ice at -50°C. The distribution for H_2O at 24.5°C is broader than that for H_2O at 6°C, which, in turn, is broader than that for ice. The pattern for ice corresponds to strictly elastic scattering broadened by the instrumental resolution. It is thus the resolution function for the other patterns. The dotted lines represent the calculated inelastic scattering corresponding to a mass 18m gas, its amplitude being adjusted by fitting on the wings of the measured distribution. Brockhouse obtained results similar to those shown for several temperatures and for a few values of the angle of scattering.

The measurements were not compatible with the models of diffusion by large jumps (activation diffusion) or the models of simple diffusion, which are discussed in Sec. 2-9.

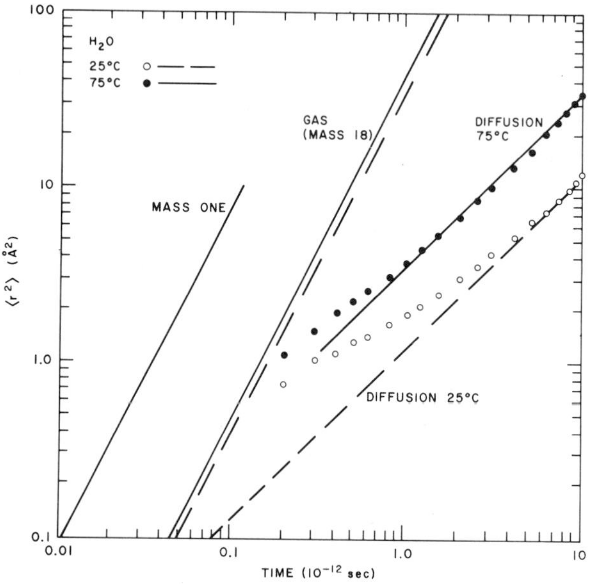

Figure 3.15 Mean square displacement of protons in water at 25° and 75°C as a function of time. (From Sakamoto et al., J. Phys. Soc. (Japan), Ref. 50.)

The conclusions are confirmed by more recent measurements of Sakamoto et al.[50] From their experimental data, these authors calculate the mean-square displacement $\langle r^2(t) \rangle$ of the protons as a function of time. The results are shown in Fig. 3.15, which also shows calculated lines for a gas of protons at 25°C, for a monatomic gas with mass $M = 18m$ at two temperatures, and for atoms which undergo simple diffusion, the coefficients of self-diffusion being equal to those measured at the two temperatures by means of nuclear magnetic resonance techniques.[53] For t greater than a few times 10^{-12} sec, the simple diffusion model gives the observed behavior. For both temperatures simple diffusion becomes the dominating mechanism only after the atoms have experienced a mean-square displacement of about 5 Å^2, corresponding roughly to the square of the mean distance between molecules.

The results represented in Fig. 3.15 show that the motions in H_2O are not like those in a gas with mass ratio A = 18. This does not imply, however, that for neutron energies much greater than k_BT it is not a good approximation to treat the center-of-mass motion of the molecules as the rectilinear motion of a free particle with a mass equal to 18 proton masses.

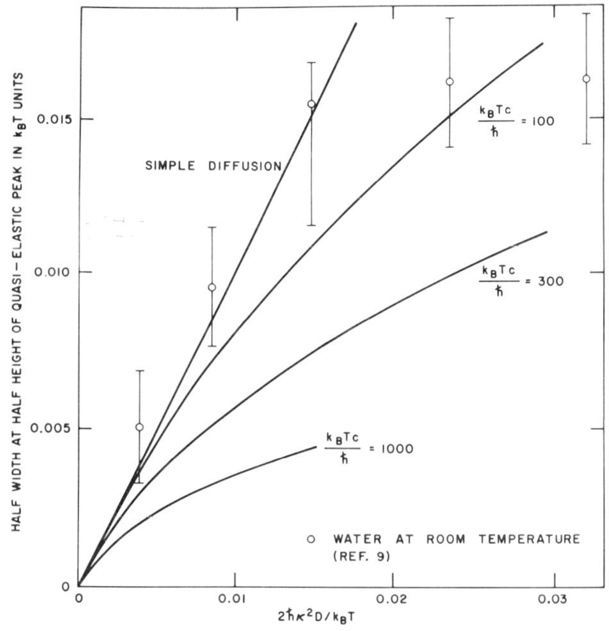

Figure 3.16 Half-width at half height of diffusion curve [Eq. (3.87)] as a function of the dimensionless variable $2\hbar\kappa^2 D/k_BT$. The departure of the curves from the simple diffusion curve is due to the sensitivity of the neutron scattering to "short-time" motions, and the comparison with experiment shows that these data are sensitive to such motions. ($2MD/\hbar$ = 0.04 for water.) (From Egelstaff and Schofield, Nucl. Sci. Eng., Ref. 23.)

Within the framework of Eq. (3.84) one can obtain a reasonable fit to the measured width of the quasi-elastic peak in H_2O as a function of κ^2. Figure 3.16 compares the half-width obtained from Eq. (3.87) with the measurements of Larsson et al. (see Ref. 23). This comparison shows that the scattering is sensitive to motions that have characteristic times comparable to or less than the characteristic time c (see Eq. (3.84) required for atoms to diffuse away from their initial positions.

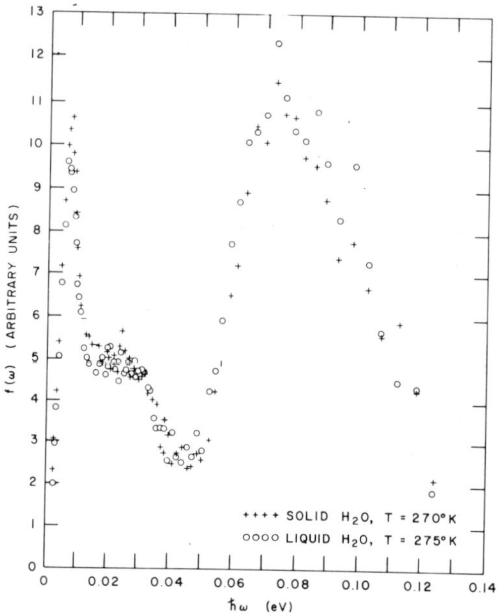

Figure 3.17 Frequency spectra derived from neutron scattering by H_2O at 2°C and ice at -3°C. The angle of observation was 30°. (From Larsson and Dahlborg, in Inelastic Scattering of Neutrons in Solids and Liquids, Ref. 32.)

The stochastic model of Rahman, Singwi, and Sjölander,[54] has been successful to some extent in explaining the scattering of cold neutrons by water. This model is physically more transparent but mathematically more complex than the one corresponding to Eq. (3.84). The stochastic model is based on the assumption that rapidly varying motions behave similarly to those in a solid, whereas slowly varying motions behave according to Langevin's equation for diffusion. For a more detailed discussion of this model and its comparison with experiment the reader is referred to Ref. 54.

In recent years much of the experimental work on H_2O and D_2O has been directed toward extracting from the measured scattering law the part of f(ω) (with ω > 0) that is associated with the hindered rotational and translational motions of these molecules.[8,32,33,35,55,56] An initial guess for f(ω) is obtained on the basis of the extrapolation procedure discussed in Sec. 3-3. The accuracy of the guess is checked by using it to calculate the neutron scattering. Then, if required, the initial guess is improved until the resulting f(ω) gives a satisfactory correlation between the observed $\bar{S}(\eta,\nu)$ and the calculated one.

In Fig. 3.17, Larsson and Dahlborg[32] compare their measured f(ω) ($\hbar\omega < 0.14$ eV) for ice at 270°K with that for water at 275°K. Their results indicate that near the melting point the motions in water resemble those in ice. Although this provides further justification for the use of a phonon expansion for the nondiffusive motions in H_2O, it does not necessarily imply the existence of excitations in liquid H_2O with well-defined energies and wave numbers.

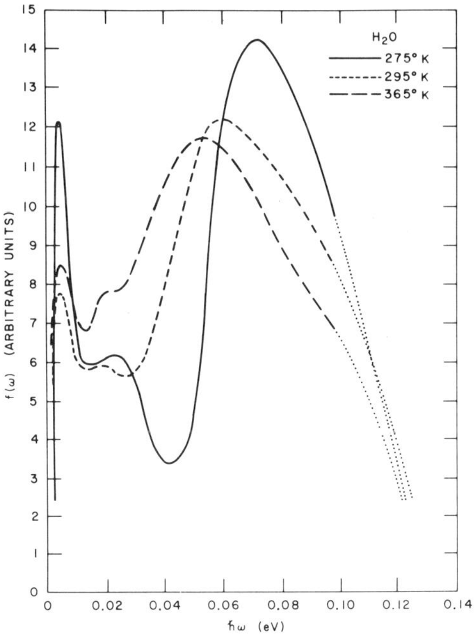

Figure 3.18 The temperature variation of f(ω) for H_2O. (From Larsson and Dahlborg, in Inelastic Scattering of Neutrons in Solids and Liquids, Ref. 32.)

As the temperature is raised, the high-frequency peak in Fig. 3.17 shifts toward lower frequencies. This is shown in Fog. 3.18, which gives only smooth-curve representations of the experimental data and indicates that the rotations are more nearly free at higher temperatures. The low-frequency peak becomes less well defined at higher temperatures, but its position does not undergo an observable shift. Egelstaff et al.[35] and Haywood and Thorson[56] obtained results that are similar to those in Fig. 3.18 but that show some differences in detail. In all of these measurements there is a strong temperature dependence of f(ω) in H_2O, thus demonstrating the absence of well-defined excitations in the sense that they are present in ice.

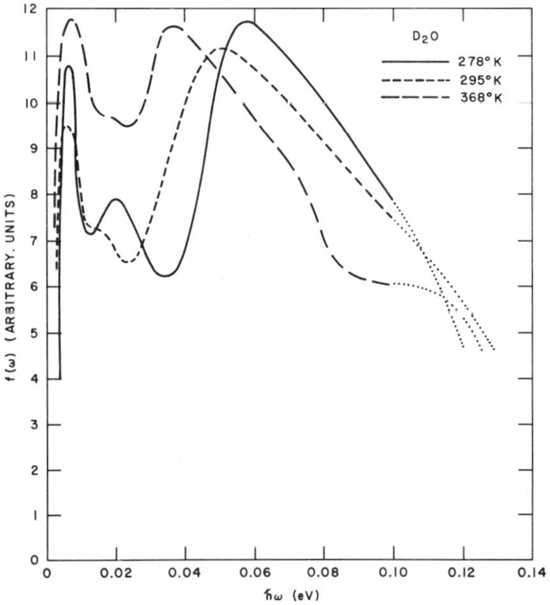

Figure 3.19 The temperature variation of f(ω) for D_2O. (From Larsson and Dahlborg, in Inelastic Scattering of Neutrons in Solids and Liquids, Ref. 32.)

For D_2O, Fig. 3.19 shows that the behavior of f(ω) for $\hbar\omega \lesssim 0.14$ eV is similar to its behavior in H_2O. The analysis of the D_2O data is complicated by the presence of interference effects and by a scattering from oxygen that is comparable to that from deuterium. Butler's calculations[40] show that the interference effect is negligible for $\nu >$ 1.4 at room temperature. Presumably, the analysis of the scattering law data in Ref. 32 is based on the assumption that the $f(\omega)$ associated with the oxygen atom in D_2O is concentrated in a small region around $\omega = 0$.

For thermalization calculations it is important to know the area under the measured part of $f(\omega)$, with $\hbar\omega \leq 0.14$ eV. So far, neutron scattering measurements have not provided an accurate determination of this area. We return to this problem after considering a simple dynamical model that has proved to be adequate for most thermalization studies in H_2O.

Brown and St. John[57] were the first to consider the effects of molecular binding on neutron thermalization by H_2O. In their calculations the energy distribution of scattered neutrons corresponds to that characteristic of a rigid, freely rotating H_2O molecule, but the neutron-proton scattering amplitude is assumed to depend on the relative velocity in such a way as to reproduce the total scattering cross section. This is a completely unphysical assumption inasmuch as the nuclear scattering amplitude is velocity independent at low energies. Further, the model does not account for the internal molecular vibrations or the hindrance of the rotations produced by molecular interactions in the liquid state.

Nelkin[58] has introduced a model for H_2O which takes into account, at least in a rough approximation, all of the principal dynamical features of the molecule in the liquid state. He assumes that the internal vibrations and the center-of-mass motion of the molecule in the liquid state are the same as in the vapor state. He approximates the hindrance of rotations by replacing the broad peak around 0.08 eV in Fig. 3.17 with a single frequency.

Stated concisely, in terms of Eq. (3.4) the model is

$$f(\omega) = \frac{M}{M_1}\,\delta(\omega) + \frac{M}{M_2}\,\delta(\omega - \omega_2) + \frac{M}{M_3}\,\delta(\omega - \omega_3) + \\ + \frac{M}{M_4}\,\delta(\omega - \omega_4) \tag{3.95}$$

with

$$\frac{M_1}{M} = 18 ,$$
$$\frac{M_2}{M} = 2.32 ,$$
$$\frac{M_3}{M} = 5.84 , \qquad (3.96)$$
$$\frac{M_4}{M} = 2.92 ,$$

and

$$\hbar\omega_2 = 0.06 \text{ eV},$$
$$\hbar\omega_3 = 0.205 \text{ eV} , \qquad (3.97)$$
$$\hbar\omega_4 = 0.481 \text{ eV} .$$

In Eq. (3.95) the symmetric and asymmetric stretching frequencies, which are very close to one another, have been combined into a single frequency, ω_4. The coefficients M/M_j are determined from simple dynamical arguments. Since it is assumed that the molecule translates freely, we must have $M/M_1 = 1/18$, that is, M_1 is the mass of the water molecule. The value $M_2/M = 2.32$, which is somewhat larger than the value that would follow from Eq. (2.239), gives the best fit of the cross sections calculated from the mass tensor approximation, Eq. (2.238), to the measured free-molecule cross section.[59] We cannot, however, explain why this value of M_2/M should apply in the liquid, where the molecular rotations are strongly hindered. To calculate the vibrational masses M_3 and M_4 we note from Eq. (2.221) that the contribution of any mode to the incoherent, vibrational, intermediate scattering function $\chi_{ss}(\kappa,t)$ depends on the square of the normal-mode amplitude for that mode. The square of this amplitude must be proportional to the inverse of the effective mass for that mode. In the approximation that the oxygen atom is infinitely heavy, the proton amplitude vectors in the symmetric stretching, the asymmetric stretching, and bending vibrations all have the same magnitude. Thus, remembering that the two stretching modes have been combined to give $f_4(\omega) = \delta(\omega - \omega_3)$, we must have $M_4^{-1} = 2M_3^{-1}$. Then, using Eq. (3.7) with

$K = 4$ and the previously given values for M_1 and M_2, we obtain the values for M/M_3 and M/M_4 given in Eq. (3.96).

By adopting the effective masses in Eq. (3.96), we assume that the fundamental internal molecular vibrations in the liquid are not very different from those in the vapor. If this is correct, then we can assign an effective mass ratio $1 - (1/5.84) - (1/2.92) = 0.488$ to that part of $f(\omega)$ with $\omega < 0.205$ eV. Thus, we may suppose that 0.488 is the area to which the $f(\omega)$ shown in Fig. 3.17 should be normalized.

The same considerations that lead to Eq. (3.95) for H_2O also apply for D_2O. Butler[40] and Honeck[60] have applied the corresponding model for D_2O to the neutron scattering problem. In Honeck's calculation

$$\begin{aligned} \hbar\omega_2 &= 0.05 \text{ eV} , \\ \hbar\omega_3 &= 0.15 \text{ eV} , \\ \hbar\omega_4 &= 0.35 \text{ eV} , \\ \frac{M_1}{M} &= \frac{20}{2} \\ \frac{M_2}{M} &= \frac{4.11}{2} \\ \frac{M_3}{M} &= \frac{29.04}{2} \\ \frac{M_4}{M} &= \frac{14.52}{2} . \end{aligned} \tag{3.98}$$

Here, $M_1?M = 20/2$ is the ratio of the mass of the D_2O molecule to the mass M of the deuteron, and M_3 is obtained from Eq. (2.239). Later, we shall consider Butler's results. Butler uses essentially the same dynamical model as Honeck but takes into account interference effects in the scattering.

The model corresponding to Eq. (3.95) has been extended to include two important effects. The extension by Egelstaff <u>et al</u>.[35] is to include the rotational band as it is measured in neutron scattering experiments, rather than having this band replaced by a single frequency. In another calculation, Koppel and Young[61] consider the non-Gaussian effects resulting from the anisotropy of the molecular

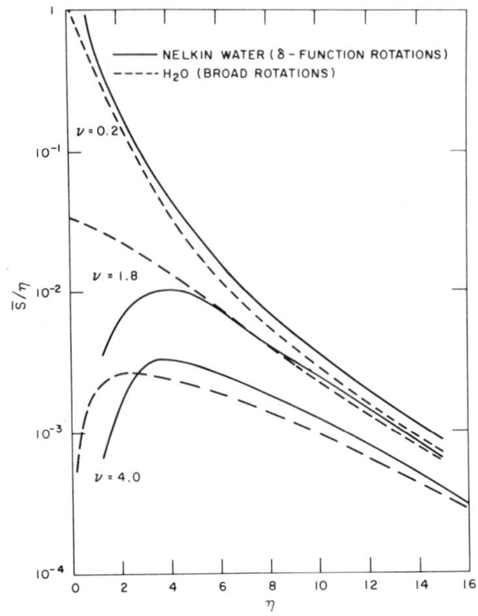

Figure 3.20 Comparison of scattering laws for different-shaped rotational components. (From Egelstaff et al., in Inelastic Scattering of Neutrons in Solids and Liquids, Ref. 35.)

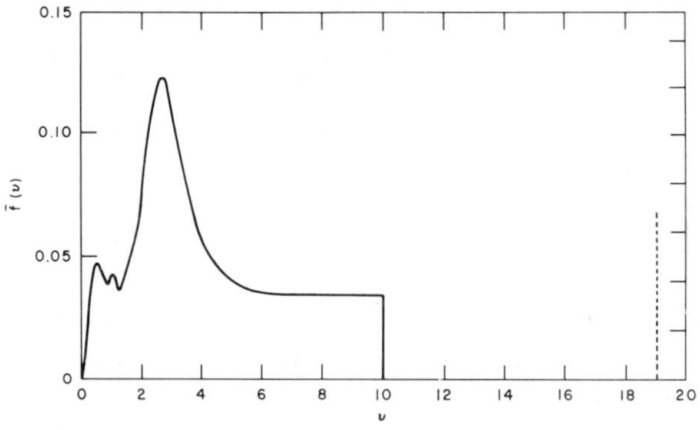

Figure 3.21 The frequency distribution for room-temperature water showing the position of the mean frequency corresponding to stretching of the OH bond; ν is the frequency in units of $k_BT/\hbar$. (From Poole et al., in Exponential and Critical Experiments, Ref. 62)

vibrations, but not the effects associated with the frequency spread of the rotational band. As shown in Fig. 3.20, Egelstaff et al. compare calculations of the scattering law based on an $f(\omega)$ like the one shown in Fig. 3.17 with calculations based on Eq. (3.95). The calculations show that the detailed shape of $f(\omega)$ is significant for $\bar{S}(\eta,\nu)$, particularly for $\eta \lesssim 4.0$.

Figure 3.21 shows an $\bar{f}(\nu)$, determined partially from scattering-law data, that has been used in neutron thermalization calculations in England.[62] Here, all motions except the high-frequency vibration at 0.48 eV are represented by a distribution of frequencies that is continuous for $\nu < 10$. At 0.48 eV, $\bar{f}(\nu)$ is a δ-function. The two effective masses associated with the continuous and discrete parts of $f(\omega)$ are such that $k_B\bar{T} = 0.155$ eV.

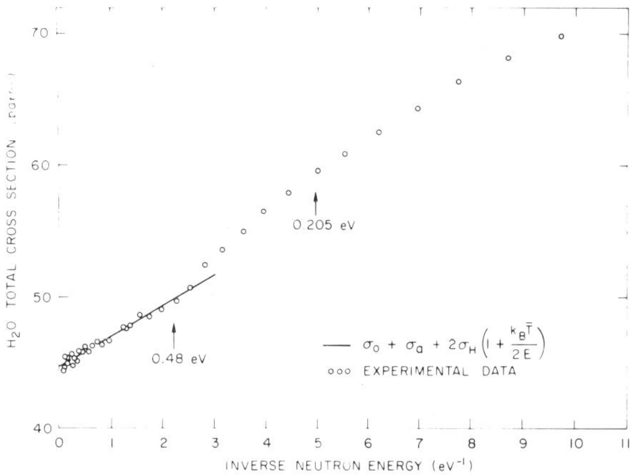

Figure 3.22 Variation of H_2O total cross section with inverse neutron energy. (From Beyster et al., USAEC Report GA-4659, Ref. 63.)

The best available evidence indicates that the effective mass associated with the 0.48-eV vibration in Fig. 3.21 is much too small. This follows from the total-cross-section curve shown in Fig. 3.22. For energies greater than about 0.5 eV we can fit Eq. (2.256) to the measured cross section. Taking into account the scattering by oxygen and the absorption by hydrogen this fit gives $k_B\bar{T} = 0.113$ eV. We compare this with the value 0.116 eV inferred from Eqs. (3.95), (3.96),

(3.97), and (3.58), and the value 0.155 eV used in England. In view of the relative ease of making total-cross-section measurements as compared with scattering-law measurements, the evidence is strong that $k_B\bar{T}$ is rather close to 0.113 eV.

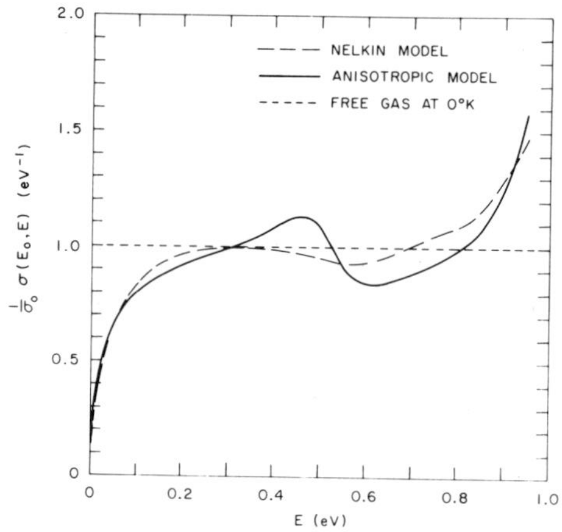

Figure 3.23 Effect of molecular anisotropy on the scattering kernel $\sigma(E_0,E)$ for $E_0 = 1.0$ eV. (From Koppel and Young, Nucl. Sci. Eng., Ref. 61.)

In their calculations, Koppel and Young[61] explore the sensitivity of the scattering kernel to the anisotropy of molecular vibrations. They state that the errors produced in $\sigma(E_0,E)$ by neglecting the anisotropy of the molecule is largest for energies above the 0.48-eV threshold for excitation of the stretching vibrations. Figure 3.23 shows the effect of anisotropies in $\sigma(E_0,E)$ for an initial energy of 1.0 eV and final energies less than 1.0 eV. The effects of the anisotropy on quantities of more direct interest in neutron thermalization are discussed in Chap. 6.

We now consider in more detail the experimental and theoretical results relating to neutron scattering by heavy water. Butler's calculations[40] for D_2O are based on a dynamical model that is analogous to the one used by Nelkin[58] for H_2O. Figure 3.24 shows a comparison of calculations with the measurements of Egelstaff et al.[8] and of Haywood and Thorson.[56] The figure shows the total $\bar{S}$, as well as the separate contributions from the direct scattering by deuterium and by oxygen and from the interference scattering by the two deuterium atoms. For the value of ν considered in Fig. 3.24, the interference term arising from the oxygen atom and either deuterium atom is negligible. For $\nu \gtrsim 1.4$, the only significant contribution to $\bar{S}(\eta,\nu)$ comes from the direct scattering by the deuterium and oxygen atoms.

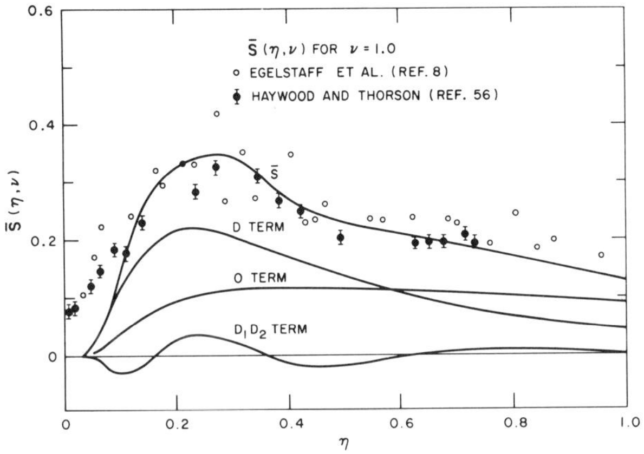

Figure 3.24 Theoretical and measured scattering laws for heavy water. (From Butler, Proc. Phys. Soc. (London), Ref. 40; the curves have been modified by Butler.)

The agreement between theory and experiment is good except for small values of . For small values of the hindrance of the molecular translations is important; to improve the theory it is necessary to account for molecular diffusion. A significant conclusion from the agreement for $\eta \gtrsim 0.2$ and $\eta \gtrsim 0.5$ is that a discrete hindered rotational level in place of the actual broad band of hindered rotational frequencies allows an accurate calculation of the scattering law.

Except for energies below about 0.01 eV, the interference scattering in D_2O does not significantly affect the total scattering cross section. This is verified in Fig. 3.25, in which calculations performed in the incoherent approximation are compared by Honeck[60] with measured values of the total scattering cross section in D_2O. The same conclusion is reached by Butler,[40] who takes account of the interference effects arising not only from atoms in the same molecule but also from atoms in different molecules. This conclusion does not apply, however, for the differential cross section $d\sigma/d\Omega$, defined by Eq. (2.72a); thus, interference scattering in D_2O may well be important for problems that involve the migration of neutrons from one point in space to another.

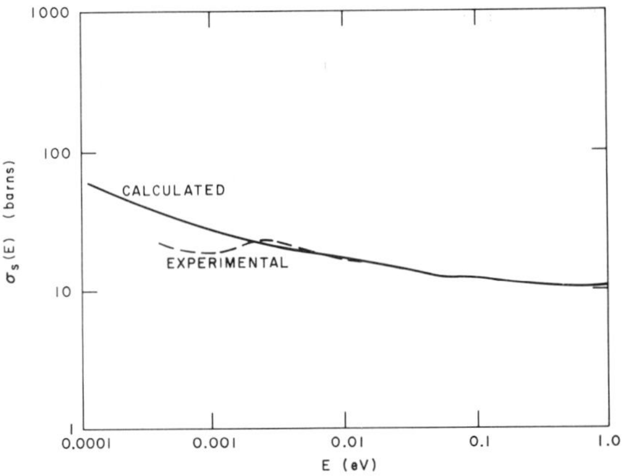

Figure 3.25 The measured and calculated total cross section of heavy water. In the calculation interference effects are neglected. (From Honeck, Trans. Am. Nucl. Soc., Ref. 60.)

3-4.2 GRAPHITE. The atoms of graphite crystallize in sheets forming a highly anisotropic structure. Within a sheet, the atoms form an array of contiguous regular hexagons, each of which is similar to the benzene ring (the carbon-carbon spacing, a, between nearest neighbors in a sheet is 1.42 Å, somewhat greater than the corresponding value of 1.39 Å in benzene). The atoms in a typical ideal sheet are located at corners of the hexagons as shown by the solid lines in Fig. 3.26. The atoms in an adjacent sheet are at the corners of the hexagons formed by the dotted lines. The distance between adjacent planes is $c/2 = 3.35$ Å. In the next-nearest sheets, a distance c from the plane of the paper, the planar coordinates of the crystal sites are the same as in Fig. 3.26.

Referring to Fig. 3.26, we may choose the basic vectors of the graphite lattice to be $\mathbf{a}_1$ and $\mathbf{a}_2$. As the third lattice vector, we choose a vector $\mathbf{a}_3$ having a magnitude c and pointing in the direction of the unit vector **k** perpendicular to the basal planes. The vectors $\mathbf{a}_1$ and $\mathbf{a}_2$ are coplanar; they form an angle of 120°, and each has a magnitude equal to $|\mathbf{a}_1| = \sqrt{3}a$. There are four atoms in each unit cell of the lattice; we choose these to be at the positions $\mathbf{r}_1$ and $\mathbf{r}_2$ in one basal plane and at $\mathbf{a}_1 + (c/2)\mathbf{k}$ and $\mathbf{r}_2 + (c/2)\mathbf{k}$ in an adjacent plane.

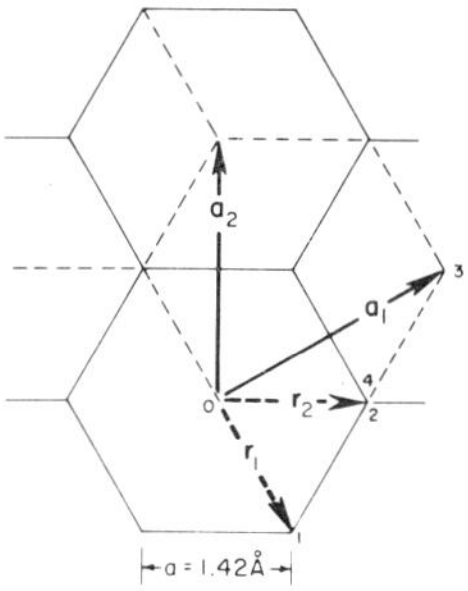

Figure 3.26. Schematic representation of the graphite lattice.

The structure of graphite suggests that the valence bond between carbon atoms acts almost completely within the separate basal planes, and thus the forces that hold the planes together are very weak compared with intraplanar forces. This idea is common to the many theoretical treatments of the lattice vibrations in graphite.[5,64-66]

The most detailed treatment is that of Yoshimori and Kitano.[5] Their theory is based on the assumption of four types of interactions between atoms. These interactions are associated with

(a) Changes in bond angles, proportional to a force constant μ;

(b) changes in the bond length between nearest neighbors in the same plane, proportional to a force constant γ;

(c) a central force resulting from changes in the bond length between nearest neighbors in two adjacent planes, proportional to a force constant γ'; and

(d) a force resulting from the bending of the planes and, with a force constant μ', proportional to the displacement perpendicular to the plane of each atom relative to the average of the normal displacements of the three neighboring atoms in the same plane.

The force constants γ and μ in graphite are assumed to be equal to the corresponding ones in benzene, namely, $\gamma = 6.71 \times 10^5$ dynes/cm and $\mu = 0.48 \times 10^5$ dynes/cm. From low-temperature specific heat data and from compressibility data for natural Ceylon graphite, Komatsu[64] determines that $\gamma' = 5.57 \times 10^3$ dynes/cm and $\mu' = 2.92 \times 10^5$ dynes/cm. The value for γ' agrees very well with that deduced from measurements of dispersion relations in pyrolitic graphite.[67] We discuss these measurements below.

There are other forces than those described above. In addition to the interaction between atoms that are nearest neighbors on adjacent planes, there are interactions between atoms and their distant neighbors in adjacent and more distant planes. These interactions give rise to additional compressional as well as shearing forces between different planes, which for $\hbar\omega/k_B \lesssim 10°K$ lead to the ω^2-dependence of the frequency distribution function. Komatsu[64] has shown, however, that the shearing forces become significant only for very low frequencies of lattice vibration, $\hbar\omega/k_B \approx 10°K$, whereas for thermalization studies at $T \gtrsim 300°K$ we are primarily interested in much higher frequencies. Yoshimori and Kitano[5] further point out that it is a good approximation to absorb the additional compressional forces into a single force of type c listed above.

We now examine some particular aspects of the problem of the normal modes of vibration of atoms in graphite. From the previous assumptions regarding the interatomic forces we see that the component of an atomic vibration that is parallel to the basal planes is uncorrelated with the motion of atoms in different planes and with the perpendicular component of the vibration of atoms in the same plane. Consequently, the polarization vectors of the normal vibrations are either parallel or perpendicular to the basal planes. Furthermore, since the parallel vibrations of atoms in a given plane are correlated only with the parallel vibrations of atoms in the same plane, the polarization vectors and frequencies associated with these vibrations depend only on $\mathbf{q}_\parallel$, the component parallel to the basal planes of the wave vector $\mathbf{q}$. Thus, in the approximation considered, the planar vibrations in graphite are completely equivalent to the planar vibrations of a two-dimensional crystal.

The one-dimensional vibration of an atom perpendicular to a basal plane is correlated with the perpendicular vibrations of atoms in the same and in neighboring planes. Consequently, the frequencies and polarizations of these vibrations depend on all three components of the propagation vector $\mathbf{q}$. If we were to neglect the forces that couple adjacent planes ($\gamma' = 0$), the normal-mode frequencies and polarizations would depend only on $\mathbf{q}_{\parallel}$ and we would be concerned with the transverse vibrations of a two-dimensional crystal. The effect of the interplanar coupling, however, is important at frequencies for which $\hbar\omega/k_B \cong 300°K$ and therefore is significant for the scattering of thermal neutrons by lattice vibrations.

The complete study of dispersion relations of graphite by means of slow-neutron scattering has been prevented by a lack of sufficiently large single-crystal specimens. Recently, however, pyrolytic graphite, a material that displays certain single-crystal characteristics, has been produced in large specimens. In this material the planes of hexagonally arranged carbon atoms have a common hexad axis, the [001] axis, but are otherwise randomly oriented. It is thus possible to measure the dispersion relations for modes in which the atoms vibrate perpendicular to the basal planes but not for modes in which the atoms vibrate parallel to these planes.

Dolling and Brockhouse[67] have measured the frequency-wave-number relation for the longitudinal acoustical vibrations of waves propagating in the [001] direction; Fig. 3.27 shows the results of these measurements. The solid curve represents the best fit to the data that can be obtained from the relation

$$4\pi^2 M\nu^2 = \gamma'\left(1 - \cos\frac{cq}{2}\right), \tag{3.99}$$

where M is the mass of the carbon atom, ν is the frequency in cycles per second, $c = 6.70$ Å, and γ' is the force constant introduced earlier. Equation (3.99) is just the expression that follows from the analysis of Yoshimori and Kitano for the case under consideration, and the value of γ' deduced from the neutron scattering data is in good agreement with the value derived by Komatsu from the observed compressibility.

In the longitudinal [001] mode of vibration, the individual sheets of carbon atoms remain undistorted while the interplanar distance is varying. The forces involved are thus interplanar and the excellent fit displayed in Fig. 3.27 indicates that these interplanar forces are almost entirely between adjacent planes.

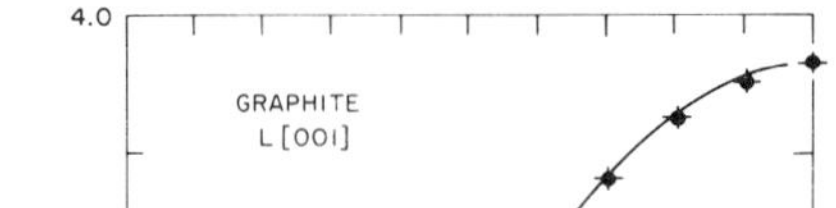

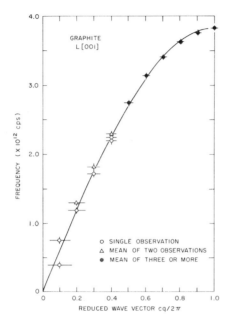

Figure 3.27 The dispersion relation for longitudinal phonons propagating in the [001] direction in graphite. (From Dolling and Brockhouse, Phys. Rev., Ref. 67.)

Yoshimori and Kitano use their model, together with the Born-von Karman approach to lattice dynamics, to calculate not only dispersion relations but also the frequency distribution functions $\rho_{\perp}(\omega)$ and $\rho_{\parallel}(\omega)$ corresponding to vibrational modes perpendicular and parallel, respectively, to the basal planes. A more accurate calculation of $\rho_{\perp}(\omega)$ and $\rho_{\parallel}(\omega)$ based on the same model and dispersion relations is the one of Young and Koppel,[68] which we show in Fig. 3.28. Figure 3.29 shows the total frequency distribution function

$$\rho(\omega) = \tfrac{1}{3}\rho_{\perp}(\omega) + \tfrac{2}{3}\rho_{\parallel}(\omega) . \tag{3.100}$$

In these calculations the value of the force constants, relative to those of Yoshimori and Kitano, are $\gamma = 0.65$, $\mu = 0.75$, $\gamma' = 1.053$, and $\mu' = 1.0$. This value of γ' is inferred from Fig. 3.27 and Eq. (3.99), and the values of γ, μ, and μ' lead to good agreement with the measured specific heat of reactor-grade graphite, at least for temperatures greater than about 200°K.

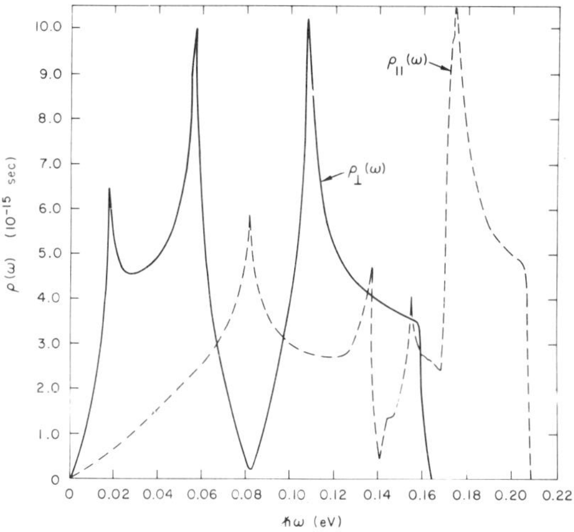

Figure 3.28 The phonon spectra for the perpendicular and parallel vibrations of the graphite lattice. (From Young and Koppel, J. Chem. Phys., Ref. 68.)

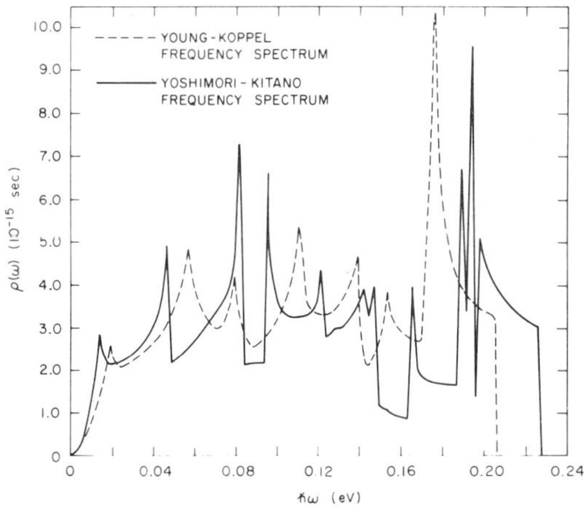

Figure 3.29 Frequency spectra of graphite. (From Young and Koppel, J. Chem. Phys., Ref. 68.)

Table 3.3

OBSERVED AND CALCULATED VALUES OF THE SPECIFIC HEAT, $C_V/3R$, OF REACTOR-TYPE GRAPHITE

(In calories/gram-atom/degree)

Temp. (°K)	Observed Value	Calculated Using Dashed Curve in Fig. 3.29	Calculated Using Solid Curve in Fig. 3.29
100	0.068613*	0.07304	0.0786
125	-----	0.1033	-----
150	0.13052*	0.1354	-----
175	-----	0.1688	-----
200	0.201*	0.2031	0.2103
225	-----	0.2379	-----
250	-----	0.2729	-----
275	-----	0.3079	-----
300	0.3439*	0.3426	0.3471
400	-----	0.4721	0.4701
500	0.582†	0.5800	0.5725
600	-----	0.6645	0.6541
700	-----	0.7292	0.7198
800	0.7817†	0.7787	0.7673
900	-----	0.8167	0.8060
1000	0.8455†	0.8462	0.8364

*From DeSorbo and Tyler, J. Chem. Phys., Ref. 70.

†From Spencer, Ind. Eng. Chem., Ref. 71.

The Yoshimori and Kitano values of γ and μ, it is recalled, lead to accurate predictions of the specific heat for natural Ceylon graphite below about 300°K. The thermalization model of Parks is based on the $\rho(\omega)$ calculated by Yoshimori and Kitano; this $\rho(\omega)$ is compared in Fig. 3.29 with that calculated by Young and Koppel.[6] In the Parks model, however, the maximum frequency is reduced from the value given by Yoshimori and Kitano to force the theoretical specific heat into agreement with values measured at temperatures greater than about 300°K[69] (see Table 3.3). Although the value $\gamma' = 1.053$ is probably not accurate for reactor-grade graphite, the lattice vibration spectrum for $\hbar\omega/k_B \gtrsim 100$°K and most quantities of interest in thermalization studies are not sensitive to small variations in γ'. The values for γ and μ cannot be determined uniquely from specific heat data, but the $\rho_\perp(\omega)$ and $\rho_\parallel(\omega)$ in Fig. 3.28, which are computed with $\gamma = 0.65$ and $\mu = 0.75$, show peaks whose positions on the frequency scale correspond with those measured in slow-neutron scattering experiments.[72] The value $\gamma = 0.65$ further leads to a high-frequency cutoff of the lattice vibration spectrum which is only five percent lower than the value measured by Egelstaff and Harris[73] by scattering cold neutrons from a hot sample (950°C) of graphite. Using the measured value of the cutoff frequency, we calculate that $\gamma \approx 0.71$ relative to the Yoshimori and Kitano value. In any case, the bond-stretching force constant γ in reactor-grade graphite is considerably lower than the value deduced from benzene vibrations.[74]

For many purposes it is convenient to have a simple form for $\rho(\omega)$. Egelstaff has suggested the form

$$\begin{aligned}
\rho(\omega) &= \left(\omega_1\omega_0^2\right)^{-1}\omega^2 && 0 < \omega < \omega_1\,, \\
&= \omega_0^{-2} && \omega_1 < \omega < \omega_2\,, \\
&= \omega_{0\;2}^{-2} && \omega_2 < \omega < \omega_3\,, \\
&= 0 && \omega > \omega_3\,.
\end{aligned} \tag{3.101}$$

For reactor-grade graphite, the best choice for the parameters is

$$\hbar\omega_0/k_B = 854°K ,$$
$$\hbar\omega_1/k_B = 9°K ,$$
$$\hbar\omega_2/k_B = 300°K , \qquad (3.102)$$
$$\hbar\omega_3/k_B = 2580°K .$$

The value of ω_1 is chosen to fit low-temperature specific heat data; however, in any case except at very small energy transfers the scattering law is not very sensitive to even fairly large variations of ω_1.[69] The values of ω_2 and ω_3 are fixed from scattering-law data and high-temperature ($T > 300°K$) specific heat data, respectively.

In Fig. 3.30 we show the measured[29] and calculated scattering laws for graphite at $T = 300°K$. The calculations are performed numerically, using the phonon expansion, Eq. (3.22), in which the anisotropy of the lattice vibrations and interference effects are neglected.[f] The legend above the figure identifies the $\rho(\omega)$ used to obtain the calculated scattering law; for the $\rho(\omega)$ given in Eq. (3.101), we use two different values of ω_3. Except for low values of η, the theory and experiment generally agree with each other to within about thirty percent. A free-

---

[f] The frequency functions $\rho(\omega)$ given in Fig. 3.29 are linear functions of ω for ω near zero. Thus, Eq. (3.8a) is not satisfied, and the phonon expansion does not exist. In reality, a small coupling between vibrations perpendicular and parallel to the basal planes of the lattice makes $\rho(\omega) \sim \omega^2$ for small ω, in which case the phonon expansion does exist. In practice, however, the convergence of the expansion is extremly slow even for neutrons with energies of the order of 0.01 eV.

Physically, the slow rate of convergence is associated with the emission by the neutron of many low-frequency phonons. This multiple phonon emission is manifested in the scattering as a large narrow peak of inelastically scattered neutron in the neighborhood of zero-energy transfer. We can overcome this slow rate of convergence by writing $\rho(\omega)$ as the sum of two parts,

$$\rho(\omega) = \frac{M}{M_1}\rho_1(\omega) + \frac{M}{M_2}\rho_2(\omega) .$$

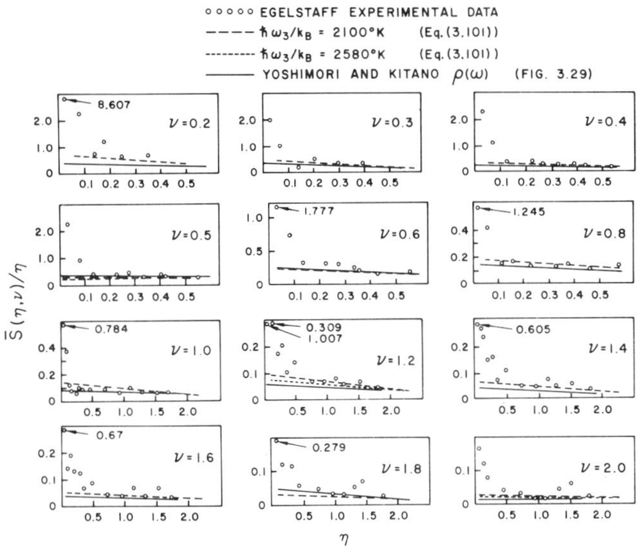

Figure 3.30 Comparison between measured and calculated scattering laws for graphite at T = 300°K. The calculated curves were obtained in the incoherent approximation. (From Wikner, Joanou, and Parks, Nucl. Sci. Eng., Ref. 69.)

carbon model or a Debye model, $\rho(\omega) = (3/\omega_D^3)\omega^2$ with a reasonably chosen ω_D, would not agree within an order of magnitude with the measured results. (Other calculations[69] show that the anisotropy of the lattice vibrations has about a ten percent effect on the theoretical results for $\eta \lesssim 2$.) Without a detailed calculation of the interference effects, however, we cannot conclude that either of the considered frequency distributions leads to a better fit of the experimental data than the other. In Fig. 3.30, the interference effects are characterized by the large values of $\bar{S}(\eta,\nu)/\eta$ in the neighborhood of $\eta = 0$, and it is not known for which, if any, of the measured values of η these effects are insignificant.

One part, $\rho_1(\omega)$, vanishes for all $\omega > \omega_c$ (say with $\hbar\omega_c/k_B \approx 60°K$) but is such that $[(M/M_1)\rho_1(\omega)]$ is very nearly equal to $\rho(\omega)$ for $\omega \leq \omega_c$. Then at room temperature the phonon expansion for $S_2(\kappa,\omega)$, corresponding to $\rho_2(\omega)$, converges rapidly except for neutron energies greater than about

The frequency distributions in Fig. 3.29 and Eq. (3.101) are essentially equivalent insofar as we are concerned with neutron spectra. This is demonstrated in Sec. 7-2.2, where we investigate the sensitivity of neutron spectra to changes in the frequency distribution. There are large differences, however, between the thermalization properties of a free gas of carbon atoms and carbon atoms bound in a graphite lattice.

So far all of our calculations of neutron scattering by graphite have been based on the incoherent approximation. As we discussed in Sec. 3-1, we expect this to be a good approximation for the thermalization properties of moderators as long as these properties depend only on the inelastic component of the scattering. The scattering, however, is completely coherent and the elastic part is dominated by interference effects. These effects have an important influence on the transport properties of the moderator, such as the total scattering cross section and the diffusion coefficient, defined in Sec. 4-6. The elastic scattering cross section is not very sensitive to the detailed lattice dynamics but depends mainly on the static structure of the crystal. We can calculate $\sigma_{el}(E_0)$ from Eq. (2.145). In Eq. (2.145), n = 4 (four atoms per unit cell), V_0 is the volume of a unit cell of the graphite lattice,

$$2W(\tau) = \frac{\hbar\tau_{\|}^2}{2M} \int \rho_{\|}(\omega) \cot h \frac{\hbar\omega}{k_B T} d\omega + + \frac{\tau_{\perp}^2}{2M} \int \rho_{\perp}(\omega) \cot h \frac{\hbar\omega}{2k_B T} d\omega , \tag{3.103}$$

0.5 eV. We can approximate the partial scattering function $S_1(\kappa,\omega)$, on the other hand, by means of the short-collision-time result, Eq. (3.61). It is $S_1(\kappa,\omega)$ that gives the small-energy-transfer peak corresponding to the multiple phonon emission of low-frequency phonons. The $S_1(\kappa,\omega)$ is appreciable only for energy transfers that are very small, and these contribute very little to the rate of neutron thermalization. In fact, we make very little error if we set $S_1(\kappa,\omega) = \delta(\omega)$, which corresponds to including this small-energy-transfer scattering in an elastic peak. In the decomposition of $\rho(\omega)$ given above, this treatment of small-energy-transfer scattering corresponds to setting $M_1 = \infty$ and $M_2 = M$.

and $\tau_{\perp}$ and $\tau_{\parallel}$ are the projections of the reciprocal lattice vector $\boldsymbol{\tau}$ perpendicular and parallel, respectively, to the basal planes. The sum in Eq. (2.145) extends over all reciprocal lattice vectors for which $\tau \leq 2k_0$. Finally, if

$$\boldsymbol{\tau} = 2\pi(\ell_1\mathbf{b}_1 + \ell_2\mathbf{b}_2 + \ell_3\mathbf{b}_3) , \tag{3.104}$$

where the $\mathbf{b}_1$, Eq. (2.133), are the reciprocal lattice basis vectors and the ℓ_i are positive or negative integers not all of which are zero, then

$$\left.\begin{aligned} |F(\boldsymbol{\tau})|^2 &= 4 \sin^2 \frac{\pi}{3}(\ell_1 + \ell_2) \qquad && \ell_3 \text{ odd} \\ &= 6 + 10 \cos^2 \frac{\pi}{3}(\ell_1 + \ell_2) \qquad && \ell_3 \text{ even} . \end{aligned}\right\} \tag{3.105}$$

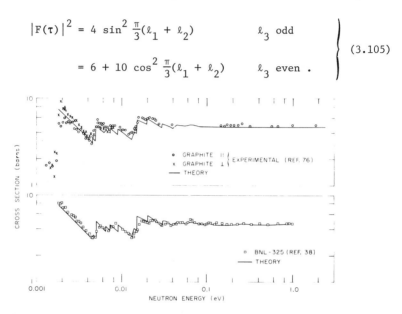

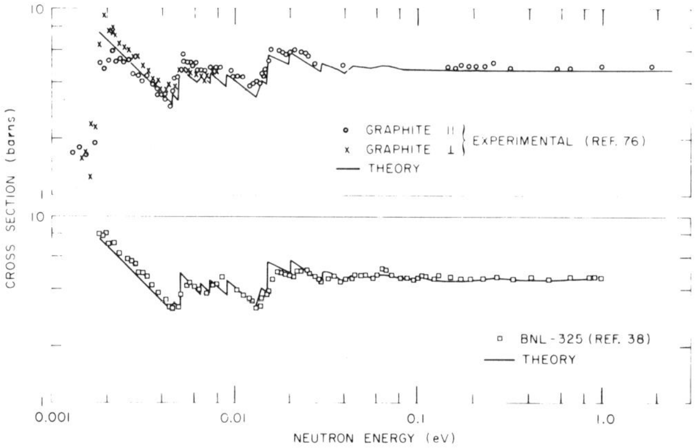

Figure 3.31. Total scattering cross section for graphite.

For the Yoshimori and Kitano model, Eq. (3.103) is an exact result if the force constant $\gamma' = 0$. For the more realistic case, where $\gamma' << \mu'$, it is an extremely good approximation, the inexactness arising from the slight difference in values of $W^s(\boldsymbol{\kappa})$ for the different atoms in a unit cell[75] (Eq. (2.194) is the exact expression from which we can calculate the elastic scattering). Figure 3.31 compares theoretical with measured values[38,76] of the total scattering cross section in polycrystalline graphite. The elastic part of the cross section is given by Eq. (2.145), while the inelastic part is calculated from the $\rho(\omega)$ of Fig.

3.29 (solid curve) using the incoherent approximation and the methods described in Sec. 3-2. Although the calculations are for randomly oriented crystals in a polycrystalline sample, the measurements are for a sample in which the microcrystals are preferentially oriented to a slight degree by the extrusion process used to produce the sample. Consequently, the upper part of Fig. 3.31 shows two sets of experimental data corresponding to the cases where the sample is oriented with the extrusion axis either parallel or perpendicular to the neutron beam. The lower part of the figure compares the theorectical cross section with measurements made by the Brookhaven group.[38] The agreement between theory and experiment is not as good as we expect it to be in the range $0.005 \text{ eV} \lesssim E \lesssim 0.02 \text{ eV}$. In this range, however, the difference between the two independently measured cross sections is greater than the difference between either measured result and the theoretical result.

3-4.3 ZIRCONIUM HYDRIDE. The ZrH_n lattice consists of face-centered-cubic units of zirconium atoms for $1.5 \lesssim n \lesssim 2.0$. (For a description of the phase relations in ZrH_n $(0 \leq n \leq 2.0)$ see Refs. 77 through 81.) Figure 3.32 shows such a unit for the stoichiometric composition ZrH_2. The zirconium atoms (light-colored) are located at the vertices of the large cube and at the centers of its faces. The hydrogen atoms (dark-colored) form the eight vertices of a second cube that is half the size of the zirconium cube and is centered inside the latter. Each hydrogen atom has four nearest zirconium neighbors at a distance $(\sqrt{3}/4)a$, where $a = 4.79$ Å is the length of the edge of the cubic unit cell shown in Fig. 3.32. These four zirconium atoms for a regular tetrahedron with the hydrogen atom at its center. The next nearest neighbors to any hydrogen atom are six other hydrogen atoms at a distance $\frac{1}{2}a$.

Actually, near $n = 2$ the ZrH_n lattice is not cubic but slightly tetragonal, i.e., the cubes are slightly deformed. As a result, the tetrahedrons enclosing the zirconium site are slightly irregular. For regular tetrahedral symmetry and in the approximation that hydrogen atoms interact only with fixed zirconium atoms that are nearest neighbors, the potential energy of interaction, neglecting cubic and higher-order terms in the displacement of the hydrogen atom from its equilibrium position, is a spherically symmetric function of that

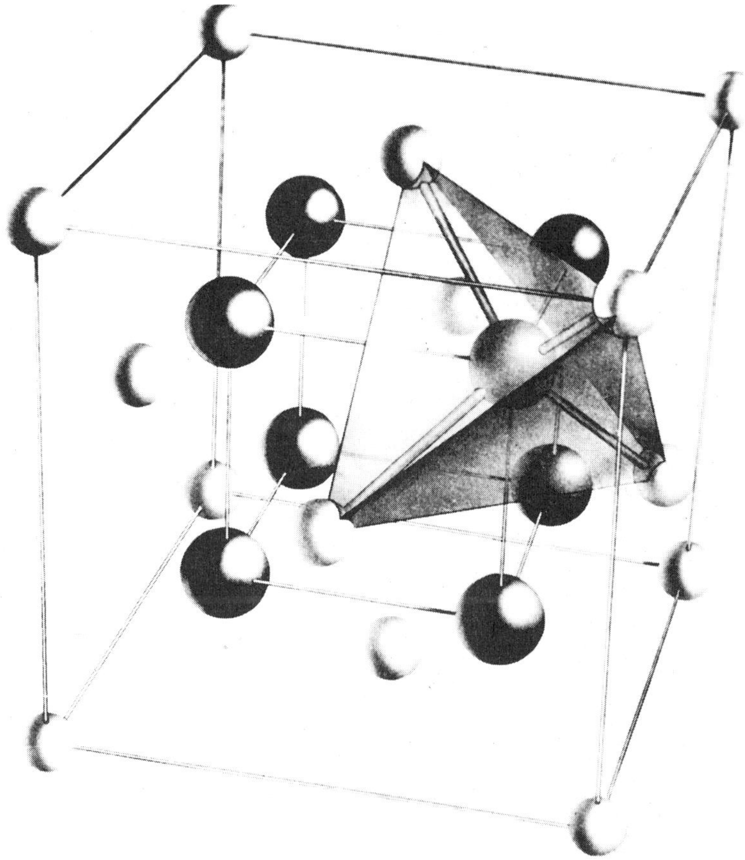

Figure 3.32 A unit cell of ZrH_2. The tetrahedral structure of four zirconium atoms (light colored) surrounding a hydrogen atom (dark colored) is emphasized for one of the hydrogen atoms. (From Technical Foundations of TRIGA, General Atomic Report GA-471, Ref. 77.)

displacement. Stated more concisely, the hydrogen atom vibrates harmonically with a single frequency ω_0 in the potential well formed by its nearest zirconium neighbors. Small departures from regular tetrahedral symmetry lead to a small splitting of this frequency. Further, a finite mass of the zirconium atoms results in a very small broadening $[\Delta\omega \sim (\omega_0/M)]$ of the frequency. Also, whenever the mass is finite there must be an acoustical branch of the vibration spectrum. In addition, the level is broadened by interaction of a hydrogen atom with other hydrogen atoms, although, in this case, not knowing the magnitude of the hydrogen-hydrogen interaction, we cannot estimate the amount of broadening.

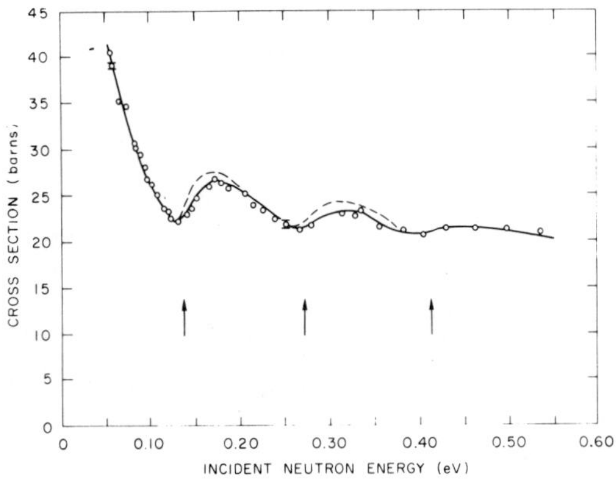

Figure 3.33 The solid curve gives the total neutron cross section of hydrogen in $ZrH_{1.5}$ obtained from the total cross section by subtracting the constant cross section of zirconium. The Fermi theoretical curve corrected for Doppler broadening has been fitted to the above data with $\hbar\omega_0 = 0.137$ eV and agrees with the experimental curve except as shown by the dashed curve. Arrows indicate excitation levels for hydrogen vibrations. (From McReynolds, et al., Proc. 2nd U.N. Conf., Ref. 82.)

Slow-neutron scattering experiments confirm that in rough approximation each hydrogen atom vibrates like a simple harmonic oscillator, not only in zirconium hydride but also in other heavy metal hydrides.[2,72,82,83] The measured total cross section of $ZrH_{1.5}$ is shown in Fig. 3.33. Figure 3.34 shows the observations of Andresen et al.[2] on the scattering of beryllium-filtered neutrons by $ZrH_{1.5}$, while Fig. 3.35 shows the results measured by Woods et al.[72] for $ZrH_{1.8}$ using the inverted beryllium filter method. In the former case, neutrons having an energy less than that corresponding to the Bragg cutoff (0.005 eV) are scattered and their energy distribution measured; in the latter case, an initially monoenergetic beam of neutrons is scattered and passed through a beryllium filter, and the intensity is measured as a

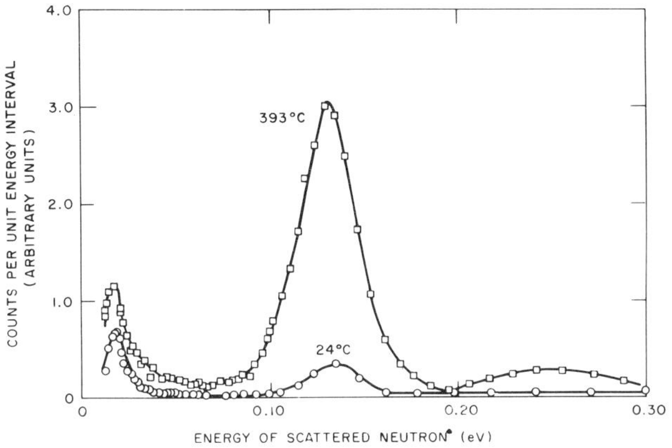

Figure 3.34 Energy distribution of neutrons of average incident energy 0.004 eV after scattering at 90° angle from $ZrH_{1.5}$. Debye spectrum is at left and optical or Einstein level is at right. In the 393°C curve the intensity from the first optical level is much higher than for the 24°C curve, and the peak arising from the second optical level becomes apparent. (From Andresen et al., Phys. Rev., Ref. 2.)

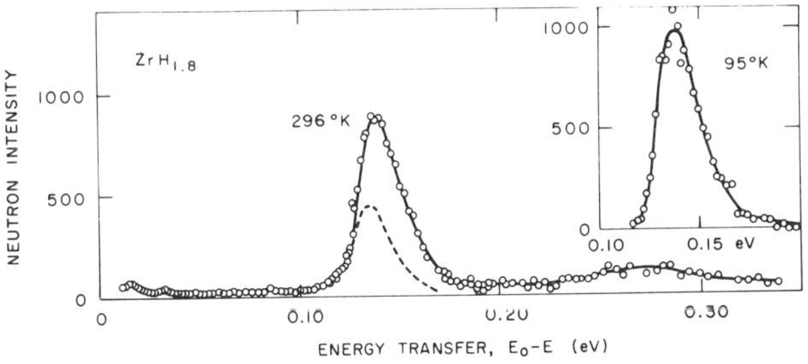

Figure 3.35 Energy distributions for zirconium hydride at 296°K. The dashed curve indicates the form of the distribution as it would be seen in an energy-gain experiment. The inset shows the 95°K distribution. (From Woods et al., in Inelastic Scattering of Neutrons in Solids and Liquids, Ref. 72.)

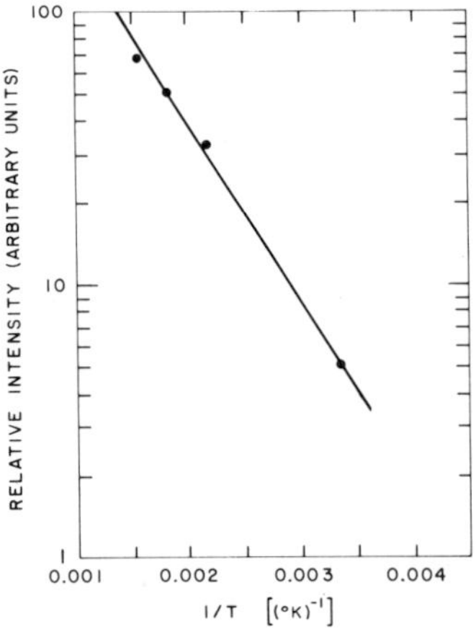

Figure 3.36 Temperature variation of integrated intensity in optical-level peak. Slope of curve gives excitation energy of 0.130 ± 0.005 eV. The size of the point includes statistical error. (From Andresen et al., Phys. Rev., Ref. 2.)

function of the initial energy. Because of the low value of the Bragg cutoff energy, the incident energy and the energy transfer in the inverted beryllium filter method are essentially the same for incident energies $\gtrsim$ 0.10 eV. Both experiments show evidence not only of "one-quantum excitation" but also of "two-quantum excitation" of the optical level. In more recent experiments, Whittemore[83] has observed "three-quantum excitation" of this level. Figure 3.36 shows the intensity of neutrons scattered from the optical level of $ZrH_{1.5}$ as a function of sample temperature. From the slope of this curve, Andresen et al.[2] conclude that the fundamental energy of oscillation is $\hbar\omega_0 = 0.13 \pm 0.005$ eV. That this is somewhat lower than the values $\hbar\omega = 0.136$ eV and $\hbar\omega = 0.140$ eV inferred from other experiments (Refs. 72 and 83, respectively) is partially accounted for by the fact that in an energy-gain experiment the Boltzmann factor changes considerably over the width of the peak.

Detailed studies of the shape of the optical level in ZrH_n and in other heavy metallic hydrides indicate that most of the breadth observed in the energy distribution in neutron scattering experiments reflects

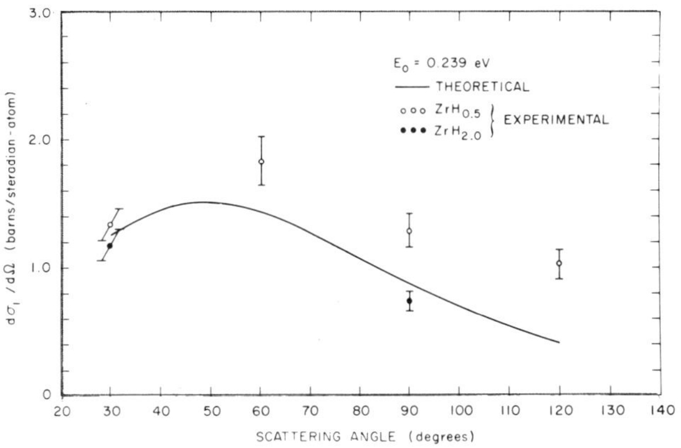

Figure 3.37 Theoretical partial differential scattering $d\sigma_1/d\Omega$ for the first excited level in ZrH, in barns per steradian per hydrogen atom. Experimental data for $ZrH_{0.5}$ and $ZrH_{2.0}$ are also presented. (From Whittemore, USAEC Report GA-5554, Ref. 85.)

intrinsic properties of these materials. In this context, for titanium hydride, Saunderson and Cocking[84] show that the line shape observed in neutron scattering cannot result from processes that involve the single excitation of an Einstein oscillator and multiple excitation of acoustical phonons. The same conclusion is reached by Woods et al.[72] for $ZrH_{1.8}$ after they observed that the width and shape of the neutron intensity versus incident energy (see Fig. 3.37) does not change very much when the temperature is changed from 95°K to 296°K. Whittemore[83] also reached this conclusion for ZrH_2 on observing only a small change in the width of the energy distribution around 0.14 eV when the scattering angle varies between 30° and 90°. The width resulting from the neutron interactions with acoustical modes completely excited by thermal agitation would be proportional to $(\kappa^2 T)^{1/2}$, contradicting the observations. Similarly, if the level width were associated predominantly with anharmonic effects or diffusion of hydrogen atoms away from their equilibrium sites, we would also expect the width to exhibit a strong temperature dependence, again contrary to observation. In first approximation, then, the hydrogen atoms in the first excited state in zirconium hydride vibrate as a superposition of harmonic oscillators having a distribution of frequencies with a width at half maximum of about 0.028 eV and a mean frequency of about 0.14 eV. Although we have still not determined which inherent property of zirconium hydride leads to the level width observed, we have already mentioned some of the possible broadening mechanisms.

For $ZrH_{0.5}$ and $ZrH_{2.0}$, Fig. 3.37 shows Whittemore's results for the angular dependence of the differential cross section $d\sigma_1/d\Omega$ corresponding to one optical phonon interacting with the neutron. For both $ZrH_{0.5}$ and ZrH_2, $\omega_0 = 0.14$ eV to within experimental error. Although the angular dependence of $d\sigma_1/d\Omega$ is considerably different for the two cases, we do not know whether this difference is associated with differences in the crystalline arrangement of zirconium atoms or with the occurrence of hydrogen-atom vacancies in the lattice.

In addition to the presence of the optical level, neutron scattering experiments show the effects associated with the acoustical vibrations in zirconium hydride. Figure 3.35, for example, gives evidence of an acoustical branch with $\omega_{max} \approx 0.02$ eV. Saunderson and Cocking[84] have made similar observations for titanium hydride and have estimated the values of M_1 and M_2 of Eq. (3.4) for K = 2, associated with the acoustical and optical parts, respectively, of $f(\omega)$. They find $M_2/M_1 = 126$, which is to be compared with 48, the titanium-hydrogen mass ratio. For zirconium hydride it is reasonable to assume that

$$\frac{M_2}{M_1} = \frac{126}{48} \times 91 \approx 239 \ .$$

From Eq. (3.7) with K = 2, it follows that

$$\left.\begin{aligned} M_1 &= \frac{1}{240} \ , \\ M_2 &= \frac{239}{240} \ , \end{aligned}\right\} \tag{3.106}$$

An $f(\omega)$ that reproduces the main features of the neutron scattering experiments discussed above is

$$f(\omega) = \frac{1}{M_1} f_1(\omega) + \frac{1}{M_2} f_2(\omega) \tag{3.107}$$

with

$$
\left.\begin{aligned}
f_1(\omega) &= \frac{3\omega^2}{\omega_{max}^3} \qquad \hbar\omega \leq \hbar\omega_{max} = 0.02 \text{ eV} \\
&= 0 \qquad \omega > \omega_{max}
\end{aligned}\right\} \tag{3.108}
$$

$$
f_2(\omega) = \frac{1}{(2\pi\sigma^2)^{\frac{1}{2}}} \exp\left[-\frac{(\omega - \omega_0)^2}{2\sigma^2}\right], \tag{3.109}
$$

where $\omega_0 = 0.14$ eV and $\sigma = 0.012$ eV. An $f(\omega)$ not very different from that defined by Eqs. (3.108) and (3.109) have been used to analyze measurements of neutron spectra in zirconium hydride. We present some of these results in Chap. 6.

3-4.4 BERYLLIUM. Beryllium possesses a hexagonal close-packed structure that is based on a hexagonal lattice with primitive lattice basis vectors $\mathbf{a}_1$, $\mathbf{a}_2$, and $\mathbf{a}_3$, as illustrated in Fig. 3.38a. Values of the lattice parameters are $|\mathbf{a}_1| = |\mathbf{a}_2| = a = 2.2856$ Å and $|\mathbf{a}_3| = c = 3.5832$ Å. The reciprocal lattice is also hexagonal with lattice basis vectors as shown in Fig. 3.38b. The primitive unit cell of beryllium contains two atoms, one positioned at the origin 0 in Fig. 3.38a, the other at $\mathbf{r} = \frac{1}{3}\mathbf{a}_1 + \frac{2}{3}\mathbf{a}_2 + \frac{1}{2}\mathbf{a}_3$.

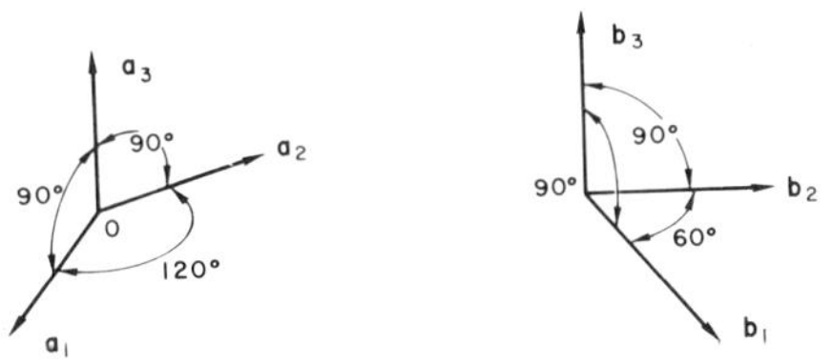

Figure 3.38 Lattice vectors for beryllium; (a) direct and (b) reciprocal.

Using the MTR phased-chopper slow-neutron velocity selector, Schmunk et al.[3] have made extensive measurements of the dispersion relations in single crystals of beryllium for waves traveling in the directions of $\mathbf{b}_3$ and $\mathbf{b}_2$, which in crystallographic notation are the

[0001] and [01$\bar{1}$0] directions, respectively; Fig. 3.39 shows the results of these measurements.

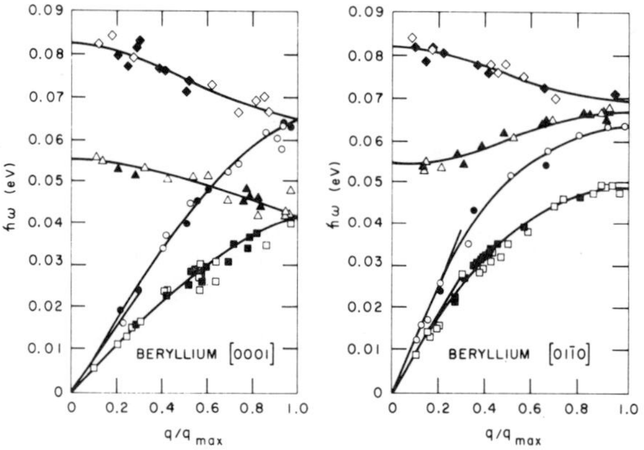

Figure 3.39 Beryllium dispersion relation for the symmetry directions [0001] and [01$\bar{1}$0]. Open points refer to experimental data that plot with the phonon wave vector within ±5° of the desired direction and solid points refer to data obtained from the scattering surface method (see Ref. 3). The character of the data is as follows: ◇ , upper optical; △, lower optical; ○ , longitudinal acoustical; □, transverse acoustical. The smooth curves were drawn through the data, and the straight lines at the origin indicate the initial slopes of the acoustical branches predicted from elastic constant data. For the direction [01$\bar{1}$0], phonon polarization for the transverse acoustical and upper optical branches is parallel to [0001]. (From Schmunk et al., Phys. Rev., Ref. 3.)

Slutsky and Garland[86] have developed a general model for the lattice dynamics of hexagonal close-packed crystals. They assume central force interactions between an atom and its first, second, and third nearest neighbors. Schmunk et al.,[3] on finding that this model did not fit their experimental data, extended the model (the extended SG model) to include interactions with fourth and fifth nearest neighbors. From the experimental data, they find force constants which, in units of

10^4 dynes/cm, are $k_1 = 1.60$, $k_2 = 2.66$, $k_3 = 0.188$, $k_4 = 0.668$, and $k_5 = 0.454$. Here k_j ($j = 1...5$) is the force constant associated with the interaction between an atom and its j*th* nearest neighbor. The extended Slutsky-Garland (SG) model, fitted to the experimental data and displayed as the dashed lines in Fig. 3.40 is compared with the solid curves that are smoothed-curve representations of the experimental points.

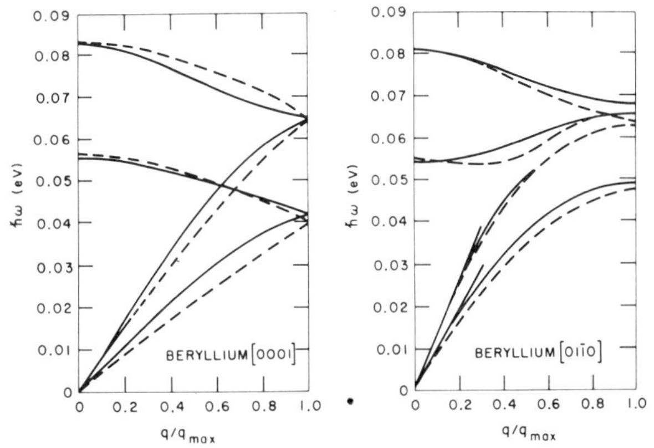

Figure 3.40 The dashed curves are for the extended model of Slutsky and Garland[86] fitted to the neutron scattering data and compared with the smoothed curves of Fig. 3.39 which were drawn through the experimental data. (From Schmunk et al., Phys. Rev., Ref. 3.)

Schmunk et al. have also attempted to fit their experimental data with the model of Begbie and Born (BB).[87] In this BB model there are no assumptions regarding the nature of the interatomic forces, but these forces are restricted to be between nearest neighbors in and out of the hexagonal plane. Figure 3.41 shows the fit of this model to the solid curve drawn through the data of Fig. 3.39. From the comparisons made in Figs. 3.40 and 3.41 a better model than the two considered previously would allow for general forces between near neighbors and central-force interactions with more distant neighbors. Nevertheless, the degree of agreement between the extended SG and BB models is considered adequate to justify the use of either one for calculating properties involving the frequency distributions of lattice vibrations.

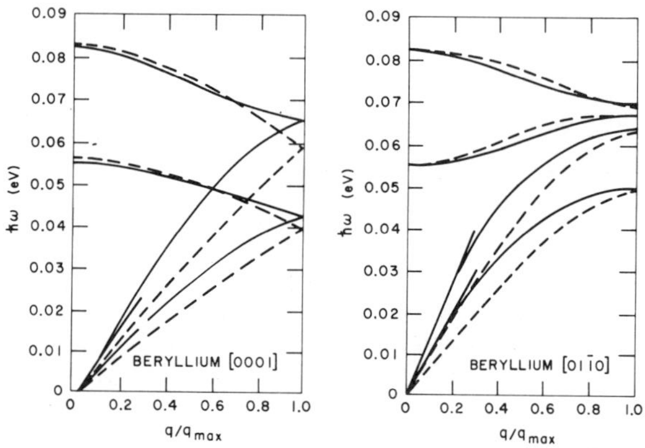

Figure 3.41 The dashed curves are for the model of Begbie and Born[87] fitted to the neutron scattering data and compared with the smoothed curves of Fig. 3.39 which were drawn through the experimental data. (From Schmunk et al., Phys. Rev., Ref. 3.)

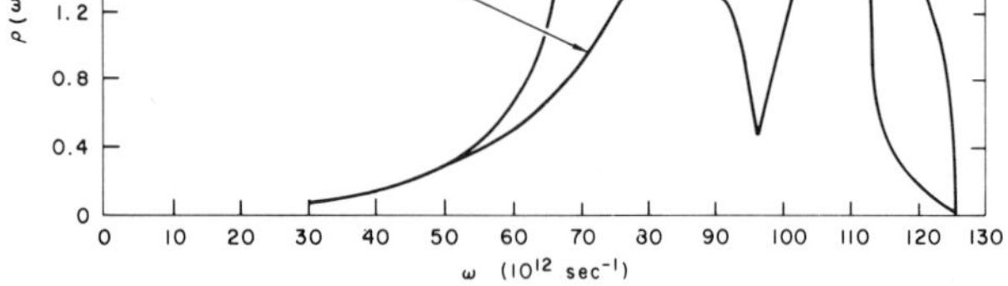

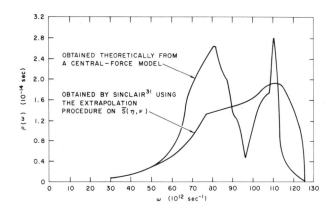

Figure 3.42 Frequency spectra of beryllium. (From Young and Koppel, Nucl. Sci. Eng., Ref. 4.)

Young and Koppel[4,88] have used the extended SG model and the force constants determined by Schmunk et al. to calculate the frequency distribution function for beryllium. A comparison of their calculated frequency distribution with that obtained by Sinclair[31] using the extrapolation procedure based on Eq. (3.91) is given in Fig. 3.42. The measured scattering law at two values of ν is compared in Fig. 3.43 with that calculated from Eq. (3.22) using the frequency distribution shown in Fig. 3.42. Because of the effects of interference on neutron scattering, the calculated frequency distribution is most probably a more accurate representation of the true distribution than the one obtained from the scattering data by the extrapolation method. In addition to yielding a reasonable fit to the measured scattering law except at low values of η where the interference effects are dominant, the theoretical $\rho(\omega)$ in Fig. 3.42 leads to measured specific heats that agree with measured values to within three percent.[4]

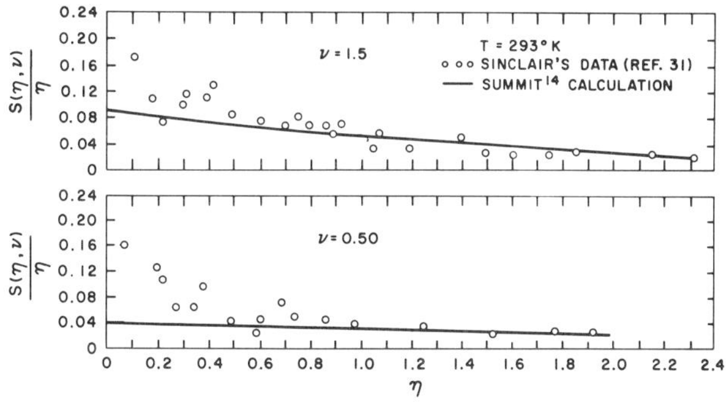

Figure 3.43 Scattering law for beryllium at two different values of ν. (From Young and Koppel, Nucl. Sci. Eng., Ref. 4.)

The difficulties inherent in the application of the extrapolation technique are directly exhibited in recent measurements by Schmunk on the scattering of neutrons from beryllium powder.[28] The results of these measurements are shown in Fig. 3.44. In the extrapolation process it is necessary to use scattering-law data from an intermediate range of κ values so that structure effects from interference scattering, which is important at low κ, are damped out, and contributions from multiphonon scattering, which is important at large κ, are insignificant. Schmunk states that at least for beryllium this extrapolation would be difficult because structure is observed over nearly the full range of the scattering-law data.

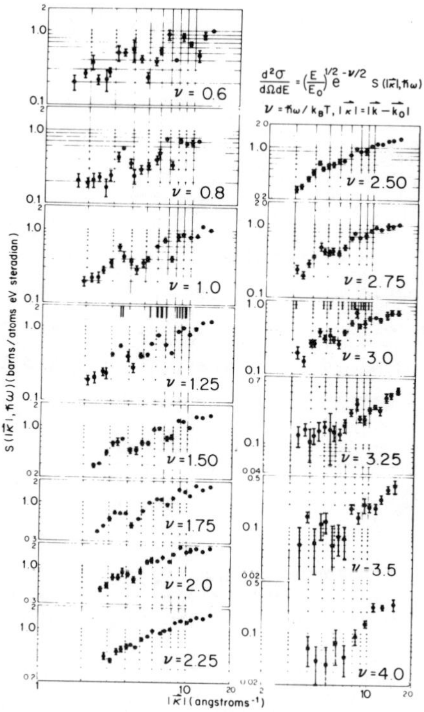

Figure 3.44 A composite of the reduced partial differential cross sections for room-temperature beryllium powder. The data have been averaged and errors evaluated as discussed in Ref. 28. Errors, where bars are not displayed, are smaller than the size of the points. (From Schmunk, Phys. Rev., Ref. 28.)

The elastic scattering of neutrons by beryllium was calculated first by Fermi et al.[89] and was repeated by Bhandari[90] and by Singwi and Kothari.[91] This cross section is given by Eq. (2.145), where now

$$V_0 = \frac{\sqrt{3}(a^2)c}{2}$$

and

$$n = 2 .$$

The expression

$$2W(\boldsymbol{\tau}) = \frac{\tau^2}{2M}\int_0^{\omega_{max}} \frac{\rho(\omega)}{\omega} \cot h \frac{\hbar\omega}{2k_BT}\, d\omega \qquad (3.110)$$

approximates Eq. (2.187) and is derived from Eq. (2.187) by averaging the latter equation over orientations and over the two atoms in the unit cell. The sum in Eq. (2.145) extends over all $\boldsymbol{\tau}$ for which $\tau \lesssim 2k$. Finally, with $\boldsymbol{\tau}$ given by Eq. (3.104),

$$|F(\boldsymbol{\tau})|^2 = 2[1 + \cos \tfrac{\pi}{3}(2\ell_1 + 4\ell_2 + 3\ell_3)] \ . \qquad (3.111)$$

Figure 3.45 compares the calculated and measured total scattering cross sections in beryllium. For these calculations beryllium is assumed to be a Debye crystal with $\hbar\omega_D/k_B = 1000°$K. To calculate the inelastic cross section, Bhandari[90] uses the incoherent approximation and neglects contributions that involve three or more phonons. Except for neutron energies between 0.04 and 0.07 eV there is good agreement between theory and experiment.

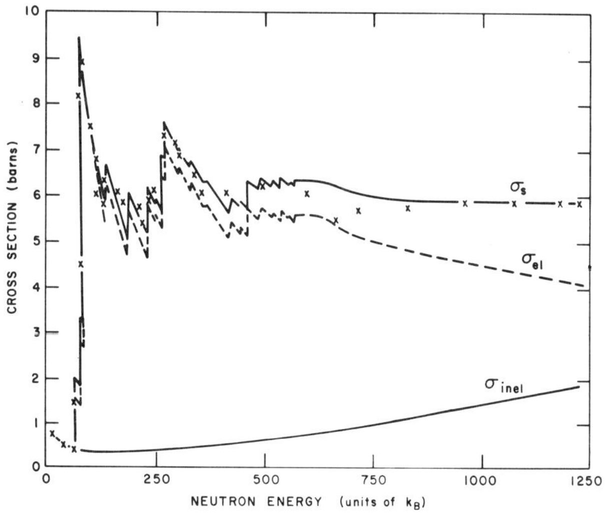

Figure 3.45 Elastic σ_{el}, inelastic σ_{inel}, and total σ_s scattering cross sections plotted as functions of neutron energy. Crosses indicate the experimental points. (From Bhandari, J. Nucl. Energy, Ref. 90.)

3-4.5 OTHER MODERATORS. There exist theoretical and experimental studies of neutron scattering by beryllium oxide,[31] polyethylene,[63,92-94] and the polyphenyls.[95,96] In particular, for polyethylene and the polyphenyls the theoretical results have been successfully applied to the calculation of thermal-neutron spectra. We refer the reader to the literature for a detailed description of these studies.

3-5 SOME IMPORTANT INTEGRATED PROPERTIES OF $\sigma(E_0, E, \theta)$

We have seen that the detailed characteristics of the dynamical and structural properties of a moderator are strongly reflected in $\sigma(E_0, E, \theta)$. Ordinarily, these characteristics are not reflected in any very definite way in problems of neutron thermalization. One important exception to this statement is provided by the heavy metallic hydrides, where, to a good approximation, the hydrogen atoms vibrate like Einstein oscillators and the energy transferred between the neutron and the moderator is quantized in multiples of the oscillator frequency. For moderation by zirconium hydride, we shall see in Chap. 6 that this quantization is manifested clearly in the neutron spectrum.

A second important exception occurs for polycrystalline moderators. The behavior of the scattering cross section as a function of energy is very different for crystalline substances that scatter neutrons coherently and for noncrystalline substances. Figure 3.45 is representative of $\sigma(E_0)$ for a polycrystal that scatters neutrons coherently. This cross section consists of elastic and inelastic parts. The former, which is responsible for the discontinuous behavior of the cross section, vanishes like E^{-1} as the neutron energy increases. The inelastic part, which makes the only contribution to $\sigma(E_0)$ for E_0 less than the Bragg cutoff energy, is generally small for $E_0 \sim k_B T$ but makes the dominant contribution to $\sigma(E_0)$ at energies greater than or approximately 0.1 eV.

The average energy exchanged between the neutron and the moderator does not exhibit a discontinuous variation with neutron energy. Figure 3.46 shows schematically the first energy-change moment,

$$\left\langle \sigma \Delta E \right\rangle = \int_0^\infty \sigma(E_0, E)(E_0 - E)\,dE \ ,$$

as a function of the initial energy E_0. For this quantity and for all energy change moments $\left\langle \sigma(\Delta E)^n \right\rangle$ the elastic scattering makes no contribution. At low energies where most probably the neutron gains energy from the moderator, $\left\langle \sigma \Delta E \right\rangle$ is negative. As E_0 increases, $\left\langle \sigma \Delta E \right\rangle$ passes through a null value and then gradually increases to its free-atom limit.

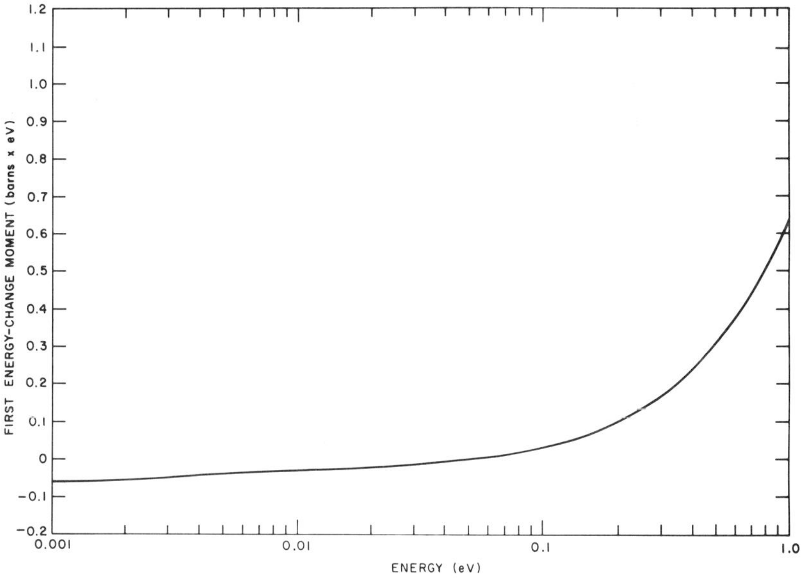

Figure 3.46 Typical curve of the first energy-change moment as a function of energy.

The cross section of a liquid exhibits an energy dependence that is entirely different from that associated with coherently scattering crystals. This is exemplified by comparing the smooth behavior of the cross section for H_2O in Fig. 3.22 with the discontinuous variation for the beryllium cross section in Fig. 3.45. The scattering by a liquid is in the strict sense completely inelastic and does not exhibit the discontinuities characteristic of Bragg scattering by a crystal. The energy-change moments for a liquid, however, are qualitatively like those for a crystal.

In the following chapters we give the explicit connection between the scattering properties of a material and the neutron balance in an assembly containing that material. We shall see that the possibly discontinuous behavior of the total cross section has an important bearing in problems involving neutron migration from one point in space to another, but that in the absence of migration the occurrence of such discontinuities is of no particular relevance.

REFERENCES

1. I. Pelah *et al.*, Detection of Optical Lattice Vibrations in Ge and ZrH by Scattering of Cold Neutrons, *Phys. Rev.*, **108**: 1091 (1957).

2. A. Andresen *et al.*, Neutron Investigation of Optical Vibration Levels in Zirconium Hydride, *Phys. Rev.*, **108**: 1092 (1957).

3. R. E. Schmunk *et al.*, Lattice Dynamics of Beryllium, *Phys. Rev.*, **128**: 562 (1962).

4. J. A. Young and J. U. Koppel, Scattering Kernel for Beryllium, *Nucl. Sci. Eng.*, **19**: 367 (1964).

5. A. Yoshimori and Y. Kitano, Theory of the Lattice Vibration of Graphite, *J. Phys. Soc. (Japan)*, **11**: 352 (1956).

6. D. E. Parks, J. R. Beyster, and N. F. Wikner, Thermal Neutron Spectra in Graphite, *Nucl. Sci. Eng.*, **13**: 306 (1962).

7. J. A. Young, N. F. Wikner, and D. E. Parks, Neutron Thermalization in Graphite. II, *Nucleonik*, **7**: 295 (1965).

8. P. A. Egelstaff *et al.*, The Thermal Neutron Scattering Law for Light and Heavy Water, in *Inelastic Scattering of Neutrons in Solids and Liquids*, Proceedings of the symposium held in Vienna, Austria, 11-14 October 1960, p. 309, International Atomic Energy Agency, Vienna, 1961.

9. K.-E. Larsson, S. Holmryd, and K. Otnes, Cold-Neutron Scattering Experiments on Light and Heavy Water, in *Inelastic Scattering of Neutrons in Solids and Liquids* (*op. cit.*, Ref. 8), p. 329.

10. G. Placzek and L. Van Hove, Interference Effects in the Total Neutron Scattering Cross-Section of Crystals, *Nuovo Cimento*, **1**: 233 (1955).

11. W. Marshall and R. N. Stuart, The Scattering of Neutrons from Polycrystalline Materials, in *Inelastic Scattering of Neutrons in Solids and Liquids* (*op. cit.*, Ref. 8), p. 75.

12. W. Marshall and R. N. Stuart, *The Scattering of Neutrons from Polycrystalline Materials*, USAEC Report UCRL-5568, University of California, Lawrence Radiation Laboratory, April 8, 1959.

13. J. D. Macdougall, Application of Scattering Law Data to the Calculation of Thermal Neutron Spectra, in *Proceedings of the Brookhaven Conference on Neutron Thermalization*, Vol. I, *The Scattering Law*, USAEC Report BNL-719, p. 121, Brookhaven National

Laboratory, 1962.

14. J. Bell, *SUMMIT, An IBM-7090 Program for the Computation of Crystalline Scattering Kernels*, USAEC Report GA-2492, General Atomic Division, General Dynamics Corporation, February 1962.

15. H. C. Honeck, *THERMOS, A Thermalization Transport Theory Code for Reactor Lattice Calculations*, USAEC Report BNL-5826, Brookhaven National Laboratory, 1961.

16. J. R. Beyster *et al.*, *Integral Neutron Thermalization, Annual Summary Report, October 1, 1963, through September 30, 1964*, USAEC Report GA-5798, p. 155, General Atomic Division, General Dynamics Corporation, October 1964

17. F. D. Federighi and D. T. Goldman, *KERNEL AND PAM--Programs for Use in the Calculation of the Thermal Scattering Matrix for Chemically Bound Systems*, USAEC Report KAPL-2225, Knolls Atomic Power Laboratory, September 17, 1962.

18. A. Sjölander, Multi-phonon Processes in Slow Neutron Scattering by Crystals, *Arkiv Fysik*, **14**: 315 (1958).

19. H. Cramér, *Mathematical Methods of Statistics*, Chap. 17, Princeton University Press, Princeton, 1946.

20. C. B. Walker, X-Ray Study of Lattice Vibrations in Aluminum, *Phys. Rev.*, **103**: 547 (1956).

21. P. Schofield and A. Hassitt, The Calculation of Thermal Neutron Spectra, in *Proceedings of the Second United Nations International Conference on the Peaceful Uses of Atomic Energy, Geneva, 1958*, Vol. 16, p. 217, United Nations, Geneva, 1958.

22. M. S. Nelkin and D. E. Parks, Effects of Chemical Binding on Nuclear Recoil, *Phys. Rev.*, **119**: 1060 (1960).

23. P. A. Egelstaff and P. Schofield, On the Evaluation of the Thermal Neutron Scattering Law, *Nucl. Sci. Eng.*, **12**: 260 (1962).

24. P. M. Morse and H. Feshbach, *Methods of Theoretical Physics*, Vol. I, p. 437, McGraw-Hill Book Company, New York, 1953.

25. A. Erdélyi *et al.*, *Higher Transcendental Functions*, Vol. 2, p. 5 (Bateman Manuscript Project, California Institute of Technology), McGraw-Hill Book Company, New York, 1953.

26. G. Placzek, The Scattering of Neutrons by Systems of Heavy Nuclei, *Phys. Rev.*, **86**: 377 (1952).

27. S. N. Purohit, Neutron Thermalization in a Crystalline Medium in Incoherent Approximation, in *Proceedings of the Brookhaven Conference on Neutron Thermalization*, Vol. I, *The Scattering Law*, USAEC Report BNL-719, p. 203, Brookhaven National Laboratory, 1962.

28. R. E. Schmunk, Slow-Neutron Inelastic Scattering from Beryllium Powder, *Phys. Rev.*, **136**: A1303 (1964).

29. P. A. Egelstaff and S. J. Cocking, The Phonon Frequency Distribution in Graphite at Several Temperatures, in *Inelastic Scattering of Neutrons in Solids and Liquids* (*op cit.*, Ref. 8), p. 569.

30. B. C. Haywood and I. M. Thorson, Frequency Distribution of Normal Modes in Graphite, in *Inelastic Scattering of Neutrons in Solids and Liquids*, Proceedings of the symposium held at Chalk River, Canada, 10-14 September 1962, Vol. II, p. 111, International Atomic Energy Agency, Vienna, 1963.

31. R. N. Sinclair, The Neutron Scattering Law and the Frequency Distribution of the Normal Modes of Beryllium and Beryllium Oxide, in *Inelastic Scattering of Neutrons in Solids and Liquids* (*op. cit.*, Ref. 30), Vol. II, p. 199.

32. K.-E. Larsson and U. Dahlborg, A Study of the Diffusive Atomic Motions in Glycerol and the Vibratory Motions in Glycerol and Light and Heavy Water by Cold Neutron Scattering, in *Inelastic Scattering of Neutrons in Solids and Liquids* (*op. cit.*, Ref. 30), Vol. I, p. 317.

33. D. A. Kittwitz and B. R. Leonard, Jr., The Scattering Law for Room Temperature Light Water, in *Inelastic Scattering of Neutrons in Solids and Liquids* (*op. cit.*, Ref. 30), Vol. I, p. 359.

34. P. A. Egelstaff, Practical Analysis of Neutron Scattering Data into Self and Interference Terms, in *Inelastic Scattering of Neutrons in Solids and Liquids* (*op. cit.*, Ref. 30), Vol. I, p. 65.

35. P. A. Egelstaff, B. C. Haywood, and I. M. Thorson, The Motion of Hydrogen in Water, in *Inelastic Scattering of Neutrons in Solids and Liquids* (*op cit.*, Ref. 30), Vol. I, p. 343.

36. P. A. Egelstaff, The Scattering of Thermal Neutrons by Moderators, *Nucl. Sci. Eng.*, **12**: 250 (1962).

37. B. N. Brockhouse, Methods for Neutron Spectrometry, in *Inelastic Scattering of Neutrons in Solids and Liquids* (*op cit.*, Ref. 8), p. 113.

38. D. J. Hughes and R. B. Schwartz, *Neutron Cross Sections*, USAEC Report BNL-325, 2nd ed., Brookhaven National Laboratory, July 1, 1958.

39. G. Herzberg, *Infrared and Raman Spectra of Polyatomic Molecules*, Vol. II of *Molecular Spectra and Molecular Structure*, D. Van Nostrand Company, Inc., New York, 1945.

40. D. Butler, The Scattering of Slow Neutrons by Heavy Water; I. Intramolecular Scattering, *Proc. Phys. Soc. (London)*, **81**: 276 (1963); and II. Intermolecular Scattering, *Proc. Phys. Soc.*

(London), **81**: 294 (1963).

41. T. Springer, Die Streuung von langsamen Neutronen an Wasser, Eis und Wasserdampf, *Nukleonik*, **3**: 110 (1961).

42. D. J. Hughes *et al.*, Atomic Motions in Water by Scattering of Cold Neutrons, *Phys. Rev.*, **119**: 872 (1960).

43. D. J. Hughes *et al.*, Atomic Motions in Water by Scattering of Cold Neutrons, *Phys. Rev. Letters*, **3**: 91 (1959).

44. G. Herzberg, *op. cit.*, Ref. 39, p. 281.

45. S. Yip and R. K. Osborn, "Hindered Rotations" in Liquids and Slow-Neutron Scattering, in *Proceedings of the Brookhaven Conference on Neutron Thermalization*, Vol. I, *The Scattering Law*, USAEC Report BNL-719, p. 289, Brookhaven National Laboratory, 1962.

46. M. Magat, Investigations of the Raman Spectrum and the Constitution of Liquid Water, *Ann. Physik*, **6**: 108 (1936).

47. H. H. Stiller and H. R. Danner, Quasi-elastic and Inelastic Scattering of Cold Neutrons from Water, in *Inelastic Scattering of Neutrons in Solids and Liquids* (*op. cit.*, Ref. 8), p. 363.

48. B. N. Brockhouse, Diffusive Motions in Liquids and Neutron Scattering, *Phys. Rev. Letters*, **2**: 287 (1959).

49. B. N. Brockhouse *et al.*, Liquid Dynamics from Neutron Spectrometry, in *Inelastic Scattering of Neutrons in Solids and Liquids* (*op. cit.*, Ref. 30), Vol. I, p. 189.

50. M. Sakamoto *et al.*, Neutron Inelastic Scattering Study of Water, in *Proceedings of the International Conference on Magnetism and Crystallography, Kyoto, 25-30 September 1961*, II. *Electron and Neutron Diffraction, J. Phys. Soc. (Japan)*, **17**: Suppl. B-II: 370 (1962).

51. D. Cribier and B. Jacrot, Diffusion Quasi-élastique des Neutron Froids par l'Eau et Coefficient d'Autodiffusion du Liquide, in *Inelastic Scattering of Neutrons in Solids and Liquids* (*op. cit.*, Ref. 8), p. 347.

52. D. A. Kottwitz, B. R. Leonard, Jr., and R. B. Smith, Quasi-elastic Scattering by Room Temperature Light Water, in *Inelastic Scattering of Neutrons in Solids and Liquids* (*op. cit.*, Ref. 30), Vol. I, p. 373.

53. J. H. Simpson and H. Y. Carr, Diffusion and Nuclear Spin Relaxation in Water, *Phys. Rev.*, **111**: 1201 (1958).

54. A. Rahman, K. S. Singwi, and A. Sjölander, Stochastic Model of a Liquid and Cold Neutron Scattering. II, *Phys. Rev.*, **126**: 997 (1962).

55. K.-E. Larsson and U. Dahlborg, Some Vibrational Properties of Solid and Liquid H_2O and D_2O Derived from Differential Cross-Section Measurements, *J. Nucl. Energy, Pt. A & B*, **16**: 81 (1962).

56. B. C. Haywood and I. M. Thorson, The Scattering Law for Light and Heavy Water at 20°C and 150°C, in *Proceedings of the Brookhaven Conference on Neutron Thermalization*, Vol. I, *The Scattering Law*, USAEC Report BNL-719, p. 26, Brookhaven National Laboratory, 1962.

57. H. D. Brown and D. S. St. John, *Neutron Energy Spectrum in* D_2O, USAEC Report DP-33, E. I. du Pont de Nemours and Company, February 1954.

58. M. S. Nelkin, Scattering of Slow Neutrons by Water, *Phys. Rev.*, **119**: 741 (1960).

59. T. J. Krieger and M. S. Nelkin, *The Scattering of Slow Neutrons by Hydrogenous Moderators*, USAEC Report KAPL-1597, Knolls Atomic Power Laboratory, August 1, 1956; and Addendum, April 1957.

60. H. C. Honeck, An Incoherent Thermal Scattering Model for Heavy Water, *Trans. Am. Nucl. Soc.*, **5**: 47 (1962).

61. J. U. Koppel and J. A. Young, Neutron Scattering by Water Taking Into Account the Anisotropy of the Molecular Vibrations, *Nucl. Sci. Eng.*, **19**: 412 (1964).

62. M. J. Poole, P. Schofield, and R. N. Sinclair, Some Measurements of Thermal-Neutron Spectra, in *Exponential and Critical Experiments*, Proceedings of the symposium on exponential and critical experiments held by the International Atomic Energy Agency in Amsterdam, Netherlands, 2-6 September 1963, Vol. III, p. 87, International Atomic Energy Agency, Vienna, 1964.

63. J. R. Beyster *et al.*, *Integral Neutron Thermalization, Annual Summary Report October 1, 1962, through September 30, 1963*, USAEC Report GA-4659, General Atomic Division, General Dynamics Corporation, January 24, 1964.

64. K. Komatsu, Theory of the Specific Heat of Graphite. II, *J. Phys. Soc. (Japan)*, **10**: 346 (1955).

65. J. Krumhansl and H. Brooks, The Lattice Vibration Specific Heat of Graphite, *J. Chem. Phys.*, **21**: 1663 (1953).

66. G. R. Baldock, Lattice Vibrations and Specific Heat of Graphite, *Phil. Mag., Ser. 8*, **1**: 789 (1956).

67. G. Dolling and B. N. Brockhouse, Lattice Vibrations in Pyrolitic Graphite, *Phys. Rev.*, **128**: 1120 (1962).

68. J. A. Young and J. U. Koppel, Phonon Spectrum of Graphite, *J. Chem. Phys.*, **42**: 357 (1965).

69. N. F. Wikner, G. D. Joanou, and D. E. Parks, Neutron Thermalization

in Graphite, *Nucl. Sci. Eng.*, **19**: 108 (1964).

70. W. DeSorbo and W. W. Tyler, The Specific Heat of Graphite from 13° to 300°K, *J. Chem. Phys.*, **21**: 1660 (1953).

71. H. M. Spencer, Empirical Heat Capacity Equations of Gases and Graphite, *Ind. Eng. Chem.*, **40**: 2152 (1948).

72. A. D. B. Woods *et al.*, Energy Distributions of Neutrons Scattered from Graphite, Light and Heavy Water, Ice, Zirconium Hydride, Lithium Hydride, Sodium Hydride, and Ammonium Chloride by the Beryllium Detector Method, in *Inelastic Scattering of Neutrons in Solids and Liquids* (*op. cit.*, Ref. 8), p. 487.

73. P. A. Egelstaff and D. H. C. Harris, High Energy Modes in the Frequency Distribution of Graphite, *Phys. Letters*, **7**: 220 (1963).

74. D. H. Saunderson, *Phonon Frequency Distribution in Graphite*, British Report AERE-M-1199, March 1963.

75. D. E. Parks, *Relation of Crystal Symmetry in Graphite to Lattice Vibrations and Their Interaction with Slow Neutrons*, USAEC Report GA-2125, General Atomic Division, General Dynamics Corporation, March 15, 1961.

76. J. R. Beyster *et al.*, Measurements of Neutron Spectra in Water, Polyethylene, and Zirconium Hydride, *Nucl. Sci. Eng.*, **9**: 168 (1961).

77. *Technical Foundations of TRIGA*, Report GA-471, General Atomic Division, General Dynamics Corporation, August 27, 1958.

78. D. A. Vaughan and J. R. Bridge, High Temperature X-Ray Diffraction Investigation of the Zr-H System, *J. Metals*, **8**: 528 (1956).

79. C. E. Ells and A. D. McQuillan, A Study of Hydrogen-Pressure Relationships in the Zirconium-Hydrogen System, *J. Inst. Metals*, **85**: 89 (1956).

80. L. D. LaGrange *et al.*, A Study of Zirconium-Hydrogen and Zirconium-Hydrogen-Uranium Systems Between 600 and 800°C, *J. Phys. Chem.*, **63**: 2035 (1959).

81. R. L. Beck, Zirconium-Hydrogen Phase System, *Trans. Am. Soc. Metals*, **55**: 542 (1962).

82. A. W. McReynolds *et al.*, Neutron Thermalization by Chemically-Bound Hydrogen and Carbon, in *Proceedings of the Second United Nations International Conference on the Peaceful Uses of Atomic Energy, Geneva, 1958*, Vol. 16, p. 297, United Nations, Geneva, 1958.

83. W. L. Whittemore, *Neutron Interactions in Zirconium Hydride*, USAEC Report GA-4490 (rev.), General Atomic Division, General Dynamics Corporation, January 6, 1964.

84. D. H. Saunderson and S. J. Cocking, Studies of Proton Vibrations in γ-Titanium Hydride, in *Inelastic Scattering of Neutrons in Solids and Liquids* (*op. cit.*, Ref. 30), Vol. II, p. 265.

85. W. L. Whittemore, *Differential Neutron Thermalization, Annual Summary Report October 1, 1963, through September 30, 1964*, USAEC Report GA-5554, General Atomic Division General Dynamics Corporation, October 20, 1964.

86. L. J. Slutsky and C. W. Garland, Lattice Dynamics of Hexagonal Close-packed Metals, *J. Chem. Phys.*, **26**: 787 (1957).

87. G. H. Begbie and M. Born, Thermal Scattering of X-Rays by Crystals. I. Dynamical Foundation, *Proc. Roy. Soc. (London)*, **A188**: 179 (1946).

88. J. A. Young and J. U. Koppel, Lattice Vibrational Spectra of Beryllium, Magnesium, and Zinc, *Phys. Rev.*, **134**: A1476 (1964).

89. E. Fermi, W. J. Sturm, and R. G. Sachs, The Transmission of Slow Neutrons through Microcrystalline Materials, *Phys. Rev.*, **71**: 589 (1947).

90. R. C. Bhandari, Scattering of Thermal Neutrons in Beryllium, *J. Nucl. Energy*, **6**: 104 (1957).

91. K. S. Singwi and L. S. Kothari, Transport Cross-Section of Thermal Neutrons in Solid Moderators, in *Proceedings of the Second United Nations International Conference on the Peaceful Uses of Atomic Energy, Geneva, 1958*, Vol. 16, p. 325, United Nations, Geneva, 1958.

92. D. T. Goldman and F. D. Federighi, Calculation of Thermal Neutron Flux Spectra in an Infinite Polyethylene Moderated Medium with Varying Amounts of Absorption, in *Proceedings of the Brookhaven Conference on Neutron Thermalization*, Vol. I, *The Scattering Law*, USAEC Report BNL-719, p. 100, Brookhaven National Laboratory, 1962.

93. D. T. Goldman and F. D. Federighi, Calculation of Thermal Neutron Flux Spectra in a Polyethylene Moderated Medium, *Nucl. Sci. Eng.*, **16**: 165 (1963).

94. J. U. Koppel and J. A. Young, Neutron Scattering by Polyethylene, *Nucl. Sci. Eng.*, **21**: 268 (1965).

95. V. C. Boffi, V. G. Molinari, and D. E. Parks, Slow Neutron Scattering by Benzene, in *Inelastic Scattering of Neutrons in Solids and Liquids* (*op cit.*, Ref. 30), Vol. I, p. 285.

96. V. C. Boffi, V. G. Molinari, and D. E. Parks, Slow Neutron Scattering and Thermalization by Benzene and Other Polyphenyls, in *Proceedings of the Brookhaven Conference on Neutron Thermalization*, Vol. I, *The Scattering Law*, USAEC Report BNL-719, p. 69, Brookhaven National Laboratory, 1962.

Chapter 4

THERMAL-NEUTRON TRANSPORT THEORY

4-1 ASSUMPTIONS AND LIMITATIONS

The migration of neutrons in macroscopic media is accurately described by a linear Boltzmann equation. This equation gives the rate of change of the average number of neutrons in an element of phase space as the sum of three terms. The first term describes the streaming in position space associated with spatial inhomogeneities in the neutron distribution, and the second describes the transitions in velocity space due to collisions of the neutrons with the nuclei of the medium. The third term represents external sources of neutrons. The theory based on this linear Boltzmann equation, with cross sections for the individual collisions defined by the single scattering experiments of the preceding chapter, is known as neutron transport theory. This theory is quite thoroughly covered in several books,[1-3] and we do not repeat any substantial fraction of its development here. We do, however, state the basic assumptions underlying the theory and indicate their limits of valicity. The primary emphasis in this chapter is on those aspects of neutron transport theory that are most directly connected with the energy distribution of neutrons near thermal energies. In particular, we extend those concepts, such as thermal diffusion length and extrapolation distance, which are well understood for neutrons diffusing with a single speed, to a theory that includes energy exchange between thermal neutrons and the moderator.

The assumptions of neutron transport theory have been summarized in an elegant article by Wigner.[4] We repeat parts of his discussion here with some additions. The basic quantity in neutron transport theory is $f_B(\mathbf{r},\mathbf{v},t)$, the neutron density in phase space,[a] where $f_B(\mathbf{r},\mathbf{v},t)d\mathbf{r}d\mathbf{v}$ is the average number of neutrons in the phase space element $d\mathbf{r}d\mathbf{v}$ at the time t. The first assumption of transport theory is that this function can be considered as a continuous function rather than as a random variable describing the stochastic process of neutron migration. The Boltzmann equation thus does not apply to the fluctuations in neutron flux, which are significant in a reactor operating at a low power level.[b] A closely related property of the transport equation is that $f_B(\mathbf{r},\mathbf{v},t)$ is determined by a self-contained equation that makes no reference to correlations in position between pairs of neutrons. The Boltzmann equation (linear or nonlinear) is a "kinetic" equation involving only the single-particle distribution function $f_B(\mathbf{r},\mathbf{v},t)$. There are position correlations between two neutrons originating in the same fission. The Boltzmann equation has nothing to say about these correlations. A more general theory giving a space- and energy-dependent description of correlations and fluctuations does not yet exist in a practically useful form, but preliminary considerations of the problem have made some formal progress.[5]

The second essential assumption of the theory is that the Boltzmann equation is linear. This linearity requires the neglect of neutron-neutron collisions, which is always justified. It also requires the neglect of any effects the neutrons have on the medium. The latter requirement is much more restrictive, since heat production by the

---

[a] The subscript B indicates that we are using Boltzmann's form for f. We reserve the symbol f without a subscript for the angular flux $f(\mathbf{r},E,\mathbf{\Omega},t)$ as used by Weinberg and Wigner.[3]

[b] That this is so is clear from the fact that the Boltzmann equation contains the average number of neutrons per fission which is not an integer. The low-power fluctuations depend on the probability distribution of the number of neutrons per fission. There is, however, a large class of reactor noise experiments where the Boltzmann equation does apply. These involve the response to a random source, and are equivalent to determining the impulse response function of the reactor.

fission process can be large. It should be emphasized, however, that the nonlinear form of the Boltzmann equation, in which the feedback between the neutron flux and material temperatures is included, remains an accurate description of the neutron distribution in a reactor. Furthermore, for a reactor operating at a steady power, the neutron distribution is determined in terms of a given temperature distribution by a linear equation. A self-consistent determination of the neutron and temperature distributions, however, requires the solution of the Boltzmann equation coupled to the heat-transfer equations for the reactor. For reactor statics, this coupling does not usually play an essential role; static problems can thus be considered solely in terms of a detailed solution of the linear Boltzmann equation. For high-power reactor dynamics, however, the coupled equations must be considered, and then the limitations of computer capacity require gross simplifications in the neutronic aspects of the problem. These simplifications usually take the form of so-called "point reactor kinetics" equations in which the space and energy dependence of the neutron distribution are completely ignored.[c]

In this book we assume a known moderator temperature as a function of position, and deal only with a linear Boltzmann equation for the space and energy distribution of the neutrons. As we have already implicitly assumed, we do require that the moderator be in local thermodynamic equilibrium. Although we make no assumptions about "neutron temperatures," we do assume that at every position the energy-transfer cross sections are defined in terms of the moderator temperature at that position; i.e., we calculate the distribution of neutron energies as a function of position in terms of a known distribution of moderator energy states as a function of position. This

---

[c] For the special case where all the heat is stored in a solid moderator and only prompt neutrons are considered, considerable progress has been made in deriving the point reactor kinetics equations from the Boltzmann equation and defining precisely the parameters that appear therein.[6-8] This case is particularly simple, however, and it is not likely that a comparable approach will work in the near future for more complicated problems.

assumption is expected to be very good for all cases of interest for this book. If it were not, we would have to consider the problems of irreversible statistical mechanics in the moderator. There is some question as to whether these problems are solved in principle, and they are certainly too difficult to solve in connection with the neutronic behavior of a reactor.

Finally, we consider the limitations imposed by the wave-mechanical nature of the neutron.[9] The Boltzmann equation for thermal neutrons is based on an essentially classical description of neutron motion. Quantum mechanics enters only in the determination of the transition probability per unit time for a neutron at a given position with a given momentum to be scattered with a given change in velocity or to be absorbed. Since the wave nature of the neutron requires a fully quantum-mechanical description for problems of neutron diffraction, it is necessary to clarify the use of the Boltzmann equation for the calculation of the neutron distribution in macroscopic moderating samples. We do not attempt to give a mathematical justification; we merely discuss the appropriate physical limitations on a semiquantative basis.

The Boltzmann equation describes the interaction of slow neutrons and moderator in terms of multiple random collisions. The de Broglie wavelength of a thermal neutron is comparable to interatomic spacings; therefore, the neutrons cannot be considered to interact with individual nuclei. The validity of the Boltzmann equation depends on the size of the aggregates with which the neutron interacts, as well as on the spatial resolution with which the neutron distribution is described.

In a fluid, the range of the atomic order is microscopic and comparable to the neutron wavelength, but the mean free path for a neutron collision is of the order of centimeters. For a spatial resolution that is coarse compared with a neutron wavelength but fine compared with a mean free path, the neutron distribution can be accurately described by a linear Boltzmann equation of the form we consider. We note that the large ratio of mean free path to interatomic spacing implies an effectively weak interaction between slow neutrons and moderator. This effective interaction is the Fermi pseudopotential discussed in Sec. 2-2. The weakness of this interaction is a necessary but not a sufficient condition for the derivation of a Boltzmann equation, because it implies that the neutrons propagate through the medium essentially as free wave packets except for occasional collisions.

In a crystalline medium, the applicability of the Boltzmann equation is more limited. Large single-crystals with order extending over distances comparable to a neutron mean free path are often encountered. The elastic scattering of neutrons by these crystals must be described in terms of wave propagation in a refractive medium. It should be noted, however, that the multiple inelastic scattering in such a crystal can still be well described by a Boltzmann equation. This follows from the microscopic range of the dynamical order (as manifested by the phonon mean free path arising from the anharmonic terms in the energy of lattice vibrations).

For polycrystalline solids, however, the crystallite size is usually of the order of microns and is small compared with a neutron mean free path. If one has a random orientation of microcrystals and is interested in a description with spatial resolution large compared with the crystallite size, then the Boltzmann equation is an accurate description even for the elastic scattering. In this description the basic scattering unit must be considered as the microcrystal and not as a truly microscopic aggregate of nuclei, as in a fluid.

We conclude then that the Boltzmann equation accurately describes the neutron distribution in a fluid or in a polycrystalline solid if the neutron distribution is considered with a spatial resolution that is coarse compared both with the range of order in the scatterer and with the neutron wavelength. The spatial resolution, which can still be fine compared with a scattering mean free path, allows the construction of neutron wave packets that are adequately localized in position and at the same time have an uncertainty in velocity compared with thermal velocities. The Boltzmann equation then applies to the essentially classical motion described by these wave packets.

4-2 THE TRANSPORT EQUATION

The neutron density in phase space changes in time as the result of three mechanisms. The linear Boltzmann equation describing this change can be written as

$$\frac{\partial f_B}{\partial t} = \left(\frac{\partial f_B}{\partial t}\right)_{\text{streaming}} + \left(\frac{\partial f_B}{\partial t}\right)_{\text{collisions}} + S(\mathbf{r},\mathbf{v},t) \ . \tag{4.1}$$

The streaming term represents the net flow from adjacent elements in phase space. The flow in velocity space is associated with the accelerations resulting from the average force acting on the neutron. This term is important for charged particles in electromagnetic fields but is negligible for neutrons. The streaming in position space is associated with spatial gradients and is given by

$$\left(\frac{\partial f_B}{\partial t}\right)_{\text{streaming}} = -(\mathbf{v} \cdot \nabla) f_B(\mathbf{r},\mathbf{v},t) \ . \tag{4.2}$$

where the gradient is with respect to the $\mathbf{r}$ dependence of f_B, and the quantity $(\mathbf{v} \cdot \nabla)f_B$ is the product of the neutron speed and the directional derivative of f_B in the direction of the neutron motion. It is important to note that Eq. (4.2) is not an approximation valid only for slow spatial variation, but is an exact result.

The source of neutrons $S(\mathbf{r},\mathbf{v},t)$ may be a radioactive source, a pulsed neutron source from an accelerator, or the fast neutrons arising from fission. In the last case, it appears as a homogeneous rather than an inhomogeneous term in the transport equation, but this is not of importance except when considering eigenvalue problems associated with criticality. In most of this book we either consider the source as an inhomogeneous term in the equation or consider source-free problems in which the source occurs implicitly as an initial condition or boundary condition.

Since the dynamics of the neutron interaction with the system requires a quantum-mechanical description, a derivation of the Boltzmann equation for neutron migration must start from the quantum-mechanical form of Liouville's theorem, since neutrons can be "destroyed" by nuclear absorption and "created" by fission. The basic approximation of the Boltzmann equation is that the rate of change of the distribution function due to dynamical interactions is determined by the probability of a single collision as given by the double differential cross section $\sigma(E_0,E,\theta)$ of Chaps. 2 and 3. This approximation assumes free-particle streaming between discrete random-collision events, and, classically, requires that the duration of a collision be short compared with the time between collisions. This requirement is well satisfied in practice, since the collision time of Chap. 2 is of the order of $(1/\omega_D) \cong 10^{-12}$ sec, where ω_D is a typical lattice vibration frequency. The time between collisions is of the order of $(v\Sigma)^{-1} \cong 10^{-5}$ sec. The further limitations imposed by quantum mechanics have been discussed qualitatively in Sec. 4-1.

The dynamical interaction between the neutrons and the medium is thus described entirely in terms of discrete collisions in which a neutron at a definite point in space has a finite change in its (vector) velocity. We must also include the process of nuclear absorption of the neutrons. The collision term in the transport equation is given by

$$\left(\frac{\partial f_B}{\partial t}\right)_{\text{collision}} = -v\Sigma(v) f_B(\mathbf{r},\mathbf{v},t) + \int W(\mathbf{v}',\mathbf{v}) f_B(\mathbf{r},\mathbf{v}',t)\, d\mathbf{v}' \ , \tag{4.3}$$

where $\Sigma(v)$ is the probability per unit length for a neutron of speed v to make a collision, and $W(\mathbf{v},\mathbf{v}')d\mathbf{v}'$ is the transition probability per unit time for a neutron of velocity $\mathbf{v}$ to scatter to the velocity interval $d\mathbf{v}'$ at velocity $\mathbf{v}'$.

The quantity $\Sigma(v)$ is known as the macroscopic total cross section and is the reciprocal of the mean free path. It is determined experimentally from a "good geometry" transmission experiment. If a collimated neutron beam of velocity $\mathbf{v}$ is incident on a sample of thickness d, the fraction of the neutrons transmitted with no collisions is $\exp[-\Sigma(v)d]$. The term $\Sigma(v)$ is the sum of a macroscopic scattering cross section $\Sigma_s(v)$ and a macroscopic absorption cross section $\Sigma_a(v)$. The absorption cross section includes fission as well as radiative capture, $(n\text{-}\alpha)$ processes, etc. The fast neutrons born in fission are considered separately as part of the source term. For a material that contains N_i nuclei per cubic centimeter of nuclide i with microscopic cross sections $\sigma_s^i(v)$ and $\sigma_a^i(v)$, the macroscopic cross sections are given by

$$\Sigma_s(v) = \sum_i N_i \sigma_s^i(v) ,$$
$$\Sigma_a(v) = \sum_i N_i \sigma_a^i(v) . \qquad (4.4)$$

The transition probability $W(\mathbf{v},\mathbf{v}')$ has the property that

$$v\Sigma_s(v) = \int W(\mathbf{v},\mathbf{v}')d\mathbf{v}' , \qquad (4.5)$$

where $v\Sigma_s(v)$ is the scattering probability per unit time for a neutron with speed v. The transition probability $W(\mathbf{v},\mathbf{v}')$ is also directly measurable from a single scattering experiment of the scattering-law type discussed in the preceding two chapters. For a single scattering species with N nuclei per cubic centimeter, it is given by

$$W(\mathbf{v},\mathbf{v}') = (Nmv/v')\sigma(E,E',\theta) , \qquad (4.6)$$

where $\sigma(E,E',\theta)$ is as defined in Sec. 2-2 and $\cos\theta = (\underset{\sim}{v} \cdot \underset{\sim}{v}'/vv')$. We frequently find it convenient to use a mixed notation in which both the neutron speed v and the kinetic energy E appear; it is to be understood that these are not independent variables but are always related through $E = \frac{1}{2}mv^2$.

Combining Eqs. (4.1) through (4.3) the Boltzmann equation becomes

$$\frac{\partial f_B(\mathbf{r},\mathbf{v},t)}{\partial t} = -[(\mathbf{v} \cdot \nabla) + v\Sigma_a(v) + v\Sigma_s(v)]f_B(\mathbf{r},\mathbf{v},t) + $$
$$+ \int W(\mathbf{v}',\mathbf{v})f_B(\mathbf{r},\mathbf{v}',t)d\mathbf{v}' + S(\mathbf{r},\mathbf{v},t) . \qquad (4.7)$$

With the aid of Eq. (4.5), the condition of over-all neutron balance can be made more explicit if we write Eq. (4.7) in the form

$$\frac{\partial f_B(\mathbf{r},\mathbf{v},t)}{\partial t} = -[(\underset{\sim}{v} \cdot \underset{\sim}{\nabla}) + v\Sigma_a(v)]f_B(\mathbf{r},\mathbf{v},t) + S(\mathbf{r},\mathbf{v},t) + $$
$$+ \int [W(\mathbf{v}',\mathbf{v})f_B(\mathbf{r},\mathbf{v}',t) - $$
$$- W(\mathbf{v},\mathbf{v}')f_B(\mathbf{r},\mathbf{v},t)]d\mathbf{v}' . \qquad (4.8)$$

If we introduce the total neutron density

$$n(\mathbf{r},t) = \int f_B(\mathbf{r},\mathbf{v},t)d\mathbf{v} \; , \tag{4.9}$$

the current

$$\mathbf{J}(\mathbf{r},t) = \int \mathbf{v}f_B(\mathbf{r},\mathbf{v},t)d\mathbf{v} \; , \tag{4.10}$$

and the source strength

$$Q(\mathbf{r},t) = \int S(\mathbf{r},\mathbf{v},t)d\mathbf{v} \; , \tag{4.11}$$

and integrate Eq. (4.8) over velocity space, the scattering terms on the right-hand side cancel identically to give

$$\frac{\partial n(\mathbf{r},t)}{\partial t} = -\nabla \cdot \mathbf{J}(\mathbf{r},t) + Q(\mathbf{r},t) - \int v\Sigma_a(v)f_B(\mathbf{r},\mathbf{v},t)d\mathbf{v} \; . \tag{4.12}$$

In the special case of a 1/v absorption cross section

$$\Sigma_a(v) = \lambda_a/v \; , \tag{4.13}$$

the absorption probability per unit time λ_a is independent of neutron energy, and Eq. (4.12) simplifies to

$$\frac{\partial n(\mathbf{r},t)}{\partial t} = -\nabla \cdot \mathbf{J}(\mathbf{r},t) + Q(\mathbf{r},t) - \lambda_a n(\mathbf{r},t) \; . \tag{4.14}$$

The continuity equation (4.14) is much simpler than the original Boltzmann equation, but it does not by itself determine either $n(\mathbf{r},t)$ or $\mathbf{J}(\mathbf{r},t)$; a further relation between current and density is needed. Such a relation is given by Fick's law, which is the frequently used approximation

$$\mathbf{J}(\mathbf{r},t) \approx -D_0\nabla n(\mathbf{r},t) \; . \tag{4.15}$$

The quantity D_0 is called the diffusion coefficient. Substitution of Eq. (4.15) into Eq. (4.14) leads to the familiar diffusion equation for thermal neutrons

$$\frac{\partial n(\mathbf{r},t)}{\partial t} = \nabla \cdot D_0\nabla n(\mathbf{r},t) + Q(\mathbf{r},t) - \lambda_a n(\mathbf{r},t) \; . \tag{4.16}$$

In Eq. (4.16) we have allowed for the possibility that D_0 depends on position. The Boltzmann equation (4.7) or (4.8) remains valid when all the transition probabilities $\Sigma_s(v)$, $\Sigma_a(v)$, and $W(\mathbf{v},\mathbf{v}')$ depend on position, but we have not explicitly indicated this position dependence. This dependence on position can arise from variation in composition with position or from variation in temperature with position. The microscopic cross sections $\sigma_a(E)$, $\sigma_s(E)$, and $\sigma(E,E',\theta)$ depend on temperature as do the atomic densities. It is important, however, to distinguish the variation of $\sigma_a(E)$ with temperature from the variation of a spectrum-averaged cross section with temperature. The former variation arises only from the Doppler broadening of absorption resonances and is normally negligible at thermal energies. The calculation of the latter variation is one of the principal objectives of thermalization theory. It arises from the temperature dependence of the neutron energy distribution, which, in turn, arises from the temperature dependence of the energy-transfer cross sections in the Boltzmann equation.

The diffusion equation (4.16) is a partial differential equation for a single scalar function of position and time. In neutron transport theory we consider it as a first approximation to the Boltzmann equation (4.8), applicable when the phase-space distribution is almost isotropic in velocity space and has an energy distribution that is independent of position and time. Even when this considerably simplified description is applicable, we must still use the Boltzmann equation to calculate D_0 in terms of the cross sections for individual collisions. For the simplest model of a reactor, a bare homogeneous mixture of fuel, absorber, and moderator, the calculation of D_0 depends on an accurate solution of the thermalization problem, since D_0 is sensitive to the neutron energy spectrum. This sensitivity is most clearly seen in the consideration of the criticality of bare homogeneous graphite-moderated assemblies in Chap. 7.

It is interesting to compare the roles of the diffusion equation and the Boltzmann equation for neutron diffusion with their roles for heat diffusion. In the former case, the Boltzmann equation gives an accurate and reasonably tractable description. When the diffusion equation is applicable, it is as a systematic first approximation to the

Boltzmann equation, and the diffusion coefficient D_0 is obtainable from a well-defined calculation. In heat diffusion, on the other hand, a scalar function $T(\mathbf{r},t)$, giving the local temperature, is almost always well-defined, and the heat flow is accurately given by

$$\mathbf{J}(\mathbf{r},t) = -\kappa\nabla T(\mathbf{r},t) \;, \tag{4.17}$$

where κ is the heat conductivity which may depend on T as well as on the material. Only in very special cases, however, is it possible to calculate κ from a Boltzmann equation. This can be done for a dilute gas by the methods of kinetic theory,[10] but for dense fluids a description in terms of binary collisions between molecules is inadequte. We therefore have a situation where the diffusion equation is an essentially exact description and the Boltzmann equation is a first approximation from which the heat conductivity can be estimated. (The development of a more accurate microscopic description of transport processes in fluids[11] is an active field of research in irreversible statistical mechanics which is beyond the scope of this book.) We have then an accurate description of the diffusion of heat by using Eq. (4.17) to obtain

$$\frac{\partial T(\mathbf{r},t)}{\partial t} = \nabla \cdot \kappa\nabla T(\mathbf{r},t) + Q(\mathbf{r},t) \;, \tag{4.18}$$

but we have no generally valid microscopic description of the dynamics of individual collisions from which we can calculate κ.

The situation appropriate to heat diffusion is far more common in physics than the one appropriate to neutron diffusion. Consider, for example, the hydrodynamic description of sound propagation. For a dilute gas, this can be derived from the Boltzmann equation, but the resulting hydrodynamic equations have a much wider range of validity than the Boltzmann equation from which they were "derived." This is not the case for neutron diffusion. The linear Boltzmann equation (4.8) is an essentially exact description of the migration of neutrons through macroscopic media. The diffusion equation (4.16) is a useful and frequently applicable approximation, but there are many practical situations where it is inadequate. These situations present no problems of principle, but only problems of increased complexity, since the transport equation (4.8) is usually difficult to solve.

4-3 THE APPROACH TO EQUILIBRIUM

The most characteristic problem of neutron thermalization theory is the approach of the neutron velocity distribution to thermal equilibrium with the moderator. In its emphasis on the velocity distribution rather than the spatial distribution, the physical content of the theory is close to that of the Boltzmann equation in the kinetic theory of gases. This is in contrast to the usual development of neutron transport theory where the emphasis on the spatial distribution is in close analogy to the theory of radiative transfer in stellar atmospheres. In our choice of notation we have stressed the similarity to kinetic theory by working with the neutron density in phase space and the transition probability per unit time. We will eventually find it preferable to work with a notation using the angular flux and energy-transfer cross sections, which are more convenient for reactor theory.

Before making this change, we consider the time evolution of the neutron velocity distribution in a uniform infinite homogeneous medium. The medium is assumed to be at a uniform temperature and to have a 1/v absorption cross section. We note first that the solution in the presence of a 1/v absorber can be written as

$$f_B(\mathbf{r},\mathbf{v},t) = \exp(-\lambda_a t) f_0(\mathbf{r},\mathbf{v},t) , \tag{4.19}$$

where $f_0(\mathbf{r},\mathbf{v},t)$ is the solution of Eq. (4.8) for zero absorption and $\lambda_a = v\Sigma_a$. This result, which can be verified by substitution of Eq. (4.19) into Eq. (4.8), is of fundamental importance in reactor theory. In the present case, it allows us to consider the approach to equilibrium without absorption and to carry over the general features of the result to the thermalization of neutrons in an absorbing system.

Let $n_0(\mathbf{v})$ be the integral over all space of the initial distribution $f_0(\mathbf{r},\mathbf{v},0)$. If this quantity is finite, then it is reasonable to expect that $f_0(\mathbf{r},\mathbf{v},t)$ also has an integral, $f_0(\mathbf{v},t)$, over all space, and that $f_0(\mathbf{r},\mathbf{v},t)$ approaches zero as $\mathbf{r}$ approaches infinity. An integration of Eq. (4.8) over all space, with f_B replaced by f_0 and Σ_a set equal to zero, then gives

$$\frac{\partial f_0(v,t)}{\partial t} = \int [W(\mathbf{v}',\mathbf{v}) f_0(\mathbf{v}',t) - W(\mathbf{v},\mathbf{v}') f_0(\mathbf{v},t)] d\mathbf{v}' . \tag{4.20}$$

Moreover,

$$f_0(\mathbf{v},0) = n_0(\mathbf{v}) \ , \tag{4.21}$$

where $n_0(\mathbf{v})$ is the (vector) velocity distribution of the source. The source may be localized in space, and the distribution $f_0(\mathbf{r},\mathbf{v},t)$ may have a spatial dependence, but the uniformity of the medium guarantees that the velocity distribution integrated over all space is the same as would obtain if there were no spatial dependence. Integration of Eq. (4.20) over all velocities gives the condition of over-all neutron conservation

$$N_0(t) = \int f_0(\mathbf{v},t)d\mathbf{v} = N_0 = \int n_0(\mathbf{v})d\mathbf{v} \ . \tag{4.22}$$

From Eqs. (4.22) and (4.19) it follows that the total number of neutrons in the presence of a 1/v absorber is given by

$$N(t) = \iint f_B(\mathbf{r},\mathbf{v},t)d\mathbf{r}d\mathbf{v} = N_0 \ \exp(-\lambda_a t) \ . \tag{4.23}$$

Since the absorption probability per unit time is independent of velocity, the neutron intensity decays exponentially independently of the way in which the velocity distribution changes in time. This simplifying feature is lost when the absorption cross section deviates from 1/v behavior, but it is a good first approximation in a reactor to assume that the absorption of neutrons and the approach to equilibrium occur independently. (In actual reactor calculations this assumption is not made, as it does not simplify the numerical calculations needed. We make it now to obtain some physical insight into the thermalization process.)

Equation (4.20) describes the approach of $f_0(\mathbf{v},t)$ to an equilibrium distribution through successive random collisions. For a dilute neutron gas, the equilibrium distribution is the Maxwellian

$$f_M(\mathbf{v}) = (m/2\pi k_B T)^{3/2} \ \exp(-mv^2/2k_B T) \tag{4.24}$$

at the moderator temperature T. That Eq. (4.20) leads to this as the equilibrium distribution follows directly from the detailed balance condition

$$W(\mathbf{v},\mathbf{v}')f_M(\mathbf{v}) = W(\mathbf{v}',\mathbf{v})f_M(\mathbf{v}') \tag{4.25}$$

satisfied by the transition probabilities for a single collision. Equation (4.25) can be derived from Eq. (4.6) and the detailed balance condition (2.79) that is satisfied by the microscopic differential-energy-transfer cross section. This condition arises from the thermodynamic equilibrium of the scattering system and from the microscopic reversibility of the interaction leading to the neutron scattering.

An interesting question is raised in considering the role of time reversibility in Eqs. (4.20) and (4.25). In order to obtain Eq. (4.20) from the original dynamical equations, which are invariant under time reversal, a statistical assumption must somewhere be made to derive the Boltzmann equation, which is not invariant under a change in the sign of time. This was the source of a great controversy in 19*th* century physics in which Boltzmann was violently attacked, but was proved to be essentially correct as the nature of the statistical assumption became better understood. There is a vast literature on this subject, both in classical and quantum mechanics. The classic work on the subject by the Ehrenfests[12] is still of interest. We note, however, that in order to get an approach to equilibrium from the irreversible classical Boltzmann equation (4.20) we must use the detailed balance condition (4.25) that arises from the reversible nature of the quantum-mechanical single-collision process. This problem arises in the mixed classical and quantum-mechanical description appropriate to neutron thermalization, and we do not pretend to understand it in any depth. As in the theory of transport phenomena in solids, a fundamental understanding depends on a clear statement of the underlying statistical assumption within a fully quantum-mechanical description. In the remainder of this book we remain within the framework defined by Eqs. (4.20) and (4.25) and do not return to the much more difficult question of the underlying physical basis of the Boltzmann equation for neutron migration.

From Eq. (4.25) it follows directly that a distribution which is initially Maxwellian at the moderator temperature will not change in time. We have the characteristic feature of statistical mechanics, i.e., the equilibrium distribution is independent of the mechanism for reaching equilibrium. The solution of Eq. (4.20) for the particular initial condition $n_0(\mathbf{v}) = N_0 f_M(\mathbf{v})$ is $f_0(\mathbf{v},t) = N_0 f_M(\mathbf{v})$, and this solution does not depend on $W(\mathbf{v},\mathbf{v}')$. To show that Eqs. (4.20) and (4.25) imply an approach to equilibrium for an arbitrary initial velocity distribution requires more explicit consideration. To do this, we follow a discussion given by Van Kampen[13] in a related context, and show that the entropy of the neutron distribution increases monotonically in time and is maximum for a Maxwellian distribution. If the transition probability $W(\mathbf{v},\mathbf{v}')$ is such that it is possible to get from the initial distribution to the equilibrium distribution, then it follows that the equilibrium distribution is approached at long times. The detailed time evolution of the distribution is, of course, quantitatively dependent on $W(\mathbf{v},\mathbf{v}')$ and on the initial conditions.

We define the entropy of the neutron distribution by

$$S(t) = -\int f_0(\mathbf{v},t)\ln\left[\frac{f_0(\mathbf{v},t)}{N_0 f_M(\mathbf{v})}\right]d\mathbf{v} \ . \tag{4.26}$$

Differentiating Eq. (4.26) with respect to time and using Eq. (4.20), we obtain

$$\frac{dS}{dt} = \iint [W(\mathbf{v},\mathbf{v}')f_0(\mathbf{v},t) - W(\mathbf{v}',\mathbf{v})f_0(\mathbf{v}',t)] \times \times \{1 + \ln[\rho(\mathbf{v},t)]\}d\mathbf{v}d\mathbf{v}' \ , \tag{4.27}$$

where

$$\rho(\mathbf{v},t) = \frac{f_0(\mathbf{v},t)}{N_0 f_M(\mathbf{v})} \ . \tag{4.28}$$

By using the detailed balance condition (4.25) and the conservation condition (4.22) we can rewrite Eq. (4.27) in the form

$$\frac{dS}{dt} = \tfrac{1}{2}\iint W(\mathbf{v},\mathbf{v}')N_0 f_M(\mathbf{v})\{[\rho(\mathbf{v}',t) - \rho(\mathbf{v},t)] \times \times \ln[\rho(\mathbf{v}',t)/\rho(\mathbf{v},t)]\}d\mathbf{v}d\mathbf{v}' \ . \tag{4.29}$$

Since $W(\mathbf{v},\mathbf{v}')$, $f_M(\mathbf{v})$, and the term in braces in Eq. (4.29) are all non-negative, it follows that

$$\frac{dS}{dt} \geq 0 \ . \tag{4.30}$$

This is Boltzmann's famous H-theorem for our special case. If $W(\mathbf{v},\mathbf{v}')$ is a sufficiently smooth function, dS/dt will be equal to zero only if the term in braces in Eq. (4.29) is identically zero. This happens only for

$$\rho(\mathbf{v},t) = 1 \quad \text{or} \quad f_0(\mathbf{v},t) = N_0 f_M(\mathbf{v}) ,$$

i.e., only when the neutron distribution is an equilibrium Maxwellian. For any actual moderator, it is plausible but not proved that $W(\mathbf{v},\mathbf{v}')$ is sufficiently smooth that the equilibrium distribution is eventually reached. (In the singular case of an isotropic harmonic oscillator, however, where the energy transfer is quantized in multiples of $\hbar\omega_0$, the neutron distribution does not approach a true equilibrium, and the asymptotic behavior of $f_0(\mathbf{v},t)$ at long times depends on the initial velocity distribution.)

To show that the equilibrium Maxwellian distribution maximizes the entropy, we use Eqs. (4.22) and (4.28) and the normalization of $f_M(\mathbf{v})$ to rewrite Eq. (4.26) in the form

$$S(t) = N_0 \int f_M(\mathbf{v})\{\rho(\mathbf{v},t) - 1 - \rho(\mathbf{v},t)\ \ln[\rho(\mathbf{v},t)]\}d\mathbf{v} . \qquad (4.31)$$

The function $(\rho - 1 - \rho \ln \rho)$ is negative for all positive of ρ except $\rho = 1$, and is zero for this value. The integrand of Eq. (4.31) is thus zero for an equilibrium Maxwellian distribution and negative otherwise.

Finally, consider the steady-state velocity distribution in the presence of absorption. It is generally true that the distribution for a steady source is the integral over time of the distribution for a pulsed source. This exact relation is the basis of the Poole technique for measuring reactor spectra discussed in Chap. 6. For the special case of 1/v absorption, we obtain

$$\Psi(\mathbf{v},\lambda_a) = \iint f_B(\mathbf{r},\mathbf{v},t)d\mathbf{r}dt = \int_0^\infty \exp(-\lambda_a t) f_0(\mathbf{v},t)dt . \qquad (4.32)$$

We are frequently interested in "thermal" reactors where the absorption time is long compared with the thermalization time. For this case, we can expand in powers of λ_a if we take care to note that the leading term is of order of $1/\lambda_a$, and that this term, which is Maxwellian, must be integrated over time before the weak-absorption limit is taken. We thus obtain

$$\lambda_a \Psi(\mathbf{v},\lambda_a) = \lambda_a \int_0^\infty \exp(-\lambda_a t)\{N_0 f_M(\mathbf{v}) + [f_0(\mathbf{v},t) - N_0 f_M(\mathbf{v})]\}dt$$

$$= N_0 f_M(\mathbf{v}) + \lambda_a \int_0^\infty [f_0(\mathbf{v},t) - N_0 f_M(\mathbf{v})]dt + O(\lambda_a^2) \ . \quad (4.33)$$

In a reactor the source of fission neutrons is isotropic so that we need consider only the approach of the energy distribution to equilibrium. The energy distribution in the weak-absorption limit is discussed in detail in Sec. 5-2. In the present discussion, we have also included the approach to isotropy of the angular distribution. In problems of interest in neutron transport theory, this effect is usually closely coupled to the spatial dependence of the neutron distribution, so that the additional generality of the present discussion is of little practical importance.

This completes our general discussion of the approach to equilibrium. We return to this subject in a more specific context in Chap. 8. At this point, we find it convenient to make a change in notation to conform more closely to the usual notation of nuclear reactor theory.

4-4 THE CONSTANT-CROSS-SECTION APPROXIMATION

We have noted the simplification that occurs in time-dependent problems when the absorption cross section varies inversely as the neutron speed. The total neutron density then has a time-dependence that is independent of the way in which the neutron energy spectrum varies with time. A similar simplification occurs in space-dependent problems when all cross sections are independent of neutron energy. The dependence of neutron flux on position and on direction of motion is then determined independently of the variation of the neutron energy spectrum with position. The total neutron flux integrated over energy satisfies the same transport equation that describes the migration of monoenergetic neutrons with no energy transfer. This enables a useful contact to be made between thermalization theory and the extensively developed theory of monoenergetic neutron transport.

For these and subsequent considerations, it is more convenient to work with the neutron energy E and direction of motion $\boldsymbol{\Omega}$ as independent variables, rather than with the vector velocity **v**. Following Weinberg and Wigner,[3] we introduce the angular flux $f(\mathbf{r},E,\boldsymbol{\Omega},t)$, which is related to the phase-space density $f_B(\mathbf{r},\mathbf{v},t)$ through

$$f(\mathbf{r},E,\boldsymbol{\Omega},t) = (v^2/m)f_B(\mathbf{r},\mathbf{v},t) \;. \tag{4.34}$$

The angular flux (4.34) is defined such that $f(r,E,\Omega,t)drd\boldsymbol{\Omega}dE$ is the neutron speed, $v = (2E/m)^{\frac{1}{2}}$, times the number of neutrons in the volume element $d\mathbf{r}$ at $\mathbf{r}$ and in the energy interval dE at E whose directions of motion lie in the solid angle $d\Omega$ around Ω. This angular flux gives the number of neutrons in $d\mathbf{r}$ and dE which cross the element $d\Omega$ on the unit sphere per second. The element is normal to the direction $\boldsymbol{\Omega}$.

The neutron density (4.9) is given by

$$n(\mathbf{r},t) = \iint (1/v)f(\mathbf{r},E,\boldsymbol{\Omega},t)dEd\boldsymbol{\Omega} \;, \tag{4.35}$$

and the current (4.10) is given by

$$\mathbf{J}(\mathbf{r},t) = \iint \boldsymbol{\Omega}f(\mathbf{r},E,\boldsymbol{\Omega},t)dEd\boldsymbol{\Omega} \;. \tag{4.36}$$

New quantities of interest are the total angular flux

$$f(\mathbf{r},\boldsymbol{\Omega},t) = \int_0^\infty f(\mathbf{r},E,\boldsymbol{\Omega},t)dE \tag{4.37}$$

and the total scalar flux

$$\Phi(\mathbf{r},t) = \int f(\mathbf{r},\boldsymbol{\Omega},t)d\boldsymbol{\Omega} = \iint f(\mathbf{r},E,\boldsymbol{\Omega},t)dEd\boldsymbol{\Omega} \;. \tag{4.38}$$

The total scalar flux is related to the neutron density through the average neutron speed:

$$\Phi(\mathbf{r},t) = \hat{v}(\mathbf{r},t)n(\mathbf{r},t) \;, \tag{4.39}$$

where

$$\hat{v}(\mathbf{r},t) = \frac{\iint f(\mathbf{r},E,\boldsymbol{\Omega},t)\,dE\,d\boldsymbol{\Omega}}{\iint (1/v) f(\mathbf{r},E,\boldsymbol{\Omega},t)\,dE\,d\boldsymbol{\Omega}}$$

$$= \frac{\iint v f_B(\mathbf{r},\mathbf{v},t)\,d\mathbf{v}}{\iint f_B(\mathbf{r},\mathbf{v},t)\,d\mathbf{v}} \tag{4.39a}$$

A frequently used form of the diffusion approximation is

$$\mathbf{J}(\mathbf{r},t) \approx -D\nabla\Phi(\mathbf{r},t). \tag{4.40}$$

If $\hat{v}(\mathbf{r},t)$ is independent of position, this is equivalent to the approximation (4.15) with

$$D_0 = D\hat{v} . \tag{4.41}$$

If $\hat{v}$ depends on position, Eqs. (4.15) and (4.40) are not equivalent and neither one is usually a good approximation. In Sec. 4-6 we develop a generalization of Eq. (4.40) which is often applicable even when the neutron energy spectrum is changing with position. Before considering new approximations, however, we restate the Boltzmann equation in our new notation.

The Boltzmann equation (4.7) for the angular flux (4.34) is

$$\frac{1}{v}\frac{\partial f(\mathbf{r},E,\boldsymbol{\Omega},t)}{\partial t} = -[(\boldsymbol{\Omega}\cdot\nabla) + \Sigma_a(E,\mathbf{r}) + \Sigma_s(E,\mathbf{r})]f(\mathbf{r},E,\boldsymbol{\Omega},t)$$
$$+ S(\mathbf{r},E,\boldsymbol{\Omega},t) + \iint \Sigma(\mathbf{r},E' \to E,\boldsymbol{\Omega}' \to \boldsymbol{\Omega}) \times$$
$$\times f(\mathbf{r},E',\boldsymbol{\Omega}',t)\,dE'\,d\boldsymbol{\Omega}' . \tag{4.42}$$

We refer to this equation as the transport equation. In Eq. (4.42), $S(\mathbf{r},E,\boldsymbol{\Omega},t)$ is the source strength per unit energy and per unit solid angle. The macroscopic energy-transfer cross section is defined by

$$\Sigma(\mathbf{r},E' \to E,\boldsymbol{\Omega}' \to \boldsymbol{\Omega}) = \sum_i N_i(\mathbf{r})\sigma_i(E',E,\theta) , \tag{4.43}$$

where $\sigma_i(E',E,\theta)$ is the double differential cross section (defined in Sec. 2-2) and the atomic concentration of the *ith* constituent is $N_i(\mathbf{r})$. We assume randomly-oriented polycrystalline material; thus this cross section depends only on $\cos\theta = \mathbf{\Omega}' \cdot \mathbf{\Omega}$ and not separately on the initial and final directions of motion. (This assumption is not always accurately justified for reactor graphite, but we do not consider the effects of anisotropic diffusion which occur in this case.)

The macroscopic scattering cross section is related to the energy-transfer cross section through

$$\Sigma_s(E,\mathbf{r}) = \iint \Sigma(\mathbf{r},E \to E',\mathbf{\Omega} \to \mathbf{\Omega}')de'd\mathbf{\Omega}' \ . \qquad (4.44)$$

Under the assumption of random orientation, $\Sigma_s(E,\mathbf{r})$ defined by Eq. (4.44) is independent of $\mathbf{\Omega}$. It is often more useful to rewrite Eq. (4.42) in a form similar to that of Eq. (4.8), in which the scattering into and out of a given energy and directional interval are grouped together. Making use of Eq. (4.44) to do this, we obtain

$$\frac{1}{v}\frac{\partial f}{\partial t} = -[(\mathbf{\Omega} \cdot \nabla) + \Sigma_a(\mathbf{r},E)]f(\mathbf{r},E,\mathbf{\Omega},t) + S(\mathbf{r},E,\mathbf{\Omega},t) + $$
$$+ \iint [\Sigma(\mathbf{r},E' \to E,\mathbf{\Omega}' \to \mathbf{\Omega})f(\mathbf{r},E',\mathbf{\Omega}',t) - $$
$$- \Sigma(\mathbf{r},E \to E',\mathbf{\Omega} \to \mathbf{\Omega}')f(\mathbf{r},E,\mathbf{\Omega},t)]dE'd\mathbf{\Omega}' \ . \qquad (4.45)$$

For time-independent problems, a considerably simpler but still self-contained equation can be obtained if the macroscopic differential cross section

$$\Sigma(\mathbf{r},E',\mathbf{\Omega} \cdot \mathbf{\Omega}') = \int_0^\infty \Sigma(\mathbf{r},E' \to E,\mathbf{\Omega}' \to \mathbf{\Omega})dE \qquad (4.46)$$

and the absorption cross section $\Sigma_a(\mathbf{r},E)$ are independent of the initial neutron energy. We can then integrate Eq. (4.42) over energy to get

$$[(\mathbf{\Omega} \cdot \nabla) + \Sigma_a(\mathbf{r}) + \Sigma_s(\mathbf{r})]f(\mathbf{r},\mathbf{\Omega})$$
$$= \int \Sigma(\mathbf{r},\mathbf{\Omega}' \cdot \mathbf{\Omega})f(\mathbf{r},\mathbf{\Omega}')d\mathbf{\Omega}' + S(\mathbf{r},\mathbf{\Omega}) \ . \qquad (4.47)$$

where

$$S(\mathbf{r},\Omega) = \int_0^\infty S(\mathbf{r},E,\Omega)dE \ .$$

We thus have an equation for the total angular flux (4.37) which does not depend on the energy-transfer properties of the moderator. The neutron energy spectrum may vary with position, but the energy integral of the angular flux can be calculated from Eq. (4.47) independently of this variation.

The transport equation (4.47) for the migration of monoenergetic neutrons as the equation governing the steady-state total flux of polyenergetic neutrons in the special case of constant cross sections is thus the same as the transport equation for the flux of monoenergetic neutrons which diffuse without changing their energy. This point of view toward the transport equation (4.47) has been emphasized by Davison and Sykes,[1] and it is very useful in making contact with the extensively developed techniques of monoenergetic transport theory. It is important, however, to recognize the severely limited applicability of this description for actual problems of thermal-neutron migration.

We note first the restriction to time-independent problems. When Eq. (4.42) is integrated over energy, the time-derivative term becomes $\partial n(\mathbf{r},t)\partial t$, which is related to the flux through the average speed $\hat{v}(\mathbf{r},t)$, which in turn requires the full energy-dependent transport equation (4.42) for its determination. The time scale for a collision is $(v\Sigma)^{-1}$, while the distance scale for a collision is Σ^{-1}; these cannot both be energy-independent. If the former is constant, we get a simplified description of the total neutron density in space-independent problems; if the latter is constant, we get a simplified description of the total neutron flux in time-independent problems. Although the monoenergetic transport equation is often used to describe space- and time-dependent problems, it applies only to those time-dependent problems in which energy exchange between neutrons and the scattering medium is negligible. It is not a useful approximation for calculating the time-dependence of neutron migration in moderating materials.

Even for time-independent problems, integrating Eq. (4.42) over energy only eliminates the neutron energy when all cross sections are constant. They are not constant at thermal energies where absorption cross sections typically vary as 1/v and where scattering cross sections can also be strongly energy dependent. Even if the cross sections are constant, Eq. (4.47) gives only the spatial distribution of the energy-integrated flux. The spatial distribution of neutron "flux" is usually measured by activation of foils. In order to measure the quantity determined by Eq. (4.47), a foil material with an energy-independent activation cross section is required. For the usual 1/v activation cross section, it is the neutron density $n(\mathbf{r})$ that is measured. If the average neutron speed $\hat{v}(\mathbf{r})$, defined by Eq. (4.39a), is changing with position, the full energy-dependent transport equation (4.42) must be solved in order to calculate the "flux" that is measured by foil activation. The actual energy-integrated neutron flux $\Phi(\mathbf{r})$ could be measured by determining the scattering rate of an "ideal" scatterer with constant isotropic cross section. Such techniques have proved feasible for measuring the energy spectrum of the scalar flux and are discussed in Chap. 6; however, for spatial distributions the burden is on the theory to calculate the neutron densities measured by foil activation, since these are, in any case, more closely related to the reaction rates of primary interest in a reactor. The constant-cross-section approximation and its associated reduced description, Eq. (4.47), are thus primarily useful in furnishing a limited description in which analytic results are known. At least for problems of thermal-neutron migration, they are not a practical calculational tool.

4-2 SPHERICAL HARMONICS AND THE P_1 APPROXIMATION

The last two terms on the right-hand side of Eq. (4.42) are usually called the "birth-rate density." For thermal-neutron problems these terms are almost always nearly isotropic, even when the angular flux itself is strongly anisotropic. Because of this fact and the dependence of the energy-transfer cross section only on $\boldsymbol{\Omega} \cdot \boldsymbol{\Omega}'$, it is convenient to expand the birth-rate density in spherical harmonics. If we introduce the spherical-harmonics expansion of the source

$$S(\mathbf{r},E,\Omega,t) = \sum_{\ell=0}^{\infty} \sum_{m=-\ell}^{\ell} S_{\ell m}(\mathbf{r},E,t) P_{\ell m}(\Omega) \tag{4.48}$$

and of the angular flux

$$f(\mathbf{r},E,\Omega,t) = \sum_{\ell=0}^{\infty} \sum_{m=-\ell}^{\ell} f_{\ell m}(\mathbf{r},E,t) P_{\ell m}(\Omega) , \tag{4.49}$$

and the Legendre polynomial expansion of the energy-transfer cross section

$$\Sigma(\mathbf{r},E \to E',\Omega \to \Omega') = \sum_{\ell=0}^{\infty} \frac{2\ell + 1}{4\pi} \Sigma_{\ell}(\mathbf{r},E,E') P_{\ell}(\Omega \cdot \Omega') , \tag{4.50}$$

the birth-rate density becomes

$$\begin{aligned} h(\mathbf{r},E,\Omega,t) &\equiv S(\mathbf{r},E,\Omega,t) + \\ &+ \iint \Sigma(\mathbf{r},E' \to E,\Omega' \to \Omega) f(\mathbf{r},E',\Omega',t) dE' d\Omega' \\ &= \sum_{\ell,m} [S_{\ell m}(\mathbf{r},E,t) + \\ &+ \int_0^{\infty} \Sigma_{\ell}(\mathbf{r},E',E) f_{\ell m}(\mathbf{r},E',t) dE'] P_{\ell m}(\Omega) . \end{aligned} \tag{4.51}$$

The spherical harmonics used here are defined by Eq. (9.4a) of Weinberg and Wigner,[3] and the derivation of Eq. (4.51) uses the spherical harmonics addition theorem

$$P_{\ell}(\Omega \cdot \Omega') = \sum_{m=-\ell}^{\ell} P_{\ell m}(\Omega) P^{*}_{\ell m}(\Omega') . \tag{4.52}$$

The notation in this section follows Weinberg and Wigner except that $\Sigma_{\ell}(\mathbf{r},E,E')$ defined by Eq. (4.50) $[4\pi/(2\ell + 1)]$ times $s_{\ell}(E,E')$ defined by Weinberg and Wigner. For completeness, we summarize in Appendix IV some of the more useful properties of the spherical harmonics used here.

We note that the right-hand side of Eq. (4.51) is nearly isotropic (for a nearly isotropic source) if either the flux or the cross section is nearly isotropic. For thermal-neutron problems, the cross section can almost always be well represented by a few terms in the expansion (4.50) so that a few terms in the expansion (4.51) are sufficient. If, in addition, the space and time variations of the flux are sufficiently slow, the expansion (4.49) is also rapidly convergent. If we neglect all terms for $\ell > L$ in Eq. (4.49), we obtain the familiar P_L approximation to transport theory. The P_1 approximation is particularly attractive physically since the P_0 and P_1 components of the angular flux are the scalar flux and the current, respectively. It is well known[3] that this approximation is not identical to diffusion theory for energy-dependent (and/or time-dependent) problems, and it is helpful to consider this circumstance explicitly for thermal-neutron migration.

Introducing the flux per unit energy

$$\phi(\mathbf{r},E,t) = \int f(\mathbf{r},E,\boldsymbol{\Omega},t)d\boldsymbol{\Omega} \tag{4.53}$$

and the current per unit energy

$$\mathbf{j}(\mathbf{r},E,t) = \int \boldsymbol{\Omega} f(\mathbf{r},E,\boldsymbol{\Omega},t)d\boldsymbol{\Omega} \ , \tag{4.54}$$

we find that the P_1 approximation takes the form

$$f(\mathbf{r},E,\boldsymbol{\Omega},t) \approx \frac{1}{4\pi} [\phi(\mathbf{r},E,t) + 3\boldsymbol{\Omega} \cdot \mathbf{j}(\mathbf{r},E,t)] \ . \tag{4.55}$$

This approximation leads to two coupled equations for $\mathbf{j}(\mathbf{r},E,t)$ and $\phi(\mathbf{r},E,t)$ which are usually called the consistent P_1 equations. The first equation is exact and is obtained by integrating Eq. (4.35) over solid angle to give

$$\begin{aligned} \frac{1}{v}\frac{\partial\phi(\mathbf{r},E,t)}{\partial t} = {} & -\Sigma_a(\mathbf{r},E)\phi(\mathbf{r},E,t) - \nabla \cdot \mathbf{j}(\mathbf{r},E) + \\ & + 4\pi S_{00}(\mathbf{r},E,t) + \int_0^\infty [\Sigma_0(\mathbf{r},E',E)\phi(\mathbf{r},E',t) - \\ & - \Sigma_0(\mathbf{r},E,E')\phi(\mathbf{r},E,t)]dE' \ . \end{aligned} \tag{4.56}$$

If Eq. (4.56) is integrated over energy, the scattering terms cancel and the equation of continuity, Eq. (4.12) is obtained. It is convenient throughout much of this book to have a shorthand for energy-exchange integrals such as those appearing in Eq. (4.56). We introduce the notation

$$R_\ell \phi(\mathbf{r},E,t) \equiv \int_0^{E_m} [\Sigma_\ell(\mathbf{r},E',E)\phi(\mathbf{r},E',t) - - \Sigma_\ell(\mathbf{r},E,E')\phi(\mathbf{r},E,t)]dE' \tag{4.57}$$

in terms of which Eq. (4.56) becomes

$$\frac{1}{v}\frac{\partial\phi}{\partial t} = -\Sigma_a\phi - \nabla \cdot \mathbf{j} + 4\pi S_{00} + R_0\phi \ . \tag{4.58}$$

In Eq. (4.57) E_m is an arbitrary maximum thermal-neutron energy, which can be allowed to approach infinity in some problems but is best kept finite in others. This point is discussed further in Chaps. 5 and 7.

The second equation coupling $\mathbf{j}$ and ϕ is obtained by multiplying Eq. (4.45) by $\mathbf{\Omega}$, integrating over solid angle, and neglecting those terms proportional to the P_2 components of the angular flux. If we add and subtract a term $\Sigma_1(\mathbf{r},E)\mathbf{j}(\mathbf{r},E,t)$, where

$$\Sigma_\ell(\mathbf{r},E) \equiv \int_0^{E_m} \Sigma_\ell(\mathbf{r},E,E')dE' \ , \tag{4.59}$$

we obtain

$$\frac{1}{v}\frac{\partial \mathbf{j}(\mathbf{r},E,t)}{\partial t} = -[\Sigma_a(\mathbf{r},E) + \Sigma_s(\mathbf{r},E) - \Sigma_1(\mathbf{r},E)]\mathbf{j}(\mathbf{r},E,t) + + R_1\mathbf{j}(\mathbf{r},E,t) - \frac{1}{3}\nabla\phi(\mathbf{r},E,t) \ . \tag{4.60}$$

The bracketed term on the right-hand side of Eq. (4.60) is the macroscopic transport cross section $\Sigma_{tr}(\mathbf{r},E)$, and its inverse is the transport mean free path. This is in accord with the conventional definition in the P_1 approximation, since $\Sigma_1(\mathbf{r},E)$ can be written as

$$\Sigma_1(\mathbf{r},E) = \Sigma_s(\mathbf{r},E)\bar{\mu}(\mathbf{r},E) \ ,$$

where $\bar{\mu}(\mathbf{r},E)$ is the average value of the cosine of the scattering angle in the laboratory system for a neutron of energy E making a collision with a material of composition appropriate to the position **r**.

We have then two coupled equations, (4.58) and (4.60), for the flux and current per unit energy in the P_1 approximation. (Actually Eq. (4.60) is a vector equation and represents three scalar equations.) These equations differ from diffusion theory because of the presence of the energy-exchange term $R_0\phi$, the anisotropic energy-transfer term $R_1\mathbf{j}$, and the time-derivative term $(1/v)\partial\mathbf{j}/\partial t$. If the last two terms are neglected, a modified form of the diffusion approximation results, which we call energy-dependent diffusion theory. This approximation is comparable in accuracy to the consistent P_1 approximation for thermal-neutron diffusion in reactors and is conveniently adapted to the usual multigroup diffusion equations used in computer calculations of criticality. Much of the succeeding discussion will be in terms of this approximation. We note that the energy-dependent diffusion approximation allows for an energy spectrum that depends on position and/or time. When the energy spectrum is separable, the ordinary diffusion equation results. We now proceed to develop these approximations in more detail.

4-6 DIFFUSION THEORY FOR THERMAL NEUTRONS

The time-derivative term on the left-hand side of Eq. (4.60) is negligible unless the current varies appreciably over a time $(v\Sigma_{tr})^{-1}$. If the current varies significantly over the time between collisions, the P_1 approximation is not quantitatively applicable in any case; thus it is reasonable to neglect this term. Its inclusion in the monoenergetic case, discussed by Weinberg and Wigner,[3] leads to a qualitative understanding of the "wave front" associated with a localized source of neutrons; however, a quantitative treatment of this phenomenon requires a more accurate approximation to the angular flux. (A comparable argument cannot be made concerning the left-hand side of Eq. (4.58), since the scattering terms on the right-hand side largely cancel, and the absorption, if $1/v$, can be taken out of the problem through use of Eq. (4.19).)

The anisotropic energy-transfer kernel $\Sigma_1(E,E')$ is weighted toward much smaller energy transfers than is the isotropic term $\Sigma_0(E,E')$. Furthermore, the cross section does not vary greatly over the average energy change in a collision. For these two reasons, the term $R_1\mathbf{j}(E)$ in Eq. (4.60) tends to be small. This term is important in reactor theory only for slowing down by hydrogen, and there it is significant primarily because of the rapid variation of cross section with energy in the meV region. If we neglect this term also, we obtain the energy-dependent diffusion approximation

$$\mathbf{j}(\mathbf{r},E,t) \approx -D(\mathbf{r},E)\nabla\phi(\mathbf{r},E,t) , \tag{4.62}$$

where

$$D(\mathbf{r},E) = [3\Sigma_{tr}(\mathbf{r},E)]^{-1} , \tag{4.63}$$

and $\Sigma_{tr}(E)$ is the bracketed term on the right-hand side of Eq. (4.60). Substituting Eq. (4.62) into Eq. (4.58) gives the energy-dependent diffusion equation

$$\frac{1}{v}\frac{\partial\phi(\mathbf{r},E,t)}{\partial t} = -\Sigma_a(\mathbf{r},E)\phi(\mathbf{r},E,t) + \nabla\cdot D(\mathbf{r},E)\nabla\phi(\mathbf{r},E,t) + 4\pi S_{00}(\mathbf{r},E,t) + R_0\phi(\mathbf{r},E,t) . \tag{4.64}$$

The accuracy with which this equation approximates the consistent P_1 equations has been checked only for a few special cases, which are discussed in Chap. 8. For these cases, the error in replacing Eq. (4.60) by Eq. (4.62) is of the order of 1% for hydrogen and less for heavier moderators.

Equation (4.64) is a convenient starting point for many discussions of neutron thermalization and thermal-neutron diffusion. It includes, to a good approximation, the approach to equilibrium in moderating samples that are large compared with a mean free path and the change of the neutron energy spectrum with position in adjacent homogeneous media. It is not adequate to describe the space and energy distribution of thermal neutrons in heterogeneous reactors. Such a description requires more accurate approximations for the angular flux than does the P_1 approximation. Except for the eigenvalue problems discussed in Chap. 8, which describe relaxation lengths and relaxation times in infinite homogeneous media, the more accurate solutions to the transport equation employ exclusively the numerical methods used in Chap. 9.

The approximate relation (4.62) between current and flux eliminates the direction of motion Ω an an independent variable in the neutron-balance equation, but the resulting diffusion equation (4.64) is still an integral equation in energy as well as a partial differential equation in position and time. It remains too complicated to solve for a general space, time, and energy dependence, but we can gain some insight by considering special problems in which the energy dependence is separable. For these problems, it is important to note two general properties of the integral operator R_ℓ defined by Eq. (4.57). The first of these is the property of neutron conservation, which follows directly from the definition. This is conveniently written as

$$\int_0^{E_m} R_\ell \phi(E) dE = 0 , \tag{4.65}$$

where $\phi(E)$ is an arbitrary function. The second property is a restatement of the detailed balance condition (4.25) in the form

$$\Sigma_\ell(E',E)\phi_M(E') = \Sigma_\ell(E,E')\phi_M(E) , \tag{4.66}$$

which implies

$$R_\ell \phi_M(E) = 0 , \tag{4.66a}$$

where

$$\phi_M(E) = \beta^2 E \exp(-\beta E) \tag{4.67}$$

is the flux per unit energy corresponding to the equilibrium Maxwellian distribution $f_M(\mathbf{v})$ given by Eq. (4.25). In Eq. (4.67), $\beta = (1/k_B T)$ and the normalization is chosen so that

$$\int_0^{\infty} \phi_M(E) dE = 1 . \tag{4.68}$$

To solve Eq. (4.64) in any finite geometry, we must specify spatial boundary conditions. Before considering these questions, we consider the steady-state problem with a plane isotropic source in an infinite homogeneous medium. This problem has been studied extensively in monoenergetic neutron transport theory. In the monoenergetic case, the diffusion and P_1 approximations are identical and are relatively trivial mathematically. This is no longer the case when the thermalization aspects of the problem are included. Even in diffusion theory the determination of the diffusion length requires the solution of a nontrivial eigenvalue problem, whereas the mean-square distance traveled from birth to absorption is the solution to a different problem. By comparing the diffusion length as an eigenvalue with the mean-square distance traveled from birth to absorption, we can obtain considerable insight into the expanded physical content of neutron diffusion when the energy exchange between neutrons and moderator is included.

4-7 THE DIFFUSION LENGTH

In monoenergetic neutron transport theory, the neutron flux at large distances from a plane source asymptotically approaches an exponential spatial behavior,

$$\Phi(z) = \exp(-z/L) , \tag{4.69}$$

where L is known as the diffusion length. (For other geometries the result generalizes to that solution of $(L^2\nabla^2 - 1)\Phi = 0$ having the appropriate symmetry, but this does not affect the physical content of the present discussion.) The calculation of the diffusion length as a function of the ratio of absorption to scattering is a classic problem of neutron transport theory. In the limit of weak absorption, the diffusion length is accurately given by the P_1 approximation result:

$$L^2 \approx L_0^2 = (3\Sigma_a\Sigma_{tr})^{-1} = (D/\Sigma_a) . \tag{4.70}$$

In this weak-absorption limit the asymptotic solution (4.69) is dominant, except within a few mean free paths of the source, where transients associated with a more accurate treatment of the angular flux becomes important. As the absorption becomes stronger, the asymptotic solution retains the form of Eq. (4.69), but the distance from the source at which it becomes dominant increases. The diffusion length approaches the mean free path Σ^{-1}, and Eq. (4.70) becomes a poor approximation.

As noted in Sec. 4-4, this description applies directly to the flux of thermal neutrons integrated over energy if all cross sections are independent of energy. For the energy-dependent cross sections appropriate to thermal-neutron diffusion, however, the situation is appreciably different. For sufficiently weak absorption, the asymptotic flux retains the form of Eq. (4.69) and is associated with a characteristic energy spectrum that is independent of position. The diffusion length becomes an eigenvalue of an integral equation in the energy variable, and the energy spectrum is the associated positive eigenfunction. In the limit of weak absorption, the diffusion length is of the form (4.70), with Σ_a and Σ_{tr} replaced by appropriate averages over a Maxwellian energy spectrum. As the absorption is increased, the energy spectrum departs from Maxwellian and corrections to the P_1 approximation become important. In Chap. 8, a systematic study of the eigenvalue problem defining the diffusion length as a function of absorber concentration is presented. This study shows that departures from a Maxwellian energy spectrum and departures from the P_1 approximation are of the same order, and are very nearly separable. It also shows that the diffusion length in the P_1 approximation differs from that in energy-dependent diffusion theory even in the weak-absorption limit, but that this difference is very small in practice. The dominant correction to the diffusion length, however, is found to be due to the "diffusion heating" of the energy spectrum which occurs in the energy-dependent diffusion approximation (Eq. (4.72). Our analysis gives the correct prescription for averaging the energy-dependent cross sections to obtain the diffusion length as well as a good estimate of the second-order term in the expansion of $1/L^2$ in powers of Σ_a. It is therefore sufficient for the interpretation of the measured diffusion lengths in pure moderating materials. The measurement of diffusion lengths and analysis of the measurement on a more systematic basis are discussed in detail in Chap. 8.

We consider an infinite homogeneous medium with a plane isotropic time-independent source at z = 0. The source-free equation for $z \neq 0$ is

$$\left[\Sigma_a(E) - D(E)\frac{\partial^2}{\partial z^2}\right]\phi(z,E) - R_0\phi(z,E) = 0 . \tag{4.71}$$

We look for solutions of the form

$$\phi_\kappa(E)\ \exp(\pm\kappa z)$$

which define the eigenvalue problem

$$[\Sigma_a(E) - D(E)\kappa^2]\phi_\kappa(E) - R_0\phi_\kappa(E) = 0\ . \tag{4.72}$$

If E_m in Eq. (4.57) goes to infinity, Eq. (4.72) becomes a singular homogeneous Fredholm integral equation of the second kind. In the absence of a slowing-down source, we must choose those solutions of Eq. (4.72) which fall off exponentially at high energy. Then there is no difficulty with the singular nature of the equation. Physically, we are interested in the diffusion of the fully thermalized neutrons far from the source, so that the characteristic spectrum does not have a 1/E tail at high energy but behaves qualitatively like a Maxwellian. (This is to be contrasted with the spectrum in a reactor core that is fed by a slowing-down source. For this latter case, the singularity of the integral equation at $E = \infty$ is important. We return to this question in Sec. 4-8 and in Chap. 5.) We consider only the lowest eigenvalue $\kappa_0^2 = 1/L_0^2$ and the associated positive eigenfunction $\phi_0(E)$. We assume that the remaining eigenvalues are separated from κ_0 by a finite amount. If this is so, the asymptotic solution far from the source is of the form

$$\phi_0(E)\ \exp(\pm\kappa_0 z) \tag{4.72a}$$

(with the minus sign for positive z and the plus sign for negative z) and is independent of the energy spectrum of the source. The complete solution of Eq. (4.71) also contains contributions from the higher eigenfunctions of Eq. (4.72), which represent the transient behavior of $\phi(z,E)$. The amplitudes of these higher eigenfunctions depend on the source energy spectrum. The existence of these "energy transients" is a new feature of the energy-dependent problem that is already present in the diffusion approximation. We recall that there are no transients in the diffusion approximation to monoenergetic transport theory. It therefore follows that there are no transients in the total flux $\Phi(z)$ for constant cross section. For constant cross section, Eq. (4.71) has a solution in which the total flux decays exponentially with the relaxation length given by Eq. (4.70), but the energy spectrum changes with position in a way that is determined by the complete solution of Eq. (4.72) subject to the boundary condition specified by the source energy spectrum. For energy-dependent cross sections, there are transients in the total flux as well. The quantitative discussion of the transient behavior depends on the eigenvalue spectrum of Eq. (4.72) and, more generally, on the spectrum of relaxation lengths for the homogeneous transport equation. We assume now that the lowest eigenvalu eigenvalue of Eq. (4.72) exists and that the asymptotic solution to Eq. (4.71) is of the form (4.72a).

Before considering the solution of Eq. (4.72), we integrate it over energy and use Eq. (4.65) to obtain

$$L_0^2 = \kappa_0^{-2} = \frac{\int_0^\infty D(E)\phi_0(E)dE}{\int_0^\infty \Sigma_a(E)\phi_0(E)dE} . \tag{4.73}$$

We find that the appropriate generalization of Eq. (4.70) is

$$L_0^2 = \bar{D}/\bar{\Sigma}_a , \tag{4.74}$$

where $\bar{\Sigma}_a$ and $\bar{D}$ are defined by

$$\bar{\Sigma}_a = \frac{\int_0^\infty \Sigma_a(E)\phi_0(E)dE}{\int_0^\infty \phi_0(E)dE} , \tag{4.75}$$

and

$$\bar{D} = \frac{\int_0^\infty D(E)\phi_0(E)dE}{\int_0^\infty \phi_0(E)dE} , \tag{4.76}$$

We have thus derived the important result that the diffusion length for energy-dependent cross sections depends on the flux-averaged absorption cross section and on the flux average of the transport mean free path. It is often convenient to rewrite Eq. (4.73) in terms of the number of neutrons per unit energy, $n_0(E) = (1/v)\phi_0(E)$, to give

$$L_0^2 = \hat{D}_0 T_0 \,, \qquad (4.77)$$

where

$$\hat{D}_0 = \frac{\int_0^\infty D(E) v n_0(E)\, dE}{\int_0^\infty n_0(E)\, dE} = \bar{D}\hat{v} \,, \qquad (4.78)$$

and

$$T_0 = \frac{\int_0^\infty n_0(E)\, dE}{\int_0^\infty v\Sigma_a(E) n_0(E)\, dE} = (\hat{v}\bar{\Sigma}_a)^{-1} \qquad (4.79)$$

is the neutron lifetime for absorption. We note that $\bar{D}$ given by Eq. (4.76) is the same as D defined by Eq. (4.40), and that $\hat{D}_0$ given by Eq. (4.78) is the same as D_0 defined by Eq. (4.15). As noted in Sec. 4-4, these two quantities are related through Eq. (4.41) when the neutron energy spectrum is independent of position. In our later discussions of the rate at which neutrons leak from a finite moderating sample, the notation of Eqs. (4.77) through (4.79) is more convenient. To clarify any possible ambiguity in Eqs. (4.78) and (4.79), we recall that $\hat{v}$, as defined by Eq. (4.39a), is the average of the speed over n(E) and is therefore the reciprocal of the average of (1/v) over $\phi(E)$. Stated in other words, the flux average of a 1/v absorption cross section,

$$\Sigma_a(E) = (vT_0)^{-1} \,, \qquad (4.80)$$

is given by

$$\bar{\Sigma}_a = (\hat{v}T_0)^{-1} \,. \qquad (4.81)$$

In the special case of a Maxwellian distribution, this yields the familiar results

$$\left(\bar{\Sigma}_a\right)_M = \frac{\pi^{\frac{1}{2}}}{2}\,\Sigma_a(k_BT)\ , \qquad (\hat{v})_M = \left(\frac{8k_BT}{\pi m}\right)^{\frac{1}{2}} . \tag{4.82}$$

In general, we cannot immediately calculate L_0^2 from Eq. (4.73) or the subsequent equivalent expressions since they contain the unknown eigenfunction $\phi_0(E)$ associated with the desired eigenvalue. In the special case that $D(E)$ and $\Sigma_a(E)$ have the same energy dependence, the eigenfunction is the Maxwellian (4.67); thus the averages in Eqs. (4.75) and (4.76) are over a Maxwellian flux spectrum. (When we go beyond diffusion theory, the asymptotic solution is Maxwellian only for the more restricted case of constant cross section.) For actual materials, $\Sigma_a(E)$ is a decreasing function of E, while $D(E)$ is either approximately constant or an increasing function of E. (An exception is crystalline material at low temperature, where $D(E)$ drops sharply as the energy is raised above the Bragg cutoff. If the temperature is not too low, however, the energy region below the Bragg cutoff will have a negligible effect on the diffusion length.) Thus, the absorption term tends to remove neutrons at a lower average energy than the average energy at which they are replenished by the negative leakage term $-D(E)\kappa_0^2$. The net effect is a "diffusion heating" of the energy spectrum which is resisted by the tendence of the thermalization operator R_0 to maintain the equilibrium Maxwellian distribution.

In the limit that Σ_a is much less than $\xi\Sigma_s$, we expect on physical grounds that the diffusion-heating effect is negligible, and that the diffusion length is calculable from Eq. (4.73) by replacing $\phi_0(E)$ with the equilibrium Maxwellian distribution $\phi_M(E)$. Mathematically, it follows from Eq. (4.73) that a determination of $\phi_0(E)$ in the limit of zero absorption gives κ_0^2 to first order in Σ_a. Since both the absorption and leakage terms in Eq. (4.72) are of first order in Σ_a, the equation for $\phi_0(E)$ in the limit of zero absorption is

$$R_0\phi_0(E) = 0 \ .$$

It follows from the detailed balance condition (4.66) that one solution of this is the equilibrium Maxwellian $\phi_M(E)$. We assume on physical ground that this is the only solution that is everywhere positive. Thus, to lowest order in Σ_a, the diffusion-length eigenvalue is associated with a Maxwellian energy spectrum at the moderator temperature. This is in accord with our intuitive picture of the diffusion of thermal neutrons in a pure moderator, and gives the first term in an expansion of κ_0^2 in powers of Σ_a.

In order to calculate κ_0^2 to second order in Σ_a, we need to know the diffusion-heating correction to the energy spectrum to first order in Σ_a. This requires a solution of Eq. (4.72) which depends on the particular model used for the thermalization operator R_0. We can obtain some insight, however, by examining Eq. (4.73). As the absorption is increased, there is a shift of $\phi_0(E)$ to higher energies. This shift is of order $\Sigma_a(k_BT)$. If D(E) and $\Sigma_a(E)$ have the assumed energy-dependence, this in turn leads to an increase in $\bar{D}$ and a decrease in $\bar{\Sigma}_a$. As a function of $\Sigma_a(k_BT)$, κ_0^2 goes through the origin with a slope determined by the Maxwell average of the transport mean free path, and bends down with increasing Σ_a. For 1/v absorption, the result is of the form

$$(\bar{D})_M(\hat{v})_M\kappa_0^2 = \frac{1}{T_0} - \frac{H}{T_0^2} + \ldots, \tag{4.83}$$

where the subscript M indicates an average over a Maxwellian flux spectrum at the moderator temperature, and H is a positive number known as the diffusion-heating coefficient. We defer the quantitative consideration of this coefficient until Chap. 8. For the present, we merely note that monoenergetic neutron transport theory predicts a curvature in the κ_0^2 versus $1/T_0$ plot in the opposite direction from that in Eq. (4.83).

Measurements of the asymptotic distribution far from a thermal source have been carried out for most moderators. For light water, these measurements have been performed as a function of boric acid concentration, and the downward curvature in the plot of κ_0^2 versus $\Sigma_a(k_BT)$ has been clearly observed and quantitatively interpreted. The measurement of diffusion length as a function of poison concentration is a valuable quantitative check on the scattering law one obtains from differential experiments and/or theory. The discussion given here is a good illustration of the way in which thermalization effects enter into even the simplest problems of thermal-neutron diffusion. In the actual analysis of diffusion-length experiments for thermal neutrons, we find that the dominant corrections arise from a shift in the thermal-neutron spectrum which is calculable in energy-dependent diffusion theory. A naive application of monoenergetic transport theory is not only wrong quantitatively but also gives a correction of the wrong sign. In Chap. 8 we carry out a systematic quantitative analysis of diffusion-length experiments using the full transport equation. We find there that even the question of the existence of a diffusion-length eigenvalue is affected by the energy-dependence of the cross sections, but that the measured diffusion lengths can be quantitatively understood in terms of the differential cross sections and explicit solutions of the eigenvalue problem using these data.

4-8 THE MIGRATION AREA

The diffusion length for thermal neutrons is most conveniently measured as the relaxation length of the asymptotic distribution far from a spatially localized source. A different quantity, which is also often called the "thermal-neutron diffusion length," can be defined in terms of the thermal-neutron contribution to the migration area. It is this latter quantity which is of greater practical importance in determining the leakage probability for thermal neutrons in a reactor. An understanding of the relation between these two diffusion lengths is an important task of thermalization theory. In this section we demonstrate this relation, at least within the limitations of the energy-dependent diffusion approximation, Eq. (4.71).

We consider the steady-state spatial distribution of neutron density $n(z)$ in an infinite homogeneous medium with a plane isotropic source at $z = 0$, and we define the migration area by

$$M^2 = \tfrac{1}{2}\overline{z^2} = \tfrac{1}{2}\frac{\int_0^\infty z^2 n(z)\,dz}{\int_0^\infty n(z)\,dz} . \tag{4.84}$$

This is conventionally separated into two terms:

$$M^2 = L^2 + \tau , \tag{4.85}$$

where $\tau^{1/2}$ is the slowing-down length from source energy to thermal energies, and L is the thermal-neutron diffusion length. Although this separation has an intuitive physical meaning in terms of the distances traveled by a neutron while slowing down and after thermalization, it is difficult to make this separation in a precise way. In the final analysis this separation is arbitrary, since it depends on the way one chooses to define a "thermal" neutron. This point becomes clear as the mathematical development proceeds.

We recall that in monoenergetic transport theory the migration area defined by Eq. (4.84) is given exactly by Eq. (4.70). This simple result arises from the well-known[3,14] property of the transport equation that the second spatial moment is given exactly in the P_1 approximation (and the 2*nth* moment is given exactly in the P_n approximation). This simplifying feature remains true for energy-dependent problems. The migration area, $\frac{1}{2}\overline{z^2}$, is determined by a relatively simple system of equations, whereas the exact determination of the asymptotic relaxation length requires the solution of a complicated eigenvalue problem. In the P_1 approximation the migration area and the asymptotic relaxation length are both given by Eq. (4.70) in monoenergetic transport theory. This is no longer true for the energy-dependent problem. The asymptotic relaxation length and its associated energy spectrum are independent of the energy of the source neutrons. By contrast, the mean-square displacement of a neutron of energy E depends both on E and on the energy of the source neutrons. We find, however, that the mean-square distance traveled from birth to absorption is given by a result formally identical with Eq. (4.73), except that $\phi_0(E)$ is replaced by an energy spectrum that is distorted from a Maxwellian in a quite different way. The flux spectrum is now the solution of an inhomogeneous integral equation, and the departure from equilibrium arises from absorption of the neutrons while they are slowing down from source energies and before they have been fully thermalized. In the limit of weak absorption, the flux spectrum again approaches a Maxwellian, and the migration area to absorption is the same as the asymptotic relaxation length. Both are given by

$$L_M^2 = (\bar{D})_M / (\bar{\Sigma}_a)_M \ . \tag{4.86}$$

The corrections of first order in Σ_a to the flux spectrum are quite different in the two cases. In the migration area problem, their effect can be interpreted in terms of a separation of the form of Eq. (4.85), with L^2 as given by Eq. (4.86) and $\tau^{\frac{1}{2}}$ as the slowing-down length to a well-defined transition energy between the slowing-down and thermal energy regions.

The appropriate formalism for this discussion is the moments form of the transport equation as given in Chap. 11 of Weinberg and Wigner.[3] In plane geometry the angular flux is given as a function of one space coordinate, z, and one angular coordinate, μ, the cosine of the angle between the direction of motion and the z-direction. The angular flux integrated over azimuth can be written in the form

$$f(z,E,\mu) = \sum_{\ell=0}^{\infty} f_\ell(z,E) P_\ell(\mu) \ , \tag{4.87}$$

in which only the m = 0 components of Eq. (4.49) appear. The transport equation (4.45) becomes

$$\left[\mu \frac{\partial}{\partial z} + \Sigma_a(E) + \Sigma_s(E)\right] f(z,E,\mu)$$

$$= S(z,E,\mu) + \sum_{\ell=0}^{\infty} \int \Sigma_\ell(E',E) f_\ell(z,E')\,dE' P_\ell(\mu) \ . \tag{4.88}$$

We introduce the spatial moments

$$M_{n\ell}(E) = \frac{1}{n!} \int_{-\infty}^{\infty} z^n f_\ell(z,E)\,dz \tag{4.89}$$

and specialize to the case of a plane isotropic source at z = 0 with source energy spectrum S(E). Using Eq. (4.89) and transforming the $\mu(\partial/\partial z)$ term by partial integration, we obtain the hierarchy of moments equations (Weinberg's and Wigner's Eq. (11.45a)). We are concerned here only with the second moment of the scalar flux

$$\tfrac{1}{2}\overline{z^2}(E) = M_{20}(E)/M_{00}(E) \ , \tag{4.90}$$

which requires the sequential solution of the three equations

$$\Sigma_a(E)M_{00}(E) = R_0M_{00}(E) + S(E) \ , \tag{4.91}$$

$$\Sigma_{tr}(E)M_{11}(E) = R_1M_{11}(E) + M_{00}(E) \ , \tag{4.92}$$

$$\Sigma_a(E)M_{20}(E) = R_0M_{00}(E) + \frac{1}{3}M_{11}(E) \ , \tag{4.93}$$

which are the first three of the hierarchy of moments equations. The term $M_{00}(E)$ determined by Eq. (4.91) is the flux energy spectrum integrated over all space. This is identical to the spatially-independent steady-state energy spectrum that would exist in an infinite homogeneous medium of the same composition with a spatially uniform source of the same energy spectrum. This idealized problem is closely related to the problem of the spectrum in a homogeneous reactor core, and is the focal point of Chaps. 5 and 7. In particular, Chap. 5 is devoted primarily to the study of Eq. (4.91) and its solution for some idealized models for the thermalization operator R_0.

The term $M_{11}(E)$ is simply related to $M_{00}(E)$ if we neglect the term $R_1M_{11}(E)$ on the right-hand side of Eq. (4.92). This neglect reduces the level of description from the consistent P_1 approximation of Sec. 4-5 to the energy-dependent diffusion approximation of Secs. 4-6 and 4-7. It is familiar in slowing-down theory as the "Selengut-Goertzel approximation."[3] Although this approximation introduces considerable error in calculations of the slowing-down length for fission neutrons in water, it is expected to introduce very little error for energies near thermal. To our knowledte, however, a quantitative check of this approximation for the migration area at thermal energies has not been made. If we neglect $R_1M_{11}(E)$, Eq. (4.92) becomes

$$M_{11}(E) = M_{00}(E)/\Sigma_{tr}(E) = 3D(E)M_{00}(E) \ . \tag{4.94}$$

Substitution of Eq. (4.94) into Eq. (4.93) gives

$$\Sigma_a(E)M_{20}(E) = R_0M_{20}(E) + D(E)M_{00}(E) \ . \tag{4.95}$$

To obtain $M_{20}(E)$, we must solve the integral equation (4.95) numerically for some model of the thermalization operator R_0. Such solutions are discussed for the heavy-gas model in Chap. 5. A considerable simplification occurs, however, if we consider only the migration area to absorption

$$M^2 = \frac{\int \Sigma_a(E) M_{20}(E)\, dE}{\int \Sigma_a(E) M_{00}(E)\, dE} . \tag{4.96}$$

For 1/v absorption, this definition is equivalent to Eq. (4.84). For non-1/v absorption, we consider Eq. (4.96) to replace Eq. (4.84) as the definition of the migration area. The migration area to absorption can be obtained directly without solving for $M_{20}(E)$ by integrating Eq. (4.93) or Eq. (4.95) over energy to give

$$M^2 = \frac{1}{3} \frac{\int M_{11}(E)\, dE}{\Sigma_a(E) M_{00}(E)\, dE} \tag{4.97}$$

or

$$M^2 \approx \frac{\int D(E) M_{00}(E)\, dE}{\int \Sigma_a(E) M_{00}(E)\, dE} . \tag{4.98}$$

We note that Eq. (4.98) is formally identical with Eq. (4.73), with $\phi_0(E)$ replaced by $M_{00}(E)$.

The integrals in Eqs. (4.96) through (4.98) extend from zero to infinity and are effectively cut off at high energy by the maximum energy of the source spectrum. Despite their formal similarity, Eqs. (4.73) and (4.98) are significantly different physically, since the flux spectra $\phi_0(E)$ and $M_{00}(E)$ are quite different. The former is an eigenfunction of the source-free diffusion equation and falls off exponentially above thermal energies. Although the diffusion-heating effect may shift $\phi_0(E)$ to higher energies than the equilibrium Maxwellian, the thermal component of the spectrum is shifted without the addition of any epithermal component. For the source of fission neutrons appropriate to reactor considerations, $M_{00}(E)$, on the other hand, has a 1/E behavior in the (epithermal) energy region above thermal energies but below source energies. Although this component of the spectrum will be small compared with the Maxwellian component for weak absorption, the range of energy over which it contributes to the numerator of Eq. (4.98) is very large. It is necessary to include this correction even in pure moderators.

In order to calculate M^2 accurately from Eq. (4.98), we must first study the infinite-medium flux spectrum $M_{00}(E)$. Since this is done in detail in Chap. 5, we defer a quantitative calculation of M^2 until Sec. 5-8. We can obtain a physical understanding, however, from a very crude model of the spectrum in a medium where all the absorption are assumed to take place at thermal energies. Assume that $M_{00}(E)$ has some characteristic behavior $\phi(E)$ below a transition energy E^* and that $\Sigma_a(E)$ is zero above E^*. For a monoenergetic source at energy E_s, neglecting transients near the source energy as well as thermalization effects just above E^*, we have

$$\begin{aligned} M_{00}(E) &= \phi(E) \qquad & 0 < E < E^* \\ &= C/E\xi\Sigma_s(18) \qquad & E^* < E < E_s \end{aligned} , \tag{4.99}$$

where C is determined by neutron conservation to satisfy

$$C = \int_0^{E^*} \Sigma_a(E)\phi(E)dE . \tag{4.100}$$

Substituting Eqs. (4.99) and (4.100) into Eq. (4.98), we obtain

$$M^2 = \frac{\int_0^{E^*} D(E)\phi(E)dE}{\int_0^{E^*} \Sigma_a(E)\phi(E)dE} + \int_{E^*}^{E_s} \frac{D(E)}{\xi\Sigma_s(E)} \frac{dE}{E} . \tag{4.101}$$

In the limit of weak absorption, $\phi(E)$ in Eq. (4.101) is Maxwellian, and the first term in Eq. (4.101) becomes identical with Eq. (4.86), since the appropriate value of E^* will be much greater than k_BT (see Sec. 5-8). Since E_s can be arbitrarily large, the second term in Eq. (4.101) can be comparable to the first, even for weak absorption. This second term is just the Fermi age in the age-diffusion approximation to slowing-down theory. This term depends logarithmically on E^*, and a more accurate calculation of $M_{00}(E)$ is necessary before E^* can be properly specified. In Sec. 5-8, we discuss a calculation that gives correctly the contributions to M^2 of order $1/\Sigma_a$ and of order unity. This calculation gives a precise definition of E^*. For the present, we see crudely that the migration area from birth to absorption splits up in the form (4.85), and that this split is directly associated with the split of the energy spectrum into a thermal and an epithermal part.

4-9 BOUNDARY CONDITIONS AND INTEGRAL TRANSPORT THEORY

Since the transport equation (4.45) is a first-order partial differential equation in the space and time variables, the specification of spatial boundary conditions and of an initial distribution is required in order to determine its solution. In practice, we must deal with two different kinds of spatial boundaries. We have internal boundaries associated with discontinuous changes in the macroscopic cross sections. Familiar examples in reactors are the interface between core and reflector, and the boundary between a fuel rod and the moderator. A somewhat different internal boundary of particular interest in the thermalization problem is the interface between two media of the same composition but at different temperatures. For these internal boundaries we must specify some type of continuity condition for the angular flux. The external boundary of the system requires special consideration. For those directions which are directed out of the system, the same continuity condition holds as at an internal boundary. In addition, however, we must specify explicitly the flux entering the system at an external boundary.

The appropriate boundary condition at an internal boundary is the continuity of $f(\mathbf{r},E,\mathbf{\Omega},t)$ across the boundary in the direction $\mathbf{\Omega}$; that is, we require that

$$f(\mathbf{r} + \varepsilon\mathbf{\Omega},E,\mathbf{\Omega},t) = f(\mathbf{r},E,\mathbf{\Omega},t) , \qquad (4.102)$$

where ε is an infinitesimal quantity such that $\mathbf{r}$ lies on one side of the boundary and $\mathbf{r} + \varepsilon\mathbf{\Omega}$ lies on the other side. This boundary condition is a direct statement of neutron conservation. It should be noted[1] that we have only required continuity in $\mathbf{r}$ in the direction $\mathbf{\Omega}$. It will usually be true that $f(\mathbf{r},E,\mathbf{\Omega},t)$ is continuous in $\mathbf{r}$ in all directions, but specifying this can overdetermine the problem.

At the external boundary of the system, we will normally apply the "vacuum boundary condition"

$$f(\mathbf{R},E,\Omega_i,t) = 0 \ , \tag{4.103}$$

where the points $\mathbf{R}$ define a closed convex surface and Ω_i denotes any direction heading into the system at the point $\mathbf{R}$. We do not consider the additional complications associated with re-entrant bodies (i.e., bodies where a neutron can leave the system and re-enter it with no collisions). This vacuum boundary condition (4.103) is a mathematical statement of the physical condition that no neutrons enter the system from the outside. There are some experimental situations, however, where this phsyical condition does not apply. In some of the linear accelerator experiments discussed in later chapters, the source of neutrons is located outside the system of interest. In this situation, we must specify the boundary source

$$f(\mathbf{R},E,\Omega_i,t) = S_B(\mathbf{R},E,\Omega_i,t) \tag{4.104}$$

due to the neutrons streaming into the system from the external source.

Finally, there is a third condition that is frequently used as an external boundary condition. In heterogeneous reactors, the fuel elements are usually arranged on a regular lattice. It is customary to consider separately the effects of the heterogeneity and of the finite size of the core. The heterogeneity is treated by assuming a flux that is strictly periodic with the period of the lattice. With this assumption, the problem is reduced to that of a single unit cell of fuel surrounded by moderator with a "cell boundary condition" at the outside of the moderator region. This boundary condition uses the symmetry of the lattice to replace the continuity condition (4.102) on the flux leaving the cell by a condition equating this flux to the flux entering the cell in the appropriate direction from an adjacent cell. We can state this condition in the form

$$f(\mathbf{R},E,\Omega_o,t) = f(\mathbf{R},E,\Omega_s,t) \ , \tag{4.105}$$

where $\mathbf{R}$ is any point on the boundary of the unit cell, and Ω_s is the ingoing direction obtained from the outgoing direction Ω_o by specular reflection from the cell boundary. In practice this cell boundary is either square or hexagonal. This unit-cell description is exact except for effects of the finite size of the reactor; however, it defines a problem in two spatial variables, two angular variables, and an energy variable. This problem requires a prohibitively large amount of computer time to be solved accurately on a routine basis for design calculations. A commonly used approximation is obtained by reducing the problem to one spatial variable by replacing the actual cell with a cylinder of equal area. This is the so-called "Wigner-Seitz cell approximation" since it is analogous to the sphericizing of the first Brillouin zone in crystal lattices.[15] The solution for the cylindrical unit cell is no longer exact for the actual cell. It is in fact seriously in error for moderator regions as thin as one mean free path. We are not concerned with such geometrical points in the present volume, except where necessary for an understanding of experiments in which thermalization effects are also important. For further discussion of the cylindrical-cell approximation, the reader is referred to the work of Honeck.[16]

The integro-differential equation (4.42) with the boundary conditions (4.102) and (4.103) can be transformed into an integral equation with a direct physical interpretation. The angular flux $f(\mathbf{r},E,\Omega,t)$ can be written as a line integral along the ray $(\mathbf{r} - s\Omega)$ of the birth-rate density, Eq. (4.51), times the probability of reaching the point $\mathbf{r}$ without collision. The resulting integral equation is[3]

$$f(\mathbf{r},E,\Omega,t) = \int_0^{R_0} \exp[-P(\mathbf{r},E,\Omega,s)] \times$$

$$\times\, h(\mathbf{r} - s\Omega,E,\Omega,t - sv^{-1})ds \;, \tag{4.106}$$

where $h(\mathbf{r},E,\Omega,t)$ is the birth-rate density (4.51), and the "optical path" P is defined by

$$P(\mathbf{r},E,\Omega,s) = \int_0^{s} \Sigma(\mathbf{r} - x\Omega,E)dx \;, \tag{4.107}$$

and

$$\Sigma(\mathbf{r},E) = \Sigma_s(\mathbf{r},E) + \Sigma_a(\mathbf{r},E) \tag{4.108}$$

is the total macroscopic cross section. The term $R_0(\mathbf{r},\mathbf{\Omega})$ is the distance from the point $\mathbf{r}$ to the external boundary of the system along the direction $-\mathbf{\Omega}$. In Eq. (4.106) the birth-rate density h gives the rate at which neutrons have entered the energy interval dE and solid angle $d\mathbf{\Omega}$ either as source neutrons or by collision. Since they entered this interval at a distance s from the point under study, they must have done so at an earlier time $t - sv^{-1}$, where $v = (2E/m)^{\frac{1}{2}}$. To obtain the angular flux at a given point, we must add up all the events that have sent neutrons in the right direction with the right energy, but we must weight these events with the probability, exp(-P), that they will reach the point in question without another collision changing their energy or direction.

The continuity condition (4.102) at internal boundaries is included in the integral equation (4.106) through the definition (4.107) of the optical path. In the integral-equation form of transport theory, we need make no distinction between a continuous variation of $\Sigma(\mathbf{r},E)$ with position and a discontinuity in composition and/or temperature. The integral in Eq. (4.107) is well defined in either case. The vacuum boundary condition (4.103) enters the integral equation through the upper limit $R_0(\mathbf{r},\mathbf{\Omega})$ of the integral in Eq. (4.106). If a boundary source were present, it could be included by adding a term

$$S_B(\mathbf{r} - R_0\mathbf{\Omega},E,t - R_0v^{-1})\ \exp[-P(\mathbf{r},\mathbf{\Omega},E,R_0)] \qquad (4.109)$$

to the right-hand side of (4.106). The use of the cell boundary condition (4.105), rather than the vacuum boundary condition (4.103), at the external boundary of the system would lead to a more complicated problem, which we do not discuss here. For the use of the integral formulation of transport theory in cell problems, the reader is referred to the work of Honeck.[17]

The integral-equation form of transport theory is simplified considerably if the birth-rate density is isotropic. This implies the rather mild restriction of isotropic sources and the more severe restriction of isotropic scattering in the laboratory coordinate system. We can then carry out the integral over solid angle to obtain an integral equation for the scalar flux $\phi(\mathbf{r},E,t)$. This is most easily done by introducing the new variable $\mathbf{r}' = \mathbf{r} - s\mathbf{\Omega}$ and noting that

$$d\Omega ds = \frac{d\mathbf{r}'}{|\mathbf{r} - \mathbf{r}'|^2} \tag{4.110}$$

with the integral over $d\mathbf{r}'$ extending over the volume V of the system. The resulting integral equation for the scalar flux is

$$\phi(\mathbf{r},E,t) = \int_V \frac{d\mathbf{r}'}{|\mathbf{r} - \mathbf{r}'|^2} \times$$

$$\times \exp[-P(\mathbf{r},\mathbf{r}',E)]\, h(\mathbf{r}',E,t - |\mathbf{r} - \mathbf{r}'|v^{-1}) , \tag{4.111}$$

where

$$h(\mathbf{r},E,t) = 4\pi S_{00}(\mathbf{r},E,t) + \int_0^{\infty} \Sigma_0(E',E)\phi(\mathbf{r},E',t)dE' , \tag{4.112}$$

and the optical path $P(\mathbf{r},\mathbf{r}',E)$ is defined by Eq. (4.107), with the vector variable $\mathbf{r}'$ replacing the variables s and Ω. For the special case of a single medium of uniform composition, Eq. (4.107) reduces to

$$P(\mathbf{r},\mathbf{r}',E) = \Sigma(E)s = \Sigma(E)|\mathbf{r} - \mathbf{r}'| . \tag{4.113}$$

As in the case of diffusion theory, we again have a single equation for the scalar flux in which the angular variable Ω has been eliminated. The integral equation (4.111) does not, however, depend on any approximation to the angular distribution of the flux; it is exact if the scattering and source are isotropic. Equation (4.111) applies to small systems where the angular flux is strongly anisotropic as well as to large systems where the angular flux is almost isotropic. Once the solution for the scalar flux is known, the angular flux can be obtained by quadrature from Eq. (4.106). We find the integral-equation form of transport theory very useful both when we discuss the effect of a vacuum boundary in the following two sections and again in Chap. 8 when we consider the diffusion-length problem more rigorously than in Sec. 4-7.

In many problems of interest, energy-dependent diffusion theory is an accurate approximation except within a few mean free paths of an internal boundary or an external vacuum boundary. Although it is not possible to give an accurate description of the neutron flux near the boundary using diffusion theory, we can specify boundary conditions such that diffusion theory correctly gives the effects of the boundary on the flux distribution at distances greater than a few mean free paths from the boundary. If the system under study is large compared to a mean free path, the over-all neutron balance is accurately described by diffusion theory with the appropriate boundary conditions.

Since the diffusion approximation gives a second-order partial differential equation in the spatial variables, boundary conditions must be specified for both the scalar flux and its normal derivative. The usual choice at an internal boundary is that

$$\phi(\mathbf{r},E,t) \quad \text{is continuous} \tag{4.114}$$

and

$$D(\mathbf{r},E)\hat{\mathbf{n}} \cdot \nabla\phi(\mathbf{r},E,t) \quad \text{is continuous,} \tag{4.115}$$

where $\hat{\mathbf{n}}$ is a unit vector normal to the boundary. Only in the special case of one-dimensional geometry with piecewise constant variation of $D(\mathbf{r},E)$ has this choice been shown to be completely satisfactory. As discussed by Wilkins,[18] there is a need for further investigation to determine the most appropriate boundary conditions in general.

The boundary condition (4.115) fortunately can be applied with correct results even to thin regions if these regions do not absorb many neutrons. This boundary condition leads to the correct result that the flux is approximately constant in such regions although diffusion theory does not give correctly the small variations from constancy. For practical problems, such as the power distribution near the interface between core and reflector, the energy-dependent diffusion approximation is a significant improvement over ordinary diffusion theory in which the variation of energy spectrum with position is ignored. In energy-dependent diffusion theory, the flux spectrum is continuous across the boundary, and the transient behavior of the spectrum is described to a good approximation if the spectrum varies slowly in a mean free path, as it does, for example, in graphite-moderated reactors.

A problem requiring special attention is the choice of diffusion-theory boundary condition at the external vacuum boundary. In this case, diffusion theory applies far from the boundary only on one side of the boundary. This problem is familiar in monoenergetic transport theory where it leads to the well-known result[1,2] that the flux should be set equal to zero at the extrapolated boundary. For a plane boundary with a nonabsorbing material, this extrapolated boundary is 0.71 transport mean free paths outside the actual boundary.

In the next section we consider in detail the extension of this result to the case in which the cross section is energy dependent.

4-10 MILNE'S PROBLEM FOR THERMAL NEUTRONS

We want to determine the effect of a vacuum boundary on the neutron distribution far from the boundary. In order to isolate the effect due to the boundary, we consider first the simple example of a uniform source-free half space with isotropic scattering and no absorption. The properties of the medium which fills the region $0 \leq z < \infty$ are determined entirely by the energy-dependent mean free path

$$\ell(E) = 1/\Sigma_s(E) \tag{4.116}$$

and the energy-transfer kernel $\Sigma_0(E',E)$ which is normalized by

$$\Sigma_s(E) = \int_0^\infty \Sigma_0(E,E')dE' \tag{4.117}$$

and satisfies the detailed balance condition (4.66). The differential form of the steady-state transport equation for this problem is

$$\mu \frac{\partial}{\partial z} + \Sigma_s(E)\, f(z,E,\mu) =$$

$$= \tfrac{1}{2} \int_0^\infty \Sigma_0(E',E) \int_{-1}^{1} f(z,E',\mu')d\mu' dE' \ , \tag{4.118}$$

and the vacuum boundary condition is

$$f(0,E,\mu) = 0 \qquad \mu \geq 0 \ . \tag{4.119}$$

This is a special case of the plane-geometry form (4.88) of the transport equation. The angular flux is assumed to be independent of x and y as well as of the azimuth around the z-direction.

The integral equation (4.106) for the angular flux in this case is

$$f(z,E,\mu) = -(2\mu)^{-1} \int_0^\infty \int_z^\infty \Sigma_0(E',E) \exp[(z' - z)/\mu \ell(E)] \times$$

$$\times \phi(z',E')dz'dE' \qquad \mu < 0 , \tag{4.120}$$

and

$$f(z,E,\mu) = (2\mu)^{-1} \int_0^\infty \int_0^z \Sigma_0(E',E) \exp[(z' - z)/\mu \ell(E)] \times$$

$$\times \phi(z',E)dz'dE' \qquad \mu > 0 , \tag{4.121}$$

where the scalar flux $\phi(z,E)$ is defined by

$$\phi(z,E) = \int_{-1}^{1} f(z,E,\mu)d\mu . \tag{4.122}$$

Integrating Eq. (4.120) over μ from -1 to 0 and Eq. (4.121) over μ from 0 to 1, and adding the results, we obtain the integral equation (4.111) for the scalar flux specialized to the present case:

$$\phi(z,E) = \tfrac{1}{2} \int_0^\infty \int_0^\infty \Sigma_0(E',E)E_1 \times$$

$$\times [|z - z'|/\ell(E)]\phi(z',E')dE'dz' . \tag{4.123}$$

The functions $E_n(x)$ are defined by

$$E_n(x) = \int_1^\infty e^{-xu}u^{-n}du \tag{4.124}$$

and are tabulated in Ref. 2.

If the medium were uniform throughout all space, the integral over z in Eq. (4.123) would extend from $-\infty$ to ∞, and the integral over z' in Eq. (4.121) would extend from $-\infty$ to z. The most general solution to this problem is

$$\phi_0(z,E) = A(z + z_0)\phi_M(E) , \tag{4.125}$$

where A and z_0 are arbitrary constants, and $\phi_M(E)$ is the equilibrium Maxwellian flux spectrum (4.67). The term proportional to z in Eq.

(4.125) represents a constant current of neutrons from a source at $z = \infty$. We note that the infinite-medium solution (4.125) exactly satisfies energy-dependent diffusion theory. Inserting Eq. (4.125) back into Eqs. (4.120) and (4.121), we obtain for the angular flux

$$f(z,E,\mu) = \tfrac{1}{2}A\phi_M(E)[z + z_0 - \mu\ell(E)] \ , \tag{4.126}$$

which gives a current per unit energy

$$j(z,E) = \int_{-1}^{1} \mu f(z,E,\mu)d\mu \tag{4.127}$$

$$j(z,E) = -\frac{1}{3}A\ell(E)\phi_M(E) = \frac{1}{3}\ell(E)\frac{\partial}{\partial z}\phi(z,E) \ . \tag{4.128}$$

The total current is given by

$$J = \int_0^{\infty} j(z,E)dE = -\frac{1}{3}A(\bar{\ell})_M \ , \tag{4.129}$$

where $(\bar{\ell})_M$ is the Maxwellian flux average of $\ell(E)$.

With a vacuum boundary at $z = 0$, the asymptotic solution for large z is still Eq. (4.125), but the presence of the boundary determines the quantity z_0. Since the asymptotic solution extrapolates to zero at $z = -z_0$, then z_0 is known as the extrapolation length. If we use diffusion theory with the boundary condition that the flux vanish at $z = -z_0$, we obtain the correct solution far from the boundary.

The determination of z_0 is therefore the objective of our calculation. Since the asymptotic solution is separable in position and energy, there is a single value of z_0, and not an energy-dependent boundary condition. The use of the boundary condition

$$\phi(-z_0,E) = 0 \tag{4.130}$$

in conjunction with energy-dependent diffusion theory gives the correct asymptotic behavior but does not describe the spectrum transients near the vacuum boundary since it yields Eq. (4.125) for all values of z. This is in contrast to the findings of Sec. 4-9, where our diffusion theory boundary conditions did describe the change in energy spectrum near a boundary between two media when diffusion theory was applicable in both media. The problem of specifying an energy-dependent boundary condition that can be combined with energy-dependent diffusion theory to study the energy spectrum near (but not so near as a few mean free paths) a vacuum boundary has not, to our knowledge, been systematically studied.

In order to calculate z_0, it is convenient to write it as an integral over the emergent angular distribution at $z = 0$, which in turn can be written as an integral over $\phi(z,E)$ by using Eq. (4.120). To do this, we go back to the differential form of the transport equation (4.118) and introduce some auxiliary quantities. Integrating Eq. (4.118) over energy and angle gives the neutron conservation condition

$$\frac{dJ}{dz} = 0 \ , \tag{4.131}$$

where J is as defined by Eqs. (4.127) and (4.129). Multiplying Eq. (4.118) by μ and integrating over energy and angle, we get

$$\frac{dK}{dz} = -J \quad \text{or} \quad K(z) = -J(z + c) \ , \tag{4.132}$$

where K(z) is defined by

$$K(z) = \int_{-1}^{1} \mu^2 \int_0^\infty \ell(E) f(z,E,\mu) dE d\mu \ . \tag{4.133}$$

For large z, $f(z,E,\mu)$ is given by Eq. (4.126). The function K(z) is given by Eq. (4.132) for all z, however, so we must have

$$c = z_0 \quad \text{and} \quad J = -\frac{1}{3} A(\bar{\ell})_M \ . \tag{4.134}$$

In particular, we have

$$K(0) = \frac{1}{3} A(\bar{\ell})_M z_0 \ . \tag{4.135}$$

Evaluating Eq. (4.133) for $z = 0$, using Eq. (4.119), we get the desired expression for z_0 in terms of the emergent angular distribution. Normalizing the current so that $A = 1$, combining Eqs. (4.133), (4.135), and (4.120), and integrating over μ, we get

$$z_0 = \frac{3}{2(\bar{\ell})_M} \int_0^\infty \int_0^\infty \int_0^\infty \ell(E) \Sigma_0(E',E) E_3[z/\ell(E)] \phi(z,E') dz dE dE' . \tag{4.136}$$

Before considering approximate calculations of z_0 when $\ell(E)$ depends on energy, we briefly review the results for constant cross section. In this case, we can use the detailed balance condition (4.66) to show that

$$\phi(z,E) = \Phi(z)\phi_M(E) \tag{4.137}$$

is a solution of Eq. (4.123) if $\Phi(z)$ satisfies

$$\Phi(z) = \frac{1}{2\ell}\int_0^\infty E_1(|z - z'|/\ell)\Phi(z')dz' \ . \tag{4.138}$$

Equation (4.138) is known as Milne's integral equation and was first considered for the closely related problem of radiative transfer in stellar atmospheres. It was solved analytically by Wiener and Hopf in 1931 by a technique that has several applications. This technique was applied to problems of interest in neutron diffusion by Placzek and Seidel.[19] The results are extensively discussed in books on neutron transport theory[1,2] and radiative transfer.[20]

Of particular interest to us here is the result for the extrapolation distance

$$\left.\begin{aligned} z_0 &= \frac{\ell}{\pi}\int_0^\infty \left[\frac{3}{t^2} - \frac{(1 + t^2)^{-1}}{(1 - t^{-1}\tan^{-1} t)}\right] dt \\ &= 0.71044609\ell \ . \end{aligned}\right\} \tag{4.139}$$

The derivation of Eq. (4.139) is available in many places[1,2,19] and so is not given here.

For constant cross sections, the energy exchange between neutrons and moderator does not enter the problem. By restricting our attention to the case of zero absorption, we have an asymptotic solution for which the scalar flux spectrum is Maxwellian. (Since a current is flowing, the asymptotic solution does not, however, represent full thermal equilibrium between neutrons and moderator.) For constant mean free path, there is no preferential leakage for any neutron energy and thus no distortion of the Maxwellian energy spectrum for the scalar flux even near the boundary. For an energy-dependent mean free path, however, there is preferential leakage for those energies corresponding to longer mean free paths. This tends to deplete the spectrum inside the medium at those energies, but this depletion is resisted by the tendency of the energy-transfer kernel to maintain thermal equilibrium. Since the asymptotic form of the scalar flux remains separable, a single value of z_0 is still defined, but the solution is no longer of the form (4.137) except for large z, and the value of z_0 depends both on the energy-dependence of $\ell(E)$ and on $\Sigma_0(E',E)$. The approximate solutions that have been obtained, however, show that z_0 is remarkably independent of the energy-transfer kernel and depends almost entirely on $\ell(E)$.

A useful approximate procedure for calculating z_0 is a generalization of the variational method introduced by LeCaine[21] in the case of constant cross section. In this method it is convenient to first separate out the term proportional to z in $\phi(z,E)$ by introducing

$$q(z,E) = \phi(z,E) - z\phi_M(E) \ . \tag{4.140}$$

Substitution of Eq. (4.140) into Eq. (4.123) leads to an integral equation for q(z,E). The spatial integrals arising from this substitution, and from later manipulations, are the same as in the monoenergetic case, and have been tabulated by LeCaine.[22]

$$q(z,E) = \tfrac{1}{2}\ell(E)\phi_M(E)E_3[z/\ell(E)] + \tfrac{1}{2}\int_0^\infty\int_0^\infty \times$$

$$\times\ \Sigma_0(E',E)E_1[|z - z'|/\ell(E)]q(z',E')dz'dE' \ . \tag{4.141}$$

Examination of Eq. (4.141) verifies the assertion that $\phi(z,E)$ has the asymptotic form of Eq. (4.125) for large z. Substitution of Eq. (4.140) into Eq. (4.136) gives an expression for z_0 in terms of q(z,E):

$$z_0 = \frac{3}{2(\bar{\ell})_M}\left[\frac{(\overline{\ell^2})_M}{4}\right] + I_1 \ , \tag{4.142}$$

where

$$I_1 = \int_0^\infty\int_0^\infty\int_0^\infty \ell(E)\Sigma_0(E',E)E_3[z/\ell(E)]q(z,E')dzdEdE' \ . \tag{4.143}$$

To calculate z_0, we can use a variational expression for I_1 which is a generalization of LeCaine's[21] result for constant cross sections.

The result is[23]

$$z_0 = \frac{3}{2(\bar{\ell})_M}\left[\frac{\overline{(\ell^2)}_M}{4} + \frac{\tilde{I}_1^2}{2\tilde{I}_2}\right], \tag{4.144}$$

where $\tilde{I}_1$ is given by (4.143) with q(z,E) replaced by the trial function $\tilde{q}(z,E)$, and

$$\tilde{I}_2 = \int_0^\infty \int_0^\infty \int_0^\infty \frac{\tilde{q}(z,E)\Sigma_0(E',E)}{\phi_M(E)} \left\{ \tilde{q}(z,E') - \tfrac{1}{2} \int_0^\infty \int_0^\infty \times \right.$$

$$\left. \times\, E_1\left[\frac{|z - z'|}{\ell(E')}\right] \Sigma_0(E'',E')\tilde{q}(z',E'')dz'dE'' \right\} dzdEdE' . \tag{4.145}$$

For a trial function $\tilde{q}(z,E)$ satisfying Eq. (4.141), Eq. (4.144) reduces to Eq. (4.142). This choice of $\tilde{q}(z,E)$ maximizes z_0. Any other choice gives a smaller value of z_0, but the error in z_0 is of second order in the error in $\tilde{q}(z,E)$. A trial function that gives fairly simple results is the asymptotic solution

$$\tilde{q}(z,E) \approx c\phi_M(E) . \tag{4.146}$$

Substituting Eq. (4.146) into Eq. (4.144), using detailed balance, and evaluating the same integrals over z that occur for constant cross sections, we obtain

$$z_0 \approx \frac{3}{8}\frac{\overline{(\ell^2)}_M}{(\bar{\ell})_M} + \frac{1}{3}(\bar{\ell})_M . \tag{4.147}$$

For constant cross sections, this reduces to

$$z_0 \approx \frac{17}{24}\ell = 0.7083\ell , \tag{4.148}$$

which is only 0.3% smaller than the exact result (4.139). For a 1/v cross section, Eq. (4.147) becomes

$$z_0 = \frac{17}{24}(\bar{\ell})_M \left[\frac{8}{17} + \frac{9}{17}\left(\frac{32}{9\pi}\right)\right] = 1.07 \times 0.71(\bar{\ell})_M . \tag{4.149}$$

The extrapolation distance is thus 7% greater than would be obtained from the naive assumption that one should average the mean free path over the flux spectrum as in the case of the diffusion coefficient.

This result is approximately valid for hydrogenous moderators.

Extension of the result (4.147) to more complicated trial functions is possible but cumbersome.

Another development of interest is a different variational expression for z_0 than Eq. (4.144). Schofield[24] has obtained interesting results from such a development. A different point of view is to work with approximate forms for the energy-transfer kernel. An elegant result has been obtained in this way by Williams.[25] Williams worked with a separable kernel

$$\Sigma_0(E',E) = \frac{\Sigma_s(E)\Sigma_s(E')\phi_M(E)}{\int_0^\infty \Sigma_s(E'')\phi_M(E'')dE''} \tag{4.150}$$

which satisfies detailed balance and gives the correct total cross section as a function of energy. This kernel, which has the maximum thermalization rate consistent with a given $\Sigma_s(E)$, was introduced by Corngold, Michael, and Wollman[26] in a study of the decay-time eigenvalues in the thermalization problem. This kernel is a useful artifice in analytical studies of thermalization and is used again in Chap. 8. Using the kernel (4.150), Williams generalized the result of Placzek and Seidel[19] to obtain an exact Wiener-Hopf solution. His result for z_0 is

$$z_0 = \frac{1}{\pi}\int_0^\infty \left[\frac{3}{\tau^2} - \frac{\int_0^\infty \Sigma(E)\phi_M(E)[\tau^2 + \Sigma^2(E)]^{-1}dE}{\left\{(\bar{\Sigma})_M - \int_0^\infty \Sigma^2(E)\phi_M(E)\tau^{-1}\tan^{-1}[\tau/\Sigma(E)]dE\right\}}\right]d\tau \;. \tag{4.151}$$

In the limit of constant cross sections, this reduces to Eq. (4.139), with the substitutions

$$\Sigma(E) = (\bar{\Sigma})_M = \ell^{-1}, \qquad \tau = t/\ell \;,$$

since $\phi_M(E)$ is normalized to unity through Eq. (4.68). The integral has been evaluated numerically for a monatomic ideal gaseous moderator with

$\Sigma(E)$ given by Eq. (2.100) and the results compared with the variational result (4.147) and with a numerical solution of Eq. (4.123) using a modified form of the THERMOS code.[17] These results are given in Table 4.1 for hydrogen. For heavier gaseous moderators $\Sigma(E)$ is more nearly constant, and the differences among different approximate results are smaller.

Table 4.1

EXTRAPOLATION DISTANCE FOR MONATOMIC HYDROGEN IN UNITS OF THE EPITHERMAL MEAN FREE PATH

Method of Calculation	Kernel	$z_0/\ell(\infty)$
Wiener-Hopf, Eq. (4.151)	Separable, Eq. (4.150)	0.5378
Variational, Eq. (4.147)	-----	0.5354
$0.7104(\bar{\ell})_M$	-----	0.5278
Numerical solution of Eq. (4.123)	Hydrogen gas, Eq. (2.99)	0.5375
Numerical solution of Eq. (4.123)	Separable, Eq. (4.150)	0.5383

The first and last lines of Table 4.1 should agree exactly. Their disagreement is a measure of the numerical accuracy of the calculation. In more recent work Williams[27] has used the separable kernel as the first step in a perturbation procedure, and has explicitly calculated corrections due to the difference between $\Sigma_0(E',E)$ and the separable kernel (4.150). It should be noted that we have examined a favorable case for both the variational and separable kernel methods. Both methods should work well for strong energy transfer between neutrons and moderator. An example of an unfavorable case is that of a heavy crystalline moderator at low temperature. Bragg scattering causes strong fluctuations in the mean free path as a function of energy, and the weak energy transfer between neutrons and moderator results in these fluctuations having a large effect on the spectrum. This effect extends many mean free paths from the boundary; therefore the asymptotic solution is not a good trial function to use in Eq. (4.144). In this case, Schofield's improved variational principle,[24] which is exact in the limit of no energy transfer, is more accurate.

Although the problem we have discussed is highly idealized, it has been treated in some detail as an extension to thermal-neutron problems of a classic problem in monoenergetic neutron transport theory. The energy dependence of the total cross section is most important for hydrogenous moderators. In this case, it is also important to include the effect of anisotropic scattering. This has been done by Kladnik and Kuščer,[28] who have extended the variational result (4.147).

In the presence of absorption, the asymptotic flux is of the form

$$\phi_{\kappa_0}(E)\ \exp(\kappa_0 z)\ ,$$

where κ_0 is the smallest solution of the diffusion-length eigenvalue problem of Secs. 4-7 and 8-1. For constant cross sections, the flux spectrum remains Maxwellian, and the spatial problem has been extensively studied.[1,2] The extension to energy-dependent thermal cross sections has not been carried out. In the limit of weak absorption, the asymptotic flux is almost linear in z in the sense that $\kappa_0 \ell \ll 1$, and the asymptotic spectrum is almost Maxwellian. Under these conditions, the zero absorption limit that we have considered in detail is applicable. A good example is the flux at the outer face of a thermal column. A careful comparison of theory and experiment for the energy and angular distribution of the emergent flux, and for the spatial distribution of the scalar flux inside the medium, would be of considerable interest. To our knowledge no such investigation has been performed.

Our analysis also applies to a good approximation to the decay of a thermalized-neutron pulse in a finite moderating sample that is many mean free paths in extent. The use of such pulsed-neutron techniques to measure the thermal-neutron diffusion parameters of moderators has many advantages over the diffusion-length measurement of Sec. 4-7. Now that we have an understanding of the boundary conditions to be used with diffusion theory at a vacuum boundary, we can proceed to the analysis of these experiments. As in Sec. 4-7, we formulate the problem in energy-dependent diffusion theory and examine the principal physical effects which are present. A quantitative analysis of the experiments and a more systematic development of the theory are presented in Chap. 8.

4-11 THE NEUTRON LIFETIME IN A FINITE MODERATOR

If one waits a sufficiently long time after a pulse of neutron has excited a finite moderating sample, the space and time distribution of the neutron density is determined by the leakage and absorption of the thermalized neutrons. For a sample that is large compared to a mean free path, energy-dependent diffusion theory with the boundary condition that the flux vanish at the extrapolated boundary is an accurate description. In a large sample, the neutron spectrum is nearly Maxwellian, and the extrapolation distance is nearly equal to the z_0 of the thermal-neutron Milne problem of Sec. 4-10.

The basic experimental technique is to wait for a fundamental mode to be established and to measure the lifetime associated with the decay of this fundamental mode. The fundamental-mode decay constant is measured as a function of sample size, and the result is analyzed in terms of the thermal-neutron diffusion properties of the moderator. In this section, we show that this experiment, when analyzed in the energy-dependent diffusion approximation (4.64) has the same content as the measurement of the diffusion length.

The equation we must solve is the special case of Eq. (4.64):

$$\frac{1}{v}\frac{\partial \phi(\mathbf{r},E,t)}{\partial t} = -[\Sigma_a(E) - D(E)\nabla^2 - R_0]\phi(\mathbf{r},E,t) \tag{4.152}$$

subject to the boundary condition

$$\phi(\mathbf{R}_e,E,t) = 0 \tag{4.153}$$

and the initial condition

$$\phi(\mathbf{r},E,0) = \phi_i(\mathbf{r},E) \ . \tag{4.154}$$

The problem is greatly simplified if we suppose that the extrapolated boundary surface $\mathbf{R}_e$ is independent of E and t. With the assumption (4.153), we can write

$$\phi(\mathbf{r},E,t) = \sum_{n=1}^{\infty} \phi_n(E,t)\chi(r) , \qquad (4.155)$$

where the $\chi_n(r)$ are a complete set of orthonormalized eigenfunctions of the scalar Helmholtz equation

$$(\nabla^2 + B_n^2)\chi_n(\mathbf{r}) = 0 \qquad (4.156)$$

subject to the boundary condition

$$\chi_n(\mathbf{R}_e) = 0 . \qquad (4.157)$$

The $\phi_n(E,t)$ satisfy

$$\frac{1}{v}\frac{\partial\phi_n(E,t)}{\partial t} = -[\Sigma_a(E) + D(E)B_n^2 - R_0]\phi_n(E,t) , \qquad (4.158)$$

with

$$\phi_n(E,0) = \int_V \chi_n(\mathbf{r})\phi_i(\mathbf{r},E)d\mathbf{r} . \qquad (4.159)$$

The scalar Helmholtz equation (4.156) with the boundary condition (4.157) has a complete set of eigenfunctions with discrete positive eigenvalues B_n^2. The smallest B^2 eigenvalue is associated with an eigenfunction that is positive everywhere inside the extrapolated boundary of the system.[29] This value of B^2 is known in reactor theory as the geometric buckling of the system, and the associated eigenfunction is the fundamental-mode flux distribution. It is important to realize that these quantities depend only on the geometry of the system and not on the composition. For completeness, in Table 4.2 we recall the geometric buckling $B^2 = B_1^2$ and fundamental-mode shape $\chi(\mathbf{r})$ for a few simple geometries. By making the modal expansion (4.155) we have removed the finite size of the moderator from further consideration. Each spatial mode is associated with an energy spectrum that is the solution of Eq. (4.152) in an infinite homogeneous medium with a cos $B_n z$ spatial distribution. The size and shape of the

Table 4.2

GEOMETRIC BUCKLING AND FUNDAMENTAL-MODE SHAPE FOR SIMPLE GEOMETRIES

Shape	$\chi(\mathbf{r})$	B^2
Slab of thickness a	$\sin\left(\frac{\pi z}{a}\right)$ for $0 \le z \le a$	$\left(\frac{\pi}{a}\right)^2$
Sphere of radius R	$\frac{1}{r}\sin\left(\frac{\pi r}{R}\right)$ for $0 \le r \le R$	$\left(\frac{\pi}{R}\right)^2$
Cylinder of radius ρ and height H	$J_0\left(\frac{2.405\,\mathbf{r}}{\rho}\right)\sin\left(\frac{\pi z}{H}\right)$ for $0 \le z \le H$, $0 \le r \le \rho$	$\left(\frac{2.405}{\rho}\right)^2 + \left(\frac{\pi}{H}\right)^2$

moderator enter only in the values of B_n^2 that must be considered. In each spatial mode the energy spectrum approaches an asymptotic form while the neutron density in this mode is decaying. If both $\Sigma_a(E)$ and D(E) are proportional to 1/v, we can integrate Eq. (4.158) over energy to obtain

$$\frac{dn_j(t)}{dt} = -(v\Sigma_a + vDB_j^2)n_j(t) \,, \tag{4.160}$$

where

$$n_j(t) = \int_0^\infty \frac{1}{v}\,\phi_j(E,t)dE \,. \tag{4.161}$$

In this case, the neutron density in each mode decays exponentially. The fundamental spatial mode decays more slowly than the higher modes, since it has the smallest value of B^2 and thus the slowest leakage rate. If one waits long enough (and the initial distribution is not orthogonal to the fundamental mode), the spatial distribution will approach its fundamental mode. This waiting time is sufficiently long (see Chap. 8) that the energy spectrum in the fundamental mode will also have reached its asymptotic form. The asymptotic solution is of the form

$$\phi_{as}(\mathbf{r},E,t) = \chi_1(\mathbf{r})\phi_0(E)T(t) \ . \qquad (4.162)$$

Substitution of Eq. (4.162) into Eq. (4.152) leads to

$$\frac{dT}{dt} = -\lambda T \quad \text{or} \quad T(t) = \exp(-\lambda t) \ , \qquad (4.163)$$

where the decay constant λ is determined by solving the eigenvalue problem

$$\left[\frac{\lambda}{v} - \Sigma_a(E) - D(E)B^2\right]\phi_0(E) + R_0\phi_0(E) = 0 \ . \qquad (4.164)$$

Introducing the geometric buckling B^2 converts the problem of the asymptotic decay constant in a finite medium to an equivalent infinite-medium problem where the decay constant is an eigenvalue and the energy spectrum is the associated eigenfunction. The accuracy of this equivalence is subject to both experimental and theoretical study. Experimentally, it implies that two different-shaped samples of the same material, but with the same geometric buckling, have the same decay constant. Theoretically, there are two levels of improvement over Eq. (4.164) that are possible. The first is to extend the study of an infinite medium with a cos Bz spatial distribution to larger values of B^2 where nondiffusion corrections become important. This has been done systematically and is discussed in Chap. 8. The second level is to consider more carefully the relation between the infinite-medium and finite-medium problems for the fundamental-mode decay constant. This is not yet so well understood, but the present state of the art is also summarized in Chap. 8. For the remainder of this chapter, we accept without further discussion the assignment of a geometric buckling to the sample, and discuss the problem of calculating λ versus B^2 in the energy-dependent diffusion approximation (4.164).

We first note that for 1/v absorption

$$\lambda = \alpha + v\Sigma_a \ , \qquad (4.165)$$

where α is the decay constant for zero absorption and satisfies

$$[\alpha/v - D(E)B^2]\phi_0(E) + R_0\phi_0(E) = 0 \ . \qquad (4.166)$$

This is a special case of the more general result (4.19). We restrict our attention to 1/v absorption for the remainder of this section. With this restriction, the eigenvalue problem (4.166) becomes essentially identical to the diffusion-length eigenvalue problem (4.72); the only difference is that the signs of the absorption and leakage terms are interchanged. We can interpret Eq. (4.166) as defining the negative macroscopic absorption cross section $-\alpha/v$, which gives an imaginary diffusion length $L = (i/B)$. The formal appearance of the eigenvalue in the absorption term rather than in the leakage term makes no substantive difference. The situation is illustrated in Fig. 4.1. We note that the region of positive absorption between zero and the pure moderator absorption cross section is not accessible experimentally. Diffusion theory is a reliable guide to the behavior of the λ versus B^2 curve for small B^2. For large B^2 or large Σ_a, we see in Chap. 8 that the transport equation implies that the curve in Fig. 4.1 terminates at certain finite limit points. For most of the available experiments, however, the analysis of Sec. 4-7 and of this section is quantitatively applicable.

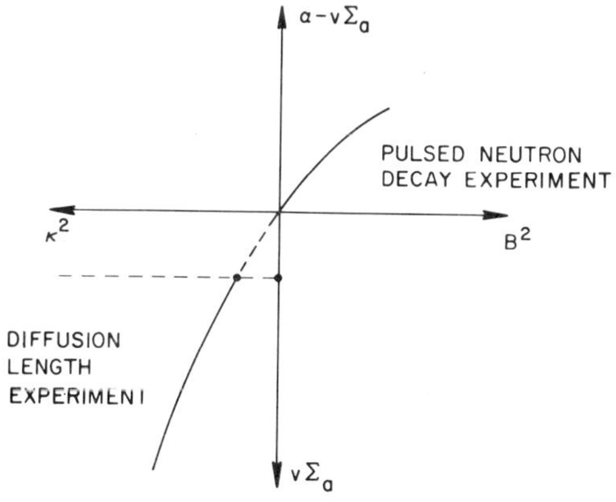

Figure 4.1 The equivalence of decay constant versus buckling to diffusion length versus poison concentration for 1/v absorption.

As with Eq. (4.72), we can integrate Eq. (4.166) over energy to get

$$\alpha = \hat{v}\bar{D}B^2 , \tag{4.167}$$

where $\hat{v}$ is the average speed in the sense of Eq. (4.39) and $\bar{D}$ is defined by Eq. (4.76). The averages are over the unknown flux spectrum $\phi_0(E)$. In the pulse-decay problem, the neutrons are removed by the leakage term $D(E)B^2$ at a higher average energy than that at which they are inserted by the equivalent negative absorption $-\alpha/v$. The asymptotic spectrum thus undergoes a "diffusion cooling" that is proportional to B^2. The quantities $\hat{v}$ and $\bar{D}$ in Eq. (4.167) are therefore dependent on B^2. Formal expansion in powers of B^2 gives

$$\lambda = (v\Sigma_a) + D_0B^2 - CB^4 + \dots , \tag{4.168}$$

where

$$D_0 = (\hat{v})_M(\bar{D})_M \tag{4.169}$$

and the diffusion cooling coefficient C is related to the diffusion heating coefficient H defined by Eq. (4.83) through

$$C = D_0^2H . \tag{4.170}$$

The relation between C and H follows from equating the expansion of α versus B^2 to the expansion of κ^2 versus Σ_a on a term by term basis. This is allowed, since these two curves define a single function. That this function is analytic for small B^2 has been assumed but not proven.

The pulsed-source neutron technique for measuring thermal-neutron diffusion parameters was first introduced by von Dardel[30] and by Antonov et al.[31] It has proven very useful as a check on theoretical models for thermal-neutron diffusion and neutron thermalization. As we see in Chap. 8, there are considerable difficulties in making the λ versus B^2 curve into a precisely definable method of analyzing experimental data. The concepts remain, however, both qualitatively and quantitatively useful. The intercept of the λ versus B^2 curve for zero B^2 can yield a precise measurement of the absorption cross section if sufficient care is taken to eliminate higher spatial modes.[32] The slope of the λ versus B^2 curve at $B^2 = 0$ is accurately measurable and gives the diffusion coefficient D_0 in a Maxwellian spectrum. (We have used the notation D_0 in Eq. (4.168) to be consistent with the usual convention in analyzing experimental data.) The diffusion cooling coefficient C is measurable with fair accuracy[33] from the diffusion-length experiment of Sec. 4-7, but there are difficulties in unambiguously determining it from pulsed-source neutron measurements.

4-12 MILNE'S PROBLEM FOR A CRITICAL SYSTEM

In Sec. 4-10 we considered the generalization of Milne's problem to a thermal-neutron problem with energy-dependent cross sections. For constant cross sections, monoenergetic transport theory applies not only to thermal-neutron problems but also to multiplying systems, which have a quite different energy spectrum. From the results of Secs. 4-7 and 4-8, we suspect that the generalization to energy-dependent cross sections can give quite different results, depending on the characteristic energy spectrum of the system.

The appropriate genralization of the case considered in Sec. 4-10 is to an infinite homogeneous half-space with a composition that is critical. We again consider a steady-state problem with isotropic scattering. Since the system is infinite, the criticality condition is that appropriate to an infinite homogeneous medium, and is not affected by the boundary. In the present case, the differential form of the transport equation is

$$\left[\mu \frac{\partial}{\partial z} + \Sigma(E)\right] f(z,E,\mu) = \tfrac{1}{2} \int_0^\infty \tilde{\Sigma}(E',E)\phi(z,E')dE' \qquad (4.170)$$

and the vacuum boundary condition is again (4.119). In Eq. (4.170)

$$\Sigma(E) = \Sigma_a(E) + \Sigma_s(E) \qquad (4.171)$$

is the total cross section and

$$\tilde{\Sigma}(E',E) = \Sigma_0(E',E) + \chi(E)\nu(E')\Sigma_f(E') \ , \qquad (4.172)$$

where $\Sigma_f(E')$ is the macroscopic fission cross section, $\nu(E')$ is the average number of neutrons resulting from the fission of a nucleus by a neutron of energy E', and $\chi(E)$ is the spectrum of neutrons liberated in the fission process. The normalization of $\chi(E)$ is

$$\int_0^\infty \chi(E)dE = 1 . \tag{4.173}$$

The integral equation analogous to Eq. (4.123) for the present problem is

$$\phi(z,E) = \tfrac{1}{2}\int_0^\infty \int_0^\infty \tilde{\Sigma}(E',E)E_1 \times \times [|z - z'|/\ell(E)]\phi(z',E')dE'dz' \tag{4.174}$$

with

$$\ell(E) = 1/\Sigma(E) .$$

The asymptotic solution for large z is again of the form of Eq. (4.125). The problem of determining z_0 and the asymptotic spectrum, which we denote by $\phi_r(E)$, is formally similar to the corresponding thermal-neutron problem. Now, however, $\phi_r(E)$ has a physical content which differs from that of the corresponding quantity $\phi_M(E)$ in Sec. 4-10. The asymptotic flux spectrum satisfies the equation

$$[\Sigma_a(E) + \Sigma_s(E)]\phi_r(E) = \int_0^\infty \Sigma_0(E',E)\phi_r(E')dE' + + \chi(E)\int_0^\infty \nu(E')\Sigma_f(E')\phi_r(E')dE' \tag{4.175}$$

for the spectrum in a critical, infinite homogeneous medium.

Consider first Eq. (4.174) when Σ_a, Σ_s, and $\nu\Sigma_f$ are constants. In this case, there is a separable solution of the form (4.137). Now, however, $\phi_M(E)$ is replaced by the solution of Eq. (4.175) and $\Phi(z)$ satisfies

$$\Phi(z) = \tfrac{1}{2}\int_0^\infty (\Sigma_s + \nu\Sigma_f)E_1(|z - z'|/\ell)\Phi(z')dz' = \frac{c}{2\ell}\int_0^\infty E_1(|z - z'|/\ell)\Phi(z')dz' , \tag{4.176}$$

where

$$c = \frac{\Sigma_s + \nu\Sigma_f}{\Sigma_s + \Sigma_a} \tag{4.177}$$

is the mean number of secondary neutrons per collision. For a critical system, c = 1. The constant c is familiar in the monoenergetic transport theory of regenerative systems.[1]

The discussion of the constant-cross-section limit following Eq. (4.139) applies here except for the identification of the asymptotic scalar flux spectrum as a Maxwellian. A direct implication of the property of separability is the validity of monoenergetic transport theory for the extrapolation distance in a nuclear reactor with constant cross sections. The conditions for which separability applies, however, are not satisfied by the actual cross sections of importance to reactor design. In most practical problems, these conditions are not even satisfied to a good approximation.

We now return to the approximate calculation of z_0 for a critical half-space. This problem is treated in detail in Ref. 34. We develop an approximate expression for z_0 which is formally analogous to Eq. (4.147). For this purpose, we require the analogies of Eqs. (4.131) through (4.136).

The problem simplifies in the special case where $\nu(E)\Sigma_f(E)/\Sigma_a(E)$ is independent of E. In this case, the quantity

$$C(E) = \frac{\Sigma_s(E) + \nu(E)\Sigma_f(E)}{\Sigma_s(E) + \Sigma_a(E)}$$

is independent of energy. The adjoint function $\phi_r^\dagger(E)$, which satisfies the adjoint equation to (4.175),

$$[\Sigma_a(E) + \Sigma_s(E)]\phi_r^\dagger(E) = \int \tilde{\Sigma}(E,E')\phi_r^\dagger(E')dE' , \tag{4.178}$$

is constant in this special case. This constancy indicates that the probability for a neutron to cause a fission is independent of the neutron energy. In this special case, which is crudely applicable for a thermal reactor, Eqs. (4.131) through (4.136) and Eq. (4.147) apply with only the modifications that $\Sigma_0(E',E)$ is replaced by $\tilde{\Sigma}(E',E)$ and the averages of $\ell(E)$ and $\ell^2(E)$ over $\phi_M(E)$ are replaced by averages over

$\phi_r(E)$.

For the more general case, the non-self-adjoint nature of the problem is more explicitly manifested by the appearance of a $\phi_r^\dagger(E)$ that is not constant. We begin by defining

$$(z) = \int_{-1}^{1} \mu \int_{0}^{\infty} \phi_r^\dagger(E) f(z,E,\mu) dE d\mu \tag{4.179}$$

and

$$(z) = \int_{-1}^{1} \mu^2 \int_{0}^{\infty} \ell(E) \phi_r^\dagger(E) f(z,E,\mu) dE d\mu \ . \tag{4.180}$$

Equations (4.179) and (4.180) are generalizations of Eqs. (4.129) and (4.133), respectively, and are identical to the latter equations when $\phi_r^\dagger(E)$ is a constant. This is the case when $\nu(E)\Sigma_f(E)$ and $\Sigma_a(E)$ have the same energy dependence or when they both vanish as in Sec. 4-10. Multiplying Eq. (4.170) by $\phi_r^\dagger(E)$, integrating over E and μ, and using Eq. (4.178), we obtain

$$\frac{d}{dz} = 0 \ . \tag{4.181}$$

Similarly, multiplying Eq. (4.170) by $\mu\phi_r^\dagger(E)$ and integrating over E and μ, we obtain

$$\frac{d}{dz} = - \qquad \text{or} \qquad (z) = - \quad (z + c) \ . \tag{4.182}$$

The constants and c are determined by arguments like those which lead to Eqs. (4.134) and (4.135). We have

$$= -\frac{1}{3}\ell_{av} \ , \tag{4.183}$$

$$(0) = \frac{1}{3}\ell_{av} z_0 \ , \qquad (4.184$$

where

$$\ell_{av} = \int_0^\infty \ell(E)\phi_r(E)\phi_r^\dagger(E) dE \ , \tag{4.185}$$

and the adjoint function $\phi_r^\dagger(E)$ is so normalized that

$$\int_0^\infty \phi_r(E)\phi_r^\dagger(E)dE = 1 \,. \tag{4.186}$$

Equation (4.120) still applies for the angular flux spectrum if therein we replace $\Sigma_0(E',E)$ by $\tilde{\Sigma}(E',E)$. Combining this result with Eqs. (4.180) and (4.184) gives

$$z_0 = \frac{3}{2\ell_{av}}\int_0^\infty\int_0^\infty\int_0^\infty \ell(E)\tilde{\Sigma}(E',E)E_3[z/\ell(E)] \times$$
$$\times\ \phi(z,E')\phi_r^\dagger(E)dzdEdE' \,. \tag{4.187}$$

Inserting

$$q(z,E) = \phi(z,E) - z\phi_r(E) \tag{4.188}$$

into Eq. (4.174), we find

$$q(z,E) = \tfrac{1}{2}\,\ell(E)\phi_r(E)E_3[z/\ell(E)] + \tfrac{1}{2}\int_0^\infty\int_0^\infty \times$$
$$\times\ \tilde{\Sigma}(E',E)E_1[|z - z'|/\ell(E)]q(z',E')dz'dE' \,, \tag{4.189}$$

whereupon

$$z_0 = \frac{3}{\ell_{av}}\left[\frac{(\ell^2)_{av}}{8} + \int_0^\infty\int_0^\infty\int_0^\infty \tfrac{1}{2}\ell(E)\tilde{\Sigma}(E',E)E_3[z/\ell(E)] \times \right.$$
$$\left. \times\ \phi_r^\dagger(E)q(z,E')dzdEdE'\right] , \tag{4.190}$$

with

$$(\ell^2)_{av} = \int_0^\infty \ell^2(E)\phi_r^\dagger(E)\phi_r(E)dE \,. \tag{4.191}$$

A variational expression for z_0 can be written conveniently in terms of the functions

$$\Psi(z,E) = \int_0^\infty \tilde{\Sigma}(E',E)q(z,E')dE' \tag{4.192}$$

$$g(z,E) = \tfrac{1}{2}\int_0^\infty \ell(E')\phi_r(E')E_3[z/\ell(E')]\tilde{\Sigma}(E',E)dE' \ , \tag{4.192a}$$

where $\Psi(z,E)$ satisfies the integral equation

$$\Psi(z,E) = \tfrac{1}{2}\int_0^\infty \int_0^\infty \tilde{\Sigma}(E',E)E_1\left[\frac{|z - z'|}{\ell(E)}\right] \times$$

$$\times \Psi(z',E')dz'dE' + g(z,E) \ . \tag{4.193}$$

Since Eq. (4.193) is an inhomogeneous integral equation, the adjoint equation is not unique but depends on the choice of adjoint source $g^\dagger(z,E)$. For a given adjoint source, the adjoint function $\Psi^\dagger(z,E)$ satisfies

$$\Psi^\dagger(z,E) = \tfrac{1}{2}\int_0^\infty \int_0^\infty \tilde{\Sigma}(E,E')E_1\left[\frac{|z - z'|}{\ell(E)}\right] \times$$

$$\times \Psi^\dagger(z',E')dz'dE' + g^\dagger(z,E) \ . \tag{4.194}$$

The adjoint source will be specified so that the stationary quantity is the desired one. Consider now the functional

$$I(\tilde{\Psi},\tilde{\Psi}^\dagger) = \frac{(\tilde{\Psi},g^\dagger)(\tilde{\Psi}^\dagger,g)}{(\tilde{\Psi}^\dagger,\tilde{\Psi} - O\tilde{\Psi})} \ , \tag{4.195}$$

where the operator O is defined by

$$O\tilde{\Psi} = \tfrac{1}{2}\int_0^\infty \int_0^\infty \tilde{\Sigma}(E',E)E_1\left[\frac{|z - z'|}{\ell(E')}\right]\tilde{\Psi}(z',E')dz'dE' \tag{4.196}$$

and the notation (A,B) means

$$(A,B) = \int_0^\infty \int_0^\infty A(z,E)B(z,E)dzdE \ . \tag{4.197}$$

One can verify that I is stationary with respect to small independent variations of $\tilde{\Psi}$ and $\tilde{\Psi}^{\dagger}$ about Ψ and $\Psi^{\dagger}$, and that the stationary value is

$$I(\Psi,\Psi^{\dagger}) = (\Psi,g^{\dagger}) = (\Psi^{\dagger},g) \ . \qquad (4.198)$$

Unlike the thermal-neutron case, however, it has not been shown that this stationary value is either a maximum or a minimum.

By choosing

$$g^{\dagger}(z,E) = \tfrac{1}{2}\,\ell(E)E_3\left[\frac{z}{\ell(E)}\right]\phi_r^{\dagger}(E) \qquad (4.199)$$

and using Eq. (4.198), we find that Eq. (4.195) is a stationary expression for the triple integral in Eq. (4.190). If we choose the asymptotic solutions

$$\tilde{\Psi}(z,E) = \phi_r(E)/\ell(E) \qquad (4.200)$$

$$\tilde{\Psi}^{\dagger}(z,E) = \phi_r^{\dagger}(E) \qquad (4.201$$

as trial functions, we obtain

$$z_0 \approx \frac{3}{8}\,\frac{(\ell^2)_{av}}{\ell_{av}} + \frac{1}{3}\,\ell_{av} \ . \qquad (4.202)$$

This result is formally similar to the result (4.147). The only difference is in the manner in which the averages are to be taken. In the limit of vanishing absorption and fission cross sections, (4.202) and (4.147) are identical since the adjoint is then constant and the flux spectrum is Maxwellian.

In the presence of absorption and fission, the physical content of Eq. (4.202) differs considerably from that of Eq. (4.147). This difference is contained in the different spectra over which the mean free path and its square are averaged. In Eq. (4.202), $\phi_r(E)$ satisfies Eq. (4.175) and contains appreciable numbers of fast neutrons as well as thermal neutrons. This, coupled with the relatively long mean free path of fission neutrons, leads to extrapolation distances for multiplying assemblies which can be considerably greater than for nonmultiplying assemblies with the same moderator. This fact affects the interpretation of pulsed-source neutron experiments.

The flux spectrum $\phi_r(E)$ in an infinite homogeneous medium with a steady-state source of fission neutrons is of great importance in reactor theory. Its determination in the energy range from zero to a few electron volts is one of the principal problems of neutron thermalization. This problem is considered in detail, both theoretically and experimentally, in the next three chapters.

REFERENCES

1. B. Davison and J. B. Sykes, *Neutron Transport Theory*, Clarendon Press, Oxford, 1957.

2. K. M. Case, F. de Hoffmann, and G. Placzek, *Introduction to the Theory of Neutron Diffusion*, Vol. I, Superintendent of Documents, U.S. Government Printing Office, Washington 25, D.C., 1953.

3. A. M. Weinberg and E. P. Wigner, *The Physical Theory of Neutron Chain Reactors*, University of Chicago Press, Chicago, 1958.

4. E. P. Wigner, Mathematical Problems of Nuclear Reactor Theory, in *Nuclear Reactor Theory*, Proceedings of Symposia in Applied Mathematics, Vol. XI, pp. 89-104, American Mathematical Society, Providence, R.I., 1961.

5. R. K. Osborn and S. Yip, Theory of Fluctuations and Correlations in Neutron Distributions, *Trans. Am. Nucl. Soc.*, **6**: 6 (1963).

6. B. W. Hapke, *Problems Connected with the Use of Pulsed Reactors to Attain Very High Neutron Fluxes*, Doctoral Dissertation, Cornell University, 1962.

7. R. Scalletar, Space-Dependent Corrections to the Fuchs-Nordheim Model, in *Reactor Kinetics and Control*, Proceedings of a Symposium held at the University of Arizona, March 25-27, 1963, p. 253, USAEC Report TID-7662, April 1964.

8. A. Ghatak and M. Nelkin, Microscopic Prompt Neutron Kinetics of the TREAT Reactor, *Trans. Am. Nucl. Soc.*, **6**: 110 (1963).

9. M. Nelkin, Asymptotic Solutions of the Transport Equation for Thermal Neutrons, *Physica*, **29**: 261 (April 1963).

10. S. Chapman and T. G. Cowling, *The Mathematical Theory of Non-uniform Gases*, 1st ed., Cambridge University Press, 1939.

11. S. A. Rice and H. L. Frisch, Some Aspects of the Statistical Theory of Transport, *Ann. Rev. Phys. Chem.*, **11**: 187 (1960).

12. P. Ehrenfest and T. Ehrenfest, *The Conceptual Foundations of the Statistical Approach in Mechanics*, a translation by M. Moravcsik of the 1912 German original, Cornell University Press, Ithaca, N.Y., 1959.

13. N. G. Van Kampen, Fundamental Problems in Statistical Mechanics of Irreversible Processes, in *Fundamental Problems in Statistical Mechanics: Proceedings of the NUFFIC International Summer Course*

in Science at Nijenrode Castle, The Netherlands, August, 1961, p. 173, compiled by E. G. D. Cohen, North Holland Publishing Company, Amsterdam, 1962.

14. H. Goldstein, *Fundamental Aspects of Reactor Shielding*, p. 174, Addison-Wesley Publishing Company, Inc., Reading, Mass., 1959.

15. E. Wigner and F. Seitz, On the Constitution of Metallic Sodium, *Phys. Rev.*, **43**: 804 (1933).

16. H. C. Honeck, Some Method for Improving the Cylindrical Reflecting Boundary Condition in Cell Calculations of the Thermal Neutron Flux, *Trans. Am. Nucl. Soc.*, **5**: 350 (1962).

17. H. C. Honeck, The Distribution of Thermal Neutrons in Space and Energy in Reactor Lattices; Part I: Theory, *Nucl. Sci. Eng.*, **8**: 193 (1960).

18. J. Ernest Wilkins, Jr., Diffusion Approximation to the Transport Equation, in *Nuclear Reactor Theory*, Proceedings of Symposia in Applied Mathematics, Vol. XI, pp. 113-114, American Mathematical Society, Providence, R.I., 1961.

19. G. Placzek and W. Seidel, Milne's Problem in Transport Theory, *Phys. Rev.*, **72**: 550 (1947); also see Refs. 1 and 2.

20. V. Kourganoff, *Basic Methods in Transfer Problems; Radiative Equilibrium and Neutron Diffusion*, Clarendon Press, Oxford, 1952.

21. J. LeCaine, Application of a Variational Method to Milne's Problem, *Phys. Rev.*, **72**: 564 (1947).

22. J. LeCaine, *A Table of Integrals Involving the Functions $E_n(x)$*, Canadian Report MT-131 (NRC-1553), National Research Council of Canada, 1945.

23. M. Nelkin, Milne's Problem for a Velocity-Dependent Mean Free Path, *Nucl. Sci. Eng.*, **7**: 552 (1960).

24. P. Schofield, *The Thermal Neutron Milne Problem*, British Report AERE-TP-99, March 1963.

25. M. M. R. Williams, The Exact Solution of the Velocity Dependent Milne Problem Using a Simple Scattering Kernel, *Trans. Am. Nucl. Soc.*, **6**: 27 (1963).

26. N. Corngold, P. Michael, and W. Wollman, The Time Decay Constants in Neutron Thermalization, *Nucl. Sci. Eng.*, **15**: 13 (1963).

27. M. M. R. Williams, University of Birmingham, England, private communication.

28. R. Kladnik and I. Kuščer, Milne's Problem for Thermal Neutrons in a Nonabsorbing Medium, *Nucl. Sci. Eng.*, **11**: 116 (1961).

29. R. Courant and D. Hilbert, *Methoden der Mathematischen Physik*, Vol. 2, Chap. 7, Sec. 3, pp. 488-495, Springer Verlag, Berlin, 1937.

30. G. F. von Dardel, The Interaction of Neutrons with Matter Studied with a Pulsed Neutron Source, *Trans. Roy. Inst. Technol., Stockholm*, **75**: 1954; and G. F. von Dardel and N. G. Sjöstrand, Diffusion Parameters of Thermal Neutrons in Water, *Phys. Rev.*, **96**: 1245 (1954).

31. A. V. Antonov *et al.*, A Study of Neutron Diffusion in Beryllium, Graphite, and Water by the Impulse Method, in *Proceedings of the First United Nations International Conference on the Peaceful Uses of Atomic Energy, Geneva, 1955*, Vol. 5, p. 3, United Nations, New York, 1956.

32. W. M. Lopez and J. R. Beyster, Measurement of Neutron Diffusion Parameters in Water by the Pulsed Neutron Method, *Nucl. Sci. Eng.*, **12**: 190 (1962).

33. E. Starr and J. Koppel, Determination of Diffusion Hardening in Water, *Nucl. Sci. Eng.*, **14**: 224 (1962).

34. V. Boffi, V. Molinari, and D. E. Parks, On the Decay of Neutrons in a Subcritical Assembly, *J. Nuclear Energy, Pt. A & B*, **16**: 395 (1962).

Chapter 5

THE THERMAL-NEUTRON SPECTRUM IN A UNIFORM HOMOGENEOUS MEDIUM

We wish to calculate the neutron flux spectrum $\phi(E)$ in the energy region from zero up to some maximum energy E_m. We call this energy region the "thermal group" since it plays this role in multigroup reactor calculations. The energy E_m can be chosen by various criteria, as discussed in Chap. 7, but we normally consider an extended thermal group covering the entire energy region in which the effects of chemical binding and atomic motion are significant. The energy region above E_m can then be treated by the well-developed theory of neutron slowing down by free atoms at rest. A typical choice of E_m for reactors that are not at very high temperatures would be the indium resonance energy of 1.46 eV. This gives a natural separation, in terms of foil-activation experiments, between slowing down and thermal diffusion of neutrons. In this extended thermal-energy group we thus include not only the quasi-Maxwellian portion of the energy spectrum but also the transition region between thermal and epithermal flux spectra. We can then appropriately distinguish between these two portions of our extended thermal group for the sake of physical clarity; however, for practical calculations this distinction is usually not useful.

Consider the neutron balance in an energy interval dE at the energy E. Neutrons are removed from this energy interval by absorption and inelastic scattering, and enter it as a result of scattering from other energies. Any purely elastic scattering from crystalline moderators cancels out of the equations, since it only changes the neutron direction of motion, which does not concern us at present. It is convenient to distinguish between those neutrons that enter the energy interval dE as a result of scattering from other energies less than E_m, and those that enter as a result of slowing down from epithermal energies greater than E_m. The resulting neutron-balance equation is

$$[\Sigma_a(E) + \Sigma_s(E)]\phi(E) = S(E) + \int_0^{E_m} \Sigma_0(E',E)\phi(E')dE' , \qquad (5.1)$$

where the slowing down from epithermal energies is represented as an effective external source

$$S(E) = \int_{E_m}^{\infty} \Sigma_0(E',E)\phi_e(E')dE' , \qquad (5.2)$$

and the epithermal flux $\phi_e(E)$ is assumed known from a calculation of the slowing down from fission energies.

Equation (5.1) is formally identical to the governing integral equation in the theory of neutron slowing down, but the physical content is quite different because of the different properties of the energy-transfer kernel. The most fundamental difference is that neutrons can both gain and lose energy in a collision. If the absorption is sufficiently weak that on the average a neutron makes many scattering collisions at thermal energy, then the neutron spectrum approximates a Maxwellian energy distribution in thermal equilibrium with the moderator. This distribution extends approximately up to an energy where the Maxwellian and epithermal flux spectra are comparable. This energy is only a few times k_BT, except for extremely weak absorption, and is thus well within the boundary of our extended thermal-energy group. There is a substantial energy region in which the Maxwellian and epithermal flux spectra join smoothly. The Maxwellian distribution in the neighborhood of k_BT is also shifted to higher energies because of the incomplete thermalization of the neutrons slowing down from high energies before they are absorbed. In order to have a physical understanding of the general features of the thermal-neutron spectrum, we study the general properties of Eq. (5.1) in some detail and consider some simplified models of the thermalization process.

5-1 GENERAL PROPERTIES OF THE NEUTRON-BALANCE EQUATION

There are several general properties of the flux spectrum $\phi(E)$ arising from the nature of Eq. (5.1) as a neutron-balance equation and from the general properties of the cross sections appearing in the equation. We describe these properties here, and the approximate solutions (to be calculated later) are always constrained to have these properties. The first important constraint is that of neutron conservation, which states that every neutron slowing down into the thermal group is absorbed there. We have chosen E_m sufficiently large that energy-gain scattering out of the thermal group can be neglected. This corresponds to the definition of $\Sigma_s(E)$ in Eq. (5.1) as

$$\Sigma_s(E) = \int_0^{E_m} \Sigma_0(E,E')dE' . \tag{5.3}$$

With this definition, the scattering terms cancel when we integrate (5.1) over energy from zero to E_m, and we obtain the statement of overall neutron conservation

$$\int_0^{E_m} \Sigma_a(E)\phi(E)dE = \int_0^{E_m} S(E)dE . \tag{5.4}$$

We note that an arbitrary term proportional to $\delta(E' - E)$ can be added to $\Sigma_0(E',E)$, as long as an elastic scattering cross section is added to $\Sigma_s(E)$ so that (5.3) remains exactly satisfied. The elastic terms exactly cancel from (5.1). We assume in the remainder of this chapter that $\Sigma_s(E)$ is defined to exclude the purely elastic component.

The physical content of (5.1) as a neutron-balance equation is most concisely stated in terms of a generalized slowing-down density

$$q(E) = \int_E^{\infty} dE_1 \int_0^{E} dE_2 \Sigma_0(E_1,E_2)\phi(E_1) -$$

$$- \int_0^{E} dE_1 \int_E^{\infty} dE_2 \Sigma_0(E_1,E_2)\phi(E_1) . \tag{5.5}$$

Thus, $q(E)$ is defined as the rate at which neutrons slow down past energy E minus the rate at which they speed up past energy E, and is a natural generalization of the slowing-down density occurring in the theory of neutron slowing down. In terms of $q(E)$, the neutron-balance equation (5.1) becomes

$$\frac{dq(E)}{dE} = \Sigma_a(E)\phi(E) , \tag{5.6}$$

and $q(E)$ satisfies the boundary condition that $q(0) = 0$ and

$$q(E_m) = \int_0^{E_m} S(E)dE = \int_0^{E_m} \Sigma_a(E)\phi(E)dE . \tag{5.7}$$

The quantity $q(E_m)$ is just the conventional "slowing-down density at thermal energies" appearing in the diffusion equation for thermal neutrons in a reactor. (In early treatments of reactor theory, the question of defining E_m often was not carefully considered, so that the slowing-down density was often evaluated at k_BT. This corresponds to counting the energy region between k_BT and E_m twice. This point is perhaps best illustrated by the discussion of the migration area given in Sec. 5-8.)

An important general property of the energy-transfer kernel is the detailed balance condition

$$\Sigma_0(E',E)\phi_M(E') = \Sigma_0(E,E')\phi_M(E) , \tag{5.8}$$

where $\phi_M(E)$ is the equilibrium Maxwellian flux spectrum (4.67) at the moderator temperature. Equation (5.8) follows directly from the detailed balance condition (2.79) satisfied by the microscopic differential-energy-transfer cross section. Its derivation requires only that the scattering system be in thermal equilibrium, and that the collision process obey microscopic reversibility. It should be emphasized that all the quantities in (5.8) depend on the moderator temperature, and that this equation is valid at all temperatures and for all cases of physical interest. The detailed balance condition states that the rate of scattering between any two energy intervals is the same in each direction when the neutrons are in thermal equilibrium with the moderator. In the context of a systematic development starting from the Boltzmann equation, the proof of (5.8) serves as a derivation of the

Maxwellian distribution as the thermal-equilibrium neutron distribution.

As we have seen in Sec. 4-3, the detailed balance condition is a necessary condition for the time-dependent neutron energy distribution to approach an equilibrium Maxwellian. In the steady-state problem defined by (5.1), Eq. (5.8) implies that $\phi_M(E)$ is a solution of the neutron-balance equation in the limit that the source and absorption terms are vanishingly small. In this limit, q(E), as defined by Eq. (5.5), is identically zero, since at equilibrium the number of neutrons speeding up past any given energy is equal to the number slowing down past that energy. In actual calculations with approximate energy-transfer kernels, we ensure that the detailed balance condition is exactly satisfied by using an approximate model only to calculate the energy-loss scattering, and by using Eq. (5.8) to calculate the energy-gain scattering from the energy-loss scattering. This has the advantage of considerably reducing the amount of numerical computation, as well as giving more accurate neutron spectra for weak absorption.

A second important general property of the scattering kernel is the approach to free-atom scattering at high energies. This has been extensively discussed in Sec. 2-8 in connection with the short-collision-time approximation. One obtains a smooth transition from the thermal-neutron spectrum to the epithermal-neutron spectrum that is determined by the well-established theory of neutron slowing down by free atoms at rest. In the epithermal-energy region, the energy-transfer kernel is given from (2.96) as

$$\left.\begin{aligned} \Sigma_0(E',E) &= \frac{\Sigma_s}{(1-\alpha)E'} && \alpha E' \le E \le E' \\ &= 0 && E > E' \quad \text{and} \quad E < \alpha E' , \end{aligned}\right\} \tag{5.9}$$

where

$$\alpha = [(A-1)/(A+1)]^2 , \tag{5.10}$$

and A is the ratio of moderator atomic mass to neutron mass. This result applies when the scattering cross section is constant and isotropic in the center-of-mass system. In this energy region, if absorption is negligible, the neutron spectrum satisfying Eq. (5.1) is the Fermi spectrum

$$\phi_e(E) = C/E . \tag{5.11}$$

The value of C can be related to the slowing-down density by carrying out the integration in Eq. (5.5) using the free-atom kernel (5.9). This gives the well-known relation

$$q(E_m) = C\Sigma_s \int_{E_m}^{E_m/\alpha} \frac{dE}{(1 - \alpha)E^2} (E_m - \alpha E) = \xi_0 \Sigma_s E_m \phi_e(E_m) , \quad (5.12)$$

where

$$\xi_0 = \frac{1}{1 - \alpha} \int_1^{I/\alpha} \frac{dx}{x^2} (1 - \alpha x) = 1 - \frac{\alpha}{1 - \alpha} \ln \frac{1}{\alpha} . \quad (5.13)$$

It should be noted that ξ_0, defined by the ratio of slowing-down density to flux, is identical to the average logarithmic energy loss per collision for elastic scattering by a free atom at rest. This is true, however, only when the scattering is isotropic in the center-of-mass system and the scattering cross section is independent of energy.

We can combine Eq. (5.12) with the neutron conservation condition (5.7) to obtain a relation between the thermal and epithermal flux spectra:

$$\xi_0 \Sigma_s E_m \phi(E_m) = \int_0^{E_m} \Sigma_a(E)\phi(E)dE . \quad (5.14)$$

This relation is independent of the energy-transfer kernel for thermal neutrons, but depends on the neglect of epithermal absorption in obtaining the Fermi spectrum (5.11). If the epithermal absorption is not negligible, more accurate solutions of the slowing-down equation can be used to relate the epithermal flux spectrum to the slowing-down density, and thus to the total number of thermal absorptions.

In Sec. 2-8 we saw that the short-collision-time approximation could be effectively employed to calculate corrections to the scattering at energies where atomic motions and chemical binding cause only a small perturbation to the cross section of a free atom at rest. We can, in fact, go one step further and use the short-collision-time expansion to directly calculate thermalization corrections to the epithermal spectrum. This procedure is discussed in detail in Sec. 5-6.

In addition to the general physical properties of the neutron-balance equation (5.1) which we have discussed, it is useful at this point to examine briefly some of its mathematical properties. Equation (5.1) is a linear integral equation that can be classified as an inhomogeneous Fredholm equation of the second kind. It is a nonsingular equation with a unique solution that is readily obtained numerically on a digital computer by dividing the energy interval from zero to E_m into a large number of subintervals. The resulting matrix equations are conveniently solved by an iterative method discussed in Sec. 7-1.2.

Although the integral equation expressing neutron balance does not have a symmetric kernel in the form (5.1), the detailed balance condition (5.8) can be used to rewrite it in a form with a symmetric kernel. We introduce the new dependent variable

$$\chi(E) = \phi(E)[\Sigma(E)/\phi_M(E)]^{\frac{1}{2}} , \tag{5.15}$$

which satisfies

$$\chi(E) = \int_0^{E_m} K(E,E')\chi(E')dE' + S(E)[\Sigma(E)\phi_M(E)]^{-\frac{1}{2}} , \tag{5.16}$$

where

$$K(E,E') = \Sigma_0(E',E)[\phi_M(E')/\phi_M(E)]^{\frac{1}{2}}[\Sigma(E)\Sigma(E')]^{-\frac{1}{2}} \tag{5.17}$$

is a symmetric function of E and E', and

$$\Sigma(E) = \Sigma_a(E) + \Sigma_s(E) .$$

It is therefore possible, in principle, to apply the general theory of nonsingular integral equations with symmetric kernels to the problem of the thermal-neutron spectrum in an infinite homogeneous medium. In practice, however, the kernels that apply to actual moderators are sufficiently complicated that numerical solutions are considered almost exclusively, and very little has been done with the analytic properties of the neutron-balance equation in the form (5.16).

The neutron-balance equation has been considered, however, from an analytic point of view in certain idealized problems in which the integral equation can be reduced to a second-order differential equation. These idealizations apply for moderation by an ideal gas of monatomic hydrogen and for the limiting case of a heavy monatomic ideal gas, and are considered in detail in Secs. 5-4 and 5-5, respectively. For these idealized cases, it is convenient to avoid explicit consideration of the thermal-group boundary E_m. We can do this by replacing the slowing-down source with a boundary condition on the spectrum at infinite energy, but we no longer have a nonsingular integral equation with a unique solution.

Although it was convenient to introduce the energy E_m separating the slowing-down and thermalization energy regions, we could, in principle, have considered the entire energy region from fission to thermal energies in a single integral equation of the form (5.1), with S(E) being replaced by the actual fission-spectrum source. The range of integration could formally be extended to infinity, but all quantities of interest at very high energies fall off rapidly enough that the formally singular nature of the resulting integral equation is of no concern. For the thermalization problem, however, there is a large energy region far above thermal energies, but far below fission energies, that is conveniently interpreted as an asymptotic region for the neutron-balance equation. We consider then an artificial problem where the true energy-transfer kernel is replaced even at very high energies by one calculated from the Fermi approximation through Eq. (2.64), so that the scattering cross section remains constant and isotropic in the center-of-mass system up to arbitrarily high energies. The actual source of neutrons is assumed to be located at arbitrarily high energies. For this problem, the neutron-balance equation is replaced by the homogeneous equation

$$[\Sigma_a(E) + \Sigma_s(E)]\phi(E) = \int_0^\infty \Sigma_0(E',E)\phi(E')dE' , \tag{5.18}$$

supplemented by the boundary condition

$$\lim_{E\to\infty} E\phi(E) = C , \tag{5.19}$$

where C is related to the epithermal slowing-down density through Eq. (5.12). We now have a singular homogeneous Fredholm integral equation of the second kind which does not have a unique solution, but which requires specification of the asymptotic behavior of $\phi(E)$ in the limit of large energy in place of the inhomogeneous source term in Eq. (5.1).

The singular equation (5.18) has the advantage that the solution satisfying Eq. (5.19) varies smoothly from the thermal into the epithermal energy region, with no cutoff energy E_m dividing these regions. (The solution does not apply, however, for energies high enough for the actual nuclear scattering to vary, since the artificial limit of high neutron energy was constructed as an extrapolation of the epithermal energy range.) The singular nature of the solution is apparent when we integrate Eq. (5.18) over energy to obtain a neutron-conservation condition. We find that the integrals of the left- and right-hand side do not separately exist. We must first subtract $\Sigma_s(E)\phi(E)$ from both sides of the equation, and note that the integral of the right-hand side is the slowing-down density defined by Eq. (5.5) in the limit of large neutron energy. A more direct manifestation of the singular character of our solution is the behavior of $\chi(E)$, the solution of the symmetrized equation (5.16). This diverges exponentially at large energies.

We emphasize that the nonexistence of the scattering rates mentioned previously is associated with locating the source at infinitely high energies and the consequent boundary conditions, Eq. (5.19), and not with the infinite limit of integration in Eq. (5.18). If we write the balance equation as an inhomogeneous equation with an explicit source S(E)

$$[\Sigma_a(E) + \Sigma_s(E)]\phi(E) = \int_0^\infty \Sigma(E',E)\phi(E')dE' + S(E) \ ,$$

then the scattering rate $\int_0^\infty \Sigma_s(E)\phi(E)dE$ does exist.

5-2 THE SPECTRUM FOR WEAK ABSORPTION

We have repeatedly noted that the spectrum in the limit of weak absorption is a Maxwellian distribution at the moderator temperature independent of the thermalization mechanism. In order to study this limit more precisely we fix the intensity of the slowing-down source, and vary the absorber concentration through a dimensionless parameter. This parameter, λ, is defined by

$$\Sigma_a(E) = \lambda \Sigma_a^0(E) , \tag{5.20}$$

where $\Sigma_a^0(E)$ is independent of absorber concentration. We see from the conservation condition (5.4) that the flux contains a component of order λ^{-1} which accounts for all the absorptions. Upon expanding ϕ in Eq. (5.1) in powers of λ, we find that this component is proportional to $\phi_M(E)$ and that the remaining terms of order unity oscillate in sign to give no net absorptions. We thus obtain a weak-absorption expansion that gives the correct thermal and epithermal flux spectra and a quantitative prescription for joining them.

We can calculate the relative amplitudes of the Maxwellian and epithermal flux spectra directly from neutron conservation without more detailed consideration. If we substitute $\Phi\phi_M(E)$ for ϕ on the right-hand side of Eq. (5.14), we obtain

$$\Phi\left(\bar{\Sigma}_a\right)_M = \xi_0 \Sigma_s E_M \phi(E_m) , \tag{5.21}$$

where $(\bar{\Sigma}_a)_M$ is the Maxwellian average of $\Sigma_a(E)$. (We have assumed E_M to be sufficiently large that $\phi_M(E)$ integrates to unity over the thermal group, so that Φ is the total thermal flux.) The ratio of the maximum value of $E\phi(E)$ to its epithermal value is thus given by

$$\frac{4}{e^2} \frac{\xi_0 \Sigma_s}{\left(\bar{\Sigma}_a\right)_M} . \tag{5.22}$$

In the limit of weak absorption that we are considering, this ratio is much greater than one. The criterion that this ratio be large is, in fact, a good criterion for defining a "thermal" reactor.

In a thermal reactor, the flux spectrum will be nearly Maxwellian up to an energy where the number of neutrons slowing down from high energy becomes comparable with the number speeding up from the peak of the Maxwellian distribution. This is roughly defined by the point at which the Maxwellian and epithermal flux spectra have equal amplitude. This occurs at an energy E_c given by

$$(\beta E_c)^2 e^{-\beta E_c} = (\bar{\Sigma}_a)_M/\xi_0\Sigma_s \,. \tag{5.23}$$

This energy depends only logarithmically on the absorption, and is $5k_BT$ for a typical value of 0.168 for $(\bar{\Sigma}_a)_M/\xi_0\Sigma_s$. Even in the weak absorption limit, however, the shape of the spectrum in the neighborhood of E_c can only be determined by a more detailed calculation. This calculation can be formally carried out by expanding the neutron-balance equation in powers of λ.

Consider the neutron-balance equation (5.1) with λ defined by Eq. (5.20). We subtract the term of order λ^{-1} by letting

$$\phi(E) = \lambda^{-1}\phi_0(E) + G(E,\lambda) \,, \tag{5.24}$$

where $G(E,\lambda)$ has a power-series expansion for small λ of the form

$$G(E,\lambda) = \phi_1(E) + \lambda\phi_2(E) + \lambda^2\phi_3(E) + \ldots \,. \tag{5.25}$$

For the subsequent manipulations, it is convenient to introduce the thermalization operator R_0 defined by (4.57). In terms of this operator, the neutron-balance equation (5.1) takes the form

$$R_0\phi(E) = \lambda\Sigma_a^0(E)\phi(E) - S(E) \,, \tag{5.26}$$

the detailed balance condition reads

$$R_0\phi_M(E) = 0 \,, \tag{5.27}$$

and the neutron-conservation condition can be written as

$$\int_0^{E_m} R_0 f(E)dE = 0 \,, \tag{5.28}$$

where $f(E)$ is an arbitrary function.

Equating terms of order λ^{-1} in Eq. (5.26), we obtain

$$R_0\phi_0(E) = 0 \,,$$

which implies that

$$\phi_0(E) = \Phi\phi_M(E) \,, \tag{5.29}$$

since $\phi_M(E)$ is the unique thermal-equilibrium distribution. The coefficient Φ is given by Eq. (5.21) approximately, and is more precisely obtained by integrating Eq. (5.26) over energy. The left-hand side gives zero, from Eq. (5.28). The right-hand side must therefore be zero to all orders in λ. The term of order unity on the right-hand side vanishes only if

$$\Phi \int_0^{E_m} \Sigma_a^0(E)\phi_M(E)\,dE = \int_0^{E_m} S(E)\,dE = q(E_m) \ . \tag{5.30}$$

The vanishing of the higher-order terms on the right-hand side implies

$$\int_0^{E_m} \Sigma_a^0(E)G(E,\lambda) = 0 \ . \tag{5.31}$$

That Eq. (5.31) must be true is easily seen by substituting Eq. (5.24) into Eq. (5.26), and using Eq. (5.29) to obtain the integral equation

$$R_0 G(E,\lambda) = \lambda\Sigma_a^0(E)G(E,\lambda) + \Phi\Sigma_a^0(E)\phi_M(E) - S(E) \tag{5.32}$$

determining $G(E,\lambda)$. Integrating this equation over energy from zero to E_m implies Eq. (5.31) by virtue of Eqs. (5.28) and (5.30).

We have subtracted a Maxwellian component of the spectrum at the moderator temperature which contains all the absorptions. The remaining part of the spectrum, $G(E,\lambda)$, obeys an integral equation of the same kind as before, but with a source term that removes as many neutrons at thermal energies as it supplies as a slowing-down source. The function $G(E,\lambda)$ is therefore necessarily negative at thermal energies and positive at higher energies, going over into the slowing-down spectrum for energies well above thermal. It contains the "hardening" of the thermal spectrum by removal of slow neutrons, as well as the smooth joining between thermal and epithermal spectra. By subtracting the large model-independent term, the resulting Eq. (5.32) for $G(E,\lambda)$ has the physical advantage that it more directly exhibits the effects of thermalization models on the spectrum and the mathematical advantage that it is not singular in the weak-absorption limit.

We have not yet, however, reduced the complexity of the necessary calculations. The spectrum still depends on the absorber concentration λ, on the energy-dependence of the absorption cross section, and on the temperature and state of chemical binding through the thermalization operator R_0. We can reduce this dependence in the case of weak absorption by expanding $G(E,\lambda)$ in a power series in λ, as given by Eq. (5.25). The leading term is determined by the integral equation

$$\Phi\Sigma_a^0(E)\phi_M(E) - S(E) = R_0\phi_1(E) , \tag{5.33}$$

and the higher-order terms by the sequence of integral equations

$$\Sigma_a^0(E)\phi_n(E) = R_0\phi_{n+1}(E) . \tag{5.34}$$

We note that Eqs. (5.33) and (5.34) do not have unique solutions. To any solution of these equations we can add an arbitrary multiple of $\phi_M(E)$ and still have a solution. This ambiguity does not occur in the original neutron-balance equation. It arises from the power-series expansion (5.25), which changes the role of the absorption term from a homogeneous term to an inhomogeneous source term in the next higher order. This ambiguity is removed by noting that the neutron-conservation condition (5.31) must be satisfied to all orders in λ. Combined with Eq. (5.25), this implies

$$\int_0^{E_m} \Sigma_a^0(E)\phi_n(E)dE = 0 \qquad n \geq 1 . \tag{5.35}$$

This condition must be satisfied to each order if the next higher-order equation in the sequence (5.34) is to have a solution.

As seen from Eqs. (5.33) and (5.34), the leading term $\phi_1(E)$ plays a special role in the expansion (5.25). For sufficiently high energies, the term in $\phi_M(E)$ on the left-hand side of Eq. (5.33) becomes negligibly small, and (5.33) goes over into the slowing-down equation in the absence of absorption. As we have noted in Sec. 5-1, this has as its solution

$$E\phi_1(E) = C = q/\xi_0\Sigma_s .$$

Terms of higher order than $\phi_1(E)$ are not fed by a slowing-down source and therefore decrease more rapidly at high energy. For weak absorption, we can neglect these terms and still have an accurate representation of the spectrum from the thermal through the epithermal energy regions.

The thermal-neutron spectrum for weak absorption has thus been written as a Maxwellian distribution at the moderator temperature, plus a correction term $\phi_1(E)$ that has the correct asymptotic behavior at epithermal energies. By requiring that neutron conservation be satisfied to every order in the absorber concentration, this correction term is uniquely defined. It is normally negative at low energies and positive at higher energies, thus acting to harden the thermal spectrum and introducing a smooth transition between the thermal and epithermal energy spectra. The concept of a "neutron temperature" greater than the moderator temperature does not appear naturally in the theory and is not particularly useful in characterizing the neutron spectrum. The function $\phi_1(E)$, on the other hand, is a soundly based and useful characterization of the neutron spectrum in the same range of weak absorption in which the "neutron temperature" concept can be given a semiquantitative meaning in terms of the peak of a thermal spectrum that is approximately Maxwellian in shape. The convergence properties of the weak-absorption expansion have not been well studied, so that its present utility as a practical tool for reactor-design calculations is limited. We therefore arbitrarily restrict the consideration of $\phi_1(E)$ to the special case of 1/v absorption. In this case, it furnishes an excellent method for studying the effects on the neutron spectrum of the thermalization model. We illustrate this in Secs. 5-4 and 5-5, where $\phi_1(E)$ for the heavy monatomic gas and for monatomic hydrogen gas are compared.

5-3 THE SPECTRUM IN MONATOMIC HYDROGEN GAS

Having discussed some of the general properties of the neutron spectrum in an infinite homogeneous medium, we now turn to the consideration of specific examples. We consider first a moderator consisting of an ideal gas of monatomic hydrogen. The thermalization problem for this case was originally studied by Wigner and Wilkins in 1944.[1] Although monatomic hydrogen gas is not a moderator of practical interest, it does constitute the simplest example that is expected to have the same general mathematical properties as an actual moderator.

The energy-transfer kernel for this case can be obtained from the general result (2.99) by setting A equal to one, to give the considerably simpler formula

$$\Sigma_0(E',E) = (\Sigma_s/E')\,\mathrm{erf}[(\beta E)^{\frac{1}{2}}] \qquad E \leq E' ,$$
$$= (\Sigma_s/E')\,\mathrm{erf}[(\beta E')^{\frac{1}{2}}]\,\exp[\beta(E' - E)] \qquad E > E' , \tag{5.36}$$

where $\beta = 1/k_B T$, and erf(x) is the normalized error function defined in Eq. (2.100a). This kernel obeys the detailed balance condition (5.8) and has a characteristic discontinuity in its derivative when E = E'. Some typical plots of the kernel (5.36) are given in Fig. 5.1 as solid curves and are compared with the corresponding slowing-down kernels for free atoms at rest, which are shown as dashed-line curves.

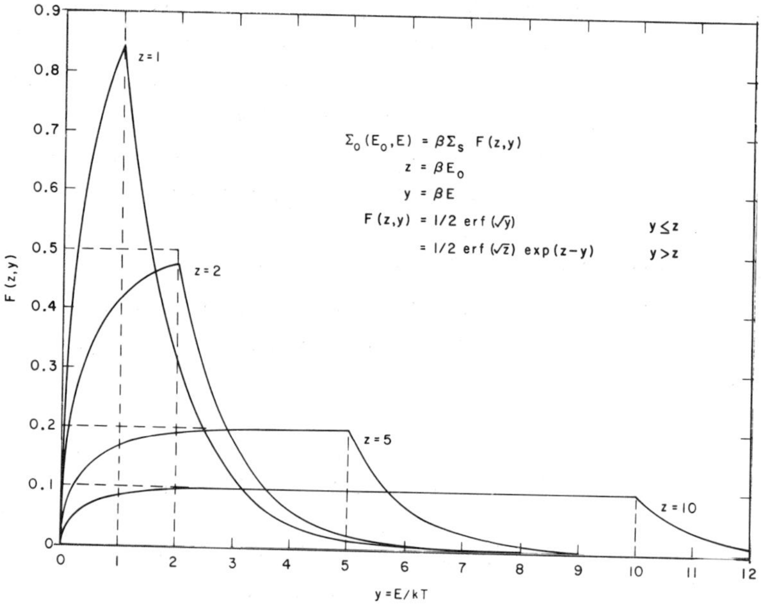

Figure 5.1 The energy-transfer kernels for monatomic hydrogen gas for several incident energies are plotted as the solid lines. The dashed lines are the corresponding slowing-down kernels for free hydrogen atoms at rest.

Even for this simple kernel, analytic solutions for the spectrum have not been found. It is possible, however, in this special case, to transform the neutron-balance equation to a second-order differential equation. This transformation is helpful in obtaining the correction term $\phi_1(E)$ for weak absorption, and is considered in Sec. 5-4. As in the case of slowing down by hydrogen, there is no particular physical simplification associated with the transformation of the neutron-balance equation to a differential equation. We therefore consider first the direct numerical solution of the integral equation (5.1) with the kernel (5.36).

In order to solve the integral equation numerically, we must first calculate the source S(E) resulting from slowing down into the thermal-energy group. Using the definition (5.2) and the kernel (5.36), we obtain

$$S(E) = \Sigma_s[\mathrm{erf}(\beta E)^{\frac{1}{2}}] \int_{E_m}^{\infty} \phi_e(E') \frac{dE'}{E'} \,. \tag{5.37}$$

This function starts from zero at E = 0 and rises smoothly to a plateau value at a few times k_BT, remaining constant for higher energies. The thermal source (5.37) has the interesting property that its dependence on energy is independent of the epithermal flux spectrum $\phi_e(E)$, and thus depends only on the moderator temperature and not on the absorption. We also note that it is the slowing-down source from free hydrogen atoms at rest, except for a depletion at energies of the order of k_BT. This depletion indicates that the neutron is inhibited from losing all of its energy in a collision by the thermal motion of the protons and is entirely kinematic in origin.

In our preceding considerations we have neglected the possibility of neutrons leaving the thermal-energy group by energy-gain scattering. A correction for this effect can be included by subtracting a term

$$\delta S(E) = \phi(E) \int_{E_m}^{\infty} \Sigma_0(E,E')dE' \tag{5.38}$$

from the source term S(E) in the neutron-balance equation. For the kernel (5.36), this correction is given by

$$\delta S(E) = \Sigma_s[\mathrm{erf}(\beta E)^{\frac{1}{2}}](\beta E)^{-1}\phi(E)\exp[\beta(E - E_m)] \ . \qquad (5.39)$$

This correction is important only within an energy interval of the order of k_BT in the neighborhood of E_m. In this neighborhood, the unknown flux spectrum $\phi(E)$ approximately equals the known epithermal flux spectrum $\phi_e(E)$, so that the correction can be made with some confidence. In addition, the magnitude of the correction, even in this energy region, decreases as $1/E_m$ as the cutoff energy E_m increases. If the correction is neglected (as is usual in practice), the only effect will be a small spurious rise in the calculated spectrum in the neighborhood of E_m. The physical reason for this rise is that the scattering cross section, Eq. (5.3), used in the neutron-balance equation is slightly smaller than the true scattering cross section, Eq. (2.100).

Having explicitly defined all the terms in the neutron-balance equation (5.1) in terms of a specific thermalization model, we can now consider the direct numerical solution of the integral equation. The resulting spectrum will depend on the amount of absorber present, on the energy-dependence of the absorption cross section, and on the temperature of the moderator. We consider initially only the dependence of the spectrum on the concentration of an absorber with a 1/v absorption cross section. Absorption resonances and non-1/v absorption are included later, when theory and experiment are compared and when actual reactor spectra are calculated. For the direct numerical solution of the integral equation, the restriction to 1/v absorption is not a simplification, but much of the essential physical content can be illustrated even with this restriction. Even in this initial highly idealized model, the quantitative calculation of the neutron spectrum requires a numerical solution of the neutron-balance equation. This is easily obtained by replacing the integrals by sums, and solving the matrix equation by an iterative procedure. This procedure is described in Sec. 7-1.2. The fact that numerical solutions are necessary even at this early stage helps to explain the orientation of this book towards digital computation rather than analytic techniques. This orientation has the advantage that more realistic models of the thermalization process can be considered in the same framework as the simple models based on an ideal monatomic gaseous moderator. The more realistic models use the same numerical techniques and require only slightly more computation.

Since the source and scattering terms in Eq. (5.1) are proportional to the hydrogen-atom concentration, the spectrum depends only on the absorption per hydrogen atom. We characterize this for 1/v absorption by the hardening parameter

$$\zeta_0 = \Sigma_a(E_0)/\Sigma_s \ , \tag{5.40}$$

where E_0 = 0.0253 eV, and ζ_0 is the ratio of the macroscopic absorption cross section at a neutron speed of 2200 m/sec to the free-atom macroscopic scattering cross section. We often find it convenient to state this parameter in barns of absorption per hydrogen atom. The two definitions are related through

$$\text{barns/H} \quad \text{atom} = 20.36\zeta_0 \ . \tag{5.41}$$

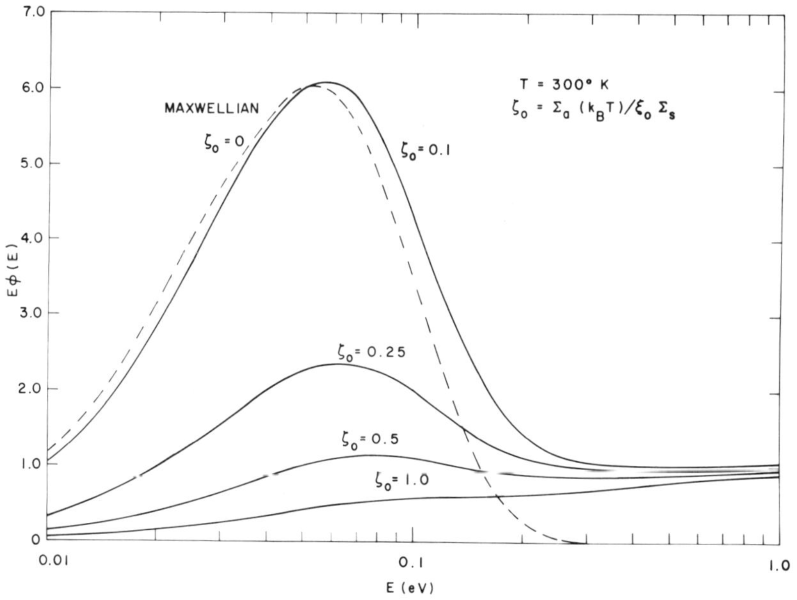

Figure 5.2 Thermal-neutron spectra in monatomic hydrogen as a function of 1/v absorber concentration. A Maxwellian at the moderator temperature is plotted as a dashed line, for comparison.

In Fig. 5.2, we show the calculated flux spectra for several values of ζ_0 at a temperature of 300°K. A Maxwellian flux spectrum is also shown for comparison. Since the total number of neutrons slowing down is proportional to the epithermal value of $E\phi(E)$, we plot the spectra by giving $E\phi(E)$ as a function of E. A logarithmic energy scale is used, since $E\phi(E)$ is the flux per unit logarithmic energy interval. The area under the curves thus represents the total flux below 1 eV. The curves are normalized to the same total number of absorptions at all energies, and thus to the same intensity of fast-neutron source. To obtain this normalization, the extrapolation above 1 eV is made using the analytic solution for free hydrogen atoms at rest, which is given by

$$E\phi(E) = \left[\Sigma_a(E) + \Sigma_s\right]^{-1}\left[\exp - \int_E^\infty \frac{dE'}{E'} \frac{\Sigma_a(E')}{\Sigma_a(E') + \Sigma_s}\right]$$

$$= \Sigma_s\left[1 + \zeta_0 (E_0/E)^{\frac{1}{2}}\right]^{-3} . \qquad (5.42)$$

Note that the Maxwellian plotted in Fig. 5.2 is

$$E\phi_M(E) = (\beta E)^2 \exp(-\beta E) ,$$

which peaks at an energy of $2k_BT$, rather than $\phi_M(E)$, which peaks at an energy of k_BT. The Maxwellian is plotted with arbitrary normalization.

For the weak absorption of $\zeta_0 = 0.1$, the thermal-flux spectrum is nearly Maxwellian, and the maximum flux per unit lethargy is much greater than the epithermal value. As the absorption is increased, the thermal-flux spectrum becomes less Maxwellian in shape and is shifted to higher energies. At the same time, the characteristic peaking in the flux per unit lethargy becomes smaller, until for $\zeta_0 = 1.0$, $E\phi(E)$ is a monotonically increasing function of energy. By plotting $E\phi(E)$ rather than $\phi(E)$, we emphasize the change in character of the spectrum with increasing absorption. The function $\phi(E)$ continues to have a maximum near thermal energy, even for strong absorption, but $E\phi(E)$ has a maximum in the thermal-energy region only when there is a significant probability of reaching an energy of the order of k_BT before absorption.

Finally, we consider the temperature-dependence of the spectrum. If we introduce the new variables

$$z = E/k_BT \quad \text{and} \quad y = E'/k_BT \qquad (5.43)$$

and the parameter

$$\zeta = \Sigma_a(k_B T)/\Sigma_s = \zeta_0(E_0/k_B T)^{\frac{1}{2}} , \tag{5.44}$$

then the spectrum as a function of z depends only on ζ, and not separately on the temperature and on the absorber concentration. This result depends, however, both on the assumption of 1/v absorption and on the restriction to an ideal monatomic gaseous moderator. For this special case, it follows from the dependence of the energy-transfer kernel on energy only through y and z, and the fact that the absorption-to-scattering ratio can be expressed as a scale factor times a function of z. In actual reactor calculations, where effects of chemical binding and absorption resonances are important, this simplification does not occur. Absorption resonances occur at definite energies, not definite values of $E/k_B T$, so that the energy scale is defined without reference to the temperature. Similarly, the chemical binding introduces characteristic energies associated with the natural frequencies of atomic vibration. We will see, in fact, that the spectrum in the presence of binding is harder at low temperatures than that for an ideal gas, but approaches the gas result at high temperatures. The spectrum is thus less dependent on temperature in the presence of chemical-binding effects.

5-4 DIFFERENTIAL EQUATION FOR MONATOMIC HYDROGEN

The original work of Wigner and Wilkins[1] was based on the reduction of the neutron-balance equation (5.18) with the kernel (5.36) to a second-order differential equation. (A corresponding reduction occurs in the calculation of the thermal-diffusion coefficient D_0 in monatomic hydrogen. This latter problem is exactly analogous to the kinetic-theory problem of calculating the coefficient of self-diffusion in a gas of hard spheres. The reduction to a differential equation for this problem was first carried out by Boltzmann. This interesting historical note is discussed further in Chap. 8.) We carry out the derivation here following the discussion by Galanin.[2] It is convenient to introduce the dimensionless variables

$$\left.\begin{aligned} \zeta(z) &= \Sigma_a(k_B Tz)/\Sigma_s , \\ w(z) &= z\Sigma_s(z)/\Sigma_s , \\ \phi(z) &= k_B T\phi(E) , \end{aligned}\right\} \qquad (5.45)$$

with z and y the same as in Eq. (5.43). In terms of these variables the neutron-balance equation for monatomic hydrogen becomes

$$[\zeta(z) + z^{-1}w(z)]\phi(z) = e^{-z}\int_0^z \phi(y)y^{-1}e^y \operatorname{erf}(y^{\frac{1}{2}})dy + $$

$$+ \operatorname{erf}(z^{\frac{1}{2}})\int_z^\infty \phi(y)y^{-1}dy . \qquad (5.46)$$

Following Galanin, we introduce the new dependent variable

$$f(z) = \int_z^\infty \phi(y)y^{-1}dy . \qquad (5.47)$$

We now multiply both sides of Eq. (5.46) by e^z, differentiate with respect to z, and replace $\phi(z)$ by $zf'(z)$ wherever it appears. The prime signifies differentiation with respect to z. This gives a second-order differential equation for f(z) that is

$$[e^z \operatorname{erf}(z^{\frac{1}{2}})]'f(z) + [e^z w(z)f'(z)]' = -[ze^z\zeta(z)f'(z)]' . \qquad (5.48)$$

This equation can be put into a more compact form by using (2.100) for A = 1 to obtain

$$w(z) = (z + \tfrac{1}{2}) \operatorname{erf}(z^{\frac{1}{2}}) + (z/\pi)^{\frac{1}{2}}e^{-z} , \qquad (5.49)$$

and noting that

$$w'(z) = e^{-z}[e^z \operatorname{erf}(z^{\frac{1}{2}})]' . \qquad (5.50)$$

With the use of Eqs. (5.49) and (5.50), Eq. (5.48) reduces to

$$\{w(z)[f(z) + f'(z)]\}' = -e^{-z}[\zeta(z)zf'(z)e^z]' . \qquad (5.51)$$

It is instructive to rewrite Eq. (5.51) in the notation used at the beginning of the chapter to obtain

$$\frac{d}{dE}\left[E\Sigma_s(E)\int_E^\infty \frac{\phi(E')}{E'}\,dE' - k_BT\Sigma_s(E)\phi(E)\right]$$

$$= \left(1 + k_BT\frac{d}{dE}\right)\Sigma_a(E)\phi(E) \ . \qquad (5.52)$$

Since Eq. (5.51) is a second-order differential equation we must specify two boundary conditions to obtain the desired solution. The first condition is Eq. (5.19) (which is already necessary to obtain the desired solution of the original singular integral equation) and corresponds to the presence of a source of neutrons at very high energy. In terms of the variable f(z) defined by Eq. (5.47), this condition becomes

$$\lim_{z\to\infty} zf(z) = C \ . \qquad (5.53)$$

The second boundary condition arises from the differentiation we have performed to obtain Eq. (5.52). We must choose that solution of the differential equation that is also a solution of the integral equation. This requires choosing that solution of the differential equation corresponding to no source or sink of neutrons at E = 0. This solution was obtained originally by Wigner and Wilkins by direct examination of the integral equation (5.46).

We take an alternate point of view and obtain the boundary condition at zero energy by considerations of neutron conservation. We first note that all cross sections go as $E^{-\frac{1}{2}}$ for sufficiently small E, as is explicitly demonstrable for the scattering cross section given by Eq. (2.100). We obtain the condition of over-all neutron conservation by integrating Eq. (5.52) over energy from zero to infinity, and making use of Eq. (5.19). If we require that

$$\lim_{E\to 0} E^{-\frac{1}{2}}\phi(E) = 0 \ , \qquad (5.54)$$

we obtain the desired condition that

$$\lim_{E\to\infty}\left[E\Sigma_s(E)\int_E^\infty \frac{\phi(E')}{E'}\,dE'\right] = C\Sigma_s = \int_0^\infty \Sigma_a(E)\phi(E)dE \ . \qquad (5.55)$$

Equation (5.54) thus serves as the boundary condition at zero energy which chooses that solution of the differential equation (5.52) which is also a solution of the integral equation (5.46).

The differential equation (5.52) also satisfies the other general properties of the neutron-balance equation discussed in Sec. 5-1. The detailed balance condition is satisfied, since the bracketed term on the left-hand side of Eq. (5.51) or Eq. (5.52) vanishes when

$$f(z) = e^{-z} \quad \text{or} \quad \phi(E) = \phi_m(E) \ . \tag{5.56}$$

We note that Eq. (5.51) has a second independent solution for zero absorption for which the bracketed term does not vanish but its derivative does. This solution does not correspond to a solution of the original integral equation and can be excluded.

The differential equation (5.52) reduces to the slowing-down equation for free hydrogen atoms at rest in the limit that E is much greater than k_BT. This can be seen in two ways: First, the differential equation was drived from the neutron-balance equation (5.18) by operating on both sides with $[1 + k_BT(d/dE)]$. In the limit of high energy, the derivative term is negligible, and the first term gives the integral form of the slowing-down equation. Second, the differential form of the slowing-down equation in hydrogen is obtained directly from Eq. (5.52) by setting $\Sigma_s(E)$ equal to its constant epithermal value, Σ_s, neglecting the terms proportional to k_BT, and carrying out the differentiation of the left-hand side.

The differential equation (5.52) is discussed in more detail in the original report by Wigner and Wilkins,[1] and in the 1955 Geneva Conference review paper by E. R. Cohen.[3] It forms the basis for the SOFOCATE code,[4] which has been extensively used to calculate average thermal-neutron cross sections in water-moderated reactors. More accurate calculations of spectra in water-moderated reactors are now available, using more realistic models for the energy-transfer kernel in water. These more realistic models do not allow an exact reduction to a second-order differential equation, so that the integral-equation formulation as discussed in Sec. 5-3 is more appropriate. For occasional calculations to compare with experiment or for design purposes, the integral-equation formulation is much more general, and its solution requires only a moderate amount of computer time. It does require, however, a rather large amount of computer storage for the energy-transfer kernel, and thus it is not very suitable for small computers or for frequent spectrum calculations to be carried out automatically within a large computer program that calculates fuel depletion in a long-burning reactor. It is therefore of interest to consider more general second-order differential equations of the Wigner-Wilkins type which include the effects of chemical binding in an approximate manner. Such differential equations are discussed in Sec. 5-7.

To conclude our discussion of the Wigner-Wilkins differential equation, we will consider its solution for weak absorption. We wish then to calculate the function $\phi_1(E)$ defined in Sec. 5-2. We do this by setting

$$f(z) = f_0(z) = ae^{-z} \tag{5.57}$$

on the right-hand side of Eq. (5.51). Specializing to the case of 1/v absorption so that

$$\zeta(z) = \zeta z^{-\frac{1}{2}} \tag{5.58}$$

and integrating once, we obtain

$$w(z)\frac{d}{dz}[e^z f_1(z)] = \frac{a\zeta\pi^{\frac{1}{2}}}{2}\operatorname{erf}(z^{\frac{1}{2}}) + C_1 . \tag{5.59}$$

The constant of integration C_1 must be set equal to zero in order that Eq. (5.54) be satisfied, i.e., in order that Eq. (5.59) give a solution to the integral equation to this order as well as to the differential equation. Integrating once more, we obtain

$$f_1(z) = \frac{\zeta\pi^{\frac{1}{2}}}{2} ae^{-z}\int_0^z \frac{e^x \operatorname{erf}(x^{\frac{1}{2}})}{w(x)}\,dx + C_2e^{-z} . \tag{5.60}$$

Using the definition (5.47), we obtain finally

$$\phi_1(z) = k_BT\phi_1(E) = zf_1'(z)$$

$$= \frac{a\zeta\pi^{\frac{1}{2}}}{2}\left[ze^{-z}\int_0^z \frac{e^x \operatorname{erf}(x^{\frac{1}{2}})}{w(x)}\,dx - \frac{z\operatorname{erf}(z^{\frac{1}{2}})}{w(z)}\right] + C_2ze^{-z} . \tag{5.61}$$

(In the derivation by Galanin, the factor e^{-z} multiplying C_2 in Eq. (5.60) is incorrectly omitted, so that the term proportional to C_2 in Eq. (5.61) is lost upon differentiation.)

As we saw in Sec. 5-2, the weak-absorption solution $\phi_1(E)$ is undetermined to within an arbitrary multiple of $\phi_M(E)$. To determine the constant C_2 we must apply the condition (5.35) that the total number of absorptions associated with $\phi_1(E)$ is zero. In the special case that we are considering, this condition becomes

$$\int_0^\infty z^{\frac{1}{2}} f_1'(z)\,dz = \int_0^\infty z^{-\frac{1}{2}} f_1(z)\,dz = 0 \ . \tag{5.62}$$

The constant C_2 is therefore given by

$$C_2 = -\tfrac{1}{2} a\zeta \int_0^\infty z^{-\frac{1}{2}} e^{-z} \int_0^z \frac{e^x \operatorname{erf}(x^{\frac{1}{2}})}{w(x)}\,dx\,dz \ . \tag{5.63}$$

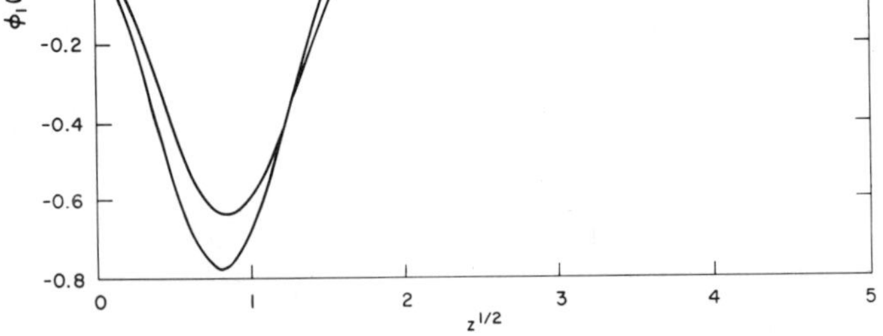

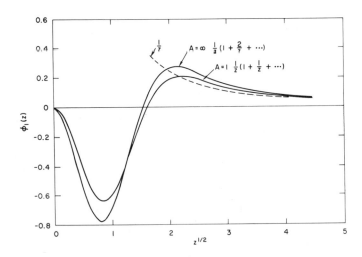

Figure 5.3 The correction functions $\phi_1(z)$ for weak 1/v absorption in monatomic hydrogen and in a heavy gas are compared. (From Cadilhac et al., BNL-719, Ref. 6.)

The calculation of $\phi_1(E)$, as determined by Eq. (5.61) was first carried out in an unpublished memorandum by Davison.[5] The calculation has also been carried out by Cadilhac et al.,[6] and compared with the corresponding result for the heavy-gas model. Their results are shown in Fig. 5.3. We defer further discussion until we give the derivation of the heavy-gas result in the following section.

5-5 THE HEAVY-GAS DIFFERENTIAL EQUATION

In the theory of neutron slowing down, there are two cases where the integral equation expressing neutron balance can be reduced to a first-order differential equation. The first of these is moderation by hydrogen, for which the reduction to a differential equation is a mathematical manipulation with no simple physical interpretation but with the result applicable to an actual moderator. The other case is slowing-down theory which can be considerably simplified is the limit of a very heavy moderator. A first-order differential equation is again obtained, now describing the limit of continuous slowing down in which collisions are very frequent but the energy change per collision is very small. This continuous slowing-down theory is known as the Fermi "age" theory, since in this limit there is a unique slowing-down time associated with each neutron energy (for a given source energy). Unlike the case of monatomic hydrogen, the age-theory differential equation does not apply rigorously to any actual moderator but only in a certain mathematical limit. Age theory must be used with some care in practice, particularly if absorption resonances are important. On the other hand, the mathematical procedures leading to this limiting description have a simple physical interpretation, and their discussion is of considerable help in understanding the slowing-down process.

An analogous situation exists in thermalization theory, at least for the special case of an ideal monatomic gaseous moderator. We have seen that the thermal-neutron spectrum for moderation by monatomic hydrogen gas is given by the solution of a second-order differential equation. This differential equation has no simple physical interpretation, but it is an exact equation for this moderator. We now consider the thermal-neutron spectrum in a monatomic ideal gaseous moderator in the limit that the atomic mass is much greater than the neutron mass. This limit, which was first considered by Wilkins[7] in

1944, also leads to a second-order differential equation for the spectrum. The resulting equation in this case has a simple physical interpretation, but it does not apply exactly to the spectrum in any actual or possible moderator. As in the case of age theory, the equation is derived by noting that the energy-transfer kernel is a highly peaked function and that the flux spectrum varies slowly over the range of the kernel. The integral equation can thus be converted to a differential equation by a Taylor-series expansion of the unknown function in the integral. When terms of first order in the mass ratio are consistently retained, a second-order differential equation results. In the limit of energies large compared with thermal, this second-order equation reduces to the first-order differential equation of age theory.

In carrying out the Taylor-series expansion mentioned above, it is convenient to first transform the neutron-balance equation (5.18) to a form where the coefficients of the expansion have a direct physical interpretation. To do this, we introduce the new dependent variable

$$\Psi(E) = \phi(E)/\phi_M(E) \tag{5.64}$$

and use the detailed balance condition (5.8) to obtain

$$[\Sigma_s(E) + \Sigma_a(E)]\Psi(E) = \int_0^\infty \Sigma_0(E,E')\Psi(E')dE' \tag{5.65}$$

as the integral equation satisfied by $\Psi(E)$. This is a singular integral equation, and we must again specify the behavior at high energies through Eq. (5.19) in order to obtain the solution corresponding to a source of neutrons at $E = \infty$. If $\Psi(E')$ in Eq. (5.65) is slowly varying over the range of the energy-transfer kernel, an approximate differential equation can be obtained by expanding $\Psi(E')$ in a Taylor-series expansion about $E' = E$ and truncating the series after a finite number of terms. The resulting equation is of the form

$$\Sigma_a(E)\Psi(E) = \Sigma_s(E)\left[A_1(E)\,\frac{d\Psi}{dE} + \frac{1}{2!}\,A_2(E)\,\frac{d^2\Psi}{dE^2} + \ldots\right]. \tag{5.66}$$

The coefficients $A_n(E)$ are given by

$$\Sigma_s(E)A_n(E) = \int_0^\infty \Sigma_0(E,E')(E' - E)^n dE' , \tag{5.67}$$

so that $A_n(E)$ is the mean value of the *nth* power of the energy transfer in a collision for which the initial neutron energy is E.

For a general moderator, the energy-transfer moments will all be of the same order, so that no consistent reduction to a differential equation can be made by truncating the expansion in Eq. (5.66). For the special case of a heavy monatomic ideal gas, however, the first two moments are of first order in $\mu = (m/M)$, while the higher moments are all of higher order.

We consider then the limit of a heavy moderator. In order that the rate of approach to equilibrium remain finite and of the same order as the absorption rate, we take the limit in which $\xi_0 = 2\mu$ approaches zero but $\xi_0\Sigma_s$ remains finite. This limit corresponds to a finite rate of energy loss but zero time between collisions, and therefore goes over into Fermi age theory for neutron energies large compared to k_BT.

To calculate the energy-transfer moments $A_n(E)$ for a heavy gaseous moderator, we can proceed in either of two ways. We can first calculate the moments analytically for an ideal gas, using techniques similar to those employed in Appendix I, and then expand in powers of the mass ratio. Analytic expressions for the first few moments have been obtained by Rajagopal.[8] The second approach, which we follow here, is to expand the energy-transfer kernel to first order in μ and then calculate the moments. This approach, which was used by Hurwitz et al.,[9] involves only simple algebraic manipulations but has the disadvantage that it involves the expansion of the nonsingular function (2.99) in terms of highly singular functions.

We start from Eq. (2.97), which gives the function $\chi(\kappa,t)$ for an ideal monatomic gas. Expanding to first order in μ and carrying out the integration over scattering angles, we obtain

$$\Sigma_0(E,E') = \Sigma_s(1 + 2\mu)\delta(E' - E) + (\Sigma_s\mu/2\pi)(E' + E)(E'/E)^{\frac{1}{2}} \times$$

$$\times \int_{-\infty}^{\infty} (it - k_BTt^2) \exp[i(E' - E)t/\ ]dt . \tag{5.68}$$

The integrals in Eq. (5.68) can be defined in terms of derivatives of the δ-function given in Chap. 2. For easy reference, we write here

$$\delta^{(n)}(x) = (2\pi)^{-1}\int_{-\infty}^{\infty} (it)^{n}\exp(itx)dt\ , \tag{5.69}$$

where these function are defined in terms of integration by

$$\int_{-\ _1}^{\ _2} dx f(x)\delta^{(n)}(x) = (-1)^{n} f^{(n)}(0)\ , \tag{5.70}$$

and $_1$ and $_2$ are any two positive numbers. To calculate the energy-transfer moments, we introduce the notation

$$\left.\begin{aligned} &\Delta = E' - E \\ \text{and}\quad &f(E,\Delta) = (E' + E)(E'/E)^{\frac{1}{2}} = 2E(1 + \Delta/E + \Delta^{2}/8E^{2} + \ldots)\ , \end{aligned}\right\} \tag{5.71}$$

in terms of which the kernel (5.68) becomes

$$\begin{aligned} \Sigma_0(E,E') &= \Sigma_s(1 + \xi_0)\delta(\Delta) + \\ &\quad + \tfrac{1}{2}\xi_0\Sigma_s f(E,\Delta)[\delta'(\Delta) + k_B T\delta''(\Delta)]\ . \end{aligned} \tag{5.72}$$

The energy-transfer moments to first order in μ are given by

$$\left.\begin{aligned} A_1(E) &= \tfrac{1}{2}\xi_0\Sigma_s[-f(E,0) + 2k_B Tf'(E,0)] = \xi_0\Sigma_s(2k_B T - E) \\ A_2(E) &= \xi_0\Sigma_s k_B Tf(E,0) = 2\xi_0\Sigma_s Ek_B T \\ A_n(E) &= 0 \qquad n \geq 3\ . \end{aligned}\right\} \tag{5.73}$$

The third and higher moments have nonvanishing contributions of higher order in μ for an ideal monatomic gas and have contributions of first order in μ for a chemically bound system. In the presence of chemical binding, the expansion of the kernel in powers of μ is not in terms of singular functions and does not convert the neutron-balance equation into a differential equation. The truncation of the expansion (5.66) after the second term is therefore only valid in the limit of a heavy moderator and in the limit of high temperature where chemical binding effects can be neglected.

Using the moments from Eqs. (5.73), we obtain a quite simple second-order differential equation

$$\Sigma_a(E)\Psi(E) = \xi_0\Sigma_s\left[(2k_BT - E)\frac{d\Psi}{dE} + Ek_BT\frac{d^2\Psi}{dE^2}\right] \tag{5.74}$$

for $\Psi(E)$, and a corresponding second-order differential equation for $\phi(E)$

$$\Sigma_a(E)\phi(E) = \xi_0\Sigma_s\left\{\frac{d}{dE}[E\phi(E)] + Ek_BT\frac{d^2\phi}{dE^2}\right\}. \tag{5.75}$$

In considering the limits of validity of these equations, we must take into account the infinite range of the energy variable and consider the dimensionless parameter E/k_BT, as well as the parameter μ. The expansion (5.68), in fact, assumes that the neutron velocity is much greater than the atomic velocities and requires not only that μ be very small but that $E/\mu k_BT$ also be very large. The differential equation is therefore not applicable in the energy region below μk_BT, but this is of no practical consequence in thermalization problems. We can use the differential equation down to zero energy so long as we choose a boundary condition that gives a physically sensible behavior of the neutron spectrum. At high energies, we note that the contribution of order μ^2E^2 to the second moment dominates the term of order μEk_BT when E is much greater than k_BT/μ. We cannot take the point of view that only energies less than k_BT/μ are of interest, inasmuch as we started with an infinite energy interval before making the expansion in powers of μ. The neglect of the terms of order μ^2 at high energies, however, is not an approximation to the thermalization problem per se, but is the approximation to the slowing-down problem which leads to Fermi age theory.

One final point concerning the derivation should be noted. Since the kernel is expanded consistently to first order in μ, the same differential equation for $\phi(E)$ is obtained whether $\phi(E)$ is expanded in the integral in Eq. (5.18) or $\Psi(E)$ is expanded in Eq. (5.65). It is not obvious physically that these expansions should be accurate under the same conditions, but, for the mathematical limiting case considered here, they give the same results. The properties of the differential equation (5.75) have been considered in more detail by E. R. Cohen,[10] and its solutions have been tabulated for a mixture of 1/v absorption and a constant absorption simulating leakage by de Sobrino and Clark.[11] In the present chapter, we consider briefly the general properties of the equation and calculate the spectrum for weak 1/v absorption.

The differential equations (5.74) and (5.75) satisfy the general properties of the neutron-balance equation discussed in Sec. 5-1. Detailed balance is satisfied, since the right-hand side of Eq. (5.74) vanishes when $\Psi(E)$ is a constant, as does the right-hand side of Eq. (5.75) when $\phi(E) = \phi_M(E)$. The approach to Fermi age theory at high energy is easily seen from Eq. (5.75) by setting k_BT equal to zero. The condition of over-all neutron conservation is obtained by integrating Eq. (5.75) over energy from zero to E_0 to give

$$\int_0^{E_0} \Sigma_a(E)\phi(E)dE = q(E_0) , \tag{5.76}$$

where

$$\begin{aligned} q(E) &= \xi_0\Sigma_s\left[(E - k_BT)\phi(E) + Ek_BT\frac{d\phi}{dE}\right] \\ &= \xi_0\Sigma_s\, Ek_BT\phi_M(E)\,\frac{d\Psi(E)}{dE} . \end{aligned} \tag{5.77}$$

To obtain Eq. (5.77), we have required $q(0) = 0$. As in the case of hydrogen moderation, this corresponds to choosing the regular solution of the differential equation in the neighborhood of $E = 0$. Unlike the case of hydrogen moderation, however, neither of the solutions of Eq. (5.75) are exact solutions of the integral equation for neutron balance, so that the criterion for choosing the appropriate solution is not as direct as in the previous section. Also unlike the case of monatomic hydrogen, the heavy monatomic gas leads to the particularly simple equation (5.77) relating the slowing-down density and the flux spectrum. In this form, the slowing-down density can be seen by inspection to vanish identically for a Maxwellian flux spectrum and to approach the Fermi age theory value in the limit of high energy. It is in this sense that we find a significant physical simplification in the heavy-gas model which is not present in other models of the thermalization process.

Finally, we calculate the neutron spectrum in the limit of weak

1/v absorption in order to compare with the results previously obtained for monatomic hydrogen. When the new variables

$$\left.\begin{aligned} z &= E/k_BT \ , \\ \phi(z) &= k_BT\phi(E) \ , \\ \text{and} \quad \zeta &= \Sigma_a(k_BT)/\xi_0\Sigma_s \end{aligned}\right\} \tag{5.78}$$

are introduced, Eq. (5.76) becomes

$$\zeta\int_0^z y^{-\frac{1}{2}}\phi(y)dy = (z-1)\phi(z) + z\frac{d\phi}{dz} \ . \tag{5.79}$$

The solution to first order in ζ is given by

$$\phi(z) = bze^{-z} + \phi_1(z) \ , \tag{5.80}$$

where b is a constant and $\phi_1(z)$ satisfies

$$\begin{aligned} b\zeta\int_0^z y^{\frac{1}{2}}e^{-y}dy &= (z-1)\phi_1(z) + z\frac{d\phi_1}{dz} \\ &= z^2e^{-z}\frac{d}{dz}\left[z^{-1}e^z\phi_1(z)\right] \ . \end{aligned} \tag{5.81}$$

Integrating Eq. (5.81), we obtain

$$\phi_1(z) = ze^{-z}\left[\Gamma + b\zeta\int_0^z y^{-2}e^y\int_0^y x^{\frac{1}{2}}e^{-x}dxdy\right] \ , \tag{5.82}$$

where the constant Γ remains to be determined. The asymptotic behavior of $\phi_1(z)$ for large z is given by

$$\lim_{z\to\infty} z\phi_1(z) = \frac{\pi^{\frac{1}{2}}}{2}b\zeta \ . \tag{5.83}$$

The constant Γ is determined by Eq. (5.35), which states that the total number of absorptions in $\phi_1(z)$ must be zero. This implies that

$$\frac{\pi^{\frac{1}{2}}}{2}\,\Gamma = -b\zeta \int_0^\infty z^{\frac{1}{2}}e^{-z} \int_0^z y^{-2}e^{y} \int_0^y x^{\frac{1}{2}}e^{-x}\,dx\,dy\,dz \;. \tag{5.84}$$

The integral in Eq. (5.84) has been evaluated by Hurwitz _et al._,[9] and is given by

$$\Gamma = -\frac{2}{\pi^{\frac{1}{2}}}\,b\zeta W \;, \tag{5.85a}$$

where

$$W = 1 + 2\sum_{n=0}^{\infty} (-1)^n (2n + 1)^{-2} = 2.831931\ldots \;. \tag{5.85b}$$

The solution for $\phi(z)$ derived here differs from that given by Hurwitz _et al._, by the constant multiplicative factor $b(1 - 2\pi^{-\frac{1}{2}}\zeta W)$. The two solutions therefore correspond to the same flux spectrum. If we wish, however, to compare various thermalization models through the function $\phi_1(z)$, it is necessary to include the subsidiary condition (5.35) so that a uniquely defined function is available for the comparison.

In Fig. 5.3, the function $\phi_1(z)$ determined by Eqs. (5.82) and (5.85) is compared with the corresponding function for monatomic hydrogen gas as determined by Eqs. (5.61) and (5.63). The figure is taken from the paper of Cahilhac _et al._,[6] and is plotted by giving $\phi_1(z)$ versus $z^{\frac{1}{2}}$. The two curves are normalized to the same asymptotic behavior, and the area under each curve is zero since

$$\int_0^\infty \phi_1(z)\,d(z^{\frac{1}{2}}) = \int_0^\infty z^{-\frac{1}{2}}\phi_1(z)\,dz = 0 \;. \tag{5.86}$$

From Fig. 5.3 we see that the thermal-neutron spectrum for weak 1/v absorption is insensitive to the mass of the moderator so long as the moderating ratio $\Sigma_a/\xi_0\Sigma_s$ is the same. Since the oscillations in the heavy-mass curve are about 20% greater than in the A = 1 curve, the hardening of the thermal spectrum is slightly greater for the heavy moderator.

This insensitivity to moderator mass has been demonstrated only in the absence of chemical binding effects, and only for weak 1/v absorption. In this limit, the epithermal spectra for these two cases are identical, so that the sensitivity of the thermal spectrum to the particular moderator is somewhat greater than in the epithermal case. We see later that in particular cases of practical importance the sensitivity of the spectrum to the model used is in fact considerable.

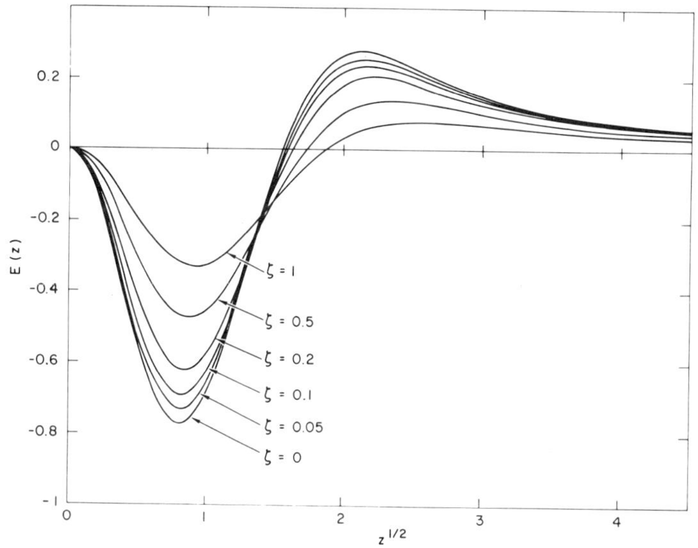

Figure 5.4 The correction functions E(z) for 1/v absorption in a heavy gas. (From Horowitz and Tretiankoff, EANDC(E)-14, Ref. 12.)

Finally, we consider the convergence of the weak-absorption expansion of Sec. 5-2, as illustrated by the heavy-gas differential equation (5.75). Figure 5.4 shows the spectrum as a function of 1/v absorption as calculated by Horowitz and Tretiakoff.[12] The spectra are plotted in the same manner as in Fig. 5.3. The ordinate is the flux spectrum $\phi(z)$ with a Maxwellian containing all the absorptions subtracted, and is given by

$$E(z) = C^{-1}\phi(z) - (2/\zeta\pi^{\frac{1}{2}})ze^{-z} .$$

The factor C^{-1} indicates a normalization to an asymptotic value of unity for $z\phi(z)$. This is the same normalization that is used in Fig. 5.3. The abscissa is $z^{\frac{1}{2}}$, so that all the curves in Fig. 5.4 satisfy

$$\int_0^\infty E(z)d(z^{1/2}) = 0 \ .$$

The curve for $\zeta = 0$ is identical to $\phi_1(z)$ for $A = \infty$ in Fig. 5.3.

From Fig. 5.4 we see that the convergence of the weak-absorption expansion is slow. For well-thermalized systems with $\zeta \lesssim 0.2$, $\phi_1(E)$ does, however, give a reasonably accurate description of the flux spectrum. It forms the basis of a prescription for effective cross sections which has a sound theoretical foundation.[12] Even in the well-thermalized case, one must use a temperature-dependent $\phi_1(E)$ that includes binding effects rather than the temperature-independent $\phi_1(z)$ of the heavy-gas model. As we see in Chap. 7, the range of spectra for applications of interest is so great that the utility of any simple prescription for effective cross sections is very limited. We therefore restrict the use of such cross sections to illustrative examples and take a straightforward numerical approach to actual reactor applications.

5-6 ASYMPTOTIC NEUTRON SPECTRA

For neutron energies greater than 0.5 to 1.0 eV the main effect of chemical binding and atomic motion is to reduce the slowing-down power of the moderator. It is a good approximation to neglect the small number of neutrons that have slowed down past this energy and have subsequently gained sufficient energy to again be above 0.5 eV. In this asymptotic energy region, one can develop useful approximate treatments that are considerably simpler than the general treatment required at lower energies. This simplification is reflected in the availability of many analytic results; some of these are given in the following discussion. The somewhat cumbersome display of these results masks the considerable simplification that has been achieved.

Two methods[13,14] exist for calculating neutron spectra in the energy region above 1 eV. The more systematic of these is due to Corngold.[13] Corngold developed an asymptotic solution of the integral equation (5.18) by applying the technique of Mellin transforms after first using the Wick short-collision-time method to expand the scattering terms. The only important physical assumption is that the inelastic scattering is accurately given by the incoherent approximation. This is true for hydrogen, and it is a very good approximation for the other moderating materials for energies where the asymptotic solution applies.

Corngold gives an explicit expansion of $\phi(E)$ in powers of $E^{-\frac{1}{2}}$ for the case of 1/v absorption and a moderator with arbitrary mass and binding characteristics. Here we present only the main result of his work. For a moderator consisting of a single atomic species,

$$E\phi(E) = \frac{1}{\xi_0 \Sigma_s E} \sum_{n=0,\frac{1}{2},1,\ldots}^{\infty} \left(\frac{k_B T}{E}\right)^n \sum_{p=0,1,\ldots}^{[n]} \tau_n^p \, -\left(\frac{\Delta}{2}\right)^{2(n-p)} , \quad (5.87)$$

where

$$\Delta = \frac{2\Sigma_a (k_B T)}{\mu \Sigma_s} \quad (5.88)$$

and $\mu = A^{-1}$. Here, [n] is the integral part of n. The coefficients τ_n^p, which characterize the moderator, are given recursively by

$$b_0(n+1)\tau_n^p + \mu b_1(N+1)\tau_{n-1}^{p-1} + \mu^2 b_2(n+1)\tau_{n-2}^{p-2} + \ldots +$$

$$+ \mu^p b_p(N+1)\tau_{n-p}^0 + \mu\tau_{n-\frac{1}{2}}^p = 0 . \quad (5.89)$$

They are defined only for $n \geq p \geq 0$. Those τ_n^p that may formally appear in Eq. (5.89) but violate this condition are taken to be zero. The construction of the τ_n^p begins with $\tau_0^0 = 1$. The $b_p(s)$ are defined by

$$\mu^p b_p(s) = \sum_{\ell=0}^{p} \frac{\mu^\ell}{(p+\ell)!} \gamma_\ell^{p+\ell} \times$$

$$\times \left[E^{p+\ell}(s) - (-1)^{p+\Omega}(1+\mu)^{p-\ell}\beta_\ell^{p+\ell} \right] , \quad (5.90)$$

where

$$E_{\ell}^{p+\ell}(s) = \frac{(1+\mu)^2}{2}(1-\mu)^{2s-p-\ell}\left\{\frac{\partial^{p+\ell}}{\partial y^{p+\ell}} \times \right.$$
$$\left. \times \int_{-1}^{1} \frac{[(y^{\frac{1}{2}}+x)^2+(1-\mu)(1-x^2)]^{\ell}}{(y^{\frac{1}{2}}+\mu x)^{2(s-p)+1}}\,dx\right\}_{y=1} . \qquad (5.91)$$

and

$$\beta_{\ell}^{p+\ell} = \frac{1}{4}\left\{\frac{\partial^{p+\ell}}{\partial y^{p+\ell}} \int_{(y^{\frac{1}{2}}-1)^2}^{(y^{\frac{1}{2}}+1)^2} x^{\ell}\,dx\right\}_{y=1} . \qquad (5.92)$$

The $\beta_{\ell}^{p+\ell}$ are universal constants independent of the atomic motions; $E^{p+\ell}$ depends on the mass ratio μ but not on other properties of the moderator. The dependence of the b_p on chemical binding is contained in the coefficients $\gamma_{\ell}^{p+\ell}$. These are defined in terms of the central moments of the scattering function given by Eq. (2.244):

$$s_n(\kappa) = \sum_{\ell=0}^{[n/2]} \gamma_{\ell}^{n}(k_B T)^{n-\ell}\left(\frac{\hbar^2\kappa^2}{2M}\right)^{\ell} . \qquad (5.93)$$

Comparing this expression with Eq. (2.249), we find for the first few γ_{ℓ}^{n},

$$\gamma_0^n = \delta_{n0} ,$$

$$\gamma_1^2 = 2\,\frac{\bar{T}}{T} , \quad \gamma_1^3 = \frac{B_{Av}}{(k_B T)^2} , \quad \gamma_1^4 = \frac{2C_{Av}}{(k_B T)^3} , \qquad (5.94)$$

$$\gamma_2^4 = \frac{16}{5}\,\frac{(K^2)_{Av}}{(k_B T)^2} .$$

Expressions for all the γ_{ℓ}^{n} have been given for a cubic crystal with one atom per unit cell. They are related to the density of modes $\rho(\omega)$ of the lattice vibration spectrum by[15]

$$
\begin{aligned}
(k_BT)^n\gamma_0^n &= {}^{n}\delta_{n0} \\
(k_BT)^{n-1}\gamma_1^n &= {}^{n-1}\left\langle \omega^{n-1} \right\rangle & n \geq 2 \\
(k_BT)^{n-2}\gamma_2^n &= \frac{1}{2!}\sum_{m=2}^{n-2} \frac{n!}{(n - m)!m!}\hbar^{n-2}\left\langle \omega^{n-m-1} \right\rangle \left\langle \omega^{m-1} \right\rangle & n \geq 4 \\
(k_BT)^{n-3}\gamma_3^n &= \frac{1}{3!} \sum_{\substack{n_1,n_2,n_3=2 \\ (n_1+n_2+n_3=n)}}^{n-4} \frac{n!}{n_1!n_2!n_3!} \, n^{-3} \times \\
&\quad \times \left\langle \omega^{n_1-1} \right\rangle \left\langle \omega^{n_2-1} \right\rangle \left\langle \omega^{n_3-1} \right\rangle & n \geq 6
\end{aligned}
\tag{5.95}
$$

and so forth, where

$$
\left.
\begin{aligned}
\omega^n &= \int_0^\infty \rho(\omega)\omega^n d\omega & n \text{ even} \\
&= \int_0^\infty \rho(\omega)\omega^n \cot h \frac{\omega}{2k_BT} d\omega \, . & n \text{ odd}
\end{aligned}
\right\}
\tag{5.96}
$$

Corngold[16] has applied Eq. (5.87) to the analysis of a neutron spectrum measured in a homogeneous water assembly.[17] The comparison of theoretical with experimental results is shown in Fig. 5.5. The bound-proton model used in the theoretical calculation is the one described in Sec. 3-4.1. To illustrate the magnitude of the chemical binding effects, the figure also shows the theoretical result for thermalization by a free-proton gas at the temperature of the moderator. Clearly, the binding effects are significant even for an energy as large as 0.5 eV.

Equation (5.87) applies only for the case of 1/v absorption. Corngold has remarked, however, that the formalism can be applied to the case where the absorption cross section can be represented as a sum of terms of the form $av^{-n} + bv^m$, where m and n are integers. In fact, Corngold and Zamick[18] have considered the case (where $a = \Sigma_a(v_0)v_0$, $n = 1$, $b = DB^2$, and $m = 0$) corresponding to a fundamental-mode spectrum in a

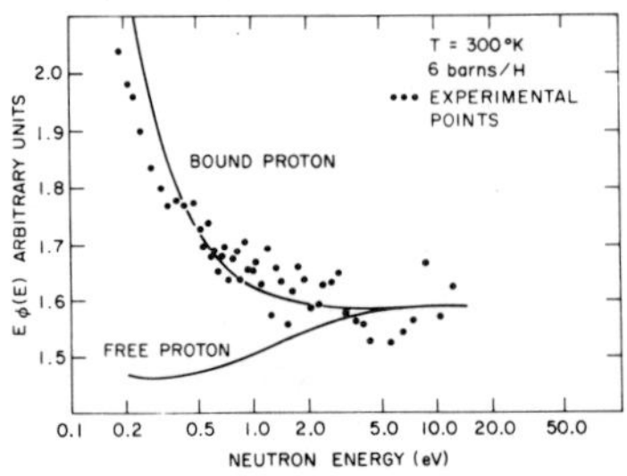

Figure 5.5 Thermalization in H_2O. Corngold's expansion is used to calculate $E\phi(E)$ for a gas of protons and for a system in which the protons are chemically bound. The model for binding takes into account the vibrational, hindered rotational, and translational modes of motion in the manner described in Chap. 3. The experimental points are due to Beyster et al.[17] (From Corngold, Ann. Phys. N.Y., Ref. 16.)

finite medium with buckling B^2 and poisoned with a 1/v absorber. In the energy region where the asymptotic series applies, they conclude that diffusive effects are, in general, as significant for the spectrum as are chemical binding effects. (For the case corresponding to Fig. 5.5, however, the diffusive effects are negligible.)

Let us now consider the second method for calculating the effects of binding on spectra. The approach is to develop an approximate first-order differential equation for the spectrum for an arbitrary dependence of the cross section on energy.[14] The result is a modified Greuling-Goertzel[19] approximation for the flux spectrum. The modification consists in including slowing-down parameters whose dependence on the state of chemical binding is expressed by means of an application of Wick's short-collision-time approximation.

For the considerations that follow, it is convenient to introduce the variable $u = \ln(k_BT/E)$ and to write the balance equation for the flux per unit lethargy $\Phi(u) = E\phi(E)$. Then, instead of Eq. (5.18), we have

$$[\Sigma_a(u) + \Sigma_s(u)]\Phi(u) = \int_{-\infty}^{\infty} \Sigma_0(u',u)\Phi(u')du' , \qquad (5.97)$$

where $\Sigma_0(u',u) = \Sigma_0(E',E)E$. At high energies $\Phi(u)$ is approximately constant, and it is thus profitable to expand $\Phi(u')$ in a Taylor series about $u' = u$. Performing the expansion, we obtain

$$[\Sigma_a(u) - \delta\Sigma(u)]\Phi(u) = -\left\langle \xi\Sigma \right\rangle \frac{d\Phi}{du} + \frac{1}{2!}\left\langle \xi^2\Sigma \right\rangle \frac{d^2\Phi}{du^2} + \ldots + \frac{(-1)^n}{n!}\left\langle \xi^n\Sigma \right\rangle \frac{d^n\Phi}{du^n} + \ldots , \tag{5.98}$$

where $\delta\Sigma(u) = \left\langle \xi^0\Sigma \right\rangle - \Sigma_s(u)$ and the moments $\left\langle \xi^n\Sigma \right\rangle$ of the distribution function $\Sigma_0(u',u)$ are defined by

$$\left\langle \xi^n\Sigma \right\rangle = (-1)^n \int_{-\infty}^{\infty} \Sigma_0(u',u)(u' - u)^n du' . \tag{5.99}$$

The expansion developed here is not in general a consistent expansion in terms of some parameter, e.g., the mass ratio μ. In order that the successive terms in the expansion given here become progressively smaller, we require both a large moderator atomic mass and a high neutron energy. It is also essential that the macroscopic absorption cross section, and hence $\Phi(u)$, are slowly varying functions of energy. This requirement rules out the possibility of directly applying the results derived below to a system containing significant amounts of material that have large resonances in the energy region $0.5 \text{ eV} \lesssim E \lesssim 2$ eV.

To calculate the moments $\left\langle \xi^n\Sigma \right\rangle$, we work with the energy variable, in terms of which

$$\left\langle \xi^n\Sigma \right\rangle = \int_0^{\infty} \Sigma_0(E_0,E) f_n\left(\frac{E_0}{E}\right) dE_0 , \tag{5.100}$$

where $f_n(x) = (1/x)(\ell n x)^n$. These moments can be expressed as series in powers of E^{-1}. This is accomplished by means of the short-collision-time approximation discussed in Sec. 2-8. The effects of thermal motion and chemical binding of the moderator atoms are contained in the coefficients of the expansion. For an isotropic scattering system consisting of a single atomic species, the scattering kernel in the incoherent approximation is

$$\Sigma_0(E_0,E) = \frac{\Sigma_s(1+\mu^2)}{4\pi\hbar}\,\frac{k}{k_0}\int\left\langle S_{inc}(\boldsymbol{\kappa},\omega)\right\rangle_R d\Omega$$

$$= \frac{\Sigma_s(1+\mu^2)}{4\pi\hbar}\,\frac{k}{k_0}\int d\Omega\,\frac{1}{2\pi}\int_{-\infty}^{\infty}\times$$

$$\times\exp\left[-i\left(\omega-\frac{\hbar\kappa^2}{2M}\right)t\right]\left\langle g(\boldsymbol{\kappa},t)\right\rangle_{T,R}dt\ ,$$

where $\omega = (E_0 - E)/$ and $g(\boldsymbol{\kappa},t)$ is given by Eq. (2.241). We now make the short-time expansion, Eq. (2.242), for $g(\boldsymbol{\kappa},t)$ and use Eq. (2.244) to obtain

$$\Sigma_0(E_0,E) = \Sigma_s\,\frac{(1+\mu^2)}{4\pi}\,\frac{k}{k_0}\int d\Omega\sum_n\frac{s_n(\kappa)}{n!}\,\delta^{(n)}\left(E-E_0+\frac{\hbar^2\kappa^2}{2M}\right). \tag{5.101}$$

Using this result, we may express Eq. (5.100) in the form

$$\left\langle\xi^n\Sigma\right\rangle = \frac{\Sigma_s(1+\mu^2)}{4\pi}\int_0^{\infty}dE_0 f_n\left(\frac{E_0}{E}\right)\left(\frac{E}{E_0}\right)^{\frac{1}{2}}\int d\Omega\ \times$$

$$\times\sum_{m=0}^{\infty}\frac{s_m(\kappa)}{m!}\,\delta^{(m)}\left(E-E_0+\frac{\hbar^2\kappa^2}{2M}\right).$$

Introducing a new variable of integration

$$x = E - E_0 + \frac{\hbar^2\kappa^2}{2M} = E - E_0 + \mu[E_0 + E - 2(EE_0)^{\frac{1}{2}}\cos\theta]$$

in place of $\cos\theta$, we find

$$\left\langle\xi^n\Sigma\right\rangle = \frac{\Sigma_s(1+\mu^2)}{4\mu}\int_0^{\infty}\frac{dE_0}{E_0}\,f_n\left(\frac{E_0}{E}\right)\int_{x_-(E,E_0)}^{x_+(E,E_0)}dx\ \times$$

$$\times\sum_{m=0}^{\infty}\frac{s_m\left\{\left[\frac{2M}{h^2}(x-E+E_0)\right]^{\frac{1}{2}}\right\}}{m!}\,\delta^{(m)}(x)\ , \tag{5.102}$$

where

$$x_{\pm}(E,E_0) = E - E_0 + \mu\left(E_0^{\frac{1}{2}} \pm E^{\frac{1}{2}}\right)^2 .$$

To evaluate the integral we first interchange the order of integration in Eq. (5.102). The region of integration in the x,E_0 plane is contained between the $x_{\pm}(E_0,E)$ curves shown in Fig. 5.6. For hydrogen ($\mu = 1$), the curve $x_+(E_0,E)$ does not cross the E_0-axis. We may treat this singular case separately or by taking the limit of Eq. (5.102) as $\mu \to 1$ after the integrations have been performed. Here, we choose to do the latter.

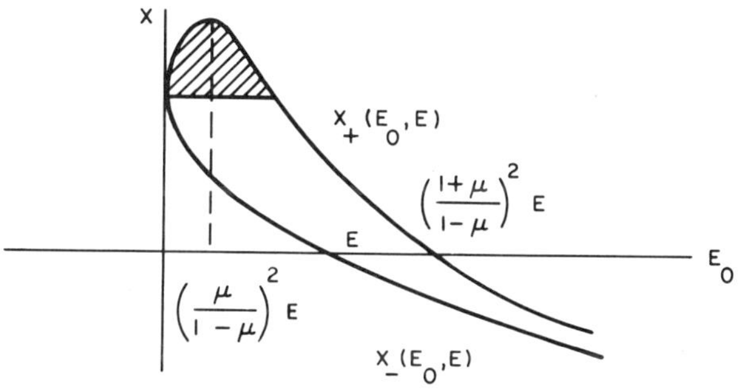

Figure 5.6

Because of the appearance of $\delta^{(m)}(x)$ in Eq. (5.102), the shaded area of Fig. 5.6 does not contribute to the integral. We may, therefore, write

$$\left\langle \xi^n \Sigma \right\rangle = \frac{\Sigma_s(1+\mu^2)}{4\mu} \sum_{m=0}^{\infty} \frac{1}{m!} \frac{\partial^m}{\partial x^m} \int_{a_-}^{a_+} \frac{dE_0}{E_0} f_n\left(\frac{E_0}{E}\right) \times$$

$$\times\, s_m \left\{ \left[\frac{(2M)}{2} (x - E + E_0) \right]^{\frac{1}{2}} \right\} \Bigg|_{x=0} ,$$

where

$$a_{\pm} = \left\{ \frac{\pm\mu + [1 - (1-\mu)(x/E)]^{\frac{1}{2}}}{1-\mu} \right\}^2 E .$$

In a slightly more concise notation this is

$$\left\langle \xi^n \Sigma \right\rangle = \frac{\Sigma_s (1 + \mu^2)}{4\mu} \sum_{m=0}^{\infty} \frac{(-1)^m (1 - \mu)^m}{m!} E^{-m} \frac{\partial^m}{\partial y^m} \times$$

$$\times \int_{b_-}^{b_+} \frac{dx}{x} f_n(x) s_m(Z) \Bigg|_{y=1} \tag{5.103}$$

where

$$Z^2 = \frac{2ME}{\hbar^2} \left(x - \frac{y - \mu}{1 - \mu} \right)$$

and

$$b_{\pm} = \left(\frac{y^{\frac{1}{2}} \pm \mu}{1 - \mu} \right)^2 .$$

Equation (5.103) is the general result from which the moments may be computed. In obtaining $\delta\Sigma$ we require the scattering cross section $\sigma(E_0)$, given by Eq. (2.256), in addition to $\left\langle \xi^0 \Sigma \right\rangle$.

The first term in Eq. (5.103) for n = 1 is $\xi_0 \Sigma_s$, where ξ_0 is given by Eq. (5.13). In general, the energy-independent term in the expansion of the n*th* moment is equal to the free-atom-at-rest value of the average of the n*th* moment of logarithmic energy loss. Except in the free-atom-limit, the quantities $\left\langle \xi^n \Sigma \right\rangle$ bear no relation to the familiar expression

$$\int_0^{\infty} \Sigma_0(E, E_0) \left(\ell n \frac{E_0}{E} \right)^n dE_0$$

for the average value of the n*th* lethargy-change moment.

The results are shown below for $\delta\Sigma$ and $\left\langle \xi^n \Sigma \right\rangle$ for arbitrary values of μ. The expression for $\delta\Sigma$ is given to terms of order E^{-2}. The moments $\left\langle \xi^n \Sigma \right\rangle$ for $n \geq 1$ are given to order E^{-1}. We have

$$\frac{\delta\Sigma}{\Sigma_s} = (1 - \alpha) \Bigg[\left(1 - \frac{\mu}{2} \right) \frac{k_B \bar{T}}{E} - \frac{1}{24} \frac{\mu^3 + 5\mu^2 - 5\mu + 15}{1 + \mu} \frac{B_{Av}}{E^2} +$$

$$+ \frac{(1 - \alpha)}{60} (\mu^3 + 6\mu^2 - 15\mu + 40) \frac{(K^2)_{Av}}{E^2} \Bigg] + 0 \left(\frac{\mu}{E^3} \right) , \tag{5.104}$$

$$\frac{\langle\xi\Sigma\rangle}{\Sigma_s} = 1 + \frac{\alpha}{1-\alpha}\,\ln\alpha + \left[\frac{1}{1-\alpha}\;\alpha(1-\alpha) - \alpha\ln\alpha + 2\alpha^{\frac{3}{2}}\ln\alpha + \right.$$

$$\left. + \alpha(1-\alpha)\left(2 + \frac{\mu}{2}\right)\ln\alpha\right]\frac{k_B\bar{T}}{E} + O\left(\frac{\mu}{E^3}\right) , \qquad (5.105)$$

and, generally,

$$(-1)^{n+1}\frac{\langle\xi^n\Sigma\rangle}{\Sigma_s} = \frac{\alpha}{1-\alpha}\left[(\ln\alpha)^n - n(\ln\alpha)^{n-1} + \right.$$

$$\left. + n(n-1)(\ln\alpha)^{n-2} + \ldots + (-1)^n n!\right] +$$

$$+ (-1)^{n+1}\frac{n!}{1-\alpha} - \frac{(1+\mu)^2}{4\mu}\times$$

$$\times\left\{\left[2\,\frac{\mu}{1+\mu} - \frac{\mu}{2}(1-\alpha) - \frac{1-\mu}{1+\mu}\right]\times\right.$$

$$\left.\times\,\alpha(\ln\alpha)^n - (1-\alpha)\alpha(\ln\alpha)^{n-1}(2\ln\alpha + n)\right\}\times$$

$$\times\frac{k_B\bar{T}}{E} + O\left(\frac{\mu}{E^2}\right) . \qquad (5.106)$$

Having calculated the slowing-down parameters, we return to the problem of finding a first-order differential equation for $\Phi(u)$. If the cross sections in Eq. (5.97), and hence $\Phi(u)$, vary slowly with lethargy, then for heavy moderators it is a good approximation to retain only a few terms on the right-hand side of Eq. (5.98). If we retain only the first term, we obtain an approximation used by Nelkin[20] to calculate the asymptotic spectrum in a heavy crystalline moderator. In his approximation the $n = 0$ and $n = 1$ moments are calculated by using the Placzek mass expansion (see Ref. 16 of Chap. 2). The spectrum $E\phi(E)$ calculated in Ref. 20 can be shown to be correct through terms of order $1/E$ by making a comparison of those calculations with the heavy-mass limit of Corngold's expansion. It is possible, however, to achieve considerably greater accuracy and still retain the simplicity of a description of the thermalization process by means of a first-order differential equation. Furthermore, an accurate description of the asymptotic spectrum by a first-order equation is possible not only in the heavy-mass limit but also for the important cases of moderation by beryllium, graphite, and hydrogen.

To obtain a useful approximate equation for the asymptotic neutron spectrum in nonhydrogenous systems, we first note that for high energies the n*th* term on the right-hand side of Eq. (5.98) is of order ξ_0^{n-1} compared with $\xi\Sigma\,(d\Phi/du)$. Hence, for all interesting moderators except hydrogen and possibly deuterium, it is a good approximation to truncate Eq. (5.98) after a few terms. Retaining terms of first order in ξ_0 gives the approximation used in Ref. 20 for the case of thermalization by a heavy crystalline moderator. Greater accuracy may be achieved, however, without increasing the order of the differential equation for Φ. To show this, we differentiate Eq. (5.98) with respect to u, obtaining

$$\frac{d}{du}\{[\Sigma_a(u) - \delta\Sigma(u)]\Phi(u)\} =$$

$$= -\left[\frac{d}{du}\left\langle\xi\Sigma\right\rangle\right]\frac{d\Phi}{du} - \left\langle\xi\Sigma\right\rangle\frac{d^2\Phi}{du^2} + O(\xi_0^2) \ . \quad (5.107)$$

If we now solve for $\left\langle\xi\Sigma\right\rangle(d^2\Phi/du^2)$ and substitute the result in Eq. (5.98), we find, after neglecting terms of order ξ_0^3 and higher, that

$$[\tilde{\Sigma}_a(u) + \gamma(u)\frac{d}{du}\tilde{\Sigma}_a]\Phi(u) =$$

$$= -[\Sigma_s(u)\xi(u) + \gamma(u)\tilde{\Sigma}_a(u) + \gamma(u)\frac{d}{du}\left\langle\xi\Sigma\right\rangle]\frac{d\Phi}{du} \ , \quad (5.108)$$

where

$$\begin{aligned} \tilde{\Sigma}_a(u) &= \Sigma_a(u) - \delta\Sigma(u) \ , \\ \Sigma_s(u)\xi(u) &= \left\langle\xi\Sigma\right\rangle \ , \\ \gamma(u) &= \frac{1}{2}\frac{\left\langle\xi^2\Sigma\right\rangle}{\left\langle\xi\Sigma\right\rangle} \ . \end{aligned} \quad (5.109)$$

Aside from an arbitrary multiplicative constant, the solution of Eq. (5.108) is given by

$$\Phi(u) \doteq \exp\left\{-\int_{-\infty}^{u}\frac{\tilde{\Sigma}_a(u') + \gamma(u')\frac{d}{du'}\tilde{\Sigma}_a(u')}{\xi(u')\Sigma_s(u') + \gamma(u')\tilde{\Sigma}_a(u') + \gamma(u')\frac{d}{du'}[\xi(u')\Sigma_s(u')]}\,du'\right\} . \quad (5.110)$$

In the limit of free atoms at rest, $\delta\Sigma$ vanishes, and $\xi(u)$, $\Sigma_s(u)$, and $\gamma(u)$ assume their free-atom values ξ_0, Σ_s, and γ_0, respectively, where ξ_0 is given by Eq. (5.13) and

$$\gamma_0 = \frac{1 - \alpha + \alpha \ell n \quad - \tfrac{1}{2}\alpha(\ell n\alpha)^2}{1 - \alpha + \alpha \ell n\alpha} . \tag{5.110a}$$

Equation (5.110) then reduces identically to the result obtained by Greuling and Goertzel:[19]

$$\Phi(u) = [\xi_0\Sigma_s + \gamma_0\Sigma_a(u)]^{-1} \exp\left[-\int_{-\infty}^{u} \frac{\Sigma_a(u')}{\xi_0\Sigma_s + \gamma_0\Sigma_a(u')}\, du'\right] ,$$

which is known to be an accurate expression for the asymptotic neutron spectrum. Equation (5.108) reduces to the usual Fermi-age equation for $\gamma = 0$. The appearance of γ in Eq. (5.108) reflects the fact that the neutrons are not continuously degraded in energy, and thus absorbers are less effective in removing neutrons. From Eq. (5.105) we see that the effective slowing-down power, $\xi\Sigma$, is reduced by the effects of thermalization. The effective absorption cross section $\tilde{\Sigma}_a$ is less than Σ_a by the quantity $\delta\Sigma$, which is positive at high energies. It is the decrease in the effective absorption rather than the decrease in $\xi\Sigma$ that gives the dominant effect of the chemical binding on the neutron spectrum at high energies.

The order of magnitude of the errors that result from truncating the series on the right-hand side of Eq. (5.98) after two terms is considered in detail in Ref. 14. Here we show that they are small by comparing the solution of Eq. (5.108) with (5.87) for the cases of moderation by a heavy gas and by graphite.

In the case of a heavy gaseous moderator, the moments are determined from Eqs. (5.104) through (5.106);

$$\begin{aligned}
\left\langle \xi^0\Sigma \right\rangle &= \mu\Sigma_s\left(1 + \frac{9}{2}\,\frac{k_BT}{E}\right) + O(\mu^2) , \\
\left\langle \xi\Sigma \right\rangle &= 2\mu\Sigma_s\left(1 - \frac{3k_BT}{E}\right) + O(\mu^2) , \\
\left\langle \xi^2\Sigma \right\rangle &= 4\mu\Sigma_s\,\frac{k_BT}{E} + O(\mu^2) , \\
\left\langle \xi^n\Sigma \right\rangle &= O(\mu^2) \quad \text{or higher} \qquad n \geq 3 .
\end{aligned} \tag{5.111}$$

Substituting Eq. (5.111) in Eq. (5.98), neglecting powers of μ higher than the first and expressing the result in terms of E and $\phi(E)$, we obtain the familiar second-order differential equation. Eq. (5.75), for the flux spectrum of neutrons thermalized by a heavy gas. For large E, the physical solution of Eq. (5.75), which corresponds to a slowing-down distribution, is given in the case of 1/v absorption by[10]

$$\Phi(u) = E\phi(E) = 1 - \frac{\Delta}{2} x^{-1} + \left(\frac{\Delta^2}{8} + 2\right) x^{-2} - \frac{\Delta(\Delta^2 + 76)}{48} x^{-3} + $$

$$+ \frac{\Delta^4 + 220\Delta^2 + 2{,}304}{384} x^{-4} - $$

$$- \frac{\Delta^5 + 500\Delta^3 + 23{,}584}{3{,}840} x^{-5} + \dots , \qquad (5.112)$$

where $x^2 = E/k_B T$ and $\Delta = [2\Sigma_a(k_B T)]/\mu\Sigma_s$. This expansion was first given by Wilkins.[7]

For the case of 1/v absorption, the approximate solution (5.110) can also be expanded in inverse powers of x. Using Eqs. (5.109), (5.110), and (5.111), together with the heavy-mass limit, we obtain an $E\phi(E)$ that is greater than the exact result given in Eq. (5.112) by the amount

$$\frac{3}{40} \Delta x^{-5} + O(x^{-6}) .$$

It is possible, therefore, to determine $\phi(E)$ with extreme accuracy, at the same time retaining the simplicity inherent in the description by a first-order differential equation.

We now apply Eq. (5.110) to the calculation of $\phi(E)$ in bound carbon. Equations (5.104) through (5.106) give

$$\frac{\delta\Sigma}{\Sigma_s} = 0.27219 \frac{k_B\bar{T}}{E} + 0.052132 \frac{(K^2)_{Av}}{E^2} - 0.16078 \frac{B_{Av}}{E^2} ,$$

$$\left\langle \frac{\xi\Sigma}{\Sigma_s} \right\rangle = 0.15773 - 0.35550 \frac{k_B\bar{T}}{E} + \dots ,$$

$$\left\langle \frac{\xi^2\Sigma}{\Sigma_s} \right\rangle = 0.034131 + 0.12044 \frac{k_B\bar{T}}{E} + \dots .$$

Table 5.1

VALUES OF $\bar{T}$, B_{Av}, AND $(K^2)_{Av}$ FOR TEMPERATURES OF 300°K AND 600°K

	Krumhansl-Brooks*			Free Gas		
Temp., T (°K)	$\frac{\bar{T}}{T}$	$\frac{(K^2)_{Av}}{(k_B T)^2}$	$\frac{B_{Av}}{(k_B T)^2}$	$\frac{\bar{T}}{T}$	$\frac{(K^2)_{Av}}{(k_B T)^2}$	$\frac{B_{Av}}{(k_B T)^2}$
300	2.363	21.63	25	1	15/4	0
600	1.432	7.794	25/4	1	15/4	0

*Values calculated from the frequency distribution of Krumhansl and Brooks, Ref. 21.

Retaining higher powers of E^{-1} in these quantities does not affect the expansion of $E\phi(E)$ to the order of accuracy given in Eq. (5.113) below.

In Table 5.1, the values of $\bar{T}$, B_{Av}, and $(K^2)_{Av}$ are listed for temperatures of 300°K and 600°K. These values follow from the vibration spectra used by Krumhansl and Brooks[21] in their investigation of the specific heat of graphite. For the purposes considered here, the frequency spectra of lattice vibrations in graphite are essentially $\rho_i = 2\theta_1^{-2}\omega$, where $\theta_1 = 1000$°K and $\theta_2 = 2500$°K are the cutoff frequencies for vibrations perpendicular and parallel, respectively, to the principal planes of the lattice.

On expanding the solution $E\phi(E)$, Eq. (5.110), in inverse powers of x to order x^{-4}, we find that the numerical coefficients differ by no more than 1.5% from those in Corngold's expansion

$$\begin{aligned} E\phi(E) = 1 &- 1.1143\left(\frac{\Delta}{2}\right)x^{-1} + \left[0.6539\left(\frac{\Delta}{2}\right)^2 + 1.917\,\frac{\bar{T}}{T}\right]x^{-2} - \\ &- \left[0.2690\left(\frac{\Delta}{2}\right)^3 + 3.334\left(\frac{\Delta}{2}\right)\frac{\bar{T}}{T}\right]x^{-3} + \\ &+ \left[0.08721\left(\frac{\Delta}{2}\right)^4 + 2.785\left(\frac{\Delta}{2}\right)^2\frac{\bar{T}}{T} + 4.996\left(\frac{\bar{T}}{T}\right)^2 + \right. \\ &\left. + 0.2028\,\frac{(K^2)_{Av}}{(k_B T)^2} - 0.6210\,\frac{B_{Av}}{(k_B T)^2}\right]x^{-4} + O(x^{-5}) \;. \qquad (5.113) \end{aligned}$$

Figures 5.7a and b show $E\phi(E)$ versus E for the Krumhansl-Brooks and free-gas models at temperatures of 300°K and 600°K; in both cases we take $\Sigma_a(2200\ \text{m/sec})/\mu\Sigma_s = 1$. The curves show that

1. The asymptotic spectrum is sensitive to the effects of binding and moderator atomic motions, even for $E > 20k_BT$, and
2. The asymptotic spectrum obtained from a tight binding model is only about half as sensitive to the temperature change from 300°K to 600°K as the spectrum of neutrons thermalized by a gas with A = 12.

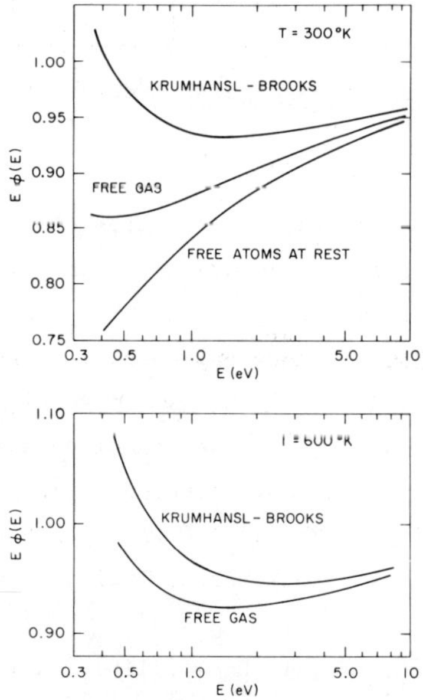

Figure 5.7 $E\phi(E)$ versus energy for bound carbon and free gas models for $\Sigma_a(2200\ \text{m/sec})/\mu\Sigma_0 = 1$; (a) at 300°K and (b) at 600°K.

The explicit results in Figs. 5.6, 5.7a, and 5.7b for $E\phi(E)$ show that the leading effect of 1/v absorption is to pull the spectrum down as the energy decreases from a high value. Since thermal motion and chemical binding reduce the rate at which neutrons slow down, the leading effect of thermalization is to hold the spectrum up at high energies. The net effect is that the curve of $E\phi(E)$ versus E has a broad minimum that is usually located at an energy between a few tenths of a volt and a few volts. The presence of a minimum in the asymptotic energy region may be precluded in some experimental situations where the effective absorption cross section is reduced by leakage into the volume element under study. For moderately strong absorption it is also possible that the minimum in $E\phi(E)$ occurs at an energy too low for Eq. (5.87) or Eq. (5.108) to apply. For very strong absorption there is no thermal peak in $E\phi(E)$, and thus $E\phi(E)$ becomes a monotonic increasing function of E.

The validity of the asymptotic series for the neutron spectrum rests on the validity of the short-collision-time approximation. The validity of the latter approximation requires that the collision times be short compared with the periods of atomic motions. Except for an isotropic harmonic oscillator with A = 1 (see Sec. 2-5), the collision time is of the order of $\hbar/E_0$, where E_0 is the energy of the incident neutron. Thus, for the asymptotic series (5.87) to be valid, we require $E_0 \gg \hbar\omega_c$, where $\hbar\omega_c$ is an average energy characteristic of atomic motions in the moderator.

In the preceding discussion we have used the term "asymptotic series" to describe the expansion of the flux spectrum in powers of $E^{-\frac{1}{2}}$. It is likely, but not proven, that (5.87) is also an asymptotic series in the precise mathematical sense.[22] An asymptotic series has the property that the error is of smaller order than the last term retained. It is not convergent, however, and the magnitude of the terms eventually increases if enough terms are included. As the energy decreases, the optimum number of terms decreases, and the minimum error increases. For a fixed number of terms the error also increases as the energy decreases. It is important to note that the accuracy of (5.87) cannot always be improved by including more terms. This is particularly true at lower energies.

We next consider the problem of neutron thermalization in a medium consisting of bound hydrogen and a heavy absorber. We recall that the validity of Eq. (5.108) depends on a rapid decrease of the quantities $\langle \xi^n \Sigma \rangle / n!$ with increasing n. This condition is not at all satisfied for moderation by hydrogen, nor is it very well satisfied for moderation by deuterium. It is well known, however, that Eq. (5.110) goes over into the exact solution for hydrogen in the limit of free atoms at rest. It also happens that for the case of 1/v absorption, the solution (5.110) agrees exactly with Corngold's expansion through terms of order E^{-2}. This agreement results from a cancellation of errors in the approximations that lead to Eq. (5.108).

To obtain the correct method for arriving at Eq. (5.108) for the case of moderation by hydrogen, we return to Eq. (5.98); now, however, the quantities $\langle \xi^n \Sigma \rangle / n! \Sigma_s$ are all of order unity at high energies. Since, in addition, all the $d^n\Phi/du^n$ are of the same order of magnitude, it is not permissible to truncate the series after a finite number of terms. There is the simplifying feature, however, that for hydrogen all moments of order $n \geq 1$ satisfy

$$\langle \xi^n \Sigma \rangle = (-1)^n n! \Sigma_s + O\left(\frac{1}{E^2}\right) \qquad n \geq 1 \ .$$

This follows from Eq. (5.106) in the limiting case $\mu = 1$, $\alpha = 0$. To obtain solutions for $E\phi(E)$ that are correct to order $1/E^2$, it is permissible, for $n \geq 1$, to neglect terms of order E^{-2} and higher in the expressions for the moments. With this approximation, Eq. (5.98) becomes

$$\frac{\Sigma_a - \delta\Sigma}{\Sigma_s} \Phi(u) = -\frac{d\Phi}{du} + \frac{d^2\Phi}{du^2} - \ldots + (-1)^n \frac{d^n\Phi}{du^n} + \ldots \ . \tag{5.114}$$

Operating on both sides of this equation with $1 + (d/du)$, we obtain

$$\frac{\Sigma_a - \delta\Sigma}{\Sigma_s} \Phi(u) + \frac{d}{du}\left[\frac{\Sigma_a - \delta\Sigma}{\Sigma_s} \Phi(u)\right] = -\frac{d\Phi(u)}{du} \ , \tag{5.115}$$

which is equivalent to Eq. (5.108) with $(d\ \xi\Sigma\ )/du = 0$. The solution of Eq. (5.115) is

$$\Phi(u) = E\phi(E) = \frac{1}{\Sigma_s + \tilde{\Sigma}_a} \exp\left(-\int_E^\infty \frac{\tilde{\Sigma}_a}{\Sigma_s + \tilde{\Sigma}_a} \frac{dE'}{E'}\right) \ . \tag{5.116}$$

To order E^{-2}, we find from Eq. (5.104) that

$$\delta\Sigma = \Sigma_s \left[\frac{1}{2} \frac{k_B \bar{T}}{E} + \frac{8}{15} \frac{(K^2)_{Av}}{E^2} - \frac{1}{3} \frac{B_{Av}}{E^2}\right] \ . \tag{5.117}$$

Substituting Eq. (5.117) in Eq. (5.116) and expanding in inverse powers of $x = (E/k_BT)^{1/2}$, we find

$$E\phi(E) = 1 - 3\left(\frac{\Delta}{2}\right)x^{-1} + \left[6\left(\frac{\Delta}{2}\right)^2 + \frac{\bar{T}}{T}\right]x^{-2} - $$
$$- \left[10\left(\frac{\Delta}{2}\right)^3 + \frac{25}{6}\left(\frac{\Delta}{2}\right)\frac{\bar{T}}{T}\right]x^{-3} + $$
$$+ \left[15\left(\frac{\Delta}{2}\right)^4 + \frac{43}{4}\left(\frac{\Delta}{2}\right)\frac{\bar{T}}{T} + \frac{3}{4}\left(\frac{\bar{T}}{T}\right)^2 + \right.$$
$$\left. + \frac{4}{5}\frac{(K^2)_{Av}}{(k_BT)^2} - \frac{1}{2}\frac{B_{Av}}{(k_BT)^2}\right]x^{-4} + O(x^{-5}) \ .$$

To the order given, this solution is in exact agreement with Corngold's expansion for $\mu = 1$. Thus, for hydrogen as well as for the heavy moderators, the initial effects of binding can be described by an equation formally similar to the Greuling-Goertzel equation of slowing-down theory.

For many purposes it is useful to have a relation between the slowing-down density q(E) and the flux per unit energy $\phi(E)$. In the Greuling-Goertzel approximation, there exists the well-known relation[19]

$$Q(E) = \int_0^E \Sigma_a(E)\phi(E)dE = (\xi_0\Sigma_s + \gamma_0\Sigma_a)E\phi(E) \qquad (5.118)$$

in which q(E) is proportional to $\phi(E)$. For the case of moderation by hydrogen this is an exact result. In the exact theory of slowing down, such a simple relation does not exist for $A > 1$.

In the presence of thermalization effects, q(E) and $\phi(E)$ are related by Eq. (5.5). In general, it is not possible to obtain a direct proportionality between q(E) and $\phi(E)$. For the asymptotic energy region we can, nevertheless, find a rather simple approximate relation that takes account of binding effects. This relation is useful for verifying whether measured thermal-neutron spectra satisfy the conditions imposed by the constraints of neutron conservation. Such verifications are carried out for some of the measured spectra shown in Chap. 6.

We start by integrating Eq. (5.6) between E and infinity to obtain

$$q(E) = q(\infty) - \int_E^\infty \Sigma_a(E')\phi(E')dE'$$

$$= \lim_{E\to\infty} \left\langle \xi\Sigma_s \right\rangle E\phi E - \int_E^\infty \Sigma_a(E')\phi(E')dE' \ . \qquad (5.119)$$

The integration of (5.108) between u and infinity gives an expression for

$$\int_E^\infty \Sigma_a(E')\phi(E')dE' \ .$$

Combining this expression with Eq. (5.119) gives

$$\int_0^E \Sigma_a(E')\phi(E')dE' = \left(\gamma\tilde{\Sigma}_a + \left\langle \xi\Sigma_s \right\rangle\right) E\phi(E) - \gamma E \left(\frac{d\left\langle \xi\Sigma_s \right\rangle}{dE}\right) \times$$

$$\times \ E\phi(E) \int_E^\infty \left[-\frac{\delta\Sigma}{E'} + \Sigma_a \frac{d\gamma}{dE'} + \frac{d\ \xi\Sigma_s}{dE'} - \right.$$

$$\left. - \frac{d}{dE'}\left(\gamma E' \frac{d}{dE'} \left\langle \xi\Sigma_s \right\rangle\right)\right] E'\phi(E')dE' \ . \qquad (5.120)$$

For hydrogen, γ and $\left\langle \xi\Sigma_s \right\rangle$ are constants, and thus

$$\int_0^E \Sigma_a(E')\phi(E')dE' = \left(\gamma\Sigma_a + \left\langle \xi\Sigma_s \right\rangle\right) E\phi(E) -$$

$$- \int_E^\infty \delta\Sigma(E')\phi(E')dE' \ . \qquad (5.121)$$

In the analysis of experiments that are designed to measure infinite-medium spectra one can use the experimental results to evaluate numerically the expressions on both sides of this equation. If the experiment approximates infinite-medium conditions very well and if $\Sigma_a(E)$ is what the experimentalist believes it to be, the two numbers that are computed should be very close to one another. Since the highest experimentally accessible energy E_e is finite, it may be necessary to estimate the contribution of the energy region $E_e < E < \infty$ to the integrals that appear on the right-hand side of Eq. (5.119).

Within the framework of energy-dependent diffusion theory the results that we have obtained also apply to a fundamental-mode spectrum in a finite medium with buckling B^2. To account for the leakage effects it is only necessary to replace $\Sigma_a(E)$ by $\Sigma_a(E) + D(E)B^2$. When this is done, we see from Eq. (5.110) that the effects of leakage on the spectrum are comparable to the effects of chemical binding at energies for which DB^2 is comparable to $\delta\Sigma$. This situation pertains to many of the practical experimental arrangements that are designed to measure infinite-medium spectra.

We have seen that the flux spectrum $\phi(E)$ can be described to a good approximation by a first-order differential equation in the energy region above 0.5 eV. The accuracy of this description is limited essentially by its ability to describe the slowing-down problem, rather than by its ability to properly include the effects of chemical binding and atomic motion. We now turn our attention to the problem of developing a similar description for the entire thermal-energy range. We find that a considerable degree of accuracy is obtainable by the use of judiciously chosen differential equation s. Since these equations now become of second order, however, their solution cannot be obtained analytically. The following discussion is thus less explicit than the one just given.

5-7 GENERALIZED DIFFERENTIAL EQUATIONS FOR THE SPECTRUM

The effects of chemical binding are usually large enough that we cannot use the differential equations of Secs. 5-4 and 5-5 to calculate reactor spectra. It is likely, however, that the full information contained in the scattering kernel $\Sigma_0(E,E')$ is not needed for most applications. It is therefore of interest to look for approximate differential equations containing corrections for chemical binding effects. Developments in this direction have been toward generalizing the heavy-gas differential equation of Sec. 5-5 and toward generalizing the Wigner-Wilkins equation of Sec. 5-4.

The most systematic development in this direction has been carried out by Cadilhac and collaborators at Saclay. We outline their method in the following discussion, but we do not give a sufficiently complete description to allow the reader to use the method. For more details, the reader is directed to Refs. 23 through 25.

First introduce the function

$$C(E) = \frac{d}{dE}\left[\frac{\phi(E)}{\phi_M(E)}\right] = \frac{d\Psi(E)}{dE} \tag{5.122}$$

representing the degree of deviation from thermal equilibrium. The slowing-down density can be written in terms of C(E) as

$$q(E) = \int_0^\infty L(E,E')C(E')dE' , \tag{5.123}$$

where

$$L(E,E') = \int_0^{E_<} \int_{E_>}^\infty \Sigma_0(E_1,E_2)\phi_M(E_1)dE_1dE_2 . \tag{5.124}$$

Here, $E_<$, $E_>$ are the lesser and greater of (E,E'), respectively. That Eqs. (5.123) and (5.124) are equivalent to Eq. (5.5) can be seen as follows: Integrating Eq. (5.123) by parts gives

$$q(E) = L(E,E')\Psi(E')\Bigg|_0^\infty - \int_0^\infty \left[\frac{d}{dE'} L(E,E')\right]\Psi(E')dE' .$$

The integrated term is zero. By differentiating Eq. (5.124) and using detailed balance, the second term gives back our original definition (5.5) for q(E).

We can write Eq. (5.123) in operator form as $q = \mathcal{L}C$. (5.125)
The operator $\mathcal{L}$ is positive definite, and self adjoint. The simplest approximation we can make for $\mathcal{L}$ is to replace it by a multiplicative operator $\mathcal{L}C = F(E)C(E)$. (5.126)

By combining Eqs. (5.122), (5.125), and (5.126), we obtain

$$q(E) = F(E)\frac{d}{dE}\left[\frac{\phi(E)}{\phi_M(E)}\right] . \tag{5.127}$$

When we compare this with Eq. (5.77) we see that Eq. (5.126) is just a generalization of the heavy-gas approximation to allow one arbitrary energy-dependent function. If we write

$$F(E) = \mathcal{S}(E) E k_B T \phi_M(E) , \tag{5.128}$$

we see that $\mathcal{S}(E)$ is the generalized slowing-down power that may be allowed to have an arbitrary energy dependence as long as

$$\mathcal{S}(E) \to \xi_0 \Sigma_s \qquad E \gg k_B T .$$

The resulting equation for the flux spectrum is, from Eqs. (5.6) and (5.127), the second-order differential equation

$$\frac{d}{dE}\left\{ F(E) \frac{d}{dE}\left[\frac{\phi(E)}{\phi_M(E)}\right]\right\} = \Sigma_a(E)\phi(E)$$

or

$$\frac{d}{dE}\left\{ \mathcal{S}(E)\left[(E - k_B T)\phi(E) + E k_B T \frac{d\phi(E)}{dE}\right]\right\} = \Sigma_a(E)\phi(E) . \tag{5.129}$$

It is natural to choose $\mathcal{S}(E)$ to give the correct spectrum for weak 1/v absorption, i.e., the function $\phi_1(E)$ discussed in Sec. 5-2. This means we choose $\mathcal{S}(E)$ so that Eq. (5.129) is satisfied with $\phi(E)$ replaced by $\phi_1(E)$ on the left-hand side and by $\phi_M(E)$ on the right-hand side. Using the fact that $q(0) = 0$, we have

$$\mathcal{S}(E)\left[(E - k_B T)\phi_1(E) + E k_B T \frac{d\phi_1(E)}{dE}\right] = \Sigma_a(E)\phi_M(E) . \tag{5.130}$$

For a given moderator at a given temperature, $\phi_1(E)$ can be calculated and then $\mathcal{S}(E)$ determined from Eq. (5.130). With this choice of $\mathcal{S}(E)$ the simplicity of a second-order differential equation for the spectrum can be retained along with a correct description of chemical binding effects for weak absorption. The spectrum for stronger absorption is given approximately by the solution of Eq. (5.129). We do not consider the accuracy of such a description, however, since it is possible to do better with no further increase in the complexity of the resulting equation.

The most important shortcoming of the generalized heavy-gas approximation is its inability to properly include the effects of resonance absorption. We know that the monatomic hydrogen differential equation of Sec. 5-4 does not have this difficulty. It is thus natural to try to generalize this description to include the effects of chemical binding. The special feature of monatomic hydrogen gas is conveniently described in terms of the operator that takes the form

$$L(E,E') = u(E_<)v(E_>) \tag{5.131}$$

and is thus Green's function of a second-order differential operator. For monatomic hydrogen u(E) and v(E) are known functions of $z = E/k_BT$:

$$u_H(z) = \frac{\Sigma_s}{\sqrt{\pi}}\left[(z - \tfrac{1}{2})\int_0^z \frac{e^{-y}}{\sqrt{y}}\,dy + \sqrt{z}e^{-z}\right]$$

$$v_H(z) = e^{-z} \ .$$

We can use Eq. (5.131) as a guide to a working approximation for an arbitrary moderator. Consider the operator

$$F = \quad {}^{-1} \qquad \text{or} \qquad C = Jq \ . \tag{5.132}$$

If J has the form

$$J = j(E) - \frac{d}{dE}\,k(E)\,\frac{d}{dE} \ , \tag{5.133}$$

then L(E,E') will be of the form (5.131), where the functions u(E) and v(E) are solutions of

$$Ju = 0 \quad \text{and} \quad Jv = 0 \ .$$

The essential physical point is that we now have two arbitrary energy-dependent functions j(E) and k(E) that correspond to definite approximations to and to $\Sigma_0(E',E)$. We need not explicitly calculate these operators, however, since Eq. (5.133) leads to a well-defined second-order differential equation for the spectrum. Combining Eqs. (5.122), (5.132), (5.133), and (5.16), we have

$$\frac{d}{dE}\left[\frac{1}{\Sigma_a(E)\phi_M(E)}\,\frac{dq(E)}{dE}\right] = j(E)q(E) - \frac{d}{dE}\,k(E)\,\frac{d}{dE}\,q(E)$$

or

$$j(E)q(E) - \frac{d}{dE}\,k_1(E)\,\frac{dq(E)}{dE} = 0 \ , \tag{5.134}$$

where

$$k_1(E) = k(E) + \frac{1}{\Sigma_a(E)\phi_M(E)} . \qquad (5.135)$$

Equation (5.134) is to be solved subject to the boundary conditions that $q(0) = 0$ and $q(\infty) = q$. The flux spectrum $\phi(E)$ can then be calculated from Eq. (5.6).

The best choice of $j(E)$ and $k(E)$ for a given moderator must still be made. This subject is under active investigation. The current status is reported in Rev. 25. We do not discuss it further here. It is worth noting the bahavior of $j(E)$ and $k(E)$ at high energy that is dictated by slowing-down theory. Consider

$$G(E) = \xi_0\Sigma_s Ej(E)\phi_M(E) , \qquad (5.136)$$

$$H(E) = \xi_0\Sigma_s k(E)\phi_M(E) . \qquad (5.137)$$

We require that $G(E) \to 1$ for $E \gg k_BT$, and that $H(E)$ become constant. In the high-energy limit it can be shown that[25]

$$\phi(E) = \frac{1}{\xi_0\Sigma_s} \quad G\,\frac{q(E)}{E} - H\,\frac{dq}{dE} \quad . \qquad (5.138)$$

As we would expect, ordinary Fermi-age theory corresponds to $G = 1$, $H = 0$. The choice of H thus determines the approximation to slowing-down theory that we are using. The choice $G = 1$, $H = \gamma_0$ (see Eq. (5.110a)) corresponds to the Greuling-Goertzel approximation.

The use of Eq. (5.134) to calculate the flux spectrum has no particular physical transparency, but it does allow a more economical description of the thermalizing properties of the moderator than is involved in a complete specification of the kernel. It also allows a simplification in computation of the spectrum through the use of a second-order linear differential equation. This procedure shows promise of being very useful for actual reactor calculations, but much work remains to be done in establishing its range of applicability and in selecting the best choice of the functions $j(E)$ and $k(E)$ for specific moderators.

5-8 THE MIGRATION AREA

Having discussed in some detail the general properties of the flux spectrum in an infinite homogeneous medium, we now turn our attention to some problems in which the spatial dependence of the neutron flux plays a role. In Sec. 4-8 we introduced the migration area to absorption and derived Eq. (4.98) in the energy-dependent diffusion approximation. The function $M_{00}(E)$ in Eq. (4.98) is the same infinite-medium flux spectrum that we have denoted by $\phi(E)$ in this chapter. Now that we have a better understanding of the properties of this function we can carry out a more refined separation of M^2 into thermal and epithermal contributions than was done in Eq. (4.101). We restrict our attention to the limit of weak absorption, and make use of the weak-absorption expansion discussed in Sec. 5-2. Keeping only terms through $\phi_1(E)$ in Eq. (5.25) and setting the parameter λ equal to one for convenience, we have

$$M_{00}(E) = \Phi\phi_M(E) + \phi_1(E) , \tag{5.139}$$

where

$$\Phi \int_0^\infty \Sigma_a(E)\phi_M(E)dE = q \tag{5.140}$$

and $\phi_1(E)$ satisfies

$$\Sigma_a(E)\Phi\phi_m(E) = R_0\phi_1(E) \tag{5.141}$$

subject to the boundary condition that

$$\lim_{E\to\infty} E\phi_1(E) = q/\xi\Sigma_s \tag{5.142}$$

and the subsidiary condition

$$\int_0^\infty \Sigma_a(E)\phi_1(E)dE = 0 . \tag{5.143}$$

If we substitute Eq. (5.139) into Eq. (4.98) and make use of Eq. (5.140) and (5.143), we obtain

$$M^2 = \frac{\int_0^\infty D(E)\phi_M(E)dE}{\int_0^\infty \Sigma_a(E)\phi_M(E)dE} + \frac{\int_0^{E_0} D(E)\phi_1(E)dE}{q} . \tag{5.144}$$

Since we are not particularly concerned with fast neutrons here, let us simplify Eq. (5.144) by choosing the source energy E_0 in the epithermal region where D(E) and $\xi\Sigma(E)$ are constant, and neglect deviations of $\phi_1(E)$ from the Fermi spectrum in the neighborhood of E_0. The first term in Eq. (5.144) is just the square of the thermal diffusion length in a purely Maxwellian spectrum. The second term represents the Fermi age from source energy to thermal energy in the limit of weak absorption; this second term includes the thermalization corrections to the migration area. The physical meaning of the second term becomes more transparent if we substitute

$$q = \xi\Sigma_s E_0 \phi_1(E_0)$$

and introduce the symbol

$$f(E) = D(E)/D \tag{5.145}$$

for the ratio of the diffusion coefficient to its epithermal value. Equation (5.144) then reads

$$M^2 = L_M^2 + \tau , \tag{5.146}$$

where L_M^2 is defined by Eq. (4.86) and

$$\tau = \frac{D}{\xi\Sigma_s} \frac{\int_0^{E_0} f(E)\phi_1(E)dE}{E_0\phi_1(E_0)} = \frac{D}{\xi\Sigma_s} \ln \frac{E_0}{E^*} , \tag{5.147}$$

a result which also defines E^*. For a particular moderator at a particular temperature, we first calculate $\phi_1(E)$ from Eq. (5.33) and then carry out the integration in Eq. (5.147). The result can be concisely stated in terms of the parameter E^*. In the limit of weak absorption, the migration area is the thermal diffusion length for a

Maxwellian spectrum plus the age from source energy to E^* calculated neglecting thermalization effects.

We illustrate by calculating the case of a heavy gaseous moderator, for which $f(E) = 1$ and $\phi_1(E)$ is given by Eq. (5.82). The result for this case was first derived in Appendix E of Ref. 9. The integrals can be done analytically, but the procedure is rather complicated; we state only the answer that

$$E^*/k_BT = 5.15 \ .$$

Consider a criticality calculation in age-diffusion theory for a reactor with a well-thermalized spectrum in which the thermal cross sections are to be calculated from averages over a Maxwellian at the moderator temperature. In such a calculation the choice of "thermal" energy to use in calculating the Fermi age should be based on obtaining the correct total migration area. From the discussion in this section we see that the proper choice is E^* as defined by Eq. (5.147). Of course, this represents a highly idealized situation. Normally one calculates the thermal spectrum in a definite energy region from zero to E_m, and the energy E_m forms an unambiguous though arbitrary division between thermal and epithermal neutrons.

The larger the value of E_m chosen, the smaller are the resulting average thermal cross sections and the larger is the resulting thermal diffusion length. This is exactly compensated, however, by a decrease in the slowing-down energy range, and thus in the age from source energies to thermal energies. The total migration area remains the same. If the epithermal and thermal energy regions are treated with the same degree of accuracy, the choice of E_m does not affect the over-all neutron balance.

5-9 GEOMETRICAL PERTURBATIONS

The desirability of obtaining measurements of the thermal-neutron spectra under the ideal conditions of an infinite homogeneous medium that is excited by a spatially flat source of fast neutrons is clear. This physical situation admits to analysis within the simple mathematical framework already described in this chapter.

In practice, however, the theoretically ideal conditions of measurement can only be approached, but not achieved. In the class of experiments under consideration here, it is the intention of the experimentalist to reduce the dependence of his measurements on phenomena of a specifically transport and geometrical nature and to emphasize, instead, the dependence of the spectrum on the scattering kernel $\Sigma_0(E',E)$.

The largest perturbations of the ideal infinite-medium conditions have origins of a geometrical nature. The effects that it is necessary to consider are associated with (1) the possibly heterogeneous composition of the moderating assembly, (2) the finite size of the assembly, and (3) the spatial variation of the neutron source. For a completely meaningful comparison of measured and calculated neutron spectra, it is advantageous that the experiment be designed in such a way that the spectral effects induced by the geometry are negligible. This condition pertains in most measurements of neutron spectra in homogeneously poisoned bare hydrogen-moderated assemblies that are very large in comparison with a mean free path of a thermal neutron. In the case of moderation by heavy nuclei such as carbon or beryllium, the spectral perturbations, though small, are significant even in quite heavily poisoned systems. Fortunately, however, the largest geometrical perturbations of the neutron spectrum are amenable to quantitative calculation by means of easily executed techniques.

In most experimental determinations of the neutron spectrum, a moderately high neutron absorption is desirable in order to emphasize the sensitivity of the spectrum to the effects of chemical binding, as well as to conform to the physical conditions encountered in practical problems of reactor design.

5-9.1 THE EFFECTS OF SMALL INHOMOGENEITIES. Ideally, in this class of experiment the absorbing material should be homogeneously distributed throughout the assembly. In systems moderated by light or heavy water, the ideal condition can be realized by forming aqueous solutions with chemical compounds that contain some neutron-absorbing element such as boron or samarium. In solid moderators, we approach the ideal condition by using thin sheets of absorbing material placed less than one mean free path apart. In general, it is not difficult to achieve a reasonable compromise between high absorption on the one hand and a low self-absorption and flux depression on the other.

Thus, the experiment is designed to render small the effects of the heterogeneous distribution of poison. Corrections for residual effects not yet accounted for must still be made, however. In most experiments involving thermalization by solid moderators, the absorber is present in the form of thin foils of thickness 2a, which are separated from one another by less than one mean free path of moderator. The macroscopic absorption cross section $\Sigma_a^F(E)$ of the foil material is usually much greater than the macroscopic scattering cross section; thus, to a good approximation, the foil may be considered as a pure absorber.

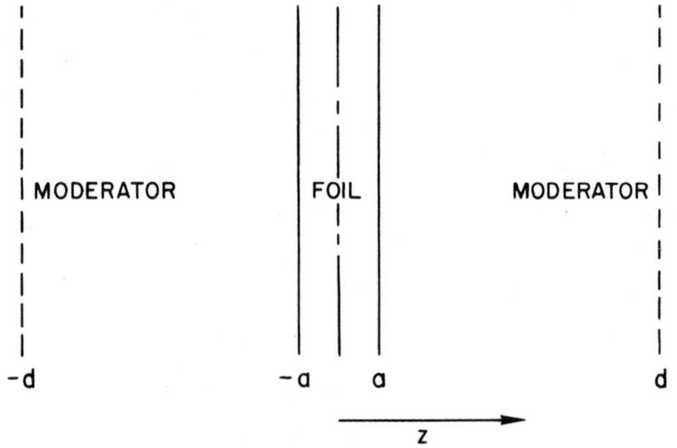

Figure 5.8. A typical cell of the infinite lattice.

To see how to estimate the effects of the heterogeneous composition of an assembly on a measured neutron spectrum, consider a one-dimensional infinite periodic array of identical thin foils whose centers are a distance 2d >> 2a apart. The space between the foils is uniformly filled with moderating material. Figure 5.8 shows a typical element or cell of this array. Assume that the foils are purely absorbing and that the moderator atoms scatter neutrons isotropically. The isotropic slowing-down source that feeds the thermal energy range is spatially flat in the moderator and, since the material in the foils does not scatter, it vanishes in the absorber.

The angular flux spectrum $f(z,E,\mu)$ in the moderator satisfies Eq. (4.88), which for the case considered here reduces to

$$\left[\mu \frac{\partial}{\partial z} + \Sigma_a(E) + \Sigma_s(E)\right] f(z,E,\mu) =$$

$$= \frac{1}{2}\int_0^\infty \Sigma_0(E',E)\phi(z,E')dE' + S(E) \ . \tag{5.148}$$

In the foil,

$$\mu \frac{\partial f}{\partial z}(z,E,\mu) + \Sigma_a^F(E)f(z,E,\mu) = 0 \ . \tag{5.149}$$

Because of the spatially periodic distribution of foils and moderator, $f(z,E,\mu)$ satisfies the conditions

$$f(d,E,\mu) = f(d,E,-\mu) \tag{5.150a}$$

$$f(0,E,\mu) = f(0,E,-\mu) \tag{5.150b}$$

$$f(z,E,\mu) = f(-z,E,-\mu) \tag{5.150c}$$

as well as the usual condition of continuity with respect to z.

To an extremely good approximation, for $a \ll d$, the quantity that is measured in the class of experiment being considered is the spatially averaged flux spectrum in the moderator, i.e.,

$$\phi_m(E) = \frac{1}{d-a}\int_a^d dz \int_{-1}^1 f(z,E,\mu)d\mu \approx \frac{1}{d}\int_0^d \phi(z,E)dz \ . \tag{5.151}$$

We can obtain an equation for $\phi_m(E)$ by integrating Eq. (5.148) over all μ and over all positions extending from a to d:

$$-\frac{1}{d-a}\int_{-1}^1 \mu[f(a,E,\mu) - f(d,E,\mu)]d\mu + [\Sigma_a(E) + \Sigma_s(E)]\phi_m(E)$$

$$= \int_0^\infty \Sigma(E',E)\phi_m(E')dE' + S(E) \ . \tag{5.152}$$

Similarly, integrating Eq. (5.149) over all angles and over the interval from zero to a gives

$$\frac{1}{a}\int_{-1}^{1} \mu[f(a,E,\mu) - f(0,E,\mu)]d\mu = \Sigma_a^F(E)\phi_i(E) \ , \qquad (5.153)$$

where

$$\phi_i(E) = \frac{1}{a}\int_0^a dz \int_{-1}^{1} f(z,E,\mu)d\mu \qquad (5.153a)$$

is the average flux per unit energy in the foil. Combining Eqs. (5.152) and (5.153), taking into account Eq. (5.150), gives

$$\frac{a}{d - a}\,\Sigma_a^F(E)\phi_i(E) + [\Sigma_a(E) + \Sigma_s(E)]\phi_m(E) =$$

$$= \int_0^{\infty} \Sigma(E',E)\phi_m(E')dE' + S(E) \ , \qquad (5.154)$$

which is just the balance equation for neutrons with energy E, regardless of their position or direction of motion.

The theoretical problem of homogenizing the heterogeneously distributed absorber throughout the volume of the assembly consists in finding an approximate relation between $\phi_i(E)$ and $\phi_m(E)$. The simplest approximation consists in setting $\phi_i(E) = \phi_m(E)$. Since the foil tends to shield its interior from neutrons at its surface, this leads to an overestimate of the rate at which neutrons are absorbed.

To estimate the effects of self-shielding, we start from Eq. (5.149). This equation admits the solution

$$f(z,E,\mu) = f(a,E,\mu)\exp\left[\frac{\Sigma_a^F(E)(a - z)}{\mu}\right] \qquad |z| < a \ . \qquad (5.155)$$

Equation (5.155) allows us to relate the absorption rate in the foil to the angular flux at its surface. As a first step in obtaining a simple connection between the $\phi_i(E)$ and $\phi_m(E)$, we assume an isotropic surface flux incident on the slab:

$$f(a,E,\mu) = \frac{\phi_-(E)}{2} \qquad \mu < 0 \ . \qquad (5.156)$$

Observing the symmetry implied by Eq. (5.150c), we find

$$a\phi_i(E) = \int_{-a}^{a} dz \int_{-1}^{0} f(z,E,\mu)d\mu$$

$$= \frac{\phi_-(E)}{2\Sigma_a^F(E)} \int_{-1}^{0} d\mu\mu \left\{ \exp\left[\frac{2\Sigma_a^F(E)a}{\mu}\right] - 1 \right\}$$

$$= \frac{\phi_-(E)}{2\Sigma_a^F(E)} \left\{ \frac{1}{2} - E_3\left[2a\Sigma_a^F(E)\right] \right\} . \qquad (5.157)$$

It still remains to relate the incident flux $\phi_-(E)$ to the average flux in the moderator. To do this we first note that

$$f(a,E,\mu) = \frac{\phi_-(E)}{2} \exp\left[-\frac{2a\Sigma_a^F(E)}{\mu}\right] \qquad \mu > 0 . \qquad (5.158)$$

The exponential factor just gives the attenuation of the flux $f(-a,E,\mu)$, for $\mu < 0$, after transmission through the slab. We now assume that the flux spectrum in the moderator is independent of position. It then follows from Eqs. (5.156) and (5.158) that

$$\phi_m(E) = \int_{-1}^{1} f(a,E,\mu)d\mu = \frac{\phi_-(E)}{2} \left\{ 1 + E_2\left[2a\Sigma_a^F(E)\right] \right\} . \qquad (5.159)$$

Using this result together with Eq. (5.157) in Eq. (5.154), we find

$$[\Sigma_a^{eff}(E) + \Sigma_s(E)]\phi_m(E) = \int_0^{\infty} \Sigma_0(E',E)\phi_m(E')dE' + S(E) , \qquad (5.160)$$

where the cross section that takes into account the self-shielding effect is given by

$$\Sigma_a^{eff}(E) = \Sigma_a(E) + \frac{a}{d-a} \Sigma_a^F(E) K\left[2a\Sigma_a^F(E)\right] , \qquad (5.161)$$

with

$$K(x) = \left[\frac{1 - 2E_3(x)}{2x}\right] \frac{2}{1 + E_2(x)} \qquad (5.162)$$

and $E_2(x)$ given by Eq. (4.124).

Equation (5.162) is a result already obtained by Bengston[26] and other authors.[27] The bracketed term in Eq. (5.162) is the self-shielding factor given by Case, de Hoffmann, and Placzek[28] for the case of an absorbing foil in an isotropic neutron bath. The remaining factor is a correction for the effect of the anisotropy of the flux at the surface of the foil.

Equation (5.162) gives a reliable estimate of the self-shielding effect as long as the scale of the heterogeneity is small compared with a diffusion length in the poisoned system, and as long as the ratio $\phi_{-}(E)/\phi_m(E)$ is not very different from unity.

5-9.2 THE EFFECTS OF FINITE GEOMETRY AND SPATIAL NONUNIFORMITY OF THE SOURCE. Next we consider the perturbations of the neutron spectrum arising from the finite size of the experimental assembly and from the nonuniform spatial distribution of the source that excites the thermal neutrons in the assembly. In general, these geometrical factors significantly perturb the infinite-medium spectrum. Fortunately, even if the effects associated with the experimental geometry cannot be made negligible, they can be theoretically accounted for to quite a high degree of accuracy. In the following, we present two different mathematical methods for calculating geometry or source-induced perturbations of the infinite-medium spectrum. Each of the methods has the properties that it can be easily executed and that it leads to quite definite and accurate estimates of the effects it is intended to describe. In addition, one of the methods gives a precise quantitative expression to some of the considerations that strongly influence the experimental design.

For both methods of calculation, the problem to be solved is formulated in energy-dependent diffusion theory. We consider the region of neutron energy from zero to the indium resonance energy, $E_m = 1.46$ eV. The important approximation that we require is that the source of thermal neutrons slowing down past indium resonance have a spatial distribution independent of the energy with which the neutrons first enter the thermal-energy region, $0 \leq E \leq E_m$. To the extent that diffusion theory is valid, this approximation is exact for moderation by hydrogen and is extremely accurate for moderation by heavy nuclei. To see this, we consider the thermal source

$$4\pi S_{00}(\mathbf{r},E) = \int_{E_m}^{E/\alpha} \frac{\Sigma_0^f(E',E)}{E'} [E'\phi(\mathbf{r},E')]dE' \qquad \alpha E_m \le E \le E_m$$

$$= 0 \qquad E \le \alpha E_m ,$$

where $\Sigma_0^f(E',E)$, the kernel for scattering by moderator atoms that are free and at rest, is given by Eq. (5.9). For slowing down by hydrogen, $\alpha = 0$, and we find

$$4\pi S_{00}(\mathbf{r},E) = \frac{\Sigma_s}{E_m} Q(\mathbf{r}) ,$$

where

$$Q(\mathbf{r}) = E_m \int_{E_m}^{\infty} \frac{\phi(\mathbf{r},E')}{E'} dE' \qquad \text{(hydrogen)}$$

$$\approx E_m \phi(\mathbf{r},E_m) .$$

In the case of slowing down by heavy nuclei, $E'\phi(\mathbf{r},E')$ varies slowly with energy in the interval $E_m/\alpha < E' < E_m$. Thus it is a good approximation to extract $E'\phi(\mathbf{r},E') \approx E_m\phi(\mathbf{r},E_m)$ from the integral and write, for heavy moderators,

$$4\pi S_{00}(\mathbf{r},E) = f(E)Q(\mathbf{r}) \qquad \alpha E_m \le E \le E_m$$

$$= 0 \qquad E \le \alpha E_m ,$$

where

$$f(E) = \frac{\Sigma_s}{1-\alpha}\left(\frac{1}{E_m} - \frac{\alpha}{E}\right) .$$

Then for a homogeneous system in a steady state, Eq. (4.64) becomes

$$-D(E)\nabla^2\phi(\mathbf{r},E) + [\Sigma_a(E) - R_0]\phi(\mathbf{r},E) = f(E)Q(\mathbf{r}) , \qquad (5.163)$$

which applies to moderation both by hydrogen and by heavy nuclei. By means of Eq. (5.163) we can make accurate predictions of the neutron-spectrum $\phi(\mathbf{r},E)$ provided that $Q(\mathbf{r})$ is known. The latter can be determined experimentally as the spatial distribution of the activation of cadmium-covered indium foils, or it can be calculated theoretically by the methods of slowing-down theory. In heavy moderators, an excellent approach is to make an age-theory fit to the measured spatial dependence of the activation of cadmium-covered indium foils.

The methods that we use to solve Eq. (5.163) are (1) the localized buckling method and (2) the method of modal expansion. The method of localized buckling is particularly useful for the case of excitation by a source of fast neutrons (~1 meV) localized about a point within or near a nonmultiplying assembly whose dimensions are greater than or comparable to $\sqrt{\tau_m}$, where τ_m is the migration area from the initial source energies to the thermal cutoff energy E_m.

In the localized buckling method, the approximation employed is based on the smallness of the thermal migration area L^2 as compared with τ. Under these conditions the thermal flux tends to follow the thermal source, and the thermal spectrum is approximately independent of position. The method of approximation results in a solution expressed as a series in powers of L^2/τ_m, the corrections to the first approximation beginning with terms of order L^4/τ_m^2. This method is of particular value not only because it allows one to make definite predictions of the spectral perturbation due to spatial effects but also because it gives quantitative expression to a principal requirement in the design of the class of experiment under consideration, namely that L^2 be small in comparison with τ_m.

We begin by assuming a solution of Eq. (5.163) of the form

$$\phi(\mathbf{r},E) = \phi_0(E)Q(\mathbf{r}) + \Psi(\mathbf{r},E) \tag{5.164}$$

and choose $\phi_0(E)$ to satisfy the infinite-medium equation

$$DB_0^2\phi_0(E) + [\Sigma_a(E) - R_0]\phi_0(E) = f(E) , \tag{5.165}$$

where

$$B^2(\mathbf{r}) = -\frac{\nabla^2 Q(\mathbf{r})}{Q(\mathbf{r})} = B_0^2 + \delta B^2(\mathbf{r}) \tag{5.166}$$

and B_0^2 is the value of $B^2(\mathbf{r})$ at the point of observation, $\mathbf{r} = r_0$. Thus $\Psi(\mathbf{r},E)$ satisfies

$$-D(E)\nabla^2\Psi(\mathbf{r},E) + [\Sigma_a(E) - R_0]\Psi(\mathbf{r},E) = \bar{\Sigma}_a\phi_0(E)Q_1(\mathbf{r}) \ , \qquad (5.167)$$

where

$$Q_1(\mathbf{r}) = -\frac{D}{\bar{\Sigma}_a}\,\delta B^2(\mathbf{r})Q(\mathbf{r}) \ . \qquad (5.168)$$

So far $\bar{\Sigma}_a$ is arbitrary, but we provide a convenient definition for it later. Equation (5.167) is similar to Eq. (5.163), and the source terms in both equations are separable in energy and position. An essential difference is that the source term is now well-thermalized compared with the slowing-down source.

The purpose of our subsequent procedure is to show that $\Psi(\mathbf{r},E)$ is calculable and small. To find Ψ, we expand it in the series

$$\Psi(\mathbf{r},E) = Q_1\Psi_0(E) + L^2\nabla^2 Q_1\Psi_1(E) + \ldots + L^{2n}\nabla^{2n}Q_1\Psi_n(E) + \ldots \ , \qquad (5.169)$$

where $L^2 = D/\bar{\Sigma}_a$. If we choose $\Psi_n(E)$ to satisfy

$$\left.\begin{aligned}
&[\Sigma_a(E) - R_0]\Psi_0(E) = \bar{\Sigma}_a\phi_0(E) \ , \\
&[\Sigma_a(E) - R_0]\Psi_n(E) = \bar{\Sigma}_a\Psi_{n-1}(E) \ ,
\end{aligned}\right\} \qquad (5.170)$$

then the series in Eq. (5.169) is a formal solution of Eq. (5.167). Furthermore, the sequence of Eqs. (5.170) defines an iterative procedure for computing the eigenfunction associated with the lowest eigenvalue of the operator $[\Sigma_a(E) - R_0]$. Since each of Eqs. (5.170) is an equation of the form of Eq. (5.1), the method of Sec. 7-1.2 can be used to compute the $\Psi_n(E)$. If $\bar{\Sigma}_a$ is the eigenvalue of $\Sigma_a(E) - R_0$, then L^2 is the thermal diffusion length in the medium under investigation. Thus, since

$$Q_1/Q = O\left(\frac{L^2}{\tau_m}\right)$$

and

$$\frac{L^{2n}\nabla^{2n}Q_1}{Q} = O\left(\frac{L^{2(n+1)}}{\tau_m^{n+1}}\right) \ ,$$

the right-hand side of Eq. (5.169) is a series whose terms decrease in order of magnitude like the sequence of positive powers of L^2/τ_m. If $L^2 << \tau_m$, the correction term Ψ is small and at the point of observation is of the order of L^4/τ_m^2 compared with $\phi_0(E)Q(\mathbf{r}_0)$.

In general, the convergence of the series in Eq. (5.169) is rapid. It may happen, however, that the correction term Ψ is small but that the series in Eq. (5.169) converges slowly. Then it is convenient to sum the series approximately. To show that this is possible, we note that for n greater than or equal to some integer N it is a good approximation to write

$$\Psi_n = \Psi_{n-1} \; . \tag{5.171}$$

Thus,

$$\Psi(\mathbf{r},E) = Q_1\Psi_0 + L^2\nabla^2 Q_1\Psi_1 + \ldots + L^{2N-4}\nabla^{2N-4}Q_1\Psi_{N-2} + $$
$$+ \Psi_{N-1}L^{2N-2}\nabla^{2N-2}[Q_1 + L^2\nabla^2 Q_1 + \ldots] \; . \tag{5.172}$$

The quantity in square brackets,

$$Q_1 + L^2\nabla^2 Q_1 + \ldots = F \; , \tag{5.173}$$

satisfies the inhomogeneous equation governing the diffusion of thermal neutrons:

$$-L^2\nabla^2 F + F = Q_1 \; . \tag{5.174}$$

Equation (5.174), together with the condition that F vanish at the extrapolated boundary of the system, determines F completely. Over the range of energies for which Ψ_n has an appreciable magnitude, and for the practical cases to be considered in Sec. 6-4, the approximation Eq. (5.171) is a very good one for N as low as two or three.

We now turn to the method of analysis by Fourier series. This method is useful when the thermal source $Q(\mathbf{r})$ has a shape that is near to a fundamental mode. The method proceeds by expanding both $Q(\mathbf{r})$ and $\phi(\mathbf{r},E)$ in a series of orthogonal functions $\chi_n(\mathbf{r})$ that satisfy Eq. (4.156) and vanish on the extrapolated boundary of the moderating assembly. Thus,

$$Q(\mathbf{r},E) = \sum_{n=0}^{\infty} A_n f(E) \chi_n(\mathbf{r}) , \tag{5.175}$$

and

$$\phi(\mathbf{r},E) = \sum_{n=0}^{\infty} A_n \phi_n(E) \chi_n(\mathbf{r}) . \tag{5.176}$$

Here we assume a modal buckling B_n^2 and an extrapolation distance that are independent of energy. Substituting Eqs. (5.175) and (5.176) into Eq. (5.163) and equating the coefficients of $\chi_n(\mathbf{r})$ gives

$$DB_n^2 \phi_n(E) + [\Sigma_a(E) - R_0]\phi_n(E) = f(E) . \tag{5.177}$$

These equations have the form of Eq. (5.1) which pertains in the case of an infinite medium and a spatially flat source. They can be solved conveniently by means of the numerical techniques discussed in Sec. 7-1.2.

To illustrate the conditions under which the method of modal analysis can be conveniently applied, we consider a nonmultiplying assembly excited by a point source of fast neutrons that are moderated by heavy nuclei, such as those of beryllium or carbon. For convenience, we assume that the moderating assembly has the form of a cube and that the fast source is located at the center $\mathbf{r}_0$ of the cube. We assume the validity of Fermi age theory; thus, Q satisfies

$$\nabla^2 Q(\mathbf{r},\tau) = \frac{\partial Q}{\partial \tau} (\mathbf{r},\tau) \tag{5.178}$$

and

$$Q(\mathbf{r},0) = \delta(\mathbf{r} - \mathbf{r}_0) \tag{5.179}$$

and vanishes on the extrapolated boundary of the cube. The solution that satisfies the prescribed boundary conditions is

$$Q(\mathbf{r},\tau) = \left(\frac{2}{a}\right)^3 \sum_{\ell,m,n=0}^{\infty} e^{-B_{\ell mn}^2 \tau} \times$$

$$\times \sin\left[\frac{(2\ell+1)\pi x}{a}\right] \sin\left[\frac{(2m+1)\pi y}{a}\right] \sin\left[\frac{(2n+1)\pi z}{a}\right] , \tag{5.180}$$

where a is the length of a cube edge, including the extrapolation distance, and

$$B_{\ell mn}^2 = \left(\frac{\pi}{a}\right)^2 [(2\ell + 1)^2 + (2m + 1)^2 + (2n + 1)^2] . \qquad (5.181)$$

The origin of coordinates 0 is at a corner of the cube, and the three edges that intersect at 0 form the coordinate axes. The series (5.180) converges rapidly if $(\pi/a)^2\tau >> 1$. More generally in the case of arbitrary geometry and for an arbitrarily located localized source of fast neutrons, the modal expansion converges rapidly if the longest dimension through the assembly is smaller than the square root of the migration area from the initial source energies to E_m. In the case of an experimental design that conforms to this criterion, the perturbation of the infinite-medium spectrum due to the finite size of the medium is of the order of $B_{000}^2 L_a^2$, where L_a^2 is the thermal migration area from the thermal cutoff energy E_m to absorption. The magnitude of this perturbation may be considerable in a practical case. Nevertheless, $\phi(E)$ satisfies an equation of the infinite-medium type and can be easily calculated. The corrections arising from the higher harmonics in the series (5.176) are small and can be accurately and conveniently calculated provided that the convergence of the series is sufficiently rapid.

Many experiments have been designed to measure the energy spectrum of thermal neutrons under nearly infinite-medium conditions. Experimentalists have made extensive studies of spectra of thermal neutrons that have been moderated by hydrogen-containing materials, including light water, and by the heavy crystalline moderators, graphite and beryllium oxide. In Chap. 6 we apply some of the results of this and previous sections to the analysis of some of these measured spectra.

REFERENCES

1. E. P. Wigner and J. E. Wilkins, Jr., *Effect of the Temperature of the Moderator on the Velocity Distribution of Neutrons with Numerical Calculations for H as Moderator*, USAEC Report AECD-2275, Oak Ridge National Laboratory, September 1944.

2. A. D. Galanin, *Thermal Reactor Theory*, 2nd ed., p. 347 ff, translated from the Russian by J. B. Sykes, Pergamon Press, New York, 1958.

3. E. Richard Cohen, A Survey of Neutron Thermalization Theory, in *Proceedings of the First United Nations International Conference on the Peaceful Uses of Atomic Energy, Geneva, 1955*, Vol. 5, p. 405, United Nations, New York, 1956.

4. Harvey Amster and Roland Suarez, *The Calculation of Thermal Constants Averaged Over a Wigner-Wilkins Flux Spectrum; Description of the SOFOCATE Code*, USAEC Report WAPD-TM-39, Westinghouse Electric Corporation, January 1957.

5. B. Davison, Harwell research memorandum, UKAEA, circa 1950, unpublished.

6. M. Cadilhac, J. Horowitz, J. L. Soulé, and O. Tretiakoff, Some Mathematical and Physical Remarks on Neutron Thermalization in Infinite Homogeneous Systems, in *Proceedings of the Brookhaven Conference on Neutron Thermalization*, Vol. II, *Neutron Spectra in Lattices and Infinite Media*, BNL-719, p. 439, 1962.

7. J. E. Wilkins, Jr., *Effect of the Temperature of the Moderator on the Velocity Distribution of Neutrons for a Heavy Moderator*, USAEC Report CP-2481, November 1944.

8. A. K. Rajagopal, *Neutron Thermalization on Doppler Model*, Brookhaven National Laboratory Memorandum, September 1961. Results are given in S. N. Purohit and A. K. Rajagopal, Scattering of Thermal Neutrons in the Doppler Approximation, Appendix III, Energy Transfer Moments for the Monatomic Gas Model, *Nucl. Sci. Eng.*, **13**: 250 (1962).

9. M. Hurwitz, Jr., M. S. Nelkin, and G. J. Habetler, Neutron Thermalization. I. Heavy Gaseous Moderator, *Nucl. Sci. Eng.*, **1**: 280 (1956).

10. E. Richard Cohen, The Neutron Velocity Spectrum in a Heavy Moderator, *Nucl. Sci. Eng.*, **2**: 227 (1957).

11. Luis de Sobrino and Melville Clark, Jr., A Study of Wilkins

Equation, *Nucl. Sci. Eng.*, **10**: 388 (1961).

12. J. Horowitz and O. Tretiakoff, *Effective Cross Sections for Thermal Reactors*, presented at Oak Ridge Meeting, November 14, 1960, European-American Nuclear Data Committee Report EANDC(E)-14, May 1962.

13. N. Corngold, Thermalization of Neutrons in Infinite Homogeneous Systems, *Ann. Phys. (N.Y.)*, **6**: 368 (1959).

14. D. E. Parks, The Effects of Atomic Motions on the Moderation of Neutrons, *Nucl. Sci. Eng.*, **9**: 430 (1961).

15. N. Corngold, Chemical Binding Effects in Neutron Thermalization, in *Inelastic Scattering of Neutrons in Solids and Liquids: Proceedings of the Symposium held at Vienna, Austria, October 11-14, 1960*, p. 607, International Atomic Energy Agency, Vienna, 1961.

16. N. Corngold, Chemical Binding Effects in the Thermalization of Neutrons, *Ann. Phys. (N.Y.)*, **11**: 338 (1960).

17. J. R. Beyster, J. L. Wood, W. M. Lopez, and R. B. Walton, Measurements of Neutron Spectra in Water, Polyethylene, and Zirconium Hydride, *Nucl. Sci. Eng.*, **9**: 168 (1961).

18. N. Corngold and L. Zamick, The Effect of Diffusion Upon the Initial Phases of the Thermalization of Neutrons, *Nucl. Sci. Eng.*, **9**: 367 (1961).

19. S. Glasstone and M. C. Edlund, *The Elements of Nuclear Reactor Theory*, p. 168, Van Nostrand, New York, 1952.

20. M. Nelkin, Neutron Thermalization. II. Heavy Crystalline Moderator, *Nucl. Sci. Eng.*, **2**: 199 (1957).

21. J. Krumhansl and H. Brooks, The Lattice Vibration Specific Heat of Graphite, *J. Chem. Phys.*, **21**: 1663 (1953).

22. P. M. Morse and H. Feshbach, *Methods of Theoretical Physics*, p. 434 ff, McGraw-Hill Book Company, New York, 1953.

23. M. Cadilhac, *Simple Thermalizer Models*, Special Report S-9, November-December 1962 (a translation), EURAEC-778, 1963.

24. M. Cadilhac, *Theoretical Methods for the Study of Neutron Thermalization in Infinite and Homogeneous Absorbing Media*, Special Report No. 26 (a translation), EURAEC-1203, 1963.

25. M. Cadilhac, J.-L. Soulé, and O. Tretiakoff, Thermalization and Neutron Spectra, in *Proceedings of the Third International Conference on the Peaceful Uses of Atomic Energy, Geneva, 1964*, Vol. 2, p. 153, United Nations, New York, 1965.

26. J. Bengston, *Neutron Self-Shielding of a Plane-Absorbing Foil*, Report CF-56-3-170, Oak Ridge National Laboratory, March 1956.

27. J. S. Martinez, T. H. Pigford, and A. J. Kirschbaum, Neutron Self-Shielding in Slab Absorbers, *Trans. Am. Nucl. Soc.*, 4: 156 (1961).

28. K. M. Case, F. de Hoffmann, and G. Placzek, *Introduction to the Theory of Neutron Diffusion*, Vol. I, Superintendent of Documents, U.S. Government Printing Office, Washington 25, D.C., 1953.

Chapter 6

THE MEASUREMENT OF NEUTRON SPECTRA

A main objective in the study of neutron thermalization is to develop theoretical methods for determining thermal-neutron spectra from first principles. The calculation of thermal-neutron scattering kernels and the development of analytical and numerical methods for solving the transport equation for systems of varying geometric complexity are essential to this development. To the extent that it is possible, thermal-neutron scattering cross-section calculations should be compared with the corresponding measurements, as discussed, for example, in Chap. 3. Because of the difficulty in obtaining complete scattering cross-section data experimentally, however, it is simpler to ascertain that the theoretically determined scattering kernels lead to calculated spectra that agree with measured spectra in systems involving a minimum of geometric complexity.

The ideal experimental system of this sort is an infinite homogeneous system with a spatially uniform source of fast monoenergetic neutrons. Although it is impossible to realize this ideal system, it is possible to construct an assembly in such a way that the spectrum in certain of its regions accurately approximates that in the ideal system, except possibly at energies near those of the source neutrons. We refer to each such region as a quasi-infinite medium and to the corresponding spectrum as a quasi-infinite-medium spectrum, inasmuch as to a high degree of accuracy this spectrum satisfies an equation of the form of Eq. (5.1). More specifically, for energies below the indium resonance energy (1.46 eV) or some other suitably chosen cutoff energy, the quasi-infinite-medium spectrum is to a good approximation the solution of Eq. (5.165); the term DB_0^2 describes the presumably small effects of neutron leakage at the point of observation. If the quasi-infinite medium is homogeneous, $\Sigma_a(E)$ is its macroscopic absorption cross section; if it is inhomogeneous, $\Sigma_a(E)$ is that cross section modified in the manner described, for example, in Eq. (5.161) to include self-shielding effects. It is only after we have demonstrated that our theoretical models are capable of yielding thermal-neutron spectra in infinite media that we can proceed with confidence to verify experimentally that our knowledge of thermal-neutron scattering and our methods for solving the transport equation are compatible with measured spectra in more complex systems.

After the construction of the first multiplying assemblies, many techniques for measuring neutron spectra were developed. The earliest spectral studies were performed in reactor environments by means of resonance and thermal activation of foils and by beam-absorption techniques. For most purposes, however, we require greater precision in spectral determinations than we can conveniently provide by using integral techniques. This is especially true of spectra that we measure for the express purpose of detailed comparisons with calculated spectra. In this chapter we stress measurement of spectra by neutron time-of-flight techniques.

There are two different approaches to the time-of-flight method for the measurement of spectra. They are

1) The direct, pulsed-source time-of-flight spectrum experiment, and
2) The chopped-beam, steady-source time-of-flight spectrum experiment.

These techniques permit study of neutron spectra with good resolution from subthermal energies to 14 meV; however, in accordance with the subject of this book, we emphasize the energy region between 0.001 eV and 10 eV.

6-1 DESIGN OF SPECTRUM EXPERIMENTS

6-1.1 GENERAL CONSIDERATIONS. The experimental arrangements for the two types of time-of-flight measurements are shown schematically in Figs. 6.1a and 6.1b. With the aid of Fig. 6.1a, we can explain the general principles involved in the measurement of the energy spectrum of neutrons in the moderator by the pulsed-source time-of-flight technique.

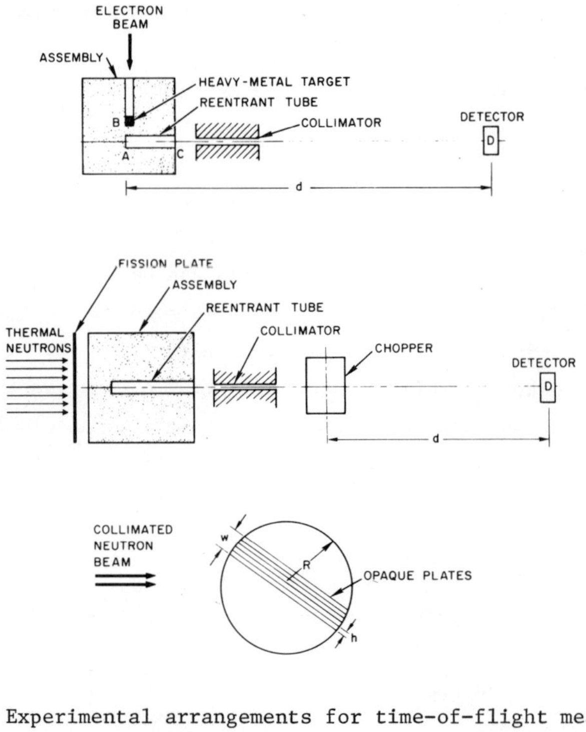

Figure 6.1. Experimental arrangements for time-of-flight measurements.

At an instant that we denote by $t = 0$, the electron beam from a pulsed high-intensity linear accelerator has been directed through a small tube and into an assembly containing the moderator to be studied, where it strikes a heavy-metal target, B, producing fast neutrons. The neutrons slow down by collisions with the nuclei of the moderator and then thermalize in a time t_{th} that ranges from a few to a few hundred microseconds. A small re-entrant tube, A-C in Fig. 6.1a, enters the assembly, and a collimator is located so that we view only neutrons issuing from the base, A, of this tube. These neutrons travel a distance d from A to the detector bank, D. The counting rate produced by all neutrons reaching the detector bank is measured as a function of the elapsed time t_e between the accelerator pulse and the time of detection. If on the average a neutron of energy E spends a much shorter time $t_a(E)$ in the assembly than in the flight path, the flight time t_f from A to D is very nearly equal to t_e. We convert the counting rate to a function of the neutron energy E by using

$$E = \frac{m}{2} \frac{d^2}{t_e^2} , \qquad (6.1)$$

where m is the neutron mass. In Sec. 6-2.1 we show that the spectrum measured by this method is equivalent to the angular-flux spectrum produced at A by a steady source having the same dependence on energy, angle, and position as the time integral of the pulsed source.

The principal requirement for the application of the pulsed-source time-of-flight measurement is that the average time that neutrons spend in the assembly, $t_a(E)$ be very short compared with their flight time $t_f(E)$; otherwise, the instant at which neutrons begin their flight from A to the detector D is not well defined and the dependence of the counting rate on t_e is not simply related to the spectrum of neutrons in the assembly. To fulfill this requirement, we can either employ a long flight path or we can reduce the neutron lifetime by poisoning the assembly with a thermal absorber. In general, $t_a(E)$ includes the time required to slow down to the thermal energy region. Because of this, the light moderators, which have short slowing-down times, lend themselves most naturally to study by the pulsed-source time-of-flight method.

Figure 6.1b illustrates the essentials of the method used in the production of chopped neutron beams. The continuous neutron beam from a steady source is collimated to have a width equal to w (see Fig. 6.1c). The chopper plates, which for simplicity we assume to be perfectly opaque to neutrons, are separated from one another by spacers of thickness $h \ll R$. Ideally, these spacers are perfectly transparent to neutrons. Every half-revolution the chopper transmits neutrons for only a time

$$\Delta t = \frac{h}{2\pi Rf}, \tag{6.2}$$

where f is the frequency of rotation of the chopper. Only those neutrons with a speed greater than the chopper cutoff velocity

$$v_c = \frac{2R}{\Delta t} \tag{6.3}$$

are transmitted. The chopper thus produces a chopped beam that contains all neutrons in the original continuous beam with velocities that are greater than v_c. The neutrons in the chopped beam begin their flight from the chopper to the detector within an interval Δt about some well-defined instant of time. Provided that Δt is much less than the flight time over the distance d shown in Fig. 6.1b, the experiment satisfies the conditions required for a time-of-flight measurement.

The chopper experiment is usually performed with a steady-state neutron source, such as that obtained by placing a fission plate in the thermal column of a reactor, as shown in Fig. 6.1b. Fast neutrons from this source slow down in the moderator, and there is established a steady-state spectrum consisting of thermal and epithermal neutrons. Beams may be extracted and collimated by procedures similar to those used in pulsed-source experiments. Here, however, since the extracted neutron beam is chopped by rotating slits, the neutron flight time from the slits to the detectors is not made uncertain by the finite time that neutrons spend in the moderating assembly, and therefore we do not require a longer flight path for lightly poisoned assemblies than we do for heavily poisoned ones. The lack of intensity and difficulties in attaining good energy resolution, however, to some extent limit the range of neutron energies for which the neutron spectrum can be accurately determined by the chopper technique.

Many considerations relevant to the precise measurements of thermal-neutron spectra pertain equally to both time-of-flight techniques. The specification of the assembly containing the moderator to be studied, i.e., its poison content, geometry, and, to a large extent, the neutron detector system and the experimental energy resolution, are all factors that we can consider regardless of which time-of-flight technique we use. Other considerations that are special to either the pulsed-source or the chopped-beam measurements are discussed in Secs. 6-2 and 6-3, respectively.

Poison concentrations can usually be determined to an accuracy of 1%, but it is important that the particle size and the spatial distribution of the particles of poisoning material be such as to keep self-shielding effects small. For all common absorbers, particles of micron and submicron dimensions give self-absorption effects of less than 5%.

Normally, to be certain that agreements between theoretical and experimental spectra are not fortuitous, we should measure spectra in any moderator for several poison concentrations. In addition, we should use absorbers with radically different absorption properties in order to emphasize thermalization effects in one neutron-energy region relative to those in another.

A further consideration in either the pulsed-source or chopped-beam technique is the determination and control of temperature. Since differential spectral measurements require high-intensity fluxes, source-heating of the experimental assembly is often significant. However, one may expect to measure and control the temperature of an experimental assembly to within 1°K.

6-1.2 GEOMETRICAL CONSIDERATIONS. Geometrical arrangements depend largely on the experimental objectives. These may be either (a) physics objectives or (b) engineering objectives.

To emphasize the sensitivity of spectra to chemical-binding effects, it is desirable first to minimize transport effects that are difficult to calculate and which, at the same time, produce significant departures from a Maxwellian spectrum. Having accomplished this, one can then use absorbers with resonances at thermal energies to emphasize the spectral effects of the scattering mechanism. On the other hand, for one-dimensional geometries--say for slabs and lattices--it may be desirable to emphasize the effects of gradients and perhaps of anisotropic scattering on the spectrum. In this case, the well-designed experiment provides, for example, a test of our ability to calculate currents into thick control rods.

Engineering studies of neutron spectra are generally motivated by a need for specific answers. Usually they are performed in mockups of a practical reactor configuration. Measured spectra are simply a part of the design information.

From the physics standpoint, we should first measure the neutron spectrum in a quasi-infinite medium. With the assurance that the calculational methods and experimental techniques are adequate for this relatively simple case, an excellent basis exists from which to proceed to the experimental investigations of transport problems involving spatial variations of flux spectra in one and in two dimensions.

One procedure for obtaining a quasi-infinite-medium spectrum is to utilize a multiplying assembly. This is done at the expense of experimental flexibility. Actually, quasi-infinite-medium conditions can be attained by a nonmultiplying, finite homogeneous assembly with a source of high-energy neutrons produced, for example, in a manner illustrated in Fig. 6.1a or 6.1b. To show that this is possible, we consider the steady-state form of the diffusion equation (4.64) for the thermal-neutron flux spectrum $\phi(\mathbf{r},E)$. Two requirements must be satisfied if an equation of the form given in Eq. (5.1), for example, Eq. (5.165), is to be appropriate to the experimental conditions:

1) The energy-dependence of the thermal source must vary only slightly with position, so that it is approximately of the form $f(E)Q(\mathbf{r})$.
2) Either $D(E)[\nabla^2\phi(\mathbf{r},E)/Q(\mathbf{r})]$ must be nearly independent of position or $|D(E)[\nabla^2\phi(\mathbf{r},E)/\phi(\mathbf{r},E]|$ must be much less than $\Sigma_a(E)$ for all energies less than the thermal-group cutoff energy E_m.

If these two conditions are satisfied, then to a good approximation the flux can be written in the form $\phi(\mathbf{r},E) = \phi_0(E)Q(\mathbf{r})$, where $\phi_0(E)$ satisfies Eq. (5.165). If the second condition is not satisfied, then possibly we can calculate corrections to this result by one of the methods given in Sec. 5-9. Experimentally, we can verify whether a quasi-infinite medium is realized by comparing measurements of spectra at different points in the assembly.

To satisfy the first condition, the energies of neutrons produced by the initially localized source (lead target or fission plate) must be large compared with E_m. Furthermore, the spectrum must be measured in the medium at a distance from boundaries that is large compared with a mean free path. Normally, the linear dimensions of the assembly are chosen to be a few times the slowing-down length from the initial source energies to E_m.

To satisfy the second condition, it is generally necessary to introduce a relatively high concentration of poison into the assembly. The required concentration of poison can often be reduced by properly choosing the regions in which the spectrum is measured and in which the source is placed. For the external-source arrangement shown in Fig. 6.1b, the buckling in the two transverse dimensions (perpendicular to the axis of the re-entrant tube) is positive while at a properly chosen point the buckling in the axial direction (parallel to the axis of the re-entrant tube) may be negative. A cancellation of the different components of the leakage is thus possible. If, on the other hand, the source is located at the center of the assembly, as in Fig. 6.1a, no such cancellation is likely at any point in the assembly where the spectrum approximates an infinite-medium spectrum. Except in small assemblies where $B^2\tau \gg 1$, for excitation by a point source it is unlikely that $D(E)[\nabla^2\phi(\mathbf{r},E)/\phi(\mathbf{r},E)]$, if large, is spatially independent.

The interior of a homogeneous reactor with a multiplication greater than about three constitutes a quasi-infinite medium since the flux is very nearly a normal-mode flux regardless of the source position. Even if the effects from higher modes are not negligible, they can be accounted for by the method of harmonic analysis.

In the application of a linear accelerator to the study of spectra in homogeneous systems moderated by beryllium or beryllium oxide, a promising method for satisfying the second condition mentioned above is to use photo-induced reactions to distribute the fast-neutron source throughout the volume of the assembly.

Any experiment designed to measure a neutron spectrum should provide precisely known boundary conditions for thermal neutrons. This can be done by covering the walls of the assembly with cadmium, borated polyethylene, or, preferably, boron carbide. Since the dimensions of the assembly must be precisely known, the absorbing container walls must be as thin as possible. Furthermore, in order to avoid backscattering from the container, poisons should be placed on the inside walls of the container; in addition, the room in which the experiments are performed should be designed for minimal scattered room-return to the assembly. For pulsed-source spectral measurements, we can easily minimize room-return by building a cubical test enclosure, measuring at least 8 ft on a side, of heavily borated paraffin. For experiments performed with a reactor source, where the test area is normally located in the shield or thermal column of the installation, it is more difficult to reduce the room-return neutrons to acceptable proportions. In any event, mapping the thermal and epithermal flux in the test geometry gives a measure of the extent to which a quasi-infinite-medium environment is realized.

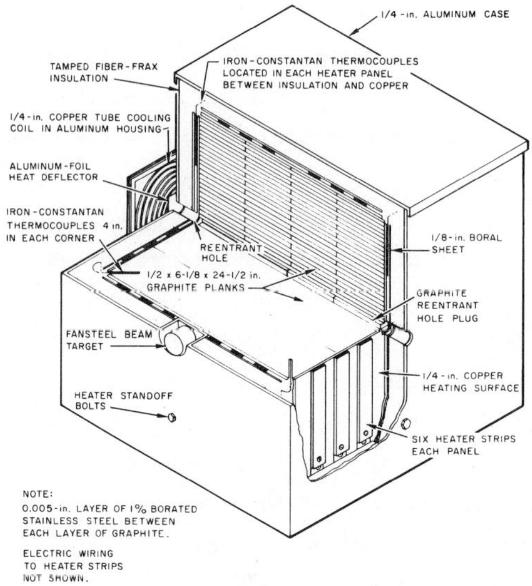

Figure 6.2 Cut-away view of a 2-ft-cube experimental graphite assembly showing the heating arrangement. (From Parks, Beyster, and Wikner, Nucl. Sci. Eng., Ref. 1.)

For most moderators, we want to measure the thermal-neutron spectrum at different temperatures. A graphite assembly that can be heated is shown in Fig. 6.2. The entire assembly is contained in a cubical aluminum case. The strip heaters are mounted on copper plates that press firmly against the poisoned graphite stack on all sides. Boral plates behind the heaters establish the boundaries of the assembly, and thermal insulation separates the aluminum case from the boral sheets. The strip heaters can heat the assembly to 800°K. Temperature profiles taken during the spectrum measurements indicate a temperature that is spatially uniform to within approximately ±5°C throughout the assembly. Thermal neutrons are extracted from a re-entrant tube terminating at the center of the assembly. The central location for the re-entrant tube in the assembly minimizes the influence on the neutron spectrum of even harmonics in the transverse flux distribution. A configuration similar to that shown in Fig. 6.2 was used for the study of a moderator of ZrH_2 powder poisoned with boron carbide.

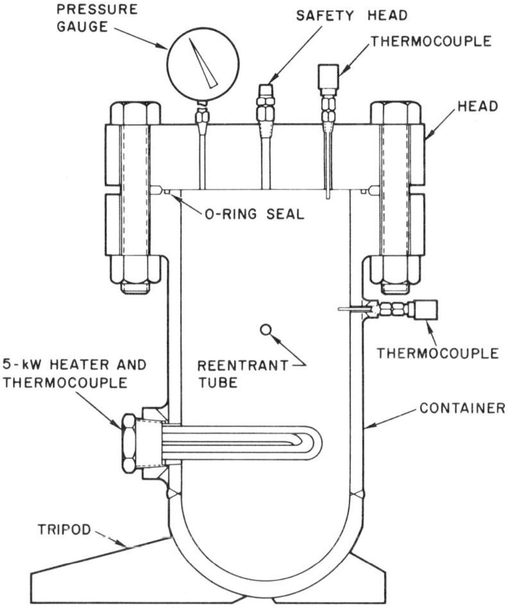

Figure 6.3 Schematic diagram of high-temperature pressure vessel for pulsed-source measurements in a homogeneously poisoned water moderator. (From Beyster et al., USAEC Report GA-3542, Ref. 2.)

Figure 6.3 is a diagram of a 2000-psi pressure vessel for making high-temperature measurements in a homogeneously poisoned water moderator by means of the pulsed-source technique. Investigators at the Knolls Atomic Power Laboratory Thermal Test Reactor[3] (TTR) have used chopper techniques to measure neutron spectra in a pressurized water-moderated metal fuel assembly.

Let us now consider the measurement of spectra in those cases where we wish to emphasize the spectral effects of geometry and anisotropic scattering. To facilitate theoretical analysis, it is important that we keep the geometry as simple as possible: we can simulate a one-dimensional, infinite-slab geometry by means of a rectangular parallelepiped slab, a cross section of which is shown in Fig. 6.4. If we excite the assembly by means of a point source of neutrons, we must be sure that the thermal-neutron spectrum in the assembly behaves very much as if the assembly were excited by a plane source. It is easy to verify that this is the case. Experimental studies of the effect on spectra of point sources placed outside an assembly show that the spectrum at one or two mean free paths into the assembly is little affected by source location, even if the source is close to the assembly.

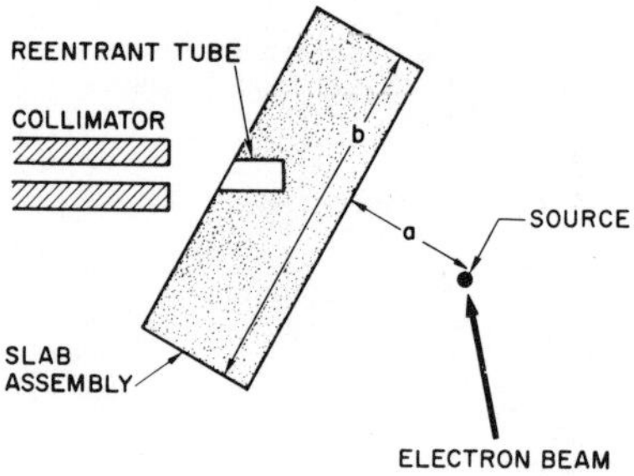

Figure 6.4 An arrangement for measuring neutron spectra in slab geometry.

Placing the external source in such a way that a = 0.3b, as shown in Fib. 6.4, gives an approximate cosine shape for the indium-resonance flux along all lines parallel to either of the two longer edges of the slab (transverse directions). A lead scattering wall between the fast-neutron point source and the assembly helps to establish the cosine flux shape in the transverse dimensions. If this flux shape does pertain, we can calculate the spatial dependence of the flux in the thin (axial) dimension of the slab by means of the one-dimensional transport equation, Eq. (4.88), with the transverse leakage rate appearing only as an additional absorption term.

To study the angular variation of $f(z,E,\mu)$ in slab assemblies,[a] it is convenient to be able to move the slab and the source without disturbing the remainder of the experimental assembly. (In general, the flight path and collimator are too bulky to be moved.) In pulsed-source experiments, for example, we can move the source and direct the electron beam to it as shown in Fig. 6.4. For stationary reactor sources, this is difficult to do.

Scattered room-return neutrons can significantly perturb one-dimensional angular-flux measurements, especially near slab surfaces that are remote from the neutron source. To measure this perturbation, we first observe the spectrum at the slab surface farthest from the source and then repeat the measurement with a thick absorbing shield adjacent to this surface. The shield prevents surrounding materials from scattering neutrons into the assembly. Another determination of room-return may be made by placing foils containing a resonance absorber at the surface of the assembly. For example, with cadmium-covered indium foils stacked against the surface the ratio of the indium-resonance activation in the foil next to the assembly to that in the outermost foil is a direct measure of the room-return background.[4]

There are a few recent experimental studies of neutron spectra in two dimensions. Neutron spectra in graphite-moderated reactor cells have been measured in the ZENITH lattice[5] and in the Calder lattice.[6] Bunch _et al._,[7] Mostovoi _et al._,[8] and Campbell _et al._[9] have studied spectra in two-dimensional water-moderated lattices. Unfortunately, considerable ambiguity exists in the interpretation of these experiments. In these complicated geometries it is useful to supplement spatially dependent spectral studies with extensive measurements of the spatial distribution of thermal and epithermal fluxes.

6-1.3 BEAM EXTRACTION. We must extract a beam of neutrons from the neighborhood of a point in an assembly in such a way that the spectrum of neutrons incident on the detector is identical with that

---

[a] For the convenience of the reader, we recall that $f(z,E,\mu)d\mu$ is the flux per unit energy of neutrons whose velocities make angles with the z-direction, the cosines of which lie between μ and $\mu + d\mu$.

which results from the angular or scalar flux that we intend to measure. To do this, it is necessary (1) to make negligible the spectral perturbations produced by the beam-extraction tube and (2) to properly orient the beam-extraction tube with respect to the assembly.

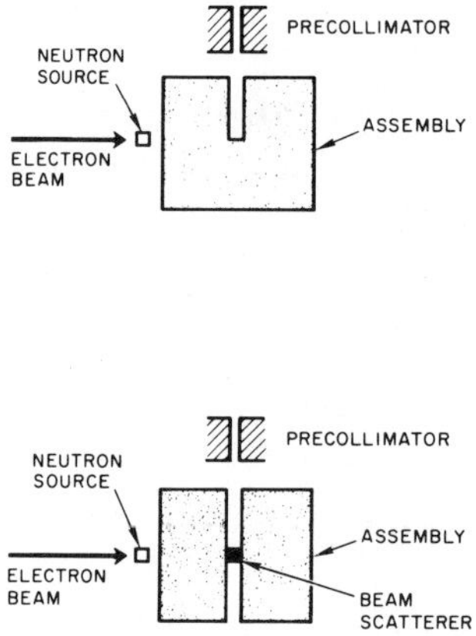

Figure 6.5 Experimental arrangements for measurement of scalar flux.

The re-entrant probe is one of the most commonly used devices for beam extraction. Figure 6.5a illustrates the use of a re-entrant probe to measure the scalar flux. External collimators are arranged so that the detectors intercept neutrons coming from the base of the re-entrant tube but not from its walls. To investigate the spectra in liquid moderators, we must construct the walls of the re-entrant tube so that they absorb very few neutrons; we can construct the walls so that they transmit at least 95% of a normally incident beam of thermal neutrons.

There is considerable latitude in the size of re-entrant holes that can be used to extract beams from homogeneous assemblies. Re-entrant tubes with cross sections as large as 2 in. by 2 in. give beams from honogeneously poisoned and unpoisoned water- and polyethylene-moderated assemblies without producing observable spectral perturbations. The time-of-flight spectra of neutrons extracted from a pure polyethylene assembly by means of 1/2 inc. by 1/2 inc. and by means of 2 in. by 2 in. re-entrant holes are compared in Fig. 6.6.

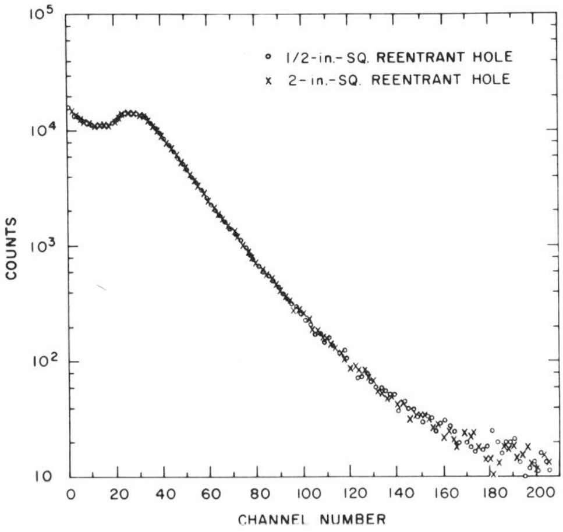

Figure 6.6 Time-of-flight neutron spectra for two sizes of re-entrant hole. (From Beyster et al., USAEC Report GA-2544, Ref. 10.)

Another method for investigating the effect of the size of the re-entrant hole consists in activating natural lutetium with neutrons.[11] For thermal neutrons, the 97%-abundant isotrope Lu^{175} is primarily a 1/v absorber, whereas the 2.6%-abundant isotope^{176} is primarily a resonance absorber (0.14 eV). The ratio of the two saturated activities is a spectral index of the low-energy spectrum (i.e., it is sensitive to the thermal-neutron spectrum). The ratios measured in an open 1-in.-square re-entrant hole and in a re-entrant hole filled with the assembly moderator (polyethylene) agree to within the experimental error of 1%. In heterogeneous assemblies the re-entrant tube must be considerably smaller than the cell size; if it is not, streaming effects would significantly perturb the spectrum and flux shape in the cell.

Even if the re-entrant tube does not significantly perturb the neutron spectra in the medium in which it is located, it is still necessary to orient the re-entrant tube properly. To measure scalar flux spectra we must orient the probe tube so that the observed angular flux is very nearly proportional to the scalar flux. Thus, if simple diffusion theory applies, the angular flux is given by

$$f(\mathbf{r},E,\boldsymbol{\Omega}) = \frac{1}{4\pi}\left[\phi(\mathbf{r},E) + 3\boldsymbol{\Omega}\cdot\mathbf{j}(\mathbf{r},E)\right] , \qquad (6.4)$$

where $\mathbf{j}(\mathbf{r},E) = D(E)\nabla\phi(\mathbf{r},E)$. In a direction such that $\boldsymbol{\Omega}\cdot\mathbf{j}$ is zero,

$f(\mathbf{r},E,\Omega)$ is nearly equal to the scalar flux. If $\Omega \cdot \mathbf{j}$ is not equal to zero, we can obtain ϕ from f and Eq. (6.4) only if $\mathbf{j}(\mathbf{r},E)$ can be accurately related to $\phi(\mathbf{r},E)$.

In general, for a particular orientation of the re-entrant tube, we measure the spectrum of the angular flux in a given direction. Since a long-flight-path collimating system designed to extract an essentially parallel neutron beam subtends the same solid angle everywhere in an assembly, it is easy to directly compare the spectra measured at different space points.

Johansson[12] and Honeck[13] have introduced a technique of beam extraction which enables us to measure scalar flux spectra directly. This technique consists in penetrating through an assembly with a through-tube, into which a small beam scatterer is placed (see Fig. 6.5b). We thus measure

$$\iint f(\mathbf{r},E',\Omega')\Sigma(E' \rightarrow E,\Omega' \rightarrow \Omega)dE'd\Omega' , \tag{6.5}$$

where $\Sigma(E' \rightarrow E,\Omega' \rightarrow \Omega)$ is the differential-energy-transfer cross section for the beam scatterer. If this technique is to give the scalar flux

$$\phi(\mathbf{r},E) = \int f(\mathbf{r},E,\Omega)d\Omega , \tag{6.6}$$

the scatterer must have

1) A very low absorption cross section,
2) A total scattering cross section Σ_s that is not a function of energy,
3) A small volume so that multiple scattering effects are negligible, and
4) A $\Sigma(E' \rightarrow E,\Omega' \rightarrow \Omega)$ equal to $1/4\pi[\Sigma_s\delta(E - E')]$, that is, the scattering must be isotropic and elastic in the laboratory system of coordinates.

When these conditions apply, the scatterer is called an ideal one. Zirconium is very nearly an ideal scatterer except at energies below about 0.015 eV, where pronounced coherent effects make the scattering very anisotropic.

There exist a number of tests of the reliability of the beam-scatterer technique.[2] The beam-scatterer technique and the standard re-entrant-tube-extraction techniques have given the same result to within the experimental errors for the case where no flux gradient exists. In another test, scatterers (lead and zirconium) placed at the position of a strong flux gradient in a poisoned polyethylene assembly gave the same neutron spectrum. The results of these measurements are given in Fig. 6.7. To within a few percent, the same spectrum was observed from 10 eV to 0.02 eV. Empirically, therefore, except for energies less than about 0.02 eV, we can be confident of spectral data obtained by the beam-scatterer technique. Since Pb^{206} appears to be a nearly ideal scatterer up to 100-keV neutron energy, this technique may be very useful even for high-energy beam extraction.

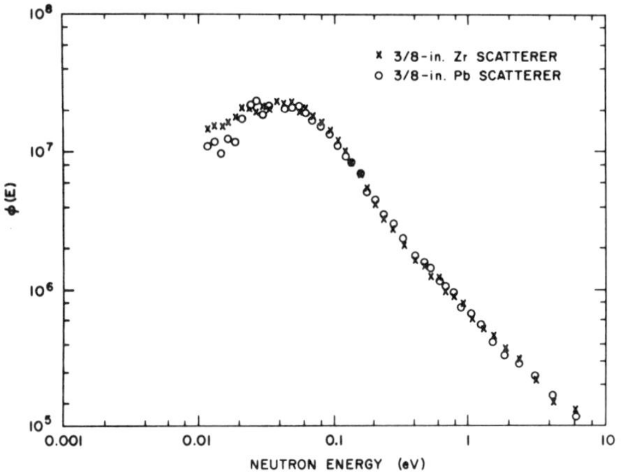

Figure 6.7 Comparison of spectra measured with a zirconium and with a lead scatterer. (From Beyster et al., USAEC Report GA-3542, Ref. 2.)

Since we can readily move the small beam scatterer from place to place in the through-tube, it is easy to measure the spatial dependence of the scalar flux spectrum. Since the collimator must not view the walls of the through-tube, however, collimation criteria are more severe than with the re-entrant-hole studies, and the through-tube should be backed up outside the assembly with a B^{10} plug to eliminate extraneous room-background that might leak into the tube. Further, a long tube through a thick assembly may not give an accurate representation of the flux because of possible large streaming from regions of high flux to regions of low flux.

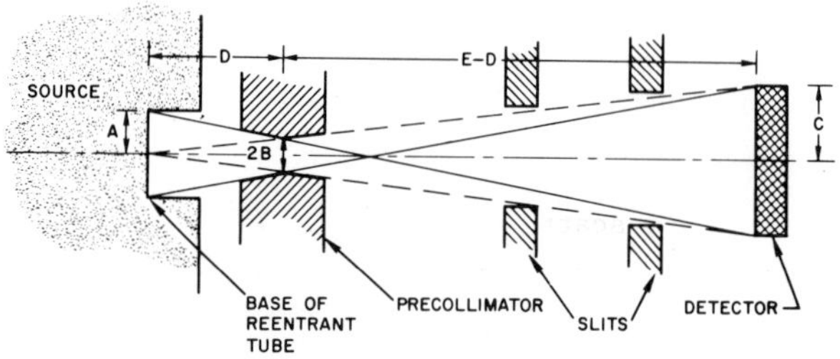

Figure 6.8 Ray diagram for a typical flight-path collimation system.

The flight-path collimation system is a critical element in the design of spectrum experiments. The detailed design characteristics depend on the experiment to be performed, but all experimental arrangements have certain common features. Since one is viewing a surface with a source strength equal to an extremely small fraction of that of the fast pulsed source, background problems can be serious. In general, we require a heavily shielded flight path in order to exclude background neutrons. The main collimation of the neutron beam usually occurs near the assembly in a B^{10} or B_4C precollimator. A ray diagram for a typical precollimator is shown in Fig. 6.8. The precollimator is situated so that the detector views only the desired surface area of the source. This position is calculated from

$$D = \frac{(A - B)E}{A + C} . \tag{6.7}$$

The neutrons issuing from the edges of the re-entrant tube base just see the detector bank. The design of the precollimator aperture is not as critical as one might expect. Long cylindrical precollimators appear to work as well as the tailored one illustrated in Fig. 6.8. In addition to the precollimator, it is often desirable to utilize oversized slits throughout the flight path so as not to detect neutrons scattered into the main collimator or from its walls.

Many materials are used in the fabrication of collimators. Often, heavily borated paraffin or polyethylene covers the main flight path. Very dense B_4C can be used to form the slits. The precollimator normally consists of B^{10} or B_4C; however, brass or iron precollimators do not distort the spectrum in many experiments with good geometry. In addition, lead shot is often put into precollimators in order to suppress gamma flash at the neutron detectors.

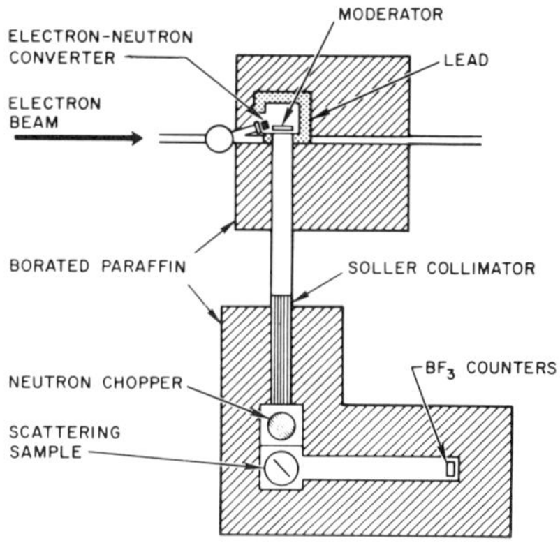

Figure 6.9 Schematic drawing of a neutron scattering experiment showing the use of a Soller collimator. (From Whittemore, USAEC Report GA-3409, Ref. 14.)

The Soller slit shown in Fig. 6.9 is a collimator system often used in chopper experiments.[14,15] It guarantees that a beam of neutrons with very small angular divergence is incident on the chopper. Because of its many openings, it also gives relatively high neutron transmission. Soller slits, although they have never been used in pulsed-source spectral investigations, may find several applications, especially in studies of spectra in heterogeneous arrays. It seems particularly promising to utilize this type of slit system for measurements of the average spectrum in a reactor cell.

Finally, for low-energy spectral studies we must eliminate neutron absorption in the air contained in the flight path. In doing this, we also eliminate scattering by the water vapor in the air. The use of helium- or argon-filled flight paths provides a practical method for avoiding these difficulties.

6-1.4 NEUTRON DETECTION. Thermal-neutron detection is based on secondary effects resulting from a neutron interaction. The absorption of a neutron by a nuclide may result in the prompt emission of a fast charged particle and/or the formation of a radioactive nuclide. For time-of-flight spectral studies we must employ the former of these reactions. Description of the principles involved in the construction and operation of neutron detectors based on the prompt emission of particles can be found elsewhere.[16] For our purposes it suffices to mention that the reactions $B^{10}(n,\alpha)Li^7$, $He^3(n,p)H^3$, $Li^6(n,\alpha)H^3$, and $U^{235}(n,f)$ occur promptly and with high probability in thermal-neutron fluxes.

A neutron detection system for low-energy spectrum measurements should have the following properties:

1) High efficiency for all energies less than 10 to 100 eV.
2) Long-term stability against electronic gain and temperature variation.
3) Large-diameter capability.
4) Short length.
5) Minimum dead space.
6) Proper wall materials.
7) An energy sensitivity function that varies smoothly with neutron energy.
8) Stability under irradiation.
9) A capability for easy calibration.

There are a large number of detection systems that satisfy all or many of the above criteria, depending on the exact application. These systems can be divided naturally into (1) gas counters and (2) scintillators. For most time-of-flight investigations of neutron spectra, gas counters are used, but in many other thermal-neutron problems scintillation detectors are essential. This is particularly true in those applications where the length of the gas-filled proportional counter limits the time and energy resolution. The most common proportional counters are filled with either BF_3 or He^3. They may be obtained commercially in a wide variety of sizes and shapes. Scintillation detectors for thermal neutrons can be prepared by including the neutron absorber in special scintillating glasses, dispersing it in scintillating plastic, or combining it chemically in an inorganic scintillator such as LiI.[17] The resulting detectors invariably possess some sensitivity to gamma radiation. Usually, a means of minimizing or eliminating this effect must be found. Techniques such as pulse-shape discrimination are proving to be very successful.

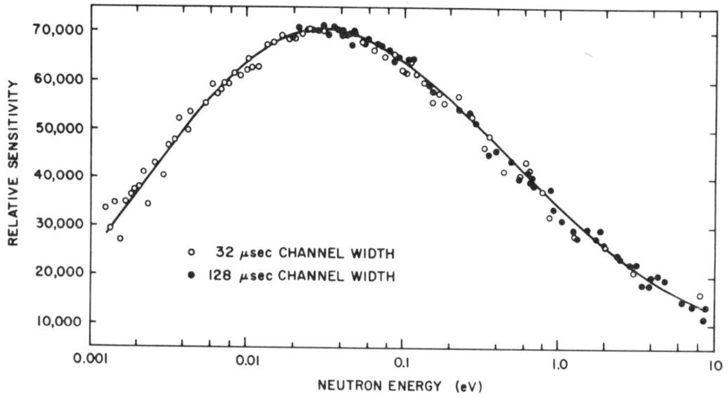

Figure 6.10 Relative energy sensitivity s(E) of the neutron detector bank measured for two time-of-flight channel widths. (From Beyster et al., USAEC Report GA-2544, Ref. 10.)

The high-pressure BF_3 counter is the most commonly used detector for thermal-neutron spectral studies. For most spectral experiments the efficiency of a low-pressure 1/v detector system is too low. Normally, 1-in.- or 2-in.-diameter tubes filled to one or two atmospheres with $B^{10}F_3$ (93% enriched) give an excellent voltage plateau and are essentially black at thermal energies. A stack of six to twelve such tubes arranged coaxially with the incident neutron beam yield a counting efficiency of 10%, even at 10 eV. Figure 6.10 gives the relative energy energy-dependent sensitivity of one such detector bank. In addition, BF_3 detectors have been arranged in banks to give large sensitive areas. A bank measuring three feet in diameter is common.

Wall materials and dead space can strongly affect the sensitivity of a detector bank. To be able to determine the energy-dependent detector sensitivity accurately, we must avoid even small insensitive regions in the detector. For a BF_3 detector an insensitive region is one in which the assembly gas is present but, because of a lack of sufficient collecting potential, all the alpha-pulses from capture events are not detected.

BF_3 detectors, although quite gamma-insensitive, are still subject to gamma-induced pile-up effects. Intense gamma irradiation also breaks down counter gas, poisoning the counting mixture. Also, in some detectors intense ion bombardment of walls can free occluded gas. These effects usually destroy or shift the counter operating plateau. Counters with walls made of aluminum are more susceptible to these effects than are counters made of OFHC outgassed copper. Nevertheless, carefully designed BF_3 detector systems can be used for spectral investigations in almost any practical environment.

It is usually difficult to calculate the sensitivity of high-pressure enriched BF_3 systems because even the best detectors do not have ideal geometries (for example, small dead spaces occur near insulators). More frequently, high-efficiency detectors are calibrated experimentally. Such a calibration can be made by using the detector to measure a known spectrum. For example, geometric conditions can be set up both in water and in graphite so as to obtain a Maxwellian thermal-neutron spectrum $f_M(v)$, at least for energies less than about 0.3 eV. The sensitivity function $\eta(v) = s(E)$ is then related to the observed velocity distribution $f_{obs}(v)$ by

$$\eta(v) = \frac{f_{obs}(v)}{f_M(v)} . \tag{6.8}$$

A second method for calibrating an efficient BF_3 counter consists in comparing the spectrum measured by this detector with that observed by a black detector located subsequently in the same spectrum. A thick sample of B^{10} constitutes a flat-response neutron detector that is black to at least 20 eV. With the sample in the neutron beam, the 0.511-keV capture gamma-rays from the (n,γ) reaction in boron are counted with a sodium iodide scintillation detector that is placed as close as possible to the sample in order to achieve maximum over-all gamma-detection efficiency. The pulse-height window is adjusted to bracket only the 0.511-keV capture radiation from boron. For intensity reasons, it is usually out of the question to use the B^{10}-NaI system directly in spectral work. Although its response is known, the absolute over-all efficiency of the B^{10}-NaI system is at best 20%, because of poor efficiency in the collection of capture gamma-rays.

Another way to produce a flat-response detector in the thermal energy region is to use a Li^6 glass scintillation detector. It has a very nearly black response below 1 eV and a calculable response above that energy.

It is possible, in principle, to calibrate thermal-neutron detectors by measuring their neutron transmission T(E) in a typical flight path. The efficiency is then

$$s(E) = [1 - T(E)] . \qquad (6.9)$$

There are at least two reasons why this method does not yield the precision desired for thermalization studies. Neutron detectors contain appreciable quantities of materials other than those producing the desired nuclear reaction. These materials may scatter and absorb neutrons. Thus they may affect the result of a transmission measurement appreciably but not necessarily contribute to the effective sensitivity. Further, for high energies, where the transmission is almost unity, the efficiency is inaccurately determined as the difference between two nearly equal numbers.

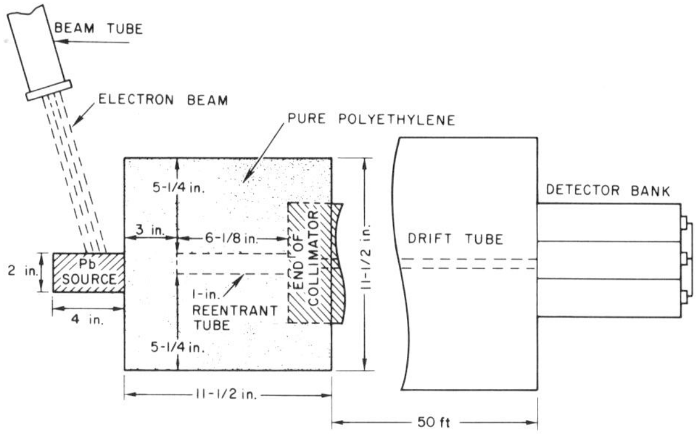

Figure 6.11 Standard polyethylene geometry for measurement of the relative energy sensitivity of the neutron detector bank. (Beyster et al., USAEC Report GA-2544, Ref. 10.)

Finally, we consider detector calibration accomplished by comparing the response of an efficient detector bank with that of a 1/v detector[18] (low-pressure, unenriched BF_3 counter). In this experiment, it is important that the neutrons entering each of the efficient detectors in a large bank have the same spectrum. Figure 6.11 shows the geometry of a polyethylene-moderated assembly that accomplishes this objective.

In general, care in the measurement of the sensitivity function makes it possible to establish the relative energy-dependent sensitivity of a detector system to within ±3% over 5 decades of energy, i.e., from 0.001 eV to 100 eV.

6-1.5 RESOLUTION. In time-of-flight measurements, the imperfect resolution produces a distortion of the time-of-flight spectrum N(t). The imperfect resolution results from the uncertainties in the flight-path length, the finite width of the time-analyzer channels, and from the imprecise determination of the instant when neutrons begin their traversal of the flight path. In the case of a pulsed source, this lack of precision in the definition of a zero time comes from the finite time that neutrons spend in the moderating assembly, whereas in the case of a chopped beam it comes from the finite time-width of the burst produced by the chopper. In Secs. 6-2 and 6-3, respectively, we give a description of the spectral distortions in pulsed-source and chopped-beam experiments.

A rough estimate of the spectral distortion corresponding to a finite uncertainty ΔT in the neutron flight time is given by the difference between the average value of N(t) over the interval ΔT and its value $N(T_0)$ at the center T_0 of the interval. Dividing this difference by $N(T_0)$ gives the relative error

$$\delta = \frac{\dfrac{1}{\Delta T}\displaystyle\int_{T_0-(\Delta T/2)}^{T_0+(\Delta T/2)} N(t)\,dt - N(T_0)}{N(T_0)} .$$

$$\approx \frac{1}{24}\,\frac{(\Delta T)^2}{N(T_0)}\,\frac{d^2N(t)}{dt^2}\bigg|_{t=T_0} \tag{6.10}$$

(Actually, this is the spectral distortion that obtains in the case where the dominant contribution to ΔT comes from the width of the time-analyzer channels, assuming that the detector is equally sensitive to neutrons of all energies that are counted in this time interval. Equation (6.10) is not a rigorously correct expression for the spectral distortion arising from other sources of uncertainty in the neutron time-of-flight.)

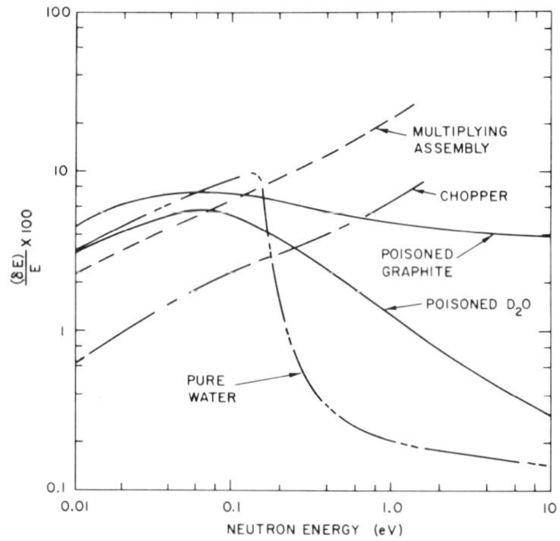

Figure 6.12 Energy resolution used in spectral measurements. (From Beyster, in USAEC Report BNL-719, Ref. 19.)

The quantity δ should be at least as small as the uncertainty in the relative energy sensitivity of the detector, which we saw in the previous section to be of the order of a few percent. Usually, the above correction can be made rather small both for the pulsed-source and chopped-beam time-of-flight experiments by proper experimental design. The resolution shown in Fig. 6.12, which is typical for both types of experiments, has not produced significant spectral distortions below a neutron energy of 1 eV.

By definition, imperfect resolution is equivalent to an uncertainty in the energy, or time-of-flight, of a detected neutron. Some experimental factors that give imperfect resolution for pulsed-source experiments are the pulse-width of the accelerator and the finite time that neutrons spend in the moderating assembly. The chopper's speed of rotation and the window width and diameter, as well as the finite detector length, have a significant effect on resolution in experiments involving steady sources.

The finite diameter d_r of the chopper rotor introduces an uncertainty in the neutron flight time: $\delta t_f = t_f(d_r/d)$. The time spent in the assembly by original-source neutrons and by other fission-generated neutrons plays no part in the determination of resolution. For this reason, it has been natural to adopt chopped-beam techniques for measuring spectra in multiplying assemblies. On the other hand, the average transmission of choppers is quite low (10^{-5}). We thus expect a low counting rate of detected neutrons in high-resolution experiments. With the source strengths now available, however, Poole[20] concludes that it is practical to utilize resolutions of 1.5 to 5 μsec per meter for chopper work. More detailed consideration of chopper resolution appears in Sec. 6-3.

The ultimate limitation on the resolution in pulsed-source spectral studies is usually set by the average time $t_a(E)$ spent by neutrons of energy E in the moderating assembly. The slowing-down time and thermal lifetime of neutrons in moderators broaden the neutron burst, thereby limiting experimental energy resolution. This is especially true for multiplying assemblies. We can nevertheless study fueled assemblies provided we use a long flight path and maintain a short generation time by heavily poisoning the assembly. Depending on the extent to which the moderator can be poisoned, resolution from 0.2 to 20 μsec per meter is readily achievable in the thermal energy region in pulsed-source experiments. We further discuss the inherent resolution limitations in pulsed-source experiments in Sec. 6-2.

In both pulsed-source and chopper experiments the absence of a unique distance into the detector at which all neutron absorptions occur results in an imprecise determination of the neutron energy. If the detector is aligned coaxially with the incident neutron beam, the mean distance into the detector at which neutrons of energy E are absorbed is

$$\bar{X}(E) = \frac{\int_0^L x e^{-x/\lambda(E)} dx}{\int_0^L e^{-x/\lambda(E)} dx} = \lambda(E) - L\left[e^{L/\lambda(E)} - 1\right]^{-1}, \qquad (6.11)$$

where L is the length of the detector and $\lambda(E)$ is the energy-dependent absorption mean free path of neutrons in the detector. We can include the distance $\bar{X}(E)$ in the length d of the flight path, but there is still a distribution of absorptions around $\bar{X}(E)$ which introduces an uncertainty δd in d. For a low-efficiency detector, $\delta d = L$; for a thick counter, $\delta d = \lambda(E)$.

The energy resolution arising from the finite channel width and from the uncertainties in flight-path length and in the time at which flight begins is

$$\frac{\delta E}{E} \cong \left[\left(\frac{2\delta t_f}{t_f}\right)^2 + \left(\frac{2\delta d}{d}\right)^2 + \left(\frac{2\delta t_c}{t_f}\right)^2\right]^{\frac{1}{2}}, \qquad (6.12)$$

where δt_c is the width of the analyzer time channel corresponding to the flight time $t_f(E)$. For the pulsed-source experiment, $\delta t_f = t_a(E)$, and for the chopper experiment $\delta t_f = t_f(d_r/d)$.

From Eq. (6.12) it is clear that we do not improve the resolution very much by making the analyzer channel width δt_c much less than the chopper rotor width, or the accelerator or moderator pulse width. Usually the δt_c of the analyzer is made equal to δt_f.

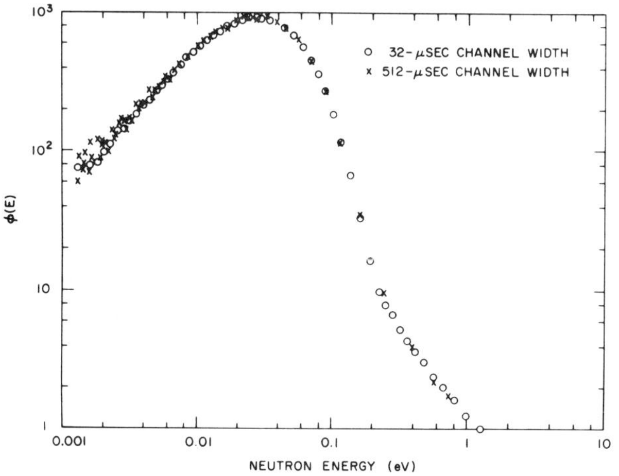

Figure 6.13 Neutron-spectrum measurements in unpoisoned polyethylene for two different analyzer channel widths. (From Beyster et al., USAEC Report GA-2544, Ref. 10.)

We can often determine experimentally the adequacy of the resolution used in a given measurement. Figure 6.13 shows a thermal-neutron spectrum measured with two very different analyzer channel widths. In this case, a 512-μsec channel width permits an accurate measurement of the spectrum. By changing the chopper rotor speed, the accelerator pulse width, etc., we can also determine empirically the importance of the resolution effects arising from these various factors.

Unfortunately, most time analyzers do not permit the selection of more than one channel width. Thus, for example, if δt_c is chosen to give good resolution at neutron energies of the order of 1 eV, the resolution at 0.01 eV may be ten time better than that needed. In practice, it is common to use a large (1000-channel) analyzer, which permits the measurement of an entire thermal-neutron spectrum in one experiment, and then to obtain the desired resolution and statistical counting accuracy by grouping channels together in the data reduction process.

In many experiments it is useful to employ more than one time analyzer. For example, one analyzer is arranged to measure the entire spectrum from 0.005 eV to 2.0 eV with good resolution, while a second is set for the particularly high resolution required to resolve the spectral effects of thermal-energy resonances.

High-resolution experiments, which require a large number of time channels and perhaps more than one analyzer, result in a great amount of experimental data. The reduction of this data to accurate and meaningful experimental results in turn requires a rather elaborate automatic data processing system. The requirements mentioned here apply not only for pulsed-source experiments but also for the chopped-beam measurements.

6-1.6 NEUTRON SOURCES. Neutrons that excite a moderating assembly in which the thermal spectrum is to be measured can be obtained from artificial sources or reactors, or they can be produced indirectly by means of positive-ion or electron accelerators. With the possible exception of neutrons coming directly from reactors, practically all of these source neutrons are produced within or near the moderating assembly with energies near to or higher than those of the fission spectrum. These fast neutrons are moderated and become slow neutrons.

The spectrum of the slow neutrons is, for all practical purposes, completely insensitive to the spectrum of the fast source neutrons. Consequently, the rate of production of fast neutrons and not their spectrum is the essential factor involved in the selection of a neutron source for studies of thermal spectra.

The strength of a fast-neutron source necessary for chopper spectral studies is of the order of 10^{10} neutrons/sec or greater if a viewing area of a few square centimeters is used. For thermal spectral studies by the pulsed-source technique, peak source intensities during the burst of 10^{15} neutrons/sec or greater are required for comparable source areas. It is therefore clear that artificial neutrons sources are not strong enough for definitive spectral studies. Radium beryllium, polonium beryllium, plutonium beryllium, antimony beryllium, and other (γ,n) sources yield less than 10^8 neutrons/sec.

The positive-ion accelerator most commonly used to produce neutrons is the low-voltage deuteron accelerator (usually a Cockcroft-Walton machine). Machines of this class have been used in time-of-flight measurements of spectra in water-moderated system. They utilize either the $T^3(d,n)He^4$ of $D(d,n)He^3$ reaction, both of which are exothermic. Charged-particle bombarding energies of about 150 keV or greater are required to give appreciable yields. Steady-state neutron yields given by Cockcroft-Walton accelerators are around 10^{11} neutrons/sec. A typical 150-keV installation is shown in Fig. 6.14. A particularly perplexing experimental problem that arises in attempting to optimize yields from (d,d) and (d,t) sources is that a short target life results from the limitation of power dissipation to the target (normally zirconium or titanium deuteride or tritide).

With positive-ion accelerators, pulsing can be accomplished by either postacceleration deflection or terminal pulsing of the beam. It is easy to obtain pulse widths and repetition rates ranging from 2×10^{-3} to 10^3 μsec and from 1 pulse/sec to 10^6 pulses/sec, respectively. Terminal pulsing is particularly useful since background sources of neutrons formed at the final slit system of the accelerator are eliminated. In addition, terminal pulsers generally accommodate higher instantaneous currents than do steady-state ion sources.

A second class of positive-ion accelerators includes those which produce charged particles with energies in the meV range. Of these, the

Figure 6.14 A typical 150-keV neutron generator. (From Beyster et al., USAEC Report GA-3542, Ref. 2.)

van de Graff and Dynamitron are most easily adapted to spectral investigations. For ion currents of 1 ma, neutron yields from the exothermic $Be^9(d,n)B^{10}$ reaction at 3 meV deuteron energy are in excess of 10^{11} neutrons/sec. These yields are adequate for thermalization studies and are somewhat greater than those attainable from the frequently used (d,d) and (d,t) reactions. Also, a greater source stability obtains for the $Be^9(d,n)B^{10}$ reaction.

Electron linear accelerators can provide one of the most intense pulsed sources of neutrons. Electrons, in the field of the heavy target nucleus, convert about half of their energy into high-energy X-radiation. This in turn produces neutrons by means of (γ,n), (γ,pn), (γ,fission), (γ,2n), and other reactions in the target. Although the residual gamma-radiation is highly directional, the neutrons have a nearly isotropic distribution. Yields for lead and uranium targets of various thicknesses are shown in Fig. 6.15. The target thicknesses in integral radiation lengths are indicated by the Roman numeral appearing after the symbol for the element.

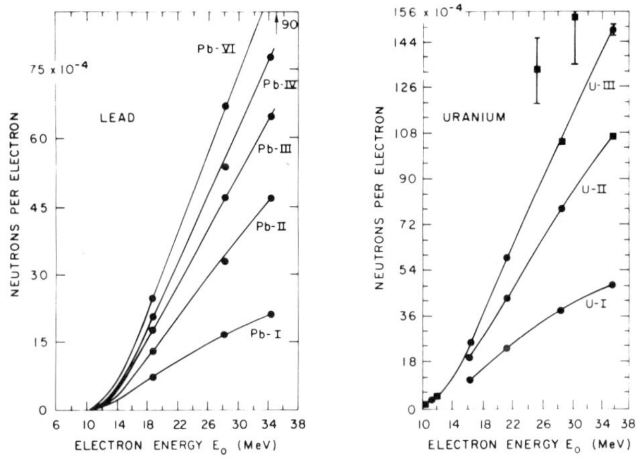

Figure 6.15 Yields of neutrons per incident electron as a function of initial electron energy for targets of various thicknesses: (a) lead targets; here the number 90 refers to the value of the VI curve at the indicated energy; (b) uranium targets; here the results of calculations by Biram are shown by solid squares. (From Barber and George, Phys. Rev., Ref. 21.)

The experimental flexibility accruing to the General Atomic linear accelerator installation, which is used for neutron spectral studies, may be seen by the over-all facility plan shown in Fig. 6.16. This and similar accelerators have been used in both pulsed-source and chopped-beam time-of-flight studies of neutron spectra. Reactor sources are also frequently used in the study of thermal-neutron spectra.

There are many ways to introduce the neutrons from a reactor into an experimental assembly in which we desire to measure the spectrum. For example we may excite the assembly with

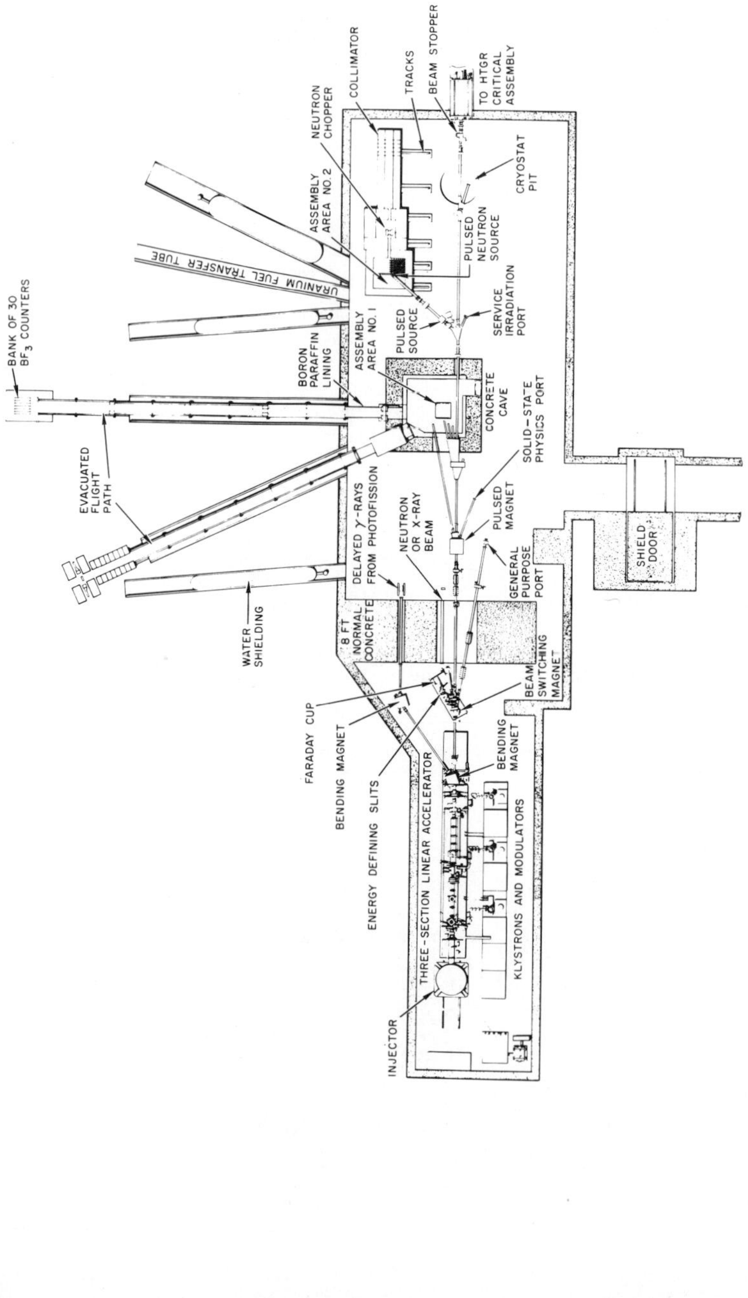

Figure 6.16. General Atomic electron linear accelerator experimental facility.

(a) neutrons leaking from a reactor core,

(b) fast neutrons produced in a fission plate placed at the interface between the assembly and a reactor thermal column,

(c) thermal neutrons from a reactor thermal column, or

(d) neutrons from a repeatedly pulsed fast reactor.[22]

We must use the chopped-beam technique to measure spectra in assemblies excited by the methods (a), (b), and (c), but not in the case of excitation by method (d).

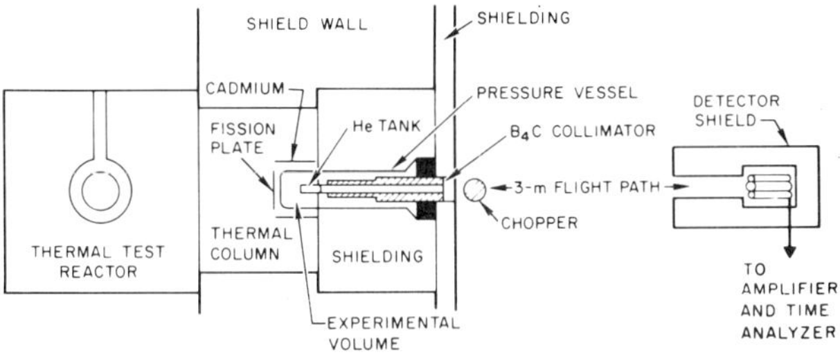

Figure 6.17 Experimental arrangement used to study spectra excited by means of a fission plate placed at the end of a thermal column. (From Stone and Slovacek, Nucl. Sci. Eng., Ref. 3.)

Methods (a) and (b) are the ones most commonly used to excite an experimental assembly. Slovacek[3] has utilized both techniques at the Thermal Test Reactor (TTR) installation of the Knolls Atomic Power Laboratory (KAPL). Figure 6.17 illustrates schematically the KAPL experimental arrangement used to study spectra excited by means of a fission plate placed at the end of a thermal column. For this method to give a fast-source strength sufficient for high-resolution studies of thermal-neutron spectra, approximately 10^{10} thermal neutrons/sec must be incident on the fission plate. Most present-day research reactors are capable of providing this required number of neutrons.

In method (a), the reactor feeds slow and epithermal, as well as fast neutrons, into the assembly in which the spectrum is to be measured. The thermal and epithermal neutrons from the reactor may result in a large rate of leakage of thermal neutrons at the point where we want to measure the thermal spectrum. In addition, the source feeding the thermal energy range may not be separable in the manner indicated in Eq. (5.163). Thus, the two conditions cited in Sec. 6-1.2 for simulating the spectrum in an infinite reactor may not be satisfied. If, however, the assembly is poisoned so that only neutrons with energies greater than the thermal cutoff energy can migrate very far, then the aforementioned conditions are satisfied.

It is feasible to use method (d) for pulsed-source time-of-flight measurements of thermal-neutron spectra. The Dubna fast burst reactor (IBR) is one that provides enough neutrons for such measurements.[22] The pulse width is 36 μsec and the peak neutron yield is ~ 10^{17} neutrons/sec. The roughly 4×10^{12} neutrons/pulse level is greater than can be achieved with most accelerator pulsed sources. With this source intensity one can use a 50-meter flight path and measure spectra with less than 10 percent resolution to energies as high as 10 eV.

6-2 PULSED-SOURCE MEASUREMENTS OF NEUTRON SPECTRA

6-2.1 GENERAL CONSIDERATIONS. As reported early in 1954 by Sykes,[23] P. A. Egelstaff at Harwell suggested "that it might be possible to examine the thermal neutron spectrum in water by a time-of-flight method." In 1955, M. J. Poole introduced the pulsed-source method for measuring thermal-neutron spectra. He tested the method on a 12-meter flight path, using the 13-meV Harwell electron linear accelerator to generate the fast-neutron source. Sykes showed that the spectrum measured by this method is equivalent to a steady-state spectrum.[23] Qualitatively, this follows since the only individual neutron histories that are possible for a steady-state source $S(\mathbf{r},E,\Omega)$ are possible for a pulsed source $S(\mathbf{r},E,\Omega)\delta(t)$, and conversely. More quantitatively, we can prove that the pulsed-source method described in Sec. 6-1.1 does indeed yield the desired steady-state spectrum by considering the time-dependent transport equation (4.42) for the flux $f(\mathbf{r},E,\Omega,t)$. In a pulsed-source experiment with perfect resolution, each data channel (time channel) corresponds to one and only one energy and collects counts from all neutrons of this energy. We thus measure

$$\int_0^\infty f(\mathbf{r},E,\Omega,t)dt = F(\mathbf{r},E,\Omega) \quad . \tag{6.13}$$

Integrating Eq. (4.42) over all t and utilizing the fact that f vanishes at t = 0 and at t = ∞, we find that F satisfies

$$\boldsymbol{\Omega} \cdot \nabla F(\mathbf{r},E,\boldsymbol{\Omega}) + [\Sigma_a(E) + \Sigma_s(E)]F(\mathbf{r},E,\boldsymbol{\Omega}) =$$

$$= \int \Sigma(\mathbf{r},E' \rightarrow E,\boldsymbol{\Omega}' \rightarrow \boldsymbol{\Omega})F(\mathbf{r},E',\boldsymbol{\Omega}')dE'd\boldsymbol{\Omega}' + S(\mathbf{r},E,\boldsymbol{\Omega}) \ , \qquad (6.14)$$

where

$$S(\mathbf{r},E,\boldsymbol{\Omega}) = \int_0^\infty S(\mathbf{r},E,\boldsymbol{\Omega};\ t)dt \ .$$

This is the equation for the steady-state angular flux spectrum excited by the time-independent source $S(\mathbf{r},E,\boldsymbol{\Omega})$. Finally, since the boundary conditions on F follow from Eq. (6.13), F is the angular flux spectrum induced by the source $S(\mathbf{r},E,\boldsymbol{\Omega})$.

The main requirement that we must satisfy in pulsed-source experiments is that the mean time after the accelerator pulse for the emission of neutrons from the assembly be much less than the flight time $t_f(E)$. If this requirement is satisfied, it is reasonable to correct $t_f(E)$ for the mean time $t_a(E)$ that neutrons of energy E spend in the moderating assembly.

In Sec. 6-1.5 we mentioned the spectral distortion resulting from imperfect resolution, which in the present context is associated with finite mean emission times. In the pulsed-source technique, the extent of the spectral distortion (as roughly estimated from Eq. (6.10), for example) produced by the finite emission times depends strongly on the energy dependence of the flux being measured. Generalizations of what moderators can or cannot be studied by this technique tend to be misleading. We can study all moderators by the pulsed-source method; however, the concentration of neutron absorbers required to give a sufficiently small $t_a(E)$ depends strongly on the moderating material. The situation is further complicated in pulsed-source measurements of spectra in multiplying assemblies, where given neutrons and their progeny reside for long times before escaping or being absorbed. In any case, a reasonable guide to experimental design is to limit the die-away time of 0.025-eV neutrons in an assembly to roughly 2 percent of their flight time.

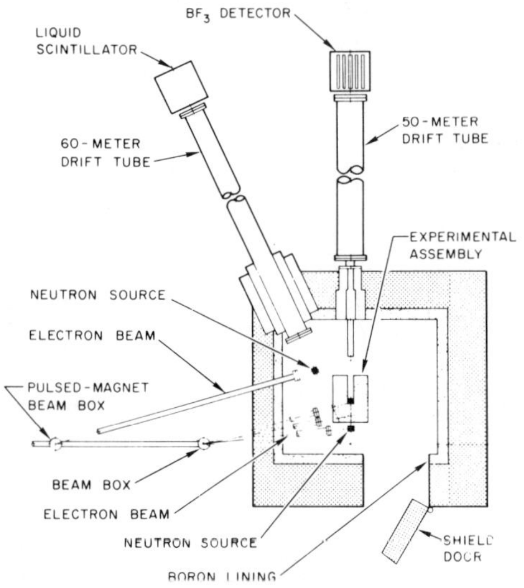

Figure 6.18 Schematic diagram of the low-room-return neutron cave. (From Parks, Beyster, and Wikner, Nucl. Sci. Eng., Ref. 1.)

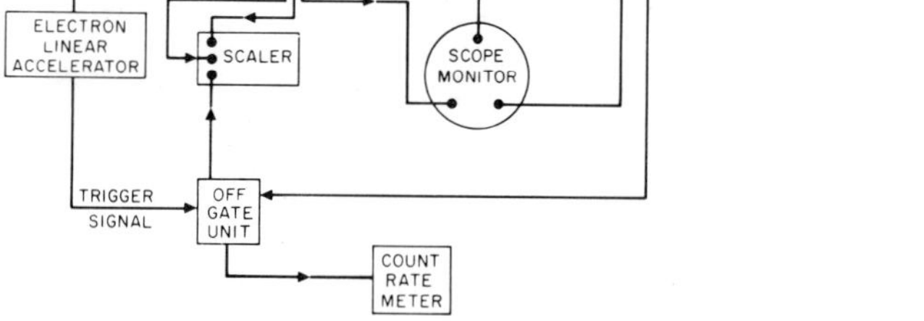

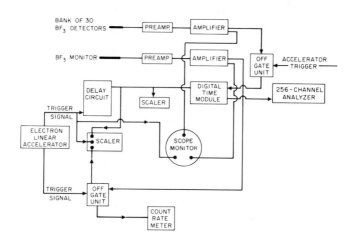

Figure 6.19 Instrumentation for pulsed-beam time-of-flight measurements. (From Beyster et al., USAEC Report GA-2544, Ref. 10.)

The general layout of a typical pulsed-source spectrum experiment using a linear accelerator is shown in Fig. 6.18. The experiment is performed in a heavily shielded, low-room-return neutron cave whose walls, floor, and ceiling are loaded with boron. To prevent neutrons from leaking into the flight-path system, it is heavily shielded on the sides with borated paraffin. Figure 6.19 is a block diagram of the electronics system required to collect the experimental data. A master trigger from the linear accelerator turns off all counting and monitoring channels during the accelerator gamma-flash pulse. A master timing signal, however, is obtained from a photomultiplier-phosphor combination located near the fast source. This signal, after a suitable delay, activates the time analyzers.

The source can be monitored by any one of several methods. For example, it may be viewed from another flight path containing a small, standard moderating assembly to spread out the detection of source neutrons over the resolvling time of the experimental electronics. By far the best procedure for monitoring is to activate thermal-neutron-absorbing foils located in the assembly near the point of measurement.

The accelerator-pulse repetition rate must be adjusted so that thermal neutrons from one pulse do not arrive at the detector during or after the initiation of the next pulse. Neutrons with energies lower than 0.001 eV usually need not be considered in establishing the pulse repetition rate. The width of the accelerator pulse is not an important factor as long as it is in the microsecond range inasmuch as resolution in pulsed-source experiments is, in general, dominated by the analyzer channel width and the residence time of neutrons in the moderating assembly.

In the remainder of Sec. 6-2 we discuss in more detail the sources of error usually encountered in pulsed-source spectral measurements and the methods for correcting them. Over-all precision in pulsed measurements can be extremely high. Resolution, counting loss, and background effects are not in general the major factors that limit this precision. The major determinant of experimental accuracy for most pulsed-source experiments is the relative precision of the determination of detector sensitivity.

6-2.2 DATA CORRECTIONS. In a pulsed-source measurement, the relative neutron flux spectrum $\phi(E_i)$ is related to the measured time-of-flight spectrum by[b]

$$\phi(E_i) = \frac{N(T_i)}{\Delta E_i T_f(E_i) s(E_i)} , \tag{6.15}$$

where $N(T_i)$ is the counting rate, corrected for counting losses, in the time channel i, ΔE_i is the channel width in energy units, $T_f(E_i)$ is the flight-path transmission, and $s(E_i)$ is the energy-dependent sensitivity of the detector (discussed in Sec. 6-1.4). The counting-loss correction for $N(T_i)$ depends on the time-analyzer logic. We emphasize the importance of utilizing an analyzer system with simple logic, inasmuch as counting rates in pulsed-source spectrum experiments are high and good experimental accuracy requires that we understand this logic. With currently available time analyzers, counting losses can be kept to within 10 to 20 percent while accepting instantaneous counting rates of 10 to 10^6 counts/sec. We discuss below three systems that are currently used for spectral measurements and the counting-loss correction required for each.

(a) A-1 Logic (One Count per Fixed Time Interval). The first instrument we consider consists of a fast clock memory address generator and typical analyzer core storage.[24] The time interval that usually limits the acceptance rate of counting data is the memory-cycle time τ_m (storage time) of the analyzer. During this cycle (say 10 to 16 μsec), only one event can be counted. If C_i is the observed counting rate in the *ith* channel, ρ the number of accelerator pulses per second, and $\omega = \delta t_c$ the channel width, and if only losses occurring in any one τ_m affect the losses in the *ith* channel, then

$$N(T_i) = \frac{1}{\omega} \frac{\omega\rho}{\omega\rho - \tau_m C_i} . \tag{6.16}$$

(b) A-2 Logic (Resolving Time Due to Input Circuitry). There is available time-of-flight instrumentation with logic ideally arranged to

---

[b] For convenience, we suppress all variables, except the energy, on which the neutron flux spectrum depends.

minimize counting-loss corrections. The analyzer works as follows: When a time channel is active, the number of counts it can accept is limited only by the two-ulse resolving time of the input circuitry. After the channel becomes inactive, data acceptance is stopped, and all previously collected information is loaded into the memory. In this case

$$N(T_i) = \frac{C_i}{\omega\rho - [C_i\omega\tau/(\frac{1}{2}\tau + \omega)]} . \qquad (6.17)$$

(c) A-3 Logic (Short Time-Channels). Especially in short-channel work, the counting loss is often determined by the probability of having received information in a previous time interval that was composed of perhaps many channels. For example, suppose that the memory cycle time τ_m of the analyzer is 10 μsec and that we are using 1-μsec channel widths; then the number of counts missed in any given channel is equal to the number of counts in any of the previous 10 channels. In general, the corrected counting rate is

$$N(T_i) = C_i\left(1 + \sum_{i-(\tau/\omega)}^{i-1} \frac{C_n}{\rho}\right) . \qquad (6.18)$$

(d) Monitor Count Correction. The monitor counts C_m obtained during a pulsed-source experiment from the auxiliary monitor must often be corrected for counting losses. For most monitoring systems the corrections discussed in the preceding paragraphs do not apply. For the large counting rates following a beam pulse that decays exponentially with some time constant β that is long compared with the two-pulse resolving time τ of the counting circuitry (usually 1 to 2 μsec), the corrected monitor count is

$$C_m = \frac{C_{obs}}{\rho}\left(1 - \frac{kC_{obs}}{\rho}\right)^{-1} , \qquad (6.19)$$

where

$$k = \frac{1 - e^{-\tau/\beta}}{2} .$$

6-2.3 BACKGROUND. Some mention of background problems in spectral studies has been made in Sec. 6-1. In general, background is measured in pulsed-source spectral work by placing a B^{10} plug at the point at which the spectrum is being measured. This removes the primary spectrum, leaving only background sources. Backgrounds range from less than 1 percent to as much as 30 percent of the desired signal, depending on many factors that include the cross-sectional area of the re-entrant tube and the poison concentration in the moderator.

Ambient backgrounds are quite low and come principally from fast neutrons which, on leaking through the massive shielding surrounding source and assembly, slow down in the shielding system of the collimator or detector.

Spurious noise, especially with linear accelerator sources, leads to background that can be eliminated by proper grounding and r-f protection. Detector after-pulsing due to charging of insulators by the intense gamma-flash is a typical source of high background, especially with certain types of BF_3 chambers.

6-2.4 EMISSION TIME. In Eq. (6.15), for each time channel one must identify both an energy width ΔE_i and an energy E_i. The simple conversion from time to energy is normally performed after subtracting from the channel time the mean time $t_a(E_i)$ that a neutron of energy E_i spends in the moderating assembly.

We develop below a method for computing $t_a(E)$ for the case of an infinite homogeneous assembly excited by a spatially flat pulse of neutrons.[1] It is easy to generalize the formalism for spatially dependent problems. The time-dependent flux spectrum $\phi(E,t)$ satisfies the equation

$$-\frac{1}{v}\frac{\partial\phi}{\partial t}(E,t) = [\Sigma_a(E) + \Sigma_s(E)]\phi(E,t) - \int_0^\infty \Sigma_0(E',E)\phi(E',t)dE' - S(E)\delta(t) \ . \qquad (6.20)$$

The quantity to be computed is

$$t_a(E) = \frac{\Psi(E)}{\phi(E)} \ , \qquad (6.21)$$

where

$$\Psi(E) = \int_0^\infty \phi(E,t)tdt \ ,$$

and

$$\phi(E) = \int_0^\infty \phi(E,t)dt \ .$$

The function $\phi(E)$ satisfies the equation for the infinite-medium steady-state spectrum

$$[\Sigma_a(E) + \Sigma_s(E)]\phi(E) = \int_0^\infty \Sigma_0(E',E)\phi(E')dE' + S(E) \ . \qquad (6.22)$$

To obtain an equation for $\Psi(E)$, we multiply Eq. (6.20) by t, integrate over all times, and use the fact that $t\phi(E,t)$ vanishes for t = 0 and t = ∞. The resulting equation is

$$[\Sigma_a(E) + \Sigma_s(E)]\Psi(E) = \int_0^\infty \Sigma_0(E',E)\Psi(E')dE' + \frac{\phi(E)}{v} \ . \qquad (6.23)$$

This is a problem of the infinite-medium type with a source equal to $\phi(E)/v$. First, Eq. (6.22) is solved for $\phi(E)$, and then the result is used to find $\Psi(E)$ from Eq. (6.23). Finally, $t_a(E)$ is obtained from Eq. (6.21). In practice, we must use numerical methods (see Sec. 7-1.2) to solve Eqs. (6.22) and (6.23). Figure 6.20 gives calculated mean-emission-time curves as a function of neutron flight time in a 16-meter flight path. These curves are for assemblies of polyethylene having 50-, 100-, and 150-μsec thermal die-away times.

6-2.5 SPECTRAL DISTORTIONS IN PULSED-SOURCE EXPERIMENTS. The uncertainties in time-of-flight and in flight distance in the pulsed-source spectrum experiment not only produce uncertainties in the energy scale but also produce spectral distortions that may be large if the spectrum changes rapidly with energy. We may choose the analyzer channel width so that the distortion of the spectrum given by Eq. (6.10) is below an acceptable value. That neutrons with different energies can

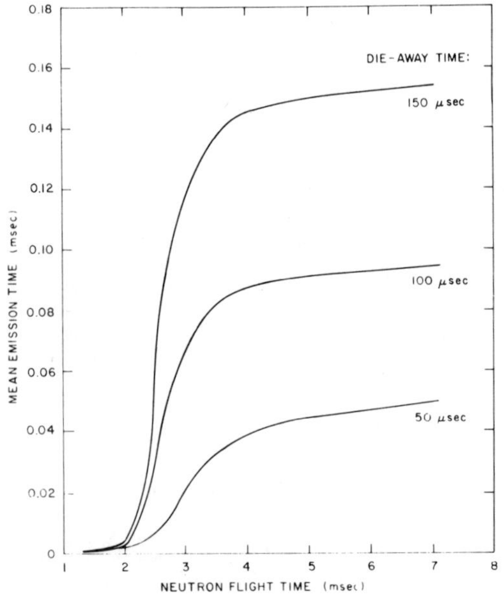

Figure 6.20 Mean-emission-time curves for polyethylene. (From Beyster et al., USAEC Report GA-2544, Ref. 10.)

arrive at the detector at the same time is another factor that distorts the observed spectrum. This can occur, for example, when a neutron is emitted from the assembly with a velocity just sufficient to overtake a previously emitted slower neutron at the detector. Koppel has considered this problem in some detail.[25] The observed counting rate at time T_0 in a perfectly resolving time analyzer is

$$\frac{dC_0}{dT_0}(T_0) = \int_0^{\infty} \phi(v,t)\eta(v)dv , \tag{6.24}$$

where

$$t = T_0 - \frac{d}{v} , \tag{6.25}$$

$\phi(v,t)$ is the velocity spectrum in the moderating assembly at time t after the pulse, and $\eta(v)$ is the velocity-dependent efficiency of the detector system. For the hypothetical case in which all neutrons of a given velocity are emitted from the infinite assembly at the same instant,

$$\phi(v,t) = \phi(v)\delta[t - \tau(v)] , \qquad (6.26)$$

where $\tau(v) = t_a(E)$ is the emission time. When we substitute Eqs. (6.26) and (6.25) into Eq. (6.24) and integrate, we find that

$$\frac{dC_0}{dT_0} = \frac{v_0^2}{d} \phi(v_0)\eta(v_0)\left[1 - \frac{v_0^2}{d}\left(\frac{d\tau}{dv}\right)_{v=v_0}\right]^{-1} , \qquad (6.27)$$

where

$$v_0 = \frac{d}{T_0 - \tau(v_0)} = \frac{d + v_0\tau(v_0)}{T_0} \qquad (6.28)$$

includes the correction for the effect of the finite emission time. The bracketed term in Eq. (6.27) represents the spectral distortion owing to the finite emission time. By setting T equal to d/v, we can write this term in the form

$$\left[1 + \left(\frac{d\tau}{dT}\right)_{T=T_0}\right]^{-1} . \qquad (6.29)$$

From the mean-emission-time curve in Fig. 6.20 corresponding to a 150-μsec die-away time, we can deduce that this expression (6.29) differs from unity by as much as 0.2 for the indicated range of flight times. In most experiments, however, the die-away time is much shorter than 150 μsec and the spectral distortion is very small.

6-3 CHOPPED-BEAM MEASUREMENTS OF NEUTRON SPECTRA

6-3.1 CHOPPER DESIGN. Reviews of the application of chopper to neutron research are given in Refs. 26 and 27. In this section we describe chopper techniques only insofar as they have proven useful in measurements of low-energy neutron spectra.

One of the best known slow-neutron choppers is the parallel-plate Fermi-type cadmium rotor.[28] Figure 6.21 shows a typical Fermi-type chopper arrangement that has been used at Harwell.[29] With such a device we cannot measure spectra above the cutoff energy in the cadmium cross section, which is at about 0.3 eV. This limits the utility of the device for thermalization studies since measurements should extend to at least 1 eV, and preferably to 10 eV. Nevertheless, spectra at low energy (≲ 0.4 eV) have been measured at many laboratories with these instruments.[30,31]

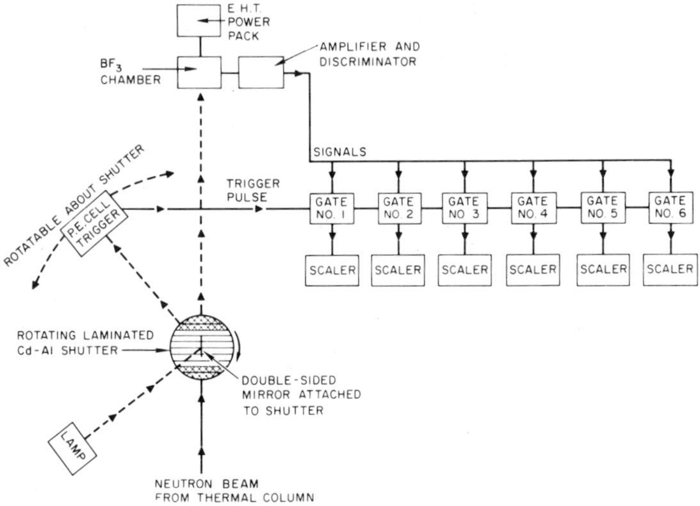

Figure 6.21 Schematic diagram of a neutron velocity selector. (From Egelstaff, J. Nucl. Energy, Ref. 29.)

The Harwell nickel-alloy rotor[32] and the borated plastic rotor[33] are two other chopper configurations extensively used for thermal-spectrum measurements. They can be used at all neutron energies below about 10 eV. The Harwell rotor has been used for spectral studies in LIDO, in Calder-Hall lattices, and in Zenith. The chopper has parallel plates of K Monel. It is 7 in. in diameter and can be rotated to angular velocities as high as 10^4 rpm by a hysteresis motor. Its energy resolution at 10 eV is 5%.

The borated plastic rotor was designed by Slovacek and Stone[33] for use in measurements of thermal-neutron spectra. Their rotor was made from boron-loaded paper-base phenolic. Figure 6.22 shows a design similar to the original one; here, however, heavily borated epoxy plates replace the phenolic. The 6-in.-diameter rotor contains sixteen 1/16-

Figure 6.22 Chopper rotor and rotor containment jacket. (From Beyster et al., USAEC Report GA-2544, Ref. 10.)

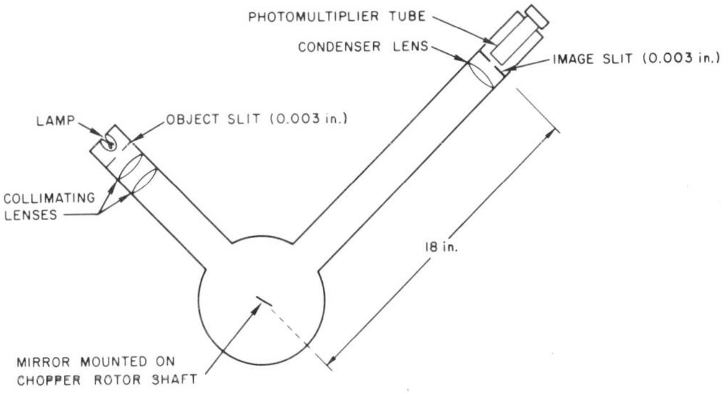

Figure 6.23 Diagram of the light-pickup arm of the chopper assembly. (From Beyster et al., USAEC Report GA-2544, Ref. 10.)

in. slits. Hollowed aluminum spacers separate the absorbing plates. The rotor, which is jacketed in thin stainless steel to reduce air drag and buffeting, is capable of reaching 10^4 rpm in air and can produce pulses with a half-width of about 8 μsec. A typical optical system for picking up a timing pulse from a rotating chopper is shown in Fig. 6.23.

At the present time, however, magnetic pickup devices are more frequently used; they are more simply applied to chopper work than are optical devices and, further, they permit more nearly jitter-free operation than most optical systems permit.

6-3.2 DESIGN OF CHOPPER EXPERIMENTS. With steady-state neutron sources (reactors) the chopped-beam technique is normally used in making measurements of neutron spectra. One possible experimental configuration is shown in Fig. 6.17. We select a flight-path length such that a high counting rate and good resolution are achieved. As in the pulsed-source experiment, one must arrange that the slow neutrons from one burst do not overlap neutron signals at the detector from a subsequent burst. The absence of such overlap signals is achieved if

$$d = \frac{v_c}{2f}, \tag{6.30}$$

where d is the flight path length, f is the frequency of chopper rotation, and v_c is the chopper cutoff velocity (Eq. (6.3)).

The geometrical criteria necessary to cbtain a meaningful measurement of neutron spectra with the chopped-beam technique are the same as those for pulsed-source experiments, as discussed in Sec. 6-1.2. Neutron detectors for chopper experiments are usually gas-filled proportional counters, and time-of-flight procedures are very similar to those used in pulsed-source spectrum studies. In the chopped-beam experiments, the neutron flight time is the interval between a reference signal from the chopper and the instant of neutron detection.

A chopper may also be used in conjunction with a high-intensity pulsed neutron source such as is afforded by a present-day linear accelerator or pulsed positive-ion machine. Figure 6.24 is a diagram of an experiment in which the technique is applied to the study of the spatial dependence of the neutron spectrum in the vicinity of an interface separating a hot region from a cold region. We may simulate steady-state neutron spectra either (1) by operating the chopper and accelerator in a completely uncorrelated time relation or (2) by systematically varying the phase between them. In the first case, the accelerator may be operated at a high pulse-repetition rate in order to more nearly represent a steady-state source. Except for gating off the detectors during the beam flash, the spectrum measurement proceeds in the same way as in the case of a reactor source.

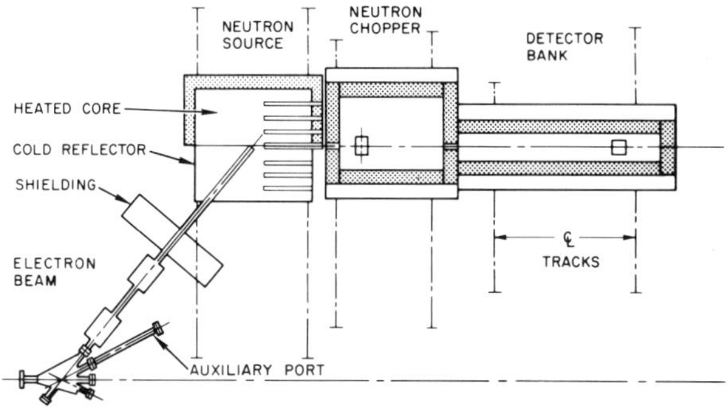

Figure 6.24 Diagram of the neutron chopper facility. (From Beyster *et al.*, USAEC Report GA-2544, Ref. 10.)

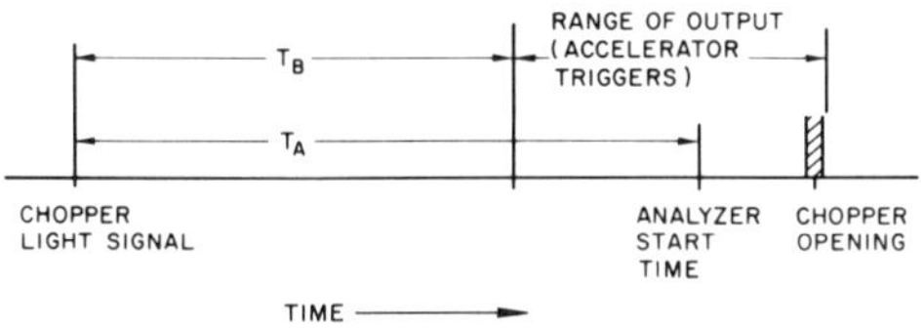

Figure 6.25 Time sequencing used to accomplish time integration with accelerator and chopper.

In a second method of simulating a steady-state spectrum measurement, the timing signals from the chopper provide trigger signals for the accelerator. These signals are accurately time-phased so that the spectrum measured in this manner is the same as would be measured by the pulsed-source method in which one integrates the spectrum over the entire decay time of neutrons in the assembly. (One can measure the time-dependent neutron spectrum by stopping the time sampling.) To accomplish the time integration with an accelerator and chopper, the signal from the chopper optical system may be fed into a dynamic digital delay device;[34] the output of this device is used to trigger the linear accelerator. The time sequencing is indicated in Fig. 6.25. The chopper opens at a fixed time after the timing signal. The delays for the activation of the time-of-flight analyzer and the accelerator trigger are T_A and T_B, respectively. The dynamic digital delay device varies the accelerator trigger from T_B to a time just beyond the instant of the chopper opening. T_B is adjusted so that the dynamic time range is equal to about three thermal die-away times plus the time that is required for the slowest neutron of interest to travel from the experimental assembly to the chopper. The whole die-away of the pulse is thus scanned by the chopper.

A third method of measuring steady-state spectra utilizing both a pulsed accelerator and chopper consists in measuring the asymptotic spectrum in a pulsed multiplying assembly.[35] The assembly is pulsed with a burst of fast neutrons generated by electrons from the linear accelerator. These neutrons thermalize and eventually establish an asymptotic distribution that is characteristic of the properties of the assembly. The chopper then is phased with respect to the accelerator burst to measure this distribution. In the asymptotic distribution the effective absorption present in the assembly is given by $\Sigma_a - (\alpha/v)$, where α is the fundamental-mode decay constant of the assembly. An advantage of this type of measurement is that most spatial flux transients have died out and the analysis of the experiment is in some cases simplified.

6-3.3 DATA CORRECTIONS. In a chopper experiment, as in the pulsed-source measurements, the measured data must be corrected for a number of characteristic effects. The flux per unit energy may be calculated from

$$\phi(E_i) = \frac{N(T_i)}{\Delta E_i s(E_i) T_c(E_i) T_f(E_i)} , \tag{6.31}$$

where $N(T_i)$ is the measured time-of-flight distribution corrected for counting losses, as discussed in Sec. 6-2.2; ΔE_i is the energy width of the channel; $s(E_i)$ is the detector sensitivity function; and T_c and T_f are the chopper transmission and flight-path transmission, respectively. The chopper transmission T_c is one of the main sources of uncertainty in spectral determinations by this technique. The neutron beam that traverses the chopper is transmitted in a manner that depends on neutron velocity. A neutron with velocity less than the cutoff velocity v_c is intercepted by the chopper blades, whereas a neutron with velocity many times greater than v_c is always transmitted. Stone and Slovacek[36] have derived the following time-averaged relative transmission functions for flat-plate choppers:

$$\left.\begin{aligned} T_c &= \frac{h}{2\pi R}\left(1 - \frac{8}{3}\beta^2\right) & \beta &< \frac{1}{4}, \\ T_c &= \frac{h}{2\pi R}\left(\frac{8}{3}\beta^2 - 8\beta + \frac{16}{3}\beta^{1/2}\right) & \frac{1}{4} &\le \beta \le 1, \end{aligned}\right\} \tag{6.32}$$

where

$$\beta = \frac{v_c}{v} = \frac{\omega R^2}{hv} .$$

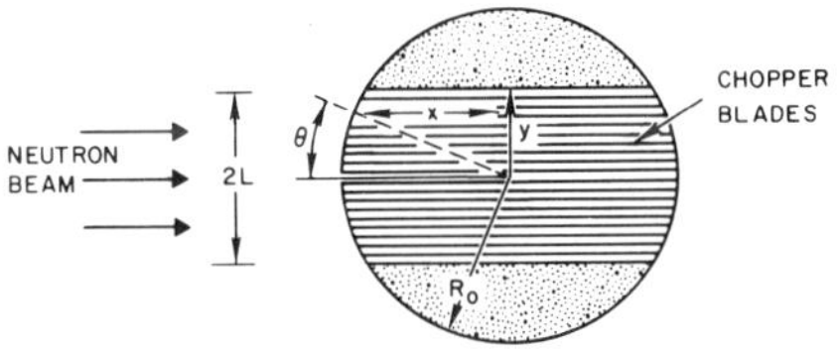

Figure 6.26. Determination of average length of chopper blades.

The factor $h/2\pi R$ appearing in Eqs. (6.32) is the transmission for neutrons with infinite velocity. Since the off-axis blades in the chopper are shorter in the beam direction than the chopper diameter, 2R, it is more accurate to use the average dimension of the blades in Eqs. (6.32) for transmission calculations. Suppose that a neutron beam of height 2L is incident on the chopper as indicated in Fig. 6.26. Then in place of R we use

$$\bar{R} = \frac{\int_0^L x\,dy}{\int_0^L dy} = \frac{R^2}{2L}\left(\theta_0 + \frac{\sin 2\theta_0}{2}\right), \tag{6.33}$$

where

$$\theta_0 = \arcsin \frac{L}{R} . \tag{6.34}$$

For many experiments L is much smaller than R and this correction is negligible.

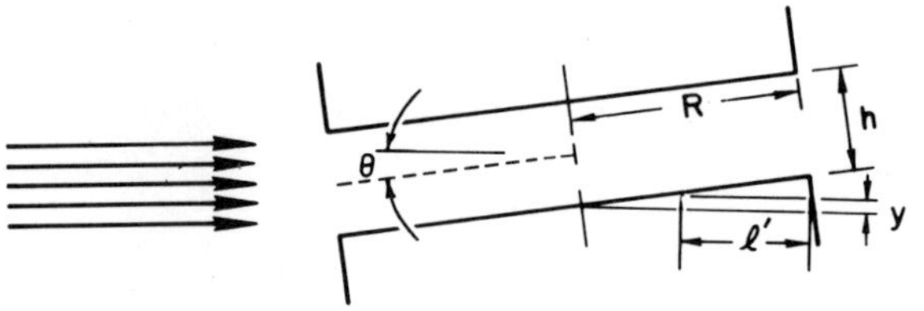

Figure 6.27. Determination of transmission through chopper blades.

In deriving transmission equations, Stone and Slovacek[36] assume that chopper blades are perfectly opaque. In practice, chopper blades are not completely opaque and some neutrons are transmitted through the blade edges. In order to take this effect into account we find the dimensions of an equivalent opaque chopper blade. Suppose, as shown in Fig. 6.27, that a slit opening (between collimating blades) is exposed to a parallel beam of neutrons when the vanes are oriented at an angle θ to the beam direction. The static transmission for this orientation relative to that for θ = 0 is

$$T_r(\theta) = 1 - \frac{2R\theta}{h} + \frac{2}{h}\int_0^{R\theta} \exp[-N\sigma\ell'(y)]dy \qquad \theta < \frac{h}{2R} ,$$

$$= 0 \qquad \theta \geq \frac{h}{2R} \tag{6.35}$$

where $\ell'(y)$ is as shown in Fig. 6.27 and $N\sigma$ is the total macroscopic cross section of the materials used in the chopper blades.

The term $[1 - (2R\theta/h)]$ is the expression for the transmission in the case of perfectly opaque blades, while the remaining term gives the transmission of neutrons through the blade edges. Using the approximation that $\ell'(y) = R - (y/\theta)$ for a small angle,

$$T = \frac{h}{2\pi R} T_r(\theta) = (\pi R)^{-1}\left[\left(\frac{h}{2} - R\theta\right) + \frac{\theta}{N\sigma}\left(1 - e^{-N\sigma R}\right)\right] . \tag{6.36}$$

Equating this transmission to that obtained with perfectly opaque vanes having a radius $R_0 = R - \Delta R$ yields

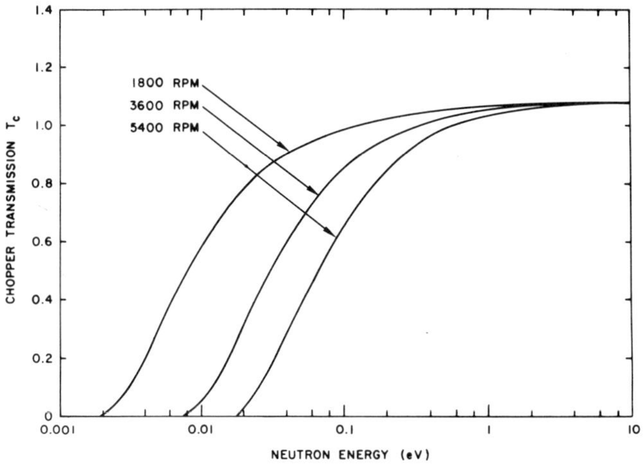

Figure 6.28 Dynamic transmission of 6-in.-diameter borated epoxy chopper with 1/16-in. slits. (From Beyster et al., USAEC Report GA-2544, Ref. 10.)

$$\frac{h}{2} - R\theta + \Delta R\theta = \frac{h}{2} - R\theta + \frac{\theta}{N\sigma}\left(1 - e^{-N\sigma R}\right) . \tag{6.37}$$

Usually we may neglect $e^{-N\sigma R}$. Thus

$$\Delta R = \frac{1}{N\sigma} = \Sigma^{-1} ,$$
$$R_0 = R - \Sigma^{-1} . \tag{6.38}$$

With corrections for the velocity-dependent radius, the neutron-chopper-transmission functions are still given by Eqs. (6.32), but with R replaced by R_0 and β replaced by

$$\beta_c = \frac{v_c}{v}\frac{R_0^2}{R^2} = \beta\frac{R_0^2}{R^2} .$$

This correction turns out to be rather large even for poisoned rotors; it is largest at neutron energies near cutoff. To make the corrections, we must measure $N\sigma$ for the rotor material as a function of energy. This can be obtained for a transmission measurement of the chopper-blade material. These procedures lead to the typical calculated transmission function shown in Fig. 6.28. The value of these functions at high energies is greater than unity because T_c is given in units of $h/2\pi R$ instead of $h/2\pi R_0$.

In addition to the dynamic transmission of the chopper, the final transmission used in reduction of chopper data must include the transmission of the casing that neutrons traverse in passing through a chopper slit (opening) and the transmission of the T_f flight path between the base of the re-entrant hole and the neutron detector. These transmissions are calculated using total-cross-section data and the known thicknesses of materials.

To further check the soundness of chopper construction and to lend confidence to the calculation of chopper dynamic transmission functions, we can compare the calculated and measured static angular resolution functions for the chopper-detector system. We usually measure the angular resolution by determining the neutron transmission as a function of angle of rotation of the chopper. Stone and Slovacek[36] have derived an expression for the static angular resolution function $H(\theta)$ of a chopper system. This function is the convolution of the triangular chopper transmission function T_r (see Eq. (6.36)) and the rectangular-detector angular response function S. It is given by

$$H(\theta) = \int_{-\phi}^{\phi} T_r(\theta' - \theta)S(\theta')d\theta' , \tag{6.39}$$

where for opaque chopper blades $T_r(\theta) = 1 - \frac{\theta}{\alpha}$, with $\alpha = \frac{h}{2R}$,

ϕ = angle subtended at the center of the chopper by one half the detector bank,

$S(\theta) = 0$ when $\theta < -\phi$,

$S(\theta) = 1$ when $-\phi < \theta < \phi$, and

$S(\theta) = 0$ when $\theta > \phi$.

Performing the integration with these functions, we get

$$H(\theta) = 1 + \frac{4\theta_m^3}{\pi^3\alpha^2\phi}\left[\sin\frac{\pi\phi}{\theta_m}\left(1 - \cos\frac{\pi\alpha}{\theta_m}\right)\cos\frac{\pi\theta}{\theta_m} + \frac{1}{8}\sin\frac{2\pi\phi}{\theta_m}\left(1 - \cos\frac{2\pi\alpha}{\theta_m}\right)\cos\frac{2\pi\theta}{\theta_m} + \frac{1}{27}\sin\frac{3\pi\phi}{\theta_m}\left(1 - \cos\frac{3\pi\alpha}{\theta_m}\right)\cos\frac{3\pi\theta}{\theta_m} + \cdots\right], \tag{6.39a}$$

where $\theta_m = \alpha + \phi$.

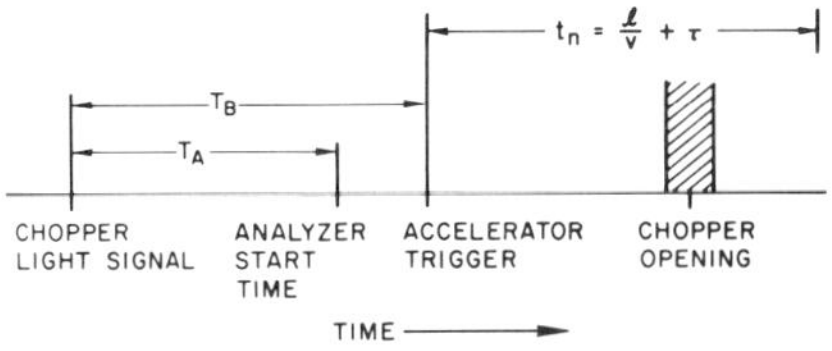

Figure 6.29 Time sequence of events in the determination of the interval between the light or magnetic timing signal and the instant the chopper window opens.

A key problem in chopper studies is the precise determination of the difference between the time of the light or magnetic timing signal and the time the chopper window opens. There is a pulsed-source technique for measuring this experimentally as well as for verifying the static angular resolution function given by Eq. (6.39a). Figure 6.29 shows the time sequence of events. $t_n(v)$ is the total time for a neutron with a velocity v to slow down and travel to the center of the chopper; it consists of the slowing-down time τ and the flight time ℓ/v. The neutrons we are considering all have energies much greater than k_BT; if the energies are lower, the time of arrival of the neutron at the chopper after the accelerator pulse is not sharply defined. If one selects a particular detection time after the accelerator pulse (thereby defining the neutron energy) and examines the counting rate of the detector as a function of T_B, the bell-shaped static angular resolution curve is measured. The time at which the chopper is fully open relative to the chopper timing signal is

$$\tau_c = T_B(v_0) + \frac{\ell}{v_0} + \tau(v_0) ,$$

where $T_B(v_0)$ is the delay from the chopper light signal to the accelerator pulse that is associated with the maximum transmission by the chopper for neutrons of velocity v_0. In Fig. 6.30 we show, as a function of T_B, measured and calculated angular resolution functions for a 6-in.-diameter borated plastic rotor running at 1800 rpm. The calculation and measurement are in good agreement.

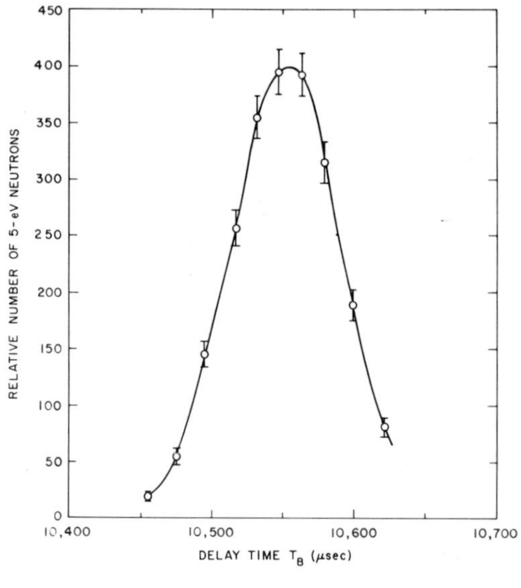

Figure 6.30 Dynamic measurement of the angular resolution of a 6-in.-diameter borated plastic chopper running at 1800 rpm. (From Beyster et al., USAEC Report GA-2544, Ref. 10.)

The range of neutron energies for which a particular chopper rotor speed is useful is limited at low energies by the cutoff of the transmission function and at high energies by the effect of the chopper burst width on the energy resolution of the measurement. The energy ranges that may be covered with the various chopper speeds for a 6-in.-diameter borated plastic chopper are depicted in Fig. 6.31. We choose the low-energy limit for each chopper speed as the neutron energy corresponding to the velocity $3v_c$. We choose this velocity because the

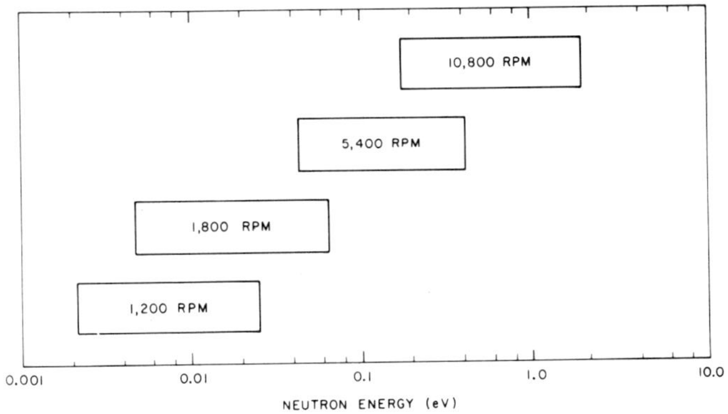

Figure 6.31 Energy range for 6-in.-diameter borated plastic chopper. (From Beyster et al., USAEC Report GA-2544, Ref. 10.)

error in the chopper transmission calculated from Eqs. (6.32) is very large when the velocity is close to the cutoff velocity, even when we account for transmission by the chopper blades. The upper energy limit for a given chopper speed is arbitrarily chosen by requiring that the energy resolution of the system be better than 50%. A resolution of 50%, though not desirable, is adequate for spectrum measurements as long as the spectrum does not vary rapidly with energy. This is the case in the 1/E portion of the spectrum. Corrections to time-of-flight spectra from the time resolution of the measuring system can be performed quantitatively when Eq. (6.10) indicates they may be required. The correction requires a knowledge of the spectrometer time-resolution function R. The observed time-of-flight spectrum O(t) is then related to the time-of-flight spectrum N(t) by the following equation:[37]

$$O(t) = \int_{-\infty}^{\infty} R(\lambda)N(t + \lambda)d\lambda \ . \tag{6.40}$$

After a Taylor series expansion of $N(t + \lambda)$ about $\lambda = 0$, we solve for the undistorted time-of-flight spectrum N(t) and obtain

$$N(T) = O(t)\left[1 - \frac{1}{2O(T)}\frac{d^2O(t)}{dt^2}\int_{-\infty}^{\infty} R(\lambda)\lambda^2 d\lambda + \right.$$
$$+\left\{\frac{1}{4}\left[\int_{-\infty}^{\infty}\lambda^2 R(\lambda)d\lambda\right]^2 - \right.$$
$$\left.\left.-\frac{1}{24}\int_{-\infty}^{\infty}\lambda^4 R(\lambda)d\lambda\right\}\frac{1}{O}\frac{d^4O(t)}{dt^2}\right] . \qquad (6.41)$$

Here we have assumed that $R(\lambda) = R(-\lambda)$. The quantity $[N(t)/O(t)]$ has been shown to differ from unity by as much as 20 percent for thermal spectra measured by the chopper technique.[3]

The background sources that exist in pulsed-source spectrum measurements also exist in measurements with choppers. In addition, the chopper rotor and associated equipment, unless they are heavily poisoned and shielded, provide a source of background. A further difficulty is that much of the background may originate at positions on a line with the chopper and detector. The resulting time-dependent background flux at the detectors depends on chopper orientation. It is common practice to utilize half the time between two consecutive chopper pulses for accumulating signal data and the other half for accumulating background. In this procedure, the chopper orientation at the time of each background count is the same as during a prior count of signal plus background, and thus the corresponding background count can be subtracted from the total count. In addition to this background subtraction procedure, it is desirable to measure the background in the neutron detectors with the direct beam absorbed at the base of the re-entrant tube. The background in reactor-source, chopper experiments is usually relatively high, especially below 0.01-eV neutron energy. In the application involving a phased chopper and a pulsed source, this problem disappears and background are greatly reduced. The reason for this is that neutrons are produced in the assembly only at times such that they are capable of being transmitted by the chopper, not at all times as in reactor-source experiments.

6-4 COMPARISONS OF MEASURED AND CALCULATED NEUTRON SPECTRA

The methods discussed in this chapter have been applied extensively to the measurement of spectra of low-energy neutrons in almost all of the common moderators. Many of the experiments were designed to provide as closely as possible a quasi-infinite medium in which the environment for thermal neutrons is essentially identical to that in a homogeneous reactor. It is for hydrogen-moderated assemblies that we can most easily arrange the experiment so that to a good approximation the measured spectrum is a solution of Eq. (5.165). For nonhydrogeneous moderators, on the other hand, the differences between the solution of Eq. (5.165) and the measured spectrum are small but usually greater than in hydrogeneous systems. We can nevertheless account for these small differences by the methods discussed in Sec. 5-9, or by solving numerically either the diffusion or the transport equation. Actually, the adaptation of high-speed computers to the solution of the one-dimensional transport equation is now so well-developed that the computers provide not only a possible but also a convenient means for multigroup (as high as thiry or forty groups) calculations of thermal-neutron spectra. The disadvantage with this approach is that it does not allow a separation of effects associated with the energy-transfer properties of the moderator from those of a specifically transport nature.

In the remainder of this chapter we discuss the results of some measurements of neutron spectra, emphasizing in particular the spectra of neutrons thermalized by light water, by zirconium hydride, and by graphite. In addition, we give results for heavy water, polyethylene, and benzene, the latter being considered as a prototype of the organic moderators.

For water systems, we consider spectra measured in thin slabs as well as in large assemblies. The purpose of the thin-slab measurements is to determine the importance to neutron spectra of effects associated with the thin geometry of the assembly and with anisotropic scattering.

We reserve until Chap. 9 the discussion of the spatial variation of the neutron spectrum in a region that is characterized by extremely different absorbing and thermalizing properties. In that chapter we discuss in particular the variation of the neutron spectrum with the distance from a plane boundary that separates a hot, heavily poisoned, graphite region from a cold, pure, graphite region.

6-4.1 HYDROGENEOUS MODERATORS. The earliest measurements of neutron spectra were made in water-moderated assemblies.

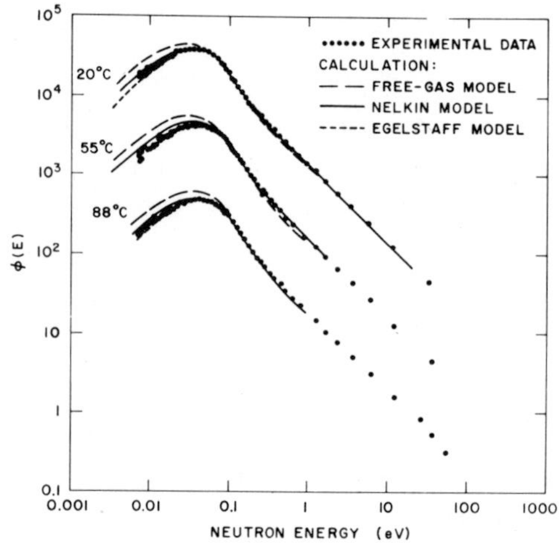

Figure 6.32 Neutron spectra in water poisoned with boron at various temperatures (0.025-eV macroscopic absorption cross section divided by the number density of protons, 7 barns/H atom). (From Poole et al., in Exponential and Critical Experiments, Ref. 38.)

(a) Light Water. Measurements of neutron spectra in homogeneous H_2O assemblies have been made by Poole,[18,38] by Beyster et al.,[39] by Young et al.,[40] by Burkart and Reichardt[41] using a pulsed source, and by Stone and Slovacek[3] and Mostovoi et al.[42] using a chopper with a reactor source. In Fig. 6.32 Poole's measured results at several temperatures are compared with calculations using the free-gas model, the Nelkin model, and the kernel based on the $f(\omega)$ shown in Fig. 3.21 (the Egelstaff model). Figures 6.33 and 6.34 show spectra measured by Beyster et al. in aqueous solutions of boric acid and cadmium sulfate, respectively. In these two figures each curve is labeled with the 0.025-eV macroscopic absorption cross section divided by the number density of protons. The experimental errors in the measured spectra are

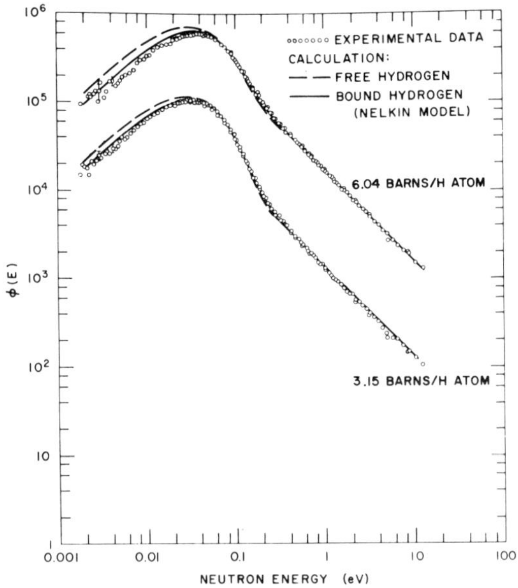

Figure 6.33 Neutron spectra in water poisoned with 1/v absorber (boric acid). (From Beyster et al., Nucl. Sci. Eng., Ref. 39.)

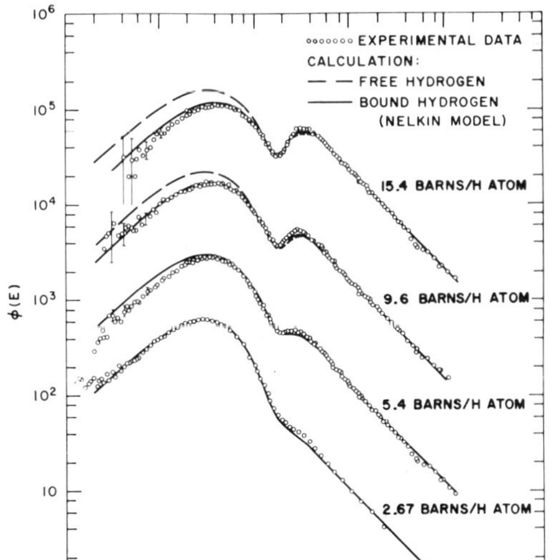

Figure 6.34 Neutron spectra in water poisoned with cadmium sulfate. (From Beyster et al., Nucl. Sci. Eng., Ref. 39.)

shown at a few low values of the neutron energy; at higher energies, the plotted "points" mask the experimental error. The experimental error results primarily from the counting statistics and from the error in the measured sensitivity of the detector bank.

The dashed curves in Figs. 6.33 and 6.34 are calculated assuming the hydrogen atoms in water are free while the solid curves are obtained on the basis of the water model described by Eqs. (3.95) through (3.97). In each case, the calculated spectra are obtained as numerical solutions of Eq. (5.165).

We can see from Fig. 6.33 that spectra measured in water poisoned with a 1/v absorber agree much better with predictions of the bound-hydrogen model than with those of the free-gas model. The same conclusion was obtained by Poole[38] from recent experiments he performed (Fig. 6.32). Some early experimental results[18] are inconsistent with this conclusion. For example, Poole's first measurements using boric acid solutions differ from Amster's free-hydrogen calculations[43] only in the energy region between about 0.1 eV and 0.5 eV. The spectra measured by Stone and Slovacek[3] are in even closer agreement with calculations based on the free-hydrogen model. Instead, in Fig. 6.33 the differences between measured results and free-hydrogen predictions are significant, being about a factor of three greater than the experimental error at low energies. The results in Fig. 6.33 and Poole's more recent experiments tend to verify Poole's early results in the joining region; however, the more recent results extend to considerably lower energies than do the early results obtained by Poole, and it is at these lower energies that the difficulties with the free-gas model are more apparent.

Figure 6.34, which shows the effect of the thermal-energy resonance in cadmium, further emphasizes the differences between the thermalization properties of free protons and protons bound in water. The conclusions drawn from Figs. 6.33 and 6.34 are further verified by the results shown in Fig. 6.35. Table 6.1 lists the concentrations of the various poison materials that were present in the experiments. In addition, since the results of the quasi-infinite medium measurements must satisfy neutron conservation, the ratio r of the absorption rate to the slowing-down rate is also given for each experiment. To obtain these ratios the experimental data were substituted into Eq. (5.121). Except for the case of gadolinium, the ratio calculated from the data is sufficiently close to unity to justify the interpretation of the measured spectrum as an infinite-medium spectrum. Inasmuch as r for the case of gadolinium differs from unity by 9 percent, we cannot be certain of the significance of the rather large differences between theory and experiment. These differences, however, may be seen to be significant in curve B of Fig. 6.35 and confirm the existence of a small but real difference between theory and experiment.

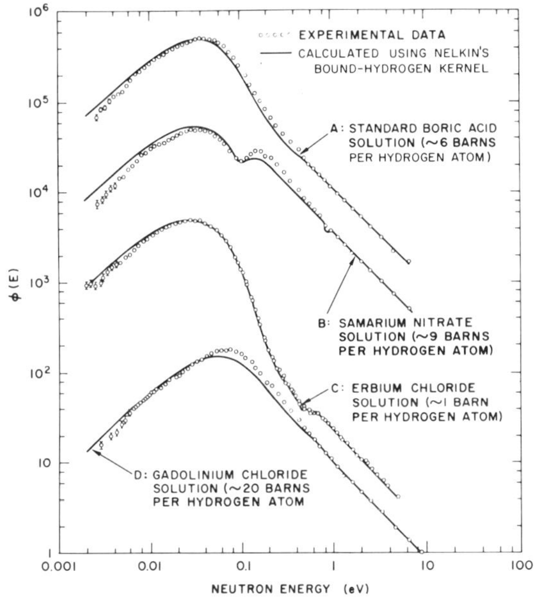

Figure 6.35 Neutron spectra in poisoned water solutions. (From Young et al., Nucl. Sci. Eng., Ref. 40.)

Neutron spectra at four different temperatures in water solution containing $N_B = 4.2118 \times 10^{20}$ atoms/cm^3 of boron are shown in Figs. 6.36a, b, c, and d. The insert in each figure shows the ratio of the experimental spectrum to the spectrum calculated from the bound-hydrogen model. The value of r calculated from the experimental data for each temperature is about 0.95. The differences, ranging from about 10% to 30% at low energies, are significant and consistent with previously presented comparisons between theory and experiment in water.

In the Nelkin model, used to calculate the spectra shown in Figs.

Table 6.1

ATOM DENSITIES AND NEUTRON CONSERVATION FOR WATER-MODERATED SPECTRUM EXPERIMENTS

Experiment	Concentration (atoms/barns-cm)							Neutron Conservation (Ratio of Absorption Rate to Slowing-down Rate, r)
	H	B	Sm	Er	Cl	Gd	O	
$H_2O(H_3BO_3)$	6.66×10^{-2}	4.26×10^{-4}					3.39×10^{-2}	1.002
$H_2O(Sm(NO_3)_3)$	6.61×10^{-2}		9.45×10^{-5}				3.39×10^{-2}	1.015
$H_2O(ErCl_3)$	6.63×10^{-2}			1.99×10^{-4}	5.96×10^{-4}		3.31×10^{-2}	1.023
$H_2O(GdCl_3)$	6.68×10^{-2}				8.46×10^{-4}	2.82×10^{-5}	3.34×10^{-2}	1.090

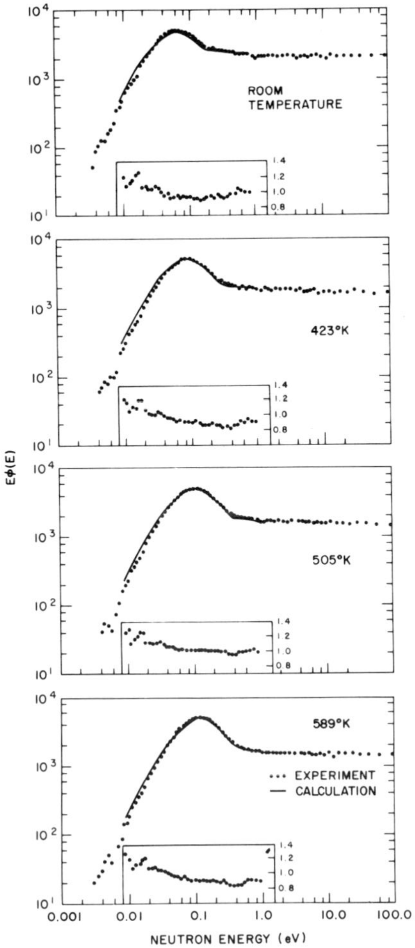

Figure 6.36 Neutron spectra in borated water in pressure vessel: (a) at room temperature; (b) at 423°K; (c) at 505°K; and (d) at 589°K. The inserts in each part of this figure show the ratio of the experimental spectrum to the calculated spectrum based on the bound-hydrogen model. (From Young, USAEC Report GA-5319, Ref. 44.)

6.33 through 6.35, there are a number of oversimplifications concerning the atomic motions in water which can affect the calculated spectra. Correcting for these oversimplifications may bring measured and calculated spectra into better agreement. Two obvious over-simplifications are (1) using a single frequency to replace the broad band of hindered rotational frequencies in water and (2) neglecting the anisotropy of molecular vibrations. In Egelstaff's model, which is one of the models used to calculate spectra shown in Fig. 6.32, the hindered rotational frequencies and the lowest vibrational level are treated as a band rather than as a single frequency. Since, however, the effective mass associated with the highest vibrational level differs from that in the Nelkin model, we cannot yet decide the importance for thermal-neutron spectra of having a broad band of rotational frequencies instead of a single frequency.

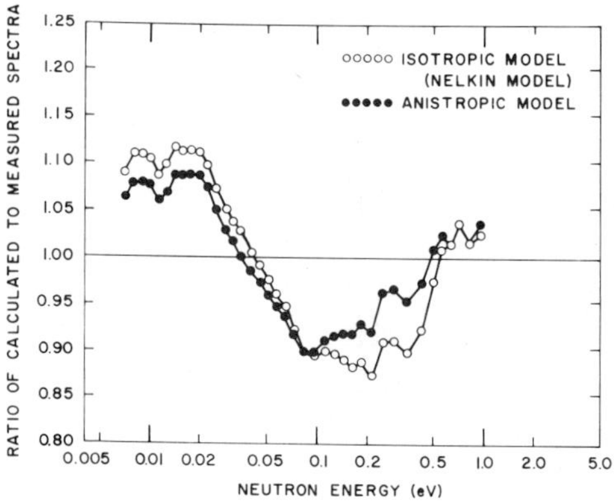

Figure 6.37 Ratio of calculated to measured spectrum in 43.7 g/liter boric acid-water solution. (From Koppel and Young, Nucl. Sci. Eng., Ref. 45.)

The effect of the anisotropy of the molecular vibrations has been considered by Koppel and Young.[45] For water that is heavily poisoned with boron, when we take account of this effect the difference between theoretical and experimental spectra is reduced in the way shown in Fig. 6.37.

There remain residual discrepancies of about 10 percent between theory and experiment. The calculated spectra are always better thermalized than the experimental ones. Spectra calculated from a free-hydrogen model are thermalized even more and typically deviate from experimental spectra by more than 30 percent. The source of this residual discrepancy is not known. It may be due to the oversimplified treatment of molecular rotations used in the model or to geometrical effects that have not been accurately included in the analysis. Given the uncertainties in detector sensitivity and other possible small systematic experimental errors, agreement to within better than 5 percent would not be meaningful.

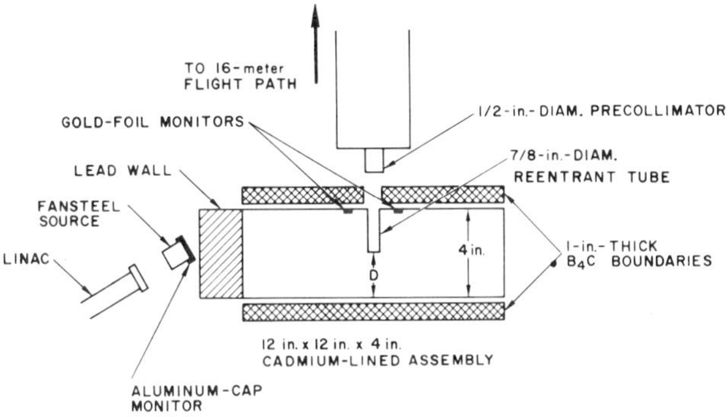

Figure 6.38 Geometrical arrangement used for measuring neutron spectra in water. (From Young et al., Nucl. Sci. Eng., Ref. 46.)

The previous experiments have all been performed in quasi-infinite media. It is possible to obtain an additional integral check of the procedures used to describe thermal-neutron scattering by performing measurements of angular flux spectra in thin-slab, one-dimensional geometry. An experiment for studying spectra in the presence of strong one-dimensional flux gradients is shown in Fig. 6.38. The strong flux gradients are produced by utilizing a relatively thin slab tank with highly absorbing walls. The base of the re-entrant tube is successively placed at several locations within the assembly, and the relative spatially dependent spectra are measured for comparison with the predictions of one-dimensional transport theory. The external location of the source is extremely important for the performance of this experiment. For a nonmultiplying system the arrangement shown in Fig.

6.38 is preferable to locating the source in line with the axis of the re-entrant tube. The latter source location produces a difficult physical situation in which the thermal source is highly anisotropic; we must solve a complicated slowing-down problem before we can calculate the thermal-neutron spectrum. With the source located as in Fig. 6.38, to a good approximation the thermal source has only P_0 and P_1 Legendre components. The P_1 component can be determined in the diffusion approximation from spectral flux plots. This procedure is adequate if the geometry is such that the P_1 component of the source does not have a large effect on the measured spectra, as is the case for the geometry shown in Fig. 6.38. Measured spectra for this geometry are shown in Fig. 6.39, together with P_0 and P_1 transport calculations using the Nelkin bound-hydrogen model for water, Sec. 3-4.1, to calculate the P_0 and P_1 components of the scattering kernel. The results show that at the surface of the assembly the scattering anisotropy has considerable effect on the spectrum.

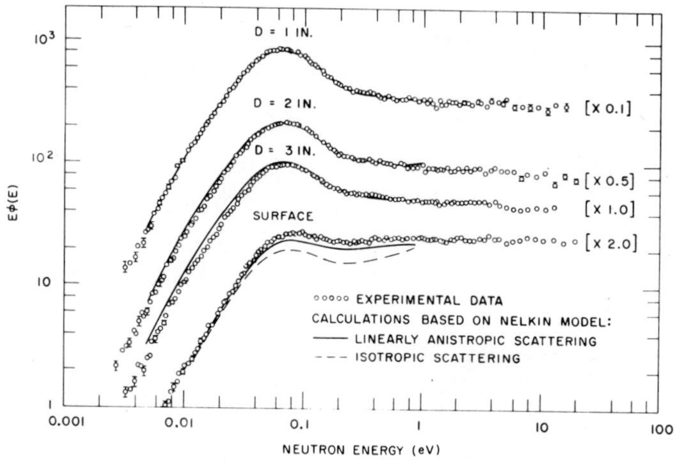

Figure 6.39 Neutron spectra in 4-in. water slab. To obtain the angular flux spectra at each position relative to the spectrum at D = 3 in. (D is as defined on Fig. 6.38), the calculated and measured results at D = 1 in., D = 2 in., and D = 4 in. (surface) must be multiplied by 0.1, 0.5, and 2.0, respectively, as shown by the factor in square brackets. (From Young et al., Nucl. Sci. Eng., Ref. 46.)

In summary, we conclude that the binding of protons in water significantly affects the thermal-neutron spectrum. The simple model described by Eqs. (3.95) through (3.97) accounts for most of the binding effect on spectra and, as we see in Chap. 9, this effect is accounted for with sufficient accuracy to permit a precise determination of the reaction rates that are important in the design of water-moderated reactors.

(b) Zirconium Hydride. Neutron spectra measured by Young *et al.*[47] at several temperatures in an assembly containing $ZrH_{1.75}$ and poisoned homogeneously with boron are shown in Figs. 640a, b, c, and d. The insert in each figure gives the ratio of calculated to measured spectra. The values of r calculated from the experimental data are listed in Table 6.2. The scattering kernel used to compute the theoretical curves is calculated numerically by the methods in Chap. 3, using the $f(\omega)$ given in Eqs. (3.107) through (3.109), but with $M_1/M_2 = 359$. This $f(\omega)$, which has a narrow band of frequencies around 0.14 eV and only a small fraction of low-frequency components, corresponds very nearly to an Einstein oscillator with $\hbar\omega_0 = 0.14$ eV.

Table 6.2

RATIO OF ABSORPTION RATE TO SLOWING-DOWN RATE IN ZIRCONIUM HYDRIDE SPECTRUM MEASUREMENTS

$$N_H = 3.799 \times 10^{22}$$

$$N_B = 1.480 \times 10^{20}$$

$$N_{Zr} = 2.176 \times 10^{22}$$

Temperature (°K)	Ratio, r
296	1.067
423	1.004
589	1.034
741	1.039

Except at the highest temperatures, the spectral effects of the nearly quantized energy transfers in zirconium hydride are clearly

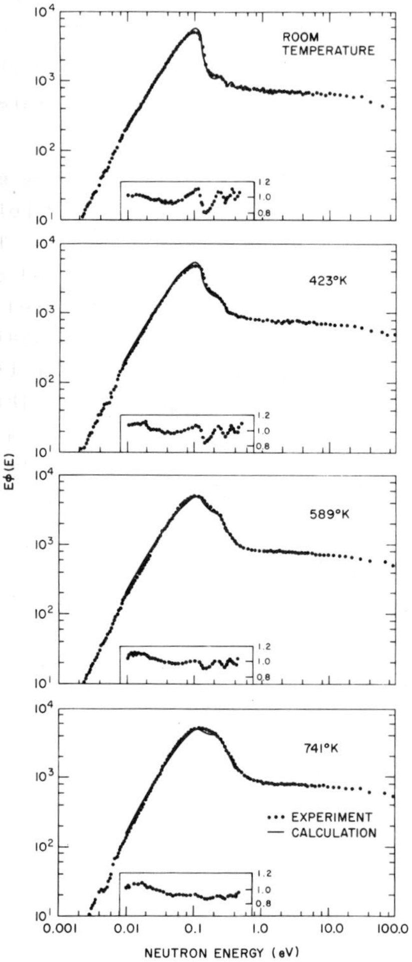

Figure 6.40 Neutron spectra in borated zirconium hydride: (a) at room temperature; (b) at 423°K; (c) at 589°K; and (c) at 741°K. The inserts in each part of this figure show the ratio of the experimental spectrum to the calculated spectrum based on the bound-hydrogen model. (From Young, USAEC Report GA-5319, Ref. 44.)

manifest. These effects are more pronounced in the theoretical curves, suggesting perhaps that the width of the optical level near 0.14 eV (see Eq. (3.109)) is less than it should be and/or that the value M_1 = 360 m is too large. The most important conclusion nevertheless is that the neutron spectra in poisoned zirconium hydride shift markedly with temperature and in very nearly the manner predicted by using the model corresponding to Eqs. (3.107) through (3.109).

(c) Polyethylene. A comparison of spectra measured in polyethylene with those calculated by Goldman and Federighi[48] and more recently by Young and Koppel[49] is shown in Fig. 6.41. The value of r is calculated from the experimental data to be 1.10; thus the small difference between the measured spectrum and spectra computed from Eq. (5.1) is not entirely meaningful. Goldman and Federighi use a discrete oscillator model similar to that given by Eq. (3.95) but with fundamental frequencies and effective masses that are supposedly appropriate for polyethylene. Actually, these frequencies are not completely consistent with those observed in the infrared absorption spectrum of polyethylene.[49] In their calculation Young and Koppel[49] employ a discrete oscillator model but with frequencies that are compatible with those measured by the methods of infrared spectroscopy.

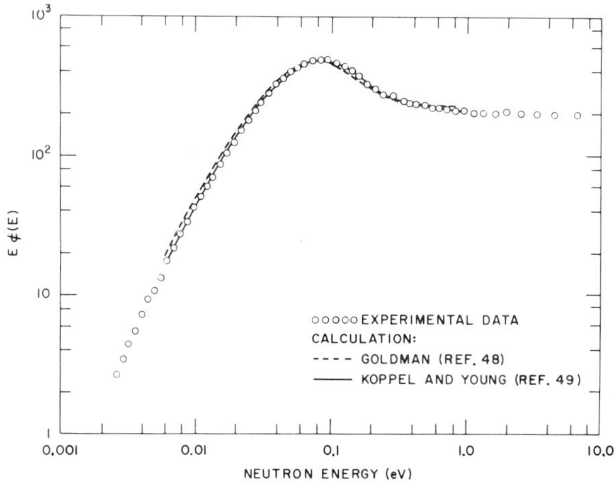

Figure 6.41 Neutron spectrum in 1% borated polyethylene. (From Koppel and Young, Nucl. Sci. Eng., Ref. 49.)

The use of a discrete oscillator model is dictated more by the structure of the available programs for calculating neutron scattering than by physical considerations. A more realistic theory of neutron scattering by polyethylene will take account of the distributed nature of $f(\omega)$ of polyethylene.

(d) Benzene. In Fig. 6.42 the measured[40] spectrum in benzene is compared with the calculation of Boffi et al.[50,51] Here, $r = 0.96$. The results for benzene are interesting because we expect that its gross thermalization properties do not differ significantly from those of the polyphenyls that are used as coolants and moderators in some reactors.

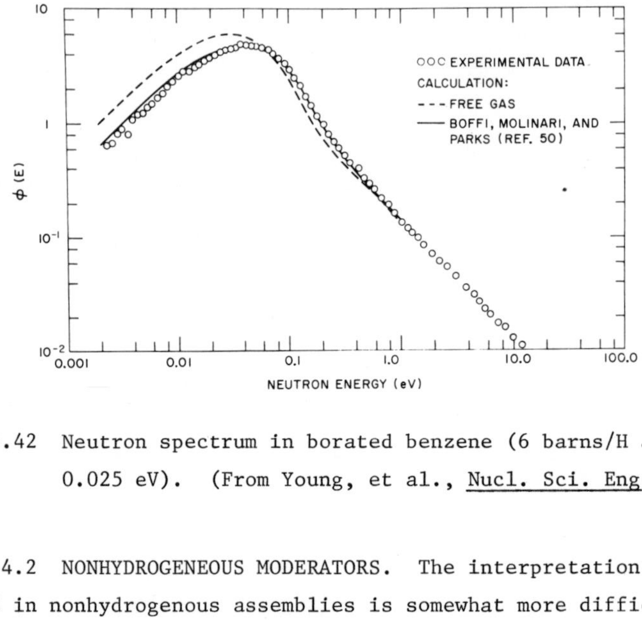

Figure 6.42 Neutron spectrum in borated benzene (6 barns/H atom at 0.025 eV). (From Young, et al., Nucl. Sci. Eng., Ref. 40.)

6-4.2 NONHYDROGENEOUS MODERATORS. The interpretation of spectra measured in nonhydrogenous assemblies is somewhat more difficult than it is in hydrogenous assemblies. We must correct for the self-shielding effects resulting from an inhomogeneous distribution of neutron-absorbing materials. At least equally important are the deviations from quasi-infinite-medium conditions produced by neutron diffusion. These deviations, though small, are usually greater than in water assemblies, and a precise comparison between theory and experiment requires that we account for the effects of neutron diffusion. We can do this either by one of the methods described in Sec. 5-9 or by numerically solving either the diffusion or transport equation with the prescribed source distribution and boundary conditions.

(a) Graphite. There are a number of recent measurements of thermal-neutron spectra in graphite.[1,38,52] Figures 6.43 through 6.47 are spectra measured in the graphite assembly shown in Fig. 6.2. The temperature of the experimental assembly was varied from 323°K to 810°K. The boron-containing foils are 0.005 in. thick and are separated by 0.525 in. of graphite. This yields a volume-averaged absorption cross section of about 0.4 barn per carbon atom at a neutron energy of 0.025 eV. In the interpretation of these experimental results, corrections were made for the self-shielding effects in the foils and for the effects of neutron diffusion. The latter corrections were calculated by the method of localized buckling described in Sec. 5-9. For this purpose it was necessary to know the spatial dependence $Q(\mathbf{r})$ of the source of thermal neutrons slowing down past indium resonances. For the assembly shown in Fig. 6.2, with the source located on one side, $Q(\mathbf{r})$ was determined by an age-theory fit to the measured spatial dependence of the activations of cadmium-covered indium foils. The theoretical spectra were computed using (1) a free-gas kernel and (2) a kernel that is calculated from the Yoshimori and Kitano frequency distribution, shown in Fig. 3. , by the methods described in Chap. 3.

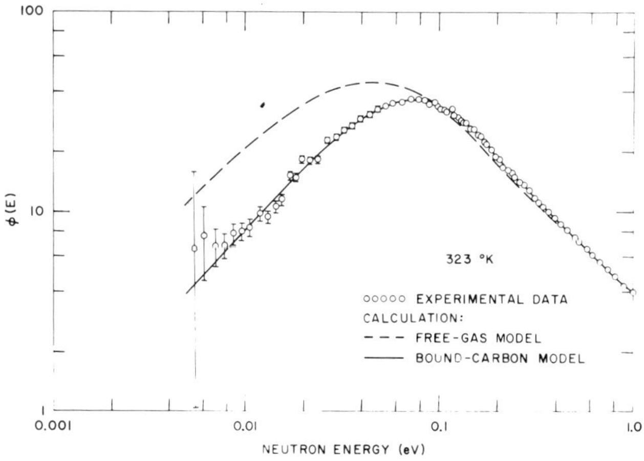

Figure 6.43 Neutron spectrum in boron-poisoned graphite at 323°K. (From Parks, Beyster, and Wikner, Nucl. Sci. Eng., Ref. 1.)

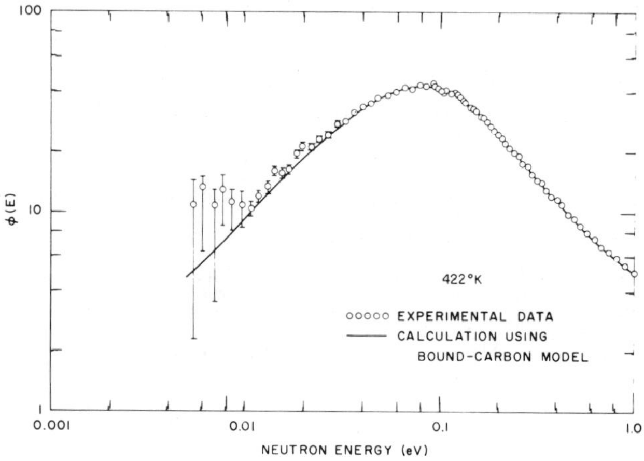

Figure 6.44 Neutron spectrum in boron-poisoned graphite at 422°K. (From Parks, Beyster, and Wikner, Nucl. Sci. Eng., Ref. 1.)

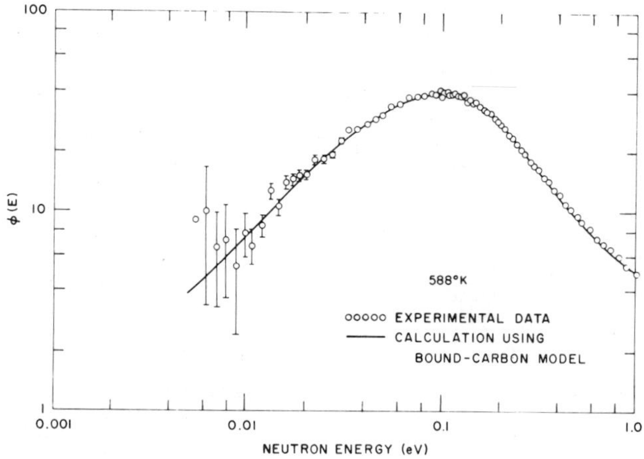

Figure 6.45 Neutron spectrum in boron-poisoned graphite at 588°K. (From Parks, Beyster, and Wikner, Nucl. Sci. Eng., Ref. 1.)

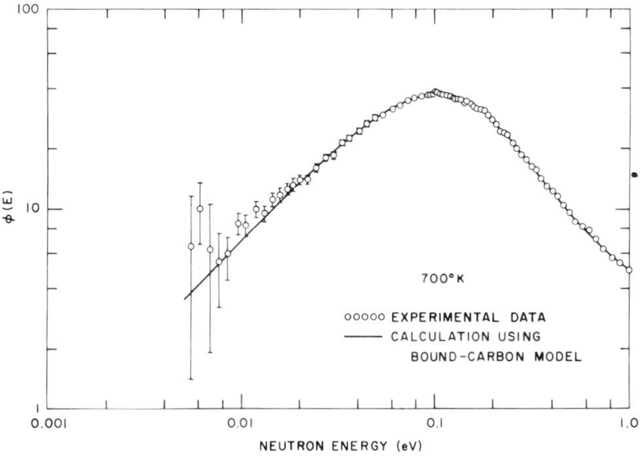

Figure 6.46 Neutron spectrum in boron-poisoned graphite at 700°K. (From Parks, Beyster, and Wikner, Nucl. Sci. Eng., Ref. 1.)

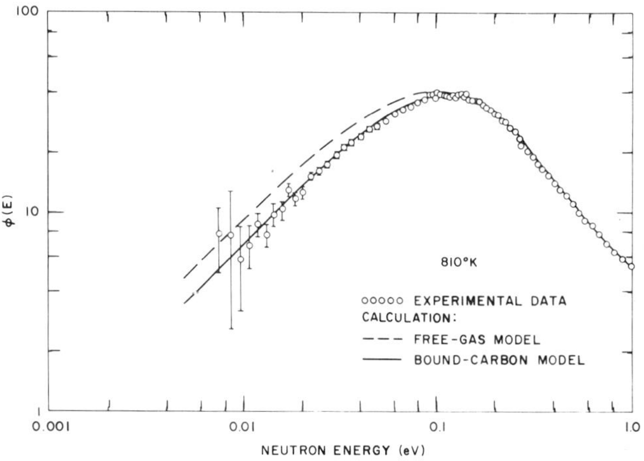

Figure 6.47 Neutron spectrum in boron-poisoned graphite at 810°K. (From Parks, Beyster, and Wikner, Nucl. Sci. Eng., Ref. 1.)

The spectra obtained from the bound-carbon model agree extremely well with the experimental results. In contrast, predictions based on the free-gas model are considerably different from the measured spectra, especially at 323°K. Even at 810°K, the free-gas result differs significantly from the experimental one.

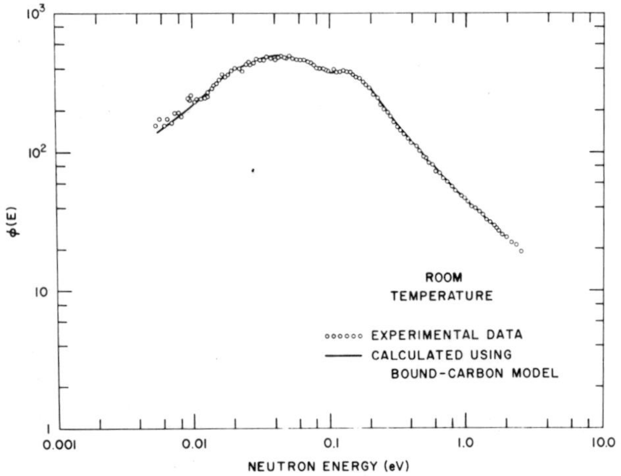

Figure 6.48 Neutron spectrum in samarium-poisoned graphite at room temperature. (From Bardes et al., Trans. Am. Nucl. Soc., Ref. 52.)

Figures 6.48 through 6.50 show spectra at 80°K, 300°K, and 600°K measured by Bardes et al.[52] in a graphite assembly poisoned with closely spaced foils containing samarium. The geometry of the assembly is essentially the same as the one shown in Fig. 6.2; again, the fast-source neutrons are produced at the center of one of the cube faces. The theoretical spectra are obtained from a 30-thermal-energy-group solution of a one-dimensional multigroup transport theory code.[53] Corrections for foil self-shielding effects are included in the calculation. There is excellent agreement between theory and experiment at 300°K, but at 80°K and 600°K the differences become as large as 15 percent. In the absence of a relation between the rate of absorption and the rate of slowing down from epithermal energies, it is difficult to determine the significance of these discrepancies.

There does exist an investigation of the sensitivity of calculated

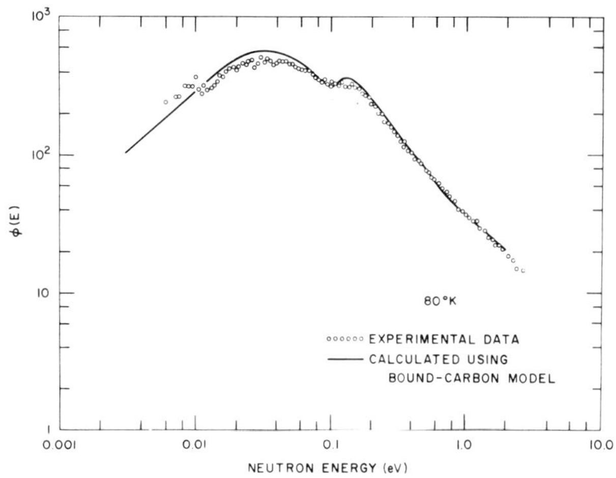

Figure 6.49 Neutron spectrum in samarium-poisoned graphite at 80°K. (From Bardes et al., Trans. Am. Nucl. Soc., Ref. 52.)

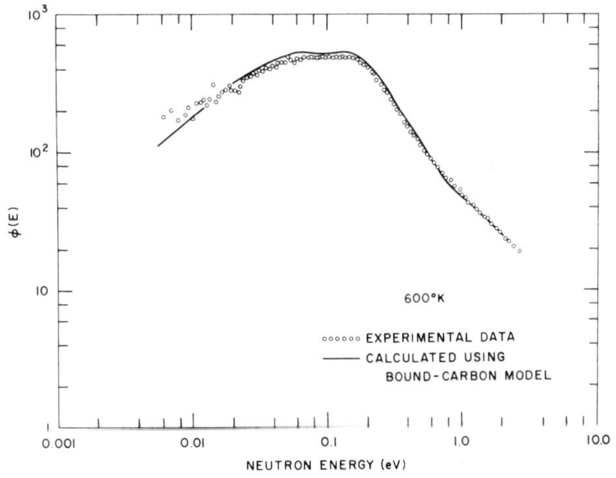

Figure 6.50 Neutron spectrum in samarium-poisoned graphite at 600°K. (From Bardes et al., Trans. Am. Nucl. Soc., Ref. 52.)

neutron spectra in graphite to the theoretical assumptions on which the scattering law calculations are based.[54] This investigation, which is discussed in Chap. 7, shows that, first, the anisotropy of carbon-atom vibrations has a negligible effect on the spectrum and, second, that the frequency spectra of Eqs. (3.101) and (3.102) and those in Fig. 3.29

lead to essentially the same neutron spectrum. Neither of these results is surprising, the second merely demonstrating that the neutron spectrum is not sensitive to the detailed features of the lattice-vibration spectrum.

(b) Heavy Water. Neutron spectra in D_2O systems have been measured by both the pulsed-source[55] and chopped-beam[12] techniques. The pulsed-source measurements were made at three positions within a $45 \times 45 \times 25\ cm^3$ rectangular assembly in which the neutrons are moderated by high-purity D_2O. A high purity must be achieved in order to reduce to acceptable propertions the considerable rate of neutron thermalization that is associated with even small concentrations of hydrogen atoms.

The assembly is poisoned with boron-aluminum foils whose dimensions are $45 \times 45 \times 0.05\ cm^3$; these foils are spaced every 0.32 cm and are placed so that the 45 cm × 45 cm face is perpendicular to the axis of the re-entrant tube. The base of the latter terminates at the midplane between the two assembly sides having the largest area, and the axis of the tube is always perpendicular to the direction of the flux gradient. The spectra were measured at 15.24, 22.86, and 30.48 cm from the source, which is located at the midpoint of a 45 cm × 25 cm side of the assembly.

Theoretical spectra are obtained as numerical solutions of the one-dimensional transport equation by means of the computer code GAPLSN.[53] The calculation takes account of the anisotropy of the scattering by D_2O. The P_0 and P_1 components of $\sigma(E_0,E,\theta)$ are obtained in the incoherent approximation from the dynamical model described in Sec. 3-4.1. The position dependence of the thermal-neutron source was determined by measuring the activation of cadmium-covered indium foils. The measured indium-resonance flux has a cosine distribution in each of the two principal assembly directions that are perpendicular to the line determined by the source and the base of the re-entrant tube. The rate of neutron leakage in these two directions is approximated as an additional absorption term in the one-dimensional transport equation.

Figure 6.51 compares the calculated and measured spectra for heavy water. In this figure we see the calculated variation of neutron flux with both energy and position. The experimental data are normalized to the calculational result at an energy of 1 eV and a position of 30.48 cm. The relative intensities at the other two positions were measured with sulfur and gold foil monitors.

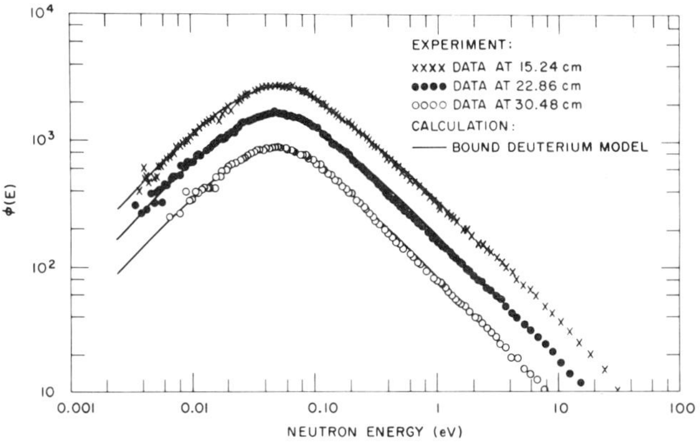

Figure 6.51 Neutron spectra in borated D_2O at various distances from the source. From Beyster et al., USAEC Report GA-5176, Ref. 55.)

(c) Beryllium and Beryllium Oxide. There exist measurements of thermal-neutron spectra in homogeneous beryllium and beryllium oxide[56] assemblies. However, inasmuch as the results have not been subjected to a detailed theoretical analysis, they are not discussed here.

REFERENCES

1. D. E. Parks, J. R. Beyster, and N. F. Wikner, Thermal Neutron Spectra in Graphite, *Nucl. Sci. Eng.*, **13**: 306 (1962).

2. J. R. Beyster *et al.*, *Integral Neutron Thermalization, Annual Summary Report, October 1, 1961, through September 30, 1962*, USAEC Report GA-3542, General Atomic Division, General Dynamics Corporation, 1962.

3. R. S. Stone and R. E. Slovacek, Neutron Spectra Measurements, *Nucl. Sci. Eng.*, **6**: 466 (1959).

4. G. A. Price, Brookhaven National Laboratory, private communication.

5. F. R. Barclay *et al.*, Low Energy Neutron Spectra in the ZENITH Heated Graphite Moderated Reactor, in *Proceedings of the Brookhaven Conference on Neutron Thermalization*, Vol. II, *Neutron Spectra in Lattices and Infinite Media*, USAEC Report BNL-719, p. 359, Brookhaven National Laboratory, 1962.

6. M. S. Coates and D. B. Gayther, *Time of Flight Measurements of Neutron Spectra in a Graphite Uranium Lattice at Different Temperatures*, British Report AERE-R-3829, September, 1961.

7. S. I. Bunch, J. R. Roesser, and R. E. Slovacek, Low Energy Measurements of the Neutron Spectrum, in *Naval Reactors Physics Handbook*, Vol. I, *Selected Basic Techniques*, A. Radkowsky, Ed., p. 1357, U.S. Atomic Energy Commission, Superintendent of Documents, U.S. Government Printing Office, Washington 25, D.C., 1964.

8. V. I. Mostovoi *et al.*, Neutron Spectrum Measurements in Uranium-Water Lattice, in *Proceedings of the Second United Nations Conference on the Peaceful Uses of Atomic Energy, Geneva, 1958*, Vol. 16, p. 254, United Nations, Geneva, 1958.

9. C. G. Campbell *et al.*, Measurements of Reactor Spectra by Time-of-Flight and Integral Measurements, in *Proceedings of the Second United Nations Conference on the Peaceful Uses of Atomic Energy, Geneva, 1958*, Vol. 16, p. 233, United Nations, Geneva, 1958.

10. J. R. Beyster *et al.*, *Integral Neutron Thermalization, Annual Summary Report, October 1, 1960, through September 30, 1961*, USAEC Report GA-2544, General Atomic Division, General Dynamics Corporation, 1961.

11. J. E. Faulkner and R. E. Heineman, Hanford Laboratories, private communication, 1960.

12. E. Johansson, E. Lampa, and N. G. Sjöstrand, A Fast Chopper and Its Use in the Measurement of Neutron Spectra, *Arkiv Fysik*, **18**: 513 (1961).

13. H. C. Honeck, Brookhaven National Laboratory, private communication.

14. W. L. Whittemore, *Differential Neutron Thermalization, Annual Summary Report, October 1, 1961, through September 30, 1962,* USAEC Report GA-3409, General Atomic Division, General Dynamics Corporation, 1962.

15. G. Caglioti and F. F. Casali, Rubber Soller Slit Collimators for Neutron Spectrometry, *Rev. Sci. Instr.*, **33**: 1103 (1962).

16. W. J. Price, *Nuclear Radiation Detectors*, McGraw-Hill Book Company, New York, 1958.

17. L. A. Wraight, D. H. C. Harris, and P. A. Egelstaff, *Improvement in Thermal Neutron Scintillator Detectors for Time of Flight Studies*, British Report AERE-R-4591, October, 1964.

18. M. J. Poole, Measurement of Neutron Spectra in Moderators and Reactor Lattices--I, Aqueous Moderators, *J. Nucl. Energy*, **5**: 325 (1957).

19. J. R. Beyster, Status of Neutron Spectra Measurements, in *Proceedings of the Brookhaven Conference on Neutron Thermalization*, Vol. II, *Neutron Spectra in Lattices and Infinite Media*, USAEC Report BNL-719, p. RC-1, Brookhaven National Laboratory, 1962.

20. M. J. Poole, General Survey on the Use of Time-of-Flight Methods for Reactor Spectrum Measurements, in *Neutron Time-of-Flight Methods*, Proceedings of a Symposium Organized by European-American Nuclear Data Committee in Collaboration with Centre d'Etudes Nucleaires de Saclay (France) and Institut National des Sciences et Techniques Nucleaires, Saclay, 24-27 July 1961, J. Spaepen, Ed., p. 221, European Atomic Energy Community (EURATOM), Brussels, September, 1961.

21. W. C. Barber and W. D. George, Neutron Yields from Targets Bombarded by Electrons, *Phys. Rev.*, **116**: 1550 (1959); some of the data appearing in Fig. 6.15b are from M. B. Biram, *Neutron Production by Electron Bombardment of Uranium*, British Report AERE-T/R-1523, October, 1954.

22. B. M. Bunin *et al.*, Operating Experience with IBR Reactor, Its Use for Neutron Investigations and Characteristics on Neutron Injection from a Microtron, in *Proceedings of the Third United Nations Conference on the Peaceful Uses of Atomic Energy, Geneva, 1964,* Vol. 7, p. 473, United Nations, New York, 1965.

23. J. B. Sykes, *Feasibility of a Proposed Method of Measurement of Thermal Neutron Spectra in a Thermal Pile*, British Report AERE-T/R-1367, 1954.

24. R. E. Nather, An All-Transistor Time Analyzer Attachment for Multi-Channel Pulse Height Analyzers, *Trans. Am. Nucl. Soc.*, **3**: 286 (1960).

25. J. U. Koppel, *Mean Emission Time Correction for the Measurement of Stationary Spectra by the Pulsed Neutron Source Technique*, USAEC Report GAMD-4131, General Atomic Division, General Dynamics Corporation, April 1, 1963.

26. L. M. Bollinger, R. E. Coté, and G. E. Thomas, A Critical Analysis of the Design and Application of Neutron Choppers, in *Proceedings of the Second United Nations Conference on the Peaceful Uses of Atomic Energy, Geneva, 1958*, Vol. 14, p. 239, United Nations, Geneva, 1958.

27. R. D. Lowde, The Principles of Mechanical Neutron-Velocity Selection, *J. Nucl. Energy, Pt. A*, **11**: 69 (1960).

28. E. Fermi, J. Marshall, and L. Marshall, A Thermal Neutron Velocity Selector and Its Application to the Measurement of the Cross Section of Boron, *Phys. Rev.*, **72**: 193 (1947).

29. P. A. Egelstaff, The Operation of a Thermal Neutron Time-of-Flight Spectrometer, *J. Nucl. Energy*, **1**: 57 (1954).

30. R. Ramanna *et al.*, On the Spectrum of Neutrons Emerging from Moderators, in *Proceedings of the Second United Nations Conference on the Peaceful Uses of Atomic Energy, Geneva, 1958*, Vol. 16, p. 260, United Nations, Geneva, 1958.

31. K. H. Beckurts and W. Reichardt, The Use of a $T(d,n)He^4$ Neutron Source for Slow Neutron Spectrum Measurements, in *Neutron Time-of-Flight Methods* (*op. cit.*, Ref. 20), p. 239, European Atomic Energy Community (EURATOM), Brussels, September, 1961.

32. M. S. Coates, Chopper Time-of-Flight Measurements on a Graphite Uranium Lattice at Different Temperatures, in *Neutron Time-of-Flight Methods* (*op. cit.*, Ref. 20), p. 233, European Atomic Energy Community (EURATOM), Brussels, September, 1961.

33. R. E. Slovacek and R. S. Stone, Low Energy Spectra Measurements, in *Proceedings of the Neutron Thermalization Conference held in Gatlinburg, Tennessee, April 28-30, 1958*, USAEC Report ORNL-2739, Paper IV-D, Oak Ridge National Laboratory, September, 1959.

34. W. M. Lopez *et al.*, A Dynamic Digital Delay Generator for Use in Neutron Chopper Experiments, *Trans. Am. Nucl. Soc.*, **3**: 276 (1960).

35. R. E. Slovacek *et al.*, Measurement of Asymptotic Neutron Spectra in Multiplying Assemblies containing U^{235}, *Nucl. Sci. Eng.*, **21**: 329 (1965).

36. R. S. Stone and R. E. Slovacek, *Reactor Spectra Measurements Using a Neutron Time-of-Flight Spectrometer*, USAEC Report KAPL-1499, Knolls Atomic Power Laboratory, March 15, 1956.

37. R. T. Frost, *Correction of Measured Flux Scans for Finite Resolution Effects*, USAEC Report KAPL-M-RTF-5, Knolls Atomic Power Laboratory, October 5, 1956.

38. M. J. Poole, P. Schofield, and R. N. Sinclair, Some Measurements of Thermal-Neutron Spectra, in *Exponential and Critical Experiments*, Proceedings of the Symposium on Exponential and Critical Experiments Held by the International Atomic Energy Agency in Amsterdam, Netherlands, 2-6 September 1963, Vol. III, p. 87, International Atomic Energy Agency, Vienna, 1964.

39. J. R. Beyster *et al.*, Measurements of Neutron Spectra in Water, Polyethylene, and Zirconium Hydride, *Nucl. Sci. Eng.*, **9**: 168 (1961).

40. J. C. Young *et al.*, Neutron-Spectrum Measurements in H_2O, CH_2, and C_6H_6, *Nucl. Sci. Eng.*, **18**: 376 (1964).

41. K. Burkart and W. Reichardt, Observation of Hardened Neutron Spectra in Water and Boric Acid Solutions, in *Proceedings of the Brookhaven Conference on Neutron Thermalization*, Vol. II, *Neutron Spectra in Lattices and Infinite Media*, USAEC Report BNL-719, p. 318, Brookhaven National Laboratory, 1962.

42. V. I. Mostovoi *et al.*, Experimental Studies on Neutron Thermalization, in *Proceedings of the Third United Nations International Conference on the Peaceful Uses of Atomic Energy. Geneva, 1964*, Vol. 2, p. 280, United Nations, New York, 1965.

43. H. J. Amster, The Wigner-Wilkins Calculated Thermal Neutron Spectra Compared with Measurements in a Water Moderator, *Nucl. Sci. Eng.*, **2**: 394 (1957).

44. J. C. Young, *Theoretical and Experimental Neutron Spectra*, USAEC Report GA-5319, General Atomic Division, General Dynamics Corporation, May 22, 1964.

45. J. U. Koppel and J. A. Young, Neutron Scattering by Water Taking Into Account the Anisotropy of the Molecular Vibrations, *Nucl. Sci. Eng.*, **19**: 412 (1964).

46. J. C. Young *et al.*, Neutron Spectrum Measurements in H_2O and D_2O, *Nucl. Sci. Eng.*, **23**: 34 (1965).

47. J. C. Young *et al.*, Neutron Thermalization in Zirconium Hydride, *Nucl. Sci. Eng.*, **19**: 230 (1964).

48. D. T. Goldman and F. D. Federighi, Calculation of Thermal Neutron Flux Spectra in a Polyethylene Moderated Medium, *Nucl. Sci. Eng.*, **16**: 165 (1963).

49. J. U. Koppel and J. A. Young, Neutron Scattering by Polyethylene, *Nucl. Sci. Eng.*, **21**: 268 (1965).

50. V. C. Boffi, V. C. Molinari, and D. E. Parks, Slow Neutron

Scattering and Thermalization by Benzene and Other Polyphenyls, in *Proceedings of the Brookhaven Conference on Neutron Thermalization*, Vol. I, *The Scattering Law*, USAEC Report BNL-719, p. 69, Brookhaven National Laboratory, 1962.

51. V. C. Boffi, V. G. Molinari, and D. E. Parks, Slow Neutron Scattering by Benzene, in *Proceedings of the Symposium on Inelastic Scattering of Neutrons in Solids and Liquids, held at Chalk River, Canada, 10-14 September 1962*, Vol. I, p. 285, International Atomic Energy Agency, Vienna, 1963.

52. R. G. Bardes *et al.*, Neutron Spectra in Graphite Poisoned with Samarium, *Trans. Am. Nucl. Soc.*, **6**: 295 (1963).

53. J. H. Alexander, G. W. Hinman, and J. R. Triplett, *GAPLSN, A Modified DSN Program for the Solution of the One-Dimensional Anisotropic Transport Equation*, General Atomic Report GA-4972, March 16, 1964.

54. N. F. Wikner, G. D. Joanou, and D. E. Parks, Neutron Thermalization in Graphite, *Nucl. Sci. Eng.*, **19**: 108 (1964).

55. J. R. Beyster *et al.*, *Integral Neutron Thermalization, Quarterly Progress Report for the Period Ending March 31, 1964*, USAEC Report GA-5176, p. 2, General Atomic Division, General Dynamics Corporation, April 17, 1964.

56. *Ibid.*, p. 6.

Chapter 7

REACTOR CROSS SECTIONS: SOME APPLICATIONS TO HOMOGENEOUS GRAPHITE REACTORS

As new reactor concepts develop, the quantitative calculation of neutron scattering processes in reactor systems acquires considerable engineering importance. Inasmuch as new reactor systems must have higher performance, they must be designed within close tolerances. To meet the new design demands the reactor analyst requires more detailed knowledge of the spatial and energy distribution of the neutron flux. In order to obtain this information at thermal energies, models for thermal-neutron scattering had to be developed. As discussed in Chap. 6, these models successfully explain the results of measurements of neutron spectra for a wide variety of moderators, poison concentrations, and temperatures. In this chapter we apply some of these models to a few simple reactor systems, our object being to determine the importance of thermalization effects on thermal-neutron reaction rates and on other quantities of interest in reactor design.

In the design of early reactors the requirements for accurate detail were small. In the early systems, the neutrons causing fission were well thermalized. There was little leakage from the core, and core lifetimes were so short that depletion effects (the depletion of fuel, the buildup in concentration of fission products, and the conversion of fertile material to fissionable material) other than those produced by the transient fission-product poisons could be treated as perturbations. The calculations could be performed with slide rules and desk calculators, and experiments, e.g., critical and subcritical assemblies, could simulate the time- and temperature-dependent characteristics of the reactor rather well.

In few, if any, reactors currently under design or construction are any of these simple conditions realized. The densities of the fuel, fertile material, and other absorbing materials are generally so high that a large fraction of the neutrons slowing down below a few electron volts are absorbed before they come into thermal equilibrium with the moderator. It is thus important to make an accurate evaluation of the competition at thermal energies between energy change and absorption in order to obtain adequate estimates of thermal-neutron reaction rates. In the early natural uranium-graphite reactors, light-water-moderated swimming pool reactors and in modern heavy-water-moderated reactors like CANDU, the thermal spectrum is almost Maxwellian. In these lightly poisoned systems the spectrum follows the moderator temperature and only an approximate treatment of the thermalization problem is required.

Another complication present today is the high thermal energy of new and proposed reactor systems. For the large power-producing graphite reactors, the mean temperature is approximately 1300°K (0.11 eV). For solid-core nuclear reactors used in the propulsion of rockets, the mean core temperature is probably near 2000°K (0.17 eV), while proposed gas-core reactors used for propulsion might have "core" temperatures near 10,000°K (0.84 eV). At the latter temperature, we must include the complete Doppler-broadened cross section in a description of thermal-neutron reaction rates (see Sec. 7-1).

In the design of reactors having short reactivity lifetimes we can approximate the stable-fission-product absorptions by a time-dependent absorption in a 1/v thermal poison. Long-life converters produce power at the end of core life from the fissile material produced during life, most of the original fuel being consumed. For the latter converters, we must consider the individual fission products, taking into account successive transmutations and distinguishing between thermal and epithermal absorptions. Many of the fission products produced have resonances at or near thermal energies so that their capture cross sections do not vary as 1/v. In some cases, these thermal-resonance poisons have a relatively small effect on available excess reactivity during core life, but their effect on the prompt or slow temperature coefficient can be large.

The need for higher operating temperatures and the long lifetimes of efficient reactor systems result in formidable problems for the reactor designer. It is exceedingly expensive to design critical or mockup assemblies that adequately represent the temperature- and time-dependent history of the reactor. Accurate theories and experimental data are at a premium. The spectrum experiments described in Chap. 6 are exemplary, and critical experiments, some of which we discuss below, also prove useful in checking the validity of theories and models used in the design of semihomogeneous graphite reactors. (Many semi-homogeneous critical experiments have been performed as part of the Naval Reactor Program. Inasmuch as this work is thoroughly discussed elsewhere,[1] we mention it only briefly in this chapter.)

Significant problems occur in all reactor systems where there is a strong spatial dependence of the thermal-neutron spectrum. In water-moderated systems we must be able to calculate the flux peaking in regions of pure water and the changes in these flux peaks with temperature. We can do this with confidence only by solving the velocity-dependent neutron transport problem. For solid moderators such as graphite, large variations in the thermal-neutron spectrum occur across the core-reflector interface, particularly if the core is at a much higher temperature than the reflector.

In both the water-moderated and solid-moderated systems, fueled regions near these material and temperature discontinuities experience abnormally large fissioning rates and hence high local temperatures. The desire for higher performance thus leads to complex designs for the core. In almost every system, local and gross zoning of fuel, fertile material, and lumped burnable poisons are used to maintain power distributions within specified shapes and to minimize the number of movable control elements. Occasionally, the lumped poisons are also used to enhance a negative temperature coefficient. Some of these matters are discussed further in Chap. 9. In the following sections we confine our attention to the effects on reactor design arising from differences in the model used to calculate the scattering kernel.

The discussion in this chapter cannot be a complete treatment of the thermalization problem as it applies to all moderators or to all classes of reactors. It is hoped, instead, that by treating a few problems the discussion will be of assistance to reactor physicists and cesigners as an illustration of the application of accurate thermalization considerations to the practical problems of reactor analysis and design. The next section establishes the essential points in calculating neutron reaction rates at thermal energies. In particular, we examine the low-energy absorption resonances of several isotopes, our object being to determine when it is important to account for the Doppler-broadening of these resonances.

The theoretical and experimental results of earlier chapters show that chemical binding significantly influences thermal-neutron spectra in moderating materials. In Sec. 7-2 we consider the specific case of bare, homogeneous, graphite-moderated, U^{235}-fueled critical assemblies. We find that although the thermal-neutron spectrum itself is quite sensitive to the details of the scattering kernel, for quantities such as thermal utilization and critical eigenvalue, which involve ratios of absorptions, the spectral effects tend to cancel. Other quantities, such as isotopic reaction rates, are linear functionals of the spectrum. Hence, estimates of core lifetime and of the time-dependent temperature coefficient are affected by the models used in the calculation of the thermal-neutron spectrum. An example of these considerations appears in Sec. 7-3, where we calculate the temperature coefficient of a high-temperature graphite-thorium converter reactor.

7-1 THERMAL-NEUTRON REACTION RATES

To fully understand the physical behavior of nuclear reactors we require methods for computing spectra and spectrum-averaged cross sections. Many experiments and almost all reactor systems require the solution of the space-energy problem for neutrons at thermal energies. With current computational techniques we can treat this problem with considerable accuracy provided that we are able to calculate suitably defined average cross sections. In this section we apply the theories developed in previous chapters to the calculation of thermal-neutron reaction rates and average cross sections in an arbitrary mixture of absorbing and scattering materials.

7-1.1 THE DOPPLER-BROADENED CROSS SECTION. In infinite homogeneous reactors, we obtain the competition between fission, absorption, and leakage by solving the neutron-balance equation, Eq. (5.1). The calculated neutron spectrum $\phi(E)$ depends not only on the scattering cross sections but also on the absorption cross sections of all materials contained in a given medium. Although it is not explicitly shown in the equations of Chap. 5, the absorption cross sections depend, in general, on the temperature. Because of the Doppler effect this dependence may be strong. Of particular concern to us are those materials with resonances at or near thermal energies. The Doppler effect, which arises from the thermal motion of the capturing nuclei, broadens the resonance and lowers the peak value of the cross section. In a reactor the broadening of a thermal resonance can cause either an increase or decrease in the thermal reaction rate, depending on the spectrum and the energy of the resonance. Furthermore, this increase or decrease can occur even in the limit of small absorber concentrations. Consequently, in order to make valid assessments of the reaction rates in absorbers with resonances below a few electron volts, we must first examine the Doppler-broadened neutron-nucleus cross section at thermal energies.

In the measurement of a neutron cross section, a monoenergetic beam of neutrons strikes a thin target. For each neutron velocity, we measure attenuation by the target of the beam intensity. The cross section σ is obtained from the well-known relation $I = I_0 e^{-N\sigma x}$, where I is the intensity at the detector with the target in place, I_0 is the incident beam intensity, and N is the density of nuclei in the target whose thickness is x. This cross section does not represent the two-body interaction for one relative speed; instead, it gives an integrated cross section between neutrons of a given speed and target nuclei that have a distribution of speeds. We must be able to relate the latter cross section to the former before we can obtain the cross section for the neutron-nucleus interaction, as published, for example, in BNL-325.[2]

For this purpose we define the following quantities:[a]

---

[a] This discussion follows a private communication of J. E. Wilkins, Jr.,[3] very closely; also see Refs. 4 and 5, which used this same source.

$\mathbf{v}$ = velocity of incident neutrons in the laboratory system,

$\mathbf{V}$ = velocity of target nucleus in the laboratory system,

$v_r = |\mathbf{V} - \mathbf{v}|$ = relative speed of neutron and nucleus, and

$\sigma_t(v_r)$ = cross section of the two-body neutron-nucleus interaction.

Classically, the total number of collisions per cubic centimeter per second for a neutron at velocity $\mathbf{v}$ with nuclei having a velocity distribution function M(V) is

$$\int_0^\infty M(V) v_r \sigma_t(v_r) d\mathbf{V} = v\sigma(v) . \tag{7.1}$$

The effective cross section $\sigma(v)$ is the one measured in the experiment discussed above.

To carry out the integral in Eq. (7.1) we express $d\mathbf{V}$ in terms of V, v, and v_r, and we denote the angle between $\mathbf{v}$ and $\mathbf{V}$ by θ. From

$$v_r^2 = v^2 + V^2 - 2vV \cos\theta$$

we obtain, for fixed values of v and V,

$$d(\cos\theta) = -\frac{v_r dv_r}{vV} .$$

Hence,

$$d\mathbf{V} = V^2 dV 2\pi d(\cos\theta) = -\frac{2\pi V dV v_r dv_r}{v} . \tag{7.2}$$

Inserting this result into Eq. (7.1) gives

$$\sigma(v) = \frac{2\pi}{v^2}\int_0^\infty VM(V)dV \int_{|V-v|}^{|V+v|} v_r^2 \sigma_t(v_r) dv_r . \tag{7.3}$$

For a Maxwellian distribution of velocities of the target nuclei,

$$M(V) = \left(\frac{M}{2\pi k_B T}\right)^{\frac{3}{2}} \exp\left(-\frac{M}{2k_B T} V^2\right) . \tag{7.4}$$

After changing the order of integration in Eq. (7.3) and performing the integration over V, we get

$$\sigma(v) = \left(\frac{M}{2\pi k_B T}\right)^{\frac{1}{2}} \frac{1}{v^2} \int_0^\infty dv_r \sigma_t(v_r) v_r^2 \left\{ \exp\left[-\frac{M}{2k_B T}(v - v_r)^2\right] - \right.$$

$$\left. - \exp\left[-\frac{M}{2k_B T}(v + v_r)^2\right] \right\} . \qquad (7.5)$$

Generally, cross sections are published as functions of the energy $E = mv^2/2$ of the neutron in the laboratory system. If we introduce $E_r = mv_r^2/2$, Eq. (7.5) becomes[b]

$$\sigma(E) = \left(\frac{A\beta}{\pi}\right)^{\frac{1}{2}} \frac{1}{2E} \int_0^\infty dE_r E_r^{\frac{1}{2}} \sigma_t(E_r) \{\exp[-A\beta(E^{\frac{1}{2}} - E_r^{\frac{1}{2}})^2] -$$

$$- \exp[-A\beta(E^{\frac{1}{2}} + E_r^{\frac{1}{2}})^2]\} , \qquad (7.6)$$

where $A = M/m$ is the nucleus-neutron mass ratio and $\beta = (k_B T)^{-1}$.

The use of a Maxwellian distribution of target velocities in Eq. (7.3) is a good approximation if the target material is at a temperature above the Debye temperature $_D$. For material temperatures below $_D$ it is still a good approximation to use a Maxwellian distribution of target speeds but with the effective temperature $\bar{T}$ of Eq. (2.250) in place of the thermodynamic temperature.[6,7] For the resonance absorbers uranium, UO_2, and thorium, the Debye temperatures are 200°K, 160°K, and 168°K, respectively.

For Eq. (7.6) to be useful we require an expression for the two-body cross section $\sigma_t(E_r)$. At thermal energies, the Breit-Wigner single-level s-wave expression applies for the absorption cross section. For the radiative capture cross section, σ_t is[8]

$$\sigma_{n,\gamma}(E) = \left(\sigma_{n,\gamma}\right)_0 \left(\frac{E_0}{E}\right)^{\frac{1}{2}} \frac{\Gamma^2(E_0)}{4(E - E_0)^2 + \Gamma(E)^2} , \qquad (7.7)$$

---

[b] The use of the symbol σ to denote the cross section both as a function of velocity and as a function of energy should lead to no confusion.

where E_0 is the resonance energy and

$$\left(\sigma_{n,\gamma}\right)_0 = \left(\frac{A+1}{A}\right)^2 \frac{2\pi\hbar^2}{m} g_J \frac{1}{E_0^{\frac{1}{2}}} \frac{\Gamma_n^0 \Gamma_\gamma}{[\Gamma(E_0)]^2}$$

$$= (2.60 \times 10^6 \text{ eV-barns}) \left(\frac{A+1}{A}\right)^2 g_J \frac{1}{E_0^{\frac{1}{2}}} \frac{\Gamma_n^0 \Gamma_\gamma}{[\Gamma(E_0)]^2} . \qquad (7.8)$$

Here, $\Gamma(E) = \Gamma_\gamma + \Gamma_n^0 E^{\frac{1}{2}}$ is the total width at half maximum; Γ_γ is the gamma width, an energy independent quantity; Γ_n^0 is the reduced neutron width ($\Gamma_n^0 = \Gamma_n / E^{\frac{1}{2}}$); and the energy is measured in electron volts. The statistical weight factor g_J is equal to $(2J + 1)/2(2I + 1)$, where J is the spin of the compound neutron-nucleus system and I is the spin of the target nucleus. For most resonances J is unknown; when it is unknown, it is usually assumed that $g_J = 1$ if $I = 0$ or that $g_J = 0$ if $I \neq 0$. Inserting Eq. (7.7) into Eq. (7.6) we finally obtain the exact expression for the cross section Doppler-broadened by a Maxwellian distribution of target speeds:

$$\sigma_{n,\gamma}(E) = \frac{1}{2}\left(\sigma_{n,\gamma}\right)_0 \left(\frac{E_0}{E}\right)^{\frac{1}{2}} \left(\frac{A\beta}{\pi}\right)^{\frac{1}{2}} \int_0^\infty dE_r \frac{\Gamma^2(E_0)}{\Gamma^2(E_r) + 4(E_r - E_0)^2} \times$$

$$\times E^{-\frac{1}{2}} \{\exp[-A\beta(E_r^{\frac{1}{2}} - E^{\frac{1}{2}})^2] - \exp[-A\beta(E_r^{\frac{1}{2}} + E^{\frac{1}{2}})^2] \quad . \qquad (7.9)$$

This expression reduces to a well-known result of Bethe and Placzek[8] if

1) $\Gamma(E_r) = \Gamma(E_0)$, i.e., if $\Gamma(E_r)$ does not vary appreciably over the resonances;
2) the term $\exp[-A\beta(E^{\frac{1}{2}} + E^{\frac{1}{2}})^2]$ is negligible compared with $\exp[-A\beta(E_r^{\frac{1}{2}} - E^{\frac{1}{2}})^2]$;
3) we approximate $E_r^{\frac{1}{2}}$ in the first exponential by

$$E_r^{\frac{1}{2}} = E^{\frac{1}{2}}\left(1 + \frac{E_r - E}{E}\right)^{\frac{1}{2}} \approx E^{\frac{1}{2}}\left(1 + \frac{E_r - E}{2E}\right) \approx E^{\frac{1}{2}} + \tfrac{1}{2}E^{-\frac{1}{2}}(E_r - E)$$

and the factor $E^{-\frac{1}{2}}$ just before the exponential by $E_0^{-\frac{1}{2}}$; and
4) we extend the lower limit of integration to $-\infty$.

Defining the Doppler width $\Delta = 2(E_0 k_B T/A)^{\frac{1}{2}}$ and making the above approximations, we find

$$\sigma_{n,\gamma}(E) = \frac{(\sigma_{n,\gamma})_0}{\sqrt{\pi}\Delta}\left(\frac{E_0}{E}\right)^{\frac{1}{2}} \int_{-\infty}^{\infty} \frac{\exp[-(E_r - E)^2/\Delta^2]}{[2(E_r - E_0)/\Gamma]^2 + 1}\, dE_r \ . \qquad (7.10)$$

When we introduce

$$x = \frac{2(E - E_0)}{\Gamma} ,$$

$$y = \frac{2(E_r - E_0)}{\Gamma} ,$$

$$\xi = \Gamma/\Delta ,$$

we obtain the Bethe-Placzek formula,

$$\sigma_{n,\gamma}(E,\Delta) = (\sigma_{n,\gamma})_0 \left(\frac{E_0}{E}\right)^{\frac{1}{2}} \Psi(\xi,x) , \qquad (7.11)$$

where $\Psi(\xi,x)$ is the shape function defined by

$$\Psi(\xi,x) = \frac{\xi}{2\sqrt{\pi}} \int_{-\infty}^{\infty} \frac{\exp[-\xi^2(x - y)^2/4]}{1 + y^2}\, dy \ . \qquad (7.11a)$$

Because of the approximations (2) through (4) listed above, this is an accurate result only if $E_0 >> k_B T$.

When $T \to 0$, then $\Delta \to 0$ and $\xi \to \infty$, and $\Psi(\xi,x)$ approaches the natural line shape $(1 + x^2)^{-1} = \{1 + [4(E - E_0)^2/\Gamma^2]\}^{-1}$; this is the case of weak Doppler broadening for which ζ, the ratio of the natural width to the Doppler width, is large. When $\xi \to 0$, then $\Gamma << \Delta$; thus, for $x << 1/\xi$,

$$\Psi(\xi,x) = \tfrac{1}{2}\sqrt{\pi}\xi \exp(-\tfrac{1}{4}\xi^2 x^2) = \tfrac{1}{2}\sqrt{\pi}\,\frac{\Gamma}{\Delta} \exp\left[-\frac{(E - E_0)^2}{\Delta^2}\right] .$$

This is the case of strong broadening and, except in the wings of the resonance, the line shape is a Gaussian whose width is the Doppler width Δ. For $x^2 >> 6/\xi^2$, the term $\Psi(\xi,x)$ is always asymptotic to the natural line shape.

Table 7.1

THE RATIO ξ OF THE NATURAL WIDTH Γ TO THE DOPPLER WIDTH Δ FOR THE THERMAL RESONANCES OF THE HEAVY ELEMENTS AND FISSION PRODUCTS

	In Electron Volts						ξ			
Isotope	E_0	Γ_γ	Γ_f	Γ_n	Γ	Δ (300°K)	300°K	1200°K	2400°K	10,000°K
Xe^{135}	0.084	0.0907	---	0.0257	0.1164	0.00802	14.51	7.26	5.13	2.51
Nd^{145}	4.37	0.048	---	0.00128	0.0493	0.0558	0.882	0.442	0.3124	0.153
Pm^{147}	1.04	0.080	---	0.0000055	0.080	0.0271	2.95	1.48	1.045	0.511
	5.43	0.080	---	0.033	0.113	0.0618	1.83	0.917	0.646	0.317
Sm^{149}	0.0976	0.063	---	0.00050	0.0635	0.00822	7.725	3.863	2.72	1.34
	0.873	0.060	---	0.00082	0.0608	0.0246	2.472	1.233	0.872	0.428
	4.98	0.067	---	0.00226	0.0693	0.0587	1.181	0.591	0.416	0.205
Sm^{151}	1.10	0.062	---	0.00052	0.0625	0.0274	2.281	1.142	0.8054	0.395
	1.70	0.062	---	0.00030	0.0623	0.0341	1.827	0.913	0.645	0.316
	2.04	0.062	---	0.00053	0.0625	0.0374	1.671	0.836	0.592	0.289
Eu^{153}	0.457	0.097	---	0.00001	0.097	0.0176	5.51	2.76	1.94	0.954
	1.730	0.092	---	0.000061	0.092	0.0342	2.69	1.345	0.950	0.465
	2.456	0.091	---	0.0013	0.0923	0.0414	2.23	1.12	0.788	0.379
Gd^{155}	2.01	0.110	---	0.00028	0.1103	0.0367	3.005	1.502	1.062	0.52
	2.64	0.111	---	0.00248	0.1135	0.042	2.702	1.351	0.955	0.468
Gd^{157}	0.030	0.100	---	0.00065	0.1006	0.00445	22.66	11.33	8.01	3.92
	2.90	0.097	---	0.00036	0.09736	0.0438	2.22	1.11	0.788	0.38
U^{233}	0.10	0.056	0.994	0.0000185	1.05	0.00666	157.66	78.83	55.73	27.31
	1.45	0.054	0.716	0.000183	0.770	0.02537	30.35	15.175	10.73	5.26
	1.76	0.049	0.231	0.000311	0.28	0.0279	10.04	5.02	3.55	1.74
	2.30	0.047	0.049	0.000176	0.0962	0.0320	3.01	1.51	1.064	0.521
U^{235}	0.273	0.029	0.099	0.0000029	0.128	0.01096	11.68	5.84	4.13	2.02
	1.140	0.044	0.1246	0.0000172	0.1686	0.02241	7.52	3.76	2.660	1.30
	2.035	0.035	0.012	0.0000077	0.047	0.02994	1.57	0.785	0.555	0.272
Pu^{239}	0.296	0.0386	0.0554	0.000114	0.094	0.01132	8.3	4.16	2.94	1.44
	7.9	0.038	0.042	0.00132	0.080	0.05849	1.37	0.68	0.484	0.237
Pu^{240}	1.054	0.034	0.000007	0.00236	0.03636	0.0213	1.707	0.853	0.603	0.296
Pu^{241}	0.25	0.043	0.099	0.000055	0.142	0.01034	13.73	6.86	4.85	2.378
	4.31	0.043	---	0.000955	---	0.043	---	---	---	---
Rh^{103}	1.257	0.155	---	0.00078	0.156	0.0352	4.44	2.22	1.57	0.77
Cd^{113}	0.178	0.113	---	0.00065	0.114	0.01275	8.94	4.47	3.16	1.55

*From Wikner and Jaye, USAEC Report GA-2113, Ref. 9.

Doppler broadening produces significant effects on reaction rates when $\xi \lesssim 1$, i.e., when $\Gamma \lesssim \Delta$. Table 7.1 gives values of ξ for 300°, 1200°, 2400°, and 10,000°K for the positive low-energy resonances of U^{233}, U^{235}, Pu^{240}, Pu^{241}, Rh^{103}, Cd^{113}, Xe^{135}, Sm^{149}, Sm^{151}, Gd^{155}, Gd^{157}, and Er^{167}. From this tabulation[9] we can determine those resonances below 3.0 eV for which the Doppler effect might be important for any given temperatures. For most isotopes, when the resonance energy $E_0 \lesssim 1.0$ eV, $\xi >> 1$ except at extremely high temperatures; however, there are a few isotopes with resonances at thermal energies that have very small natural widths.

For the temperatures greater than 1200°K or 2400°K there is significant Doppler broadening of the 2.3-eV resonance in U^{233}, the 2.035-eV resonance in U^{235}, the 1.054-eV resonance in Pu^{240}, all the resonances of Sm^{151}, and the resonances at a few electron volts in Nd^{145}, Pm^{147}, Sm^{149}, Eu^{153}, Gd^{155}, and Gd^{157}. For gaseous core reactors, the Doppler effect is significant for all materials with resonances at thermal energy. Only a few resonances in U^{233}, U^{235}, Pu^{241}, Cd^{113}, Xe^{135}, and Gd^{157} have a predominantly natural line shape at 10,000°K.

Near room temperature, since the natural width is much larger than the Doppler width for the low-energy resonances of most materials, Doppler broadening is not an important consideration in the thermal energy region. This is the case for cadmium and samarium--poisons that are frequently used to simulate the effects on thermal spectra of fissionable materials like Pu^{239}. For this reason, this effect was not considered in Chap. 6.

7-1.2 THE CALCULATION OF NEUTRON SPECTRA AND AVERAGE CROSS SECTIONS. We mentioned in Chap. 4 that we must distinguish the variation of $\sigma_a(E)$ with temperature from the variation of a spectrum-averaged cross section with temperature. The former variation, which arises from the Doppler effect, is often negligible at thermal energies. The latter variation is usually significant. It arises from the temperature dependence of the neutron energy distribution, which, in turn, arises from the temperature dependence of the energy-transfer cross sections in the Boltzmann equation. The determination of the temperature variation of the spectrum-averaged cross section is one of the primary objectives of the following discussion.

We can define the average of the cross section $\bar{\sigma}$ over the entire spectrum at a single position in a reactor in terms of the reaction rate associated with that cross section:[c]

$$\begin{aligned}\text{Reaction rate} &= N\bar{\sigma}\int_0^\infty vn(v)dv \\ &= N\int_0^\infty n(v)dv\int_0^\infty M(V)|\mathbf{V}-\mathbf{v}|\sigma_t(|\mathbf{V}-\mathbf{v}|)d\mathbf{V} \\ &= N\int_0^\infty vn(v)\sigma(v)dv\ ,\end{aligned} \tag{7.12}$$

where $\sigma(v)$ is the cross section defined by Eq. (7.1), N is the number of nuclei per cubic centimeter having the effective microscopic cross section σ, and n(v) is the number of neutrons per cubic centimeter having speeds between v and v plus dv. The density n(v) is related to the flux per unit energy $\phi(E)$, Eq. (4.53), by

$$n(v) = m\phi(E)\ .$$

Expressed in terms of $\phi(E)$, Eq. (7.12) reduces to

$$\bar{\sigma} = \frac{\int_0^\infty \sigma(E)\phi(E)dE}{\int_0^\infty \phi(E)dE}\ . \tag{7.13}$$

Equation (7.13) is exact if $\sigma(E)$ is the effective cross section defined by Eq. (7.1). If σ is replaced by the two-body cross section σ_t, Eq. (7.13) is not strictly valid unless the latter cross section varies as v^{-1}. Except for resonance absorbers, however, it is a good approximation for neutron energies much greater than k_BT.

---

[c] For convenience of notation, we are suppressing the space coordinates of any possibly position-dependent quantities.

To illustrate the magnitude of the error that may result at thermal energies on replacing σ by σ_t in Eq. (7.13), suppose that σ_t is a constant, σ_0, and that the thermal motion of the scattering nuclei is represented by a Maxwellian at a temperature T_M. The neutrons are assumed to have a Maxwellian distribution of speeds characterized by a different temperature T_N. From Eq. (7.13) with $\sigma = \sigma_0$ we find that $\bar{\sigma} = \sigma_0$, whereas the correct result is $\bar{\sigma} = \sigma_0[1 + T_M/AT_N)]^{\frac{1}{2}}$. The thermal motion of the scattering nuclei increases the reaction rate and hence the average cross section $\bar{\sigma}$.

It would seem that for moderating materials, particularly for hydrogen, we would make a significant error by using Eq. (7.13) to obtain spectrum-averaged scattering cross sections. This, however, is not the case since the averaging over the distribution of quantum states of the moderator to obtain calculated scattering cross sections is already performed, providing that we proceed on the basis of the rigorous theory given in Chap. 2.

Current methods for computing neutron spectra and average cross sections include the full complexities of the thermalization problem. To illustrate the formal and physical content of these methods we start with Eq. (4.88). We consider a region of space within which the cross sections are independent of position and assume that over most of this region the flux and source can be accurately approximated in terms of

$$f(z,E,\mu) = e^{iBz}f(E,\mu,B) ,$$

$$f_\ell(z,E) = e^{iBz}f_\ell(E,B) , \qquad (7.14)$$

and

$$S(z,E,\mu) = e^{iBz}S(E,\mu,B) ,$$

where B is a real constant. Equation (4.88) then reduces to

$$(\Sigma + iB\mu)f(E,\mu,B) = S(E,\mu,B) + \\ + \sum_{\ell=0}^{\infty} \int \Sigma_\ell(E',E)f_\ell(E',B)dE'P_\ell(\mu) . \qquad (7.15)$$

We now make the B_1-approximation.[d] To obtain this approximation, we retain the $\ell = 0$ and $\ell = 1$ terms on the right-hand side of Eq. (y.15). Then, after dividing both sides of Eq. (7.15) by $\Sigma(E) + iB\mu$, we multiply by $\mu^n d\mu$ (for n = 0, 1) and integrate over μ. This gives the pair of equations

$$BJ(E,B) + \Sigma(E)\phi(E,B) = Q_0(E,B) + \int_0^\infty \Sigma_0(E',E)\phi(E',B)dE' \qquad (7.16)$$

and

$$-B\phi + 3\gamma(E)\Sigma(E)J(E,B) = Q_1(E,B) + 3\int_0^\infty \Sigma_1(E',E)\phi(E',B)dE' , \qquad (7.17)$$

where

$$\phi(E,B) = \int_{-1}^{1} f(E,\mu,B)d\mu , \qquad (7.18a)$$

$$J(E,B) = i\int_{-1}^{1} \mu f(E,\mu,B)d\mu , \qquad (7.18b)$$

$$Q_0(E,B) = \int_{-1}^{1} S(E,\mu,B)d\mu , \qquad (7.18c)$$

$$Q_1 = 3i\int_{-1}^{1} \mu S(E,\mu,B)d\mu , \qquad (7.18d)$$

$$\gamma(E) = \frac{1}{3}\left(\frac{B}{\Sigma}\right)^2 \frac{A_{00}}{1 - A_{00}} \qquad (7.18e)$$

---

[d] For most thermalization problems the B_1-approximation is not essentially more accurate than its diffusion-theory counterpart. The numerical method given here, however, can also be used to calculate epithermal and fast spectra, in which case the difference between diffusion theory and the B_1-approximation can be significant.

and

$$A_{00} = \frac{\Sigma}{B} \tan^{-1}\left(\frac{B}{\Sigma}\right) . \qquad (7.18f)$$

The B_1-approximation is in general more accurate than the P_1-approximation to Eq. (7.15).[10] In particular, if $\Sigma_\ell(E',E)$ vanishes for $\ell > 1$, then within the framework of the B_1-approximation we can determine $f(E,\mu,B)$ rigorously. For $B = 0$, J and Q_1 vanish, and Eq. (7.16) reduces to the equation for the spectrum in an infinite homogeneous medium.

We bound the thermal range by an upper limit E_m and, as discussed in Sec. 5-1, we neglect the scattering from energies less than E_m to energies greater than E_m. Further, if we suppose that all neutrons with energy less than E_m have slowed down from energies greater than E_m, then $Q_0 = Q_1 = 0$, and Eqs. (7.16) and (7.17) become

$$BJ(E,B) + \Sigma(E)\phi(E,B) = \int_0^{E_m} \Sigma_0(E',E)\phi(E',B)dE' + \int_{E_m}^{\infty} \Sigma_0(E',E)\phi(E',B)dE' , \qquad (7.19)$$

and

$$-B\phi(E,B) + 3\gamma(E)\Sigma(E)J(E,B) = 3\int_0^{E_m} \Sigma_1(E',E)J(E')dE' + 3\int_{E_m}^{\infty} \Sigma_1(E',E)J(E')dE' . \qquad (7.20)$$

These equations are solved numerically by approximating the scattering integral by a trapezoidal integration quadrature; thus,

$$BJ_i + \Sigma_i\phi_i = \sum_{j=1}^{I} \Sigma_{0ji}\phi_j\Delta_j + S_{0i} , \qquad (7.21)$$

and

$$-B\phi_i + 3\gamma_i\Sigma_i J_i = 3\sum_{j=1}^{I}\Sigma_{1ji}J_j\Delta_j + S_{1i} , \tag{7.22}$$

where, for $E_{i+1} > E_i$,

$$\Delta_i = \frac{E_{i+1} - E_{i-1}}{2} \qquad i = 2,\ldots,I - 1 ,$$

$$\Delta_I = \frac{E_I - E_{I-1}}{2} \qquad E_I = E_m ,$$

$$\Delta_1 = \tfrac{1}{2}E_2 ,$$

and the source terms are as given below. We assume that the number of energy points is sufficiently large that we make only a negligible error by identifying the quantities with a subscript i with their true values at the points E_i. In particular, we note that $\Sigma_{0ji} = \Sigma_0(E_j,E_i)$ and $\Sigma_{1ji} = \Sigma_1(E_j,E_i)$.

To calculate the source S_{0i} we assume that neutrons with energies greater than E_m belong to a 1/E spectrum and that (except for hydrogen; see Sec. 5-3) they scatter against atoms that are free and at rest. Then, for a mixture of materials, where

$$S_{0i} = \sum_k s_{0i}^{(k)} \tag{7.23}$$

and

$$S_{1i} = \sum_k s_{1i}^{(k)} \tag{7.24}$$

we have

$$s_{0i}^{(k)} = \frac{\Sigma_{sk}}{1 - \alpha_k}\left(\frac{1}{E_m} - \frac{\alpha_k}{E_i}\right) , \tag{7.25}$$

where Σ_{sk} and α_k, respectively, are the values of the free-atom cross section and of α, Eq. (5.10), associated with the *kth* nucleus. For a single material, Eq. (7.25) is equivalent to f(E) in Sec. 5-9.2. To calculate $s_{1i}^{(k)}$, we apply the diffusion approximation (4.62) in the epithermal region. If the diffusion coefficient D is constant in this region, then

$$s_{1i}^{(k)} = \frac{3}{4} DB\Sigma_{sk} \frac{(A+1)^2}{AE_i} \left\{ \frac{A+1}{3} \left[\left(\frac{E_i}{E_m} \right)^{\frac{3}{2}} - \alpha^{\frac{3}{2}} \right] - \right.$$

$$\left. - (A-1) \left[\left(\frac{E_i}{E_m} \right)^{\frac{1}{2}} - \alpha^{\frac{1}{2}} \right] \right\} . \qquad (7.26)$$

The set of equations (7.21) and (7.22) is solved by using the "Normalized Extrapolated Gauss" iteration procedure,[e] which is discussed in Appendix V. Given the m*th* iterates ϕ_i^m and J_i^m, the (m+1)*th* iterates are

$$\phi_i^{m+1} = (1 - \lambda)\phi_i^m + \lambda n^{m+1}(U\phi_i^m + F_i^m) \qquad (7.27)$$

and

$$J_i^{m+1} = (1 - \lambda)J_i^m + \lambda n^{m+1}(VJ_i^m + G_i^m) , \qquad (7.28)$$

where the matrices U and V are defined by

$$U\phi_i^m \equiv \sum_j u_{ji}\phi_j^m$$

and

$$VJ_i^m \equiv \sum_j v_{ji}J_j^m ,$$

with

---

[e] The rate of convergence associated with this iteration procedure has been evaluated by Honeck.[11] This method is used in the infinite-medium routine GATHER-II,[12] and in the energy- and space-dependent route THERMOS.[13]

$$u_{ji} = \frac{\Sigma_{0ji}\Delta_j}{\Sigma_i - \Sigma_{0ii}\Delta_i} \qquad i \neq j\ ,$$

$$= 0 \qquad i = j\ ,$$

$$v_{ji} = \frac{\Sigma_{ji}\Delta_j}{\gamma_i\Sigma_i - \Sigma_{1ii}\Delta_i} \qquad i \neq j\ ,$$

$$= 0 \qquad i = j\ ,$$

$$F_i^m = \frac{S_{0i} - BJ_i^m}{\Sigma_i - \Sigma_{0ii}\Delta_i}\ ,$$

$$G_i^m = \frac{S_{1i} + B\phi_i^m}{3(\gamma_i\Sigma_i - \Sigma_{1ii}\Delta_i)}\ ,$$

and, finally,

$$n^{m+1} = \frac{\sum_j S_{0j}\Delta_j}{\sum_j \Sigma_{aj}\left(U\phi_j^m + F_j^m\right) + B\left(VJ_j^m + G_j^m\right)\Delta_j}\ . \tag{7.29}$$

The parameter λ in Eqs. (7.27) and (7.28) is the linear extrapolation factor. The value $\lambda = 1.2$ is commonly used; however, the assumption that a rapid convergence is associated with this value is supported by only a few numerical experiments. In any case, the optimum value of λ is not unique, but depends on the problem being solved.

The normalization factor n^{m+1}, as defined by Eq. (7.29), is the ratio of the total sources to the total losses and serves to impose a conservation condition on each iterate for both flux and current. It tends to the limit $n^{\infty} = 1$ for a convergent iterative process.

The iterative process is terminated when both the conditions

$$\left| \frac{\phi_i^{m+1} - \phi_i^m}{\phi_i^m} \right| \leq$$

$$\left| \frac{J_i^{m+1} - J_i^m}{J_i^m} \right| \leq \tag{7.30}$$

are satisfied for a-1 values of i. For modern high-speed computers, even the strict convergence criterion $\omega = 10^{-5}$ leads to rapid computation of accurate results.

For many calculations the thermal cutoff energy E_m is chosen at 2 or 3 eV with approximately 100 energy points used to cover the thermal energy range from 10^{-3} eV to 2.0 eV. A judicious choice of these points depends on the character of the scattering kernel, on the energy dependence of the principal absorbers, particularly if they have large resonances at thermal energies, and on questions of numerical accuracy. For the solid moderators carbon, beryllium, and beryllium oxide, if E_m is 2 or 3 eV, approximately 100 well-chosen energy points give accurate results. For water, comparable accuracy can be achieved with about fifty points. A thermal cutoff of 2 or 3 eV is very high for room-temperature problems (~100 k_BT) but much too low for moderator temperatures of 10,000°K. Generally, an E_m of 20 k_BT or 40 k_BT is adequate for thermalization problems. Computational convenience requires a fixed E_m rather than one that varies with temperature. The value of E_m is generally chosen to be suitable for the highest moderator temperature of interest in any class of problems.

After we have made an accurate calculation of the spectrum, we can use the results to calculate the group cross sections that appear in multigroup formulations of the diffusion equation, Eq. (4.64). Because of considerations of computing time and computer storage capability, the group structure in these formulations is, in general, coarse compared with that used in obtaining solutions of Eqs. (7.21) and (7.22). For any cross section $\Sigma_x(E)$, its group average Σ_{xg} for an energy group g bounded below and above by the energies E_g and E_{g+1}, respectively, is frequently defined by

$$\Sigma_{xg} = \frac{\int_{E_g}^{E_{g+1}} \Sigma_x(E)\phi(E)dE}{\int_{E_g}^{E_{g+1}} \phi(E)dE} . \tag{7.31}$$

Similarly, the group diffusion coefficient may be defined by

$$D_g = \frac{\displaystyle\int_{E_g}^{E_{g+1}} D(E)\phi(E)dE}{\displaystyle\int_{E_g}^{E_{g+1}} \phi(E)dE} \qquad (7.32)$$

The cross sections that determine the rate of scattering between groups g and k are defined by

$$\Sigma_{0g,k} = \frac{\displaystyle\int_{E_k}^{E_{k+1}} dE \int_{E_g}^{E_{g+1}} \Sigma_0(E',E)\phi(E')dE'}{\displaystyle\int_{E_g}^{E_{g+1}} \phi(E)dE} . \qquad (7.33)$$

For later considerations, we are particularly interested in the thermal group averages

$$\Sigma_x = \frac{\displaystyle\int_0^{E_m} \Sigma_x(E)\phi(E)dE}{\displaystyle\int_0^{E_m} \phi(E)dE}$$

$$D_T = \frac{\displaystyle\int_0^{E_m} D(E)\phi(E)dE}{\displaystyle\int_0^{E_m} \phi(E)dE} \qquad (7.34)$$

and

$$L^2 = \frac{D_t}{\Sigma_a} . \qquad (7.35)$$

Conventionally, L here is the thermal diffusion length. In keeping with this convention, we should remain aware of the distinction between this L and the $L = 1/\kappa_0$, of Sec. 4-7, where κ_0 is the smallest of the eigenvalues κ associated with Eq. (4.72).

Equations (7.31) through (7.33) are not necessarily the best definitions of the group cross sections and diffusion coefficient. Selengut[14] has given a multigroup formulation of diffusion theory in which the definitions of the group constants are similar to Eqs. (7.31) through (7.33) except that the product $[\phi^+(E)\phi(E)]$ appears in place of $\phi(E)$, where $\phi^+(E)$ is the solution of an equation that is adjoint to that satisfied by $\phi(E)$. The group constants reduce to the usual flux averages provided that the adjoint spectrum $\phi^+(E)$ is constant within each group. This, however, is the case only if there are no processes through which neutrons are lost from the system.

The spectrum function $\phi(E)$ that appears in Eq. (7.31) is assumed to have been calculated for each region of the reactor from some formulation, like Eqs. (7.19) and (7.20), which is appropriate for a bare reactor. In the usual recipe for calculating $\phi(E)$, the quantity B^2 for a particular region in a reactor assumes a value that is consistent with the best guess of the actual curvature of the flux in that region. Inasmuch as best estimates of the value of B^2 may vary from region to region in the reactor, the $\phi(E)$ calculated from Eqs. (7.19) and (7.20) and the group constants are often functions of position even if the material concentrations do not vary with position.

Equations (7.31) and (7.33) also define the group cross sections in multigroup formulations of the transport equation. If the scattering is isotropic, there are not particular difficulties associated with this formulation. For anisotropic scattering, the transport approximation is sometimes made. This approximation consists in neglecting the $\Sigma_\ell(E',E)$ for $\ell \geq 1$ in Eq. (4.88) and replacing the total cross section $\Sigma(E) = \Sigma_a(E) + \Sigma_s(E)$ and kernel $\Sigma_0(E',E)$ by a suitably defined transport cross section and transport kernel, respectively. Thus, the transport cross section or the corresponding mean free path is introduced to account for the anisotropy of neutron scattering, while the simpler mathematical theory associated with isotropic scattering is preserved. A thorough discussion of the transport cross section is given in an article by Rakavy and Yeivin.[15] For the present, we are concerned with systems in which diffusion theory is a good approximation and for which the definition of the transport cross section given in Sec. 4-5 is adequate.

We obtain the usual form of energy-dependent diffusion theory, Eq. (4.64), from the consistent P_1-approximation, Eqs. (4.56) and (4.60), by neglecting anisotropic energy transfer and introducing the transport cross section defined in Sec. 4-5. With that definition, as we noted in Chap. 4, the energy-dependent diffusion approximation and the consistent P_1-approximation are of comparable accuracy for thermal neutrons.

7-2 THE EFFECTS OF CHEMICAL BINDING ON THE PHYSICS OF GRAPHITE REACTORS

It is of interest to note some features of the thermal-neutron spectrum which depend on an accurate description of neutron scattering. The first feature is the hardening of the thermal portion of the spectrum, that is, its shift to higher energies because of the capture of neutrons before they can become completely thermalized. This effect is frequently described in terms of a neutron temperature that exceeds the moderator temperature, but this description must be used with some care. In particular, the less complete the thermalization, the less is the temperature dependence of the spectrum. The second important feature is the shape of the spectrum in the joining region between thermal and epithermal energies. In general, the interatomic binding reduces the rate of thermalization and therefore gives a harder spectrum with a more gradual transition to the epithermal 1/E shape. The spectrum and its temperature dependence in the joining region can only be obtained from a detailed calculation. A detailed calculation of the spectrum is particularly important if we must know the temperature dependence of reaction rates of materials having absorption resonances in the joining region.

The importance of an accurate calculation of thermalization varies with the over-all spectrum of the reactor. For a very lightly loaded, well-thermalized reactor, such as a swimming pool research reactor, the thermal-neutron spectrum is almost Maxwellian and follows the water temperature. On the other hand, a very heavily loaded reactor has very few thermal fissions and thus the thermal neutrons are not of great interest.

For the analysis of lightly loaded reactors with insignificant amounts of resonance absorption at thermal energies and in the evaluation of foil measurements, the neutron temperature concept is frequently used. To illustrate how to define a neutron temperature, we consider a method discussed by Beckurts and Wirtz.[16] Beckurts calculates $\phi(E)$ from the heavy-gas model, Eq. (5.75), as a function of the concentration of a 1/v absorber. At low energies ($E \lesssim 4k_BT$), the slope of the curve of $\ln[\phi(E)/E]$ as a function of energy is very nearly a constant, which we identify with $-1/k_BT_N$, where T_N is the neutron temperature. For $\Sigma_a(k_BT)/\xi_0\Sigma_s \lesssim 0.1$, Beckurts[16] (also see Cohen[17] and Coveyou et al.,[18]) finds that

$$T_N = T\left[1 + 1.46\,\frac{\Sigma_a(k_BT)}{\xi_0\Sigma_s}\right] . \tag{7.36}$$

In a similar way we can obtain a neutron temperature corresponding to other scattering models, but, in general, this requires that we solve Eq. (5.1).

In the application of the neutron-temperature concept to reactor problems and to the analysis of the activation of thin foils,[19] the neutron spectrum is written as

$$\phi(E) = \Phi_t\,\frac{E}{(k_BT_N)^2}\,\exp(-E/k_BT_N) + \Phi_e\,\frac{\Delta(E)}{E} , \tag{7.37}$$

where $\Delta(E)$ is called the joining function. It approaches unity for $E \gg k_BT_N$ and vanishes for $E < 4k_BT_N$. Once $\Delta(E)$ has been calculated, the neutron conservation condition, Eq. (5.14) with $E_m = \infty$, establishes the relative magnitudes of Φ_t and Φ_e. If we make the approximation that $\Delta(E) = 1$ for $E > 4k_BT_N$, we find

$$\frac{\Phi_e}{\Phi_t} = \frac{\frac{\sqrt{\pi}}{2}\,\Sigma_a(k_BT_N)}{\xi_0\Sigma_s}\;\frac{1}{1 - \dfrac{\Sigma_a(k_BT_N)}{\xi_0\Sigma_s}} . \tag{7.38}$$

To obtain a more accurate relation between Φ_t and Φ_e it is necessary to calculate $\Delta(E)$ for any given moderator. We can do this only by solving the infinite-medium spectrum, Eq. (5.1). Then the neutron-temperature concept has no greater utility than the weak-absorption limit discussed in Sec. 5-2 and in any case is valid only for the same range of weak absorptions as the description in terms of the function $\phi_1(E)$ that appears in Eq. (5.33). In particular, for the applications that we consider below there is no advantage to a description in terms of a neutron temperature; therefore, we do not pursue the matter further here.

It is for reactors wherein about two-thirds of the fissions take place below about 0.5 eV that considerations of neutron thermalization are of greatest importance. In this range the thermal-neutron distribution is quite far from Maxwellian. Most of the power reactors now being designed are in this category. Finally, there is the exceptional case of moderation by metallic hydrides where the energy transfer is nearly quantized in units of about 0.1 eV. For reactors using these moderators the thermalization must be considered in detail, even for very light fuel loadings, if the temperature dependence of the reactivity is to be calculated reasonably well.

We postpone until Chap. 9 a discussion of reactors containing H_2O. In that chapter we consider some important effects resulting from the interaction between the transport and thermalization of neutrons in heterogeneous water systems.

In this section we consider the sensitivity to the effects of chemical binding of some important reactor parameters, in particular, the effective multiplication constant and the temperature coefficient of reactivity. For this purpose, we consider small homogeneous graphite-U^{235} reactors. The dominant effect of chemical binding in such reactors is to cause a significant increase in the thermal leakage over that which occurs in the absence of binding effects.

In considering the effects of binding we proceed in a straightforward manner. We calculate thermal-neutron spectra as functions of absorber concentration and temperature, we average fission and absorption cross sections over these spectra, and then we perform multigroup calculations to determine effective multiplication constants and temperature coefficients of reactivity for completely homogeneous systems. In all cases, binding effects in graphite are described by the frequency distribution of Yoshimori and Kitano as modified to fit the specific heat of reactor-grade graphite (Parks model, see Chap. 3). Any of the other frequency distributions discussed in Sec. 3-4.3 would do just as well; in Sec. 7-2.2 we use the frequency distribution of Eq.

(3.101) to investigate the sensitivity of neutron spectra to changes in $\rho(\omega)$.

7-2.1 THE EFFECTS OF CHEMICAL BINDING IN HOMOGENEOUS GRAPHITE SYSTEMS. In Chap. 6 we demonstrated that our physical knowledge is adequate to account for the considerable influence that binding effects have on thermal-neutron spectra in poisoned graphite assemblies. Here we apply the results of the previous chapters in order to examine more closely the variation with absorption and temperature of the thermal-neutron spectrum. We consider an infinite homogeneous system containing only carbon atoms and U^{235} atoms. Figures 7.1 and 7.2 show thermal spectra, calculated from Eq. (5.1), for several $C:U^{235}$ atom ratios and for moderator temperatures of 300° and 1200°K.[20] These figures represent the flux per unit lethargy $E\phi(E)$ for thermalization by bound carbon (solid lines) and by free carbon (dashed lines). The spectra are normalized to give the same number of total absorptions below 1.0 eV. At 300°K, the effect of chemical binding is apparent for all values of the $C:U^{235}$ ratios considered. The effects of chemical binding are small for a moderator temperature of 1200°K; the same conclusion holds for

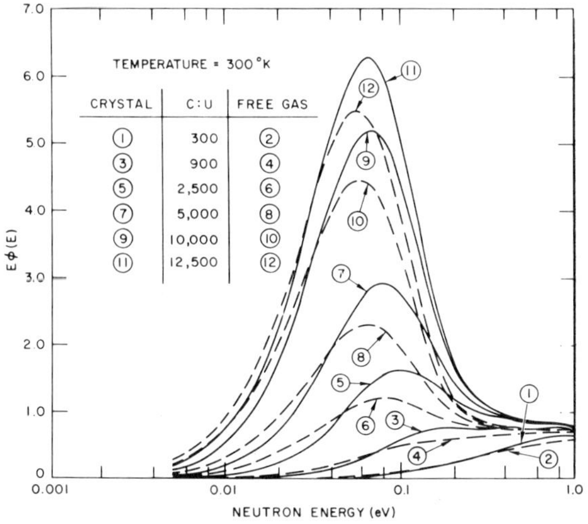

Figure 7.1 Thermal-neutron spectra at 300°K for six different $C:U^{235}$ atom ratios. (From Parks, Beyster, and Wikner, <u>Nucl. Sci. Eng.</u>, Ref. 20.)

T > 1200°K. The effects of chemical binding on thermal spectra are more strikingly exhibited by the differences in computed spectra. At T = 300°K the quantity $\delta = 100\,\{[\phi(E)_{gas} - \phi(E)_{crystal}]/\phi(E)_{gas}\}$ has a maximum value of about 50% at 0.2 eV for C:U^{235} atom ratios between 5000 and 12,5000; it is in excess of 60% at 0.01 eV for C:U^{235} atom ratios near 1000.

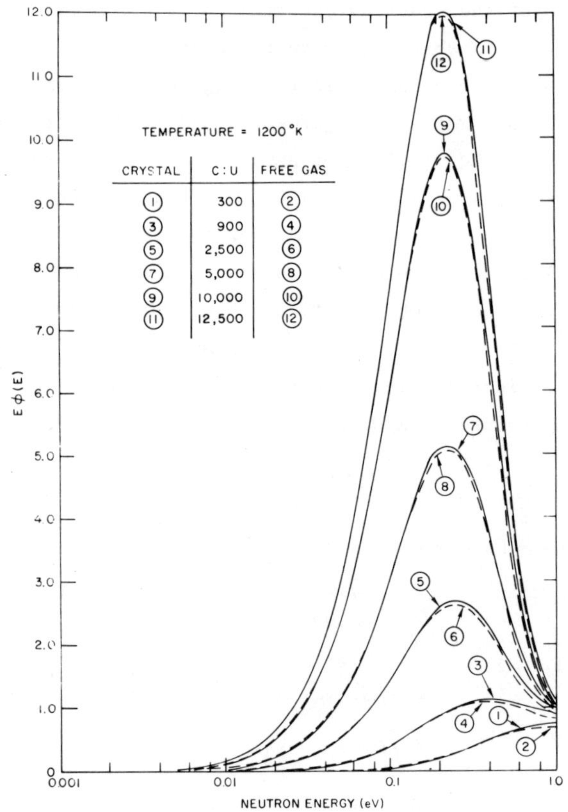

Figure 7.2 Thermal neutron spectra at 1200°K for six different C:U^{235} atom ratios. (From Parks, Beyster, and Wikner, Nucl. Sci. Eng., Ref. 20.)

7-2.2 SENSITIVITY OF NEUTRON SPECTRA TO THE FREQUENCY DISTRIBUTION OF LATTICE VIBRATIONS. The computed thermal-neutron spectra in Fig. 7.1 have been obtained on the basis of the frequency distribution represented by the solid curve in Fig. 3.29. This frequency distribution is consistent with measured scattering laws and specific heat data (see Chap. 3). Neutron spectra calculated on the basis of this frequency distribution are in good agreement with measured neutron spectra in homogeneous media (see Chap. 6). To examine the sensitivity of thermal-neutron reactor spectra to variations in the frequency distribution, we calculate infinite medium spectra from Eq. (5.1), using the frequency distribution given in Eq. (3.101) and different values of the cutoff frequency ω_3. The results[21] are shown in Fig. 7.3. In Fig. 7.4 we plot the difference between each of these spectra and the reference spectrum, computed on the basis of the $\rho(\omega)$ represented by the solid curve in Fig. 3.29. Inasmuch as for the purpose of neutron thermalization the dashed curve in Fig. 3.29 is essentially equivalent to the solid one,[22] we could just as well use the dashed curve in the calculation of the reference spectrum. The neutron spectrum obtained with ω_3 = 2580°K differs very little from the reference spectrum. For ω_3 = 2100°K and 3000°K, the predicted neutron spectra are softer and harder, respectively, than the reference spectrum by a maximum amount of about 10 percent.

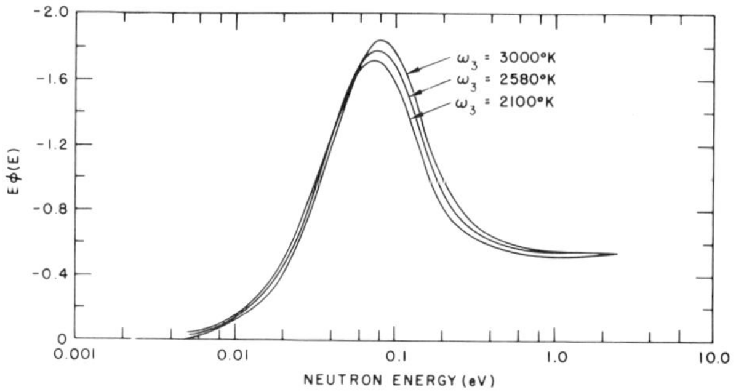

Figure 7.3 Thermal-neutron spectra at 293°K for a $C:U^{235}$ atom ratio of 5000, using the Egelstaff $\rho(\omega)$ with different high-frequency cutoff ω_3 (see Eq. (3.101)). (From Wikner, Joanou, and Parks, Nucl. Sci. Eng., Ref. 21.)

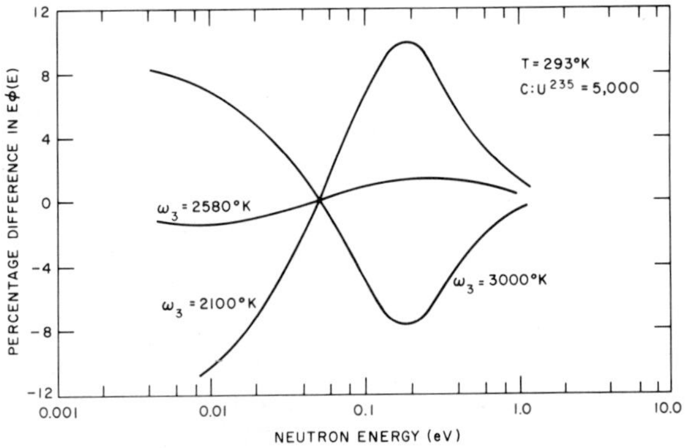

Figure 7.4 Variation of the percentage difference in thermal-neutron spectra versus energy for several different ω_3 values for the Egelstaff $\rho(\omega)$. The reference spectrum was computed using the Parks model. (From Wikner, Joanou, and Parks, Nucl. Sci. Eng., Ref. 21.)

The fact that the $\rho(\omega)$ of Eq. (3.101) can be adjusted to give agreement between calculated and measured thermal-neutron spectra is of significant practical importance. The principal result is that reactor spectra are sensitive to the gross features but not to the details of the frequency distribution. (It has also been shown that the use of the Gaussian approximation, Eq. (3.1), does not lead to significant errors in the calculation of spectra.[21])

Within the framework of Eq. (3.1), any number of judiciously chosen forms of $\rho(\omega)$ with a single variable parameter would most likely lead to scattering kernels that are sufficiently accurate for calculations of spectra. The most judicious choices, however, are the ones that are consistent with all pertinent experimental evidence. We emphasize that the form of the $\rho(\omega)$ that we choose should not be dictated by its simplicity alone. The complexity of calculating a scattering kernel is associated with the dependence of $\Sigma(E_0,E)$ on $\rho(\omega)$ and not with the dependence of $\rho(\omega)$ on ω.

To understand quantitatively the effects of binding on the behavior of homogeneous reactor systems we calculate the differences in spectrum-averaged cross sections obtained on the basis of free- and bound-carbon scattering kernels. In Table 7.2 we record spectrum-

Table 7.4

VALUES OF $[(1/k)(\partial k/\partial T)]/°C$ OBTAINED FOR FREE-GAS AND CRYSTAL SCATTERING KERNELS FOR SEVERAL $C:U^{235}$ ATOM RATIOS*

The tabulated values are obtained by taking the difference between calculated eigenvalues at 600°K and 300°K and dividing by the product of k at 300°K and the temperature difference.

$C:U^{235}$ Atom Ratio	$[-(1/k)(\partial k/\partial T)]/°C$		Difference (%)
	Free Gas	Crystal	
300	2.42×10^{-6}	6.07×10^{-7}	75
900	2.14×10^{-5}	1.07×10^{-5}	50
2,500	7.65×10^{-5}	4.91×10^{-5}	36
5,000	1.49×10^{-4}	1.07×10^{-4}	28
10,000	2.52×10^{-4}	2.04×10^{-4}	19
12,500	2.88×10^{-4}	2.41×10^{-4}	16

*From Parks, Beyster, and Wikner, Nucl. Sci. Eng., Ref. 20.

in this sense. Of interest here are the series of critical experiments performed at the University of California Lawrence Radiation Laboratory at Livermore[25] and at Argonne National Laboratory with the TREAT reactor.[26,27]

The UCRL systems were constructed to approximate closely an idealized system containing a homogeneous mixture of only moderator and fuel. Experiments accurately determined the reactivity effects of room-return (neutrons escaping an assembly are returned to it by scattering from surrounding walls), the control-rod void spaces, the nonhomogeneity of the fuel, and the porosity and poison content of the moderator blocks. These bare systems are not well thermalized, the carbon to uranium atom ratios ranging from 300 to 2500. The basic physical data for the bare UCRL assemblies are given in Table 7.5. (For completeness we note that the experiments on the effects of neutron room-return required the addition of only 0.2 in. to the critical height. The control-rod guide tube is made of aluminum; it is equivalent to a uniform distribution of 180 g of Type 2S aluminum per inch of reactor height. The amount of boron and other poisons in the graphite is equivalent to 13 ppm of boron. Additional descriptive information is available in Ref. 25.)

Table 7.5

BASIC PHYSICAL DATA FOR BARE UCRL GRAPHITE CRITICAL ASSEMBLIES*

Experiment No.	Core Size (in. ±0.03)	Critical Height (in. ±0.1)	$C:U^{235}$ Atom Ratio	Graphite Density (g/cm^3)	Foil Thickness (mils)	U^{235} Mass (kg)
U1	42.50×42.50	47.1	297	1.640	4	149.7
U2	48.50×48.50	40.1	600	1.645	2	82.26
U3	48.50×48.50	42.3	1200	1.645	2	43.57
U4	48.50×48.50	47.6	2340	1.645	1	25.14

*From Reynolds, Proc. 2nd U.N. Conf., Ref. 25.

The TREAT reactor, although a reflected system, is nearly a thermal reactor in the classic sense (see (5.22)), having a $C:U^{235}$ atom ratio of approximately 10,000, and thus nearly all fission occur below 2 eV. According to the results in Sec. 7-3.1, the maximum effect of binding on the multiplication constant occurs for this value of the $C:U^{235}$ atom ratio. In contrast to the UCRL assemblies, TREAT operates as a pulsed reactor system. It is simply constructed; the core, a right cylinder, is essentially homogeneous. The fuel element, which has a square cross section measuring 4 in. on a side, is canned in zirconium. The core is fully reflected by graphite. (A detailed description of the reactor is given in several ANL reports,[26-28] and an analysis of this reactor is given in Ref. 29.)

For the UCRL assemblies and for the TREAT reactor, the effects of chemical binding on the effective multiplication constant are determined from a series of one-dimensional calculations.[30] The results of these calculations are shows in Table 7.6. The small amount of self-shielding

Table 7.6

CALCULATED VALUES OF EFFECTIVE MULTIPLICATION CONSTANT k

Experiment No.	$C:U^{235}$ Atom Ratio	Effective Multiplication Constant k		
		Free Gas	Crystal	Difference (%)
UCRL U1	297	0.99609	0.99535	-0.073
UCRL U2	600	0.99699	0.99322	-0.38
UCRL U3	1,200	1.00621	0.99565	-1.06
UCRL U4	2,340	1.01706	0.99720	-1.99
TREAT	~10,000	1.0596	1,0350*	-2.37

*The excess reactivity is suppressed by control rods.

in the foils at thermal energies is accounted for using the procedures described in Sec. 5-9. The self-shielding in the zirconium cladding of TREAT is insignificant. Since we use only three nonthermal groups in the criticality calculations, we make a significant error in the computation of k for those cases ($C:U^{235} \lesssim 1000$) where $E\phi(E)$ varies significantly with lethargy (see Fig. 7.5). For $C:U^{235} \lesssim 1200$, calculated values of k very close to unity are the result of a cancellation of errors, one of which arises from the use of an insufficient number of epithermal groups and the other from neglecting the self-shielding effects in the fuel-bearing foils at the energies of the principal resonances in U^{235}.

Recent calculations by Stevens and Joanou[31] include an accurate treatment of phenomena in the epithermal region. The effects of chemical binding that we show in the last column of Table 7.6 are unaffected by these phenomena. The values of k computed with thermal-group parameters determined for bound graphite are lower than those obtained with the free gas parameters. The variation with $C:U^{235}$ atom ratio of the percentage difference in k is what we expect on the basis of the results given in Table 7.3.

Table 7.7

SPECTRUM-AVERAGED VALUES OF THE DIFFUSION COEFFICIENT AND DIFFUSION LENGTH FOR FREE-GAS AND BOUND CARBON VERSUS $C:U^{235}$ ATOM RATIO

$C:U^{235}$ Atom Ratio	D_T (cm)			L^2 (cm^2)		
	Free Gas	Crystal	Diff. (%)	Free Gas	Crystal	Diff. (%)
297	0.80750	0.81989	1.5	25.404	26.856	5.4
600	0.82970	0.84326	1.6	37.517	40.714	7.8
1200	0.84001	0.85809	2.1	54.770	61.640	11.2
2340	0.8468	0.87221	2.9	79.851	92.913	14.1

In Table 7.7 we compare values of D_T and L^2, computed from Eqs. (7.34) and (7.35) with $E_m = 2.0$ eV, for several $C:U^{235}$ atom ratios. In the presence of binding, the average diffusion coefficients are larger than those obtained on the basis of free-atom scattering. The increase in D_T reflects the shift in the mean energy of the thermal distribution to higher energies away from the largest Bragg peaks in the total cross section of carbon. To better understand the relevance of the information in Table 7.6 for the UCRL assemblies, we examine in Table 7.8 the sensitivity to chemical-binding effects of the intensive parameter k_∞, which is the infinite multiplication constant, and the extensive parameter P, which is the total nonleakage probability. The interatomic binding of the carbon atoms gives a pronounced effect on the nonleakage probability but a very small effect on the infinite-medium multiplication constant. We diagnose these effects further after first considering the separation of effects at thermal energies from those at nonthermal energies.

We consider the neutron balance equation in a two-group formulation. We denote the rates of leakage and absorption by L and A, respectively, and the rate at which neutrons slow down from the fast into the thermal group by D_F. In the following, attached subscripts F and T refer to fast and thermal groups, respectively. The source is normalized so that one fast neutron is produced in the core every

Table 7.8

COMPARISON OF CALCULATED VALUES OF k_∞ AND THE TOTAL NONLEAKAGE PROBABILITY P FOR THE BARE UCRL CRITICAL ASSEMBLIES

Experiment No.	k_∞			P		
	Free Gas	Crystal	Diff. (%)	Free Gas	Crystal	Diff. (%)
U1	1.608	1.6078	-0.02	0.61941	0.61909	-0.05
U2	1.673	1.6704	-0.14	0.59601	0.59458	-0.24
U3	1.766	1.76076	-0.32	0.56963	0.56547	-0.74
U4	1.822	1.8211	-0.06	0.55819	0.54738	-1.97

second:

$$1 = \int \frac{\nu(E)}{k} \Sigma_f(E)\phi(E)dE \ . \tag{7.39}$$

The balance equations for the core are

$$1 = L_F + A_F + D_F \tag{7.40}$$

and

$$D_F = L_T + A_T \ . \tag{7.41}$$

The fast and thermal nonleakage probabilities are given by

$$P_F = 1 - L_F = D_F + A_F \ , \tag{7.42}$$

and

$$P_T = 1 - \frac{L_T}{D_F} = \frac{D_F - L_T}{D_F} = \frac{A_T}{D_F} \ . \tag{7.43}$$

The thermal nonleakage probability is simply the ratio of absorptions in the core to the number of neutrons that reach thermal energies in the core. For a reflected system the nonleakage probability P_T may be greater than one, inasmuch as $L_T < 0$ if there is a net diffusion of thermal neutrons into the core from the reflector.

We define the fast absorption escape probability as the probability for a neutron to slow down into the thermal group, given that it does not leak as a fast neutron;

$$p = \frac{D_F}{1 - L_F} = 1 - \frac{A_F}{1 - L_F} . \quad (7.44)$$

We note that the product $P_F P_T$ is not equal to $1 - (L_F + L_T)$ and so is not equivalent to an over-all nonleakage probability. The product $pP_F P_T = [D_F/(1 - L_F)](1 - L_F)(A_T/D_F) = A_T$ is the total rate of thermal absorptions.

We further define

η = number of fast neutrons produced per thermal neutron absorbed in the fuel,

f = thermal utilization = fraction of thermal-neutron absorptions that occur in the fuel, and

ω = ratio of the number of fast neutrons produced by both fast and thermal fission to the number produced by thermal fission only.

The average number of neutrons k produced by each neutron through fission is related to previously defined quantities

$$k = \omega\eta f \; pP_T P_F . \quad (7.45)$$

Our definition of ω is at variance with general practice. Usually, ω denotes the total number of neutrons slowed down below the U^{238} fission threshold per neutron produced in a thermal fission. Further, with our definitions, the four-factor quantity $\eta f \omega p$ is not an intensive parameter of the reactor as it is for more conventional definitions of ω and p. The definitions that we have adopted are better suited to our purpose of expressing the separation between thermal and epithermal effects, but we must emphasize that ω is neither a specifically thermal nor a specifically epithermal quantity.

To express the separate contributions from thermal and epithermal neutrons, we arbitrarily select the upper bound of the thermal group at $E_m = 2.0$ eV. From the results of the criticality calculations, the

Table 7.9

THE EFFECTS OF CHEMICAL BINDING ON THE NEUTRON BALANCE IN SMALL GRAPHITE REACTORS

1	2	3			4			5			6	7			8	9
		$\underline{f}$			η			$\underline{P}_{\underline{T}}$				ϵ				
Experiment No.	$C:U^{235}$ Atom Ratio	Free Gas	Crystal	$(\Delta \underline{f}/\underline{f}) \times 100$	Free Gas	Crystal	$(\Delta\eta/\eta) \times 100$	Free Gas	Crystal	$(\Delta \underline{P}_{\underline{T}}/\underline{P}_{\underline{T}}) \times 100$	$[(\Delta \underline{f}/\underline{f}) + (\Delta\eta/\eta) + (\Delta \underline{P}_{\underline{T}}/\underline{P}_{\underline{T}})] \times 100$	Free Gas	Crystal	$(\Delta\epsilon/\epsilon) \times 100$	$[(\Delta \underline{f}/\underline{f}) + (\Delta\eta/\eta) + (\Delta \underline{P}_{\underline{T}}/\underline{P}_{\underline{T}}) - (\Delta\epsilon/\epsilon)] \times 100$	From Table 7.6 $(\Delta k/k) \times 100$
UCRL U1	247	0.99136	0.99126	-0.010	2.02121	2.02017	-0.051	0.94598	0.94315	-0.300	-0.361	5.0173	5.0316	+0.285	-0.08	-0.07
UCRL U2	600	0.98465	0.98450	-0.018	2.0333	2.02929	-0.222	0.92684	0.92140	-0.590	-0.830	2.2100	2.2200	+0.45	-0.38	-0.38
UCRL U3	1,200	0.971696	0.971302	-0.041	2.04652	2.04113	-0.264	0.90086	0.88949	-1.278	-1.583	1.4884	1.4961	+0.52	-1.063	-1.06
UCRL U4	2,340	0.949114	0.94818	-0.098	2.05579	2.05139	-0.214	0.87170	0.85350	-2.132	-2.448	1.2336	1.2394	+0.47	-1.98	-1.99
TREAT	~10,000	0.8224	0.8195	-0.35	2.0670	2.0656	-0.067	0.9891	0.9716	-1.80	-2.220	1.0496	1.0508	+0.11	---	-2.37

values of the quantities appearing in Eq. (7.45) are then as given in Table 7.9

The quantity $\Delta f/f$ is obtained by subtracting the bound-atom value of f from its free-atom value and dividing the result by the free-atom value. The $\Delta\eta/\eta$, etc., are defined in a similar manner.

The sum $\Delta f/f + \Delta\eta/\eta + \Delta P_T/P_T$, given in column 6, is more negative than the corresponding quantity given in Table 7.6. The difference between the two is removed when the positive contribution for ω (column 7) is included.

We ascertain from Table 7.9 that the dominant effect of binding occurs in the thermal nonleakage probability; $100(\Delta P_T/P_T)$ increases from 0.3% for the intermediate assembly U1 to 2.1% for the more thermal system U4. For U4, the value of ω is 1.24, that is, 20% of all fissions are produced by neutrons with energies greater than 2.0 eV. Inasmuch as the cross sections for U^{235} are nearly 1/v, the effects of binding on η and f are 10 to 50 times smaller, respectively, than they are on the nonleakage probability P_T.

Table 7.9 also shows results from the analysis of the TREAT reactor.[29] Because of the much softer spectrum, the non-1/v nature of the U^{235} absorption cross section is relatively more important, and thus the effect of binding on the thermal utilization of the TREAT reactor is larger than for the UCRL assemblies. Since the TREAT reactor has a reflector, the magnitude of the change caused by binding effects is not as marked as that predicted for the bare reactors (see Table 7.6), although it is still substantial. As expected, the effect on ω is reduced for the TREAT reactor. The difference between column 9 and column 6 is due to the fast nonleakage probability, which is slightly negative for this reflected system.

We conclude that the effects of chemical binding on the effective multiplication constant are significant for small graphite-U^{235} systems near room temperature. These results can be of importance in the control of small graphite power-reactor or propulsion systems. Frequently, these small systems have control rods or drums located in the side reflector.[32] The worth of an individual control element is usually small, and so is the total amount of reactivity that it can control. Chemical-binding effects reduce the magnitude of the reactivity loss that occurs when the reactor is brought to temperature (the temperature defect), thus ameliorating control requirements.

7-3 THE TEMPERATURE COEFFICIENT IN HOMOGENEOUS GRAPHITE REACTORS

The temperature coefficient of reactivity (1/k)(dk/dT) plays an important role in problems of reactor design. In water-cooled reactors the principal part of the temperature coefficient arises from density changes of the water moderator. In solid-moderator reactors, the density change of the moderator is small and spectral effects dominate the temperature coefficient. Typical of this class of reactor are the sodium-cooled graphite reactor and the various types of gas-cooled, graphite-moderated reactors. The gas-cooled reactors include both heterogeneous reactors, such as the Calder Hall type and Advanced Gas-Cooled Reactor, and the homogeneous or semihomogeneous types, such as the high-temperature gas-cooled reactors (HTR's). In the HTR's currently under construction or just beginning operation in England and the United States, the core, which is cooled by helium, will be at a mean temperature of about 1300°K. For our examples, we focus on the HTR, primarily because of the importance of neutron thermalization for its temperature coefficient.[f]

One practical HTR fuel element is illustrated in Fig. 7.6. It consists of a 1-e/r-in.-diameter central graphite spine surrounded by a 1/2-in.-thick annular region that contains fissile and fertile materials, uniformly mixed with moderator. The spine and fuel body are located in an impermeable graphite sleeve having a 3.5-in. outside diameter. The fuel elements are placed on a triangular pitch in the core. The scale of the heterogeneity is not large in comparison with a mean free path, and there is very little depression of the flux in passing from fuel to nonfuel regions. In this sense, the core of an HTR is to a good approximation homogeneous.

In general, the kinetic stability of a reactor will depend on the sign, magnitude, and time dependence of the temperature coefficient.

---

[f] The choice of this example is to some extent dictated by the authors' first-hand knowledge of these systems.

7-3.1 TEMPERATURE COEFFICIENTS OF η, f, AND P_T. We consider first the temperature coefficients of η, f, and P_T for an arbitrary homogeneous mixture of absorbers and fissionable nuclides. We establish our notation after introducing

$$\eta_j = \frac{\nu_j \Sigma_{fj}}{\Sigma_{aj}} = \nu_j \frac{\sigma_{fj}}{\sigma_{aj}} . \tag{7.47}$$

Here, ν_j is the average number of neutrons produced in the thermal-neutron fission of a nucleus of the *jth* nuclear species.[h] The quantities Σ_{fj} and Σ_{aj} (σ_{fj} and σ_{aj}) are the thermal-group (for $0 < E < E_m$) averages of the macroscopic (microscopic) fission and absorption cross sections, respectively, of the *jth* nuclear species. The Σ_{fj} and Σ_{aj} are related to the corresponding energy-dependent cross sections by the first of Eqs. (7.34). For convenience in notation, we have chosen not to label the average cross sections with some subscript denoting the thermal group. Letting N_j be the number of nuclei of the *jth* species per cubic centimeter, we have

$$\Sigma_{fj} = N_n \sigma_{fj} \tag{7.48}$$

and

$$\Sigma_{aj} = N_j \sigma_{aj} . \tag{7.49}$$

We adopt the convention that in summations over nuclides the index j enumerates only fissionable species of nuclei, whereas i enumerates all species present in the reactor. With this convention, the total macroscopic absorption and fission cross sections are given by

$$\Sigma_a = \sum_i \Sigma_{ai} \tag{7.50}$$

and

$$\Sigma_f = \sum_j \Sigma_{fj} , \qquad (7.51$$

---

[h] The quantity $\nu_j(E)$, the average number of neutrons produced when a neutron of energy E splits a nucleus of the *jth* species, is essentially independent of E near thermal energies.

respectively. Similarly, the macroscopic cross section for absorption by fuel nuclei is

$$\Sigma_{aF} = \sum_j \Sigma_{aj} \; . \tag{7.52}$$

The quantities f and η are given by

$$f = \frac{\Sigma_{aF}}{\Sigma_a} \tag{7.53}$$

$$\eta = \sum_j \frac{\eta_j \Sigma_{aj}}{\Sigma_{aF}} = \sum_j \frac{\nu_j \Sigma_{fj}}{\Sigma_{aF}} \; . \tag{7.54}$$

These never appear separately but always as the product

$$\eta f = \sum_j \frac{\nu_j \Sigma_{fj}}{\Sigma_a} \; . \tag{7.55}$$

If we specialize to the case of a single fissionable nuclear species, the temperature coefficient of ηf is

$$\frac{1}{\eta f}\frac{d}{dT}(\eta f) = \alpha_\eta + \alpha_f = \frac{1}{\nu\Sigma_f}\frac{d(\nu\Sigma_f)}{dT} - \frac{1}{\Sigma_a}\frac{d\Sigma_a}{dT} \; . \tag{7.56}$$

Since ν is essentially independent of T,

$$\frac{1}{\nu\Sigma_f}\frac{d\nu\Sigma_f}{dT} = \frac{1}{\Sigma_f}\frac{d\Sigma_f}{dT} = \frac{1}{N}\frac{dN}{dT} + \frac{1}{\sigma_f}\frac{d\sigma_f}{dT} = -\beta_V + \frac{1}{\sigma_f}\frac{d\sigma_f}{dT} \; , \tag{7.57}$$

where $\beta_V = -N^{-1}dN/dT$ is the volume coefficient of expansion of the fuel material. The temperature coefficient of Σ_f separates into two parts. The thermal expansion of the fuel material results in a decrease in the mean free path for fission and thereby a decrease in the fission rate per unit volume of the reactor (but not in the total fission rate). The term $\sigma_f^{-1} d\sigma_f/dT$ represents a specifically thermalization effect. As T increases, the spectrum shifts toward higher energies, where σ_f is normally smaller, and thus $d\sigma_f/dT < 0$.

For the temperature coefficient of Σ_a we find

$$\frac{1}{\Sigma_a}\frac{d\Sigma_a}{dT} \doteq \sum_i x_i \; -\beta_{Vi} + \frac{1}{\sigma_{ai}}\frac{d\sigma_{ai}}{dT} \quad , \tag{7.58}$$

where

$$x_i = \frac{\Sigma_{ai}}{\Sigma_a} \tag{7.59}$$

$$\beta_{Vi} = -\frac{1}{N_i}\frac{dN_i}{dT} \; . \tag{7.60}$$

For a homogeneous system where all materials expand together, $\beta_{Vi} = \beta_V$ is independent of i, and Eq. (7.58) reduces to

$$\frac{1}{\Sigma_a}\frac{d\Sigma_a}{dT} = -\beta_V + \sum_i x_i \frac{1}{\sigma_{ai}}\frac{d\sigma_{ai}}{dT} \; . \tag{7.61}$$

The contribution of the *ith* nuclear species to the spectral component of this temperature coefficient is proportional to x_i, the fraction of the total absorptions by that species.

If we combine Eqs. (7.57) and (7.61), the density effects cancel, and

$$\frac{1}{\eta f}\frac{d}{dT}(\eta f) = \frac{1}{\sigma_f}\frac{d\sigma_f}{dT} - \underset{i}{} x_i \frac{1}{\sigma_{ai}}\frac{d\sigma_{ai}}{dT} \tag{7.62}$$

is determined entirely by the energy dependence of the nuclear cross sections and the temperature dependence of the neutron spectrum. The separate η and f coefficients are given by

$$\alpha_\eta = \frac{1}{\sigma_f}\frac{d\sigma_f}{dT} - \frac{1}{\sigma_{aF}}\frac{d\sigma_{aF}}{dT} \tag{7.63}$$

$$\alpha_f = \frac{1}{\sigma_{ag}}\frac{d\sigma_{af}}{dT} - \sum_i x_i \frac{1}{\sigma_{ai}}\frac{d\sigma_{ai}}{dT} \; . \tag{7.64}$$

A given isotrope can significantly affect α_f (and for a mixture of fuel species, α_η) without greatly affecting the neutron balance. If x_i is small, say of the order of a few hundredths, but $\sigma_{ai}^{-1} d\sigma_{ai}/dT$ is large, the effect on α_f can be considerable (see Sec. 7-3.5).

We now generalize the preceding results to the case of a mixture of fissionable materials. We introduce m_j, the ratio of the number of neutrons produced by the thermal neutron fission of a fuel nucleus of type j to the total number of neutrons produced by thermal-neutron fission:

$$m_j = \frac{\nu_j \Sigma_{fj}}{\nu \Sigma_f} = \frac{\eta_j \Sigma_{aj}}{\eta \Sigma_{aF}}, \tag{7.65}$$

where

$$\nu = \frac{\sum_j \nu_j \Sigma_{fj}}{\Sigma_f} \tag{7.66}$$

We also introduce h_j, the fraction of fuel absorptions that occur in the j*th* species of fuel nuclide:

$$h_j = \frac{\Sigma_{aj}}{\Sigma_{aF}} = \frac{x_j}{f}. \tag{7.67}$$

Thus, for a mixture of fissionable materials,

$$\frac{1}{\eta f} \frac{d(\eta f)}{dT} = \sum_j m_j \frac{1}{\sigma_{fj}} \frac{d\sigma_{fj}}{dT} - \sum_i x_i \frac{1}{\sigma_{ai}} \frac{d\sigma_{ai}}{dT} \tag{7.68}$$

and

$$\alpha_f = \frac{1}{f} \frac{df}{dT} = \sum_j h_j \frac{1}{\sigma_{aj}} \frac{d\sigma_{aj}}{dT} - \sum_i x_i \frac{1}{\sigma_{ai}} \frac{d\sigma_{ai}}{dT}. \tag{7.69}$$

Subtracting Eq. (7.69) from Eq. (7.68), we obtain

$$\alpha_\eta = \sum_j m_j \frac{1}{\sigma_{fj}} \frac{d\sigma_{fj}}{dT} - \sum_j h_j \frac{1}{\sigma_{aj}} \frac{d\sigma_{aj}}{dT}. \tag{7.70}$$

For some considerations it is more convenient to use the alternative forms

$$\alpha_\eta = \sum_j m_j \frac{1}{\eta_j} \frac{d\eta_j}{dT} + \sum_j (m_j - h_j) \frac{1}{\sigma_{aj}} \frac{d\sigma_{aj}}{dT} \tag{7.71}$$

$$= \sum_j h_j \frac{1}{\eta_j} \frac{d\eta_j}{dT} + \sum_j (m_j - h_j) \frac{1}{\sigma_{fj}} \frac{d\sigma_{fj}}{dT}. \tag{7.72}$$

It is interesting to note the similarity of the first sum in Eq. (7.72) and

$$\eta = \sum h_j \eta_j \ . \tag{7.73}$$

In the second sum the quantity $(m_j - h_j) = \Sigma_{aj}(\eta_j - \eta)/\eta\Sigma_{aF}$ is negative for $\eta_j < \eta$ and positive for $\eta_j > \eta$. In a U^{235}-fueled thorium converter, $(m_j - h_j)$ is negative for U^{235} and positive for U^{233}. After the conversion of a sufficient amount of thorium to U^{233}, α_η may become positive.

We now consider the thermal nonleakage coefficient α_{P_T} for a bare homogeneous reactor. Within the framework of energy-dependent diffusion theory (Sec. 4-6), the thermal nonleakage probability is given by

$$P_T = (1 + B^2L^2)^{-1} \ , \tag{7.74}$$

where L^2 is as defined by Eq. (7.35) and B^2 is the geometric buckling of the reactor. From Eq. (7.74),

$$\alpha_{P_T} = \frac{1}{P_T}\frac{dP_T}{dT} = -\frac{B^2L^2}{1 + B^2L^2}\left(\frac{1}{L^2}\frac{dL^2}{dT}\right) \ , \tag{7.75}$$

with

$$\frac{1}{L^2}\frac{dL^2}{dT} = \frac{1}{D}\frac{dD}{dT} - \frac{1}{\Sigma_a}\frac{d\Sigma_a}{dT} \ . \tag{7.76}$$

In contrast to α_η and α_f, we cannot in general separate the spectral and density effects in $(1/D)(dD/dT)$ and $(1/L^2)(dL^2/dT)$. Where $\Sigma_a \ll \Sigma_s$, in which case D is inversely proportional to the number density N_m of moderator atoms, this separation does occur. If we set

$$D = \frac{1}{3} N_m^{-1}\left(\sigma_{tr}^{-1}\right)_{Av} \ , \tag{7.77}$$

where $\sigma_{tr\ Av}^{-1}$ is the thermal-group average of the reciprocal microscopic transport cross section associated with the moderator atoms, then

$$\frac{1}{D}\frac{dD}{dT} = \beta_V + \frac{1}{\left(\sigma_{tr}^{-1}\right)_{Av}}\frac{d\left(\sigma_{tr}^{-1}\right)_{Av}}{dT} \tag{7.78}$$

reactor[36] are shown in Fig. 7.8. For comparison, a cross section that varies as 1/v is also shown in this figure. The fission cross section of U^{235} decreases more rapidly than 1/v with increasing energy; the opposite is true for U^{233}. The positive energy resonances in U^{233} dominate the low-energy behavior of its fission cross section, whereas a negative energy resonance dominates the low-energy behavior of $\sigma_f(E)$ for U^{235}. For the accurate calculation of temperature coefficients induced by spectral shifts, it is necessary to know the energy dependence of the relevant cross sections with the greatest possible accuracy (~1%); for computational reasons the data must be represented by smooth curves.

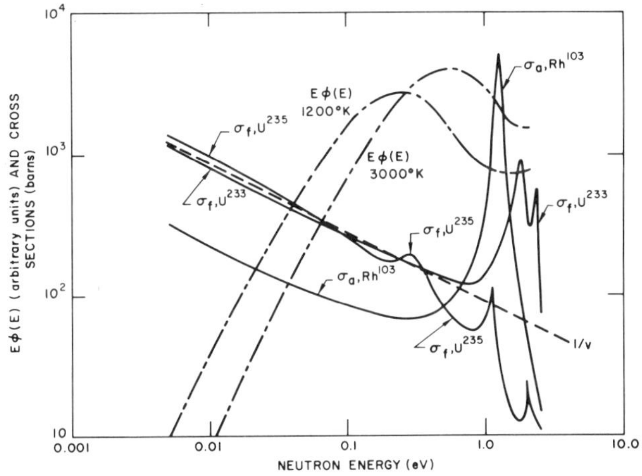

Figure 7.8 Fission cross sections of U^{233} and U^{235}, absorption cross section of Rh^{103}, and neutron flux $E\phi(E)$.

In Figs. 7.9 and 7.10 we show average cross sections of Xe^{135}, Rh^{103}, U^{233}, U^{235}, Pu^{239}, and Pu^{240} as a function of temperature. We also show the temperature variation that the average absorption cross sections of these materials would exhibit if the corresponding energy-dependent cross sections varied as 1/v. With these figures it is easy to convince ourselves of the importance of accurate calculations of average cross sections in high-temperature reactors.

For high-temperature graphite systems, each of the plutonium isotopes produced from neutron reactions with U^{238}, which constitutes 7%

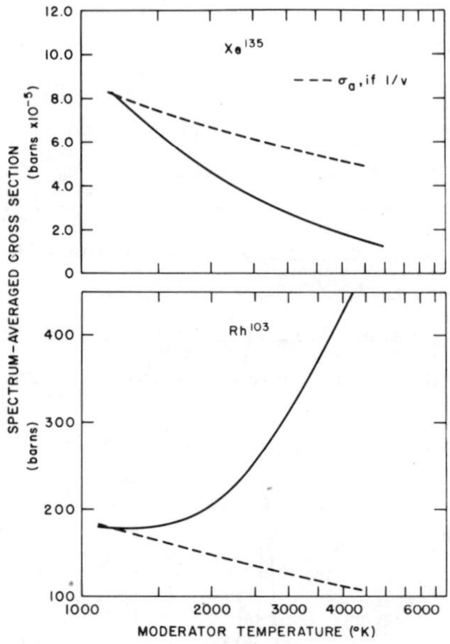

Figure 7.9 Temperature variation of absorption cross section versus moderator temperature for Xe^{135} and Rh^{103}, with $E_m = 2.1$ eV and thermal spectra appropriate to a small HTF. (From Fischer and Wikner, USAEC Report GA-2307, Ref. 37.)

of the highly U^{235}-enriched fuel, make negative contributions to the utilization coefficient α_f. At low temperatures, for example those appropriate to Hanford-type reactors, Pu^{239} makes positive contributions to α_f. For some reactors, the contributions from plutonium isotopes dominate the slow temperature coefficient,[i] and these reactors require a control system to compensate for this effect. It is not always possible possible, however, to devise such a system.[38] Where possible, it is much more advantageous to provide an inherently nuclear means to assure an over-all negative temperature coefficient.

Finally, in Fig. 7.11 we show the temperature variation of $(1/\sigma_a)(d\sigma_a/dT)$ for U^{233}, U^{235}, Rh^{103}, Xe^{135}, Sm^{149}, Pu^{240} and boron

---

[i] We distinguish between prompt and slow temperature coefficients in Sec. 7-3.3.

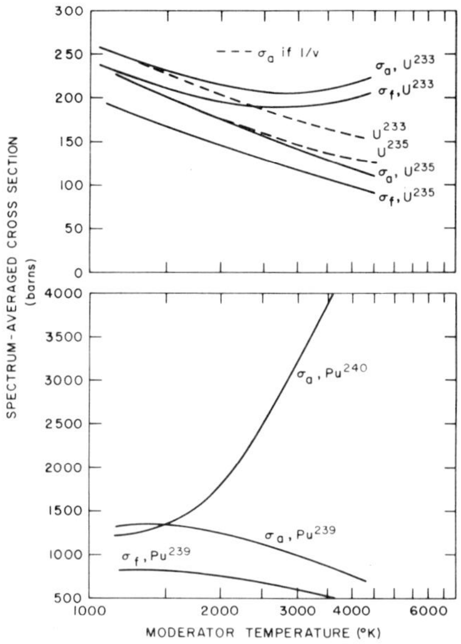

Figure 7.10 Temperature variation of absorption and fission cross sections versus moderator temperatures for U^{233} and U^{235} and for Pu^{239} and Pu^{240}, with $E_m = 2.1$ eV and thermal spectra appropriate to a small HTF. (From Fischer and Wikner, USAEC Report GA-2307, Ref. 37.)

obtained for spectra in an infinite medium containing carbon and U^{235} with $C:U^{235}$ atom ratios of 2820, 5000, and 7500.[9] The logarithmic temperature derivative of σ_a for Rh^{103} is positive and increases with increasing temperature; however, for all of the $C:U^{235}$ ratios considered the magnitude of $(1/\sigma_a)(d\sigma_a/dT)$ for Pu^{240} decreases with increasing temperature after reaching a maximum between 1900° and 2100°K. These two isotopes make strong negative contributions to α_f not only because of $(1/\sigma_a)(d\sigma_a/dT)$ but also because their fractional absorptions x_i increase with temperature. The fission products Xe^{135} and Sm^{149}, on the other hand, make positive contributions to α_f.

For the fuels U^{233} and U^{235}, $(1/\sigma_a)(d\sigma_a/dT)$ is negative for the range of temperatures shown in Fig. 7.11; however, from Fig. 7.10 it follows that $(1/\sigma_a)(d\sigma_a/dT)$ for U^{233} becomes positive at about 2800°K, a temperature considerably above the normal moderator temperature of

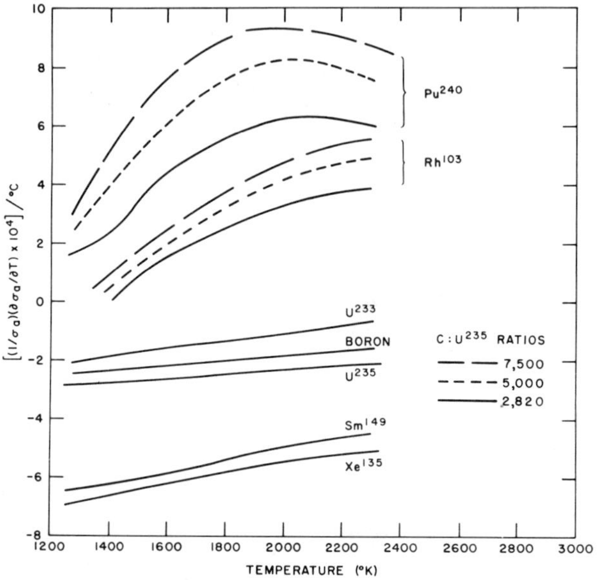

Figure 7.11 Values of the logarithmic temperature derivative of the spectrum-averaged absorption cross section for several C:U^{235} atom ratios.

1200°K but not impossible for an accident condition.

One of the important characteristics of an HTR system is the excellent neutron economy that results from using the uranium-thorium fuel cycle, since η for U^{233} is greater than η for either U^{235} or Pu^{239} in a thermal or slightly intermediate reactor spectrum. The high conversion ratios in these systems, as well as possible breeding of fuel, are achieved by recycling the U^{233} that is produced from the thorium. The high concentration of U^{233} in large HTR systems, however, and possibly even the small amount of U^{233} present in small systems, at high temperatures (>1200°K) could lead to positive temperature coefficients affecting the inherent kinetic stability of the reactor. If it is desired to assure a negative temperature coefficient under all conditions, including postulated accident conditions, we must compensate the possibly positive value of $(1/k)(dk/dT)$ at high temperatures. We can do this by taking advantage of the inherently negative spectral contributions produced by certain isotopes. Following the next section, we undertake the precise calculation of the importance of various isotopes for the temperature coefficient of semihomogeneous HTR's.

7-3.3 THE TEMPERATURE COEFFICIENT FOR SEMIHOMOGENEOUS HIGH-TEMPERATURE GRAPHITE REACTORS. In the fuel-bearing part of the semihomogeneous fuel element shown in Fig. 7.6 the fuel and admixed moderator have a common temperature at all times. The boundaries between different regions of the fuel element, however, inhibit the transfer of heat from the fuel annulus to the central spine and to the outer sleeve. Because of the inhibited heat flow and the large heat capacity of graphite, the time required to establish an essentially uniform temperature throughout the element may be long compared with the time that is characteristic of changes brought about by a rapid insertion of reactivity. In this case, the kinetic response of the reactor is determined predominantly by the changes in reactivity that are associated with temperature changes of only the fuel-bearing portion of the fuel element. On the other hand, at long times after the insertion of reactivity the kinetic response is determined by the reactivity changes associated with uniform heating of the entire element. To determine the kinetic response of the reactor for both long and short times we thus must distinguish the prompt from the slow temperature coefficient. To calculate the prompt temperature coefficient, we neglect the heat transfer from the fuel-bearing annulus to other regions of the fuel element, whereas to calculate the slow temperature coefficient, we assume that the temperature is uniform throughout the fuel element.

In the fuel elements considered here, the thorium is homogeneously mixed with the fuel. The Doppler effect, which arises from the resonance absorption by Th^{232}, therefore contributes to the prompt component of the temperature coefficient. The Doppler coefficient α_p is negative at all temperatures, but its magnitude approaches zero with increasing temperature.

The magnitude of the Doppler coefficient can be changed considerably by varying the concentration of materials in semi-homogeneous fuel elements (or, in heterogeneous systems, by changing the diameter of the rods containing the fuel and fertile material). To illustrate the case for semihomogeneous systems of the HTR type, in

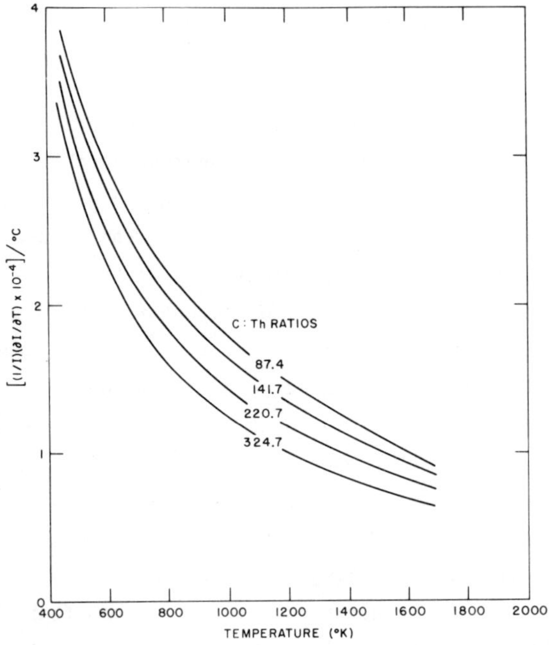

Figure 7.12 Temperature variation of (1/I)(∂I/∂T) for four carbon-to-thorium atom ratios appropriate to HTR-type reactors. (From Fischer and Wikner, USAEC Report GA-2307, Ref. 37.)

Figs. 7.12 and 7.13 we show (1/I)(∂I/∂T) and (1/p)(∂p/∂T), respectively, as functions of temperature and of the carbon-to-thorium atom ratio in the fuel body. In the relation $(1/p)(\partial p/\partial T) = (\ell np)(1/I)(\partial I/\partial T)$, the value of ℓnp varies much more rapidly with T than the remaining factor. A two-fold decrease in the carbon-to-thorium atom ratio gives a two-fold increase in α_p, although (1/I)(∂I/∂T) increases by only 25%. A significant increase in the Th^{232} concentration, however, requires a considerable increase in the U^{235} loading and can have an adverse effect on the economics of the plant. The thorium content of the reactor thus should not be entirely dictated by its influence on the temperature coefficient.

Inasmuch as in the semihomogeneous systems that we are considering a fraction of the moderator is mixed with the fuel, thermalization effects contribute not only to the slow but also to the prompt temperature coefficient. For systems containing U^{233}, the coefficients

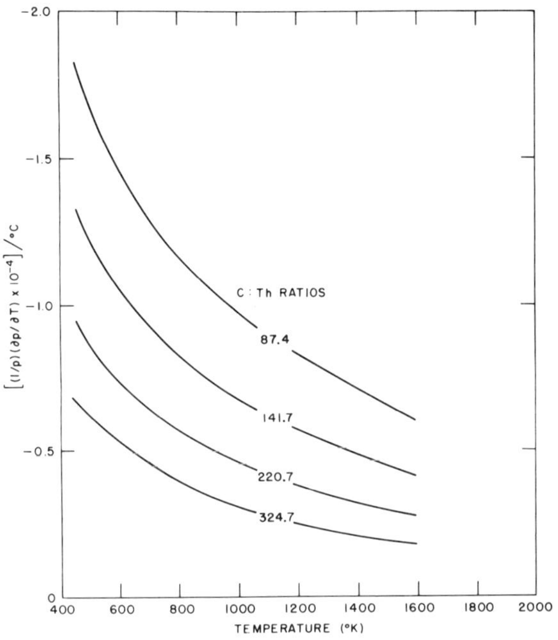

Figure 7.13 Temperature variation of the Doppler coefficient $(1/p)(\partial p/\partial T)$ for four carbon-to-thorium atom ratios appropriate to HTR-type reactors. (From Fischer and Wikner, USAEC Report GA-2307, Ref. 37.)

α_η and α_f are frequently positive at high temperatures and of the same general magnitude as the Doppler coefficient. The temperature coefficient of HTR systems is thus determined by two large[j] components of opposite sign, and these components can have different time responses. When the positive component is sufficiently delayed in time relative to a prompt negative component, a reactor can have an acceptable kinetic behavior even if it has a net positive temperature coefficient. The control-rod drive system can, of course, be designed to provide the necessary stability for normal and abnormal operating conditions. It is desirable, however, to rely as much as possible on an inherent nuclear mechanism rather than the operation of a control

[j] Compared with those in water systems, the temperature coefficients in HTR's are very small.

system. Normally, we try to find a temperature coefficient that can compensate a reactivity insertion Δk of one or two percent. Thus, dk/dT must be such that

$$\Delta k \leq \int_{T_{op}}^{T_{max}} - \frac{\partial k}{\partial T} \, dT \ , \tag{7.82}$$

where T_{op} is the average core temperature during normal operations and T_{max} is the maximum temperature reached during a power excursion. Obviously, $\partial k/\partial T$ must be negative. In the remainder of this chapter we consider how we make $\partial k/\partial T$ negative for a specific small prototype HTR.

7-3.4 THE TEMPERATURE COEFFICIENT FOR A SMALL HTR. Fischer and Wikner[37] have analyzed the temperature coefficient of an early version of a small high-temperature reactor that does not contain Rh^{103} at the beginning-of-life. They show that the slow component of the utilization coefficient at the end-of-life is positive over a large temperature range. Figure 7.14 shows $\alpha_f(T)$ and the contributions to it from Sm^{149}, Xe^{135}, Rh^{103}, and Pu^{240}, and the composite contribution from all materials having cross sections that vary as $1/v$. In plotting Fig. 7.14, we have divided all cross sections by the macroscopic fuel cross section Σ_{aF}. To explain precisely what we have plotted, we write

$$\alpha_f = -\frac{\Sigma_{aF}}{\Sigma_a}\left[\frac{d}{dT}\left(\frac{\Sigma_a}{\Sigma_{aF}}\right)\right] = -\frac{\Sigma_{aF}}{\Sigma_a}\left[\frac{d}{dT}\left(\frac{\Sigma_{aF}}{\Sigma_{aF}}\right)\right] + \sum_k -\frac{\Sigma_{aF}}{\Sigma_a}\left[\frac{d}{dT}\left(\frac{\Sigma_{ak}}{\Sigma_{aF}}\right)\right] , \tag{7.83}$$

the sum extending over nonfissionable isotopes. The separate contributions of each term on the right-hand side of Eq. (7.83) are shown in Fig. 7.14. With the chosen normalization of all cross sections to Σ_{aF}, the first term, which represents the combined contribution of all fuel isotopes, vanishes. Xe^{135}, Sm^{149}, and the "1/v materials" make positive contributions to α_f, whereas the fission-product Rh^{103} makes a large negative contribution. Nuclear reactions in the small amount of U^{238} present in the initially highly enriched U^{235} create enough Pu^{240} to more than compensate the positive contribution from

Sm^{149}. The combined effect of all materials for this early version of an HTR gives an α_f that is positive from 1200°K to 2000°K and from 3500°K to 4000°K, the approximate sublimation temperature of graphite.

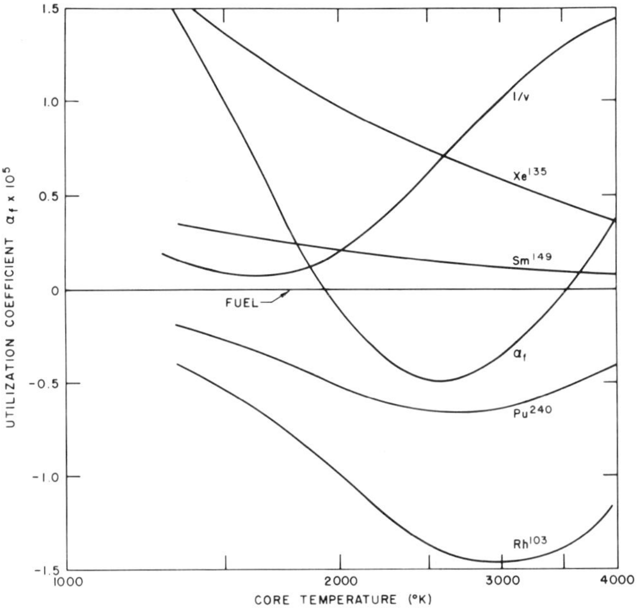

Figure 7.14 Variation of the logarithmic temperature derivative of the thermal utilization with temperature for end-of-life conditions with only fission-product Rh^{103} present. (From Fischer and Wikner, USAEC Report GA-2307, Ref. 37.)

Calculations indicate that the over-all temperature coefficient can be made strongly negative at all temperatures by adding a sufficient amount of Rh^{103} to the reactor at the beginning-of-life (Rh^{103} constitutes 100 percent of natural rhodium). The presence of this material at the beginning-of-life requires only a small increase of the original U^{235} loading to maintain the same reactivity lifetime. Rhodium has a high degree of chemical stability in a high-temperature (2000°K) graphite matrix.[39] Furthermore, the thermal-group average of its microscopic absorption cross section is smaller than those of U^{235} and boron so that any initial concentration of Rh^{103} does not burn out as rapidly as U^{235} or boron. (Rhodium can also be useful in helping to

Table 7.11

HOMOGENIZED CONCENTRATIONS OF THE FISSILE AND FISSION PRODUCTS OF INTEREST FOR THE BEGINNING-OF-LIFE* AND END-OF-LIFE OF A TYPICAL SMALL HTR†

Isotope	Concentration (atoms/barns-cm)	
	Beginning-of-Life	End-of-Life
U^{233}	0	8.038×10^{-6}
U^{235}	3.146×10^{-5}	1.133×10^{-5}
Pu^{239}	0	5.192×10^{-8}
Th^{232}	3.688×10^{-4}	3.540×10^{-4}
U^{234}	0	6.988×10^{-7}
U^{236}	0	3.820×10^{-6}
U^{238}	2.279×10^{-6}	1.967×10^{-6}
Pu^{240}	0	1.908×10^{-8}
Rh^{103}	2.093×10^{-6}	1.328×10^{-6}
Xe^{135}	3.359×10^{-10}	2.322×10^{-10}
Sm^{149}	3.138×10^{-9}	1.683×10^{-9}
Sm^{151}	0	9.453×10^{-9}
Boron	4.100×10^{-6}	2.686×10^{-8}
Carbon	7.918×10^{-2}	7.918×10^{-2}

*Beginning-of-life is defined as the time when the Xe^{135} reaches an equilibrium concentration following the start of full-power operation.

†From Fischer and Wikner, USAEC Report GA-2307, Ref. 37.

shape the power distribution in a long-lived reactor.)

Table 7.11 shows the concentration of the important fissile materials and fission products at the beginning- and end-of-life. By incorporating rhodium with the initial fuel, we increase the fractional absorption in Rh^{103} at the end-of-life from less than 1% to approximately 3%. As a result, the utilization coefficient, shown in Fig. 7.15, is positive at operating temperature but becomes negative near 1500°K and is increasingly negative with increasing temperature. Although it is not shown in Fig. 7.15, α_f reaches a minimum near 3000°K and increases to -4×10^{-5}/°K at 4000°K.

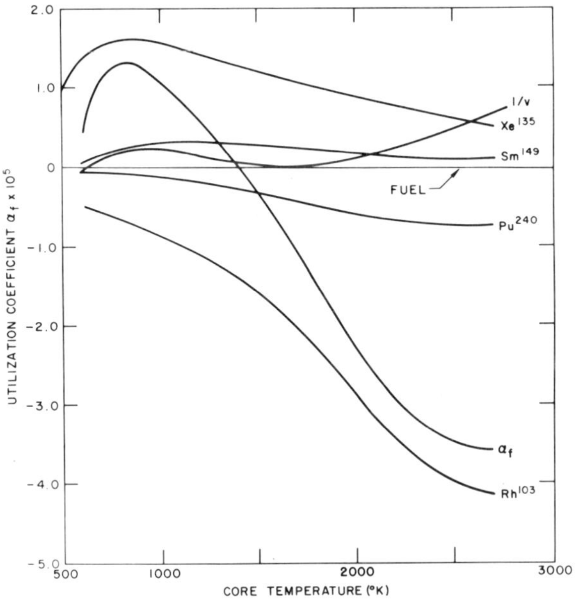

Figure 7.15 Variation of the logarithmic temperature derivative of the thermal utilization with temperature in a small HTR where Rh^{103} is introduced at the beginning-of-life. (From Fischer and Wikner, USAEC Report GA-2307, Ref. 37.)

The change that introduction of an initial concentration of Rh^{103} produces in the total slow temperature coefficient $[(1/k)(dk/dT)]_s$ is shown in Fig. 7.16. The magnitude of the temperature coefficient in the reactor without an initial concentration of rhodium decreases with temperature and probably approaches zero near 4000°K. When Rh^{103} is added (see Table 7.11), $[(1/k)(\partial k/\partial T)]_s$ remains negative and relatively large even at very high temperatures. The effect on the temperature coefficient of removing all the Xe^{135} is also shown.

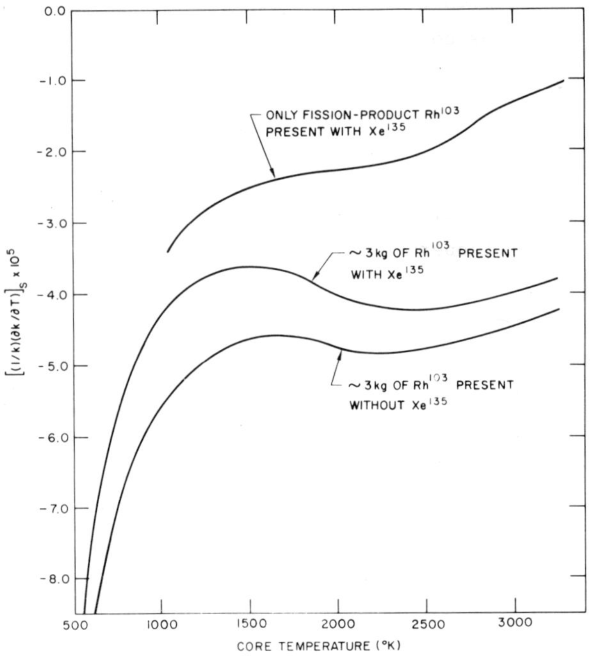

Figure 7.16 Temperature variation of the slow temperature coefficient of a typical small HTR at the end-of-life. (From Fischer and Wikner, USAEC Report GA-2307, Ref. 37.)

Finally, the temperature variations of α_η, α_f, α_ω, α_p, α_{P_F}, and α_{P_T} for the material concentrations in Table 7.11 are given in Fig. 7.17. The slow temperature coefficient is negative and it is nearly constant from 1200°K to 3000°K. For $T \gtrsim 2000$°K, its magnitude is more than twice that of α_p. Except for α_p and α_f, the magnitudes of all components of $[(1/k)(dk/dT)]_s$ are small.

So far we have given results for the slow temperature coefficient only. The Doppler coefficient α_p is the principal contributor to the prompt coefficient. The contribution of thermalization effects to the

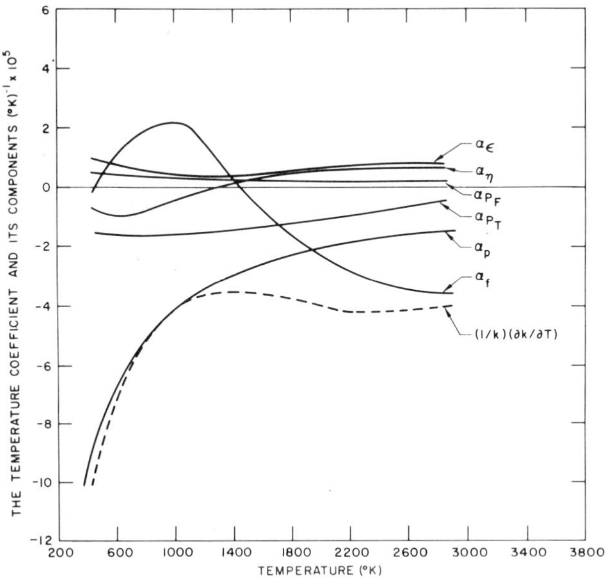

Figure 7.17 Values of the temperature coefficient at the end-of-life versus temperature.

prompt, or fuel, coefficient is significant only if f_m, the fraction of moderator atoms that are mixed with the fuel, is not very much less than unity. For the reactor considered here $f_m = 0.37$.

The slow temperature coefficient $[(1/k)(dk/dT)]_s$ is associated with a spatially uniform heating of the fuel element. The Doppler coefficient depends on the temperature of the fuel-bearing region but not on the temperature of nonfuel regions. The prompt and slow α_p are thus the same for a given fuel temperature.

To obtain the prompt η- and f-coefficients, we first note that the neutron spectrum does not change greatly with position over the fuel element and that the depression of the thermal flux in the fuel region is negligible. For spectrum calculations it is thus a good approximation to homogenize the materials in the fuel element without making corrections for self-shielding effects. The thermal spectrum then satisfies Eq. (5.1), but with a kernel given by

$$\Sigma_0(E',E) = N_m \frac{a - a_F}{a} \sigma_0(E',E,T_m) + N_{mF} \frac{a_F}{a} \sigma_0(E',E,T_F) , \quad (7.84)$$

where N_m and N_{mF} are the number densities of moderator atoms in the nonfuel and fuel regions, respectively, and a and a_F are the total cross-sectional areas of the fuel element and fuel annulus, respectively (see Fig. 7.6). The temperatures T_{mF} and T_m are average temperatures of fuel and nonfuel regions. From the solution of Eq. (5.1) with the kernel given in Eq. (7.84), we can calculate η and f, as well as the corresponding temperature coefficients. The results that we obtain for the prompt coefficient, however, do not differ significantly from

$$\left(\frac{1}{d}\frac{dk}{dT}\right)_{pr} = \left(\frac{1}{k}\frac{dk}{dT}\right)_{s} - f_m \alpha_m , \tag{7.85}$$

where

$$\alpha_m = \left(\frac{1}{k}\frac{dk}{dT}\right)_{s} - \alpha_p , \tag{7.86}$$

$[(1/k)(dk/dT)]_s$ is the slow temperature coefficient corresponding to a uniform heating of the fuel element, and

$$f_m = \frac{N_{mF} a_F}{[N_m(a - a_F)] + N_{mF} a_F} . \tag{7.87}$$

According to Fig. 7.18, the prompt coefficient for the reactor with the material composition given in Table 7.11 is negative. The addition of rhodium makes the prompt coefficient more negative, but the effect of this element for $[(1/k)(dk/dT)]_{pr}$ is only one-third as great as for $[(1/k)(dk/dT)]_s$.

From Fig. 7.16 we can estimate to what extent an initial concentration of 3 kg of Rh^{103} affects the slow temperature rise of the fuel element following the prompt excursion initiated by a reactivity insertion at the end-of-life. For this purpose, for T > 1000°K, we assume that with Rh^{103}

$$\frac{dk}{dT} = -4.0 \times 10^{-5}/°K \tag{7.88a}$$

and without Rh^{103}

$$\frac{dk}{dT} = -3.0 \times 10^{-5}\left[1 - \frac{\ln(T/1000)}{\ln 4}\right] . \tag{7.88b}$$

Equation (7.88b) approximates the upper curve in Fig. 7.16 by a straight

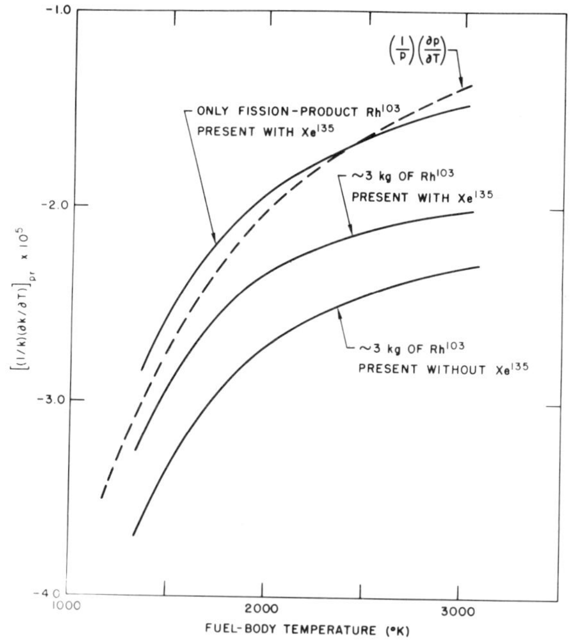

Figure 7.18 Temperature variation of the prompt temperature coefficient of a typical small HTR at the end-of-life. (From Fischer and Wikner, USAEC Report GA-2307, Ref. 37.)

line that passes through (dk/dT) = 0 at 4000°K. For an operating temperature T_{op} of 1500°K, assuming no coolant flow, we calculate the temperature rise ΔT that is required for the temperature coefficient to compensate the reactivity insertion Δk. As shown in Table 7.12, the presence of 3 kg of Rh^{103} at the beginning-of-life reduces the temperature increase in a slow transient by more than a factor of two.

Although we have restricted our attention to a small HTR, similar considerations apply also for large HTR's. For the large systems ($B^2L^2 << 1$), the nonleakage temperature coefficients α_{P_F} AND α_{P_T} are negligible and only the temperature dependence of k_∞ is of importance. Baxter and Fischer[40] show for the large HTR systems that the slow temperature coefficient can be positive for temperatures above the operating temperature, but that the use of Rh^{103} or Pu^{240} yields a slow temperature coefficient that is negative at all times and temperatures of interest.

Table 7.12

THE EFFECT OF Rh^{103} ON THE TEMPERATURE RISE ΔT
IN A SLOW TRANSIENT

Reactivity Insertion Δk	With 3 kg Rh^{103} Initially Present	With Only Fission-Product Rh^{103} Present
0.009	225	500
0.017	435	1000

The discussion in this section shows the relevance of thermalization effects in assuring a safe kinetic response of a modern reactor system. The HTR systems are particularly amenable to exploitation of the thermalization effects.

REFERENCES

1. A. Radkowsky (Ed.), *Naval Reactor Physics Handbook*, Vol. I, *Selected Basic Techniques*, U.S. Atomic Energy Commission, Superintendent of Documents, U.S. Government Printing Office, Washington 25, D.C., 1964.

2. D. J. Hughes and R. B. Schwartz, *Neutron Cross Sections*, USAEC Report BNL-325, 2nd ed., Brookhaven National Laboratory, July 1, 1958; and J. R. Stehn *et al.*, *Neutron Cross Sections*, USAEC Report BNL-325, 2nd ed., Suppl. 2, Brookhaven National Laboratory, May 1964.

3. J. E. Wilkins, Jr., General Atomic, private communication.

4. G. W. Hinman *et al.*, *Accurate Computation of Doppler Broadened Absorption Cross Sections*, General Atomic Report GA-3522, September 20, 1962.

5. G. W. Hinman *et al.*, Accurate Doppler Broadened Absorption, *Nucl. Sci. Eng.*, **16**: 202 (1963); and J. B. Sampson and G. W. Hinman, Comment on Doppler Broadened Absorption, *Nucl. Sci. Eng.*, **18**: 531 (1964).

6. W. E. Lamb, Jr., Capture of Neutrons by Atoms in a Crystal, *Phys. Rev.*, **55**: 190 (1939).

7. M. S. Nelkin and D. E. Parks, Effects of Chemical Binding on Nuclear Recoil, *Phys. Rev.*, **119**: 1060 (1960).

8. H. A. Bethe, Nuclear Physics: B. Nuclear Dynamics, Theoretical, *Revs. Mod. Phys.*, **9**: 69 (1937).

9. N. F. Wikner and S. Jaye, *Energy Dependent and Spectrum Averaged Thermal Cross Sections for the Heavy Elements and Fission Products for Various Temperatures and C:U^{235} Atomic Ratios*, USAEC Report GA-2113, General Atomic Division, General Dynamics Corporation, June 16, 1961.

10. A. M. Weinberg and E. P. Wigner, *The Physical Theory of Neutron Chain Reactors*, p. 371, University of Chicago Press, Chicago, 1958.

11. H. C. Honeck, The Distribution of Thermal Neutrons in Space and Energy in Reactor Lattices. Part I: Theory, *Nucl. Sci. Eng.*, **8**: 193 (1960).

12. G. D. Joanou, C. V. Smith and H. A. Vieweg, *GATHER-II, An IBM-7090 FORTRAN-II Program for the Computation of Thermal-Neutron Spectra and Associated Multigroup Cross Sections*, General Atomic Report

GA-4132, July 8, 1963.

13. H. C. Honeck, *THERMOS, a Thermalization Transport Theory Code for Reactor Lattice Calculations*, USAEC Report BNL-5826, Brookhaven National Laboratory, 1961.

14. D. S. Selengut, Variational Analysis of Multi-Dimensional Systems, in *Nuclear Physics Research Quarterly Report October-December, 1958*, p. 89, USAEC Report HW-59126, Hanford Laboratories, January 1959.

15. G. Rakavy and Y. Yeivin, The Transport Approximation of the Energy-Dependent Boltzmann Equation, *Nucl. Sci. Eng.*, **15**: 158 (1963).

16. K. H. Beckurts and K. Wirtz, *Neutron Physics*, p. 204 ff, Springer-Verlag, Berlin, New York, 1964.

17. E. R. Cohen, The Neutron Velocity Spectrum in a Heavy Moderator, *Nucl. Sci. Eng.*, **2**: 227 (1957).

18. R. F. Coveyou, R. R. Bate, and R. K. Osborn, *Effect of Moderator Temperature Upon Neutron Flux in Infinite, Capturing Medium*, USAEC Report ORNL-1958, Oak Ridge National Laboratory, October 20, 1955.

19. C. H. Westcott, *Effective Cross Section Values for Well-Moderated Thermal Reactor Spectra*, 3rd ed., Canadian Report AECL-1101 (CRRP-960), September 1, 1960.

20. D. E. Parks, J. R. Beyster, and N. F. Wikner, Thermal Neutron Spectra in Graphite, *Nucl. Sci. Eng.*, **13**: 306 (1962).

21. N. F. Wikner, G. D. Joanou, and D. E. Parks, Neutron Thermalization in Graphite, *Nucl. Sci. Eng.*, **19**: 108 (1964).

22. J. A. Young, N. F. Wikner, and D. E. Parks, Neutron Thermalization in Graphite, II, *Nukleonik* (to be published, 1965).

23. B. Carlson, C. Lee, and J. Woolton, *The DSN and TDC Neutron Transport Codes*, USAEC Report LAMS-2346, Los Alamos Scientific Laboratory, October 16, 1959.

24. G. D. Joanou and J. S. Dudek, *GAM-I, A Consistent P_1 Multigroup Code for the Calculation of Fast Neutron Spectra and Multigroup Constants*, USAEC Report GA-1850, General Atomic Division, General Dynamics Corporation, June 28, 1961.

25. H. L. Reynolds, Critical Measurements and Calculations for Enriched-Uranium Graphite-Moderated Systems, in *Proceedings of the Second United Nations International Conference on the Peaceful Uses of Atomic Energy, Geneva, 1958*, Vol. 12, p. 632, United Nations, Geneva, 1958.

26. D. Okrent *et al.*, *The Reactor Kinetics of the Transient Reactor Test Facility (TREAT)*, USAEC Report ANL-6174, Argonne National Laboratory, September 1960.

27. I. Kirn *et al.*, *Reactor Physics Measurements in TREAT*, USAEC Report ANL-6173, Argonne National Laboratory, October 1960.

28. G. Freund *et al.*, *Design Summary Report on the Transient Reactor Test Facility TREAT*, USAEC Report ANL-6034, Argonne National Laboratory, June 1960.

29. A. Ghatak, *Neutronic Analysis of TREAT*, General Atomic Report GA-3575, October 11, 1962.

30. G. D. Joanou and N. F. Wikner, General Atomic, private communication.

31. C. A. Stevens and G. D. Joanou, Investigation of the Flux Recovery Assumption in Resonance Computations, to be presented at the Eleventh Annual Meeting of the American Nuclear Society, Gatlinburg, Tennessee, June 1965.

32. K. J. Mitchell and R. A. Beary, The High Temperature Zero Energy Reactor ZENITH, in *Proceedings of the Second United Nations International Conference on the Peaceful Uses of Atomic Energy, Geneva, 1958*, Vol. 9, p. 310, United Nations, Geneva, 1958.

33. J. A. Larrimore, *Temperature Coefficients of Reactivity in Homogenized Thermal Nuclear Reactors*, Doctoral Dissertation, Massachusetts Institute of Technology, 1962.

34. J. D. Garrison and B. Roos, Fission-Product Capture Cross Sections, *Nucl. Sci. Eng.*, **12**: 115 (1962); and private communication.

35. M. M. Levine, Equal Charge Displacement Rule in Fission Product Poisoning, *Nucl. Sci. Eng.*, **9**: 495 (1961).

36. P. Fortescue *et al.*, HTGR-Underlying Principles and Design, *Nucleonics*, **18** (No. 1): 86 (1960).

37. P. U. Fischer and N. F. Wikner, *An Interim Report on the Temperature Coefficient of the 40-MW(e) HTGR*, USAEC Report GA-2307, General Atomic Division, General Dynamics Corporation, July 14, 1961.

38. G. Stuart, General Atomic, private communication.

39. C. J. Orth, Diffusion of Lanthanides and Actinides from Graphite at High Temperatures, *Nucl. Sci. Eng.*, **9**: 417 (1961).

40. A. M. Baxter and P. U. Fischer, General Atomic, unpublished work.

Chapter 8

CHARACTERISTIC LENGTHS AND TIMES IN NEUTRON THERMALIZATION

We have so far been primarily concerned with the thermal-neutron energy spectrum and the inelastic-scattering mechanism that establishes it. We now turn our attention to the explicit consideration of space- and time-dependent problems. In Chap. 4 we developed the general theory needed but did not consider the systematic application of the theory to thermal-neutron migration in actual moderators. We attempt this in the present chapter and give a comparison with selected experimental results.

We begin with a systematic consideration of the diffusion length, considered as the eigenvalue defining the asymptotic relaxation length of the thermal-neutron flux far from sources. This asymptotic situation can be realized experimentally to a good approximation so that a meaningful comparison between theory and experiment can be given. The same is true for the asymptotic relaxation time for the flux in a finite moderating sample, although the theory here usually deals with too highly idealized a problem to make a precise comparison with experiment.

We also consider the higher eigenvalues describing the space and time variation of the neutron spectrum. Although these are important to the physical understanding of the theory, they are not readily measurable. The comparison between theory and experiment in the area of spatial and temporal variations of the spectrum relies almost entirely on direct numerical solutions of the transport equation. Discussion of such solutions is deferred to Chap. 9, where we consider the spatially dependent spectrum in a reactor.

8-1 THE DIFFUSION LENGTH

In Sec. 4-7 we considered the exponential relaxation of the thermal-neutron flux far from a plane source in energy-dependent diffusion theory. We now reformulate this problem in transport theory and consider its solution more quantitatively. We start from the plane-geometry form of the transport equation, Eq. (4.88), and look for a solution of the form

$$f(z,E,\mu) = e^{-\kappa z} f(E,\mu) = e^{-\kappa z} \sum_{n=0}^{\infty} f_n(E) P_n(\mu) \tag{8.1}$$

to the source-free equation. This defines the eigenvalue problem

$$[\Sigma_a(E) + \Sigma_s(E) - \mu\kappa]\, f(E,\mu) =$$

$$\sum_{n=0}^{\infty} P_n(\mu) \int_0^{\infty} \Sigma_n(E',E) f_n(E') dE' \; . \tag{8.2}$$

We want to calculate the smallest eigenvalue κ_0 and the associated positive eigenfunction $\underline{f}_0(\underline{E},\mu)$. We assume that this solution exists and that κ_0 is separated from the next eigenvalue κ_1 by a finite amount. If this is the case, the solution far from the source is of the form (8.1), and κ_0 can be determined from the measured spatial dependence of the flux.

Since both $\underline{f}_0(\underline{E},\mu)$ and the right-hand side of Eq. (8.2) must be positive for the asymptotic solution, the bracketed term on the left-hand side must also be positive everywhere in the range $-1 \leq \mu \leq 1$. This implies that

$$\kappa_0 \leq \mathrm{Min}[\Sigma(E)] \; . \tag{8.3}$$

Equation (8.3) states that no separable function of position and energy can decay with a relaxation length shorter than the longest mean free path for any neutron energy. We return in Sec. 8-1.5 to a discussion of Eq. (8.3).

8-1.1 Isotropic Scattering. In studying Eq. (8.2) we consider first the simpler problem of isotropic scattering, in which only the $\ell = 0$ term is present on the right-hand side. Assuming Eq. (8.3) to hold, we divide through by $[\Sigma(E) - \kappa_0\mu]$ and integrate over μ from -1 to 1. The resulting equation is

$$\phi_0(E) = \frac{1}{2\kappa_0} \ln\left[\frac{\Sigma(E) + \kappa_0}{\Sigma(E) - \kappa_0}\right] \int_0^\infty \Sigma_0(E',E)\phi_0(E')dE', \quad (8.4)$$

where

$$\phi_0(E) = \int_{-1}^{1} f_0(E,\mu)d\mu \ . \quad (8.5)$$

In the special case of constant cross sections, we can solve Eq. (8.4) directly. The detailed balance condition

$$\Sigma_\ell(E,E')\phi_M(E) = \Sigma_\ell(E',E)\phi_M(E') \quad (8.6)$$

implies that $\phi_M(E)$ is a solution of Eq. (8.4) if κ_0 satisfies

$$1 = \frac{\Sigma_s}{2\kappa_0} \ln\left(\frac{\Sigma_s + \Sigma_a + \kappa_0}{\Sigma_s + \Sigma_a - \kappa_0}\right) \ . \quad (8.7)$$

Equation (8.7) is familiar from monoenergetic transport theory, where it has been extensively studied.[1] Consider κ_0 as a function of Σ_a for fixed Σ_s. In the weak absorption limit, the solution of Eq. (8.7) is

$$\kappa_0^2 = 3\Sigma_a\Sigma_s\left(1 + \frac{1}{5}\frac{\Sigma_a}{\Sigma_s} + \cdots\right) . \tag{8.8}$$

In the limit that $\Sigma_a \gg \Sigma_s$, there is still a solution to Eq. (8.7) with the property

$$\kappa_0 \approx \Sigma_a \ . \tag{8.9}$$

(For strong absorption the asymptotic solution exists, but is not easily separated from the transient solution; see Ref. 1, Chap. 4, for details.)

We saw in Sec. 4-7 that for energy-dependent cross sections the situation is quite different even for weak absorption due to the "diffusion heating" of the energy spectrum. To combine this effect with a correct treatment of the correction term in Eq. (8.8) we must carry out a consistent expansion of κ_0^2 to second order in Σ_a. To do this, we divide Eq. (8.4) by the coefficient of the integral on the right-hand side and subtract a term $\Sigma_s(E)\phi_0(E)$ from both sides to give

$$G(E,\kappa_0)\phi_0(E) = R_0\ \phi_0(E) \ , \tag{8.10}$$

where R_0 is the operator defined by Eq. (4.57) with $\ell = 0$ and $E_m = \infty$,

$$R_0\,\phi(E) \equiv \int_0^\infty [\Sigma_0(E',E)\phi(E') - \Sigma_0(E,E')\phi(E)]dE',$$

and

$$G(E,\kappa) = \frac{2\kappa}{\ell n\,\dfrac{\Sigma(E)+\kappa}{\Sigma(E)-\kappa}} - \Sigma_s(E)\ . \qquad (8.10a)$$

Keeping only terms up to order κ^2 in Eq. (8.10a), we obtain the energy-dependent diffusion equation (4.72) with

$$D(E) = \frac{1}{3\Sigma(E)} \approx \frac{1}{3\Sigma_s(E)}\ .$$

In order to have a convenient expansion parameter, we let

$$\Sigma_a(E) \to \varepsilon\,\Sigma_a(E) \qquad (8.11)$$

and allow ε to equal one at the end of the calculation. If we set

$$\kappa_0^2 = \varepsilon A_0 + \varepsilon^2 A_1 + \varepsilon^3 A_2 + \cdots \qquad (8.12)$$

and expand $\underline{G}(\underline{E},\kappa_0)$ to order ε^2, we obtain

$$G(E,K_0) = \varepsilon\left[\Sigma_a(E) - \frac{A_0}{3\Sigma_s(E)}\right] + \varepsilon^2\left[\frac{1}{3}\,\frac{\Sigma a(E)A_0}{\Sigma_s^2(E)} - \frac{4}{45}\,\frac{A_0^2}{\Sigma_s^3(E)} - \frac{A_1}{3\Sigma_s(E)}\right] + O(\varepsilon^3)\ . \qquad (8.13)$$

We must also expand the spectrum $\phi_0(\underline{E})$ in powers of ε:

$$\phi_0(E) = \phi_M(E) + \varepsilon\phi_1(E) + \varepsilon^2\phi_2(E) + \ldots \quad . \tag{8.14}$$

Equation (8.10) must be satisfied to each order in ε. To lowest order we have

$$R_0 \, \phi_M(E) = 0;$$

thus we were justified in assuming that the leading term in Eq. (8.14) is $\phi_{\underline{M}}(\underline{E})$. To first order in ε we have

$$\left[\Sigma_a(E) - \frac{A_0}{3\Sigma_s(E)}\right]\phi_M(E) = R_0 \, \phi_1(E) \ . \tag{8.15}$$

If we integrate Eq. (8.15) over energy, the right-hand side gives zero by virtue of Eq. (4.65) so that

$$A_0 = \frac{\int_0^\infty \Sigma_a(E)\phi_M(E)dE}{\int_0^\infty [3\Sigma_s(E)]^{-1}\phi_M(E)dE} \quad . \tag{8.16}$$

In the limit of weak absorption, the diffusion length for isotropic scattering is the same as we calculated in Chap. 4. If we let

$$G(E,\kappa_0) = \varepsilon G_1 + \varepsilon^2 G_2 + \ldots ,$$

the terms of order ε^2 in Eq. (8.10) give

$$G_1 \, \phi_1(E) + G_2 \, \phi_M(E) = R_0 \, \phi_2(E) \; . \qquad (8.17)$$

Integrating Eq. (8.17) over energy, we have

$$A_1 \int_0^\infty [3\Sigma_s(E)]^{-1} \phi_M(E) dE = - \int_0^\infty \left[\frac{A_0}{3\Sigma_s(E)} - \Sigma_a(E) \right] \phi_1(E) dE$$

$$+ \int_0^\infty \frac{1}{3} \left[\frac{\Sigma a A_0}{\Sigma_s^2(E)} - \frac{4}{45} \, \frac{A_0^2}{\Sigma_s^3(E)} \right] \phi_M(E) dE \, . \qquad (8.18)$$

Everything on the right-hand side of Eq. (8.18) is known except $\phi_1(\underline{E})$, which is the solution of Eq. (8.15) and depends on the thermalization model chosen. The right-hand side of Eq. (8.18) has an interesting structure. The first integral is exactly the same as we would get from the energy-dependent diffusion equation (4.72) if we expanded to order ε^2. It thus gives exactly the same diffusion heating coefficient as we would calculate from Eq. (4.72). For the reasons discussed in Chap. 4, this contribution to $\underline{A}_1$ is negative. The second integral on the right-hand side of Eq. (8.18) does not depend on the thermalization model but involves only averages over a Maxwellian flux spectrum. We thus find that to second order in $\Sigma_{\underline{a}}$ we can calculate diffusion heating and transport corrections to κ_0^2 separately and add the two corrections together. (We will see that this is not exact for anisotropic scattering, but it remains a good approximation.)

Before going on to the case of anisotropic scattering we mention two further properties of Eq. (8.18). First, Eq. (8.15) does not uniquely determine $\phi_1(\underline{E})$. We can add any multiple of $\phi_{\underline{M}}(\underline{E})$ to the solution and still have a solution. This does not affect our calculation of $\underline{A}_1$, however, since any such addition to $\phi_1(\underline{E})$ does not contribute to $\underline{A}_1$ by virtue of Eq. (8.18). Finally, consider the case of constant cross sections, where

$$A_0 = 3\Sigma_a\Sigma_s$$

and the left-hand side of Eq. (8.15) vanishes identically. The diffusion heating contribution to $\underline{A}_1$ also vanishes, and the transport contribution becomes

$$A_1 = \frac{3}{5}\Sigma_a^2$$

so that κ_0^2 is given by Eq. (8.8), as expected.

Our discussion for isotropic scattering has shown the essential structure of the problem. In order to compare theory and experiment, however, we must include the anisotropic scattering that occurs in actual moderators.

8-1.2 Anisotropic Scattering. Returning to Eq. (8.2), we consider the improved approximation of retaining $\underline{N} + 1$ terms on the right-hand side. In place of the single integral equation (8.4) for $\phi_0(\underline{E})$ we then have $\underline{N} + 1$ coupled integral for the quantities

$$\phi_\ell(E) = \frac{2}{2\ell + 1} f_\ell(E) \qquad \ell = 0 \ldots N . \tag{8.19}$$

Dividing Eq. (8.2) by $[\Sigma(\underline{E}) - \kappa\mu]$, multiplying by $\underline{P}_\ell(\mu)$, and integrating over μ from -1 to 1 we obtain

$$\phi_\ell(E) = \sum_{n=0}^{N} Q_{\ell n}(\kappa_0, E) \int_0^\infty \Sigma_n(E', E)\phi_n(E')dE', \tag{8.20}$$

where

$$Q_{\ell n}(\kappa_0, E) = \frac{2n + 1}{2\Sigma(E)} F_{\ell n}\left[\frac{\kappa_0}{\Sigma(E)}\right] \tag{8.21}$$

and

$$F_{\ell n}(x) = \int_{-1}^{1} \frac{P_\ell(\mu)\ P_n(\mu)\ d\mu}{1 - x\mu} . \tag{8.22}$$

The direct numerical solution of the set of equations (8.20) has been carried through by Honeck[2] for the case of H_2O with $\underline{N} = 3$. We discuss the results after first considering the limit of weak absorption.

Returning to Eq. (8.2), we multiply by $\underline{P}_\ell(\mu)$ and integrate over μ from -1 to 1. This gives the standard spherical-harmonics form of the transport equation for this problem. For convenience, we subtract a term $\Sigma_{\underline{\ell}}(\underline{E})\underline{f}_{\underline{\ell}}(\underline{E})$ from both sides of the resulting equation to give

$$[\Sigma_a(E) + \Sigma_s(E) - \Sigma_\ell(E)]f_\ell(E) -$$

$$\kappa_0\left[\frac{\ell}{2\ell - 1} f_{\ell-1}(E) + \frac{\ell + 1}{2\ell + 3} f_{\ell+1}(E)\right] = R_\ell f_\ell(E). \quad (8.23)$$

We recall that $\Sigma_{\underline{\ell}}(\underline{E})$ is defined by Eq. (4.59) and $\underline{R}_{\underline{\ell}}$ by Eq. (4.57).

We can obtain a set of equations that can be solved sequentially by carrying out a systematic expansion in powers of the absorption as we did for isotropic scattering. The functions $\underline{f}_{\underline{\ell}}(\underline{E})$ will have the same type of power-series expansion as in the isotropic-scattering case. The reader can verify that this is of the form

$$f_\ell(E) = \varepsilon^{\ell/2}[f_\ell^0 + \varepsilon f_\ell^1 + \varepsilon^2 f_\ell^2 + \ldots] . \quad (8.24)$$

Letting

$$\kappa_0 = \varepsilon^{1/2}[b_0 + \varepsilon b_1 + \varepsilon^2 b_2 + \ldots] \quad (8.25)$$

and substituting Eqs. (8.11a), (8.24), and (8.25) into Eq. (8.23), we obtain

$$\Sigma_a f_\ell^{n-1} + (\Sigma_s - \Sigma_\ell) f_\ell^n - \frac{\ell}{2\ell - 1}\left[b_0 f_{\ell-1}^n + b_1 f_{\ell-1}^{n-1} + \dots\right]$$

$$- \frac{\ell + 1}{2\ell + 3}\left[b_0 f_{\ell+1}^{n-1} + b_1 f_{\ell+1}^{n-2} + \dots\right] = R_\ell f_\ell^n \quad , \tag{8.26}$$

where $f_\ell^n \equiv 0$ if $\ell < 0$ or $n < 0$ and the energy dependence of Σ_a, Σ_s, Σ_ℓ, and f_ℓ^n has been suppressed.

For notational convenience let

$$\Sigma_s(E) - \Sigma_\ell(E) = \Sigma_s(E) c_\ell(E) \ . \tag{8.27}$$

In particular, $c_0 = 0$ and $c_1(E) = 1 - \overline{\mu}(E)$.

We now examine the set of equations (8.26). Starting with $\ell = 0$ and $n = 0$,

$$R_0 f_0^0 = 0, \tag{8.28}$$

which has the solution

$$f_0^0 = \phi_M(E) \ . \tag{8.29}$$

To calculate b_0 we need

$$\Sigma_a f_0^0 - \tfrac{1}{3} b_0 f_1^0 = R_0 f_0^1 \quad , \tag{8.30}$$

which is Eq. (8.26) with $\ell = 0$ and $n = 1$. Integrating Eq. (8.30) over energy, we obtain

$$\frac{1}{3}b_0 \int_0^\infty f_1^0(E)dE = \int_0^\infty \Sigma_a(E)\phi_M(E)dE . \tag{8.31}$$

To calculate $\underline{b}_0$ we thus need $\underline{f}_1^0(\underline{E})$, which is the solution of

$$\Sigma_s c_1 f_1^0 - b_0 f_0^0 = R_1 f_1^0 . \tag{8.32}$$

In the limiting case of isotropic scattering this reduces to

$$f_1^0 = \frac{b_0}{\Sigma_s(E)} \phi_M(E) ,$$

which gives Eq. (8.16), inasmuch as $\underline{b}_0^2 = \underline{A}_0$ by comparison of Eqs. (8.12) and (8.25). For anisotropic scattering we already have an integral equation to solve at this early stage. For thermal-neutron scattering, however, it is a good approximation to neglect the term $\underline{R}_1\underline{f}_1^0(\underline{E})$ on the right-hand side of Eq. (8.32). We thus have the approximate solution

$$f_1^0 \approx \frac{b_0\phi_M(E)}{\Sigma_s(E)c_1(E)} , \tag{8.33}$$

which gives

$$b_0^2 \int_0^\infty \frac{\phi_M(E)dE}{3\Sigma_s(E)\ c_1(E)} = \int_0^\infty \Sigma_a(E)\phi_M(E)dE . \tag{8.34}$$

This is the same as Eq. (8.16) except that Σ_s is replaced by $\Sigma_s c_1$. This corresponds to the use of energy-dependent diffusion theory instead of the consistent P_1-approximation. To obtain a more accurate solution, we can substitute Eq. (8.33) into the right-hand side of Eq. (8.32). The improved result for b_0 is given by

$$\frac{1}{3}b_0^2 \int_0^\infty \left[\frac{\phi_M}{\Sigma_s c_1} + \frac{1}{\Sigma_s c_1} R_1 \left(\frac{1}{\Sigma_s c_1}\right) \phi_M\right] dE = \int_0^\infty \Sigma_a(E)\phi_M(E)dE \, . \tag{8.35}$$

To calculate b_1 we must consider

$$\Sigma_a f_0^1 - \frac{1}{3}b_0 f_1^1 - \frac{1}{3}b_1 f_1^0 = R_0 f_0^2 \, . \tag{8.36}$$

Integrating Eq. (8.36) over energy, we have

$$\frac{1}{3}b_1 \int f_1^0 \, dE = \int \Sigma_a f_0^1 \, dE - \frac{1}{3}b_0 \int f_1^1 \, dE \, . \tag{8.37}$$

We therefore need f_1^1 that satisfies

$$\Sigma_a f_1^0 + \Sigma_s c_1 f_1^1 - b_0 f_0^1 - b_1 f_0^0 - \frac{2}{5}b_0 f_2^0 = R_1 f_1^1 \tag{8.38}$$

and also f_2^0 that satisfies

$$\Sigma_s c_2 f_2^0 - \frac{2}{3}b_0 f_1^0 = R_2 f_2^0 \, . \tag{8.39}$$

Table 8.1

DIFFUSION PARAMETERS FOR H_2O

(See Eq. (8.43))

	α_1 (cm^{-1})	α_2	$(\alpha_2)_{\underline{d}}$	$(\alpha_2)_{\underline{t}}$	α_3 (cm)
Theory (I)	5.87	2.65	2.81	-0.16	1.42
Theory (II)	5.86	2.87	----	----	1.35
Experiment	6.14 ±0.02	3.05 ±0.37	----	----	----

I - Weak-absorption expansion neglecting $\underline{R}_1$, $\underline{R}_2$, and $\underline{R}_3$.

II - Numerical solution of Eq. (8.20).

(8.35), but this has not been checked. Thus α_1 is essentially a measure of an appropriately averaged transport cross section. The agreement of theory and experiment within 5% is satisfactory, considering the crude way in which the theory treats the hindered molecular rotations and the anisotropy of vibrations in water.

For α_2 we find agreement of theory and experiment within the 12% experimental error. The approximation of isotropic energy transfer is not as good as for α_1. We find that within this approximation the transport contribution is very small. If we examine Eq. (8.41), we find that this contribution comes from the integral over $\phi_{\underline{M}}(\underline{E})$ on the right-hand side. This integral is a small difference between two large numbers, each of which depends sensitively on the energy dependence of the cross sections. The small value of $(\alpha_2)_{\underline{t}}$ is thus without quantitative significance. Again, the agreement between theory and experiment is satisfactory.

8-1.4 Discussion. We have treated the problem of the diffusion length, as defined by Eq. (8.2), in considerable detail in this chapter and in Chap. 4. This detailed treatment has two rather distinct objectives. The first is to exhibit the structure of the transport equation for thermal neutrons and to generalize a classic problem of monoenergetic transport theory so that it can be applied to experiment. Most of the algebraic manipulation that we have presented has been connected with this first objective.

The second objective is to present another clean integral experiment where our knowledge of the fundamental cross sections can be tested. This is in much the same spirit as the comparison of theory and experiment for the infinite-medium spectrum that we presented in Chap. 6. In order to be able to achieve this objective, we require that the equations connecting the basic cross sections with the experimental measurements be well understood and that the geometrical complexities of the problem be sufficiently unimportant that we can correct for them with confidence. For this second objective, we could have started directly from the numerical solution of the set of equations (8.20) and compared theory and experiment entirely by means of Fig. 8.1. This would be useful, but it would not tell us what aspects of our scattering model were being tested by experiment. From the more analytical approach in which we expand in powers of $\underline{\Sigma_a(k_B T)}$, the parameters α_1 and α_2 give us considerable physical information. The parameter

α_1 gives the Maxwellian average of the reciprocal of the energy-dependent transport cross section. The parameter α_2 depends both on the energy dependence of the cross sections and on a spectrum that is sensitive to the thermalization model. It is a very convenient quantity to use as a check point between theory and experiment in theories of neutron thermalization.

In the diffusion length experiment of Starr and Koppel,[3] the exponential relaxation of the thermal-neutron flux is measured by foil activation in a cylindrical tank of boric acid placed at the end of a reactor thermal column. Except for a small correction for neutron leakage in the radial direction, the idealized conditions of Eq. (8.2) are well fulfilled. The flux is measured to have the shape

$$e^{-\gamma z} J_0(qr)$$

and the eigenvalue κ_0 is identified as

$$\kappa_0^2 = \gamma^2 - q^2 \quad .$$

Whenever the flux distribution far from the source in an infinite medium corresponds to the exponential decay of a fully thermalized flux distribution, the theory of the present section applies. When this is true, a clean

The basic idea is to expand $f(z, E, \mu)$ in terms of the eigenfunctions of the scattering-in operator on the right-hand side of Eq. (8.44). Consider the eigenvalue problem

$$\int_0^\infty \Sigma_0(E',E)\phi_j(E')dE' = \alpha_j \Sigma_s \phi_j(E) \; . \tag{8.45}$$

It is somewhat more convenient to work with the symmetrized form of this problem. Using detailed balance, we can rewrite Eq. (8.45) in the form

$$\int_0^\infty K(E',E)\chi_j(E')dE' = \alpha_j \, \chi_j(E) \; , \tag{8.46}$$

where

$$\chi_j(E) = \phi_j(E)[\phi_M(E)]^{-1/2} \quad , \tag{8.47}$$

and

$$K(E',E) = \Sigma_0(E',E)\Sigma_s^{-1}\left[\frac{\phi_M(E')}{\phi_M(E)}\right]^{1/2} \tag{8.48}$$

is a symmetric function of E and E'. The structure of the eignevalue spectrum of Eq. (8.46) has been studied in a recent paper by Kuščer and Corngold.[12] The kernal $K(E',E)$ is symmetric, nondegenerate, and positive definite. Kuščer and Corngold show that it is also square integrable in the sense that

$$\int_0^\infty \int_0^\infty K^2(E,E')dEdE' < \infty$$

for a solid or a gas but is not square integrable for a liquid. It therefore follows for a solid or a gas that there is an infinite set of eigenfunctions $\chi_i(\underline{E})$ corresponding to positive eigenvalues α_i. The eigenvalues are bounded as follows:

$$\alpha_0 > \alpha_1 > \alpha_2 \ldots > 0 ;$$

they accumulate only at zero. From detailed balance we know that there is a solution corresponding to $\alpha_0 = 1$ with $\chi_0(\underline{E}) = [\phi_{\underline{M}}(\underline{E})]^{1/2}$. We have already stated that this is the only solution in which scattering in and scattering out balance exactly at all energies. We have stated this on physical grounds, but the work of Kuščer and Corngold allows a mathematical proof.

The eigenfunctions $\chi_i(\underline{E})$ are orthogonal, and we choose their normalization so that

$$\int_0^\infty \chi_j(E)\chi_k(E)dE = \int_0^\infty \frac{\phi_j(E)\phi_k(E)}{\phi_M(E)} dE = \delta_{jk} \quad . \tag{8.49}$$

Since the eigenfunctions $\phi_i(\underline{E})$ form a complete set, we expand $\underline{f}(\underline{z},\underline{E},\mu)$ in the form

$$f(z,E,\mu) = \sum_{n=0}^{\infty} \psi_n(z,\mu)\, \phi_n(E) . \tag{8.50}$$

We substitute Eq. (8.50) into Eq. (8.44), multiply by $\phi_j(E)/\phi_M(E)$, and integrate over energy. Making use of Eqs. (8.45) and (8.49), we obtain

$$\left(\Sigma + \mu \frac{\partial}{\partial z}\right)\psi_j(z,\mu) = \frac{1}{2}\,\alpha_j \Sigma_s \int_{-1}^{1} \psi_j(z,\mu')d\mu' + \frac{\phi_j(E_0)}{\phi_M(E_0)}\,\delta(z) . \tag{8.51}$$

The expansion of Eq. (8.50) is constructed so that different values of j are not coupled on the right-hand side of Eq. (8.51). Because of the assumption of constant cross sections, there is also no coupling on the left-hand side. We thus have an infinite set of uncoupled one-speed transport equations for the functions $\psi_j(z,\mu)$. The corresponding coupled equations when the cross sections are not constant have been considered by Leonard and Ferziger.[13] This is a much more complicated problem and will not be considered here.

Equation (8.51) is a standard problem in neutron transport theory, which is solved, for example, in Ref. 1. Consider first the total flux. Integrating Eq. (8.50) over energy, and using Eq. (8.49), we have

$$\int_0^\infty f(z,E,\mu)dE = \psi_0(z,\mu) .$$

We thus recover the one-speed transport equation for the total angular flux. We saw in Sec. 4-4 that this is the case for constant cross sections on more general grounds. The terms in Eq. (8.50) with $j \neq 0$ thus do not contribute to the total flux but describe the change of the energy spectrum with position. Since the eigenvalue α_j form a decreasing sequence, Eq. (8.51) for $j \neq 0$ and for $j = 0$ differs only in that the ratio of scattering cross section to total cross section is reduced by a factor α_j. If we introduce the new variables $c = \Sigma_s/\Sigma$ and $y = \Sigma z$, we can write Eq. (8.51) in the form

$$\left(1 + \mu\frac{\partial}{\partial y}\right)\psi_j(y,\mu) = \frac{1}{2}c\alpha_j \int_{-1}^{1} \psi_j(y,\mu')d\mu' +$$

$$\frac{\phi_j(E_0)}{\phi_M(E_0)} \delta(y) . \tag{8.52}$$

The scalar flux associated with $\psi_j(y,\mu)$ is given in Ref. 1 as

$$\rho_j(y) = \int_{-1}^{1} \psi_j(y,\mu)d\mu$$

$$= \frac{\phi_j(E_0)}{\phi_M(E)}\left[A(c\alpha_j)\exp\left(-\frac{|y|}{\nu_j}\right) + \right.$$

$$\left.\frac{1}{2}\int_0^1 g(c\alpha_j,\nu)\exp\left(-\frac{|y|}{\nu}\right)\nu^{-1}\, d\nu\right] . \tag{8.53}$$

The relaxation length ν_i is the solution of the familiar transcendental equation

$$1 = \frac{1}{2}c\alpha_j\nu_j \ln\left(\frac{\nu_j + 1}{\nu_j - 1}\right) \equiv c\alpha_j\nu_j \tanh^{-1}\left(\frac{1}{\nu_j}\right) . \qquad (8.54)$$

The function $\underline{g}(\underline{c},\nu)$ is given by

$$g(c,\nu) = \left[\left(1 - c\nu \tanh^{-1}\nu\right)^2 + \left(\frac{\pi}{2}c\nu\right)^2\right]^{-1} ,$$

and

$$A(c\alpha_j) = \frac{1}{\nu_j^2}\frac{\partial\nu_j}{\partial(c\alpha_j)} .$$

We note that Eq. (8.54) is the same as Eq. (8.7) except that the left-hand side of Eq. (8.7) is replaced by α_i^{-1}. If we had started with Eq. (8.4) and looked for the higher eigenvalues, the eigenvalue condition would be Eq. (8.54) with

$$\nu_j \rightarrow \frac{\Sigma}{\kappa_j} .$$

By starting with Eq. (8.44) we are able to exhibit the full transient behavior of the energy-dependent flux instead of just the discrete relaxation lengths for the source-free transport equation. If we have done all the algebra explicitly, this would have been a much more difficult task, but by making use of the results of Ref. 1 we get the full solution with no greater effort. In the paper of Ferziger and Leonard[11] the reduction of the energy-dependent problem to a set of uncoupled one-speed problems was combined with

the solution of the one-speed problem using the elegant reformulation of one-speed transport theory due to Case[14]. The use of Case's method in the particular problem under discussion here is desirable, but not essential, inasmuch as we are dealing with a standard problem that has already been solved by older more cumbersome methods. The essential point of the procedure is that we calculate the full spatial dependence associated with each energy eigenfunction. In addition to the discrete term with relaxation length ν_i/Σ, where $\nu_i > 1$, we have the more rapidly decaying contribution of the integral in Eq. (8.53). The complete solution for the flux is given by

$$\phi(z,E) = \sum_{j=0}^{\infty} \frac{\phi_j(E_0)\phi_j(E)}{\phi_M(E_0)} \left[A(\alpha_j c) \exp\left(-\frac{\Sigma|z|}{\nu_j}\right) + \int_0^1 g(c\alpha_j,\nu) \exp\left(-\frac{\Sigma|z|}{\nu}\right) \frac{d}{\nu} \right]. \quad (8.55)$$

Let us now examine the discrete relaxation lengths ν_i. For $i = 0$, ν_0/Σ is the usual diffusion length for constant cross section and isotropic scattering. As we have discussed in detail in Sec. 8-1, the constant-cross-section approximation is not very good for this quantity because the $1/\underline{v}$ nature of the thermal absorption plays an important role. For $\underline{i} \neq 0$, $\nu_{\underline{i}}$ has the same dependence on $\alpha_{\underline{i}}\underline{c}$ as does ν_0 on $\underline{c}$. Thus the $\nu_{\underline{i}}$ satisfy

$$\nu_0 > \nu_1 > \nu_2 > \ldots > 1 .$$

The actual values of $\nu_{\underline{i}}$ are of interest only when they are appreciably greater than one, since only then do a finite number of discrete relaxation lengths approximately determine the actual behavior of the space-energy distribution. For the special case of a heavy moderator, a small value of $\underline{i}$, and weak absorption, this will be the case. In this situation the constant-cross-section approximation should be fairly accurate.

As an example, consider the limiting case of the heavy monatomic gaseous moderator. The eigenvalue equation (8.45) for this case takes the form

$$\xi_0\Sigma_s \left\{ \frac{d}{dE} [E\,\phi(E)] + Ek_BT \frac{d^2\phi}{dE^2} \right\} = (\alpha - 1)\Sigma_s\phi(E) . \qquad (8.56)$$

In terms of the physical context of Sec. 5-5, Eq. (8.56) is just the differential equation for the flux spectrum in an infinite homogeneous medium with a constant absorption cross section of magnitude $(\alpha - 1)\Sigma_s$. If we let $\varepsilon = E/k_BT$ and

$$g(\varepsilon) = \frac{\phi(E)}{\phi_M(E)} = \varepsilon^{-1} e^{\varepsilon} \phi ,$$

Eq. (8.56) becomes a confluent hypergeometric equation,

$$\varepsilon g'' + (2 - \varepsilon)g' + kg = 0 , \tag{8.57}$$

where

$$k = \frac{1 - \alpha}{\xi_0} .$$

As we discussed in Chap. 5, we must choose that solution of Eq. (8.56) or (8.57) which corresponds to no net source of neutrons at $E = 0$. The possibility of solutions with the wrong behavior at small energies was introduced in going to the differential equation from the integral equation. If we exclude such solutions, the solution of Eq. (8.57) is a confluent hypergeometric function[15]

$$g(\varepsilon) = \Phi(-k|2|\varepsilon).$$

For negative values of k, we have the flux spectrum for a case equivalent to a positive absorption cross section and a neutron source at high energies. Physically, we want to exclude neutron sources at high energy in the present case. Mathematically, we are interested only in solutions where the $\chi_i(E)$ of Eq. (8.47) are square-integrable and belong to the Hilbert space of the operator $\underline{K}$ appearing in Eq. (8.46). The only solutions to Eq. (8.57) with satisfactory behavior as E approaches infinity occur for k equal to zero or a positive integer. For these values of k, the function $\Phi(-k|2|\varepsilon)$ becomes polynomial--specifically, an associated Laguerre polynomial of the first kind. The desired solutions of Eq. (8.56) are thus

$$\phi_j(E) = \phi_M(E)\, L_j^1\left(\frac{E}{k_B T}\right) , \tag{8.58}$$

with α_i taking on the values

$$\alpha_j = 1 - \xi_0 j \qquad j = 0,1,2, \ldots \quad . \tag{8.59}$$

To be more explicit, the Laguerre polynomials in Eq. (8.58) are defined by

$$(m + 1)^{1/2}\, L_m^1(\varepsilon) = \sum_{n=0}^{m} \binom{m + 1}{n + 1} \frac{(-\varepsilon)^n}{n!} \tag{8.60}$$

and satisfy the orthonormality condition

$$\int_0^{\infty} \varepsilon\, e^{-\varepsilon}\, L_m^1(\varepsilon)\, L_n^1(\varepsilon)\, d\varepsilon = \delta_{nm} \ . \tag{8.61}$$

The heavy-gas approximation can only be consistently applied in the limit that ξ_0 approaches zero. As in Chap. 5, we assume that $\gamma = \Sigma_{\underline{a}}/\xi_0\Sigma_{\underline{s}}$ remains finite in this limit. In this limit $\alpha_{\underline{j}}\underline{c}$ in Eq. (8.54) is nearly equal to one, so that $\nu_{\underline{j}}$ is much greater than one. Expanding the logarithm in Eq. (8.54) gives

$$\frac{1}{\alpha_j c} = \frac{1}{2}\nu_j \ \ln\left(\frac{\nu_j + 1}{\nu_j - 1}\right) \approx 1 + \frac{1}{3\nu_j^2} \quad ,$$

so that $\nu_{\underline{j}}$ is given by its diffusion-theory value

$$\frac{1}{\nu_j^2} = 3(j + \gamma)\xi_0 \approx 3j\xi_0 + 3(1 - c) \ , \tag{8.62}$$

the last form holding when $\underline{c} \approx 1$. Equation (8.62) could have been obtained somewhat more directly if we had started with energy-dependent diffusion theory. The eigenvalue problem we want to solve would then be

$$R_0\phi_j = (\Sigma_a - D\kappa_j^2)\phi_j \approx \left(\xi_0\Sigma_s\gamma - \frac{\Sigma}{3\nu_j^2}\right)\phi_j \ .$$

Using Eq. (8.45) and the definition of $\underline{R}_0$, this becomes

$$-\Sigma_s(1 - \alpha_j)\phi_j = \left(\xi_0\Sigma_s\gamma - \frac{\Sigma_s}{3\nu_j^2}\right)\phi_j ,$$

which is equivalent to Eq. (8.62) when α_i is given by Eq. (8.59).

The combination of the heavy-gas model and diffusion theory has been fruitful in yielding a semiquantitative understanding of space-dependent neutron thermalization, largely because the energy eigenfunctions and eigenvalues are known in closed form. This approach was first discussed in a Geneva Conference paper by Karaenovsky et al. in 1958.[16] It was applied to the space-energy distribution in an infinite medium by Michael[17] and to the problem of the spectrum in the neighborhood of a temperature discontinuity by Kottwitz.[18] Many authors have extended this approach as a basis for more quantitative approximate calculations, but experience indicates that this has not been particularly successful. We give several examples later in this chapter.

If the moderator is very heavy, we see from Eq. (8.62) that the relaxation lengths for the spectrum are long compared with a mean free path; it is thus consistent to use diffusion theory. The lengths ν_i depend, however, on the mass ratio only as the inverse square root; therefore, it

is not clear whether diffusion theory is reasonable for the practically interesting heavy moderators, beryllium and graphite. It turns out in fact that it is not reasonable. In Fig. 8.2 we show the results of Ferziger and Leonard for $\underline{i} = 0$, 1, and 2 and $\xi_0 = 0.2$. The solid lines give $1/\nu_i^2$ versus $\underline{c}$ as obtained from Eqs. (8.54) and (8.59). The dashed lines give the diffusion theory results from Eq. (8.62). We see that diffusion theory is not very accurate even for $\underline{i} = 1$ and zero absorption and that it is very bad for $\underline{i} = 2$. In fact, Eq. (8.62) gives a value of ν_2 less than one for all values of γ if $\xi_0 > 1/6$.

In combining Eqs. (8.54) and (8.59) we have not been mathematically consistent inasmuch as Eq. (8.59) is correct only to first order in the mass ratio. If we consider ξ_0 as being moderately large, we should calculate α_i correctly and should include the effects of the energy dependence and anistropy of the cross sections. It is often the case in calculations of neutron migration that diffusion theory works better than it should, and this could also be the case for the space dependence of the thermal spectrum. From the limited analytic information we have at present, however, it appears that the relaxation lengths for changes in the thermal spectrum are sufficiently small for all

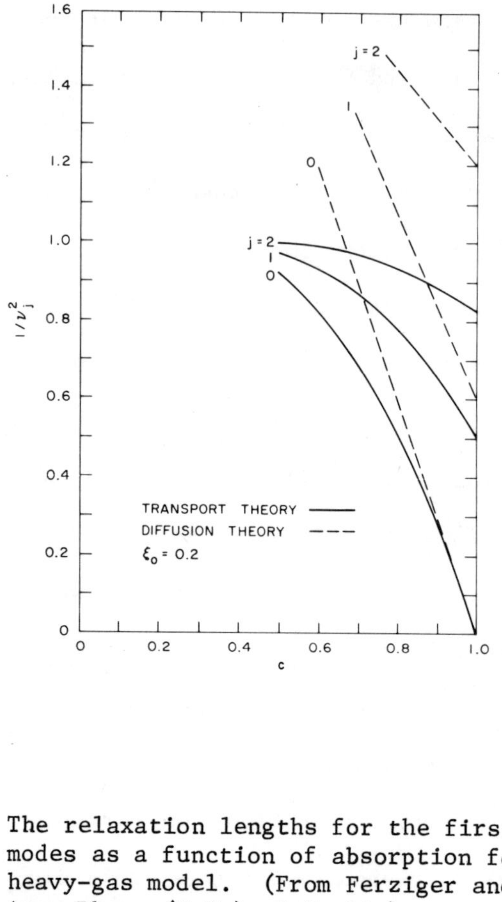

Figure 8-2 The relaxation lengths for the first few energy modes as a function of absorption for the heavy-gas model. (From Ferziger and Leonard, Ann. Phys. (N.Y.), Ref. 11.)

moderators of interest that a transport-theory treatment is needed. It should be emphasized that the transport-theory treatment we have given is not adequate for actual calculation but is merely illustrative of the structure of the solution. The only conclusion we can draw is that a more accurate theoretical description is needed. Such a description can only be given numerically. Some examples of numerical calculations of space-dependent spectra are given in Chap. 9.

8-3 THE DIFFUSION OF THERMAL NEUTRONS

In Sec. 8-1 we considered the diffusion of fully thermalized neutrons in terms of the weak-absorption limit of the diffusion length. Certain aspects of thermal-neutron diffusion are more transparent physically if we consider explicitly the case of a nonabsorbing medium. The results obtained are instructive and of some historical interest. They also are necessary preliminaries to the consideration of the problem of two nonabsorbing media at different temperatures.

We consider the solution to the transport equation in an infinite nonabsorbing medium with uniform temperature. Considering variation in one dimension only, the appropriate form of the transport equation is

$$\left[\Sigma_s(E) + \mu \frac{\partial}{\partial z}\right] f(z,E,\mu) = \sum_{\ell=0}^{\infty} \int \Sigma_\ell(E',E) f_\ell(E',z) dE' P_\ell(\mu). \tag{8.63}$$

We expect the solution of this equation far from boundaries to be an appropriate generalization of Eq. (4.126) to the case of anisotropic scattering. Following the original discussion of this problem by Kuščer and Ribarič,[19] we try a solution of the form

$$f(z,E,\mu) = \tfrac{1}{2}A\ \phi_M(E)[z + z_0 - \mu\ U(E)]\ . \tag{8.64}$$

Substituting Eq. (8.64) into the left-hand side of Eq. (8.63) and the equivalent expressions

$$f_0(z,E) = \tfrac{1}{2}A\ \ \phi_M(E)\ \ (z + z_0)$$

$$f_1(z,E) = \tfrac{-1}{2}A\ U(E)\ \phi_M(E)$$

into the right-hand side of Eq. (8.63), we find that Eq. (8.64) is a solution if

$$\Sigma_s(E)\phi_M(E) = \int_0^\infty \Sigma_0(E',E)\phi_M(E')dE' \tag{8.65}$$

and if U(E) satisfies the integral equation

$$\Sigma_s(E)U(E)\phi_M(E) = \phi_M(E) + \int_0^\infty \Sigma_1(E',E)U(E')\phi_M(E')dE'. \tag{8.66}$$

Equation (8.65) is automatically satisfied by virtue of the detailed balance condition (8.6) and the definition of $\underline{\Sigma_s(E)}$ as the integral of $\Sigma_0(\underline{E},\underline{E}')$ over all final energies $\underline{E}'$. Equation (8.66) defines the function $\underline{U}(\underline{E})$ that appears in Eq. (8.64). We can put this equation in a somewhat more transparent form by using detailed balance and extracting a common factor $\phi_{\underline{M}}(\underline{E})$ to give

$$\Sigma_s(E)\, U(E) = 1 + \int_0^\infty \Sigma_1(E,E')\, U(E')dE' \tag{8.67}$$

Alternatively, we can subtract

$$U(E)\, \phi_M(E) \int_0^\infty \Sigma_1(E,E')dE'$$

from both sides of Eq. (8.66) to give

$$\Sigma_s(E)\, c_1(E)\, U(E)\phi_M(E) = \phi_M(E) + R_1\, U(E)\phi_M(E)\,. \tag{8.68}$$

Comparing Eq. (8.68) with Eq. (8.32) we see that they define the same integral equation, and thus

$$f_1^0(E) = b_0\, U(E)\phi_M(E)\,. \tag{8.69}$$

To first approximation we neglect $\underline{R}_1$ in Eq. (8.68). An equivalent approximation is to replace $\underline{U}(\underline{E}')$ by $\underline{U}(\underline{E})$ in the integral on the right-hand side of Eq. (8.67). In this approximation we have

$$U(E) \approx [\Sigma_s(E)\, c_1(E)]^{-1} = \ell_{tr}(E) . \tag{8.70}$$

Going back to the solution (8.64), we see that it implies

$$j(z,E) = -\frac{1}{3}U(E)\,\frac{\partial\phi(z,E)}{\partial z} . \tag{8.71}$$

We can therefore think of U(E) as an effective transport mean free path. The diffusion coefficient for thermal neutrons is defined by

$$J = -D\frac{d\Phi}{dz} . \tag{8.72}$$

Again going back to Eq. (8.64), we see that this is given in a nonabsorbing medium by

$$D = \frac{1}{3}(\overline{U})_M = \frac{1}{3}\int U(E)\phi_M(E)\, dE . \tag{8.73}$$

We expect the diffusion length in the limit of weak absorption to be defined by the same diffusion coefficient. We can verify this by substituting Eq. (8.69) into Eq. (8.31) to give

$$b_0^2 = \frac{(\overline{\Sigma}_a)_M}{D}$$

where $\underline{D}$ is given by Eq. (8.73).

We consider now the accuracy of the approximate solution (8.70) to Eq. (8.67). For neutron scattering by water we have a comparison of $(\overline{\underline{U}})_{\underline{M}}$ with $(\overline{\underline{\ell}}_{\underline{tr}})_{\underline{M}}$ from the values of α_1 in the first two lines of Table 8.1. We see that for this case Eq. (8.70) is extremely accurate, at least for the Maxwellian average. For heavy moderators Eq. (8.70) should be an even better approximation. The integral on the right-hand side of Eq. (8.67) makes a smaller fractional contribution to $\underline{U}(\underline{E})$. For this reason $\underline{U}(\underline{E})$ is a more slowly varying function. Furthermore, $\Sigma_1(\underline{E},E')$ has a narrower range of energy over which it is nonvanishing. Thus, the integral on the right-hand side of Eq. (8.67) is not only smaller but more accurately approximated by setting $\underline{U}(\underline{E}') = \underline{U}(\underline{E})$.

The case for which $\underline{U}(\underline{E})$ has been most extensively studied is that of neutron diffusion in monatomic hydrogen gas. The actual problem that has been studied here is the analogous problem in gas kinetic theory of the self-diffusion coefficient in a dilute gas of rigid spheres. The adaptation to neutron diffusion was made by Kuščer and Ribarič.[19] The function $\underline{U}(\underline{E})$ for this case has been calculated exactly by Pekeris[20] by reducing the integral equation for $\underline{U}(\underline{E})$ to a second-order differential equation. The exact value of

$(\overline{U})_M$ thus obtained is 1.27325 in units of the epithermal mean free path. The Maxwellian average of the transport mean free path is 1.29, in the same units. The history of this problem is very interesting. The integral equation (8.67) was first derived and solved numerically by Pidduck in 1916 (see Ref. 20). After Pekeris derived and solved the equivalent second-order differential equation, he found that Boltzmann had also derived this differential equation and had spent 67 pages discussing the properties of its solution but had not obtained a full numerical solution. The interested reader should consult Ref. 20 for additional historical details.

If the infinite nonabsorbing medium that we consider has a boundary with another medium with different scattering properties, the solution to the transport equation will be of the form (8.64) only in the asymptotic region many mean free paths from the boundary. There are, however, two integrals of the transport equation which have a simple form everywhere in each region. From over-all neutron conservation the total current must be the same everywhere. We can prove this by integrating Eq. (8.63) over angle and energy to give

$$\int_0^\infty dE \int_{-1}^{1} d\mu\, \mu \frac{\partial}{\partial z} f(z,E,\mu) = \frac{d}{dz}\int_0^\infty j(z,E)\, dE = \frac{dJ}{dz} = 0 \;. \tag{8.74}$$

We can also generalize the K-integral of Eq. (4.133) to the case of anisotropic scattering. If we multiply Eq. (8.63) by μ $U(E)$ and integrate over E and μ, we obtain

$$\frac{d}{dz}\int_0^\infty U(E)\, dE \int_{-1}^{1} \mu^2 f(z,E,\mu) d\mu = \int_0^\infty dE \Big[-\Sigma_s(E)\, U(E)\, j(z,E) + U(E) \int_0^\infty \Sigma_1(E',E)\, j(z,E')dE'\Big] . \tag{8.75}$$

In writing Eq. (8.75) we have used

$$\int_{-1}^{1} \mu\, f(z,E,\mu) \quad d\mu = j(z,E) = \frac{2}{3} f_1(z,E) .$$

In the double integral on the right-hand side of Eq. (8.75) we can use Eq. (8.67) to carry out the integral over E. If we define

$$K(z) \equiv \int_0^\infty U(E) \int_{-1}^{1} \mu^2\, f(z,E,\mu)\, d\mu dE , \tag{8.76}$$

we obtain

$$\frac{dK(z)}{dz} = - \int_0^\infty U(E)\Sigma_s(E) j(z,E)dE + \int_0^\infty j(z,E')\Big[\Sigma_s(E')U(E') - 1\Big]dE' = - \int_0^\infty j(z,E')dE ,$$

or

$$\frac{dK}{dz} = -J . \tag{8.77}$$

With these preliminaries we can now proceed to the discussion of two adjacent nonabsorbing half-spaces at different temperatures. We also have the necessary results to consider the extension of the thermal-neutron Milne problem to the case of anisotropic scattering. We consider the former problem in the following section. The latter problem has been treated by Kladnik and Kuščer,[21] and the reader is referred to their paper on this subject.

8-4 THERMAL DIFFUSION OF NEUTRONS

It is of some interest to study the effects of a nonuniform moderator temperature on the neutron distribution. In order to isolate these effects, we consider first the example of an infinite nonabsorbing medium that is uniform except for the variation of moderator temperature with position. We consider the two extreme limits of a temperature discontinuity and of a temperature that changes very slowly with position. In the latter case, the neutron spectrum can be assumed to be Maxwellian at the local moderator temperature. The only new phenomenon is the flow of neutrons induced by the temperature gradient. This is analogous to the phenomenon of thermal diffusion in kinetic theory. This analogy has been exploited by Kuščer;[22] his paper forms the basis for most of the discussion in this section.

The problem of a moderator-temperature discontinuity is also of considerable practical importance. An example, the interface between a hot reactor core and a cold reflector, is given in Chap. 9.

We consider first the problem of a temperature discontinuity. Let the half-sapce for $z < 0$ be at a temperature T_1 and the half-space for $z > 0$ be at a temperature T_2 that is greater than T_1. For convenience, we choose a boundary condition at large distances so that no net current is flowing. The asymptotic solutions of Eq. (8.64) degenerate in this case to

$$\left.\begin{aligned} \lim_{z\to-\infty} f(z,E,\mu) &= \tfrac{1}{2}A_1\, \phi_M^{(1)}(E) \\ \lim_{z\to\infty} f(z,E,\mu) &= \tfrac{1}{2}A_2\, \phi_M^{(2)}(E), \end{aligned}\right\} \tag{8.78}$$

where

$$\phi_M^{(j)}(E) = (k_BT_j)^{-2}\, E \exp(-E/k_BT_j) \tag{8.79}$$

is the Maxwellian flux distribution at the temperature $\underline{T}_1$ and is normalized to unit total flux. The total scalar flux at large negative $\underline{z}$ approaches $\underline{A}_1$ and at large positive $\underline{z}$ approaches $\underline{A}_2$. We will see shortly that $\underline{A}_1$ and $\underline{A}_2$ are in general not equal.

Consider first the case of constant cross sections. In this case the scalar flux $\Phi(\underline{z})$ is constant, and thus $\underline{A}_1$ and $\underline{A}_2$ are equal. If we integrate the transport equation over energy in the case of constant cross sections, recalling that the energy-transfer properties of the moderator cancel, we obtain a one-speed transport equation for the total angular flux. In the present problem the only difference between the two half-spaces is in the energy-transfer properties of the moderator; thus, the total angular flux for no net current takes the trivially simple form

$$\int_0^\infty f(z,E,\mu)\, dE = f(z,\mu) = \frac{1}{2}A_1 = \frac{1}{2}A_2 .$$

The energy dependence of the scalar or angular flux in the neighborhood of the temperature discontinuity is of course not trivial and is the main object of study. We have illustrated in Sec. 8-2 how to construct the general solution in each half-space for the special case of isotropic scattering. If we have the general solution in each medium, it remains to match the boundary condition that $\underline{f}(\underline{z},\underline{E},\mu)$ be continuous at $\underline{z} = 0$. This approach has been carried through only for the special case of the heavy gaseous moderator in the diffusion approximation.[18] We give the results of this calculation by Kottwitz[18] in order to have a definite example.

Following Kottwitz we introduce the dimensionless variables

$$\left.\begin{aligned} \varepsilon_1 &= \frac{E}{k_B T_1}\,, \\ \varepsilon_2 &= \frac{E}{k_B T_2}\,, \end{aligned}\right\} \tag{8.80}$$

and

$$y = (3\xi_0)^{1/2}\Sigma_s\, z\,.$$

Making use of the results obtained in Sec. 8-2, in particular of Eqs. (8.58) and (8.62), we can write the general solution to our present problem in the form

$$\phi(y,E) = \left\{\begin{array}{ll} \varepsilon_1\, e^{-\varepsilon_1} \displaystyle\sum_{m=0}^{\infty} \frac{1}{T_1} A_m\, L_m^1(\varepsilon_1) e^{y\sqrt{m}} & y \le 0\,, \\ \\ \varepsilon_2\, e^{-\varepsilon_2} \displaystyle\sum_{m=0}^{\infty} \frac{1}{T_2} B_m\, L_m^1(\varepsilon_2) e^{-y\sqrt{m}} & y \ge 0\,. \end{array}\right\} \tag{8.81}$$

The constants $\underline{A}_m$ and $\underline{B}_m$ remain to be determined from the boundary conditions. In diffusion theory, with constant cross sections, these boundary conditions are that $\phi(\underline{y},\underline{E})$ and $[\partial\phi(\underline{y},\underline{E})]/\partial\underline{y}$ be continuous at $\underline{y} = 0$. Kottwitz has shown that these boundary conditions imply that

$$A_n = C_n\left(1 - \frac{T_2}{T_1}\right)^n$$

and (8.82)

$$B_n = C_n\left(1 - \frac{T_1}{T_2}\right)^n$$

where the $\underline{C}_n$ are a set of universal constants defined by the set of equations

$$\sum_{q=0}^{m}\binom{m}{q}\left(\frac{q}{m}\right)^{1/2} + 1\Bigg](-1)^q C_q = 0 \qquad (8.83)$$

The values of the first twenty one of these constants are given in Table 8.2.

Kottwitz finds that the series in Eq. (8.81) is always convergent in the high-temperature half-space and converges in the low-temperature half-space if $\underline{T}_2/\underline{T}_1 < 2$. When $\underline{T}_2/\underline{T}_1 > 2$, the series can be replaced by a new series that converges. The transformation that accomplishes this is given by

Table 8.2

CONSTANTS THAT OCCUR IN THE TEMPERATURE DISCONTINUITY PROBLEM*

n	C_n	n	C_n
0	1.0000 0000 00	11	0.1234 4237 11
1	0.5000 0000 00	12	0.1166 3381 97
2	0.3535 5339 06	13	0.1106 9794 47
3	0.2803 3008 59	14	0.1054 6964 17
4	0.2355 4595 17	15	0.1008 2373 87
5	0.2049 8140 26	16	0.0966 6349 71
6	0.1826 2060 27	17	0.0929 1289 26
7	0.1654 5563 36	18	0.0895 1129 64
8	0.1518 0577 74	19	0.0864 0972 46
9	0.1406 5391 58	20	0.0835 6814 09
10	0.1313 4628 83		

*From Kottwitz, Nucl. Sci. Eng., Ref. 18.

$$\sum_{n=0}^{\infty} G_n \left(1 - \frac{T_2}{T_1}\right)^n \rightarrow \frac{T_1}{T_2} \sum_{n=0}^{\infty} \left(1 - \frac{T_1}{T_2}\right)^n \left[\sum_{q=0}^{n} (-1)^q \binom{n}{q} G_q\right],$$

where

$$G_n = C_n L_n^1(\varepsilon_1) e^{n\sqrt{y}} .$$

Kottwitz has obtained numerical results for a temperature ratio $T_2/T_1 = 2$. Rather than display these here, we give the analytic results arising from two simple approximations that fit the numerical results remarkably well. The first approximation is a "neutron-temperature" approximation in which it is assumed that

$$\phi(y,E) \approx A\, E[t(y)]^{-2} \exp\left[-\frac{E}{t(y)}\right], \tag{8.84}$$

where $t(y)$ is chosen to give the correct average neutron energy as a function of y:

$$A\, t(y) = \frac{1}{2} \int_0^{\infty} E\, \phi(y,E)\, dE . \tag{8.85}$$

The function $t(y)$ defined by Eq. (8.85) is easily calculated. Going back to the energy-dependent diffusion equation for this problem, multiplying by E, and integrating over energy, one obtains a second-order differential equation for $t(y)$. Integrating this equation, Kottwitz obtains

$$t(y) = \begin{cases} \frac{1}{2}[T_1 + T_2 + (T_1 - T_2)(1-e^{y})] & y \leq 0, \\ \\ \frac{1}{2}[T_1 + T_2 + (T_2 - T_1)(1 - e^{-y})] & y \geq 0. \end{cases} \tag{8.86}$$

Equation (8.84) with $t(y)$ given by Eq. (8.86) gives a good fit to the exact solution of Eq. (8.81), as illustrated by the flux spectrum at various positions or by the position dependence of the total neutron density.

The neutron-temperature approximation is clearly limited, however, to the case where the asymptotic spectra far from the boundary are Maxwellian. A more readily generalized approximation that works equally well for the present problem is the overlapping-group method introduced by Selengut,[23] and discussed in Chap. 9. In the application to the two-region problem, two thermal groups are chosen to overlap in energy so that

$$\phi(z,E) \approx \phi_1(E)F_1(z) + \phi_2(E)F_2(z), \tag{8.87}$$

where $\phi_1(E)$ and $\phi_2(E)$ are the asymptotic flux spectra far from the boundary on the left- and right-hand sides, respectively. The functions $F_1(z)$ and $F_2(z)$ are assumed, in general, to satisfy two-group diffusion equations of the form

$$\left.\begin{aligned} -D_1 \frac{d^2F_1}{dz^2} + (\Sigma_a)_1 F_1 &= \Sigma_{21} F_2 , \\ -D_2 \frac{d^2F_2}{dz^2} + (\Sigma_a)_2 F_2 &= \Sigma_{12} F_1 . \end{aligned}\right\} \tag{8.88}$$

In the present example, the two diffusion coefficients are equal, and the absorption coefficients are zero. The transfer cross sections Σ_{12} and Σ_{21} can be chosen to give the correct average neutron energy as a function of position (i.e., the $\underline{t}(\underline{y})$ of Eq. (8.86)). The resulting expression for $\phi(\underline{y},\underline{E})$ is

$$\phi(y,E) = \left\{\begin{array}{ll} A\left[\phi_M^{(1)}(E)(1 - \frac{1}{2}e^{y}) + \frac{1}{2}\phi_M^{(2)}(E)e^{y}\right] & y \leq 0, \\ A\left[\frac{1}{2}\phi_M^{(1)}(E)e^{-y} + \phi_M^{(2)}(E)(1 - \frac{1}{2}e^{-y})\right] & y \geq 0. \end{array}\right\} \tag{8.89}$$

Equation (8.89) agrees well with the exact solution (8.81) in the example considered by Kottwitz.

In the special case of the heavy gas, Σ_{12} and Σ_{21} turn out to be equal to each other and independent of temperature This is closely related to the fact that the relaxation lengths for the spectrum are given by

$$\frac{\nu_m}{\Sigma_s} = \frac{(3m\xi_0)^{1/2}}{\Sigma_s}$$

for the heavy gas and do not depend on temperature. More generally, the relaxation lengths on the hot and cold sides will be different and Σ_{21} will be different from Σ_{12}. These transfer cross sections are sometimes called "rethermalization cross sections" and have been extensively used in phenomenological applications of the two-thermal-group method to problems with a temperature discontinuity. The reader is referred to the papers of Bennett,[24] Lindenmeier,[25] and Pearce and Kennedy[26] for a detailed discussion of the method and its application. It has been established that the two-thermal-group picture is a good approximation to the space-energy distribution within energy-dependent diffusion theory for monatomic gaseous moderators of arbitrary mass.[26] It has also been established that the method is useful for the phenomenological analysis of measured neutron densities in the neighborhood of temperature discontinuities.[24,25] As we have seen in Sec. 8-2, however, the relaxation lengths for the spectrum in moderators as light as graphite are sufficiently short that the use of diffusion theory as a starting point is open to some doubt. Our understanding of the space dependence of the neutron spectrum in the neighborhood of a temperature discontinuity thus is not yet on a sound theoretical basis for moderators of practical interest. The calculation of Kottwitz remains useful as a consistent calculation in the heavy-mass limit. It should apply quantitatively in a very heavy material such as lead; it would be interesting if experiments were carried out in such a material to check the quantitative predictions of the theory. For lighter moderators, more work with transport theory is needed to understand when the two-thermal-group diffusion method can be applied with confidence. Such work necessarily must be primarily numerical but could be profitably applied to the analysis of the idealized problem that we consider here.

If we consider the temperature-discontinuity problem for an arbitrary moderator, the asymptotic scalar flux $\underline{A}_1$ on the cold side will, in general, be different than the corresponding flux $\underline{A}_2$ on the hot side. Although we cannot say much about the behavior of $\phi(\underline{z},\underline{E})$ in the general case, we can, following Kuščer,[22] estimate the ratio $\underline{A}_2/\underline{A}_1$ by making use of the $\underline{K}$-integral defined by Eq. (8.76). In the present problem $\underline{J} = 0$ so that $\underline{K}(\underline{z})$ is constant in each half-space. Its value, $\underline{K}_1$, in the left half-space can be calculated from Eq. (8.76) for any value of $\underline{z} < 0$. Setting the values thus calculated for $\underline{z} = -\infty$ and for $\underline{z} = 0$ equal to each other gives

$$K_1 = \tfrac{1}{3}A_1\overline{U}_1 = \int_0^\infty dE\ U_1(E) \int_{-1}^1 d\mu\ \mu^2\ f(0,E\mu) \ . \qquad (8.90)$$

In obtaining Eq. (8.90) we have used Eq. (8.78) and have introduced the symbol $\overline{U}_1$ to represent the average of $U_1(E)$ over a Maxwellian flux spectrum at temperature T_1. We can write a similar expression for K_2 with $\overline{U}_2$ defined as the average of $U_2(E)$ over a Maxwellian flux spectrum at temperature T_2:

$$K_2 = \frac{1}{3}A_2\overline{U}_2 = \int_0^\infty dE\; U_2(E) \int_{-1}^{1} d\mu\; \mu^2\; f(0,E,\mu)\,. \tag{8.91}$$

We note that the effective transport mean free path $U(E)$ is a function of energy and is in general a different function of energy on the two sides of the temperature discontinuity. If we divide Eq. (8.91) by Eq. (8.90), we obtain

$$\frac{A_2}{A_1} = \frac{\overline{U}_1\left\langle U_2\right\rangle_0}{\overline{U}_2\left\langle U_1\right\rangle_0}\,, \tag{8.92}$$

where $\left\langle U_1\right\rangle_0$ is the average of $U_1(E)$ over an appropriate flux spectrum at $z = 0$, and $\left\langle U_2\right\rangle_0$ is the average of $U_2(E)$ over the same flux spectrum.

In the cases of practical interest, $U(E)$ is an increasing function of E. The flux spectrum becomes harder with increasing z as one moves from the cold to the hot half-space. Thus the factors $\overline{U}_1/\left\langle U_1\right\rangle_0$ and $\left\langle U_2\right\rangle_0/\overline{U}_2$ in Eq. (8.92) are each less than one. The total flux is therefore smaller on the hot side than on the cold side.

If the temperature difference between the two sides is small, we can make a more quantitative estimate. It is reasonable to assume that the flux spectrum at $z = 0$ is an equal mixture of Maxwellian flux spectra at temperatures T_1 and T_2. With this assumption, Eq. (8.92) reduces to

$$\frac{A_2}{A_1} = \frac{\overline{U}_2 + \langle U_2 \rangle_1}{\overline{U}_1 + \langle U_1 \rangle_2} \frac{\overline{U}_1}{\overline{U}_2} ,$$

where

$$\langle U_1 \rangle_2 = \int_0^\infty U_1(E)\phi_M^{(2)}(E)dE$$

and $\langle U_2 \rangle_1$ is similarly the average of $U_2(E)$ over a Maxwellian at temperature T_1. If the temperature difference is small, we can expand $\phi_M^{(2)}(E)$ in a Taylor series expansion about $\phi_M^{(1)}(E)$ to give

$$\langle U_1 \rangle_2 \approx \overline{U}\left[1 + 2\,\frac{\Delta T}{T}\left(\frac{\overline{\overline{U}}}{\overline{U}} - 1\right)\right] , \qquad (8.93)$$

where

$$\overline{U} = (2k_BT)^{-1} \int_0^\infty E\, U(E)\phi_M(E)dE ,$$

$$T_1 \approx T_2 = T ,$$

and we have assumed that the composition at the medium is uniform. With a similar expression for $\langle \underline{U}_2 \rangle_1$ one obtains finally

$$\frac{A_2}{A_1} = \left(\frac{T_1}{T_2} \right)^{1/2 - \alpha} , \tag{8.94}$$

where

$$\alpha = \frac{5}{2} - \frac{2\bar{\bar{U}}}{\bar{U}} . \tag{8.95}$$

In the limit of constant cross sections, $\underline{\bar{\bar{U}}} = \underline{\bar{U}}$ so that $\alpha = \frac{1}{2}$ and the flux ratio $\underline{A}_2/\underline{A}_1 = 1$, as it should.

We now turn our attention from the case of a temperature discontinuity to the case of a small temperature gradient. We assume that the temperature changes very little in a characteristic relaxation length for the spectrum. We thus expect the flux spectrum to be very nearly Maxwellian at the local moderator temperature:

$$\phi(z,E) = \beta^2(z)\ E \exp[-\beta(z)E]\Phi(z) \equiv \phi_M(E,z)\Phi(z) , \tag{8.96}$$

where $\beta(\underline{z}) = [\underline{k}_B \underline{T}(\underline{z})]^{-1}$. In this case the $\underline{P}_1$-approximation applies with good accuracy. To illustrate the physical content, consider first the energy-dependent diffusion approximation

$$j(z,E) = -D(z,E)\frac{\partial\phi(z,E)}{\partial z} \quad . \tag{8.97}$$

If we substitute Eq. (8.96) into Eq. (8.97), we obtain

$$j(z,E) = -D(z,E)\phi_M(E,z)\{\frac{d\Phi}{dz} + [\beta(z)E - 2]\,\frac{\Phi}{T}\frac{dT}{dz}\} \quad . \tag{8.98}$$

There is thus one contribution to the current from the flux gradient and another from the temperature gradient. If we integrate Eq. (8.98) over energy, we obtain

$$J = -\bar{D}(z)\left\{\frac{d\Phi}{dz} + [\frac{1}{2} - \alpha'(z)]\,\frac{\Phi}{T}\frac{dT}{dz}\right\} , \tag{8.99}$$

where

$$\alpha'(z) = \frac{5}{2} - \frac{2\bar{\bar{D}}(z)}{\bar{D}(z)} \quad . \tag{8.100}$$

$\bar{D}(z)$ is the average of $D(z,E)$ over a local Maxwellian at temperature $T(z)$, and $\bar{\bar{D}}(z)$ is given by

$$\bar{\bar{D}}(z) = \frac{1}{2}\beta(z)\int_0^\infty E\,D(z,E)\phi_M(E,z)dE \ .$$

In the limit of constant cross sections $\alpha' = \frac{1}{2}$ and therefore there is no term in Eq. (8.99) proportional to the temperature gradient.

If we now go to the consistent $\underline{P}_1$-approximation, the results are remarkably similar. Equation (8.97) is replaced by

$$-\frac{1}{3}\frac{\partial\phi(z,E)}{\partial z} = \Sigma_s(z,E)j(z,E) - \int_0^\infty \Sigma_1(z,E',E)j(z,E')dE' . \tag{8.101}$$

If we again assume Eq. (8.96) to hold and make use of the exactly satisfied local detailed-balance condition

$$\Sigma_1(z,E',E)\phi_M(E',z) = \Sigma_1(z,E,E')\phi_M(E,z)$$

it can easily be shown that

$$j(z,E) = -\frac{1}{3}\phi_M(E,z)\{U(z,E)\frac{d\Phi}{dz} + [V(z,E) - 2U(z,E)]\frac{\Phi}{T}\frac{dT}{dz}\} , \tag{8.102}$$

where $\underline{U}(z,\underline{E})$ is the solution of the integral equation

$$\Sigma_s(z,E)U(z,E) - \int_0^\infty \Sigma_1(z,E,E')U(z,E')dE' = 1 , \tag{8.103}$$

and $\underline{V}(\underline{z},\underline{E})$ is the solution of the very similar equation

$$\Sigma_s(z,E)V(z,E) - \int_0^\infty \Sigma_1(z,E,E')V(z,E')dE' = \beta(z)E . \tag{8.104}$$

Equation (8.103) is identical to Eq. (8.67) except for the $\underline{z}$-dependence of the cross sections.

If we integrate Eq. (8.102) over energy, we obtain

$$J = -\frac{1}{3}\bar{U}(z)\,\frac{d\Phi}{dz} - \frac{1}{3}[\bar{V}(z) - 2\bar{U}(z)]\,\frac{\Phi}{T}\,\frac{dT}{dz}\,, \qquad (8.105)$$

where $\bar{U}(z)$ and $\bar{V}(z)$ are the averages of $U(z,E)$ and $V(z,E)$ over the local Maxwellian $\phi_M(E,z)$. Equation (8.105) can be reduced to a form very similar to Eq. (8.99) if we note that

$$\left.\begin{aligned} &\int_0^\infty V(z,E)\phi_M(E,z)dE = \int_0^\infty \beta(z)\,E\,U(z,E)\phi_M(E,z)dE \\ &\text{or} \\ &\bar{V}(z) = 2\bar{\bar{U}}(z)\,. \end{aligned}\right\} \qquad (8.106)$$

To derive Eq. (8.106), we multiply Eq. (8.104) by $U(z,E)\phi_M(E,z)$ and integrate over energy, making use of Eq. (8.103). The derivation is very similar to the derivation of Eq. (8.77). We thus finally obtain

$$J = -\frac{1}{3}\bar{U}(z)\left\{\frac{d\Phi}{dz} + \left[\frac{1}{2} - \alpha(z)\right]\frac{\Phi}{T}\,\frac{d\Phi}{dz}\right\}\,, \qquad (8.107)$$

where

$$\alpha(z) = \frac{5}{2} - \frac{2\bar{\bar{U}}(z)}{\bar{U}(z)} \qquad (8.108)$$

is exactly analogous to the quantity α in Eq. (8.95). If we consider the special case in which $J = 0$,

$$\frac{1}{\Phi}\frac{d\Phi}{dz} = -[\frac{1}{2} - \alpha(z)]\,\frac{1}{T}\frac{dT}{dz} \;.$$

For a small temperature difference, $\alpha(\underline{z})$ is approximately constant so that

$$\frac{\Phi(\infty)}{\Phi(-\infty)} = \left[\frac{T(-\infty)}{T(\infty)}\right]^{\frac{1}{2} - \alpha} , \qquad (8.109)$$

in agreement with Eq. (8.94). Thus the change in total flux associated with a change in temperature is the same whether the temperature change occurs discontinuously or is spread out over many mean free paths.

8-5 THE APPROACH TO EQUILIBRIUM

In Sec. 4-3 we gave a semiquantitative description of the approach of the neutron velocity distribution to thermal equilibrium with the moderator. We return now to a more quantitative discussion of this problem, which has played an important role in leading to a sound basic understanding of neutron thermalization. In Sec. 4-3 we dealt with the neutron distribution in velocity space without separately considering the speed and direction of motion. The approach of the angular distribution to isotropy and of the speed distribution to a Maxwellian were thus treated together.

In the absence of any spatial variation of the neutron distribution these two aspects are in fact to a large extent independent. If we make the spherical harmonics expansion (4.49) of the angular flux, the transport equation becomes

$$\frac{1}{v}\frac{\partial f_{\ell m}(E,t)}{\partial t} = -\Sigma_s(E)f_{\ell m}(E,t) + \int_0^\infty \Sigma_\ell(E',E)f_{\ell m}(E',t)dE'. \tag{8.110}$$

In Eq. (8.110) we have assumed that there are no sources and no absorption. As we have discovered previously, a $1/v$ absorption cross section just gives a multiplicative factor $\exp(-v\Sigma_a t)$ in the angular flux and does not affect the time-dependent problem in an essential way.

The solution of Eq. (8.110) is qualitatively different for $\ell = 0$ and for $\ell \neq 0$. Consider first the case of $\ell \neq 0$ that describes the approach of the angular distribution to isotropy. If we add $\Sigma_\ell(E)\, f_{\ell m}(E,t)$ to the first term on the right-hand side of Eq. (8.110) and subtract it from the second term, we obtain

$$\frac{1}{v}\frac{\partial f_{\ell m}(E,t)}{\partial t} = -[\Sigma_s(E) - \Sigma_\ell(E)]f_{\ell m}(E,t) + R_\ell\, f_{\ell m}(E,t)\,, \tag{8.111}$$

where $\Sigma_\ell(E)$ is defined by Eq. (4.59) and the integral operator R_ℓ is defined by Eq. (4.57). In Sec. 8-1, we have found it a good approximation to neglect the operator

$\underline{R}_\ell$ for $\underline{\ell} \neq 0$. If we do this in Eq. (8.111) and define an effective collision frequency

$$\alpha_\ell(E) = v[\Sigma_s(E) - \Sigma_\ell(E)] ,\qquad(8.112)$$

we obtain the approximate solution

$$f_{\ell m}(E,t) \approx f_{\ell m}(E,0) \exp[-\alpha_\ell(E)t] .\qquad(8.113)$$

The flux anisotropy decays to zero with a continuous range of decay constants ranging over the values taken on by the functions $\alpha_\ell(\underline{E})$. We are thus led to examine the energy independent so that $\alpha_\ell(\underline{E})$ increases linearly with the neutron speed. In the limit that $\underline{E} = 0$, the cross section is proportional to $1/\underline{v}$ and is isotropic, and thus

$$\alpha_\ell(0) = \lim_{E\to0} v\, \Sigma_s(E) \equiv \lambda^* .$$

We now turn our attention to the case $\underline{\ell} = 0$, $\underline{m} = 0$, which we consider in much more detail. Letting

$$n(E,t) = (1/v)f_{00}(E,t) ,\qquad(8.115)$$

Eq. (8.111) for $\underline{\ell} = 0$, $\underline{m} = 0$ takes the form

$$\frac{\partial n(E,t)}{\partial t} = -v\, \Sigma_s(E)n(E,t) + \int_0^\infty v'\Sigma_0(E',E)n(E',t)dE' .$$

$$(8.116)$$

We look for solutions to Eq. (8.116) of the form

$$n(E,t) = n_\lambda(E)\ \exp(-\lambda t)\ , \tag{8.117}$$

thus defining the eigenvalue problem

$$[v\ \Sigma_s(E) - \lambda]\ n_\lambda(E) = \int_0^\infty v'\Sigma_0(E',E)n_\lambda(E')dE'\ . \tag{8.118}$$

From detailed balance we know that there is a solution

$$\left.\begin{aligned} \lambda &= 0 \\ n_0(E) &= \frac{1}{v}\phi_M(E) \equiv n_M(E) \end{aligned}\right\} \tag{8.119}$$

corresponding to thermal equilibrium between neutrons and moderator. We see in Sec. 8-6 that there are also singular solutions to Eq. (8.118) for all λ greater than λ^*. If we start with an arbitrary angular distribution, these singular solutions will be important for the same range of times as will the decay of the $\underline{\ell} \neq 0$ components of the angular flux. For times greater than a few times $(\lambda^*)^{-1}$, the neutron distribution will be isotropic, and the approach to equilibrium will be governed by the eigenvalues $\lambda < \lambda^*$ that satisfy Eq. (8.118), if such eigenvalues exist. In particular, the smallest nonzero eigenvalue λ_1 determines the rate of approach to equilibrium at long times and is a useful characterization of the thermalizing power of a moderator. We are thus led to consider the quantitative calculation of the first few eigenvalues of Eq. (8.118).

In order to obtain some physical insight into the factors determining λ_1, we consider first an approximate analytical calculation based on a variational expression. To obtain this expression we rewrite Eq. (8.118) so as to symmetrize the kernel of the integral equation. If we let

$$g(E) = n(E)[n_M(E)]^{-\frac{1}{2}}$$

and

$$S(E,E') = (vv')^{\frac{1}{2}}\Sigma_0(E',E)\left[\frac{\phi_M(E')}{\phi_M(E)}\right]^{\frac{1}{2}} ,$$

Eq. (8.118) becomes

$$\lambda g(E) = v\,\Sigma_s(E)\,g(E) - \int_0^\infty S(E,E')\,g(E')\,dE' . \qquad (8.120)$$

Since the kernal $\underline{S}(\underline{E},\underline{E}')$ is symmetric by virtue of detailed balance, we can apply the Rayleigh-Ritz variational principle[27] to Eq. (8.120). Multiplying by $\underline{g}(\underline{E})$ and integrating over energy gives

$$\lambda = \frac{\int_0^\infty v\Sigma_s(E)g^2(E)dE - \int_0^\infty \int_0^\infty g(E)S(E,E')g(E')dEdE'}{\int_0^\infty g^2(E)dE} \quad (8.121)$$

If the exact solution for $\underline{g}(\underline{E})$ is replaced by a trial solution $\tilde{\underline{g}}(\underline{E})$ that is orthogonal to the eigenfunction $\underline{g}_0(\underline{E})$. corresponding to $\lambda = 0$, then the approximate expression calculated from Eq. (8.121) will always be an overestimate of the exact eigenvalue. Furthermore, the error in the eigenvalue is of second order compared with the error in the eigenfunction. The orthogonality condition that we require is

$$\int_0^\infty \tilde{g}(E)[n_M(E)]^{\frac{1}{2}} dE = 0 .$$

The symmetrized formulation of Eqs. (8.120) and (8.121) is helpful in deriving the variational expression, but it is inconvenient for actual calculation. A more convenient form of Eq. (8.121) for actual calculation can be written in terms of

$$\tilde{\Phi}(E) = v\,\tilde{n}(E) = v\,\tilde{g}(E)[n_M(E)]^{\frac{1}{2}} . \quad (8.122)$$

The resulting expression is

$$\lambda = N/D , \quad (8.123)$$

where

$$N = \int_0^\infty \int_0^\infty \Sigma_0(E,E')\tilde{\phi}(E)\left[\frac{\tilde{\phi}(E)}{\phi_M(E)} - \frac{\tilde{\phi}(E')}{\phi_M(E')}\right] dE dE' , \qquad (8.124)$$

and

$$D = \int_0^\infty \frac{1}{v} \frac{\tilde{\phi}^2(E)}{\phi_M(E)} dE \ . \qquad (8.125)$$

The required orthogonality condition becomes

$$\int_0^\infty \frac{1}{v} \tilde{\phi}(E) dE = \int_0^\infty \tilde{n}(E)\, dE = 0 \qquad (8.126)$$

and expresses the fact that the total neutron density is independent of time and is therefore contained entirely in the asymptotic Maxwellian distribution. This can be seen most directly by integrating Eq. (8.118) over energy to give

$$\lambda \int_0^\infty n_\lambda(E) dE = 0$$

so that Eq. (8.126) must be satisfied unless $\lambda = 0$.

A simple trial function satisfying Eq. (8.126) is

$$\tilde{\phi}(E) = \phi_M(E)\left(\frac{3}{2} - \frac{E}{k_B T}\right). \qquad (8.127)$$

Substituting Eq. (8.127) into Eq. (8.124) gives, after

some algebra and using detailed balance,

$$\lambda_1 \approx \frac{2}{3\sqrt{\pi}} v_0 M_2 \quad , \tag{8.128}$$

where

$$v_0 = \left(\frac{2k_B T}{m} \right)^{\frac{1}{2}} \tag{8.129}$$

and

$$M_2 = (k_B T)^{-2} \int_0^\infty \int_0^\infty \Sigma_0(E,E')\phi_M(E)(E' - E)^2 \, dEdE'. \tag{8.130}$$

The quantity M_2 has been frequently employed in the literature as a convenient measure of the rate of energy transfer between thermal neutrons and moderator. It was first introduced in a variational calculation of the diffusion cooling coefficient.[28] In an Appendix of Ref. 28 it is shown that a neutron distribution characterized by neutron temperature T_n would relax to equilibrium according to

$$\frac{T_n}{T} - 1 = \exp(-\lambda t) \quad ,$$

with λ given by Eq. (8.128). This is a generalization to include chemical binding of a result originally obtained by von Dardel[29] for the "heat transfer" coefficient between a neutron gas and an ideal monatomic gaseous moderator. For an ideal monatomic gas the integrals in Eq. (8.130) can be done analytically to give

$$M_2 = 8\mu\Sigma_s (1 + \mu)^{-\frac{3}{2}} \tag{8.131}$$

where μ is the ratio of neutron mass to moderator atom mass and Σ_s is the free-atom epithermal scattering cross section. Substituting Eq. (8.131) into Eq. (8.128) gives the expression obtained by von Dardel for the relaxation of the neutron temperature to the moderator temperature.

To obtain a more accurate result for λ_1 we need a better trial function than Eq. (8.127). A natural choice is

$$\tilde{\phi}(E) = \phi_M(E) \sum_{n=0}^{N} a_n L_n^1\left(\frac{E}{k_B T}\right) . \tag{8.132}$$

If we substitute Eq. (8.132) into Eq. (8.123) and minimize with respect each of the coefficients a_n, we obtain a set of $N + 1$ linear algebraic equations for the unknown coefficients a_0, a_1 ... a_N. As pointed out by Shapiro,[30] the same set of equations is obtained by substituting

$$n(E) = \frac{1}{v} \phi_M(E) \sum_{n=0}^{N} a_n L_n^1\left(\frac{E}{k_B T}\right) \tag{8.132a}$$

into Eq. (8.118), multiplying by $L_m^1(E/k_B T)$ for $m = 0, 1, 2 \ldots N$, and integrating over energy. This set of equations will give the correct lowest eigenvalue and eigenfunction (8.119) and will also satisfy the orthogonality condition (8.126) for the higher eigenfunctions. This method was used by several authors[31-33] for quantitative calculations.

The most extensive study of methods for calculating the time eigenvalues has been carried out by Shapiro and Corngold.[30,34,35] They point out that a much more rapidly convergent choice of trial function than Eq. (8.132) is

$$\tilde{\phi}(E) = \phi_M(E) \sum_{i=0}^{N} a_i \left(\frac{v}{v_0}\right)^i . \tag{8.133}$$

The same result as is obtained from the variational method can again be obtained by substituting

$$n(E) = \frac{1}{v} \phi_M(E) \sum_{j=0}^{N} a_j \left(\frac{v}{v_0}\right)^j \tag{8.133a}$$

into Eq. (8.118), multiplying by $(v/v_0)^i$ for $i = 0, 1, 2 \ldots N$, and integrating over energy. The resulting set of $N + 1$ algebraic equations is

$$-\frac{1}{2}\lambda \sum_{j=0}^{N} \Gamma\left(\frac{i + j + 3}{2}\right) a_j = \sum_{j=0}^{N} (\Sigma_{i,j} - \Sigma_{0,i+j}) a_j , \tag{8.134}$$

where $\Gamma(x)$ is the usual gamma function,[36] and the matrix elements $\Sigma_{i,j}$ are defined by

$$\Sigma_{i,j} = \int_0^\infty \int_0^\infty \Sigma_0(E',E)\phi_M(E')\left(\frac{v}{v_0}\right)^i\left(\frac{v'}{v_0}\right)^j dEdE' = \Sigma_{j,i} . \tag{8.135}$$

With the availability of digital computers, the calculation of the matrix elements of Eq. (8.135) and the solution of the set of algebraic equations (8.134) is straightforward. Shapiro and Corngold also discuss another method for the calculation of eigenvalues in which the actual kernel is replaced by an approximate degenerate kernel, and the eigenvalue problem is subsequently solved exactly. This method gives somewhat more rapid convergence than does the use of velocity polynomials, but it requires more work to use it at a level where the number of coupled equations is the same.

We turn now to a discussion of the actual numerical results[30] for the eigenvalues λ_i as a function of moderator atom mass and of strength of chemical binding. Consider first the limit of a heavy gaseous moderator. In this limit, the "$\underline{M}_2$-approximation" of Eq. (8.128), using Eq. (8.131), gives

$$\lambda_1 = \frac{16}{3\sqrt{\pi}}(\mu v_0 \Sigma_s) = 3.009(\mu v_0 \Sigma_s) .$$

The exact value in this limit is also proportional to

$\mu v_0 \Sigma_s$, but the coefficient is 2.342 instead of 3.009. In this heavy-mass limit the rate of approach to equilibrium is very slow compared with the mean time between collisions. Measured in units of $\mu v_0 \Sigma_s$, the minimum collision frequency λ^* becomes infinite in the limit that $\mu = 0$. The above information is summarized in the first line of Table 8.3.

The variation of the λ_i with the mass ratio μ is plotted in Fig. 8.3. The ratio $\lambda_i/\mu v_0 \Sigma_s$ is plotted as a function of μ^{-1}. The extension to values of μ^{-1} less than one is of no interest for neutron moderation but is relevant for the type of diffusion process encountered in the theory of Brownian motion. The interested reader should consult Refs. 30 and 35 for more details. We see from Fig. 8.3 that the values of λ_i for $i \geq 2$ are close to λ^* for all values of μ that are of interest for neutron moderation, but that λ_1 is clearly resolvable.

For hydrogen ($\mu = 1$) the results of Fig. 8.3 are given in more detail on the second line of Table 8.3. The calculation has also been done for the bound-proton model of Eqs. (3.95-3.97), and the results are given on the third line of Table 8.3. In particular, we see that the value of λ_1 is considerably reduced by the effects of binding. This difference should be observable and is discussed further in Sec. 8-8.

Table 8.3

CALCULATED TIME EIGENVALUES IN UNITS OF $\mu \underline{v}_0 \Sigma_{\underline{s}}$

Moderator	Exact λ_1	Exact λ_2	"$\underline{M}_2$" Formula λ_1	"$\underline{M}_2$" Formula λ^*
Ideal gas, $\mu \to 0$	2.342	5.056	3.049	∞
Ideal gas, $\mu = 1$	0.9242	1.105	1.06	1.128
Water (Model of Eqs. (3.95-3.97))	0.546	0.852	0.82‡	0.975
Heavy Debye crystal, $\Theta_{\underline{D}}/\underline{T} = 0.2$	2.26	λ^*	3.001	2.6
Heavy Debye crystal, $\Theta_{\underline{D}}/\underline{T} = 1.0$	λ^*	λ^*	2.83	0.826

‡Based on $\underline{M}_2 = 3.2\ \text{cm}^{-1}$ taken from Beckurts' review paper.[37]

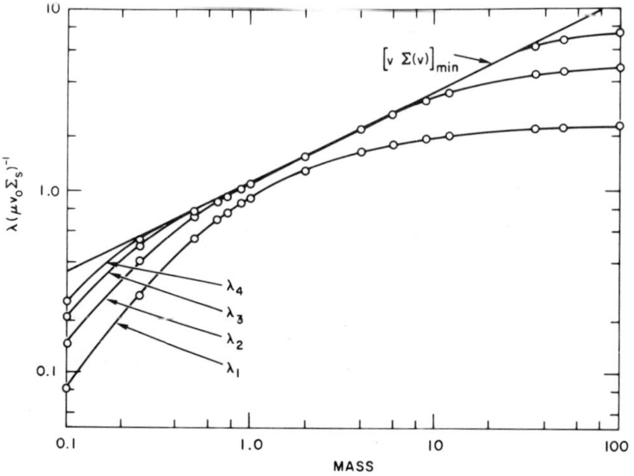

Figure 8-3 Free-gas time eigenvalues versus mass. (From Shapiro, USAEC Report BNL-8433, Ref. 30.)

In order to systematically study the effects of chemical binding, Shapiro considered the case of a heavy Debye crystal. In the presence of binding, the value of $(\lambda^*/\mu \underline{v}_0 \Sigma_{\underline{s}})$ remains finite for $\mu = 0$ and decreases rapidly as the Debye temperature $\Theta_{\underline{D}}$ is increased. For moderate values of $\Theta_{\underline{D}}/\underline{T}$, the eigenvalues λ_1 and λ_2 are insensitive to $\Theta_{\underline{D}}$. The results are illustrated in Fig. 8.4. From this figure we see that no discrete eigenvalues with values less than λ^* exist even for moderate values of $\Theta_{\underline{D}}/\underline{T}$. As $\Theta_{\underline{D}}/\underline{T}$ is increased from zero, the eigenvalues λ_1 and λ_2 decrease only slightly but λ^* decreases rapidly. When λ_1 becomes less than λ^* it can no longer describe the asymptotic approach to equilibrium of an arbitrary initial velocity dsitribution. Very slow neutrons must approach equilibrium more slowly than $\exp(-\lambda^* \underline{t})$ inasmuch as they have a mean time of $(1/\lambda^*)$ to make their first collision, after which theyare still not fully thermalized. Thus, the disappearance of the discrete eigenvalues in a crystal is related to the inability of the slowest neutrons to "keep up" with the thermalization rate typical of most of the remainder of the thermal-neutron distribution. Under these circumstances the asymptotic approach to equilibrium cannot be described by a separable function of energy and time but requires a continuous distribution of decay

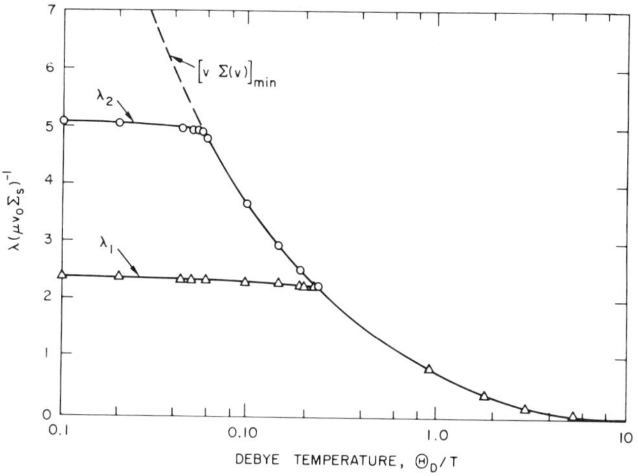

Figure 8-4 Time eigenvalues for the heavy Debye crystal versus Debye temperature. (From Shapiro, USAEC Report BNL-8433, Ref. 30.)

constants. The difference in the eigenvalue structure between water and a heavy crystal is clearly reflected in the time dependence of the average neutron velocity in water as compared with graphite. We discuss this further in Sec. 8-8. Before proceeding to a comparison with experiment, however, we turn to a more careful mathematical discussion of the eigenvalue spectrum since this has been the subject of much interesting recent work.

8-6 THE SPECTRUM OF THE COLLISION OPERATOR

In the last few years we have attained a greatly improved understanding of the mathematical structure of the eigenvalue problem Eq., (8.118). We summarize these recent results in this section without giving any explicit derivations.

First, consider separately the scattering-out and scattering-in operators in Eq. (8.118), or preferably in the symmetrized equivalent equation, Eq. (8.120). The spectrum of the scattering-out operator is particularly simple and is defined by

$$v\,\Sigma_s(E)\,n(E) = \lambda\,n(E). \tag{8.136}$$

Assuming that $\underline{v\,\Sigma_s(E)}$ is a monotone increasing function of energy, the solution to Eq. (8.136) is

$$n_\lambda(E) = \delta(E - E_\lambda)\,, \tag{8.137}$$

where $\underline{E}_\lambda$ is defined by

$$v\,\Sigma_s(E_\lambda) = \lambda\,. \tag{8.138}$$

There is a solution for all values of λ that are taken on as values of the function $\underline{v\,\Sigma_s(E)}$. For actual moderators this means that there is a continuum of eigenvalues extending from λ^* to infinity. The scattering-out operator has no discrete spectrum.

The scattering-in operator, on the other hand, is expected to have only a discrete spectrum. In Sec. 8-2 we noted that this had been proven by Kuščer and Corngold[12] for the case of a gas and of a solid. It is also expected to be true for a liquid, but this is more difficult to prove inasmuch as the symmetrized kernel is not square-integrable. To be more explicit, we expect the equation

$$\int S(E,E')\chi_j(E')dE' = \xi_j\chi_j(E) \tag{8.139}$$

to have an infinite number of solutions with

$$\xi_1 > \xi_2 > \xi_3 > \dots \xi_N > 0$$

and to have no other solutions. In particular, this implies that the kernel $\underline{S}(\underline{E},\underline{E}')$ can be represented as the limit of a degenerate kernel constructed from its eigenfunctions:

$$S(E,E') = \lim_{N\to\infty} \sum_{j=1}^{N} \xi_j \, \chi_j(E) \, \chi_j(E') \ . \tag{8.140}$$

An operator with these properties is known as a compact operator.

The operator whose spectrum we actually want is the difference between the scattering-out and scattering-in operators. We cannot say anything simple about the discrete eigenvalues of Eq. (8.120), but it can be rigorously proved that the continuous spectrum extends from λ^* to infinity. There is a theorem due to Weyl and Von Neumann which states that the continuous spectrum of an operator is unaffected if a compact operator is added to it.[38] In other words, the continuous eigenvalues of Eq. (8.120) are the same as those of Eq. (8.136). The associated eigenfunctions are no longer the simple δ-functions of Eq. (8.137), but they can be constructed.

It is of some interest to trace the historical development of these ideas. For a long time it was assumed as "plausible" that there should be a complete set of discrete eigenvalues satisfying Eq. (8.118). That this was not true was proved by Corngold, Michael, and Wollman[39] by careful examination of the second-order differential equation applicable to thermalization in monatomic hydrogen gas. They showed that there are an infinite number of discrete eigenvalues accumulating at λ^*, plus a continuum from λ^* to infinity. The next contribution was made by Koppel,[40] who showed how the continuum eigenfunctions could be explicitly constructed if one worked with the approximation of an $\underline{N}$-term degenerate kernel. Sacrificing some mathematical rigor, Koppel's demonstration gives us the essence of the Weyl-Von Neumann theorem as we use it here. The proof that the scattering-in operator is compact is essentially equivalent to proving that the representation of this operator in terms of an $\underline{N}$-term degenerate kernel converges to the true operator in the limit that $\underline{N}$ approaches infinity.

The continuum eigenfunctions can be constructed for the range $\lambda^* < \lambda < \infty$ for any value of $\underline{N}$. If the convergence of the $\underline{N}$-term degenerate kernel to the true kernel is sufficiently "well-behaved", Koppel's demonstration can then be extended to the true kernel.

Since the scattering-in operator is compact, it can be usefully approximated by a degenerate kernel. Such an approximate kernel can be constructed in a variety of ways. Perhaps the most physically reasonable is the form used by Shapiro[30] and by Shapiro and Corngold,[35] which gives the exactly correct values for the first N-velocity moments of the true kernel. Shapiro's degenerate kernel is given by

$$\Sigma_0(E',E) = \phi_M(E) \sum_{i,j,=0}^{N} A_{ij}\, \Sigma_i(E')\Sigma_j(E) ,$$

where the matrix $\underline{A}$ is the inverse of the matrix whose matrix elements are defined by Eq. (8.135), and

$$\Sigma_i(E') = \int_0^\infty \Sigma_0(E',E)\left(\frac{v}{v_0}\right)^i dE .$$

Shapiro has used this form of kernel for numerical calculation of the discrete eigenvalues of Eq. (8.118), but its full utility as an approximation method in neutron thermalization problems remains to be explored.

To complete the discussion of the mathematical aspects of the problem, consider the question of the number of discrete eigenvalues. This was the primary objective of the work of Kuscer and Corngold[12] to which we have referred in another context. They studied the behavior of the modified eigenvalue problem

$$c\,\psi_\lambda(E) = \int_0^\infty K_\lambda(E,E')\psi_\lambda(E')dE' ,$$

where

$$\psi_\lambda(E) = n(E)\left[\frac{v\,\Sigma_s(E) - \lambda}{n_M(E)}\right]^{\frac{1}{2}}$$

and

$$K_\lambda(E,E') = \frac{vv'}{\{[v\,\Sigma_s(E) - \lambda][v'\,\Sigma_s(E') - \lambda]\}^{\frac{1}{2}}}\;\Sigma_0(E',E)\left[\frac{\phi_M(E')}{\phi_M(E)}\right]^{\frac{1}{2}} .$$

By examining the behavior of eigenvalues $\underline{c_i}(\lambda)$ as λ approaches λ^* they conclude that there are an infinite number of discrete eigenvalues below λ^* for a gas of arbitrary mass, and that the number of eigenvalues for a crystal is finite. Under certain conditions the only discrete eigenvalue for a crystal is $\lambda = 0$ with the associated Maxwellian eigenfunctions.

Given our knowledge of the spectrum of eigenvalues that satisfy Eq. (8.118), the general solution to Eq. (8.116) for the time-dependent spectrum takes the form

$$n(E,t) = N\, n_M(E) + \sum_j A_j\, n_j(E) \exp(-\lambda_j t) +$$

$$\int_{\lambda*}^{\infty} A(\lambda)\, n_\lambda(E) \exp(-\lambda t) d\ .$$

In the sum over *j*, all of the λ_j are less than λ^*. The number of terms in the sum is infinite for a monatomic gas and finite (perhaps zero) for a crystalline moderator. The coefficients A_j and the function $A(\lambda)$ can together be chosen to satisfy any arbitrary initial condition.

In actual numerical calculations of $n(E,t)$ it is usually most convenient to work with a multigroup or discrete energy representation. The resulting set of *N* first-order differential equations takes the form

$$\frac{dn_j}{dt} = - V_j n_j + \sum_{i=1}^{N} C_{ij} n_i\ , \tag{8.141}$$

where

$$n_j = n(E_j, t)$$

and

$$V_j = v\, \Sigma_s(E_j) = \sum_{i=1}^{N} C_{ji}\ .$$

The detailed balance condition becomes

$$C_{ji}\, n_M(E_j) = C_{ij}\, n_M(E_i) \ .$$

As emphasized by Ohanian and Daitch,[41] the N x N matrix form of the equation has, at most N, distinct eigenvalues so that the complete solution to Eq. (8.141) takes the form

$$n_j(t) = \sum_{m=1}^{N} A_m N_{jm} \exp(-\lambda_m t) \ , \qquad (8.142)$$

where

$$\lambda_m N_{jm} = V_j N_{jm} - \sum_{i=1}^{N} C_{ij} N_{im} \qquad (8.143)$$

and the coefficients A_m can be chosen to fit the initial conditions. Ohanian and Daitch have demonstrated that Eq. (8.142) gives the same numerical results as a direct numerical integration of Eq. (8.141).

Although there are only discrete eigenvalues in the multigroup approximation, it has been pointed out by Ohanian and Daitch that the eigenfunctions that belong to the continuum in the limit of a continuous energy description can be recognized. For those eigenfunctions N_{jm} for which λ_m is less than λ^*, the dependence on the energy E_i is smooth. One has the familiar behavior that the first eigenfunction has one node, the second has two nodes, etc.

For those eigenfunctions for which λ_m is greater than λ^*, on the other hand, the dependence on E_i is quite different, with each eigenfunction having a large peak at that energy for which λ_m is nearly equal to V_i. Thus the incipient singularity in the continuum eigenfunctions can be seen even in a discrete energy representation. As the number of energy groups is increased, the smooth eigenfunctions retain a similar energy dependence and the associated eigenvalues remain approximately the same. The sharply peaked eigenfunctions with $\lambda > \lambda^*$ become greater in number and the individual eigenvalues and eigenfunctions depend sensitively on the group structure chosen. (For the discrete eigenfunctions with λ very close to λ^*, the behavior is more complicated.) The function $n(E,t)$ can be conveniently and accurately described by Eq. (8.142) even when no physical significance can be attached to the individual eigenvalues λ_m or eigenfunctions N_{im}.

Finally, we return to the eigenvalue spectrum for $\ell \neq 0$ which we discussed briefly at the beginning of Sec. 8-5. In the approximation (8.113) in which we neglect the operator R_ℓ, there are no discrete eigenvalues and a continuum extends from the minimum value of $\alpha_\ell(E)$ to infinity. The general behavior of the $\alpha_\ell(E)$ has not been studied numerically in any detail, but for monatomic hydrogen gas[42] $\alpha_1(\underline{E})$ initially decreases from its value of λ^* at $\underline{E} = 0$ and then increases. The minimum value of $\alpha_1(\underline{E})$ occurs at $\underline{E}/\underline{k}_B\underline{T} = 0.64$ and is $0.82\ \lambda^*$.

If we consider the complete problem including the operator $\underline{R}_\ell$, there will be a continuum extending from λ^* to infinity plus the possibliity of discrete modes below λ^*. Ghatak[43] has done multigroup calculations of the eigenvalues for $\underline{\ell} \neq 0$ and finds no resolvable discrete eigenvalues except for the case of monatomic hydrogen and $\underline{\ell} = 1$. For this case he finds two resolvable discrete modes with $\lambda_1 = 0.873\ \lambda^*$ and $\lambda_2 = 0.993\ \lambda^*$. Since $\alpha_1(\underline{E})$ becomes as small as $0.82\ \lambda^*$, these modes should exist to describe the relaxation process because no true continuum modes can exist with decay constants between $0.82\ \lambda^*$ and λ^*. Presumably the lack of discrete modes for other values of $\underline{\ell}$ and other moderators is associated with the fact that the minimum of $\alpha_\ell(\underline{E})$ is either equal to or very nearly equal to λ^*. This is the case for $\underline{\ell} \geq 2$ in monatomic hydrogen.

8-7 DECAY CONSTANT VERSUS BUCKLING

The fundamental spatial mode of a finite moderating sample is conventionally described in terms of a geometric buckling in which the finite-medium problem is replaced by an infinite-medium problem with an exp(iBz) spatial dependence. This procedure can be justified in diffusion theory and has been discussed for the pulse-decay experiment in Sec. 4-11. In order to estimate transport corrections to the pulse-decay experiment it is convenient to consider the infinite-medium exp(iBz) formulation of the pulse-decay problem in transport theory. We do this in this section, deferring the problems of assigning a buckling to a finite sample until Sec. 8-8.

We look for solutions of the infinite-medium source-free transport equation of the form

$$f(z,E,\mu,t) = e^{-\lambda t} e^{iBz} f(E,\mu) \quad . \tag{8.144}$$

The function $f(E,\mu)$ satisfies

$$\left[\Sigma_a(E) - \frac{\lambda}{v} + \Sigma_s(E) - iB\mu\right] f(E,\mu) =$$

$$\sum_{n=0}^{\infty} P_n(\mu) \int \Sigma_n(E',E) f_n(E)\, dE' \ , \tag{8.145}$$

where

$$f(E,\mu) = \sum_{\ell=0}^{\infty} f_\ell(E)P_\ell(\mu) .$$

Equation (8.145) can be obtained from Eq. (8.2) by letting

$$\Sigma_a \to \left(\Sigma_a - \frac{\lambda}{v}\right)$$

and

$$\kappa \to iB .$$

If we rewrite Eq. (8.145) in a form analogous to Eq. (8.20), it can be shown that λ depends only on $\underline{B}^2$. Thus for 1/$\underline{v}$ absorption, if we let

$$\left.\begin{aligned} v\Sigma_a &\to (v\Sigma_a - \lambda) \\ \kappa^2 &\to -B^2 , \end{aligned}\right\} \tag{8.146}$$

we are dealing with a continuation of the κ^2 versus $\Sigma_{\underline{a}}$ curve to negative values of κ^2. This is illustrated in Fig. 8.1, where the diffusion length and pulse-decay data for water are plotted on the same graph.

The pulsed-neutron die-away measurements are usually analyzed by plotting the fundamental-mode decay constant as a function of the theoretically assigned geometric buckling. The measured $\lambda(\underline{B}^2)$ is then fitted to a polynomial of the form

$$\lambda(B^2) = v\Sigma_a + D_0 B^2 - CB^4 + FB^6 + \ldots \,, \qquad (8.147)$$

where $\underline{D}_0$ is the thermal-neutron diffusion coefficient and $\underline{C}$ is the so-called diffusion cooling coefficient. We could start from Eq. (8.145) and systematically expand in powers of $\underline{B}^2$ to obtain a set of equations for $\underline{D}_0$, $\underline{C}$, and higher terms in the expansion (8.147). This is in fact how the problem of including transport corrections to the pulsed-neutron experiment was first attacked.[44] In the present discussion, however, we can avoid a great deal of redundancy by noting that Eq. (8.147) is just the inverse of the expansion (8.43) of κ^2 in powers of $\Sigma_{\underline{a}}$. We thus have[2]

$$D_0 = \frac{v_0}{\alpha_1} \,,$$

$$C = \frac{v_0 \alpha_2}{\alpha_1^3} \,, \qquad\qquad (8.148)$$

$$F = \frac{v_0(2\alpha_2^2 - \alpha_1\alpha_3)}{\alpha_1^5} \,.$$

The procedure for calculating α_1, α_2, and α_3 has been discussed in Sec. 8-1. We saw there that α_1 and α_2 could be calculated rather accurately in energy-dependent diffusion theory. The calculation of the fundamental-mode decay constant in energy-dependent diffusion theory is quite similar to the calculation of λ_1, which was discussed in Sec. 8-5. In particular, the same variational principle can be applied. We will not discuss approximate semianalytic results for the diffusion cooling coefficient in any detail but will only mention the simplest and earliest of these results.[28] If we assume a transport mean free path that varies with energy as $\underline{E}^{\gamma}$, and a trial function in which only one energy polynomial is retained, the resulting approximate formula for the diffusion cooling coefficient is

$$C \approx (\gamma + \frac{1}{2})^2 \frac{\sqrt{\pi} D_0}{v_0 M_2} \gamma$$

where $\underline{M}_2$ is as defined by Eq. (8.130). The approximation is similar to that leading to Eq. (8.128). Combining these two expressions to eliminate $\underline{M}_2$ leads to an approximate relation between the diffusion cooling coefficient and the thermalization time constant:

$$\lambda_1 C \approx \frac{2}{3}(\gamma + \frac{1}{2})^2 D_0$$

As emphasized by Shapiro,[30] these $\underline{M}_2$ formulas can be extremely inaccurate for tightly bound crystals. They were useful for a semiquantitative understanding in the early days of studying neutron thermalization but are quite inadequate to our present-day quantitative study.

For larger values of B^2, the power-series expansion of Eq. (8.147) may not converge sufficiently rapidly to be useful. In that case we must solve Eq. (8.20) with the substitutions (8.146). This is essentially the same problem as calculating κ^2 versus Σ_a. The functions $F_{\ell n}(x)$ (Eq. (8.22)) for imaginary argument are real if $\ell + n$ is even and imaginary if $\ell + n$ is odd. The functions $\phi_\ell(E)$ are real for even ℓ and imaginary for odd ℓ. Equation (8.20) thus contains only real terms for even ℓ and imaginary terms for odd ℓ so that there is no additional complication introduced by the presence of a complex-valued $f(E,\mu)$ associated with an $\exp(iBz)$ spatial distribution. The actual numerical solution for negative values of κ^2 has not been carried out with the same accuracy as for positive κ^2. Quantitative studies for large B^2 have been limited to the case of isotropic scattering. This case should have the same qualitative features as a more accurate treatment.

To illustrate the behavior for large values of B^2, consider first the case of isotropic scattering and monoenergetic neutrons. The equation to be solved is

$$\left(\Sigma - \frac{\lambda}{v} + iB\mu\right) f(\mu) = \frac{1}{2}\Sigma_s \int_{-1}^{1} f(\mu')\, d\mu' \ . \tag{8.149}$$

This has the solution

$$\lambda = v\Sigma - vB \cot(B/\Sigma_s) \qquad \text{for } B \leq \frac{1}{2}\pi\Sigma_s \ . \tag{8.150}$$

When $\underline{B} = \frac{1}{2}\pi\Sigma_{\underline{s}}$, $\lambda = v$, which is its maximum allowable value for a nonsingular solution. For larger values of $\underline{B}^2$, there is no value of λ satisfying Eq. (8.149). For sufficiently large values of $\underline{B}^2$, there are thus no nonsingular solutions to the source-free transport equation of the form

$$e^{iBz}\, e^{-\lambda t}\, f(\mu)$$

Thus, if we make a spatial Fourier analysis of an initial value problem in an infinite homogeneous medium, we find that only those wave numbers less than $\frac{1}{2}\pi\Sigma_{\underline{s}}$ have an asymptotic time behavior that corresponds to a purely exponential decay.

Corngold[7] has shown that the same situation obtains in the energy-dependent problem for thermal neutrons. Making the substitutions of Eq. (8.146) into Eq. (8.4), we have

$$\phi(E) = \frac{1}{B} \tan^{-1}\left[\frac{Bv}{v\,\Sigma(E) - \lambda}\right] \int_0^\infty \Sigma_0(E',E)\phi(E')dE'. \tag{8.151}$$

Equation (8.151) has nonsingular solutions only for

$$\lambda \leq \min_{0 \leq E \leq \infty} v\,\Sigma(E) \;. \qquad (8.152)$$

It remains to show that the fundamental-mode decay constant reaches the upper bound given by Eq. (8.152) for some finite value of $\underline{B}^2$, as in the monoenergetic case. This has been shown by Corngold,[7] using only general properties of the cross sections for thermal neutrons. The result is confirmed by the numerical calculations of Shapiro.[30,35] We defer a discussion of numerical results until Sec. 8-9.

Finally, we briefly mention one other technique of measuring thermal-neutron diffusion parameters, i.e., the oscillating source, or "neutron wave," technique. In this technique the neutron source is oscillated in time, and we observe the complex spatial relaxation length. We are thus interested in solutions to the source-free transport equation of the form

$$e^{i\omega t}\, e^{-\gamma z}\, f(E,\mu) \;,$$

where γ is complex. This experiment contains essentially the same information as the diffusion-length or pulsed-neutron experiment. The theory for purely moderating samples has been discussed by Perez and Uhrig,[45] and recent experimental results using the technique have been presented by Perez and Booth.[46] This technique has the same advantages as the diffusion-length technique in that it is not sensitive to a proper treatment of the boundary between the moderating sample and vacuum. It has the further practical advantage for solid moderators that the equivalent of homogeneous $1/\underline{v}$ poison can be uniformly inserted by means of looking at a single frequency component. Furthermore, one has both an amplitude and a phase to measure at every point, and a combination of these can be found which is sensitive to thermalization effects to first order. It seems likely that further development of the "neutron wave" technique will proceed rather rapidly. To the extent that one wishes to emphasize the effects of thermalization on the measurement of thermal-neutron diffusion parameters, this technique promises to be a desirable alternative to the measurements of time decay constants. This is particularly true since the desired information can be obtained either directly from modulated source measurements or from the Fourier analysis of pulse propagation measurements.

8-8 PULSE DECAY IN FINITE MEDIA

In the measurement of the fundamental-mode decay constant of a finite moderating sample, one waits a sufficiently long time that an asymptotic solution of the form

$$f(\mathbf{r},E,\mathbf{\Omega},t) = f_0(\mathbf{r},E,\mathbf{\Omega})\ \exp(-\lambda_0 t) \qquad (8.153)$$

is established. The asymptotic solution $f_0(\mathbf{r},E,\mathbf{\Omega})$ is a positive solution of the steady-state source-free transport equation with the removal term $\Sigma_a + \Sigma_s$ replaced by $\Sigma_a + \Sigma_s - \lambda_0/v$; that is, a uniform poisoning of macroscopic cross section λ_0/v is removed from the system until a steady solution is obtained. The equation is to be solved with the boundary condition that no neutrons enter the sample.

The existence of a fundamental-mode solution $f_0(\mathbf{r},E,\mathbf{\Omega})$ does not require that this solution be separable in space and energy. For samples that are many mean free paths in size, however, it is often a good approximation that

$$\int f_0(\mathbf{r},E,\mathbf{\Omega})\ d\Omega \approx \Phi(\mathbf{r})\phi_0(E,B^2), \qquad (8.154)$$

where $\Phi(\mathbf{r})$ is a solution of

$$(\nabla^2 + B^2)\Phi(\mathbf{r}) = 0 \ .$$

Equation (8.154) is only applicable many mean free paths from the boundary of the system. If there is a region of the sample where it applies to a good approximation, we can identify $\underline{B}^2$ as the geometric buckling of the sample. We would expect that any sample of a given material that yeilds the same $\underline{B}^2$ in the sense of Eq. (8.154) will have the same fundamental-mode decay constant. The energy spectrum $\phi_0(\underline{E})$ and the decay constant λ_0, for a given value of $\underline{B}^2$, will be the same as that for an infinite medium with an $\exp(\underline{iBz})$ spatial dependence, as was discussed in Sec. 8-7.

As an illustration, consider the case of a slab of thickness $\underline{a}$ with no transverse spatial variation. The neutron flux at long times is given by

$$\phi(z,E,t) = \phi_0(z,E) \exp(-\lambda_0 t) \ .$$

The asymptotic solution $\phi_0(\underline{z},\underline{E})$ far from the boundaries of the slab is given approximately by

$$\phi_0(z,E) \approx \cos(Bz)\phi_0(E,B^2) \ , \tag{8.155a}$$

where

$$B = \frac{\pi}{a + 2z_0} \ . \tag{8.155b}$$

Equation (8.155b) can be taken as a definition of the extrapolation distance for the pulsed system. In the limit of small $\underline{B}^2$, $\underline{z}_0$ is just the extrapolation distance for the thermal-neutron Milne problem of Sec. 4-10. We recall that the solution to this problem is dependent on the energy dependence of the mean free path. For constant mean free path we have the well-known result

$$z_0 = 0.7104\lambda_{tr} \quad .$$

For the energy-dependent mean free path, the variational result of Kladnik and Kuscer[21] gives

$$z_0 = \frac{3}{8}\frac{\left(\overline{U^2}\right)_M}{\left(\overline{U}\right)_M} + \frac{1}{3}(\overline{U})_M \quad , \tag{8.156}$$

where $\underline{U}(\underline{E})$ is the effective transport mean free path discussed in Sec. 8-3. Equation (8.156) is a generalization of Eq. (4.147) to include anisotropic scattering. If $\underline{U}(\underline{E})$ is proportional to $\underline{E}^{\frac{1}{2}}$, as is approximately true for water, Eq. (8.156) gives

$$z_0 = 0.76(\overline{\lambda}_{tr})_M \quad . \tag{8.157}$$

This is consistent with numerical calculations in slab geometry carried out by Gelbard and Davis.[47]

Walker, Brown, and Wood[48] have recently measured the extrapolation distance for cubes of water. Space-time separability at long times was verified, and the spatial distribution across one dimension of the cube was measured and fitted to Eq. (8.155a). The measured value for a large cube was $z_0 = 0.35 \quad 0.02$ cm. The calculated value using Eq. (8.157) and the value of $(\bar{\lambda}_{tr})_M$ determined from the diffusion-length measurement of Starr and Koppel[3] is $z_0 = 0.33$ cm. [In terms of the parameter α_1 of Eq. (8.43) and Table 8.1, $(\bar{\lambda}_{tr})_M = 3\sqrt{\pi}/2\alpha_1$.] The agreement between theory and experiment is satisfactory.

In order to consider theoretically the dependence of z_0 on buckling, we need a more precise definition than Eq. (8.155a). This can be obtained by using Eq. (8.155b) to define $z_0(B^2)$ in terms of B^2, and defining the value of B^2 for a given slab thickness so that an infinite-medium $\exp(iBz)$ solution has the same fundamental-mode decay constant as is calculated for the finite slab. This definition has been used by Gelbard and Davis to calculate extrapolation distances for water in various finite geometries. In slab geometry they find that z_0 varies with B^2 in the manner shown in Fig. 8.5.

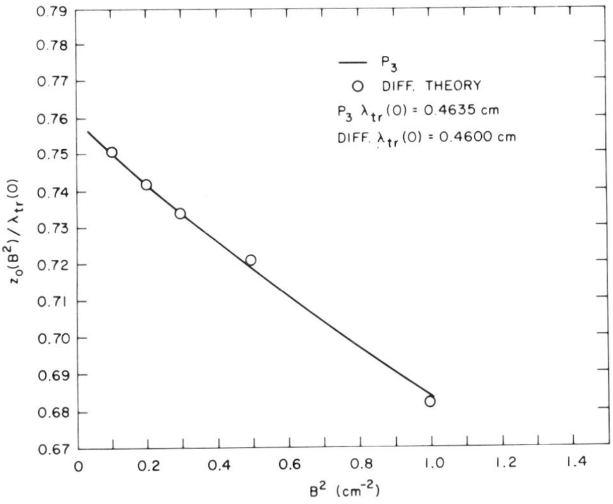

Figure 8-5 $\underline{P_3}$ and diffusion-theory extrapolation distances for infinite slabs of water. (From Gelbard and Davis, Nucl. Sci. Eng., Ref. 47.)

We show below that the extrapolation distance for a slab defined in terms of the equivalent decay constant increases with increasing buckling for monoenergetic neutrons. The decrease seen in Fig. 8.5 is therefore essentially a spectral effect. As the buckling increases, the spectrum in the interior of the slab is diffusion-cooled so that the average transport mean free path is decreased. The decrease in extrapolation distance can be calculated with reasonable accuracy from energy-dependent diffusion theory as shown in Fig. 8-5. It is to be expected that the buckling defined from the equivalent decay constant and that defined from Eq. (8.155a) in terms of the curvature of the flux near the center of the slab should agree rather well. Measurements of the extrapolation distance using Eq. (8.155a) have been carried out for small cubes of water,[48] but the accuracy was not great enough to test the variation of z_0 with buckling. The calculations of Gelbard and Davis[47] indicate that there are interesting effects of sample shape on extrapolation distance which deserve further investigation. For slab geometry, however, it appears that the decrease of z_0 with increasing B^2 shown in Fig. 8.5 can be applied with a fair degree of confidence to the analysis of pulsed-neutron decay constants in terms of a λ_0 versus B^2 curve for water.

For moderators other than hydrogen, the energy dependence of the mean free path is not so strong and thus there is less difficulty in determining an extrapolation distance. For crystalline moderators, however, there is the phenomenon of the Bragg cutoff (see Sec. 2-6). For sufficiently low energies, no elastic scattering is possible and the cross section drops to a low value that is a strongly increasing function of temperature. The energy region below the Bragg cutoff does not have a very large weight in a typical diffusion-cooled spectrum except at very low temperature. There is, however, a more fundamental difficulty. We recall that for the infinite-medium exp(iBz) solution the fundamental-mode decay constant cannot exceed

$$\lambda^* = \min[v\Sigma(E)] = \lim_{E\to 0} v\,\Sigma(E) .$$

For crystalline moderators at low temperatures λ^* is much smaller than the typical collision rates for energies above the Bragg cutoff. As pointed out by Corngold and Michael,[8] the observed decay constants for small beryllium and graphite assemblies in many cases exceed λ^*. We are thus led to examine more closely the conditions for the existence of a fundamental mode in a finite moderating sample. This turns out to involve rather subtle considerations that have become reasonably well understood only in the last few years. The basic conclusion of Corngold and Michael is found to be correct, and the observed decay constants greater than λ^* must represent a quasi-exponential decay applicable only in an intermediate time range. Experimental evidence supporting this conclusion is now available and is discussed in Sec. 8-9. The recent theoretical developments have been very nicely summarized in the introduction to a recent paper by Albertoni and Montagnini[49] and we paraphrase their discussion below.

Most of the theoretical investigations concerning the eigenvalue spectrum of the neutron transport equation originate from Lehner and Wing's fundamental papers,[50,51] where the time-dependent monoenergetic transport equation in a bare slab was first fully and rigorously treated. These authors showed that the eigenvalue spectrum of this equation (considered in a Hilbert space of square-integrable functions) is made up of the following parts: a continuous spectrum occupying the half-plane Re $\lambda > \underline{v}\Sigma$, and a point spectrum consisting of a finite (but not zero) number of real eigenvalues.

In particular, the fundamental-mode decay constant exists for any slab thickness and approaches $\underline{v}\Sigma$ as the slab thickness approaches zero. If we compare this result with the infinite-medium exp(*iBz*) case of Eq. (8.150), we see that a buckling can be chosen for any slab thickness so as to give the same fundamental-mode decay constant. In the limit of a very thin slab, we must choose $B = \frac{1}{2}\pi\Sigma_s$, which corresponds to an extrapolation distance of $1/\Sigma_s$. Thus, as previously stated, the extrapolation distance increases with increasing buckling in the case of monoenergetic neutrons, isotropic scattering, and slab geometry.

Lehner and Wing also showed that an entirely different result is achieved if a finite body (e.g., a sphere) is considered. The spectrum is then purely discrete, the eigenvalues whose real part is less than $v\Sigma$ being real. Moreover, there are eigenvalues with $\mathrm{Re}\,\lambda > v\Sigma$ which form an infinite sequence of positive real numbers without accumulation points except for $\lambda = \infty$, but it cannot be excluded that complex eigenvalues also exist.[52] The nature of the spectrum, in the slab case and in the finite-body case, does not change in an essential way when passing to N energy groups[53,54] or even to continuous energy dependence, provided that the set of admissible neutron speeds is bounded away from zero.[54]

In particular, for a sufficiently small sphere with isotropic scattering and monoenergetic neutrons, the fundamental-mode decay constant becomes greater than $v\Sigma$.

In this case it is no longer possible to define an equivalent buckling or extrapolation distance on the basis of the equal fundamental-mode decay constant inasmuch as the infinite-medium exp(iBz) solution does not allow for so large a decay constant.

The appearance of the continuous spectrum in the slab case is related to the fact that a neutron can spend an indefinitely large time in the slab without undergoing collisions or escaping, provided it moves parallel to the slab sides. These long-uncollided neutrons form an exceptional class that cannot be treated like the ordinary neutron population and cannot be expected to allow for a discrete eigenfunction expansion. Since the first investigations, it was thus clear that a neutron can take in crossing the infinite slab. A related phenomenon has proved of importance in the case of thermal neutrons. It has been noted[55] that the theory of pulsed-neutron experiments in moderators involves the solution of an initial-value problem relating to a finite sample but without any lower bound for the speed of the neutrons moving in it, inasmuch as thermal neutrons can have any energy between zero and infinity. The transit time increases indefinitely as neutrons of vanishing small energy are considered, and thus a continuous spectrum is to be expected, in analogy with the work of Lehner and Wign, for $\underline{\mathrm{Re}}\ \lambda > \lambda^*$.

For a finite body the value of λ^* is intrinsically the value of $\underline{v}\ \Sigma(\underline{E})$ for zero-energy neutrons. For all the thermalization models we have considered this is also the minimum value of $\underline{v}\ \Sigma(\underline{E})$, but the minimum value of $\underline{v}\ \Sigma(\underline{E})$ is not relevant, per se. If this minimum were to occur at some energy greater than zero, it would still be the zero-energy value of $\underline{v}\ \Sigma(\underline{E})$ that would determine the lower limit of the continuum decay constants. This is in contrast to the infinite-slab or infinite-medium $\exp(\underline{iBz})$ case where the minimum value of $\underline{v}\ \Sigma(\underline{E})$ is the intrinsically relevant parameter.

The most striking result for thermal neutrons in finite bodies, however, concerns the discrete spectrum. If the body is small enough, contrary to Lehner and Wing's case, the discrete spectrum is a void set. Fundamental decay modes thus should no longer be present in samples of very small size. This was first demonstrated in Ref. 55 for a very simple (and rather unphysical) degenerate scattering kernel. This result has been extended on a rigourous mathematical basis to more realistic scattering kernels with the same conclusion. Albertoni and Montagnini[49,56] have studied the spectrum of decay constants for the free-gas model, Eq. (2.99), of the scattering kernel. In addition to the vanishing of the point spectrum for sufficiently small samples, they have obtained rather detailed results concerning the remainder of the spectrum; we do not discuss these results here. Bednarz[57] has worked with a very weakly restricted scattering kernel and has shown that the point spectrum vanishes for a sufficiently small sample. The nonexistence of a fundamental decay mode for a sufficiently small moderating sample thus is now established on rather general grounds.

To illustrate some of the above results, consider the integral-equation from of the eigenvalue problem defining the fundamental mode in a finite sample with isotropic scattering and no absorption

$$\phi_0(\mathbf{r},E) = \int dE' \int d^3r \, \frac{\exp[-\alpha(E)|\mathbf{r} - \mathbf{r}'|]}{4\pi|\mathbf{r} - \mathbf{r}'|^2} \Sigma_0(E',E)\phi_0(\mathbf{r}',E'). \tag{8.158}$$

In Eq. (8.158)

$$\alpha(E) = \Sigma_s(E) - \frac{\lambda_0}{v}$$

and λ_0 is the fundamental-mode decay constant. The constant λ_0 is the smallest value of λ satisfying Eq. (8.158). The associated flux $\phi_0(\mathbf{r},E)$ represents the actual neutron distribution at long times and so must be positive and nonsingular.

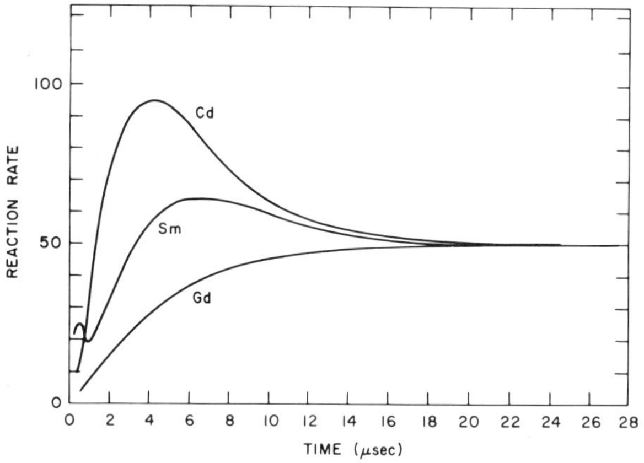

Figure 8-6 Measured reaction rate curves for cadmium, samarium, and gadolinium in water, corrected to zero absorption. (From Möller and Sjöstrand, <u>Arkiv Fysik</u>, Ref. 58.)

Ghatak and Honeck[59] have studied the eigenvalues describing the approach to equilibrium by the multigroup method. Using the same basic idea as Ohanian and Daitch,[41] discussed at the end of Sec. 8-6, they found that there are two resolvable discrete eigenvalues for the bound proton model of Eqs. (3.95-3.97). These are

$$\lambda_1 = 0.177 \ (\mu \ \text{sec})^{-1}$$

and

$$\lambda_2 = 0.276 \ (\mu \ \text{sec})^{-1} \ .$$

The approach to equilibrium will be governed by λ_1 without appreciable effects due to λ_2 only for times long compared with $(\lambda_2 - \lambda_1)^{-1} = 10 \ \mu\text{sec}$. Thus, in order to measure λ_1 one must look at times long enough that the thermalization process is almost complete. This is illustrated by a calculation of Ghatak and Honeck. They calculated $[\overline{v}(\underline{t}) - \overline{v}(\underline{T})]$, where $\underline{T}$ is the time assumed for the spectrum to become asymptotic. Fitting this to an exponential gives an effective λ_1. This effective λ_1 is plotted versus $\underline{T}$ in Fig. 8.7. It is clear from this plot that fitting any reaction rate over the range from 9 μsec to 45 μsec with an effective λ_1 will give too large a value of λ_1 and thus too short a

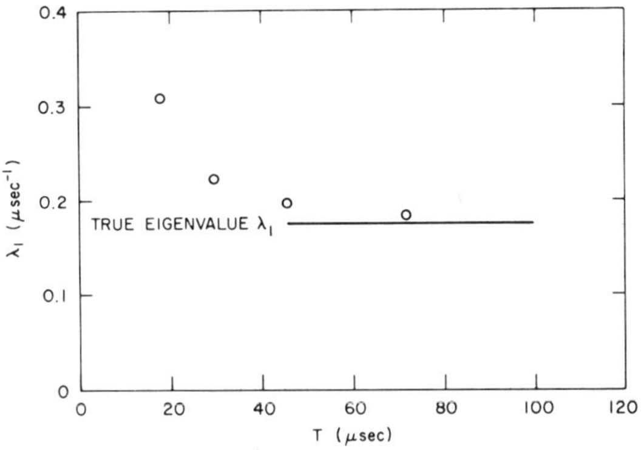

Figure 8-7 Dependence of the effective value of λ_1 in water on T, the time assumed for the spectrum to become asymptotic. (From Ghatak and Honeck, Nucl. Sci. Eng., Ref. 59.)

thermalization time constant. This is consistent with the sign of the discrepancy between the calculated λ_1 and the value inferred from experiment by Möller and Sjöstrand.

In a more complete analysis of the experiment, a thermalization model is used to compute $\phi(\underline{E},\underline{t})$, which in turn is used to compute the measured time-dependent reaction rates $\underline{R}(\underline{t})$. Such an analysis has been carried through by Purohit.[60] He finds that the bound-proton model of Eqs. (3.95-3.97) gives better agreement with experiment than does a free-proton model but that there are residual discrepancies. Purohit finds that the measured decay of the cadmium activation falls between that calculated from the free- and the bound-proton models, indicating that the bound-proton model gives too slow a thermalization rate. This is the opposite conclusion from that reached in Chap. 6 from steady-state spectra. The reason for this disagreement is not understood.

We turn now to the measurement of the fundamental-mode decay constant as a function of buckling. This method is hampered by three basic difficulties: The first of these is a purely experimental one -- the precise determination of λ_0 in the presence of transient behavior and background. The second problem is a theoretical one; it consists in the calculation of the geometrical buckling of the sample. We have discussed this problem in Sec. 8-8. Thirdly, there is the problem of analyzing the λ_0 versus $\underline{B}^2$ curve properly, keeping the right number of terms in the expansion (8.146).

For water, all of these problems occur. In the region of small bucklings the major problem is the effect of higher spatial modes. These can be eliminated by waiting a long time before beginning the measurement. Alternatively, the space-time distribution of the thermal-neutron density can be measured, and a spatial Fourier analysis can be performed to give the time dependence of the various spatial-mode amplitudes. This latter method was successfully employed by Lopez and Beyster.[4] They were able to demonstrate the exponential decay of the higher spatial modes with fair accuracy and also to measure the fundamental-mode decay with high accuracy.

For bucklings $\underline{B}^2 \lesssim 0.4\ \text{cm}^{-2}$, almost all of the recent measurements are in rather good agreement. If these are combined with the stationary measurements of Starr and Koppel[3] (see Sec. 8-1), a very accurate set of diffusion parameters can be obtained. In addition, it has been pointed out by Joest and Memmert[61] that the region of the λ_0 versus $\underline{B}^2$ curve between these two experiments can be filled in by measuring the diffusion length in the presence of a source that is exponentially decaying in time. This measurement has been performed by Arai and Küchle,[62] who have also carried out a least-squares fit of the form (8.147) to the combined data from all three types of experiments. Actually, the pulsed-neutron data used was that of Küchle,[63] but the use of the Lopez and Beyster data would not change the results appreciably if one used only data for $\underline{B}^2 \leq 0.4\ \text{cm}^{-2}$. The heaviest weight in the data analysis goes to the accurate diffusion-length measurements of Starr and Koppel[3] so that the results stated here are essentially a translation of Table 8.1 to the language of pulsed-neutron experiments, using Eq. (8.148). The experimental numbers for water at 20°C are

$$v\,\Sigma_a = 4782 \pm 15\ \text{sec}^{-1}\ ,$$

$$D_0 = 35{,}630 \pm 80\ \text{cm}^2\ \text{sec}^{-1},$$

$$C = 3420 \pm 170\ \text{cm}^4\ \text{sec}^{-1},$$

$$F = 214 \pm 140\ \text{cm}^6\ \text{sec}^{-1}\ ,$$

The theoretical numbers using the bound-proton model of Eqs. (3.95-3.97) have been calculated by Honeck[2] and are

$$D_0 = 37{,}530\ \text{cm}^2\ \text{sec}^{-1}\ ,$$

$$C = 3130\ \text{cm}^4\ \text{sec}^{-1}\ ,$$

$$F = 270\ \text{cm}^6\ \text{sec}^{-1}\ .$$

Except for the addition of the comparison of $\underline{F}$ with experiment, the same comments apply here as applied to the discussion of the diffusion length in Sec. 8-1. It is interesting to note that the correction of Koppel and Young[64] for anisotropic molecular vibrations reduces $\underline{D}_0$ by about 4%, improving the agreement with experiment.

Although all the measured decay constants in water are much less than λ^*, the experimental situation for large bucklings ($\underline{B}^2 \geq 0.4\ \text{cm}^{-2}$) is still somewhat confused. The results of Lopez and Beyster[4] for small cubes do not agree with those of Küchle[63] for flat cylinders of the same buckling. Measurements at Birmingham[65] have emphasized this discrepancy between "flat" and cubical systems of the same buckling. Part of the discrepancy could arise from the variation of extrapolation distance with shape and size of the sample, but the published differences seem rather large to be explained entirely by this cause. Recent measurements at Columbia on spherical systems[66] are in general agreement with earlier measurements on cubes but not in agreement with the data on flat systems. The possibility of a purely experimental reason for part of this discrepancy cannot be excluded.

On the theoretical side, the main interest in the measurements in small water systems is not in thermalization but rather as an experimental check on our ability to solve the transport equation in various geometries. With the exception of the work of Gelbard and Davis[47] and Paik,[67] little work has been done on this problem in a way that makes useful contact with pulsed neutron experiments. Hopefully, the recent Columbia data,[66] which extends down to 1.5-cm sphere radius, will stimulate further calculations in this area.

8-9.2 Zirconium Hydride. In hydrogenous moderators other than water, the thermalization properties vary greatly. This is clearly demonstrated in recent measurements by Kallfelz and Reichardt[68] of the time dependence of neutron energy spectra in zirconium hydride and in ice. The final approach to equilibrium in zirconium hydride is determined largely by the small energy exchange between neutrons and the acoustic vibrations of the lattice, and is consequently very slow. From the measured time-dependent spectra shown in Fig. 8.8, Kallfelz and Reichardt determine that the average neutron energy approaches its asymptotic value with a time constant

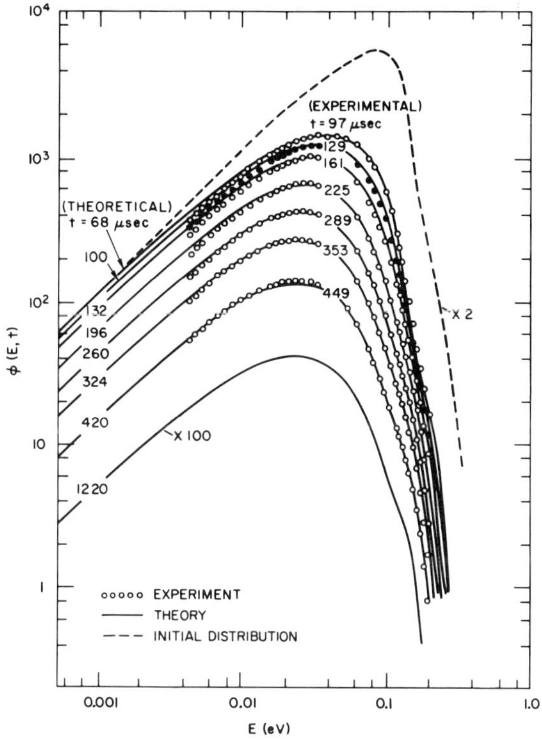

Figure 8-8 Comparison of measured and theoretical time-dependent spectra in zirconium hydride. The initial distribution is taken from measurements of Ismaev et al.[68a] The model for zirconium hydride assumes a Gaussian distribution of frequencies centered around 0.14 eV with a width of 0.028 eV and a relative weight of 0.995, and a Debye spectrum with a Debye temperature of 312°K and a relative weight of 0.005.

$$\frac{1}{\lambda_1} = 97 \pm 7 \ \mu\text{sec} \ .$$

The approach to equilibrium appears from the experiment to be governed by a discrete λ_1 with the above value. The value of λ^* for zirconium hydride is not well known, so it is not certain that the measurement corresponds to a true discrete eigenvalue.

The slow approach to equilibrium in zirconium hydride is also clear in the λ_0 versus $\underline{B}^2$ measurements of Meadows and Whalen.[69] The curvature of the λ_0 versus $\underline{B}^2$ plot is so great that the power-series expansion of Eq. (8.146) is not usefully applied. This measurement is also sensitive to the treatment of the acoustic modes, which have not been very well determined from scattering measurements, from lattice dynamical theories, or from the steady-state measurements discussed in Chap. 6. A careful analysis of the λ_0 versus $\underline{B}^2$ and time-dependent spectrum measurements in zirconium hydride would thus be of considerable interest. No such analysis has yet been carried out.

8-9.3 $\underline{D_2O}$. For heavy water, data on time-dependent spectra have recently become available[70,71] supplementing existing data on diffusion parameters from measurements of λ_0 versus

$\underline{B}^2$. Most of the analysis has been based on Honeck's bound-deuteron model (see Eqs. (3.95) and (3.98)). In general, the agreement between theory and experiment is good, but there remain discrepancies among the experiments and thus the situation is not completely clear.

The recent data of Poole and Wydler[70] on time-dependent spectra in D_2O are in good agreement with energy-dependent diffusion theory where the scattering is based on Honeck's model. The comparison of theory and experiment is shown in Fig. 8.9. The effects of including several spatial modes in the analysis were appreciable. The experiments of the Rensselaer Polytechnic Institute (RPI) group[71] carried out by very similar techniques, on the other hand, do not agree as well with a theory that is essentially the same, but which includes only the fundamental spatial mode. Since the geometries of the two sets of experiments are not identical, it is not clear whether the disagreement between the results from these two experiments is in the data or in the treatment of spatial modes in the analysis.

The measurement of λ_0 versus $\underline{B}^2$ for D_2O has been carried out by several authors, and the resulting diffusion parameters are in good agreement among the experiments and between theory and experiment. The only significant

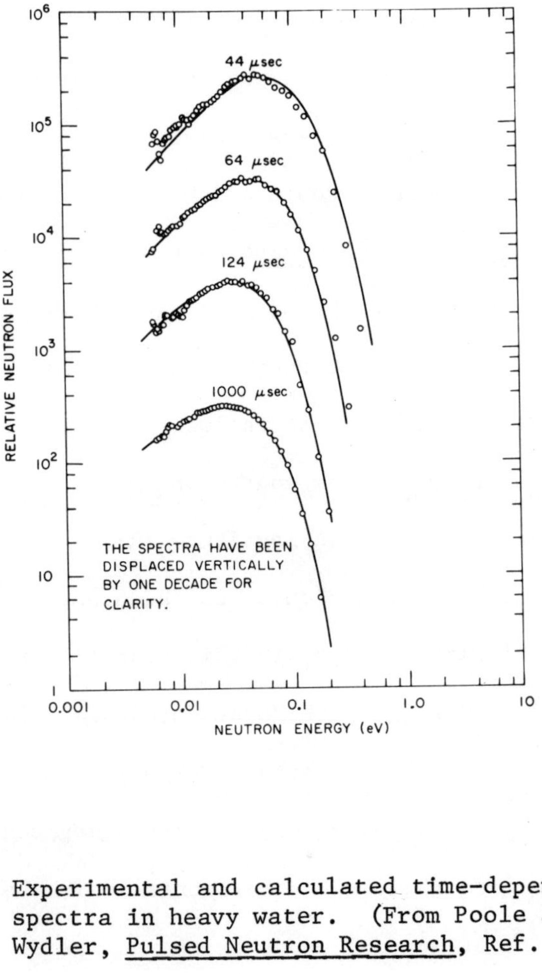

Figure 8-9 Experimental and calculated time-dependent spectra in heavy water. (From Poole and Wydler, Pulsed Neutron Research, Ref. 70.)

difficulty arises in including the correct number of terms in the expansion (8.146). This is most readily illustrated in terms of the theory. Honeck[2] calculates

$$D_0 = 2.07 \times 10^5 \text{ cm}^2 \text{ sec}^{-1} ,$$

$$C = 5.13 \times 10^5 \text{ cm}^4 \text{ sec}^{-1} ,$$

$$F = 7.39 \times 10^5 \text{ cm}^6 \text{ sec}^{-1} ,$$

When he fits the computed λ_0 versus B^2 to the function

$$A + D_0'B^2 - C'B^4$$

over the range $0 < B^2 < 0.1 \text{ cm}^{-2}$, he finds, however, that

$$D_0' = 2.11 \times 10^5 \text{ cm}^2 \text{ sec}^{-1}$$

and

$$C' = 3.65 \times 10^5 \text{ cm}^2 \text{ sec}^{-1} ,$$

indicating that the FB^6 term plays an appreciable role over this range of buckling. The tendency for the measured diffusion-cooling coefficient to decrease with an increase in the range of bucklings considered is present in the experiments. The least ambiguity occurs when a small enough buckling range is used so that the FB^6 term is negligible. This should be true for the experiment of

Kussmaul and Meister,[72] which was conducted in the range $0.013\ \text{cm}^{-2} < B^2 < 0.045\ \text{cm}^{-2}$; they obtain

$$D_0 = 2.00 \pm 0.01\ \text{cm}^2\ \text{sec}^{-1}$$

and

$$C = 5.25 \pm 0.25 \times 10^5\ \text{cm}^4\ \text{sec}^{-1}$$

The agreement between theory and experiment is comparable to that obtained for H_2O and is entirely satisfactory, considering the crudity of the theoretical model. The experiment is for 99.8% D_2O, and the theory for 100% D_2O. The correction for H_2O content will improve the agreement in D_0. The temperature dependence of D_0 and C for heavy water has recently been measured by Daughtry and Waltner[73] in the range from 25°C to 250°C. The results are in reasonable agreement with theory. A summary of the measured diffusion parameters for heavy water by various investigators is given in Refs. 37 and 73.

8-9.4 Graphite. From the results of Shapiro and Corngold[35] for a heavy crystalline moderator, we expect that the approach to equilibrium in graphite should be governed entirely by the continuum decay constants greater than λ^*. This is confirmed by the multigroup calculations of Ghatak and Honeck[59] using the thermalization model of Parks (Sec. 3-4.2) for graphite. An "effective λ_1", obtained by the analysis of the computed integral behavior, depends on the particular detector used, the type of analysis made, and how long one waited before assuming an asymptotic spectrum. In Fig. 8.10, the average neutron energy and average neutron speed computed by Ghatak and Honeck are plotted as a function of time. The spectrum is assumed to be asymptotic at 4.6 msec. It is seen that the final decay appears to be very nearly exponential but that the decay constants for the speed and for the energy are different, indicating that a single discrete eigenvalue is not dominant.

This ambiguity is reflected in the discrepancies among the experiments, but there is general qualitative agreement[37] that the effective λ_1 is of the order of 2000 sec^{-1}, giving a thermalization time constant in the neighborhood of 500 μ sec. A more quantitative comparison of theory and experiment requires a theoretical analysis of the specific integral experiment in question rather than a discussion in terms of eigenvalues.

The diffusion parameters in graphite measured from the λ_0 versus $\underline{B}^2$ curve are reasonably well understood if the range of $\underline{B}^2$ considered is sufficiently small. Analyzing only the region $B^2 < 0.006\ \text{cm}^{-2}$, Klose, Küchle, and Reichardt[74] obtain

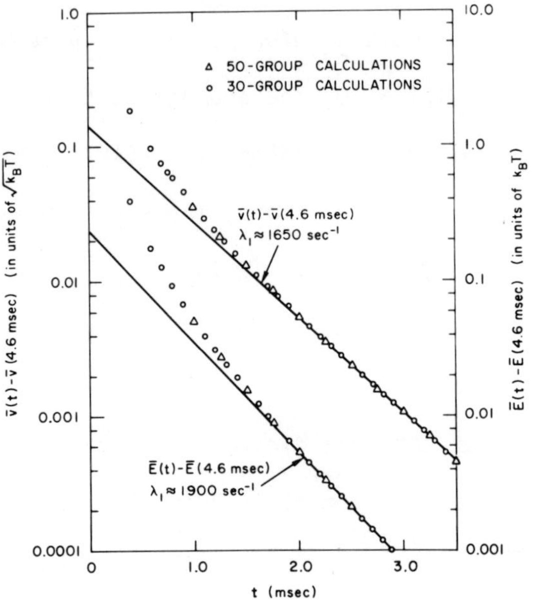

Figure 8-10 The approach to equilibrium in an infinite nonabsorbing graphite assembly using the bound-carbon atom model of Sec. 3-4.2. (From Ghatak and Honeck, Nucl. Sci. Eng., Ref. 59.)

$$D_0 = 2.13 \pm 0.02 \times 10^5 \text{ cm}^2 \text{ sec}^{-1}$$

and

$$C = 26 \pm 5 \times 10^5 \text{ cm}^4 \text{ sec}^{-1} \quad .$$

Fitting the region $\underline{B}^2 < 0.0053 \text{ cm}^{-2}$, Davis, de Juren, and Reier[75] find

$$D_0 = 2.187 \pm 0.008 \times 10^5 \text{ cm}^2 \text{ sec}^{-1}$$

and

$$C = 30 \pm 1 \times 10^5 \text{ cm}^4 \text{ sec}^{-1} \quad .$$

Other recent experimental results are summarized in Ref. 37. The above results (all results are given for a graphite density of 1.6 g/cm^3) are in good agreement with the theoretical results computed by Honeck[2] from the thermalization model of Eqs. (3.95) and (3.98):

$$D_0 = 2.178 \times 10^5 \text{ cm}^2 \text{ sec}^{-1}$$

and

$$C = 24.6 \times 10^5 \text{ cm}^4 \text{ sec}^{-1}.$$

For larger values of $\underline{B}^2$, most of the experiments remain in reasonably good agreement with respect to the value of λ_0 as a function of $\underline{B}^2$. The variations in quoted values of

$\underline{D}_0$, $\underline{C}$, and $\underline{F}$ can largely be accounted for by variations in the number of terms retained in the expansion (8.146) and in the range of $\underline{B}^2$ over which the fit is made. Despite this internal consistency, however, many of the measured values of λ_0 exceed λ^*, and thus are not true fundamental-mode decay constants. The actual value of λ^* in graphite is somewhat uncertain. The theoretical value (using the Parks model; Sec. 3-4.2) in the incoherent approximation is 1000 sec^{-1}, whereas the experimental results in BNL-325[76] give a value of 2600 sec^{-1}. If the former value were correct, there would be no fundamental mode for $\underline{B}^2 \gtrsim 0.005$ cm^{-2}. The 2600 sec^{-1} value is more likely correct and gives a maximum buckling of approximately 0.015 cm^{-2}.

The departure from a true fundamental-mode behavior at long times and large bucklings has not been clearly demonstrated experimentally in graphite, but further calculations by Ghatak and Honeck[77] illustrate the effect to be expected. They carried out multigroup diffusion-theory calculations of the time-dependent spectrum $\phi(\underline{E},\underline{t},\underline{B}^2)$. The time-dependent spectrum for $\underline{B}^2 = 0.00959$ cm^{-2} is shown in Fig. 8.11. In this calculation λ^* is 1000 sec^{-1}, so this example is considerably beyond the critical buckling for the existence of a fundamental mode. In a multigroup calculation there is always a fundamental mode, but the results of Fig. 8.11

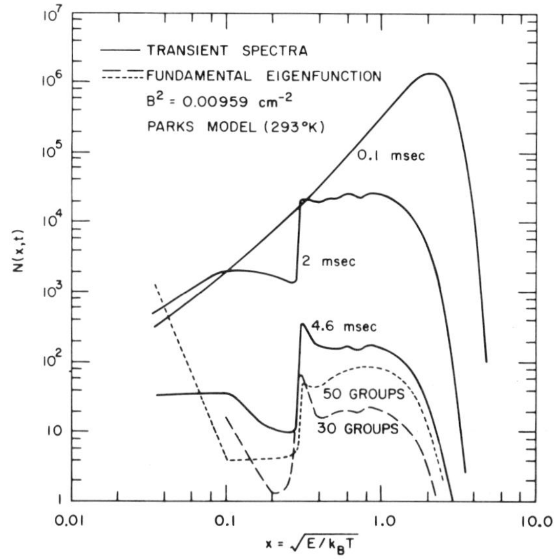

Figure 8.11 Calculated time-dependent spectra in a small graphite assembly. (From Ghatak and Honeck, J. Nuclear Energy, Pt. A & B, Ref. 77.)

show that the fundamental-mode spectrum is unstable against an increase in the number of energy groups. It is suggestive that in the limit of continuous energy variation there would be no asymptotic spectrum for this value of $\underline{B}^2$. (For sufficiently small $\underline{B}^2$, the same calculational method gives a fundamental mode that does not depend on the number of energy groups.)

8-9.5 Beryllium. We complete our discussion of time-dependent experiments by summarizing the recent experiments on small beryllium assemblies that confirm the general theoretical ideas we have discussed. We omit any discussion of thermalization times or diffusion parameters in this moderator.

The value of λ^* for beryllium at 20°C is 3800 sec^{-1}. The early experimental results indicated a fundamental-mode decay constant extending up to 8000 sec^{-1} for a buckling of 0.07 cm^{-2}. This is still quite a large assembly, and therefore it seems reasonable to characterize it by a buckling. More recently, the decay of the neutron density has been measured for $\underline{B}^2 = 0.073\ cm^{-2}$ by Fullwood, Slovacek, and Gaerttner.[78] They were able to extend their measurements to 1.2 msec, as shown in Fig. 8.12. Although the decay was very nearly exponential from 300 to 700 μsec

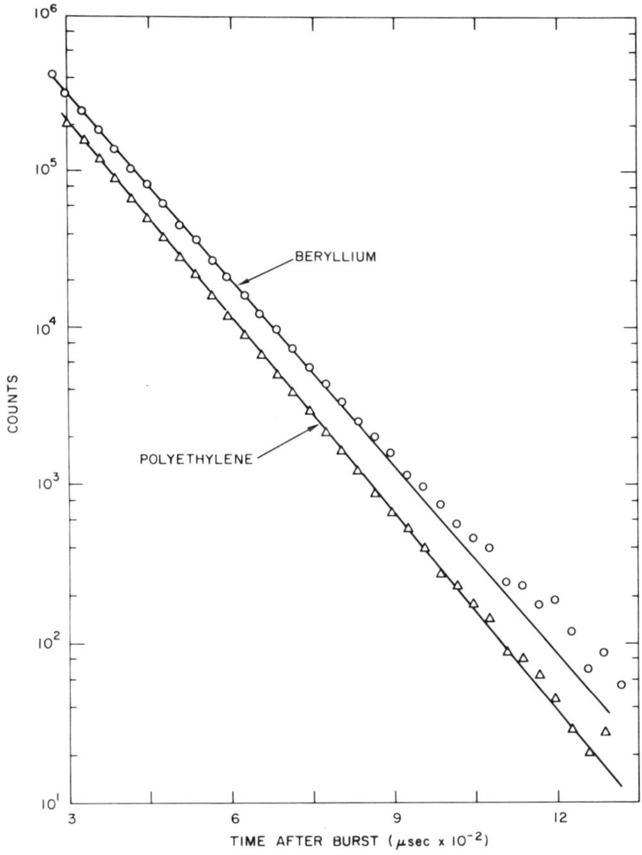

Figure 8-12 Neutron die-away in a beryllium assembly with $B^2 = 0.073\ cm^{-2}$ and in a test assembly of polyethylene selected to give approximately the same decay time. (From Fullwood et al., Nucl. Sci. Eng., Ref. 78.)

(with a decay constant of nearly 10,000 sec^{-1}), there was clear evidence of breakaway to a slower decay beyond 700 μsec. For comparison, the results for a polyethylene assembly with a similar decay constant are shown to illustrate that the nonexponential behavior at long times is not due to background effects.

One would expect that the time-dependent spectrum in such an assembly would not reach an asymptotic state but would instead become richer in very slow neutrons as time increased. This is confirmed by the recent experiments of the RPI group.[79] The time-dependent spectrum in a beryllium assembly with $\underline{B}^2 = 0.075\ cm^{-2}$ is shown in Fig. 8.13. From 277 μsec to 477 μsec the spectrum changes rather little, but at later times neutrons of energies near 0.007 eV appear to be "trapped" compared with the leakage rate of the remainder of the distribution. These neutrons are in an energy region of very large cross section, just above the Bragg cutoff. The trapping effect for these neutrons was first predicted by deSaussure[80] and is confirmed by these experiments. The trapping effect is much more pronounced for a larger value of $\underline{B}^2$, as in seen for $\underline{B}^2 = 0.12\ cm^{-2}$ in Fig. 8.14. By contrast, the spectrum for $\underline{B}^2 = 0.013\ cm^{-2}$ smoothly approaches a true fundamental mode, as shown in Fig. 8.15.

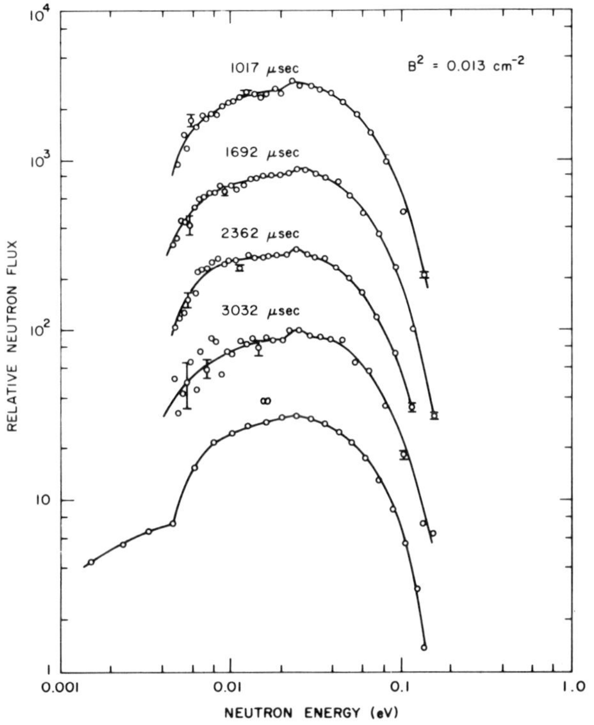

Figure 8-15 Time-dependent spectra in beryllium for $\underline{B}^2 = 0.013\ cm^{-2}$. The points on the curve labelled ∞ are theoretical results determined in Ref. 79 for the asymptotic spectrum. (From Gaerttner *et al.*, *Pulsed Neutron Research*, Ref. 79.)

In our earlier theoretical discussion we indicated that it was neutrons of very low energy which would finally be left behind after the decay of most of the energy distribution. This is still expected to be true, but will be very hard to observe. The "trapped" neutrons with energies of about 0.007 eV will also decay at a rate greater than λ^* and therefore the slower decay observed at the longest times shown in Fig. 8.12 is still more rapid than the asymptotic decay for this assembly. Whether the true asymptotic decay for this assembly will be exponential with $\lambda_0 < \lambda^*$ is not known theoretically and will be prohibitively difficult to determine experimentally in the foreseeable future. This specific point is not as important, however, as is the demonstration that the observed decay constants greater than λ^* are not true eigenvalues of the transport equation, in confirmation of theoretical predictions based on very general considerations.

8-9.6 Summary. Our survey of pulsed-neutron experiments in moderators has pointed out two general features. In those problems where the transport theory is sufficiently simple to provide a test of our models for thermal-neutron inelastic scattering in particular moderators, the models have proven adequate. Many of the experiments, however, have been extended to sufficiently small systems that they have become primarily tests of our ability to accurately solve the transport equation. In the case of crystalline moderators, the special effects of coherent scattering have led to a reexamination of the mathematical structure of the transport equation and to an emphasis on the presence of continuum as well as discrete-time decay eigenvalues. The general mathematical considerations have been given semi-quantitative support by experiment, but a quantitative comparison of theory and experiment for most of the measured time-dependent phenomena is conspicuous by its absence.

Chapter 8

REFERENCES

1. K. M. Case, F. de Hoffman, and G. Placzek, Introduction to the Theory of Neutron Diffusion, Vol. I, Superintendent of Documents, U.S. Government Printing Office, Washington 25, D. C., 1953.
2. H. C. Honeck, On the Calculation of Thermal Neutron Diffusion Parameters, in Proceedings of the Brookhaven Conference on Neutron Thermalization, Vol. IV, USAEC Report BNL-719, p. 1186, Brookhaven National Laboratory, 1962.
3. E. Starr and J. Koppel, Determination of Diffusion Hardening in Water, Nucl. Sci. Eng., 14: 224(1962).
4. W. M. Lopez and J. R. Beyster, Measurement of Neutron Diffusion Parameters in Water by the Pulsed Neutron Method, Nucl. Sci. Eng., 12: 190(1962).
5. K. S. Rockey and W. Skolnik, Measurements on the Diffusion Length of Thermal Neutrons in Water from 25° to 296°C, Nucl. Sci. Eng., 8: 62(1960).
6. M. Nelkin, Neutron Thermalization a la Mode, in Proceedings of the Brookhaven Conference on Neutron Thermalization, Vol. IV, USAEC Report BNL-719, p. RF-1, Brookhaven National Laboratory, 1962.

7. N. Corngold, Some Transient Phenomena in Thermalization, I. Theory, Nucl. Sci. Eng., 19: 80(1964).

8. N. Corngold and P. Michael, Some Transient Phenomena in Thermalization, II. Implications for Experiment, Nucl. Sci. Eng., 19: 91(1964).

9. A. Leonard, On Transient Phenomena in Thermalization, Nucl. Sci. Eng., 22: 489(1965).

10. N. Corngold, On the Maximum Absorption Theorem, Nucl. Sci. Eng., 24: 410(1966).

11. J. H. Ferziger and A. Leonard, Energy-Dependent Neutron Transport Theory, I. Constant Cross Sections, Ann, Phys. (N.Y.), 22: 192(1963).

12. Ivan Kuscer and Noel Corngold, Discrete Relaxation Times in Neutron Thermalization, Phys. Rev., 139: A981(1965).

13. A. Leonard and J. H. Ferziger, Energy-Dependent Neutron Transport Theory, II (submitted to Nucl. Sci. Eng. for publication).

14. K. M. Case, Elementary Solutions of the Transport Equation and Their Applications, Ann. Phys. (N.Y.), 9: 1 (1960).

15. P. M. Morse and H. Feshbach, Methods of Theoretical Physics, Vol. I., p. 615, McGraw-Hill Book Company, New York, 1953.

16. M. V. Kazarnovsky, A. V. Stepanov, and F. L. Shapiro, Thermalization and Diffusion of Neutrons in Heavy Media, in Proceedings of the Second United Nations International Conference on the Peaceful Uses of Atomic Energy, Geneva, 1958, Vol. 16, p. 279, United Nations, Geneva, 1958.

17. P. Michael, Thermal Neutron Flux Disbributions in Space and Energy, Nucl. Sci. Eng., 8: 426(1960).

18. D. A. Kottwitz, Thermal Neutron Flux Spectrum in a Medium with a Temperature Discontinuity, Nucl. Sci. Eng., 7: 345(1960).

19. I. Kuščer and M. Ribarič, The Diffusion Coefficient of Thermal Neutrons, Nuovo Cimento, 7: 451(1958).

20. C. L. Pekeris, Solution of The Boltzmann-Hilbert Integral Equation, Proc. Nat. Acad. Sci. U.S., 41: 661 (1955).

21. R. Kladnik and I. Kuščer, Milne's Problem for Thermal Neutrons in a Nonabsorbing Medium, Nucl. Sci. Eng., 11: 116(1961).

22. I. Kuščer, Thermal Diffusion of Neutrons, J. Nucl. Energy, Pt. A & B, 17: 49(1963).

23. D. S. Selengut, Thermal Neutron Flux in a Cell with Temperature Discontinuities, Nucl. Sci. Eng., 9: 94(1961).

24. R. A. Bennett, Neutron Rethermalization in Graphite and Water, Nucl. Sci. Eng., 17: 131(1963).

25. C. W. Lindenmeier, The Modified Heavy Gas Model and Two Thermal Group Rethermalization Theory, in Proceedings of the Brookhaven Conference on Neutron Thermalization, Vol. IV, USAEC Report BNL-719, p. 1272, Brookhaven National Laboratory, 1962.

26. R. M. Pearce and J. M. Kennedy, Thermal Neutron Spectrum in a Moderator with a Temperature Discontinuity, Nucl. Sci. Eng., 19: 102(1964).

27. P. Morse and H. Feshbach, Methods of Theoretical Physics, Vol. II, p. 1117, McGraw-Hill Book Company, New York 1953.

28. M. Nelkin, The Diffusion Cooling of Neutrons in a Finite Moderator, J. Nucl. Energy, 8: 48(1958).

29. G. F. vonDardel, The Interaction of Neutrons with Matter Studied with a Pulsed Neutron Source, Trans. Roy. Inst. Technol., Stockholm, No. 75(1954).

30. C. S. Shapiro, Time Eigenvalues and Degenerate Kernels in Neutron Thermalization, USAEC Report BNL-8433, Brookhaven National Laboratory, August 1964.

31. S. N. Purohit, Neutron Thermalization and Diffusion in Pulsed Media, Nucl. Sci. Eng., 9: 157(1961).

32. B. Baily du Bois, J. Horowitz, and C. Maurette, Thermalisation des Neutrons, French Report SPM-525, Commissariat a l'Energie Atomique, Paris, December 11, 1958.

33. H. Takahashi, Space and Time Dependent Eigenvalue Problem in Neutron Thermalization, in Proceedings of the Brookhaven Conference on Neutron Thermalization, Vol. IV, USAEC Report BNL-719, p. 1299, Brookhaven National Laboratory, 1962.

34. C. S. Shapiro and N. Corngold, Velocity Polynomials in Neutron Thermalization, Trans. Am. Nucl. Soc., 6: 26 (1963).

35. C. S. Shapiro and N. Corngold, Approach to Equilibrium of a Neutron Gas, Phys. Rev., 137: A1686(1965).

36. P. M. Morse and H. Feshbach, Methods of Theoretical Physics, Vol. I, p. 419ff, McGraw-Hill Book Company, New York, 1953.

37. K. H. Beckurts, A Review of Pulsed Neutron Experiments on Non-Multiplying Media, in Pulsed Neutron Research, Proceedings of the Symposium on Pulsed Neutron Research held at Karlsruhe, 10-14 May 1965, Vol. I, p. 3, International Atomic Energy Agency, Vienna, 1965.

38. F. Riesz and B. Sz. - Nagy, Functional Analysis, trans. L. F. Boron, Ungar, New York, 1955.

39. N. Corngold, P. Michael, and W. Wollman, The Time Decay Constants in Neutron Thermalization, Nucl. Sci. Eng., 15: 13(1963).

40. J. U. Koppel, A Method of Solving the Time Dependent Neutron Thermalization Problem, Nucl. Sci. Eng., 16: 101(1963).

41. M. J. Ohanian and P. B. Daitch, Eigenfunction Analysis of Thermal-Neutron Spectra, Nucl. Sci. Eng., 19: 343 (1964).

42. R. C. Desai, Cornell University, private communication.

43. A. K. Ghatak, On the Calculation of Time Decay Constants for Higher Angular Modes, private communication to Noel Corngold, September 1, 1964.

44. M. Nelkin, The Decay of a Thermalized Neutron Pulse, Nucl. Sci. Eng., 7: 310(1960).

45. R. B. Perez and R. E. Uhrig, Propagation of Neutron Waves in Moderating Media, Nucl. Sci. Eng., 17: 90(1963).

46. R. B. Perez and R. S. Booth, Excitation of Neutron Waves by Modulated and Pulsed Sources, in Pulsed Neutron Research, Proceedings of the Symposium on Pulsed Neutron Research held at Karlsruhe, 10-14 May 1965, Vol. II, p. 701, International Atomic Energy Agency, Vienna, 1965.

47. E. M. Gelbard and J. A. Davis, The Behavior of Extrapolation Distances in Die-Away Experiments, Nucl. Sci. Eng., 13: 237(1962).

48. J. Walker, J. B. C. Brown, and J. Wood, Extrapolation Distances for Pulsed Neutron Experiments, in Pulsed Neutron Research, Proceedings of the Symposium on Pulsed Neutron Research held at Karlsruhe, 10-14 May 1965, Vol. I, p. 49, International Atomic Energy Agency, Vienna, 1965.

49. S. Albertoni and B. Montagnini, On Some Spectral Properties of the Transport Equation and their Relevance to the Theory of Pulsed Neutron Experiments, in Pulsed Neutron Research, Proceedings of the Symposium on Pulsed Neutron Research held at Karlsruhe, 10-14 May 1965.

50. J. Lehner and G. M. Wing, On the Spectrum of an Unsymmetric Operator Arising in the Transport Theory of Neutrons, Commun. Pure Appl. Math., 8: 217(1955).

51. J. Lehner and G. M. Wing, Solution of the Linearized BoltzmannTransport Equation for the Slab Geometry, Duke Math. J., 23: 125(1956).

52. R. van Norton, The Spectrum of a Neutron Transport Operator, USAEC Report NYO-9085, New York University, Institute of Mathematical Sciences, June 1, 1960.

53. G. H. Pimbley, Jr., Solution of an Initial Value Problem for the Multi-Velocity Neutron Transport Equation with a Slab Geometry, J. Math. & Mech., 8: 837(1959).

54. K. Jörgens, An Asymptotic Expansion in the Theory of Neutron Transport, Commun. Pure Appl. Math., 11: 219 (1958).

55. M. Nelkin, Asymptotic Solutions of the Transport Equation for Thermal Neutrons, Physica, 29: 261(1963).

56. S. Albertoni and B. Montagnini, On the Spectrum of Neutron Transport Equation in Finite Bodies, J. Math. Anal. Appl., 13: 19(1966).

57. R. Bednarz, Spectrum of the BoltzmannOperator with an Isotropic Thermalization Kernel, in Pulsed Neutron Research, Proceedings of the Symposium on Pulsed Neutron Research held at Karlsruhe, 10-14 May 1965, Vol. I, p. 259, International Atomic Energy Agency, Vienna, 1965.

58. E. Möller and N. G. Sjöstrand, Measurement of the Slowing-down and Thermalization Time of Neutrons in Water, Arkiv Fysik, 27: 501(1965).

59. A. K. Ghatak and H. C. Honeck, On the Feasibility of Measuring Higher Time Decay Constants, Nucl. Sci. Eng., 21: 227(1965).

60. S. N. Purohit, Time Dependent Neutron Thermalization in Liquid Moderators H_2O and D_2O, in Pulsed Neutron Research, Proceedings of the Symposium on Pulsed Neutron Research held at Karlsruhe, 10-14 May, 1965, Vol. I, p. 273, International Atomic Energy Agency, Vienna, 1965.

61. K. -H, Joest and G. Memmert, On the Experimental Region Connecting Diffusion Length and Pulsed Neutron Experiments, in Proceedings of the Third International Conference on the Peaceful Uses of Atomic Energy, Geneva, 1964, Vol. 2, p. 345, United Nations, New York, 1965.

62. E. Arai and M. Küchle, Messung der Diffusions länge in Zeitlich abklingenden Neutronen felderen, Nukleonik, 7: 416(1965).

63. M. Küchle, Measurement of the Temperature Dependence of Neutron Diffusion in Water and Diphenyl by the Pulse Method, Nukleonik, 2: 131(1960).

64. J. U. Koppel and J. A. Young, Neutron Scattering by Water Taking into Account the Anisotropy of Molecular Vibrations, Nucl. Sci. Eng., 19: 412(1964).

65. R. S. Hall, S. A. Scott, and J. Walker, Neutron Diffusion in Small Volumes of Water: Pulsed Source Measurements, Proc. Phys, Soc. (London), 79: 257(1962).

66. E. Gon, L. Lidofsky, and H. Goldstein, Pulsed Source Measurements of Neutron-Diffusion Parameters in Water, Trans, Am. Nucl. Soc., 8: 274(1965).

67. N. C. Paik, A. Travelli, and I. Kaplan, Asymptotic Decay Constants in Small Slabs and Cubes by Spatial-Expansion Method, Trans. Am. Nucl. Soc., 8: 292(1965).

68. J. Kallfelz and W. Reichardt, Measurements of Neutron Spectra in Hydrogenous Moderators, in Pulsed Neutron Research, Proceedings of the Symposium on Pulsed Neutron Research held at Karlsruhe, 10-14 May 1965, Vol. I. p. 545, International Atomic Energy Agency, Vienna, 1965.

68a. S. N. Ismaev, V. I. Mostovoi, I. P. Sadikov, and A. A. Chernisov, "Experimental Study of the Process of Neutron Thermalization in Time in Aqueous Moderators," in Pulsed Neutron Research, Proceedings of the Symposium on Pulsed Neutron Research held at Karlsruhe, 10-14 May 1965, Vol. I. p. 643, International Atomic Energy Agency, Vienna, 1965.

69. J. W. Meadows and J. F. Whalen, Thermalization and Diffusion Parameters of Neutrons in Zirconium Hydride, Nucl. Sci. Eng., 13: 230(1961).

70. M. J. Poole and P. Wydler, Measurements of the Time-Dependent Spectrum in Heavy Water, in Pulsed Neutron Research, Proceedings of the Symposium on Pulsed Neutron Research held at Karlsruhe, 10-14 May 1965, Vol. I, p. 535, International Atomic Energy Agency, Vienna, 1965.

71. R. C. Kryter, G. P. Calame, R. R. Fullwood, and E. R. Gaerttner, Time-Dependent Thermal Neutron Spectra

in D_2O, in Pulsed Neutron Research, Proceedings of the Symposium on Pulsed Neutron Research held at Karlsruhe, 10-14 May 1965, Vol. I. p. 465, International Atomic Energy Agency, Vienna, 1965.

72. G. Kussmaul and H. Meister, Thermal Neutron Diffusion Parameters of Heavy Water, J. Nucl. Energy, Pt. A & B, 17: 411(1963).

73. J. W. Daughtry and A. W. Waltner, The Diffusion Parameters of Heavy Water, in Pulsed Neutron Research, Proceedings of the Symposium on Pulsed Neutron Research held at Karlsruhe, 10-14 May 1965, Vol. I, p. 65, International Atomic Energy Agency, Vienna, 1965.

74. H. Klose, M. Küchle, and W. Reichardt, Pulsed Neutron Measurements on Graphite, in Proceedings of the Brookhaven Conference on Neutron Thermalization, Vol. III, USAEC Report BNL-719, p. 935, Brookhaven National Laboratory, 1962.

75. S. K. Davis, J. A. de Juren, and M. Reier, Pulsed-Decay and Extrapolation-Length Measurements in Graphite, Nucl. Sci. Eng., 23: 74(1965).

76. D. J. Hughes and R. B. Schwartz, Neutron Cross Sections, USAEC Report BNL-325, 2nd ed., Brookhaven National Laboratory, July 1, 1958.

77. A. K. Ghatak and H. C. Honeck, Transient Spectra in Graphite, J. Nuclear Energy, Pt. A & B, 19: 1(1965).

78. R. R. Fullwood, R. E. Slovacek, and E. R. Gaerttner, The Effects of Coherent Scattering on the Thermalization of Neutrons in Beryllium, Nucl. Sci. Eng., 18: 138 (1964).

79. E. R. Gaerttner, P. B. Daitch, R. R. Fullwood, R. R. Lee, and R. E. Slovacek, The Effects of Coherent Scattering on the Thermalization of Neutrons in Beryllium, in Pulsed Neutron Research, Proceedings of the Symposium on Pulsed Neutron Research held at Karlsruhe, 10-14 May 1965, Vol. I, p. 483, International Atomic Energy Agency, Vienna, 1965.

80. G. deSaussure, A Note on the Measurement of Diffusion Parameters by the Pulsed-Neutron Source Technique, Nucl. Sci. Eng., 12: 433(1962.)

Chapter 9

SOME STUDIES IN NONUNIFORM REACTORS

In this chapter we discuss problems in which both the thermalization and migration of neutrons play a significant role. Such problems may be of particular significance near boundaries that separate regions with very different absorbing and moderating properties. These boundaries occur, for example, within the lattice cell of a heterogeneous system and near the core-reflector interface in a reflected system. Near such boundaries the thermal-neutron flux spectrum can vary strongly with distance. The power produced in a fueled region near a predominantly moderating region can be much higher than that produced elsewhere. The power peaking results from a migration into the fueled region of neutrons that are thermalized mostly by collisions in the weakly absorbing moderator region. The effect of these neutrons is felt to a distance of about one thermal diffusion length. Failure to remove the power peak can lead to a high local temperature, accompanied in some cases by the buckling of fuel elements and a severe nonuniformity in the depletion of fuel.

Accurate estimates of the effects of thermalization in heterogeneous systems require accurate solutions of the space-energy problem at thermal energies. There are many approaches to the problem of obtaining solutions. Some of these are based on multigroup methods, others on analytical or semi-analytic solutions of the energy-dependent transport equation.

We do not intend to give here a comprehensive review of the methods for calculating thermal-neutron spectra. (For a survey of many

related to the latter by

$$\zeta_\phi = \frac{\int_{V_m} d\mathbf{r} \int_0^{E_m} dE \int f(\mathbf{r},E,\Omega)\,d\Omega}{\int_{V_F} d\mathbf{r} \int_0^{E_m} dE \int f(\mathbf{r},E,\Omega)\,d\Omega} = \frac{\hat{v}_m}{\hat{v}_F}\,\zeta_n\ , \tag{9.4}$$

with

$$\hat{v}_m = N_m^{-1} \int_{V_m} d\mathbf{r} \int_0^{E_m} dE \int f(\mathbf{r},E,\Omega)\,d\Omega\ , \tag{9.5}$$

and a similar definition for $\hat{v}_F$.

9-1.1 SENSITIVITY TO SCATTERING MODEL. Honeck has performed a series of calculations[a] for the lattices described in Table 9.1. We consider first his calculation of ζ_n and $\hat{v}_m$ in the uranium metal slab lattices, the object being to examine the sensitivity of ζ_n to the scattering model. The dimensions and water to fuel volume ratio W/U for these slab lattices are given in Table 9.2.

Figure 9.1 shows ζ_n and $\hat{v}_m$ as functions of W/U for four models of thermal-neutron scattering by water. The four scattering models considered are: (1) the free-proton model, (2) a free gas of mass 18m (m = neutron mass), (3) the Brown-St. Joan (BSJ) model (see Sec. 3-4.1), and (4) the Nelkin model (Sec. 3-4.1). In all four cases the scattering from oxygen is approximated as that from a free gas of oxygen atoms. Similarly, it is assumed that uranium metal scatters like a free gas of U^{238} atoms.

To understand qualitatively the behavior of the disadvantage factor curves in Fig. 9.1 we make the following observations. The rate of neutron slowing down by the heavy fuel nuclei is negligible in

---

[a] In these calculations 30 groups are used in the range $0 \le E \le 0.785$ eV. In the integration over energy appearing in Eqs. (9.4) and (9.5), however, E_m is taken to be 0.632 eV.

Table 9.1

DESCRIPTION OF THE LATTICES*

Fuel	Enrichment (% U^{235})	Rod Radius (cm)	Cladding Thickness (in.) and Material	Type of Lattice	Range of Water to Fuel Volume Ratio	Fuel Density (g/cm^3)
U Metal Slabs	1.250	(†)	None	Slab	1.0 → 4.1	18.7
U Metal Rods	1.027** 1.143** 1.299**	0.3175 0.4915 0.7620	0.028 Aluminum††	Triangular	1.0 → 4.0	18.9
U Metal Rods	1.027	0.9525	0.030 Aluminum	Triangular	1.3 → 3.8	18.9
UO_2 Rods	3.00	0.6640	0.028 SS	Triangular	1.3 → 4.0	9.3

*From Honeck, Nucl. Sci. Eng., Ref. 21.

†Thickness of slab = 3.310 cm.

**Each enrichment is associated with each of the three radii listed.

††There is a 0.005-in. gap between fuel and cladding.

Table 9.2

DESCRIPTION OF UNCLAD URANIUM METAL SLAB LATTICES*

Thickness of Uranium Metal Slab† (cm)	W/U	Water Channel Thickness (cm)
0.3099	1.0245	0.3175
	1.5369	0.4763
	2.0490	0.6350
	3.0736	0.9525
	4.0981	1.2700

*From Honeck, Nucl. Sci. Eng., Ref. 21.

†The uranium is enriched to contain 1.25% U^{235} and the fuel density is 18.7 g/cm^3.

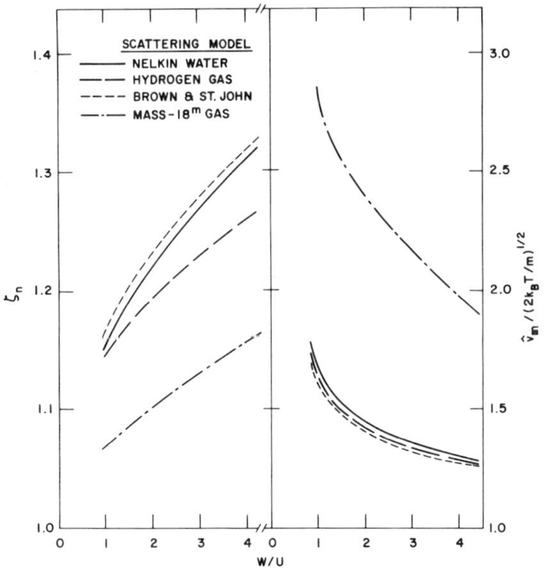

Figure 9.1 Cell disadvantage factors and average neutron speeds in the moderator for the slab lattices using various scattering kernels for water. (From Honeck, Nucl. Sci. Eng., Ref. 21.)

comparison with the rate at which neutrons, on slowing down in the moderator region, diffuse into the fuel region. The larger the average mean free path of neutrons in the moderator, the larger is this rate of diffusion, and the smaller is ζ_n. This behavior is reflected in Fig. 9.1 for $E \gtrsim 0.01$ eV, the cross sections for the BSJ and Nelkin models are about equal, and larger than those for scattering by free protons or by protons bound in a point molecule of mass 18m. The difference in slopes of the disadvantage factor curves for the BSJ, Nelkin, and free-proton models reflects the difference in the energy variation of the scattering cross section. For these models, the mean velocity curves in Fig. 9.1 and their slopes at a given value of W/U are about equal. The nearly equal cross sections $\Sigma_s(E)$ for the BSJ and Nelkin models increase more rapidly with decreasing energy than the scattering cross section for free protons. Thus, as the average spectrum in the moderator softens with increasing values of W/U, the disadvantage factor curves for the BSJ and Nelkin models increase with a greater slope than the disadvantage factor curve for the free-proton model. Similarly, the relatively small and and almost constant $\Sigma_s(E)$ for the mass-18m gas leads to a disadvantage factor that is smaller and varies less rapidly with increasing W/U than the ζ_n corresponding to the other models.

The level of the disadvantage factor curves is influenced by the mean velocity $\hat{v}_m$ as well as by the magnitude of the average scattering cross section. The Nelkin model gives a slightly greater value of $\hat{v}_m$ than the BSJ model; for the greater $\hat{v}_m$ the interior of the uranium rod is less well shielded from the neutrons thermalized in the moderator. The low value of ζ_n for the mass-18m gas is associated with both the relatively low and constant value of $\Sigma_s(E)$ and the relatively high value of $\hat{v}_m$.

Having ascertained the sensitivity of the disadvantage factor to the model for the proton dynamics in H_2O, in the rest of this section we shall present only results that are obtained from the Nelkin model. This model leads to calculated infinite-medium spectra, diffusion parameters, and a scattering cross section which are in good agreement with measured results (see Chaps. 6 and 8).

In assessing the importance of anisotropic scattering and the accuracy of the transport approximation, Honeck[21] has calculated ζ_n using, in turn, the assumptions that

curvature of the indium resonance flux as determined by activating cadmium-covered indium foils at several positions in the lattice. The source feeding the thermal energy region is taken to be independent of position. The good agreement between theory and experiment in Fig. 9.4 suggests that considerable confidence can be placed in the experimental technique, not only when applied to the lattice cell considered here but quite possibly even for cells of much smaller size.

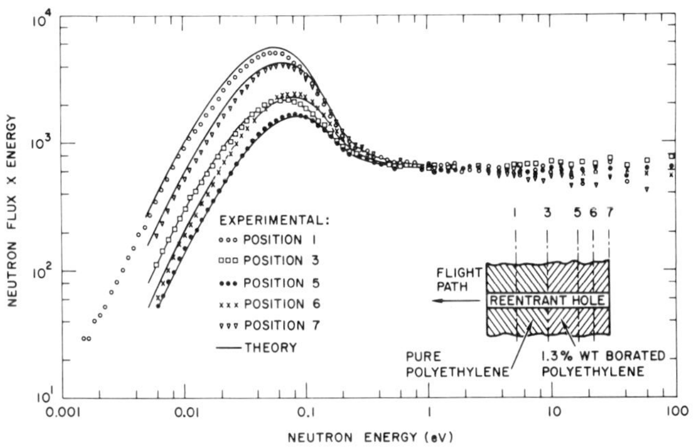

Figure 9.4 Angular flux spectra as a function of positition in a polyethylene cell. (From Beyster et al., USAEC Report GA-6323, Ref. 39.)

Most of the experimental studies of neutron distributions in lattices have been made by means of foil activation techniques. Even for activation rates, which are much less sensitive than the thermal spectrum to chemical binding effects, there are small but significant discrepancies between theory and experiment. Table 9.4 shows that the theoretical values of disadvantage factors are consistently higher than the values measured at Brookhaven (BNL) and, independently, at the Bettis Atomic Power Laboratory (BAPL).[46] The experimental techniques used at BNL have been described by Kouts et al.[47-49] A group of small (~1/16-in.-diameter, 0.007-in.-thick) foils consisting of Dy_2O_3 in either aluminum or plastic are arranged in the lattice. The activity of each foil, which includes epithermal activation, is counted and plotted as a function of position in the cell. From this function, the

Table 9.4

COMPARISON OF ζ FROM DIFFERENT FOIL TYPES AND FROM BNL AND BAPL EXPERIMENTS*

Lattice		Theory†						BNL Experiment		BAPL Experiment** (Ref. 46)	
Rod Radius (cm)	% U^{235}	W/U	ζ_n	ζ_d	ζ_{25}	ζ_n/ζ_d	ζ_{25}/ζ_d	$\zeta_{d(exp)}$	$\frac{\zeta_{d(theory)}}{\zeta_{d(exp)}}$	$\zeta_{25(exp)}$	$\frac{\zeta_{25(theory)}}{\zeta_{25(exp)}}$
0.4915	1.299	2.0	1.278	1.269	1.291	1.007	1.017	1.233	1.029	1.21	1.067
		3.0	1.318	1.310	1.332	1.006	1.017	1.229	1.066	1.31	1.017
0.7620	1.143	2.0	1.460	1.440	1.476	1.014	1.025	1.383	1.041	1.45	1.018
		3.0	1.530	1.517	1.553	1.009	1.024	1.449	1.047	1.45	0.071
0.7620	1.299	1.5	1.444	1.426	1.467	1.013	1.029	1.354	1.053	1.39	1.055
		2.0	1.495	1.477	1.518	1.012	1.028	1.427	1.035	1.45	1.047
		3.0	1.578	1.562	1.603	1.010	1.026	1.482	1.054	1.49	1.076
									1.046††		1.050††
									0.012§		0.022§

*From Honeck, Nucl. Sci. Eng., Ref. 21.

†The ζ_n and ζ_{25} are for a cutoff energy of 0.632 eV. The epithermal activity has been included in ζ_d. ζ_{25} is the disadvantage factor measured with foils containing U^{235}. ζ_d is the disadvantage factor measured with foils containing Dy_2O_3.

**The quoted uncertainty for these values is ±0.04.

††Average.

§Standard deviation.

disadvantage factor can be obtained by numerical integration. The theory predicts dysprosium disadvantage factors that are about 3% to 5% greater than the experimental ones. This systematic discrepancy is exhibited in the very large number of lattices studied.

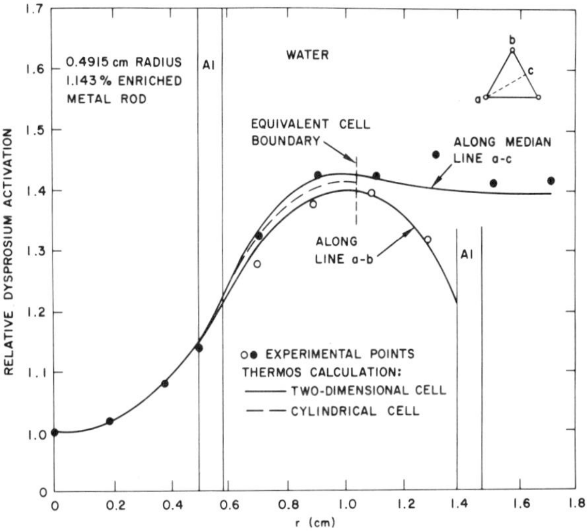

Figure 9.5 Dysprosium-foil activation distribution in a metal rod lattice. (From Honeck, Nucl. Sci. Eng., Ref. 21.)

Honeck has suggested that this discrepancy is due to fission-product contamination of the foils in the fuel. In an effort to resolve this discrepancy, the BNL group has measured ζ_d for a 0.4915 cm, 1.143% U^{235} lattice with W/U = 3.[50] Special precautions were taken to prevent fission-product contamination of the foils in the fuel. From the detailed intracell traverse shown in Fig. 9.5 one obtains a ζ_d of 1.278. This value and the detailed flux traverse compare extremely well with the corresponding theoretical results, thus confirming, but only for this case, that we are able to calculate the neutronics of the lattice accurately from first principles.

9-1.3 ONE-GROUP CALCULATIONS. With high-speed computers and for simple geometries a few calculations of thermal-neutron distributions in a cell can be performed rapidly and are not prohibitively expensive.

For complex geometries and for survey calculations, however, it is desirable to have simpler and more economical methods. (Examples are the method of Amouyal, Benoist, and Horowitz,[22] and the SPECTROX method.[23])

In recent years, hundreds of spectrum calculations have been performed for both homogeneous and lattice geometry. Honeck has found that the results of these numerical experiments can be correlated in terms of a fictitious temperature and a parameter that is a measure of the ratio of the rate of absorption to the rate of thermalization of neutrons.[21] From this correlation, prescriptions can be given for average thermal cross sections that, when used in one-group transport theory, lead to accurate results for disadvantage factors. The correlation that is made is comparable to the correlation expressed by Eq. (7.36) between the energy at which the peak in the thermal spectrum in an infinite homogeneous medium occurs and the absorption cross section $\Sigma_a(k_B T)$.

Just as for Eq. (7.36), the correlation obtained here pertains only to the case in which the absorption cross sections vary nearly as 1/v. When resonance absorbers such as plutonium are present in large concentrations, it may not be possible to provide simple prescriptions that have the wide range of applicability that is required for survey calculations; to the authors' knowledge, a study aimed at providing such prescriptions has not been attempted.

The parameters that Honeck correlates are the effective moderator "temperature," given by

$$kT_{eff} = \frac{m}{2}\left(\frac{\hat{v}_m}{1.128}\right)^2 \tag{9.7}$$

and

$$\theta = \frac{\Sigma^0_{am} N_m V_m + \Sigma^0_{ac} N_c V_c + \Sigma^0_{aF} g_a N_F V_F}{M_{2m} N_m V_m + M_{2c} N_c V_c + M_{2F} N_F V_F} . \tag{9.8}$$

Here the subscripts m, c, and F denote moderator, cladding, and fuel, respectively, Σ^0_a denotes the 2200-m/sec value of the absorption cross section, M_2 is the second moment of energy transfer defined by Eq.

(8.130), g_a is the non-1/v factor in the fuel,[b] and V is the volume of the region. For the moment we assume that the average neutron density N in each region is known. In expressing the correlation between mean neutron energy and the absorbing properties of the medium, θ is the analogue of the quantity $\Sigma_a(k_BT)/\xi\Sigma_s$ that appears in Eq. (7.36), and M_2 is the analogue of $\xi\Sigma_s$. Whereas for a given moderator $\xi\Sigma_s$ is a measure of the slowing-down rate at energies much greater than k_BT and is independent of temperature and binding effects, M_2 is a measure of the thermalization rate and depends strongly on these effects. Values of M_2 for the different thermalization models considered in this section are given in Table 9.5.

Table 9.5

VALUES OF M_2 FOR DIFFERENT MODELS OF H_2O

Kernel	M_2 (cm^{-1})
H gas	3.85
18m gas	0.67
Brown and St. John	5.23
Nelkin	3.34

The correlation between $\tau_m = T_{eff}/T$ and θ is shown in Fig. 9.6 for a large number of calculations with different scattering models in lattices and infinite media. For all geometries and scattering models (except Brown and St. John), the correlation is expressed to within about 3% by

---

[b] The non-1/v factor $g_a(T)$ is defined by[51]

$$g_a(T) = \frac{2}{\sqrt{\pi}}\left(\frac{T}{293.6}\right)^{\frac{1}{2}} \frac{1}{\Sigma_a(v_0)} \int_0^\infty \Sigma_a(E)\, \frac{E}{k_BT}\, [\exp(-E/k_BT)]\, \frac{dE}{k_BT} ,$$

where v_0 = 2200 m/sec. If $\Sigma_a(E) \sim E^{-\frac{1}{2}}$, $g_a(T) = 1$.

$$\tau_m = 1 + 8.10\,\theta\,. \tag{9.9}$$

(The larger discrepancy--7%--for the Brown and St. John model has not been explained.[21]) The value $\theta = 0.16$ corresponds to a quite strong absorption of 8 barns/H atom. The two experimental values of τ_m in Fig. 9.6 were computed from Beyster's measurements of spectra[52] in homogeneous boric acid solutions, and the M_{2_m} used to calculate θ was obtained from the Nelkin model (Sec. 3-4.1). That the experimental points lie below the line in Fig. 9.6 is consistent with the conclusion reached in Sec. 6-4.1 that the Nelkin model slightly overestimates the rate of neutron thermalization in water.[c] If the M_2 derived from the Nelkin model were reduced by 10%, the experimental points in Fig. 9.6 would fall on the line.

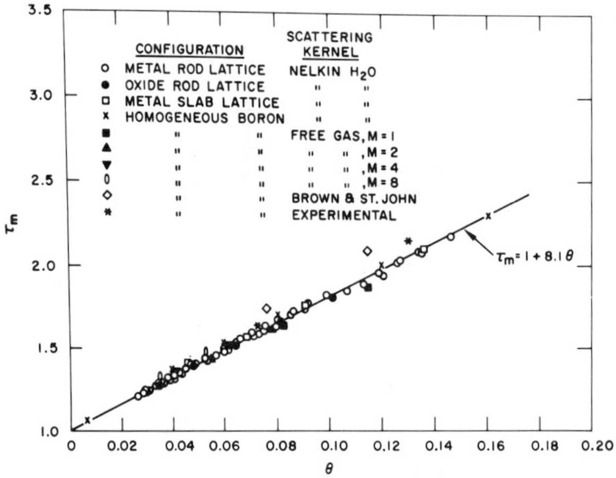

Figure 9.6 Moderator neutron-spectrum hardness parameter τ_m for all lattices and homogeneous medium. (From Honeck, Nucl. Sci. Eng., Ref. 21.)

Provided we know the neutron density in each region, we can calculate the average speed of neutrons in the moderator from Eq. (9.9). From Fig. 9.7 we can obtain the average transport cross section that

[c] This conclusion is opposite to that obtained from the analysis of measured time-dependent reaction rates (see Sec. 8-9.1).

would subsequently be used in a one-group calculation of the cell disadvantage factor. To proceed with such a one-group calculation, however, we must also give prescriptions for obtaining one-group cross sections in the fuel and cladding regions.

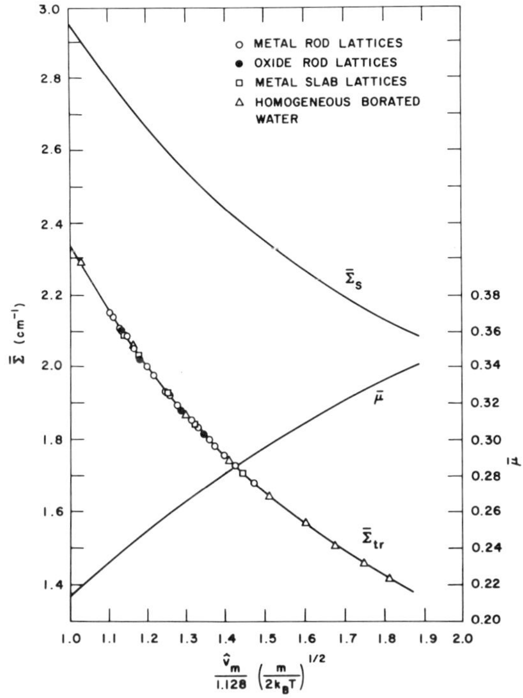

Figure 9.7 The average scattering and transport cross sections, and the average cosine of the scattering angle in H_2O as a function of the average neutron speed. (From Honeck, Nucl. Sci. Eng., Ref. 21.)

Owing to the preferential absorption at low energies of neutrons entering the fuel from the moderator, the average spectrum in the fuel is harder than that in the moderator. This increased hardening can be expressed in terms of $\hat{v}_F - \hat{v}_M$, which in first approximation is proportional to the absorption cross section and radius R of the fuel rod. Thus, for metal rods, Honeck finds

$$\hat{v}_F - \hat{v}_m = (0.366 \pm 0.003)\Sigma_a^0 R \left(\frac{2k_B T}{m}\right)^{\frac{1}{2}} , \qquad (9.10)$$

where Σ_a^0 is the 2200-m/sec value of the absorption cross section to the fuel. The 1% uncertainty corresponds to a slight trend of the multigroup results for $\hat{v}_F - \hat{v}_m$ with W/U. Probably an expression similar to Eq. (9.10) can also be found for uranium oxide rods, but there were not enough multigroup results from which to obtain it.[21] For metal slabs, $\hat{v}_F - \hat{v}_m$ varies strongly with W/U so that an expression as simple as Eq. (9.10) is not valid.

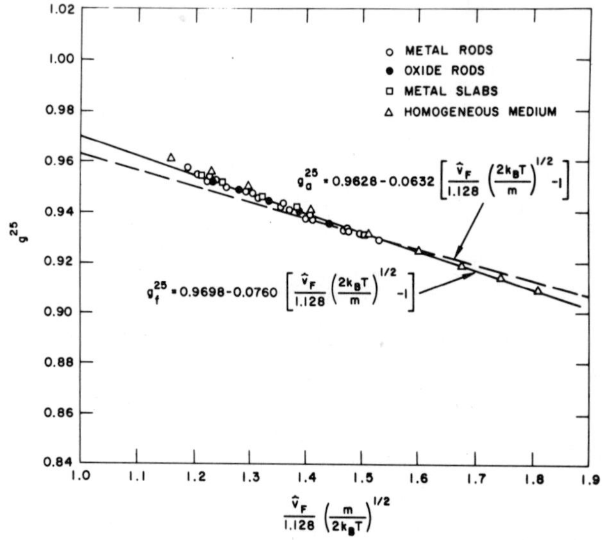

Figure 9.8 The non-1/v factors for absorption and fission, g_a and g_f, respectively, for U^{235} as a function of the average neutron speed in the fuel, $\hat{v}_F$. (From Honeck, Nucl. Sci. Eng., Ref. 21.)

For all cases studied, the expression

$$\hat{v}_c = \tfrac{1}{2}(\hat{v}_m + \hat{v}_F) \qquad (9.11)$$

gives the average neutron speed in the cladding to better than 1%. The remaining parameters needed for one-group calculations are g_a and g_f, the non-1/v absorption and fission factors for U^{235}. These quantities are plotted in Fig. 9.8 as a function of $\hat{v}_F$. Values of η derived from the curves of g_a and g_f are shown in Fig. 9.9. Finally, the scattering cross sections in the fuel and cladding regions are taken to be independent of energy.

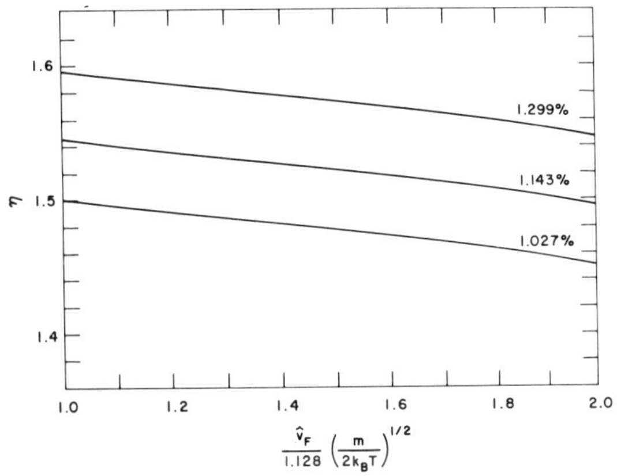

Figure 9.9 The variation of η with average neutron speed in the fuel. (From Honeck, Nucl. Sci. Eng., Ref. 21.)

The prescriptions given for obtaining one-group parameters depend on $\hat{v}_m$, which in turn depends on the average neutron densities in each region. These densities are not known a priori. Nevertheless, they can be estimated with reasonable accuracy from simple considerations, and, using the already given prescriptions, these estimates can be refined by means of a single one-group transport calculation. The improved estimates of the neutron densities in each region then allow us to obtain one-group parameters to the accuracy that is allowed by the prescriptions.

The prescriptions given above were used to obtain average cross sections and speeds which were subsequently used in a number of one-group calculations of disadvantage factors. The condition at the cell boundary corresponded in all cases to isotropic reflection from the boundary. The results are summarized in Table 9.6. The THERMOS one-group and multigroup calculations were done with identical spatial meshes so that a direct comparison of one-group and multigroup results for ζ_n is possible.

Table 9.6

NEUTRON DENSITY DISADVANTAGE FACTORS FOR THE 1.027% ENRICHED RODS*

Rod Radius (cm)	W/U	THERMOS Multi-Group	THERMOS One Group	S_4 One Group	P_3 One Group	ABH One Group
0.3175	1.0	1.127	1.126	1.118	1.070	1.125
	1.5	1.136	1.134	1.128	1.085	1.139
	2.0	1.144	1.140	1.136	1.098	1.150
	3.0	1.157	1.153	1.150	1.117	1.167
	4.0	1.171	1.166	1.162	1.132	1.181
0.4915	1.0	1.198	1.194	1.185	1.138	1.194
	1.5	1.220	1.214	1.207	1.170	1.222
	2.0	1.239	1.231	1.227	1.195	1.245
	3.0	1.272	1.262	1.259	1.233	1.282
	4.0	1.301	1.289	1.285	1.261	1.309
0.7620	1.0	1.333	1.322	1.310	1.275	1.320
	1.5	1.382	1.367	1.361	1.336	1.380
	2.0	1.424	1.406	1.404	1.384	1.426
	3.0	1.494	1.473	1.475	1.457	1.504
	4.0	1.550	1.529	1.534	1.515	1.566
0.9525	1.334	1.502	1.469	1.462	1.442	1.469
	1.584	1.537	1.501	1.497	1.479	1.516
	1.834	1.570	1.530	1.529	1.512	1.551
	2.334	1.626	1.583	1.587	1.570	1.613
	2.834	1.675	1.632	1.638	1.621	1.668
	3.834	1.759	1.714	1.725	1.707	1.762

*From Honeck, Nucl. Sci. Eng., Ref. 21.

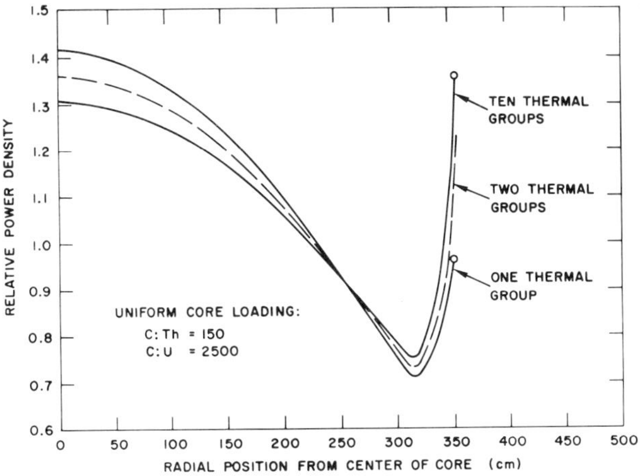

Figure 9.13 Power distribution in a high-temperature graphite reactor. (From Jaye, in USAEC Report TID-7650, Ref. 55.)

generated by thermal neutrons[d] are compared in Fig. 9.13. The curves are normalized to give the same total power

$$P = \int_{V_c} P(\mathbf{r})d\mathbf{r} \ , \tag{9.13}$$

the integral extending over the volume V_c of the core. The ratio of the power density at the center of the core to that at the core-reflector interface ranges from about 1.41 for one thermal group to about 0.93 for ten groups, the latter value being an essentially exact result.

Thus, in the multigroup scheme requirements for accurate estimates of the power peaking at the interfaces would force us to carry out

---

[d] The group boundaries are at 1.45, 1.0, 0.6, 0.4, 0.25, 0.15, 0.10, 0.07, 0.04, 0.025, and 0.00 eV for the ten-group calculations; 1.45, 0.10, and 0.00 eV for the two-group calculations; and 1.45 and 0.00 eV for the one-group calculations. Group constants in the core (reflector) are obtained by averaging over the spectrum in an infinite medium having a composition identical to that in the core (reflector). The group transfer cross sections, Eq. (7.33), include the interatomic binding effect.

calculations with many thermal groups. The excessive cost of large numbers of such calculations motivates a search for cheaper but comparably accurate methods of computation.

By studying the power distribution in several systems as functions of the composition of the core and of the temperatures of the core and reflector, we could undoubtedly construct prescriptions, at least for nearly 1/v absorption cross section, for obtaining group constants that could be employed in a one-group theory to yield accurate power distributions. Such a construction would be analogous to the one described in the previous section to provide a simple method for performing cell calculations, with the difference that here the group cross sections would necessarily be functions of position.

There are better methods for the systematic reduction of a clumsy multigroup theory to one having a simple mathematical structure. The systematic reduction of the energy-dependent transport or diffusive equations to simpler theories leads to the method of overlapping groups, a method that has been applied extensively in reactor design problems. The reduction can be carried out by means of variational methods or more directly by the method of moments[56] or, as it is sometimes called, the method of weighted residuals.

Before describing these methods we first study a relatively simple case that illustrates much of the physical content of the spectrum problem encountered in a reflected reactor. We consider two semi-infinite regions contiguous at the plane $z = 0$. Each region is uniform and moderated by the same material, but the macroscopic absorption cross section and temperature in the half space $z > 0$ may differ from those in the half space $z < 0$. We assume that the rate at which neutrons slow down from epithermal to thermal energies is the same everywhere.

We formulate the problem in energy-dependent diffusion theory. The flux $\phi(E,\mathbf{r})$ satisfies Eq. (5.163) but with a $Q(r) = Q$ that is the same everywhere and a D, Σ_a, and scattering operator R_0 (see Eq. (4.57)) that are discontinuous at $z = 0$.

In Chap. 8 we mentioned the inconsistency of applying diffusion theory to describe the spectrum within a few mean free paths of an interface between dissimilar media. We cannot completely justify diffusion theory except insofar as the spatial transients induced by the discontinuities in material properties and temperature are characterized by relaxation lengths that are long compared with a mean free path. To the extent that this is true, the leading effect of thermalization in transport theory is also present in diffusion theory, and the latter allows a simple description of the effects that dominate the spatial dependence of the spectrum. In general, however, we can gauge the accuracy of such a description only by comparing it with experiment or with a more exact theory.

Returning to the problem of two semi-infinite media, we use an obvious notation and write the solution of Eq. (5.163) as

$$\left.\begin{aligned} \phi(z,E) &= \phi_+^{(0)}(E) + \phi_+^{(t)}(z,E) \qquad z > 0 \\ &= \phi_-^{(0)}(E) + \phi_-^{(t)}(z,E) \qquad z < 0 . \end{aligned}\right\} \tag{9.14}$$

The particular solution $\phi_+^{(0)}(E)$ satisfies the inhomogeneous space-independent equations

$$\Sigma_{a+}(E)\phi_+^{(0)}(E) = R_{0+}(E)\phi_+^{(0)}(E) + f(E)Q , \tag{9.15}$$

whereas the transient spectra $\phi_+^{(t)}(z,E)$ are solutions of the homogeneous equations

$$-D_+(E)\frac{\partial^2\phi_+^{(t)}}{\partial z}(z,E) + \Sigma_{a+}(E)\phi_+^{(t)}(z,E) = R_{0+}\phi_+^{(t)}(z,E) . \tag{9.16}$$

Since similar equations hold for $z < 0$, for the sake of simplicity in notation we drop the subscript + and write equations only for $z > 0$.

We can write $\phi^{(t)}(E)$ as

$$\phi^{(t)}(E) = \sum_{n=0}^{\infty} a_n\phi_n(E)\exp(-z/L_n) , \tag{9.17}$$

where the $\phi_n(E)$, which we assume form a complete set, are solutions of the eigenvalue equations

$$-\frac{D(E)}{L_n^2}\phi_n(E) + \Sigma_a(E)\phi_n(E) = R_0\phi_n(E) . \tag{9.18}$$

The characteristic lengths L_n are determined from

$$L_n^2 = \frac{\int_0^\infty D(E)\phi_n(E)dE}{\int_0^\infty \Sigma_a(E)\phi_n(E)dE} . \tag{9.19}$$

This eigenvalue problem has been discussed in Sec. 4-7 and, more exactly, in the context of transport theory in Chap. 8. There, particular emphasis is given to the largest characteristic length L_0. Except for a rather high degree of poisoning, this length is considerably greater than the L_n for $n \geq 1$. (See Ref. 57: Calame has calculated L_0, L_1, and L_2 in poisoned H_2O using Eq. (9.18) and a free proton model for scattering. His results compare favorably with diffusion lengths measured by Wright et al.[58-61])

The modal functions $\phi_n(E)$ of Eq. (9.18) are orthonormal in the same sense as the $\phi_j(E)$ in Eq. (8.49). The function $\phi_0(E)$ is of particular interest. It determines the relaxation of the flux in an assembly at large distances from a surface adjacent to a thermal column. This function also dominates the transient behavior of the spectrum beyond a few mean free paths (more precisely, for $z \gtrsim L_1$) from an interface. It is positive at all energies and does not have an E^{-1} tail at high energies. Rather, it is a diffusion-hardened spectrum (see Sec. 4-7) that behaves qualitatively like a Maxwellian spectrum. The higher functions $\phi_n(E)$, with $n \geq 1$, which also fall off rapidly at high energies, are functions of oscillating sign. Because of this, and because the coefficient a_n in general decreases with increasing n, the contribution to neutron reaction rates and total flux from terms with $n \geq 1$ is often much less than the contribution from the leading terms, even at positions considerably less than one diffusion length from an interface. In general, however, the transient flux as measured by foil activations relaxes with a simple exponential dependence only if $z \gg L_1$ of if the contribution from the modes with $n \geq 1$ are inherently small. We shall consider two cases where the latter condition prevails.

The current $J = -D(E)(\partial/\partial z)\phi(z,E)$ for the present problem does not exhibit a $D(E)/E$ behavior at high energies; instead, it is well-thermalized compared with the infinite medium spectrum $\phi^{(0)}(E)$. This circumstance is a consequence of our assumption of a spatially

and

$$S_m(\mathbf{r}) = \int_0^{E_c} S(\mathbf{r},E)W_m(E)dE \ . \qquad (9.30)$$

Let us consider a one-dimensional system with M regions, in each of which the cross sections and temperature are independent of position. We write a trial function of the form (9.25) for each region, containing N unknown functions of the position alone, and N functions $\Psi_n(E)$, where $n = 1, \ldots N$, which are known functions of energy alone and do not vary from region to region. The method of weighted residuals gives a system of NM equations, N for each region. Since each equation is of second order in the position coordinate, we require 2NM conditions in order to obtain a unique solution. Equation (9.27) is valid everywhere and in particular at an interface. Since $\Sigma_{m,n}(z)$ and $S_m(z)$ are finite, so is

$$\sum_{n=1}^{N} \frac{\partial}{\partial z}\left[D_{m,n}(z)\frac{\partial}{\partial z}\chi_n(z)\right] \qquad 1 \le m \le N \ . \qquad (9.31)$$

This implies the continuity of

$$\chi_m(z) \qquad 1 \le m \le N \qquad (9.32)$$

and of

$$\sum_{n=1}^{N} D_{m,n}(z)\frac{\partial}{\partial z}\chi_n(z) \qquad 1 \le m \le N \ . \qquad (9.33)$$

Using these conditions to match the solutions at the (M-1) interfaces gives 2(M-1)N of the 2NM conditions required for a unique solution. The remaining 2N conditions are symmetry conditions and/or conditions at external boundaries. For example, in plane parallel geometry, with a symmetry plane at z = 0 and an extrapolated vacuum boundary at z = a,

$$\left.\begin{aligned} \chi_m(a) &= 0 \\ \left.\frac{d\chi_m}{dz}\right|_{z=0} &= 0 \end{aligned}\right\} \quad 1 \le m \le N \ . \qquad (9.34)$$

It is instructive to compare the method of weighted residuals with the variational method. We introduce an adjoint source $S^{\dagger}(\mathbf{r},E)$ and an adjoint function $\phi^{\dagger}(\mathbf{r},E)$ that is a solution of

$$H^{\dagger}\phi^{\dagger}(\mathbf{r},E) = S^{\dagger}(\mathbf{r},E) \ , \tag{9.35}$$

where $H^{\dagger}$ is the operator adjoint to H. For the diffusion equation it is given by

$$H^{\dagger}\phi^{\dagger}(\mathbf{r},E) = -\nabla \cdot D(\mathbf{r},E)\nabla\phi^{\dagger}(\mathbf{r},E) + \Sigma(\mathbf{r},E)\phi^{\dagger}(\mathbf{r},E) - \int_0^{E_c} \Sigma_0(\mathbf{r},E,E')\phi^{\dagger}(\mathbf{r},E')dE' \ . \tag{9.36}$$

Usually, we seek solutions $\phi(\mathbf{r},E)$ and $\phi^{\dagger}(\mathbf{r},E)$ that satisfy either

$$\left.\begin{aligned} \phi(\mathbf{r},E) &= 0 \\ \phi^{\dagger}(\mathbf{r},E) &= 0 \end{aligned}\right\} \tag{9.37}$$

or

$$\left.\begin{aligned} \mathbf{n} \cdot \nabla\phi(\mathbf{r},E) &= 0 \\ \mathbf{n} \cdot \nabla\phi^{\dagger}(\mathbf{r},E) &= 0 \ , \end{aligned}\right\} \tag{9.38}$$

for $\mathbf{r}$ on the surface $\mathcal{S}$ bounding the system, where $\mathbf{n}$ is a unit vector pointing along the outward to $\mathcal{S}$.

When Eq. (9.37) or (9.38) is satisfied,

$$\int_V d\mathbf{r} \int_0^{E_c} \phi^{\dagger}H\phi dE = \int_V d\mathbf{r} \int_0^{E_c} \phi H^{\dagger}\phi^{\dagger} dE \ , \tag{9.39}$$

where V is the volume bounded by $\mathcal{S}$.

We now consider the functional[e]

---

[e] For this form of variational principle, $\tilde{\phi}(\mathbf{r},E)$, $\tilde{\phi}^{\dagger}(\mathbf{r},E)$, $D(\mathbf{r})\nabla\tilde{\phi}(r,E)$, and $D(\mathbf{r})\nabla\tilde{\phi}^{\dagger}(\mathbf{r},E)$ must be continuous functions of $\mathbf{r}$ throughout the volume V. It is possible to construct variational principles that admit trial functions that are discontinuous, but we do not consider such principles here.

$$F[\tilde{\phi}(\mathbf{r},E),\ \tilde{\phi}^{\dagger}(\mathbf{r},E)] = \int_V d\mathbf{r} \int [\tilde{\phi}^{\dagger}H\tilde{\phi} - \tilde{\phi}^{\dagger}S - \tilde{\phi}S^{\dagger}]dE\ , \qquad (9.40)$$

and trial functions $\tilde{\phi}(\mathbf{r},E)$ and $\tilde{\phi}^{\dagger}(\mathbf{r},E)$ that satisfy the conditions (9.37) or (9.38). For the exact solutions $\phi^{\dagger}$ and ϕ,

$$F[\phi(\mathbf{r},E),\ \phi^{\dagger}(\mathbf{r},E)] = \int_V d\mathbf{r} \int S^{\dagger}(\mathbf{r},E)\phi(\mathbf{r},E)dE \qquad (9.41)$$

and

$$F[\phi(\mathbf{r},E),\ \phi^{\dagger}(\mathbf{r},E)] = \int_V d\mathbf{r} \int S(\mathbf{r},E)\phi^{\dagger}(\mathbf{r},E)dE\ . \qquad (9.42)$$

For small variations $\delta\phi^{\dagger}$, $\delta\phi$ about the exact solutions,

$$F[\phi + \delta\phi,\ \phi^{\dagger} + \delta\phi^{\dagger}] = \int_V d\mathbf{r} \int S^{\dagger}(\mathbf{r},E)\phi(\mathbf{r},E)dE\ +$$

$$+ \int_V d\mathbf{r} \int \delta\phi^{\dagger}H\delta\phi dE\ . \qquad (9.43)$$

Thus, first-order errors in the trial functions $\tilde{\phi}^{\dagger}$ and $\tilde{\phi}$ give a value of F which differs from that in Eq. (9.41) or (9.42) by terms of second order in the errors in the trial functions; that is, F is stationary with respect to arbitrary variations of $\tilde{\phi}$ and $\tilde{\phi}^{\dagger}$ about the exact solutions of Eqs. (9.23) and (9.35), respectively.

Variational principles derived on the basis of the functional in Eq. (9.40) have been discussed in considerable detail by Selengut.[62] This functional was first given by Roussopolos[65] as a generalization to non-self adjoint operators of a variational principle due to Kahan and Rideau.[66] As discussed by Selengut, Roussopolos' principle includes many frequently used variational principles for linear equations as special cases. We restrict our attention here, however, to variational principles based directly on Eq. (9.40).

We now use the functional (9.40) to systematically reduce the energy-dependent diffusion equation to the system of equations (9.27). We introduce as trial functions

$$\tilde{\phi}(r,E) = \sum_{n=1}^{N} \Psi_n(E)\chi_n(r) \qquad (9.44)$$

and

$$\tilde{\phi}^{\dagger}(\mathbf{r},E) = \sum_{n=1}^{N} \Psi_n^{\dagger}(E)\chi_n^{\dagger}(r) \;, \qquad (9.45)$$

where Ψ_n and $\Psi_n^{\dagger}$ are known functions of E, and $\chi_n(\mathbf{r})$ and $\chi_n^{\dagger}(\mathbf{r})$ are to be determined, subject to the same external boundary conditions as ϕ and $\phi^{\dagger}$. Inserting these trial functions into Eq. (9.40) and demanding that F be stationary with respect to small independent variations of all $\chi_n^{\dagger}$, for n = 1, ... N, subject to fixed boundary conditions, we obtain Eq. (9.27) with $D_{m,n}(\mathbf{r})$, $\Sigma_{m,n}(\mathbf{r})$ and $S_m(\mathbf{r})$ defined by Eqs. (9.28), (9.29), and (9.30), respectively, but with the weight function $W_m(E)$ replaced by $\Psi_m^{\dagger}(E)$. The value of the functional reduces to

$$F[\tilde{\phi}(\mathbf{r},E),\, \tilde{\phi}^{\dagger}(\mathbf{r},E)] = \int dr \int S^{\dagger}(r,E)dE \sum_{n=1}^{N} \Psi_n(E)\chi_n(r) \;, \qquad (9.46)$$

and this differs from the stationary value of F, Eq. (9.41), by terms that are of second order in the error in the trial function.

The accuracy that we can achieve in the calculation of any quantity, such as the power distribution, for a given class of trial functions for $\phi(\mathbf{r},E)$ depends on the choice of weight functions $W_m(E)$. If we wish to calculate to a high degree of accuracy the average over energy and volume of a particular quantity $S^{\dagger}(\mathbf{r},E)$, the best choice for $W_m(E)$ can be identified with $\tilde{\phi}_m^{\dagger}(E)$, provided we restrict ourselves to the class of adjoint trial functions given by Eq. (9.45). The $\tilde{\phi}_m^{\dagger}(E)$ are at our disposal, but they should be chosen in such a way that Eq. (9.45) allows an accurate approximation to the true adjoint flux $\phi^{\dagger}(\mathbf{r},E)$ (cf. Ref. 29).

The variational procedure given above is the generalization of the semidirect Ritz method[67] to non-self adjoint problems. In the generalization of the direct Ritz method not only are the ϕ_n and $\phi_n^{\dagger}$ known, but the χ_n and $\chi_n^{\dagger}$ are known to within multiplicative constants

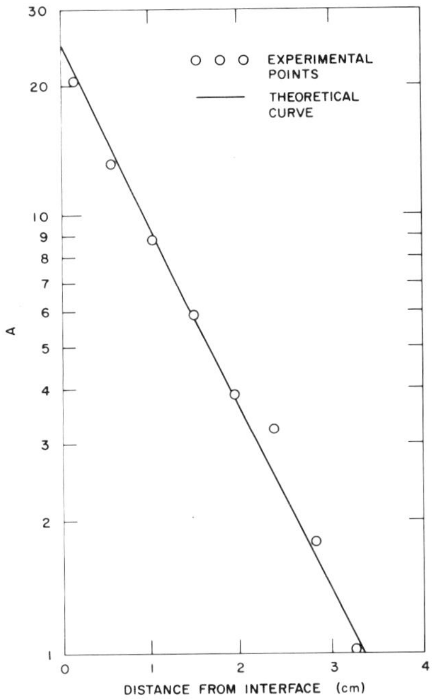

Figure 9.16 Manganese activity minus unperturbed activity near a 1.62-cm polyethylene slab in lattice L-7 (see Table 9.8). (From Calame and Federighi, Nucl. Sci. Eng., Ref. 63.)

$$W_1(E) = \phi_1^\dagger(E) = 1 \qquad (9.49)$$

$$W_2(E) = \phi_2^\dagger(E) = E\ . \qquad (9.50)$$

It is noteworthy that the calculated results are not significantly affected when $W_2(E)$ is set equal to E^2 instead of E. These results, however, are significantly affected when Eq. (9.27) does not contain the condition of neutron conservation. This condition is preserved in the formalism when one of the $W_n(E)$ is equal to unity.

The energy exchange kernel in the calculations is taken to be the free proton kernel, Eq. (5.36). The method of overlapping groups leads to agreement between theory and experiment which is considerably better than that obtained from a standard calculation using one thermal group with thermal-group parameters that are constant within a region.

In our discussion of the method of weighted residuals and of the variational method, we have restricted our attention to trial functions, of the form of Eq. (9.25), in which the energy-dependent parts $\phi_n(E)$ are known. This approach, in principle, is applicable to any space-energy problem. The postulated energy functions are generally taken to be infinite-media spectra characteristic of the various regions present in the system. The method is thus best applied when each region is large in comparison with the diffusion length in that region.[f]

Pomraning[30] has investigated an alternate method which, in principle, is also applicable to any space-energy problem but which should be best suited for small systems. In this method, the spectrum is again represented by a finite sum of separable terms, as in Eq. (9.44), but the spatial dependence is postulated and equations developed within the framework of diffusion theory for the energy-dependent coefficients. He has also applied this method to cell problems within the framework of transport theory using trial functions[30]

$$f(x,E,\mu) = \sum_{n=0}^{N} \Psi_n(E,\mu)\chi_n(z) \ . \tag{9.51}$$

With specified functions $\chi_n(z)$, the method leads to a system of equations that determine $\Psi_n(E,\mu)$.

Two authors[71,72] have shown that a polynomial spatial expansion of the solution of the transport equation leads to good results in slab lattices consisting of small cells. Corngold[73] has used an expansion in polynomials in both position and angular coordinates to study resonance absorption in slab lattices. For a detailed discussion of the methods, we refer the reader to the cited literature.

---

[f] For this reason, in the problem considered above we expect that the overlapping-group method would not give activation traverses in the perturbing gap as accurately as in the lattice region, at least for a trial function of the form of Eq. (9.47).

REFERENCES

1. H. C. Honeck, A Review of the Methods for Calculating Spatially Dependent Thermal Neutron Spectra, *Trans. Am. Nucl. Soc.*, **6**: 16 (1963).

2. G. C. Pomraning, Transport Methods for the Calculation of Spatially Dependent Thermal Spectra, in *Reactor Physics in Thermal and Resonance Regions*, Proceedings of the American Nuclear Society National Topical Meeting held in San Diego, February 7-9, 1966. MIT Press, Cambridge (to be published).

3. D. C. Leslie and M. J. Terry, A Preliminary Description of Thule, in *Heavy Water Lattices: Second Panel Report*, p. 417, Technical Reports Series No. 20, International Atomic Energy Agency, Vienna, 1963.

4. D. W. Hone *et al.*, Natural-Uranium Heavy-Water Lattices, Experiment and Theory, in *Proceedings of the Second United Nations International Conference on the Peaceful Uses of Atomic Energy, Geneva, 1958*, Vol. 12, p. 351, United Nations, Geneva, 1958.

5. Y. Girard *et al.*, Natural Uranium-Heavy Water Lattices, in *Proceedings of the Second United Nations International Conference on the Peaceful Uses of Atomic Energy, Geneva, 1958*, Vol. 12, p. 281, United Nations, Geneva, 1958.

6. A. D. Galanin *et al.*, Critical Experiments on a Heavy-Water Research Reactor, in *Proceedings of the Second United Nations International Conference on the Peaceful Uses of Atomic Energy, Geneva, 1958*, Vol. 12, p. 380, United Nations, Geneva, 1958.

7. B. Pershagen, G. Anderson, and I. Carlwik, Calculations of Lattice Parameters for Uranium Rod Clusters in Heavy Water and Correlation with Experiments, in *Proceedings of the Second United Nations International Conference on the Peaceful Uses of Atomic Energy, Geneva, 1958*, Vol. 12, p. 341, United Nations, Geneva, 1958.

8. G. Dessauer, Physics of Natural Uranium Lattices in Heavy Water, in *Proceedings of the Second United Nations International Conference on the Peaceful Uses of Atomic Energy, Geneva, 1958*, Vol. 12, p. 320, United Nations, Geneva, 1958.

9. D. Hicks, Few-Group Nuclear Design Methods for Heavy Water Reactors, in *Heavy Water Lattices: Second Panel Report*, p. 371, Technical Reports Series No. 20, International Atomic Energy Agency, Vienna, 1963.

10. B. I. Spinrad, Calculations of Lattice Parameters, in *Heavy Water*

Lattices: Second Panel Report, p. 39, Technical Reports Series No. 20, International Atomic Energy Agency, Vienna, 1963.

11. A. Radkowsky (Ed.), *Naval Reactors Physics Handbook Selected Basic Techniques*, U.S. Atomic Energy Commission, Superintendent of Documents, U.S. Government Printing Office, Washington 25, D.C., 1964.

12. D. Hicks, Light Water Lattice Calculations in the United Kingdom, in *Light Water Lattices*, p. 99, Technical Reports Series No. 12, International Atomic Energy Agency, Vienna, 1962.

13. H. C. Honeck, The Calculation of the Thermal Utilization and Disadvantage Factor in Uranium Water Lattices, in *Light Water Lattices*, p. 233, Technical Reports Series No. 12, International Atomic Energy Agency, Vienna, 1962.

14. R. L. Hellens and H. C. Honeck, A Summary and Preliminary Analysis of the BNL Slightly Enriched Uranium, Water Moderated Lattice Measurements, in *Light Water Lattices*, p. 27, Technical Reports Series No. 12, International Atomic Energy Agency, Vienna, 1962.

15. J. Chernick, The Theory of Uranium Water Lattices, in *Proceedings of the First United Nations International Conference on the Peaceful Uses of Atomic Energy*, *Geneva, 1955*, Vol. 5, p. 215, United Nations, New York, 1956.

16. R. L. Hellens and G. A. Price, Reactor Physics Data for Water-moderated Lattices of Slightly Enriched Uranium, in *Reactor Technology*, *Selected Reviews—1964*, p. 529, TID-8540, USAEC Division of Technical Information Extension, Oak Ridge, Tennessee, 1964.

17. H. C. Honeck and J. L. Crandall, The Physics of Heavy-water Lattices, in *Reactor Technology*, *Selected Reviews—1965*, p. 1, TID-8540, USAEC Division of Technical Information Extension, Oak Ridge, Tennessee, January 1966.

18. D. A. Newmarch, A Theoretical Interpretation of the Pressure-Tube, Heavy-Water, Zero-Energy Experiment in DIMPLE, in *Heavy Water Lattices: Second Panel Report*, p. 445, Technical Reports Series No. 20, International Atomic Energy Agency, Vienna, 1963.

19. G. D. Joanou and H. B. Stewart, Modern Techniques Used in Nuclear Design of Reactors, in *Annual Review of Nuclear Science*, E. Segrè (Ed.), Vol. 14, p. 259, Annual Reviews, Inc., Palo Alto, 1964.

20. M. J. Terry, *The Use of Multi-Group Transport Programmes for Thermal Spectrum Calculations on Pressure Tube Reactors*, British Report AEEW-R-186, September 1963.

21. H. C. Honeck, The Calculation of the Thermal Utilization and Disadvantage Factor in Uranium/Water Lattices, *Nucl. Sci. Eng.*, **18**: 49 (1964).

Appendix I

THE KERNEL AND CROSS SECTION FOR NEUTRON SCATTERING BY A MONATOMIC GAS

In this Appendix we shall derive the expressions (2.99) and (2.100), the scattering kernel and cross section, respectively, for the case of neutron scattering by a free gas. To obtain the scattering kernel, one can integrate the right-hand side of (2.98) over solid angles. A different procedure, which we shall follow, consists in integrating $\chi(\kappa, t)$ over solid angles in Eq. (2.88) before performing the t-integration indicated. Hence, we shall write

$$\sigma(E_0, E) = \frac{\sigma_b}{4\pi\hbar}\left(\frac{E}{E_0}\right)^{\frac{1}{2}} \frac{1}{2\pi} \int_{-\infty}^{\infty} dt \exp\left(i\frac{E - E_0}{\hbar} t\right) \int d\Omega\, \chi(\kappa, t)$$

$$= \frac{\sigma_b}{8\pi E_0}\frac{M}{m} \int_{-\infty}^{\infty} dt \frac{\exp\{i[(E - E_0)/\hbar]t\}}{it - [(k_B T/\hbar)t^2} \times$$

$$\times \left\{\exp\left[\frac{m}{M\hbar}\left(E^{\frac{1}{2}} + E_0^{\frac{1}{2}}\right)^2 \left(it - \frac{k_B T}{\hbar} t^2\right)\right] - \exp\left[\frac{m}{M\hbar}\left(E^{\frac{1}{2}} - E_0^{\frac{1}{2}}\right)^2 \left(it - \frac{k_B T}{\hbar} t^2\right)\right]\right\}. \quad \text{(I-1)}$$

Let us consider the case where $E \geq E_0$. By simple transformations of the variable of integration, we can express

$$I_+ = \int_{-\infty}^{\infty} dt \frac{\exp\left(i \frac{E - E_0}{\hbar} t\right)}{it - \frac{k_B T}{\hbar} t^2} \left\{ \exp\left[\frac{m}{M} \frac{\left(E^{\frac{1}{2}} + E_0^{\frac{1}{2}}\right)^2}{\hbar} \left(it - \frac{k_B T}{\hbar} t^2\right)\right] - \right.$$

$$\left. - \exp\left[\frac{m}{M} \frac{\left(E^{\frac{1}{2}} - E_0^{\frac{1}{2}}\right)^2}{\hbar} \left(it - \frac{k_B T}{\hbar} t^2\right)\right]\right\}$$

$$= \frac{1}{i} \left[\int_{-\infty}^{\infty} dz\, e^{-z^2 + 2i(\eta x - \rho x_0)z} \left\{ z^{-1} - \left[z - i(x+x_0)(m/M)^{\frac{1}{2}}\right]^{-1} \right\} \right.$$

$$\left. - \int_{-\infty}^{\infty} dz\, e^{-z^2 + 2i(\eta x + \rho x_0)z} \left\{ z^{-1} - \left[z - i(x-x_0)(m/M)^{\frac{1}{2}}\right]^{-1} \right\} \right]$$

$$= \frac{1}{i} \int_{-\infty}^{\infty} \frac{e^{-z^2}}{z} \left\{ e^{2i(\eta x - \rho x_0)z} - e^{2i(\eta x + \rho x_0)z} - e^{-(x^2 - x_0^2)} \left[e^{2i(\rho x - \eta x_0)z} - e^{2i(\rho x + \eta x_0)z} \right] \right\}, \tag{I-2}$$

where

$$\eta = \frac{M + m}{2(mM)^{\frac{1}{2}}} ,$$

$$\rho = \frac{M - m}{2(mM)^{\frac{1}{2}}} ,$$

$$x = (E/k_B T)^{\frac{1}{2}} ,$$

$$x_0 = (E_0/k_B T)^{\frac{1}{2}} .$$

Appendix II

ON OPERATOR IDENTITY

In Secs. 2-5 and 2-6.3 we employ the relation

$$e^{A}e^{B} = e^{A+B+\frac{1}{2}[A,B]} \tag{II-1}$$

for any operators $\underline{A}$ and $\underline{B}$ whose commutator is a multiple of the identity operator. Given

$$[A, B] = AB - BA = c1 , \tag{II-2}$$

where $\underline{c}$ is a constant and 1 is the identity operator, it is easy to show by induction that

$$AB^{n} - B^{n}A = ncB^{n-1} . \tag{II-3}$$

Now consider

$$\frac{d}{d\lambda} e^{\lambda A}e^{\lambda B} = e^{\lambda A}Ae^{\lambda B} + e^{\lambda A}e^{\lambda B}B . \tag{II 4}$$

Expanding $\underline{e}^{\lambda \underline{B}}$ in the first term on the right-hand side of this equation and using Eq. (II-3), we find

$$\frac{d}{d\lambda} e^{\lambda A}e^{\lambda B} = e^{\lambda A}e^{\lambda B}(A + B) + \lambda[A, B]\ e^{\lambda A}e^{\lambda B} . \tag{II-5}$$

Finally, integrating with respect to λ and putting λ = 1 in the result gives Eq. (II-1).

Appendix III

BLOCH'S THEOREM

If $\underline{Q}$ is any linear combination of harmonic oscillator coordinates, then

$$\langle \exp Q \rangle_T = \exp\left(\tfrac{1}{2} \langle Q^2 \rangle_T\right) \ . \qquad \text{(III-1)}$$

The following proof of this result is taken from Appendix A of the work by Zemach and Glauber.[a] We write $\underline{Q}$ as

$$Q = \lambda_1 a + \lambda_2 a^+ \ , \qquad \text{(III-2)}$$

where $\underline{a}$ and $\underline{a}^+$ are harmonic oscillator destruction and creation operators that satisfy the commutation relations

$$\left.\begin{aligned} [a, a^+] &= 1 \\ [a, a] = [a^+, a^+] &= 0 \ , \end{aligned}\right\} \qquad \text{(III-3)}$$

and λ_1 and λ_2 are complex numbers that may be time-dependent. Using Eq. (II-1), we may express $\underline{\exp Q}$ as

$$\exp Q = \exp(\lambda_2 a^+) \exp(\lambda_1 a) \exp(\tfrac{1}{2} \lambda_1 \lambda_2) \ . \qquad \text{(III-4)}$$

The convenient feature of this form for $\underline{\exp Q}$ is that in the expansion of the exponents the annihilation operators appear to the right of

[a] A. C. Zemach and R. J. Glauber, Dynamics of Neutron Scattering by Molecules, Phys. Rev., **101**: 118 (1956).

the creation operators. After expanding the first two exponential functions on the right-hand side of Eq. (III-4), we can express the diagonal element of exp Q in the nth excitation state of the oscillator by

$$\left\langle \exp Q \right\rangle_n = \sum_{p=0}^{n} \frac{(\lambda_1 \lambda_2)^p}{(p!)^2} \left\langle (a^+)^p a^p \right\rangle_n \exp(\tfrac{1}{2} \lambda_1 \lambda_2) \ . \tag{III-5}$$

Here we have used the property

$$a^p |n\rangle = 0 \tag{III-6}$$

for $p > n$. Further, since

$$\left. \begin{aligned} a|n\rangle &= n^{\frac{1}{2}} |n-1\rangle \\ a^+|n\rangle &= (n+1)^{\frac{1}{2}} |n+1\rangle \ , \end{aligned} \right\} \tag{III-7}$$

it follows that

$$\left\langle (a^+)^p a^p \right\rangle_n = \frac{n!}{(n-p)!} \ . \tag{III-8}$$

When Eq. (III-8) is substituted into Eq. (III-5), there results

$$\left\langle \exp Q \right\rangle_n = \sum_{p=0}^{n} \frac{n!}{(p!)^2 (n-p)!} (\lambda_1 \lambda_2)^p \exp(\tfrac{1}{2} \lambda_1 \lambda_2) \ . \tag{III-9}$$

We require

$$\left\langle \exp Q \right\rangle_T = \sum_{n=0}^{\infty} e^{-n\beta\hbar\omega} \frac{\left\langle \exp Q \right\rangle_n}{\sum_{m=0}^{\infty} e^{-m\beta \hbar\omega}} \ . \tag{III-10}$$

Letting $z = \exp(-\beta\hbar\omega)$ and combining Eqs. (III-9) and (III-10), we find

$$\left\langle \exp Q \right\rangle_T = (1 - z) \sum_{n=0}^{\infty} z^n \sum_{p=0}^{n} \frac{n!}{(p!)^2 (n-p)!} (\lambda_1\lambda_2)^p \exp(\tfrac{1}{2}\lambda_1\lambda_2)$$

$$= (1 - z) \sum_{p=0}^{\infty} \frac{1}{(p!)^2} (\lambda_1\lambda_2)^p z^p \frac{d^p}{dz^p} \sum_{n=0}^{\infty} z^n \exp(\tfrac{1}{2}\lambda_1\lambda_2)$$

$$= (1 - z) \sum_{p=0}^{\infty} \frac{1}{(p!)^2} (\lambda_1\lambda_2)^p z^p \frac{d^p}{dz^p} \frac{1}{1 - z} \exp(\tfrac{1}{2}\lambda_1\lambda_2)$$

$$= \sum_{p=0}^{\infty} \frac{1}{p!} (\lambda_1\lambda_2)^p \left(\frac{z}{1 - z}\right)^p \exp(\tfrac{1}{2}\lambda_1\lambda_2)$$

$$= \exp\{\lambda_1\lambda_2[\tfrac{1}{2} + z(1 - z)^{-1}]\} \ . \qquad \text{(III-11)}$$

Now we calculate

$$\left\langle Q^2 \right\rangle_T = (1 - z) \sum_{n=0}^{\infty} z^n \left\langle (\lambda_1 a + \lambda_2 a^+)^2 \right\rangle_n$$

$$= (1 - z)\, \lambda_1\lambda_2 \sum_{n=0}^{\infty} z^n \left\langle aa^+ + a^+a \right\rangle_n \ . \qquad \text{(III-12)}$$

Since

$$\left\langle aa^+ \right\rangle_n = \left\langle a^+a + 1 \right\rangle_n = n + 1 \ , \qquad \text{(III-13)}$$

Updating. As soon as each new value of an unknown is computed, it may be inserted at once into the equations that remain to be evaluated in the current iteration rather than waiting until all equations have been iterated before performing the substitutions.

Normalization. The equations usually express some conservation condition or balance between "gain" and "loss" processes. Until the true solutions are determined, this balance condition is not exactly satisfied. For the inhomogeneous systems of equations of the type being considered, however, it is possible to impose a requirement of gross conservation on each iteration step; that is, the unknowns are each multiplied by a single normalizing factor that is specified in such a way that some appropriately weighted sum of all gain terms in the entire set will be exactly balanced by a similarly defined sum of all loss terms. This normalizing factor changes from one iteration to another but tends to unity as the process converges.

The effects of these iteration procedures individually and in combination on the rate of convergence have been evaluated by Honeck.[a] If none of the three refinements is used, the method is termed the "Gauss iteration." In the Gauss-Seidel method, updating only is used, while in the Extrapolated Liebmann method both updating and extrapolation are employed. The "Normalized Gauss" method involves only normalization, while the "Normalized Extrapolated Gauss" method includes both normalization and extrapolation. The number of iterations required to converge a simple test problem using these five methods are 620, 399, 172, 32, and 17, respectively.[a] Although the test is not exhaustive, it does indicate that the last method, Normalized Extrapolated Gauss, is probably an excellent choice for thermalization problems.

---

[a] H. C. Honeck, The Distribution of Thermal Neutrons in Space and Energy in Reaction Lattices. Part I: Theory, Nucl. Sci. Eng., **8**: 193 (1960).

Index

Date Due